Geochronology and
Thermochronology

Geochronology and Thermochronology

Peter W. Reiners

University of Arizona
Tucson, AZ, USA

Richard W. Carlson

Carnegie Institution for Science
Washington, DC, USA

Paul R. Renne

Berkeley Geochronology Center
and
University of California
Berkeley, CA, USA

Kari M. Cooper

University of California, Davis
Davis, CA, USA

Darryl E. Granger

Purdue University
West Lafayette, IN, USA

Noah M. McLean

University of Kansas
Lawrence, KS, USA

Blair Schoene

Princeton University
Princeton, NJ, USA

WILEY

Contents

Preface

Geochronology, including thermochronology, is an essential component of practically all modern Earth and planetary science and provides fundamental information for many other areas, including archeology, marine sciences, and ecology. Geochronology establishes the timing of critical events ranging from the age of the Earth to stratigraphic boundaries, and it provides unique constraints on the pace and dynamics of processes ranging from condensation of the solar nebula to planetary differentiation to surface exposure to biologic evolution. Given that Earth and planetary scientists commonly seek to understand relationships between events or phenomena for which physical evidence is incomplete or ambiguous, establishing temporal relationships through geochronology often provides a substantial basis for causality arguments.

Although the concept of geochronology has existed for millennia, and the particular name has been around since 1893, most scientists would probably agree that the modern practice or discipline is based on application of radioisotopic (or cosmogenic) systems in natural materials, which has existed for only a little more than a century (or less). Even into the 20th century, the geologic timescale floated freely in time. Geologists had established sequences of evolutionary and orogenic events in the rock record, but numerical estimates ranged widely, more so further back in geologic history. Without precise dates, only poorly constrained arguments could be made about the relative durations and the time separating major events in the geologic record. Likewise, prior to radioisotopic methods, the best available estimates for the age of the Earth (and solar system) disagreed by several orders of magnitude. The rather sudden recognition of nuclear structure and radioactive decay around the beginning of the 20th century, changed Earth and planetary science fundamentally. The very first radioisotopic dates measured increased the previously deduced minimum age of the Earth by about an order of magnitude, and subsequent work, less than 100 years ago, increased it by another factor of ten.

Although the numerical age of the Earth and the temporal anchoring of the geologic timescale are of immense practical (as well as philosophical and fundamental) value, these were only a start to the revolution that radioisotopic geochronology imparted to Earth and planetary sciences. Geochronology continues to be essential in the way it was originally used, to establish formation ages of rocks, but has also evolved into a broad array of methods and approaches for providing temporal constraints on natural phenomena ranging from the pacing of orbital oscillations, rates of erosion and paleotopographic change, subsurface fluid fluxes, timescales of lithospheric recycling to the deep mantle, periodicity of continental growth, collisions in the asteroid main belt, and much more. Modern geochronology is about more than dating events—it is also about using rigorous, quantitative, and innovative approaches to measuring rates, fluxes, and timescales, and using temporal constraints to understand the processes driving natural phenomena.

Our book is intended to provide both an introduction and reference for users and innovators in geochronology. Because it is a dynamic field, many aspects of geochronology change quickly, from the atomic-scale understanding of radioisotopic decay, experimental investigations and kinetic calibrations of thermal (and other) sensitivities to daughter product retention, analytical measurement techniques, mathematical modeling and interpretational approaches, and the types of geologic or planetary questions on which applications focus. We have done our best, in the chapters of this book, to provide modern perspectives on the current state-of-the art in most of the principal areas of geochronology, while recognizing that they are changing rapidly. We intend for students and scientists to use the chapters in this book as a foundation for understanding each of the methods we cover, and for illuminating directions that we think will be important in the near future. Users of this edition of this book may wish to complement the chapters with emerging references and reviews to provide valuable perspectives on the fields and topics, as well as opportunities for important questions and problems in the near future. We have attempted to provide sufficient references to rapidly evolving topics that will enable readers to pursue future developments via citation strings in bibliometric databases.

This book attempts to present the state-of-the-art on most of the most important geochronologic methods, emphasizing fundamentals and systematics, historical perspective, analytical methods, data interpretation, and some applications chosen from the literature. The presentation is designed to be useful to students in graduate courses or to upper-level undergraduates with a solid background in mathematics, geochemistry, and geology. Although this book will be useful as a reference to users, we cannot make claims to encyclopedic coverage of all these topics. Indeed actual encyclopedias of dating techniques are available elsewhere. In addition, this book is different from the several others that do an excellent job of describing geochronology as a subfield or application of isotope geochemistry. We cannot supplant the comprehensive utility of isotope geology books, but we aim to complement them by expanding on those parts of isotope geochemistry that are concerned with dates and rates and insights into Earth and planetary science that come from temporal perspectives. We have attempted to present the fundamentals, perspectives, and opportunities in modern geochronology together in a way that we hope will inspire further innovation and creative technique development and applications.

We acknowledge helpful reviews and advice from several colleagues, particularly Rebecca Flowers and Peter Zeitler, who provided helpful feedback on the entire book. We are also grateful to reviews and advice from Willy Guenthner, Frederic Herman, Richard Ketcham, Georgina King, Larry Nittler, Stuart Thomson, Jibamitra Ganguly, and Doug Walker, and editorial assistance from Matt Dettinger, Diana Gutierrez, and especially Erin Abel. We also gratefully acknowledge the strong and vibrant community of geochronologists and thermochronologists in Earth and planetary science today, who may not always agree but who inspire our desire to contribute to the geochronologic conversation and wield the power of radioisotopic dating to gain and share real insights to nature.

Peter W. Reiners
Richard W. Carlson
Paul R. Renne
Kari M. Cooper
Darryl E. Granger
Noah M. McLean
Blair Schoene

CHAPTER 1

Introduction

Occasionally debates arise and hands are wrung about what parts of a scientific discipline really distinguish it from others. Geoscientists often find themselves trying to define the unique perspectives or essential skills at the heart of their field as if failure to properly indoctrinate students in them might put the entire profession at risk. Without commenting on the wisdom of such disciplinary exceptionalism, a reasonable person asked to engage in it could, after some thought, suggest that if there is something distinctive about Earth science, it might have something to do with time. Naturalistic thinking about the evolution and workings of the Earth have been around for centuries if not millennia, and considerations of time at scales far surpassing human experience are a central and obligatory part of any serious endeavor in this area. The facility to deal easily with enormous timescales is such an ingrained part of Earth and planetary science that occasional meditative realizations of even the most hardened scientists are sometimes required to remind them that our ability to envision geologic time accurately and precisely has been in some ways hard won. Before quantitative measurements were available of the durations of time separating events of the past from the present, and of the rates of geologic processes, practically all attempts to understand Earth were, to paraphrase a key historical figure in geochronology (Lord Kelvin), meagre and of a most unsatisfactory kind. Quantitative geochronology as a concept, and especially radioisotopic geochronology as a field in and of itself, revolutionized our understanding of the Earth and planets. More importantly, geochronology continues to be one of, if not the most, important foundation and means of exploration in modern geoscience.

The tools and applications of geochronology find use in a variety of fields besides Earth and planetary science, including archeology, evolutionary ecology, and environmental studies. But the impact of geochronology on Earth science was fundamentally transformative. For one thing, it laid out the boundary conditions for reconstructing the history of the planet and quantitative understanding of the significance of ongoing physical processes like erosion, sedimentation, magmatism, and deformation. It also established, for the first time, a realistic temporal context of existence—not just of life as we know it, but for the recognizable planetary environment that hosts life. This is because the timescales of Earth history and Earth processes (including biotic evolution at that scale) require a fundamentally different temporal perspective than human experience (much less historical records) can offer. While some important geologic and evolutionary processes happen over very short timescales and require chronometers with commensurate sensitivity, many of the most challenging and important observations we make about the Earth reflect processes that occur either very slowly or very rarely, relative to the perspective of humans as individuals, civilizations, or even species. Modern radioisotopic techniques span vast timescales from seconds to billions of years, finding application in problems ranging from the age and pace of individual volcanic eruptions to condensation of the solar nebula and ongoing planetary accretion. The transformative power of geochronology comes from its capacity to expand our understanding beyond the reach of the pathetically short timescales of intuitive human or social perspectives.

1.1 GEO AND CHRONOLOGIES

Extending the timescale of our understanding does not mean just establishing a chronology of events that occurred earlier than historical records or generational folklore allow. It goes without saying that establishing pre-historical records of changes on and in Earth and other planets is practically useful: knowing when a volcano erupted or a nearby fault last ruptured or the age of an extinction or diversification event may be important. Establishing historical chronologies of tectonic events is clearly necessary for practical purposes. But a list of dates or sequence of regional events is of limited value in and of itself, and does little to represent geochronology as way of

Geochronology and Thermochronology, First Edition. Peter W. Reiners, Richard W. Carlson, Paul R. Renne, Kari M. Cooper, Darryl E. Granger, Noah M. McLean, and Blair Schoene.
© 2018 John Wiley & Sons Ltd. Published 2018 by John Wiley & Sons Ltd.

exploring how the planet works using time as an organizing principle or mode of inquiry.

For one thing, there is the question of how to define an event. At one level the question of the age of the Earth is simple, and has been the focus of countless studies since human curiosity began. Modern perspectives on the problem however, shifted years ago from simplistic numerical answers of around 4.56 Ga, to more sophisticated ones that raise issues of how to assign a single age to a protracted evolutionary process complicated by questions of the initial uniformity of and chemical fractionation in the solar nebula, and timescales of accretion, mass loss, and differentiation. Many other questions in Earth and planetary science have evolved similarly as understanding deepened. Continuing efforts to understand the geologic record are no longer satisfied with just knowing "the age" of a particular event such as the Permo-Triassic boundary, the Paleocene–Eocene Thermal Maximum (PETM), or meltwater pulse 1A, but now we need to know the duration, pace, and number of perturbations composing an event, and the detailed sequence and timing of resulting effects. Geochronology has been central to all of these as not only the intended accuracy and precision, but also the essence of the question, changed. Geochronology shows that "events" are not only finite and messy, but manifestations of more interesting phenomena in themselves.

Also, while some scientists see geochronology as a useful tool for addressing pre-defined geologic problems, using geochronology is not the same thing as doing it. The power of geochronology arises from innovative approaches. There is no single template for this, but one could make an argument for at least two types of creative geochronology. The first is adapting new geochemical, physical, or analytical insight or technology to addressing suitable geologic problems. Fission-track dating was developed after methods for observing cosmic ray tracks in insulators were extended to tracks produced by natural radiation sources *in situ* [*Fleischer and Price*, 1964]. Inductively coupled plasma mass spectrometry and its pairing with laser ablation sample introduction both changed isotope geochemistry and geochronology in key ways [e.g., *Halliday et al.*, 1998; *Lee and Halliday*, 1995; *Kosler et al.*, 2008]. K–Ar dating was adapted into one of the most precise and powerful geochronological techniques ever developed (^{40}Ar/^{39}Ar dating) using fast neutron irradiation to create proxies of parent nuclides of the same element and chemical behavior as the daughter nuclides [*Merrihue and Turner*, 1966]. And of course the first radioisotopic date itself was calculated as a marginalia to a nuclear physics study much more concerned with "radioactive transmutations" than with determining the age of anything [*Rutherford*, 1906].

Second, and as is true in many other fields, some impactful advances in geochronology have come not from deliberate engineering but more as refusals to ignore complications. Solutions to such problems often hold potential for illuminating unknown unknowns, which may then be trained to address previously unsolvable problems. When a particular technique appears to "not work" for answering the question originally posed, it may

be time to ask why the answer is unexpected and what can be learned from it by reframing the question. Thermochronology, for example, owes a great deal of its modern utility to this sort of lemons-to-lemonade evolution, as the diffusive loss of daughter products was initially considered a debilitating limitation of noble-gas-based techniques [e.g., *Strutt*, 1906] but is now recognized as its defining strength, as increasingly complex as it appears to be [e.g., *Shuster et al.*, 2006; *Guenthner et al.*, 2013].

This is all to say that geochronology is not just a "tool" serving other fields, but is a field unto itself, and one that originates the new ideas and approaches that allow for advances in the areas to which it is applied. Geochronology generates the innovative ways to use nuclear physics and geochemistry to understand natural processes, often by using initially problematic aspects of these systems, and adapting them to questions that initially may not have been asked. It was not until long after we started wondering about the age of the Earth that we started to appreciate questions about the duration of events, stratigraphic boundaries, and diachroneity. And it was not until we developed quantitative tools (serendipitously, in many cases) for measuring dates and rates in new ways that we began to realize the value of understanding many more nuanced time-related problems, like rates of erosion, sedimentation, crystallization, or groundwater flow, the degree to which these processes are steady or episodic, and the scale at which these questions even make sense.

There is no denying geochronology's utility for addressing some of the most fundamental and, in many cases, simple questions in Earth and planetary science. This is true in both a historical sense, as geochronology provided key foundations for geoscience progress over the century, as well as in a continuing sense, as it continues to provide simple formation and cooling ages essential to many geologic studies. So it is reasonable to begin here with a review of the history of geochronology in the context of its original mothers of its necessity: the age of the Earth and, soon thereafter, ages of stratigraphic boundaries. The last part of this chapter then returns to the broader topic of geochronology—the discipline and its objectives and significance—with the hope that the perspective of the historical review drives these home.

1.2 THE AGES OF THE AGE OF THE EARTH

It is impossible to know when humans or perhaps their predecessors first started posing questions about the age of the Earth, but it seems likely that it has been a central focus of human contemplation for millennia. The scope and context of the question has likely changed, and in fact continues to evolve as our resolution of the early days of the solar system improves [*Bouvier and Wadhwa*, 2010; *Brennecka et al.*, 2010]. Ancient Greek and Hindu philosophies explained the age of the "world" in terms of infinite or cyclical ages, the latter punctuated by revolutions of destruction and rebirth, a theme that may originate from the rise and fall of human civilizations, but which may also have been inferred by observant early naturalists from the rock record's evidence for episodes of upheaval and deformation followed by quiescence and slow accumulation.

Propositions for noncyclical and finite ages have also been around since ancient times. Early estimates for a finite age of the Earth (or "world"), which typically have more religious than philosophical origins, tended to converge on timescales in the thousands of years. These included Zoroasters 17th century BCE estimate of 12,000 years, and numerous estimates based on scrutiny of details in the Christian bible. No less than eight well established bible-based estimates for the age of the Earth are known from 169 and 1650 CE, bookended by the Syrian saint Theophilus of Antioch and the famous "scholarship" of James Ussher. All of these invoke ages between 5000 and 9000 years, and all but one are within a narrower range of 5500–7500 years.

The convergence of many "world" chronologies in the range of a few thousand years ago to millennial timescales is an interesting target of speculation. The 1000-year timescales may arise from being just beyond the reach of multigenerational memory of oral histories, but not so far as to seem unreasonable or intuitively incomprehensible. This timescale also is commensurate with the rise and fall of some of the most persistent political empires and cultural dynasties, as well as the timescale of the development of recorded human history. In any case, the eventual recognition that the age of the Earth was not infinite but is actually a million times greater than a few thousand years represented a slow-moving but important change in human perception. That the planet has a deep history of such immensity that it practically challenges our ability to conceive of it, and that it predates humans' presence by more than three orders of magnitude, has been called the fourth great revolution in human cognition [*Rudwick*, 2014].

Although not scientific in modern senses, some scholars consider early attempts to estimate the age of the Earth using biblical records historically important. Many of these essentially counted the number of human generations since the birth of Abraham. Like other pre-Enlightenment scholars who mixed religious and scientific approaches, Johannes Kepler combined biblical accounts with astronomy to arrive at very similar ~6000 ka ages for the Earth as late as the early 17th century. One of the best known of these biblically based, but astronomically laced, deductions is that of the Bishop James Ussher, who in the middle of the 17th century presented the results of his scholarship proposing the beginning of the Earth to be 22 October 4004 BCE. The "9:00 a.m." often associated with Ussher's estimate actually comes from a separate but similar account from a contemporary scholar, John Lightfoot, who put the beginning at the autumnal equinox of 3928 BCE. Incidentally, these results are a good example of the difference between accuracy and precision: Lightfoot's extremely precise time-of-day estimate was a full 76 years younger than Ussher's, and both were obviously lacking much more in the way of accuracy. Readers interested in these early examples are directed to more thorough accounts in G. Brent

Dalrymple's "*The Age of the Earth*" [1994] and references therein.

Some historians of the evolution of thinking about the age of the Earth have suggested that early Christian accounts represent respectable nascent attempts to at least take the question seriously and start to frame the problem and possible solutions to it in an analytical and evidentiary way, even if the basis of the evidence was not scientific. Historian Martin Rudwick, in "*Earth's Deep History*" [2014], for example, calls Ussher's work "rigorous," and claims that it does not deserve the ridicule it commonly endures. It may indeed have been a rigorous examination of a document; less clear is the rigor of the documentation of the generations, much less their initiation as a proxy for the birth of the planet. But it may be true that Ussher's studies (and those of a few others) were not exactly sycophantic religious repetition then, but actually somewhat at odds with prevailing eternalism, the idea that the Earth has existed literally forever, at least insofar as humans are capable of understanding, and which to many seemed more reasonable and potentially reconcilable with biblical teachings. Thus Ussher's work and that of others might be considered the beginning of attempts to have a serious think about how old the Earth *could* be, using the scholarly resources available at the time. Rudwick argues that these efforts, while based largely on scripture, are continuous with later scientific attempts, which arose from the same progressive effort to understand the world. To that extent, it may be true that early studies by Ussher and others are distinguished from those of modern creationists (including proponents of intelligent design), whose absurdities are not honest attempts to comprehend anything and do not represent even primitive roots of any kind of legitimate understanding.

But before Ussher gets too much credit, his analysis was based on a religious text that represented the political, economic, and cultural authority of the time, so he probably did not lose sleep worrying whether his "rigorous" scholarship might put him in very real danger at the hands of the Christian power structure, as Galileo and others had only a few years earlier. So although Ussher's work may be detailed and arguably historically important, and may represent an early attempt to challenge the idea of equally nonscientific "eternalism," it does not rank with intellectually honest and courageous work of secular pioneers of the time who risked, and in many cases paid, the price of censure or far worse for crossing church authorities. In any case, the real challenge to human thinking (and, as it turns out, the scientific truth) though, is far different from both the relatively simple perspectives that either the Earth is eternal or its history is basically conceivable in terms of human generations. The far stranger truth is that the Earth is incredibly old, but has a finite age. Indeed it is the fact that it was born a knowable number of years ago in a relatively short period of time which we can know with somewhat startling precision that raises even more questions.

Although some 17th century scholarship on the age of the Earth mixed astronomical observations or theory with "textual" constraints, the Enlightenment brought new ideas about rates of natural processes and actual geologic observations to bear on the question. One important figure in this vein was French diplomat and amateur naturalist Benoit de Maillet. In the early 17th century he constructed a theory for the age of the Earth told through an ingenious parable designed to avoid directly antagonizing the powerful Christian church (undermining Rudwick's claims that the Christian theocracy was no barrier to free thought at the time). Speaking through his fictional Indian philosopher Telliamed (his surname backwards), he combined measurements of rates of regional sea level decline with the height of high mountains and a Cartesian assumption (common at the time) that the earliest Earth was completely covered by ocean and that water was continuously lost through Descartes' mysterious vortices. This led him to the conclusion that the highest terrain must have been covered in water more than two billion years ago, providing a minimum age for the age of the Earth. Although the initial condition and steady decline of sea level is clearly absurd, de Maillet's analysis deserves credit for combining geologic observations and uniformitarian arguments to derive an estimated duration and therefore age constraint.

Other notable 18th and early 19th century attempts to constrain the age of the Earth also followed the general approach of combining a process occurring at an assumed rate with an initial condition of some sort. A popular one was cooling of an initially molten or at least extremely hot Earth. Although as far as we know, Isaac Newton did not directly wade into the debate over the age of the Earth, in the late 17th century he calculated cooling times for planetary bodies and speculated about cooling durations of comets that passed close to the Sun. His contemporary Gottfried Leibniz also speculated about the origins of topography as resulting from differential contraction during cooling of the initially molten Earth. Neither Newton nor Leibniz used cooling timescales to actually estimate ages of planetary or solar bodies, perhaps because they recognized the potential complexities involved, this of course became a popular sport a few hundred years later, in the late 19th century. But long before the famous calculations of physicist William Thomson helped earn him the title of Lord Kelvin, similar experiments and calculations of the provocative natural historian George-Louis LeClerc helped earn him the title Comte de Buffon.

Buffon wrote his major work on the origin of the Earth, "*Époques de la Nature*," in 1778. By this time, advances in natural history had established evidence that Earth history was not static or eternal, but that the planet had changed progressively over time. This included recognition, attributed at least partially to Nicolas Steno, that sedimentary rocks lower in stratigraphic sequences, and hence older, contained macroscopic fossils that appeared to be morphologically simpler than the rocks above them. And in fact the oldest rocks contained no identifiable fossils at all. Progressive change over time, rather than strict steady-state concepts of Earth history, was an important basis of Buffon's (and others') thinking. Although he recognized uniformitarian principles, for example as represented in erosion and deposition, he did not extend these to a simple eternalist vision of the Earth as many

other contemporary thinkers, including the purported founder of modern geology, James Hutton, whose strong Christian convictions pervaded his avoidance of questions on the age of the planet, as in "[the Earth shows] no vestige of a beginning, no prospect of an end."

In fact, Buffon's publication "*Nature's Epochs*" ventured to estimate both ends of Earth's history that Hutton said were unknowable. Although many aspects of Buffon's analysis were highly speculative, such as the origin of the Earth (and other planets) by impact of a large comet with the Sun, his work was some of the first to apply basic physics and experiments to the question. Recognizing, as many did by that time, that temperatures beneath the Earth's surface generally increase with depth, Buffon combined this with the then well-accepted idea that the primeval Earth was entirely molten, and set about to experimentally determine the duration of time required to cool an Earth-size body to present surface temperatures. Using cooling times of cast iron balls of varying size at initially high temperatures, he extrapolated his experimental results to determine a minimum age for the Earth on the order of 100 ka. Buffon considered this likely far too low, for reasons that are not entirely obvious but probably related to his recognition that stratigraphic thicknesses required longer timescales if achieved by typical erosion and deposition rates, an apparent problem that was to plague the issue of the age of the Earth for the next ~150 years.

Buffon's cooling timescale experiments, which were built on those of Newton, Liebniz, and others before him, also foreshadowed some of the well-recognized thinking of one of the 19th century's most celebrated scientists, William Thomson, later named Lord Kelvin, whose influence and subsequent arguments have been documented well by *Burchfield* [1975], *Stacey* [2000], and in many other places. Beginning with the same convenient initial condition that the Earth began as a uniformly very hot sphere that cooled gradually with time, simple thermal diffusion arguments led to the basic conclusion that the current surface temperature and near-surface geothermal gradient required something on the order of 100 Ma [*Kelvin*, 1863], a number that he later revised to 20 Ma. Although it was widely recognized as heuristic, and well known that any internal advection would change the result to some degree, this estimate stood as the most reasonable and definitely the most authoritative estimate for more many decades. It also put most geologists (and the few evolutionary biologists of the time), who felt that the Earth must be far older based on observed timescales of ongoing processes, at odds with much of the scientific establishment for the next several decades.

The common account of the reason Kelvin's estimate was so far off is that it came from failure to account for the contribution of radioactive decay to Earth's internal heat. In reality this additional heat is not very significant to the basic result, and its incorporation would not have changed things significantly. The true explanation of the erroneous result is its failure to incorporate the much more effective advective, instead of conductive, transport of heat from throughout Earth's interior to the thin crustal layer where the thermal gradient used in Kelvin's calculations was measured. A far more influential reason that most of the scientific community chose to accept the physicists' estimates over the longer views of geologists came from Kelvin's work on the estimated age of the Sun. Using similar approaches, Kelvin had argued that the Sun could only contain enough heat after initial formation to remain as hot as it now is for no more than about 20 Ma. Assuming that the Earth itself was unlikely to predate the Sun, this placed a strong upper bound on the age of the Earth. Even the discovery of radioactive decay of naturally occurring nuclides near the turn of the century would not change the basics of this argument, as nuclear fusion was not recognized until the 1920s or 1930s, extending the debate and undercurrent of animosity between geologists and physics for several more decades.

Among the many geologists resistant to Kelvin's constraints was then University of Chicago professor Thomas Crowder Chamberlin. Also recognized for proposing that changing CO_2 concentrations in the atmosphere may be responsible for climate change, he suggested that Kelvin's timescale was too short to reconcile with geologic evidence and that there must be another source of heat within the Earth. From this debate comes one of his well cited quotations:

"The fascinating impressiveness of rigorous mathematical analysis, with its atmosphere of precision and elegance, should not blind us to the defects of the premise that condition the process." [*Chamberlin*, 1899]

Ironically, it would be the physicists again, including a Kiwi by the name of Ernest Rutherford known for the quote, *"All science is physics or stamp collecting,"* who would all underscore Chamberlin's quote by not only helping find the additional heat source but also creating the means for accurate and increasingly precise quantification of the real age of the Earth.

Several other approaches to estimate the age of the Earth were also taken near the end of the 19th century and beginning of the 20th. One of the more productive, at least in terms of numbers of papers, was based on ocean salinity. As described by *Dalrymple* [1994] the "salt accumulation clock" method was first proposed by Edmund Halley (of comet fame), who reasoned as early as 1715 that comparing the total salt content of the ocean to the amount delivered by rivers could provide an estimate of the age of the ocean and, to the extent that the Earth has always had an ocean (and that it began as freshwater…), the Earth itself. Between 1876 and 1909 T. Mellard Reade and later John Joly, as well as others, picked up the approach and derived estimates falling between 25 and 150 Ma, with later estimates tending to inch upwards. The fact that the approach yielded answers converging on something similar to Kelvin's calculation based on heat flow probably aided its apparent legitimacy. But as we recognize now, even if delivery rates of ions to the ocean from rivers (and groundwater, as we now also know is an important source) were to stay constant with time, the ratio of the total amount of any ion in the oceans to this rate does not necessarily produce a time that corresponds to an initial concentration of zero. Analogous to the problem of coupled production and diffusion in open-system thermochronometers, the ratio of current inventory to current rate of accumulation does not account for fluctuations in both through time. In addition to the likelihood that the ocean was not born fresh, it is also subject to loss of its dissolved load at a rate that may vary itself over time. The apparent age may therefore be better thought of as a something approximating the residence time, which for the major ions (Na, Cl, Mg, SO_4) that were the primary focus of these studies, are about 12–130 Ma.

The golden years of the late 19th century for speculative calculations bearing on the age of the Earth also saw estimates based on orbital physics. Around 1879, George Darwin, second son of Charles and most famous for the fission model of the origin of the Moon and creation of the Pacific Ocean basin, developed a complicated set of geophysical arguments involving dissipation of tidal friction and its effect on slowing Earth–Moon rotation, coming up with a minimum estimate of around 56 Ma. This line of investigation was shared by several others including Lord Kelvin himself, who constrained the problem to an age less than about 1 Ga.

Probably few approaches of the pre-radioactivity era received more attention as avenues for estimating geologic time than accumulation rates of sediments or sedimentary rocks. According to *Dunbar* [1949] the great historian Herodotus (484–425 BCE) attempted to understand durations through observing sediment deposition during flooding of the Nile. Extrapolating individual flood events to the sediment pile in the Nile delta he inferred that buildup of the sediment there must have taken thousands of years. He also discusses calculations of durations of time from a statue of Ramses II (about 3200 years old) buried beneath about 2.7 m of sediment, and the burial of a clearly much older burned brick about 12 m beneath the surface. He observed that this made sense with the observed deposition rate from the area of about 9 cm/century, suggesting a sensible uniformitarian approach could at least extend back several thousand years.

One of the most detailed and influential attempts to constrain the magnitude of geologic time was Charles Walcott's 1893 paper in the Journal of Geology. His opening lines characterize the debate at the time:

"OF ALL subjects of speculative geology few are more attractive or more uncertain in positive results than geologic time. The physicists have drawn the lines closer and closer until the geologist is told that he must bring his estimates of the age of the earth within a limit of from ten to thirty millions of years. The geologist masses his observations and replies that more time is required, and suggests to the physicist that there may be an error somewhere in his data or the method of his treatment."

Walcott divided sedimentary strata of the US Cordillera into clastic and chemical precipitated rocks (in this case limestones). But rather than use arguably more direct estimates of depositional rates extrapolated from short timescales of modern observations as Herodotus did, he employed relatively complex

Period.							Time Duration.
Cenozoic, including Pleistocene	-	-					2,900,000 years
Mesozoic	-	-	-	-	-	-	7,240,000 "
Paleozoic	-	-	-	-	-	-	17,500,000 "
Algonkian	-	-	-	-	-	-	17,500,000 "
Archean	-	-	-	-	•	-	10,000,000(?)"

Fig. 1.1. Estimated durations of time assigned to each geological era by Walcott based on stratigraphic accumulation (and erosion rate) estimates and observations from the US Cordillera. (Source: *Walcott* [1893].)

arguments about rates and areas of erosion providing the raw materials for deposition. Comparing these with thicknesses of stratigraphic units in each of the paleontologically defined eras, he came up with estimates shown in Fig. 1.1 (The Algonkian is essentially the same as the modern Proterozoic). Although the durations are obviously grossly low, it is interesting that the ratios of their apparent durations (except for the Archean and Algonkian, for which there was little sedimentary record in the region that Walcott could observe) are similar to those recognized now.

After Walcott's introduction, cited above, that contrasted geologic versus physics-based approaches, it is somewhat ironic that his estimate for the total duration of Earth history was not very different from that of Kelvin's. Many other estimates based on sediment accumulation were also published in the latest 19th and earliest 20th centuries, and although there were a few exceptions, by far most of them consistently estimated durations and total ages roughly 10 to 100 times too short. While some of this may owe to apparent legitimacy arising from similarities to physics-based methods, it is also undoubtedly an inevitable outcome of failure to properly account for unconformities, recycling, varying depositional (and erosional) rates, the fact that individual basins neither survive nor receive sediments for all of Earth history, and the increasing paucity of the preserved stratigraphic record for progressively older units. Although most of these limitations were recognized, their magnitude was obviously difficult to constrain, so when assumptions were made that yielded final results of the same order of magnitude as previous ones, those were probably considered the most reasonable.

In some ways, the relatively young field of geology of the late 19th to early 20th centuries was not held back by an inability to assign numerical ages to stratigraphic boundaries, deformation episodes, milestones of biotic evolution, or even the age of the planet. Armed with Nicolas Steno's principles of superposition and stratigraphic correlations, Cuvier's extinctions as marked by disappearances of fossil assemblages, and the ability to interpret orogenic episodes, there was a lot that could be done to interpret histories of subsidence, uplift, magmatism, deformation, and the regional extents and relationships of such processes. By 1870s, eras separated by biotic or lithologic differences were well defined (e.g., LeConte, 1879), including the Archaean (or Eozoic), Paleozoic, Mesozoic, Cenozoic, and the most recent era, the Psychozoic (which captures the defining characteristic of human degradation of our planet's habitability somewhat

more eloquently than Anthropocene). The business of reconstructing the geologic history of the planet could, apparently, go on with only relative dating and some, at least relative, sense of the amount of time represented in each era.

But pre-geochronology geology floated in time and compressed Earth history the farther back in time one looked, underestimating the true extent of the planet's age by about 100 fold. This was at least partly due to the fact that the geologic record that was interpretable without geochronology was restricted to stratigraphic correlations of sedimentary rocks, particularly those bearing fossils. Correlations and interpretations of Precambrian rocks were difficult if not impossible without geochronology, and igneous and metamorphic units lacking constraints from related sedimentary rocks could be just about any age in any part of the world. Pre-20th century geologists recognized that there was a Precambrian history to the Earth, but the early "Geologic Timescale" basically ignored it. This is somewhat ironic not only because it excludes the vast majority of Earth history but also because the Precambrian–Cambrian boundary represents one of the most significant orogenic, biotic, and sedimentary events to affect the planet. Something huge happened at this time, but there was almost no way to date it or recognize how much time lay before it. In fact, because the Precambrian–Cambrian boundary represented a very widespread if not global event of very limited duration compared to the duration represented by the younger stratified rocks above it, it is an example of one of the completely undated but well-recognized geologic "revolutions" that proved useful for regional and global correlations and establishing the floating Geologic Timescale. According to *Williams* [1893],

> "*As the period of each dynasty in ancient history is marked by continuity in the successive steps of progress of the country, of the acts of the people and of the forms of government, and the change of dynasties is marked by a breaking of this continuity, by revolutions and readjustment of affairs, so in geological history the grand systems represent periods of continuity of deposition for the regions in which they were formed, separated from one another by grand revolutions interrupting the regularity of deposition, disturbing by folding, faulting and sometimes metamorphosing the older strata upon which the following strata rest unconformably and for the beginnings of a new system.*"

Geologic revolutions of the late 19th century included the close of the "Archean" revolution, now recognized as end of the Precambrian, as well as the Appalachian revolution (also recognized in Europe by other names), the Palisades revolution of the Jurassic–Triassic, the Rocky Mountain revolution (which was extended over what might now be a disturbingly long period of time, Cretaceous to Miocene, and also apparently distance, as lumped in causally were also the Pyrenees and the Himalaya), and the Miocene Cascadian revolution in flood basalts of the Pacific Northwest. Of course all of these periods were floating in temporal space. The clear and practically instinctive association of numerical ages to geologic periods or eras that we enjoy now did not exist then, and these revolutions could have been millions to tens to hundreds of millions of years old.

By Dunbar's time in the early 20th century, after the discovery of radioactivity but before geochronology had a serious influence on mainstream geologic thinking, revolution concepts were still important bases of understanding, though they had been tweaked a bit. The Rocky Mountain disturbance had been separated into the Nevadan, peaking in the Late Jurassic, the Laramide, near the Cretaceous–Paleogene boundary, the Palisades had been relegated to a "disturbance," the Appalachian was seen to postdate two earlier disturbances, the Acadian and Taconian, and several revolutions were recognized within the Precambrian, the Penokean, Algoman, and Laurentian.

As useful as the concept of geologic revolutions were, most them were not nearly as global or even superregional as typically envisioned, and to a large degree reflected regional tectonic processes largely restricted to the stomping grounds of their investigators. Ironically, the arguably most globally preserved revolution, the Precambrian-Cambrian boundary, did not really rise to the status of a revolution because it represented such a fundamental shift in the rock record as to make it different from all the others. In any case, as late as the early 20th century there was general acceptance that very widespread "events" punctuated the stratigraphic record and therefore Earth history. Although few if any of these retain their nearly global significance, they do point out that even prior to the advent of geochronology there was a great deal known about the geologic record of the last ~542 Ma, even if it was entirely floating in time. Though reliable numerical ages were a long way off and the methods of estimating durations were primitive, some semi-quantitative constraints on the relative amounts of time represented by geologic periods were also available. J.D. Dana did this by assigning sedimentary thicknesses a standard unit of time, and proposing that an equivalent thickness of limestone (which was supposedly much slower to accumulate) represented 5 times the duration of other sedimentary rocks. Williams and LeConte extended this to the idea of the geochrone as a unit of time useful for correlations and fundamental counting tool. As an aside it was also Williams who in 1893 was the first to propose the term geochronology, with a somewhat puzzling emphasis to modern ears that the time concerned is not human-centric:

> "In all these studies in which the geological time-scale is applied to the evolution of the earth and its inhabitants, the time concerned is not human chronology but is what may be called geochronology."

In some ways the geochrone is not a complete anachronism but is still used today. But instead of referencing time to a particular thickness of sedimentary rocks (e.g., the Eocene section in a specific place, as proposed by Williams), it is the orbital period of the Earth around the Sun, 1 year, or 365.256363004 days, or in SI units, 31558149.8 s.

Geochrones, geologic revolutions, and stratigraphic correlations allowed a great deal of Earth history to be reconstructed. The fact that the resulting structure was floating in time did not seem to be of paramount importance, and in fact even the significance of assigning numerical constraints to dates and rates of geologic events and processes any older than a few thousand years was often considered (as it still is by some practically-minded stratigraphers) as wading into a kind of speculative philosophy not unlike asking what existed prior to the big bang.

As late as the 1889 fifth edition of *Elements of Geology*, the renowned Joseph LeConte wrote:

> "Previous to even the dimmest and most imperfect records of the history of the earth there is, as already said, an infinite abyss of the unrecorded. This, however, hardly belongs strictly to geology, but rather to cosmic philosophy. We approach it not by written records, but by means of more or less probable general scientific reasoning."

and

> "Thus the history of the earth, recorded in stratified rocks, stretches out in apparently endless vista. And still beyond this, beyond the recorded history, is the infinite unknown abyss of the unrecorded. The domain of Geology is nothing less than (to us) inconceivable or infinite time."

Meanwhile, a few enterprising German and French scientists in the field that had underestimated the age of the Earth for so long were busy in labs doing experiments whose sometimes serendipitous results would move these questions from philosophy to hard science and begin a scientific and in some ways cultural revolution.

1.3 RADIOACTIVITY

In a universe with only slightly different physics, a conceivable combination of circumstances like the availability of still extant radioactive parent nuclides, the happenstances of trace element partitioning in common minerals, and the achievable precision of mass spectrometers might conspire to make radioisotopic geochronology impossible or at least much more difficult than it is. An almost uncomfortably small number of parent-daughter decay systems (Table 1.1) have decay constants and parent-daughter partitioning that make them geochronologically useful. And an almost absurdly large amount of what is known about the age of the Earth and terrestrial rocks in general comes from the U–Pb system alone. The parent isotopes of this system, ^{235}U and ^{238}U, have already lost 98.5% and 50%, respectively, of their abundance since the beginning of the solar system, and the technique is most often applied to a mineral, zircon, that constitutes only a fraction of a percent in certain rock types. If it weren't for nuclear transmutation and its manifestations in minerals with particular properties, except for the distant limits from astronomers and lifetimes of main-sequence stars, we may still be arguing

Table 1.1 Geochronologically useful radioactive decay systems

Parent/daughter (system)	Reaction or key daughters	Decay constant (a^{-1})	Half-life (a)	Daughter ratio typically measured
$^{147}Sm/^{143}Nd$	$^{147}_{62}Sm \rightarrow ^{143}_{60}Nd + ^{4}_{2}He$	6.54×10^{-12}	1.06×10^{11}	$^{143}Nd/^{144}Nd$
$^{238}U/^{206}Pb$	$^{238}_{92}U \rightarrow ^{206}_{82}Pb + 8^{4}_{2}He$	1.55×10^{-10}	4.47×10^{9}	$^{206}Pb/^{204}Pb$, $^{207}Pb/^{206}Pb$
$^{235}U/^{207}Pb$	$^{235}_{92}U \rightarrow ^{207}_{82}Pb + 7^{4}_{2}He$	9.85×10^{-10}	7.07×10^{8}	$^{207}Pb/^{204}Pb$, $^{207}Pb/^{206}Pb$
$^{232}Th/^{208}Pb$	$^{232}_{91}Th \rightarrow ^{208}_{82}Pb + 6^{4}_{2}He$	4.95×10^{-11}	1.4×10^{10}	$^{208}Pb/^{204}Pb$
(U–Th–Sm)/He	Sum of four above			$^{4}He/^{3}He$
$^{87}Rb/^{87}Sr$	$^{87}_{37}Rb \rightarrow ^{87}_{38}Sr + ^{0}_{-1}\beta$	1.42×10^{-11}	4.88×10^{9}	$^{87}Sr/^{86}Sr$
$^{187}Re/^{187}Os$	$^{187}_{75}Re \rightarrow ^{187}_{76}Os + ^{0}_{-1}\beta$	1.67×10^{-11}	4.16×10^{10}	$^{187}Os/^{188}Os$
$^{40}K/^{40}Ca$	$^{40}_{19}K \rightarrow ^{40}_{20}Ca + ^{0}_{-1}\beta$	4.96×10^{-10}	1.25×10^{9}	$^{40}Ca/^{44}Ca$
$^{40}K/^{40}Ar$		0.581×10^{-10}	1.25×10^{9}	$^{40}Ar/^{36}Ar$
^{238}U/fission	Variable daughters + fission track	8.45×10^{-17}	8.20×10^{15}	Track density/^{238}U
$^{138}La/^{138}Ce$	$^{138}_{57}La \rightarrow ^{138}_{58}Ce + ^{0}_{-1}\beta$	6.80×10^{-12}	1.02×10^{11}	$^{138}Ce/^{136}Ce$
$^{138}La/^{138}Ba$	$^{138}_{57}La \rightarrow ^{138}_{56}Ba + ^{0}_{+1}\beta$	6.80×10^{-12}	1.02×10^{11}	
$^{176}Lu/^{176}Hf$	$^{176}_{57}Lu \rightarrow ^{176}_{56}Hf + ^{0}_{-1}\beta + v$	1.87×10^{-11}	3.71×10^{10}	$^{176}Lu/^{177}Hf$
^{238}U series	Commonly used daughters: ^{234}U, ^{230}Th, ^{226}Ra, ^{210}Pb, ^{210}Po	$0.01-2.82 \times 10^{-6}$	1.00×10^{2} to 2.46×10^{5}	$(^{234}U/^{238}U)$, $(^{230}Th/^{238}U)$, $(^{226}Ra/^{230}Th)$, $(^{210}Pb/^{226}Ra)$, $(^{210}Po/^{226}Ra)$
^{235}U series	Commonly used daughters: ^{231}Pa, ^{227}Ac	$0.03-2.12 \times 10^{-5}$	$22.8-3.28 \times 10^{4}$	$(^{231}Pa/^{235}U)$, $(^{227}Ac/^{231}Pa)$
^{232}Th series	Commonly used daughters: ^{228}Th, ^{228}Ra	$0.12-0.36$	$1.92-5.75$	$(^{228}Th/^{232}Th)$, $(^{228}Ra/^{232}Th)$

about the age of the Earth and solar system, or relegating such discussions to cosmic-philosophy as was common less than one hundred years ago. But fortunately, there are sufficient numbers of geochronologically useful decay systems, undecayed parents, and minerals available to us.

The first graying of the dawn of the nuclear era is usually associated with Wilhelm Conrad Roentgen's discovery of X-rays in Wurzburg, Germany in 1895. Although not due to natural radioactivity, his discovery of radiation that had the ability to penetrate most solids (c.f., the X-ray image of Roentgen's wife's hand complete with wedding ring) set the stage for an even more serendipitous and portentous discovery in the next year. In spring of 1896 Henri Becquerel performed a series of experiments in which he determined that uranium-bearing salts had the ability to darken photographic plates. His own iconic figure features a fuzzy image of two rectangular dark spots on a plate, corresponding to photographic impressions left by two plates coated in uranium-bearing salt; one of the rectangles shows a lighter region corresponding to a Maltese cross that Becquerel placed between the salt and photographic plates to demonstrate the lesser penetration through the iron (Fig. 1.2). Interestingly, very similar experiments had actually been done and observations made about 40 years before this, by a French photographic inventor with the impressively lengthy name Claude Félix Abel Niépce de Saint-Victor, who noted that uranium produces "a radiation that is invisible to our eyes." In fact, Henri Becquerel's father Edmond had written about these observations in his book about light published in 1868.

Less serendipitous and more deliberate systematic explorations into natural radioactivity and a series of foundational discoveries

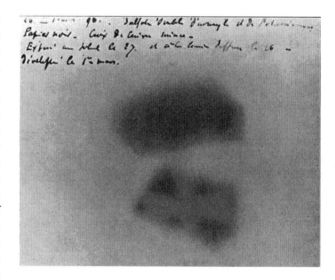

Fig. 1.2. Henri Becquerel's 1896 image of a photographic plate exposed by natural radiation from uranium-bearing salts. The lower of the two shows the shadow of an iron Maltese cross placed between the sample and the plate. (Source: https://en. wikipedia.org/wiki/Henri_Becquerel#/media/File:Becquerel_plate.jpg)

by Marie, and later Pierre, Curie soon followed Becquerel's work. Marie Curie measured electrical charge on the air surrounding uranium, and observed that its extent depended only on the amount of uranium present, leading to a hypothesis that uranium's radiation came from the atom, not molecules. She also observed that other U-bearing minerals were far more radioactive than uranium, leading to the insight that other elements must also be radioactive. Although she is often credited with discovering that Th is also radioactive, this was actually published in Berlin two months before by Gerhard Carl Schmidt. However, her more important insight, that uranium minerals contain small amounts of much more radioactive elements, soon led her and Pierre to large-scale chemical separations of constituents of uraninite (pitchblende), and the discovery of polonium and radium and recognition of their highly radioactive nature. In 1903 Marie and Pierre Curie (Fig. 1.3) and Henri Becquerel won the Nobel Prize in Physics; Marie also won a second prize in 1911 for her work on radium; their daughter Irène Joliot-Curie also won the Nobel in 1935 for discovery of artificial (neutron-bombardment induced) radioactivity.

The Curie's work on radioactivity accelerated progress on natural radioactivity at the turn of the century. In 1899 the Kiwi physicist Ernest Rutherford (Fig. 1.4) distinguished two types of radiation with different penetrating powers that he termed alpha and beta. He also discovered that thorium produced a gas, or "emanation" as he called it, that was itself radioactive, and that the activity of this gas followed a law whose differential form is $dN/dt = -\lambda N$, establishing the concept of the radioactive decay constant and half-life, which he determined for Th-emanation (now known to be ^{220}Ra) as 60 s (not far from today's accepted value of 55.6 s). In the process he also noticed that Th-emanation itself eventually produced another radioactive substance, which we now recognize as ^{212}Pb.

Working together between 1900 and 1903, *Rutherford and Soddy* [1903a,b] further characterized other intermediate daughter products of the U- and Th-series (as did several other workers of the time), suggested that He could be a decay product of radium, and they developed the "atomic theory of disintegration" that proposed radiation as a byproduct of "spontaneous transformation" of atoms of one element into those of another. They also delineated part of the first U- and Th-series decay series chain, and mathematically described its behavior (Fig. 1.5);

Fig. 1.3. The 1903 Nobel Prize winners Marie and Pierre Curie as depicted on the French 500-Franc note. (*See insert for color representation of the figure.*)

Fig. 1.4. The 1908 Nobel Prize winner Ernest Rutherford as depicted on the New Zealand 100-dollar note. (Source: https://commons.wikimedia.org/wiki/File:100Neuseeland-Dollar_vorderseite_21585256953_02d6c65788_o.jpg. Used under CC BY SA 3.0.). (*See insert for color representation of the figure.*)

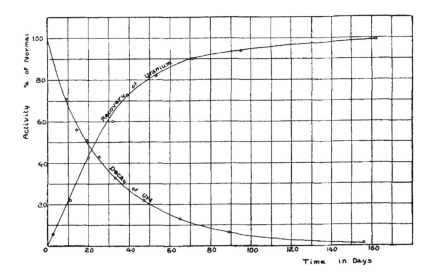

Fig. 1.5. Original figure from *Rutherford and Soddy* [1903b]. The curve showing an increase with time is the beta-activity of uranium from which a then unknown substance called uranium-X (now recognized as ^{234}Th) had been chemically removed, and the decreasing curve is the activity of the separated uranium-X. Through this experiment Rutherford and Soddy estimated the half-life of U-X (^{234}Th) as 22 days (now known to be 24.10 days) and recognized the beginning of the ^{238}U decay chain. (Source: *Rutherford* [1903]. Reproduced with permission of Taylor & Francis.)

this was followed up in a more complete way by *Harry Bateman* [1910], who laid out the differential equations and solutions for chained decay systems.

The year 1905 is sometimes referred to as the *annus mirabilis* for publication of Albert Einstein's four foundational physics papers, including the one with the famous equation relating nuclear energy, mass, and the speed of light. But it also was not a bad year for the study of radioactivity and geochronology. New Haven chemist Bertram Boltwood noted that Pb was likely a decay product of U, and in a series of lectures at the World's Fair in St. Louis and also at Yale, Rutherford presented a calculation of the first radioisotopic age. He described the calculation in a 1905 publication. Because the rate of He production directly from U was not known, but the Ra to He production was, Rutherford combined an assumption of U–Ra secular equilibrium with He and U concentrations, measured previously by William Ramsay and Morris Travers, on a sample of fergusonite, a Nb–Ta oxide, to obtain an apparent age of about 40 Ma [*Rutherford*, 1905]. In his book *Radioactive Transformations* [1906] he uses a different Ra to He production rate to obtain an age on the same sample of about 500 Ma, which he noted was likely a minimum as some of the He may have escaped. In the same work he also calculated 500 Ma for a uranium-bearing mineral from Glastonbury, Connecticut analyzed by W.F. Hillebrand. For some reason these calculations were undertaken assuming a simple linear relationship between the U (Ra) and He concentrations and Ra decay rate, even though Rutherford was aware of, and in fact derived with Soddy, the mathematical descriptions of radioactive decay and growth several years earlier.

Interestingly, in his 1906 book Rutherford also described W.F. Hillebrand's observations that U/He (though Hillebrand thought, as others prior to about 1902, that the inert gas in U-bearing samples was N_2, not He) ratios appeared to be fairly constant for "primary" minerals in certain locations but different from U/He ratios of similar minerals in other places. Given that Rutherford recognized the likelihood that some fraction of radiogenic He is lost from samples over time, the attention he pays to this observation suggests that he recognized the potential of the U–He system to represent something about regional geologic histories.

Although the accuracy, precision, and exact geologic significance of these first radioisotopic ages are not clear, their symbolic scientific importance was huge. Simple as they were, they were literally the first time humans resolved the timing of *something* in deep time using fundamental physical foundations relying only on uniformitarianism of decay constants. The ages came with a kind of cosmic insight that had been missing from centuries of "scholarly reasoning," wishful thinking, semi-quantitative heurism, and floating timescales. The fact that two almost arbitrary mineral samples from different places yielded ages about an order of magnitude older than prevailing estimates for the age of the entire Earth must have suggested to some that either the physics was missing something and the method was completely useless, or that physics had just provided something like a Promethean

lens with which to understand the Earth and universe in an entirely new and powerful way.

Also in 1905, and in rapid succession over the next few years, Robert John Strutt (later the 4th Baron Rayleigh) published a large number of analyses and calculated ages of minerals and related materials based on the relative concentrations of He, U and Th, the latter which Strutt recognized as also producing radiogenic He. Strutt noticed that (U–Th)/He ages of specimens thought to be from the same geologic stratum in different places yielded different apparent ages, and that many samples lost He at room temperatures at rates approaching, and in some cases higher than, their production rates. This He "leakage" rendered the method largely unusable for the tasks of the day, which were to establish the age of the Earth and to place reliable numerical estimates on key parts of the geologic timescale.

Around the same time, Bertram Boltwood, a prolific scientific penpal of Ernest Rutherford [*Badash*, 1969], carried out experiments showing that Pb was likely the end-product of decay of U. Using the reasoning that the number of decays of U to Ra was the same as the number of decays of Ra to Pb (i.e., secular equilibrium), Boltwood calculated an apparent decay rate of U (around 1×10^{-10} a^{-1}), and combined this with measured U/Pb ratios in a series of minerals from a variety of areas, coming up with ages ranging from 410 Ma (for a sample from Glastonbury, Connecticut) to 2.2 Ga. If Rutherford's initial ages were in essence thermochronologic ages, Boltwood's were the first real geochronologic age estimates, insofar as they came closer to estimating formation rather than potential cooling ages.

Recognizing that Pb provided a more promising daughter product than He for measuring formation ages, one of Strutt's most famous students, Arthur Holmes, began a long series of studies carefully characterizing U/Pb ratios and apparent ages of a wide variety of samples strategically chosen from various parts of the geologic timescale. Initial results, published in 1911, carried a lot of promise for the new U/Pb dating method, showing regionally and stratigraphically consistent ages (Fig. 1.6).

Somewhat surprisingly, given geologists' decades of kvetching that physicists had the age of the Earth far too young, radioisotopic geochronology did not catch on quickly in geology. Once ages of many minerals were starting to look one to two orders of magnitude older than mainstream physicists' estimates, it was the geologists who generally became skeptical of the whole approach. Many proposed variable decay rates as a most probable culprit. Two US Geological Survey geologists led the skepticism prominently. *George Becker* [1908] measured U/Pb ratios of altered uranium minerals in Texas, finding apparent ages in some cases older than 10 Ga, prompting him to reiterate earlier arguments based on oceanic sodium accumulation and terrestrial "refrigeration" (à la Kelvin) that made these new estimates essentially untenable [*Becker*, 1910]. Even as late as 1924, the famous geochemist Frank Wigglesworth Clarke (with Henry Washington) cast copious doubt on radioisotope methods because they appeared to be so discordant with earlier methods that seemed

The Association of Lead with Uranium in Rock-Minerals.

Geological period.	Pb/U.	Millions of years.
Carboniferous	0·041	340
Devonian	0·045	370
Pre-carboniferous	0·050	410
Silurian or Ordovician	0·053	430
Pre-Cambrian—		
a. Sweden {	0·125	1025
	0·155	1270
b. United States {	0·160	1310
	0·175	1435
c. Ceylon	0·20	1640

Fig. 1.6. Table of early U/Pb ages. (Source: *Holmes* [1911].)

to converge in the much more "reasonable" tens to (maximum!) hundred Ma age range.

Although dating based on radioactive decay had a long way to go before mainstream acceptance, it was pursued in the early 20th century by a number of pioneering geologists with clever ways of estimating ages from U/Pb chemical dating. In 1917, Joseph Barrell made the first real attempts at delineating the boundaries between the main geologic eras, in his poetic and prescient publication "Rhythms and the measurement of geologic time." Aside from discussing what we now call Milankovitch cycles and climatic influences on sedimentation, he came pretty darn close (especially considering the tools and data available) to the currently recognized era boundaries: Cenozoic–Mesozoic at 55–65 Ma, Mesozoic–Phanerozoic at 135–180 Ma, and Phanerozoic–Precambrian at 360–540 Ma.

Attempts to use radioactivity to hone in on the age of the Earth itself started to approach at least the right order of magnitude, for the right reasons, with Henry Russell, an astronomer, who used relative concentrations of radioactive and radiogenic elements in the Earth's crust. His best estimates came up with a maximum age of about 8 Ga from U, Th, and Pb concentrations, and a minimum of about 1.1 Ga, from the oldest U/Pb age on minerals that he considered reliable.

As the era of radioactive "chemical" dating came to a close near the end of the 1920s, the geochronologic giant Arthur Holmes redid Russell's calculations, using U, Th, and Pb concentrations in the crust to estimate a most likely age of 1–3 Ga for the Earth [*Holmes and Lawson*, 1927]. He also amassed a good deal of previous data to delineate ages for various parts of the geologic timescale, and compared estimates from various methods.

With increasing convergence of an increasing number of calculations pointing to ages in the 1–3 Ga age range, and with the official authoritative blessing of an National Research Council committee report appointed by the National Academy of Sciences in 1931, the question of the age of the Earth appeared to have been solved (though not to all geologists, by any means) to at least an order of magnitude, and probably a factor of a few.

In the middle part of the 20th century, further progress in geochronology received a huge boost from the proliferation of mass spectrometry, begetting truly radioisotopic geochronologic methods. The recognition that some elements comprised atoms with more than one mass is generally attributed to Frederick Soddy. The name isotope referred to the fact that two different types of an element occupy the same place in the periodic table, and is said to have been suggested to him by novelist and medical doctor Margaret Todd during a dinner party in Glasgow. An example of the observations behind this insight go at least as far back as Boltwood's 1906 notes that a decay product of uranium, then called ionium and now known to be ^{230}Th, was chemically identical to thorium (i.e., ^{232}Th).

Ernest Rutherford's former advisor, J.J. Thomson, is generally given credit as the first one to separate isotopes of an element by mass spectrometry, identifying ^{20}Ne and ^{22}Ne in 1913. Another student of Thomson's, Francis Aston, built a mass spectrometer for the purpose of separating nuclides based on mass-to-charge ratios, and he identified multiple isotopes of Cl, Br, and Kr, earning a Nobel Prize in 1921. By 1929 Aston had measured the isotopic composition of radiogenic Pb in a Norwegian sample of broggerite (Th-bearing uraninite) and showed it to have much higher proportions of ^{206}Pb and ^{207}Pb than common Pb. Besides attributing the ^{206}Pb to U decay, he noted that the ^{207}Pb must have come from a precursor with an atomic mass of about 231, naming this element protactinium (actinium had actually been discovered much earlier, by DeBieren in 1899). Immediately following Aston's discovery, Fenner and Piggot were the first to use radioisotopic compositions of an element to calculate ages, combining the ^{206}Pb–^{207}Pb and ^{208}Pb abundances with the U and Th contents of the same sample analyzed by Aston, to obtain apparently discordant ages of 908 and 1310 Ma for the U–Pb and Th–Pb systems, respectively.

Also in 1929, Rutherford proposed that the parent of Aston's protactinium (actino-uranium) was likely to be a uranium isotope with a mass number of 235. After estimating the decay constant of the new element, he calculated the amount of time that would be required to reduce an assumed $^{235}U/^{238}U$ value of unity (as a heuristic assumption for an initial value in the solar nebula) to the present value, which he estimated as about 0.28%. His answer, which he presumed provided a constraint on the time since the Earth separated from the Sun, was 3.4 Ga (although using the now more accurately known constants this becomes 5.9 Ga). Rutherford noted that this was approximately twice as old as previous U/Pb age determinations on any terrestrial sample. This was the first constraint on the age of the Earth from isotopic compositions. (As an aside, in the same paper, Rutherford then used a contemporary accepted age estimate for the Sun (of seven trillion years), to infer that the Sun must have been able to produce uranium at least as recently as about 4 Ga, and probably still does today...).

Dramatic improvements in mass spectrometry occurred through the 1930s and 1940s, driven largely by concerns quite different from determining the age of the Earth and its rocks, primarily nuclear physics and the Manhattan Project. Applications to Pb isotopic compositions led quickly to generally converging estimates of the age of the Earth in the 2–4 Ga range. Alfred Nier, at the University of Minnesota, measured Pb isotopic compositions of both Pb ores (low U/Pb) and uraniferous (high U/Pb) samples from a variety of locations. Besides calculating ages of many samples that supported the existence of minerals with ages older than several billion years, he proposed that Pb isotopic variations could arise from separation of materials from primordial Pb into reservoirs with a range of U/Pb ratios over time. In 1942, E.K. Gerling, who also pioneered interpretation of diffusion kinetics of noble gas thermochronometry well ahead of his time, used this approach and Nier's measurements to develop some of the first Pb-dating approaches that became widespread in subsequent decades. Using a combination of minerals dominated by radiogenic and ore Pb, he used complex but prescient arguments to establish durations of time required to generate Pb isotopic differences, including a minimum estimate of 3.94 Ga for the Earth.

Somewhat similar approaches to estimating durations required for generating Pb isotope differences were also taken, independently, by Holmes and Houtermann through the 1940s and 1950s, obtaining constraints generally pointing to 3–4 Ga ages. As technically robust as many of the analyses were, and as insightful and useful as they were for understanding common Pb behavior in the crust, they required assumptions about source homogeneity, isolation of reservoirs, and durations of "events" that separated these reservoirs that limit their ability to accurately constrain the age of the Earth beyond the prevailing convergence of approximately 3–4 Ga.

Clair Patterson's famous 1956 paper "*Age of meteorites and the Earth*," was similar but introduced a significant variation on the Pb isotope riff, introducing the age of meteorites into the question. In an elegant two-figure paper, Patterson combined primitive Pb isotopic compositions of meteoritic (Canyon Diablo) troilite with that of several other meteorites. He showed that these meteorites form a single isochron consistent with

(a)

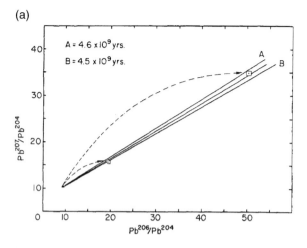

(b)

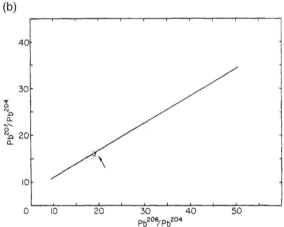

Fig. 1.7. (a) Pb-isotopic compositions of five meteorites. The least radiogenic point (near 10,10) represents troillite analyses from two different metallic meteorites; the other three points are from stony meteorites. The central line is the regression through the meteorite points; A and B represent isochrons with ages shown in legend. Curved dashed lines represent the evolution of Pb-isotopic compositions with time inferred for closed-system sources derived from the same reservoir as the troillite point. (b) The reference meteorite isochron compared with the Pb isotopic composition of oceanic sediment (unfilled circle) and a selection of recently formed galena ores (dashed field). (Source: Figures 1 and 2 from *Patterson* [1956]. Reproduced with permission of Elsevier.)

fractionation of U and Pb about 4.550 ± 0.070 Ga and closed system behavior since then (Fig. 1.7). Rather than comparing this isochron with numerous terrestrial samples, he then argued that a convenient proxy for the bulk common Pb composition of Earth could be estimated by oceanic sediment, which was rather close to many galena ores. Patterson may have been motivated to use oceanic sediment, distant from anthropogenic sources, partly because of his other work demonstrating the widespread Pb contamination of natural environments from burning of leaded gasoline. Although we now understand that even natural Pb-isotopic compositions of oceanic sediment vary more widely than the small range represented by Patterson, however fortuitous his sample choice was, the basic idea was not flawed. As long as one accepted the cogenetic nature of the Earth and the meteorites he chose for this study, and the approximation of his oceanic sediment for the Pb isotopic composition of the bulk Earth, the question of the age of the planet was at least close to settled.

Since Patterson's time, older terrestrial rocks and minerals have been discovered and dated precisely and accurately and have provided important geological understanding. Examples include early Archean units in Greenland and the Acasta gneisses of northern Canada [e.g., *Moorbath et al.*, 1975; *Bowring et al.*, 1989], and the well-known Jack Hills detrital zircons of western Australia, some of which yield concordant ages as old as 4.4 Ga [*Compston and Pidgeon*, 1986; *Wilde et al.*, 2001]. As far as determining the age of the Earth, the problem is no longer inadequate chronometers, but instead recognizing that the formation of the Earth was not an instantaneous event, not even on the scale of chronological resolution provided by radioisotopic geochronometers. The circa 4.567 Ga crystallization ages of some components of meteorites can now be determined with precisions of tens of thousands of years, but Earth growth likely took tens of millions of years. The processes involved in growing the Earth were sufficiently energetic that they continually reset the radioactive clocks in ways that are not yet well understood. As a result, most attempts to determine a singular age for the Earth, such as Patterson's, provide, at best, something approximating an average age for the interval of Earth growth. While "four and a half billion years" likely will remain a valid answer for the general age of the Earth, we can now ask this question in more detail. For example, when did Earth acquire its bulk composition, when did it form its core and atmosphere, and when did it form its first crust? Much like how the principle of superposition allowed resolution of the processes involved in the growth of sedimentary deposits, the precision obtainable with modern geochronometers is allowing the decoding of the processes involved in the growth of the Earth as a planet.

1.4 THE OBJECTIVES AND SIGNIFICANCE OF GEOCHRONOLOGY

Much of the preceding review focuses on the ability of geochronology to address ostensibly simple problems like the age of the Earth and of stratigraphic boundaries, and the history of those challenges as if they were linear pursuits with clear endings defined by sufficiently small error bars on a single number. It is true that the age of the Earth and of the punctuations in its biotic evolution are of great importance. As Martin Rudwick, and Steven J. Gould before him, suggested, the discovery of deep time and the historical evolution of the Earth and its inhabitants requires a cognitive expansion in human perspective beyond the generations, revolutions, and rises and falls of civilization, and therefore another displacement of humanity as the center of the universe, in this case with respect to time.

But as important as these questions are, geochronology is not just about determining the age of the Earth, marking precise mileposts on the geologic timescale, or even simply dating more geologic "things." If it were it would be an anachronism, minimizing its significance by its own progress—an exercise of increasingly specific, local, or minute geologic features. Simple questions posed as problems for hyperprecise dating raise the question of just how abrupt or well-defined events or processes are in the first place and so how precisely they could ever be known. Fortunately, questions that start out as easily posed usually do not end up that way. The Earth did not instantaneously appear, but was accreted over time, possibly with significant episodic mass loss, not to mention differentiation episodes; all these continue today, complicating questions of when. Similarly, stratigraphic boundaries are almost certainly diachronous to some degree, and the question of how diachronous they are may be at least as interesting as their regionally averaged age. For example, one could imagine that the K–P boundary may be diachronous over hours, whereas others defined by evidence for biostratigraphic changes that are less catastrophic may be over millions of years. The most interesting questions may have fundamental temporal aspects, but they are not solved by determining a single number with units of time.

Besides questions of instantaneousness and diachroneity, geochronology as a science in its own right comes from questions about not just dates, but also durations, rates, frequencies, and fluxes of geologic and planetary processes. Time is a fundamental characteristic of any natural process, and in many cases a timescale of some kind may be the most important part of understanding the process. For example, in struggling with the question of geochrones and stratigraphic durations, geologists of the pre-radioactivity era struggled with much more fundamental and physically enlightening questions: how long does erosion take; how long does sedimentation take? Why does it vary and what does that tell us? How long does it take to build a volcano, crystallize a pluton, or for groundwater to move through an aquifer? How steady or episodic are these processes and so over what timescales and length scales do these rates actually have a useful answer? Are these even reasonable questions with clear answers of the kind we are asking, or are these processes more complex than recognized by our simple questions? For example is groundwater transport far more complicated than can be expressed by a simple velocity, or will the attempt to answer the question lead to insights about episodicity of fluxes, mixing with ancient or multisourced fluids, dewatering and sorption with subsurface minerals, etc.? Although it is sometimes used this way, the objective of geochronology is not simply to estimate or deduce simple ages with which to label predefined geologic features as if the only thing left to understand is their birthdays. Geochronologic studies may start with simple questions, but most the power comes from harnessing the versatility of radioisotopic decay in analytically and interpretationally innovative ways. This often means resisting the temptation to declare that a geochronologic investigation "did not work." Unexpected geochronologic results are often interpreted as method failure, but the history of geochronology itself shows that these are often the most important results, leading to critical insights into the behavior of the radioisotopic systems or the geologic processes themselves.

One simple interpretation of a date is closure of the sample (a crystal, rock, fluid, etc.) to gain or loss of parent and daughter nuclides (or "daughter" damage effects). In many cases this may be reasonably argued to correspond to closure, especially for systems involving relatively immobile parent and daughter elements and single crystals or parts of crystals. Some consider this "geochronology" in the strict traditional sense—dating the age of formation of a phase. But geochronology could also refer to the collective activities of constraining other types of ages, rates, durations, and thermal histories, and processes with a key temporal aspect.

In other cases, age interpretations benefit from the consideration of open system behavior of either the parent or the daughter subsequent to formation of the sample. In most cases it is the daughter that is more easily lost, and this leads to great utility in thermochronology, U-series, and cosmogenic methods, as the chapters in this book demonstrate. Preferential loss of a daughter product is not a coincidence: typically the most useful systems are those in which the parent/daughter ratio is high, and if this is true the parent "fits" well into the crystallographic structure relative to the daughter, most likely because of its ionic radius and charge, and is therefore partitioned into it. In contrast the daughter is less welcome in the structure and more likely to be lost if opportunity or disturbance (typically thermal) arises. Therefore when we talk about closure we mostly mean cessation of loss of daughter products. But there are exceptions to this: Lu/Hf in garnet for example: although Lu partitions into garnet more than Hf, Lu has a higher diffusivity so may migrate out during high-T events.

Open-system behavior of radioisotopic systems is often associated with migration of daughter products by thermally activated diffusion (or annealing). Thermochronologic applications resulting from this have a wide range of uses in both low and high temperature settings. In some cases, geochronologic ages may be associated with neither formation nor temperature change, but other kinds of processes. Exposure or burial ages, for example, are commonly the target of cosmogenic nuclide and luminescence or ESR studies. Compositional changes, such as diagenetic uptake of parent nuclides accompanying fossilization, hydrothermal activity, or the timing of comminution are targeted by some types of U-series or ESR dating. In many cases, geochronologic approaches do not yield a simple date, but instead some other kind of temporal constraint. In many cosmogenic or low-temperature thermochronologic studies, for example, ages themselves carry little meaning other than through their relationships among samples, which can yield spatial or temporal patterns of erosion. Many U-series studies provide not dates, but minimum or maximum durations of time since material transfer, phase changes, transport, or other processes that fractionate intermediate daughter products. Some groundwater studies of He and H isotopes aim for constraints not on dates but rates of movement through underground reservoirs. Sedimentation rates have traditionally been the target of ^{230}Th excesses in deep-sea sediment.

Highly directed applications of geochronology to specific objectives have a long and successful history: e.g., determining the age of the Earth, ultraprecise stratigraphic dates, and astrochronologic calibrations. Deliberate and strategic method developments also have a distinguished track record: e.g., the engineering of ^{40}Ar/^{39}Ar dating, ^{4}He/^{3}He diffusion experiments, intercalibration of decay constants, and precise measurements of cosmogenic production rates. But some of the most important results of geochronology have been exploratory or even serendipitous, even if our professional propensity to recast our findings as resulting from carefully designed strategic plans make this hard to recognize. Few other approaches in geoscience have the ability to yield surprising results with minimal effort. Relatively straightforward techniques for measuring ages can be easily and widely applied in exploratory ways, rather than highly considered or routine conventional ways. In context, exploratory geochronology has the potential to relatively easily reveal insights that we did not know we did not know, especially when datasets are conscientiously combined, in the manner of abductive discovery advocated by *Hazen* [2014].

Discovery in geochronology also comes from exploring physical and chemical behavior of the chronometric systems. Our actual mechanistic understanding of how parent and daughter elements (or features) behave in minerals lags far behind our geologic applications. The lag is not in the basic physics of decay and decay constants, which are well known. Rather, many aspects of our understanding of the isotopic systems that we use are highly heuristic models based on relatively simple observations of complex systems. While we make many assumptions about daughter (and parent) nuclide partitioning and behavior in (usually ideal, perfect lattices of) crystals, in reality our mechanistic understanding of nuclide behavior at the atomic scale and the effects of defects are quite primitive. New discoveries of radioactive and radiogenic nuclide behavior at the atomic scale are revealing how complex these systems can be, but also what incredible archives of planetary history these complications can reveal [e.g. *Kusiak et al.*, 2015; *Valley et al.*, 2015]. Surprising but powerful insights also come from the behavior of parents and daughters in cases where interphase partitioning, intragranular media, or fluids impart unexpected behavior [e.g., *Camacho et al.*, 2005]. Innovative geochronology of the terrestrial planets, an endeavor that is evolving from analysis of accidentally launched samples [e.g. *Shuster and Weiss*, 2005; *Zhou et al.*, 2013] to more deliberate sample return and dating *in situ* [*Farley et al.*, 2014], also holds great promise for experimenting, exploring, and discovering.

This book contains chapters on many (but not all) of the tools of the trade of geochronology, a field that has become incredibly diverse and powerful since Williams proposed the term in 1893. The chapters aim to provide a blend of history, theory, nuts and bolts, and applications, all in a modern outlook that raises questions and nudges towards innovation, for the various techniques.

What ties these chapters and techniques together is the underlying question of dates and rates. As geochronology has expanded and the applications and approaches have become so varied and versatile that many modern objectives are much more nuanced than simple formation ages, the common themes of radioisotopic decay and growth (along with nucleogenic and cosmogenic production), and the desire for more precise, accurate, and innovative approaches to understanding dates and rates of natural processes have created a kind of disciplinary cohesion that underscores the fundamental importance of time in Earth and planetary science. Although the approaches and applications in these chapters may be diverse and at least superficially distinct, even besides the isotopic bases, they have in common that they are keys to discovering, quantitatively, when and how fast. To paraphrase Lord Kelvin again, when you can measure that, and express it in numbers, you know something about it. Without the temporal context of time, your knowledge is of a meagre and unsatisfactory kind.

1.5 REFERENCES

Badash, L. (1969) Rutherford and Boltwood: Letters on Radioactivity. Yale University Press, 378 pp.

Barrell, J. (1917) Rhythms and the measurements of geologic time. *Geological Society of America Bulletin* **28** (1), 745–904.

Bateman, H. (1910) The solution of a system of differential equations occurring in the theory of radioactive transformations. *Proceedings of the Cambridge Philosophical Society* **15** (part V), 423–427.

Becker, G. F. (1908) Relations of radioactivity to cosmogony and geology. *Geological Society of America Bulletin* **19** (1), 113–146.

Becker, G. F. (1910) *The Age of the Earth*, Vol. **56**. Smithsonian institution.

Bouvier, A. and Wadhwa. M. (2010) The age of the Solar System redefined by the oldest Pb–Pb age of a meteoritic inclusion. *Nature Geoscience* **3**(9), 637–641.

Bowring, S. A., Williams, I. S., and Compston, W. (1989) 3.96 Ga gneisses from the Slave Province, Northwest Territories, Canada. *Geology* **17**(11), 971–975.

Brennecka, G. A., Weyer, S., Wadhwa, M., Janney, P. E., Zipfel, J., and Anbar, A. D. (2010) ^{238}U/^{235}U variations in meteorites: extant ^{247}Cm and implications for Pb–Pb dating. *Science* **327** (5964), 449–451.

Buffon, G.-L. L. (1778) *Les Époques de la nature: 1778*. Paleo.

Burchfield, J. D. (1975) *Lord Kelvin and the Age of the Earth*. Science History, New York.

Camacho, A., Lee, J. K. W., Bastiaan J. Hensen, B. J., and Braun, J. (2005) Short-lived orogenic cycles and the eclogitization of cold crust by spasmodic hot fluids. *Nature* **435**(7046), 1191–1196.

Chamberlin, T. C. (1899) On Lord Kelvin's address on the age of the Earth as an abode fitted for life. *Annual Report*, 223–246. Smithsoniam Institution, Washington, DC.

Clarke, F. W. and Washington, H. S. (1924) *The Composition of the Earth's Crust*, Vol. **127**. US Government Printing Office, Washington, DC.

Compston, W. T. and Pidgeon, R. T. (1986) Jack Hills, evidence of more very old detrital zircons in Western Australia. *Nature* **321**(6072), 766–769.

Dalrymple, G. B. (1994) *The Age of the Earth*. Stanford University Press, 474 pp.

Dunbar, C. O. (1949) *Historical Geology*. New York, 567 pp.

Farley, K. A., Malespin, C., Mahaffy, P., *et al.* (2014) *In situ* radiometric and exposure age dating of the Martian surface. *Science* **343**(6169), 1247166.

Fleischer, R. L. and Price, P. B. (1964) Techniques for geological dating of minerals by chemical etching of fission fragment tracks. *Geochimica et Cosmochimica Acta* **28**(10), 1705–1714.

Gerling, E. K. (1942) Age of the earth according to radioactivity data. In *Doklady (Proc Russian Acad Sci)*, **34**, pp. 259–261.

Guenthner, W. R., Reiners, P. W., Ketcham, R. A., Nasdala, L., and Giester, G. (2013) Helium diffusion in natural zircon: radiation damage, anisotropy, and the interpretation of zircon (U–Th)/He thermochronology. *American Journal of Science* **313**(3), 145–198.

Halliday, A. N., Lee, D-C., Christensen, J. N., *et al.* (1998) Applications of multiple collector-ICPMS to cosmochemistry, geochemistry, and paleoceanography. *Geochimica et Cosmochimica Acta* **62**, 6 919–940.

Hazen, R. M. (2014) Data-driven abductive discovery in mineralogy. *American Mineralogist* **99**(11–12), 2165–2170.

Holmes, A. (1911) The association of lead with uranium in rock-minerals, and its application to the measurement of geological time. *Proceedings of the Royal Society of London, Series A* **85**, 248–256.

Holmes, A. and Lawson, R. W. (1927) Factors involved in the calculation of radioactive minerals. *American Journal of Science* **76**, 327–344.

Kelvin, Lord. (1863) Dynamical problem regarding elastic spheroid shell; on the rigidity of the Earth. *Philosophical Transactions of the Royal Society of London, Treatise on Natural Philosophy* **2**, 837.

Kosler, J. (2008) Laser ablation sampling strategies for concentration and isotope ratio analyses by ICP-MS. *Laser Ablation ICP-MS in the Earth Sciences: Current Practices and Outstanding Issues* **40**, 79–92.

Kusiak, M. A., Dunkley, D. J., Wirth, R., Whitehouse, M. J., Wilde, S. A., and Marquardt, K. (2015) Metallic lead nanospheres discovered in ancient zircons. *Proceedings of the National Academy of Sciences* **112**(16), 4958–4963.

Lee, D-C. and Halliday, A. N. (1995) Hafnium–tungsten chronometry and the timing of terrestrial core formation. *Geochimica et Cosmochimica Acta* **62**, 919–940.

LeConte, J. (1879) Elements of Geology: a Text-book for Colleges and for the General Reader. D. Appleton and Company, New York.

Merrihue, C. and Turner, G. (1966) Potassium-argon dating by activation with fast neutrons. *Journal of Geophysical Research* **71**(11), 2852–2857.

Moorbath, S., O'nions, R. K., and Pankhurst, R. J. (1975) The evolution of early Precambrian crustal rocks at Isua, West Greenland—geochemical and isotopic evidence. *Earth and Planetary Science Letters* **27**(2), 229–239.

Patterson, C. (1956) Age of meteorites and the Earth. *Geochimica et Cosmochimica Acta* **10**(4), 230–237.

Rudwick, M. J. S. (2014) *Earth's Deep History: How it was Discovered and Why it Matters.* University of Chicago Press.

Rutherford, E. (1905) Present problems in radioactivity. *Popular Science Monthly* **67**.

Rutherford, E. (1906) *Radioactive Transformations.* Yale University Press.

Rutherford, E. (1929) Origin of actinium and age of the Earth. *Nature* **123**, 313–314.

Rutherford, E. and Soddy, F. (1903a) Radioactive change. *Philosophical Magazine* **5**, 576–591.

Rutherford, E. and Soddy, F. (1903b) XLIII. The radioactivity of uranium. *Philosophical Magazine* **6**, 441–445.

Shuster, D. L. and Weiss, B. P. (2005) Martian surface paleotemperatures from thermochronology of meteorites. *Science* **309**(5734), 594–600.

Shuster, D. L., Flowers, R. M., and Farley, K. A. (2006) The influence of natural radiation damage on helium diffusion kinetics in apatite. *Earth and Planetary Science Letters* **249**(3) 148–161.

Stacey, F. D. (2000) Kelvin's age of the earth paradox revisited. *Journal of Geophysical Research: Solid Earth (1978–2012)* **105**(B6), 13155–13158.

Strutt, R. J. (1905) On the radio-active minerals. *Proceedings of the Royal Society of London. Series A, Containing Papers of a Mathematical and Physical Character* **76**(508), 88–101.

Strutt, R. J. (1906) On the distribution of radium in the earth's crust. *Proceedings of the Royal Society of London. Series A, Containing Papers of a Mathematical and Physical Character* **78**(522), 150–153.

Valley, J. W., Reinhard, D. A., Aaron J. Cavosie, A. J., *et al.* (2015) Presidential Address. Nano- and micro-geochronology in Hadean and Archean zircons by atom-probe tomography and SIMS: new tools for old minerals. *American Mineralogist* **100**(7), 1355–1377.

Walcott, C. D. (1893) Geologic time, as indicated by the sedimentary rocks of North America. *The Journal of Geology* **1**(7), 639–676.

Wilde, S.A., Valley, J. W., Peck, W. H., and Graham, C. M. (2001) Evidence from detrital zircons for the existence of continental crust and oceans on the Earth 4.4 Gyr ago. *Nature* **409**(6817), 175–178.

Williams, H. S. (1893) Studies for students: the elements of the geological time-scale. *The Journal of Geology* **1**(3), 283–295.

Zhou, Q., Herd, C. D. K., Yin, Q-Z., *et al.* (2013) Geochronology of the Martian meteorite Zagami revealed by U–Pb ion probe dating of accessory minerals. *Earth and Planetary Science Letters* **374**, 156–163.

Foundations of radioisotopic dating

2.1 INTRODUCTION

Radioactivity, and the geologic clock it provides, is a property of the atomic nucleus. The delineation of the fundamentals of atomic structure occurred over just a couple of decades around the transition from the 19th to 20th century [e.g., *Reed*, 2014, chapter 2]. Understanding the structure of the atom and its constituent particles was without a doubt a first-order advance in our understanding of the nature of matter. Atomic structure provided explanations for the systematic, but often mysterious, behavior of the elements that had been seen over centuries of exploitation of natural ores. While the electrons of an atom primarily control the chemical behavior of an element, the behavior of the nucleus is the foundation of radioisotope geochronology. The importance of this application of nuclear physics is reflected in the fact that the first age determination for a rock using this technique occurred only eight years after the discovery of the first atomic particle, the electron [e.g., *Radash*, 1968]. The basic understanding of the physics of the nucleus provides the tool by which geoscientists over the following century, and continuing today, are able to use naturally occurring isotopic variations in Earth and planetary materials to address a vast range of topics, including questions relating to:

the origin of the elements in the solar system;
the chronology and processes involved in planet formation;
the geologic evolution of Earth and other planets;
the rates of plate tectonics, basin subsidence, and mountain building;
absolute ages for the geologic timescale and the evolution of life;
rates of erosion and modification of the near-surface environment;
temporal changes in the composition of the atmosphere and the rise of oxygen;
paleoclimate, paleoecology, and paleogeography.

Geology is in essence the history of the Earth, and for any history dates are absolutely essential. Radioisotope geochronology provides the means to decipher the timescale and rates of all the processes that have created and modified Earth and its surface environment. Understanding the physics of the nucleus and how it leads to stable, and unstable, nuclei provides the background on both the strengths and some of the weaknesses in using radioactive decay as a chronometer.

2.2 THE DELINEATION OF NUCLEAR STRUCTURE

The first big step toward our modern understanding of atomic structure came in 1897 when Joseph (J.J.) Thomson (Fig. 2.1) discovered the electron [*Davis and Falconer*, 2005], an atomic particle characterized by a single negative electrical charge (1.60×10^{-19} coulomb) and a constant mass (9.11×10^{-28}g). Positively charged particles of considerably greater mass were known from the work of Eugen Goldstein in 1886, but these were seen to have different mass to charge ratios depending on what gas was used as a source of the particles. Goldstein's positively charged particles thus could not be characterized as a discrete particle of a constant mass and charge, analogous to the electron. The discovery of both positive and negative charged particles within an atom, however, led to the idea that an electrically neutral atom must be composed of a number of electrons, whose negative charge was balanced by a similar number of positively charged components. Ernest Rutherford in 1911 targeted a beam of high-energy positively charged particles at very thin metal foils and found that in passing through the foil, some particles were strongly deflected from their paths [*Rutherford*, 1911]. Rutherford recognized that these large scattering angles could occur only if the positively charged particles occasionally passed close to an intense positive electric field. This meant that the positive charge in the atoms in the foil must be concentrated into a very small space. Rutherford's experiments showed that all the positive charge in an atom is confined to a nucleus whose diameter is about 10,000 times smaller than that of the atom.

Geochronology and Thermochronology, First Edition. Peter W. Reiners, Richard W. Carlson, Paul R. Renne, Kari M. Cooper, Darryl E. Granger, Noah M. McLean, and Blair Schoene.

Fig. 2.1. Key players in the delineation of nuclear structure, from left to right: William Prout (By Henry Wyndham Phillips, 1820–1868 (From a miniature by Henry Wyndham Phillips)), Joseph John (J.J.) Thomson (1856–1940), Ernest Rutherford (1871–1937), and Maria Goeppert Mayer (1906–1972).

Rutherford's structural model for an atom thus has a very dense (nuclear densities of 10^{14} g/cm), small ($\sim 10^{-12}$ to 10^{-13} cm), positively charged nucleus surrounded by a diffuse cloud of negatively charged electrons that define an atomic diameter on the order of 10^{-8} cm.

The idea that different elements consist of assemblages of different integer numbers of a fundamental particle dates to well before the identification of the particles that make up the atom. Noting that the mass of many elements are integer multiples of the mass of hydrogen, William Prout in 1815 suggested that different elements reflect different numbers of hydrogen "protyle" in their constituent atoms. The detection of "hydrogen" particles released when nitrogen was bombarded with energetic positively charged particles led Rutherford in 1919 to suggest that the nuclei of all atoms did indeed contain one or more particles that have a single positive charge and a mass similar to that of the hydrogen atom. As this supported Prout's theory, Rutherford named the particle "proton". To achieve charge neutrality, the number of positively

charged protons in the nucleus is balanced by the number of negatively charged electrons that orbit the nucleus in a series of "shells" whose electron densities describe shapes that range from the spherical "s" orbitals to the dumbbell shaped lobes of "p" and "d" orbitals, as described in the quantum mechanical model of the atom developed by Niels Bohr. The number of electrons and their residence in specific orbitals is the primary feature that determines the chemical behavior of different elements.

A long-standing argument against Prout's atomic model was that not all elements have masses equal to an integer multiple of the mass of hydrogen. An answer to this valid criticism came with Thomson's discovery in 1913 that neon consists of atoms of two different masses, one with mass ≈ 20 and the other with mass ≈ 22. We now know that neon also contains a low abundance species at mass ≈ 21. Because the different mass neon atoms have essentially identical chemical properties, they must have the same number of electrons and hence the same number of protons in the nucleus. Such atomic species with identical chemical properties, but different atomic masses, were termed "isotopes" of the element by Frederick Soddy in 1913 after the Greek for "equal (iso) place (topos)," as isotopes had the same chemical properties and hence appeared in the same place in the periodic table. The discovery of isotopes required the presence in the nucleus of another, uncharged, particle of similar mass to a proton. Early experiments with radioactivity had detected electrons (β particles) released by nuclear decay, so an obvious explanation for isotopes was that the nuclei of the two isotopes consisted of different mixtures of protons plus electrons, with the mixture balanced to obtain the correct nuclear charge. Various physical properties, for example the kinetic energy, of β particles were not easily reconciled with a nuclear model of a distributed mixture of electrons and protons. These properties of β particles, however, could be explained if the nuclear electron were tightly bound to a single proton, creating an electrically neutral particle with a mass close to that of a proton. This particle, called a neutron, was eventually found by James Chadwick in 1932, who reported the discovery in a half-page letter to Nature [*Chadwick*, 1932]. The neutron was later shown to have a mass about 0.14% heavier than a proton. The delay in the discovery of the neutron reflects the fact that as an uncharged particle, neutrons cannot be accelerated, and hence separated, by electric fields. In addition, neutrons free from a nucleus decay spontaneously to a proton and electron with a mean lifetime of a bit over 880 s, so they do not stay around long enough to make their detection simple.

Although we now know that both the proton and neutron are composed of mixtures of other particles (proton = 2 up and 1 down quarks; neutron = 1 up and 2 down quarks) [*Thomas and Weise*, 2001], the model of an atom consisting of a nucleus composed of protons plus neutrons, together known as nucleons, surrounded by enough electrons to electrically balance the number of protons explains most of the first order features of all the elements in the periodic table. An individual element is thus defined by the number of protons in the nucleus. The mass (A) of any isotope of that element is the sum of the number of protons (Z) plus neutrons (N) in the nucleus. The atomic mass of an element that contains more than one isotope is then calculated from the sum of the proportion of each isotope times its mass. For example, chlorine consists of a mixture of two isotopes, both containing $Z = 17$ protons, but one with $N = 18$ neutrons (mass = 34.969 g/mol) and the other with $N = 20$ neutrons (mass = 36.966 g/mol). Isotopes conventionally are designated by the integer sum of neutrons and protons in the nucleus (A), so in the case of chlorine, the two isotopes are written as ^{35}Cl and ^{37}Cl. Naturally occurring chlorine is made up of a mixture of 75.76% ^{35}Cl and 24.23% ^{37}Cl, so the atomic mass of Cl is calculated by:

$$\text{Atomic mass (g/mol)} = 0.7576 \times 34.969 \text{ g/mol} \\ + 0.2424 \times 36.966 \text{ g/mol} \tag{2.1}$$

where a mole consists of Avogadro's number (6.0221213×10^{23}) of atoms. At 35.453 g/mol, chlorine is thus one of the elements that appears to argue against Prout's model that

every element's mass should be an integer multiple of the mass of hydrogen. This comes about because chlorine has two isotopes, and while the mass of each isotope is nearly an integer multiple of the mass of hydrogen, the mass of the element chlorine is intermediate due to it being a mixture of the two isotopes.

Because electrons are distant from, and hence relatively weakly bound to, the nucleus of their atom, they are free to exchange with nearby atoms. This is the basis of chemical reactions. Given the loosely bound nature of electrons, environmental factors, such as temperature, pressure, and the density and composition of surrounding material, readily influence the chemical reactivity of all elements. As a result, chronological techniques that rely on the chemical reactivity of elements, for instance, optically stimulated luminescence, electron spin resonance (Chapter 10), and amino acid racemization, are sensitive to the environmental conditions that a given sample has experienced. In contrast, the nucleus of an atom is protected from its surroundings by its electron cloud. For this reason, nuclear stability is to a large extent independent of environmental factors, at least up to the point where temperatures or atomic densities approach the very high values found in stellar interiors. The environmental insensitivity of nuclear decay is the primary factor that makes radioisotopic dating so useful for geochronology because the rate of nuclear decay is a constant unaffected by geologic processes. As will be seen in later chapters, the ability of radioactive decay to accurately date various geologic processes does indeed depend on the chemical behavior of the elements involved, but the rate of radioactive decay does not.

2.3 NUCLEAR STABILITY

2.3.1 Nuclear binding energy and the mass defect

An obvious question presented by Rutherford's model of the atom is the nature of the force that keeps the positively charged protons tightly bound in the nucleus rather than allowing them to fly apart due to the electrostatic repulsion of their like charges. That force is referred to as the nuclear binding energy, a residual effect of the strong nuclear force that binds quarks together in the individual nucleons (*Krane*, 1987). The binding energy present in each atom can be calculated through the energy to mass conversion that derives from Einstein's special theory of relativity:

$$E = M \times c^2 \tag{2.2}$$

where E is energy, M is mass, and c is the speed of light. The mass of a proton is 1.67243×10^{-24} g, and the speed of light is 2.9979×10^{10} cm/s. Putting these values into equation (2.2) shows that the mass of a proton corresponds to an energy of 1.503×10^{-3} ergs. An erg is not a large amount of energy, for example, the amount of energy required for a mosquito to take off is estimated at 1 erg. Binding energies are more commonly expressed as electronvolts (eV), which is the energy needed to accelerate one electron through a potential difference of 1 V: 1 eV = 1.602×10^{-12} erg.

On this scale, a proton mass corresponds to an energy of 938 million electronvolts (MeV).

Because the mass of a single atom is so small, the mass of an element or isotope is commonly expressed in units called either unified atomic mass unit (u), or Dalton (Da), that corresponds to the mass of a mole of atoms. Both scales assign the mass of the carbon isotope that has 6 neutrons and 6 protons to exactly 12 Da, or 12 g/mol. On this scale, the mass of a proton is 1.00727 Da, that of a neutron is 1.00866 Da, and an electron is 5.5×10^{-4} Da. Previously, the atomic mass scale was based on the mass of the ^{16}O isotope of oxygen being exactly 16 g/mol. This scale gave rise to the commonly used term AMU, for atomic mass unit. Although use of the ^{12}C scale is now almost universal, acceptance of "u" and "Da" as the abbreviation for the unit has been slow to catch on. As a result, many modern publications still use AMU, but use it with the mass of ^{12}C being defined as exactly 12.000 g/mol.

If the mass of carbon were simply the sum of the masses of its atomic components, then the mass of ^{12}C should be:

$$6 \times 1.00727 \,(\text{proton}) + 6 \times 1.00866 \times (\text{neutron})$$
$$+ 6 \times 0.00055 \,(\text{electron}) = 12.0989 \,\text{Da}$$

When the protons and neutrons are assembled into an atom, however, the resulting nucleus is more stable than the sum of its parts because it has turned some of the mass of the individual nucleons into nuclear binding energy. The difference in mass between the combination of its atomic constituents and that of the ^{12}C isotope, often called the mass defect, is the energy released by combining these 12 nucleons into a single nucleus. The binding energy of the ^{12}C nucleus is thus:

$$12.0989 - 12.0000 \,\text{Da} = 0.0989 \,\text{Da} \,\text{which equals} \,92.12 \,\text{MeV}$$
$$\text{or} \,7.68 \,\text{MeV per nucleon.}$$

The energy released by fusing enough hydrogen together to make one gram of ^{12}C is thus:

$$92.12 \,\text{MeV/atom} \times 6.022 \times 10^{23} \text{atoms/mol} \div 12 \,\text{g/mol}$$
$$= 4.62 \times 10^{30} \text{eV/g} \,\text{or} \,7.4 \times 10^{18} \text{ergs/g.}$$

For comparison, typical chemical reactions release of order 10^{13} ergs/g. Fusion reactions occur only at the very high atomic densities and temperatures found in the cores of stars. The much greater energy production from nuclear fusion compared to chemical reactions explains why the Sun can still be emitting so much energy, presently $\sim 4 \times 10^{33}$ erg/s, over 4.5 billion years after its formation. If solar energy were derived solely from chemical reactions, the Sun would have "burned out" long ago.

As shown in Fig. 2.2, binding energy per nucleon peaks between $A = 50$ and $A = 60$, which is in the mass range of iron and nickel. This means that nuclear fusion of light nuclei that result in a product atom near the mass of iron will release energy, whereas fusion of heavier nuclei to produce a product atom significantly higher in mass than iron requires the input of energy. In contrast, nuclear fission, or breaking apart the nucleus of a very

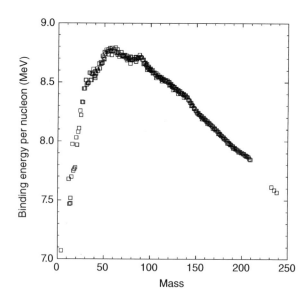

Fig. 2.2. Average binding energy per nucleon for all stable and long-lived isotopes.

heavy element (e.g. uranium) into two like-mass pieces, releases energy because both fission fragments will have nuclei with higher average binding energies than will uranium. The shape of the binding energy versus mass curve in Fig. 2.2 means that if element synthesis were taken to its lowest energy state, the universe would be composed primarily of iron and nickel. While iron and nickel indeed are more abundant in the solar system than elements of similar mass [*Lodders*, 2003], the processing of the matter in the universe is far from complete. Hydrogen and helium are still the most abundant elements in the universe by more than a factor of 10^4 over iron.

2.3.2 The liquid drop model for the nucleus

Nuclear stability is a balance between the strong nuclear force holding the nucleons together and the Coulombic repulsion of the protons due to their positive charge. George Gamow in 1928 first proposed the "liquid drop" model of the nucleus [*Stuewer*, 1997; *Basdevant et al.*, 2005, section 2.2] that treats the nucleus in an analogous way to an incompressible fluid. In this model, an empirical equation with five terms, derived by C.F. von Weizsacker in 1935, that involves only the mass (A) and number of protons (Z) in the nucleus provides a good approximation to nuclear binding energy (E) by taking into account the various forces involved in creating nuclear stability/instability:

$$E = (15.8 \,\text{MeV}) \times A - (17.8 \,\text{MeV}) \times A^{\frac{2}{3}} - (0.71 \,\text{MeV})$$
$$\times \frac{Z^2}{A^{\frac{1}{3}}} - (23.7 \,\text{MeV}) \times \frac{(A-2Z)^2}{A} \pm \frac{11.2 \,\text{MeV}}{A^{1/2}} \qquad (2.3)$$

The first three terms in this equation are concerned with the geometry of the "drop". Its volume is proportional to the total number of protons plus neutrons, A, so that its radius is proportional to $A^{1/3}$, and surface area is proportional to $A^{2/3}$. The first term, the volume term, notes that the total binding energy in the nucleus is proportional to the total number of nucleons, independent of the ratio of protons to neutrons, and reflects the short-range nature of the strong nuclear force. The second term reflects the surface energy of the "drop" and is negative because nucleons at the surface of the nucleus are not surrounded by other nucleons and thus feel less strong nuclear force from surrounding nucleons than those nucleons in the middle of the nucleus. The third term, also negative, reflects the Coulomb repulsion between protons, which will be proportional to the radius of the nucleus.

The last two terms concern the number of protons and neutrons inside the nucleus. The fourth, or asymmetry, term derives from the Pauli exclusion principle that keeps any two nucleons from occupying exactly the same quantum state in an atom. Neutrons and protons occupy different quantum states, so an equal number of protons and neutrons leads to the lowest nuclear energy state because each new nucleon added must be added to a higher energy state than the previous one. The last term, or pairing term, arises because nucleons have spin, and hence angular momentum and magnetic moments. A pair of protons, or neutrons, with opposite spin have more binding energy than a pair with the same spin. The term is written in a way that an even–even number of protons–neutrons adds to the binding energy while an odd–odd number reduces the binding energy in comparison to the baseline odd–even, or even–odd, combinations of neutrons and protons. If A is an odd number, this term is zero. The binding energies calculated as a function of number of protons and neutrons using this equation are shown in Fig. 2.3.

For an atom of any given mass, the liquid drop model equation (2.3) indicates that the binding energy is a parabolic function of the number of neutrons and protons that combine to that mass.

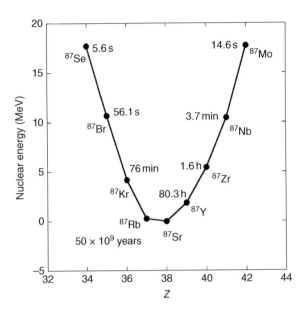

Fig. 2.4. Excess nuclear energy, in MeV, for the addition or subtraction of protons from the stable isotope at mass 87, ^{87}Sr. Each isotope is labeled along the curve with its radioactive decay half-life.

This is illustrated in Fig. 2.4 for $A = 87$. At this mass, the stable isotope is ^{87}Sr that has $Z = 38$ and $N = 49$. Swapping a neutron for a proton, or visa versa, to create an isotope with a different Z than ^{87}Sr increases the energy of the nucleus. As a result, an atom of, for example, $Z = 37$ and $N = 50$ (^{87}Rb) can reduce its nuclear energy by transforming its nucleus through radioactive decay to ^{87}Sr. The total energy released by the radioactive decay is equal to the difference in the nuclear energies of the starting and ending nucleus. In general, the farther an atom gets from the maximum nuclear binding energy for a given mass, the quicker the unstable nucleus decays to a more energetically favorable state, eventually working its way to the stable isotope of that mass, if there is one. As indicated in Fig. 2.3, the steepness of the binding energy parabolas decrease with increasing A, thus creating a broader "well" at the bottom of the parabola that allows for an increasing number of stable isotopes at any given mass as A increases.

The parabolic relationship of nuclear energy versus number of protons for any single mass isotope is the primary control over nuclear stability, but as equation (2.3) indicates, several other factors also contribute to enhancing or reducing nuclear stability. Figure 2.5 shows binding energies for elements with $A < 20$. The rapid increase with mass in binding energy per nucleon is readily apparent in these light elements, but in addition, elements with multiples of mass 4 (2 protons plus 2 neutrons, otherwise known as an α particle or ^{4}He nucleus) show an extra step in binding energy. The extra stability of the mass 4 aggregate reflects the fact that the strong nuclear force extends only over distances comparable to the "nearest neighbor" nucleon, so the grouping of 2 protons and 2 neutrons maximizes the bonding due to the strong nuclear force. The mass 4 nuclear aggregate is so stable that ^{8}Be, composed of two α particles, is unstable and

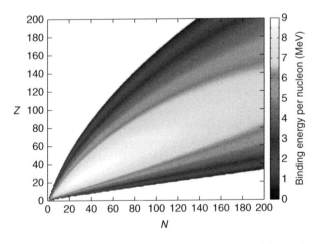

Fig. 2.3. Nuclear binding energies per nucleon as a function of the number of neutrons (N) versus protons (Z) in the nucleus. The binding energies are calculated using the Bethe–Weizsacker equation described above. (Source: Courtesy of Larry Nittler.) (*See insert for color representation of the figure.*)

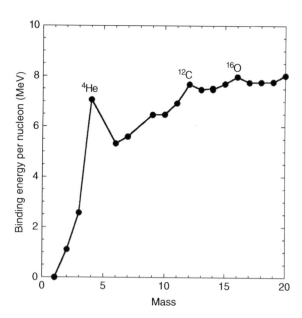

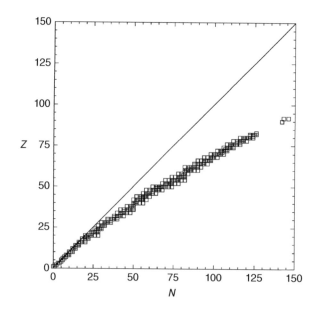

Fig. 2.5. Average binding energy per nucleon versus mass for isotopes of mass 20 or lower.

Fig. 2.6. Squares plotted for stable and long-lived isotopes comparing the number of neutrons (N) and protons (Z) in their nuclei. The line shows where $Z = N$.

decays with a half-life of about 10^{-16} seconds into two ^{4}He nuclei. The extra binding energy per nucleon at mass 8 is not yet sufficient to overwhelm the extra stability of the mass 4 aggregate, whereas it is by mass 12. Consequently, ^{12}C, ^{16}O, ^{20}Ne, ^{24}Mg, ^{28}Si, and ^{32}S are by far the most abundant isotopes in these elements, with all but ^{24}Mg constituting more than 90% of the stable isotopes of each element.

Another factor contributing to the extra stability of the ^{4}He nucleus arises from the pairing term in equation (2.3). Pairing of nucleons with opposite spins results in extra nuclear stability compared to an unpaired nucleon. For this reason, nuclei with an even number of either neutrons or protons are more stable than those with an odd number. As a result, nuclei with an even number of both protons and neutrons are abundant while isotopes with an odd number of both neutrons and protons are rare at any mass above ^{14}N. Of all the stable or long-lived isotopes, 224 are even Z, 63 are odd Z, and only 9 have both odd-Z and odd-N and of these, only 3 are found at $A > 50$.

Given the short-range nature of the strong nuclear force, the additional nuclear stability contributed by the grouping of two protons and two neutrons decreases with nuclear mass. Consequently, at higher masses, the Coulomb repulsion of the positive charge of the protons becomes a more important parameter in nuclear stability. Although the asymmetry term in equation (2.3) indicates that an equal number of protons and neutrons leads to the highest binding energy, the extra binding energy contributed by this term is overwhelmed by Coulomb repulsion as Z increases. The Coulomb repulsion acts over the whole radius of the nucleus, but is less at greater proton-to-proton distances. Below about $Z = 20$, the strong force dominates and most nuclei have an equal number of protons and neutrons due to the

asymmetry term (Fig. 2.6). Above $Z = 20$, the repulsion of protons causes the region of stable nuclei to shift to an increasing ratio of neutrons to protons. This change reflects the need to increase the nuclear radius by adding more neutrons in order to lessen the Coulomb repulsion of the protons. This competition between the strong nuclear binding force and Coulomb repulsion also explains the shape of the nuclear binding energy versus mass relationship at high Z shown in Fig. 2.2. Here, the Coulomb repulsion is sufficient to begin to reduce the binding energy per nucleon with the addition of additional protons.

2.3.3 The nuclear shell model

A final parameter that affects nuclear stability is that certain numbers of neutrons and protons provide additional stability to the nucleus. This observation was used by Maria Göppert Mayer in 1948–1950 to suggest a "shell" model for the nucleus [*Mayer*, 1950a, b], analogous to the Bohr electron shell model, where a filled nuclear shell provides additional nuclear stability. The shell model has "magic" numbers of N or Z of 2, 8, 20, 28, 50, 82, and for neutrons 126. A good example of the consequences of this effect on elements important to geochronology is seen in samarium and neodymium. Both Sm and Nd are even Z elements with 7 stable isotopes. Both have isotopes with the magic number of 82 neutrons. As a result, the lightest Nd isotope (^{142}Nd, $Z = 60$, $N = 82$) is the most abundant Nd isotope whereas for most elements in this mass range, the most abundant isotope is in the middle of the mass range for that element. Another example of the extra nuclear stability contributed by the shell model is Sm. Samarium-144 ($Z = 62$, $N = 82$) is stable, yet the next three heavier Sm isotopes are all radioactive. Stability is only reached again at ^{148}Sm.

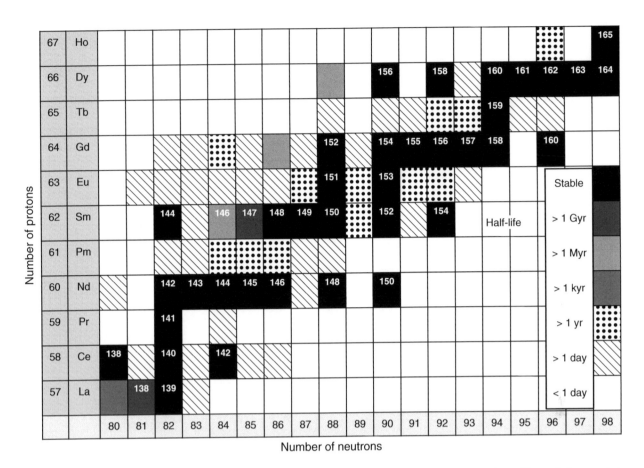

Fig. 2.7. Chart of the nuclides in the mass region of the rare earth elements. Black blocks denote isotopes with half-lives greater than 10^{12} years, dark gray are isotopes with half-lives between 10^6 and 10^{12} years, light gray are isotopes with half-lives of 1 to 10^6 years, and white blocks are isotopes with half-lives of less than a year. Numbers in the blocks give the atomic mass of the isotope. (*See insert for color representation of the figure.*)

2.3.4 Chart of the nuclides

The nuclear stability rules outlined above are well reflected in the chart of the nuclides (Fig. 2.7) that plots N on the x-axis and Z on the y-axis. A stable combination of neutrons and protons in the nucleus creates a "valley of stability", indicated by the black blocks in Fig. 2.7, within the middle of the chart of the nuclides that is coincident with the region of maximum per-nucleon nuclear binding energy shown in Fig. 2.3. For the reason depicted in Fig. 2.4, the presence of too many neutrons or protons leads to nuclei that are unstable with respect to their radioactive decay into nuclei that reside within the valley of stable nuclei. The terminology for various combinations of neutrons and protons shown on the chart of the nuclides includes:

isotopes—atoms of identical Z, but different N and A
isotones—atoms of identical N, but different Z and A
isobars—atoms of identical A, but different Z and N.

On the chart of the nuclides, isotopes define rows, isotones define columns, and isobars plot with a slope of $-45°$.

2.4 RADIOACTIVE DECAY

As shown in Fig. 2.4, an atom with an unstable number of neutrons and protons can reach a lower nuclear energy state by transforming the nucleus into a more stable combination of nucleons in the process known as radioactive decay. In radioactive decay, the radioactive isotope is often referred to as the parent isotope whereas the decay product is the daughter isotope. Depending on the mass of the radioactive nucleus, and whether it lies to the neutron-rich or neutron-poor side of the valley of nuclear stability, its transformation into a stable isotope can take many forms.

2.4.1 Fission

Given the slope of the energy per nucleon versus mass curve shown in Fig. 2.2, all nuclei heavier than about mass 100 can reach lower nuclear energy states by breaking into two fragments. This form of radioactive decay, known as fission, occurs at measureable rates only in atoms with $A > 230$. What stops lighter elements from fissioning is the need for the two fragments to

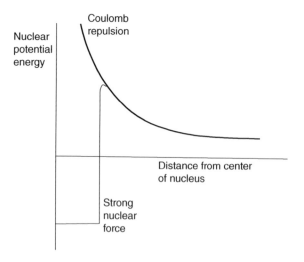

Fig. 2.8. Schematic representation of the energy balance of an atomic nucleus. Coulomb repulsion of two positively charged particles rises in proportion to the inverse of the separation distance squared, but at some distance, the attractive strong nuclear force begins to overwhelm the Coulomb repulsion. (Source: Adapted from *Friedlander et al.* [1981]. Reproduced with permission of John Wiley & Sons.)

overcome the energy barrier created by the strong nuclear force that would allow the two parts of the fissioning nucleus to get far enough apart so that Coulomb repulsive forces become dominant.

For a nucleus to be stable, one can envision a plot of nuclear potential energy versus distance from the center of the atom to look something like the curve shown in Fig. 2.8 [*Friedlander et al.*, 1981]. Two positively charged particles being brought together would encounter increasing Coulomb repulsion as their separation decreases, but at some point, the strong nuclear force must overcome the Coulomb repulsion so that the two particles can bind to form a stable nucleus. Reversing this process in order to fission, the two pieces of the nucleus would have to have enough energy to escape from the energy well created by the strong nuclear force. Given the charge and radii of two halves of a ^{238}U nucleus, the energy needed to penetrate this barrier is nearly 200 MeV. The energy released during ^{238}U fission is of order 180 MeV, so it is only through quantum mechanical effects that a uranium nucleus can occasionally "tunnel" through the energy barrier to fission. As the mass of the nucleus declines, both the Coulomb barrier and the energy released in fission decline, but the Coulomb energy declines more slowly, leading to decreasing probability of fission in lighter nuclei; of the naturally occurring elements, only ^{238}U and ^{244}Pu fission at measureable rates. Fission generally does not split the nucleus exactly in half, but instead into two daughters of unequal mass. For example, ^{238}U fission produces daughter nuclei ranging from ^{66}Mn to ^{172}Ho, but with two peaks in abundance of the daughter isotopes from $A = 87$ to 104 and $A = 132$ to 149. Because of the high neutron to proton ratio in nuclei of high mass, most of the daughter isotopes produced

by fission lie to the neutron-rich side of the valley of stability. As a result, most of the products of fission are themselves radioactive isotopes.

One way to accelerate the rate of fission is to add energy to a nucleus, for example by allowing it to capture a neutron. Neutrons can penetrate into the nucleus because they have no charge, and hence no Coulomb repulsion. Laboratory neutron-induced fission of uranium was first accomplished by Enrico Fermi in 1934, but Fermi misinterpreted the product of neutron irradiation of uranium as a heavier element. Similar experiments performed by Otto Hahn, Lise Meitner, Otto Frisch, and Fritz Strassmann proved in 1938 that one of the products of neutron irradiation of uranium was barium, an element of roughly half the mass of uranium (http://www.chemheritage.org/discover/online-resources/chemistry-in-history/themes/atomic-and-nuclear-structure/hahn-meitner-strassman.aspx).

Besides the two heavy nuclei, fission often releases α particles and free neutrons. The neutrons can then go on to induce fission in nearby uranium atoms. If the density of uranium is sufficient, a "critical mass" is achieved when the neutrons released from a fission event can induce enough additional U atoms to fission, releasing more neutrons, that the reaction becomes self-sustaining. This is the operating principle of both the fission bomb and most nuclear power plants. Nature, however, was the first to perform this experiment when at least one naturally occurring uranium deposit, Oklo in Gabon, Africa reached a sufficient uranium density to sustain a natural fission reactor. This natural reactor was in operation 1.7 billion years ago and likely ran for a few hundred thousand years. The Oklo natural reactor was discovered when it was found that the U from this mine was deficient in ^{235}U compared to most natural U, as the ^{235}U was consumed by the sustained fission [*Meshik*, 2005].

Applications of fission in geochronology include the production of xenon from the fission of U and Pu (Chapter 14), and the damage to crystals that results when the energetic fission particles rip through a crystal lattice (Chapter 10).

2.4.2 Alpha-decay

Rather than splitting into two near equal size fragments, another way for heavy elements to gain nuclear binding energy is to eject an α particle. This is a preferred decay mechanism because of the high binding energy per nucleon of the ^{4}He nucleus. The declining binding energy per nucleon versus mass at $A > 60$ means that the combined binding energy of the α particle and any nucleus with $A > 140$ that underwent α–decay would be higher per nucleon than of the predecay nuclei. As with fission, however, ejection of an α particle from the nucleus requires overcoming the strong nuclear force holding the two particles together, which leads α–decay to be a less common means of decay than expected from nuclear binding energy considerations alone. The Coulomb repulsion of a +2 charged α particle, however, is always < 13 MeV and generally ~5 MeV, compared to the > 100 MeV energies associated with fission.

The ejection of an α particle causes Z to decrease by 2 and A by 4; for example, ^{147}Sm α–decays to form ^{143}Nd. Although more common than fission, in the naturally occurring radioactive isotopes, α–decay occurs only in ^{146}Sm (Chapter 14), ^{147}Sm (Chapter 6), ^{190}Pt (Chapter 7) and Th and U (Chapter 8) and their decay chain nuclides (Chapter 12). Besides the daughter isotopes produced by α-decay, the abundance of the α particle itself, which becomes ^{4}He, is a useful chronometer (Chapter 11).

2.4.3 Beta-decay

Unlike fission and α–decay that involve interaction of the strong nuclear force and the Coulomb repulsion of two positively charged particles, β–decay reflects the transformation of a single nucleon, with the ejection of either a positive or negatively charged electron. For isotopes on the neutron-rich side of the valley of nuclear stability, β–decay involves the emission of a nuclear electron, reflecting the conversion of a neutron into a proton. This form of β–decay increases Z by 1 while reducing N by 1, with no change to A. For example, ^{87}Rb β-decays to ^{87}Sr.

An analogous decay mechanism occurs to nuclei on the proton-rich side of the valley of stability, but involves the ejection of a positron, the antimatter counterpart to the electron. Positrons have the mass of an electron, but a positive charge. Positron decay transforms a proton into a neutron and thus increases N by 1, decreases Z by 1, but keeps A constant. Both β–decay and positron-decay produce isobars of the starting atom. β-decay is the basis of many radioactive systems used in geochronology, including ^{10}Be, ^{26}Al, ^{53}Mn, ^{60}Fe, ^{87}Rb, ^{107}Pd, ^{129}I, ^{176}Lu, ^{182}Hf, ^{187}Re (Chapters 6, 7, 13, 14) and several of the members of the U and Th decay chains (Chapter 12).

2.4.4 Electron capture

A final means of radioactive decay involves the capture of an inner electron in the atom by the nucleus, transforming a proton into a neutron. Electron capture, much like positron decay, thus decreases Z by 1, increases N by 1, but leaves A unchanged. Unlike all other forms of radioactive decay, electron capture does not release a charged particle, only the energetic photon of a gamma ray if the decay occurs to an excited nuclear state of the daughter nucleus. The removal of an electron from the electron shell of the element, however, often results in emission of an X-ray photon as the atom moves an electron from an outer shell to fill the electronic state vacated by the electron consumed by the nucleus. The geochronologically most useful electron capture decay is the transformation of ^{40}K to ^{40}Ar (Chapter 9).

2.4.5 Branching decay

For a small number of isobars, the shape of the nuclear energy versus Z parabola (e.g., Fig. 2.4) allows the isotope to gain nuclear stability through more than one decay path. A good example is ^{40}K ($Z = 19$, $N = 21$), whose nuclear stability is relatively low because it has an odd number of both neutrons and

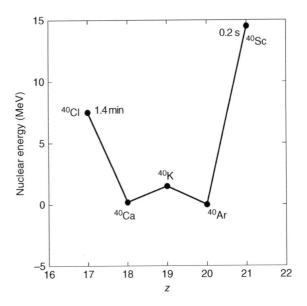

Fig. 2.9. Nuclear potential energy diagram for isobars of mass 40. Because ^{40}Ca consists of a magic number (20) of both neutrons and protons, its nuclear stability is increased enough so that it is more stable than its ^{40}K isobar. As a result, radioactive ^{40}K can decay to either ^{40}Ca or ^{40}Ar and gain nuclear stability. The decay times (half-life) of the unstable ^{40}Cl and ^{40}Sc are shown.

protons. Figure 2.9 shows the nuclear energy parabola for isobars of $A = 40$. In this mass range, ^{40}Ca ($Z = 20$, $N = 20$) is unusually stable because it has a magic number of both neutrons and protons. This feature leads the ^{40}Ca nucleus to be more stable than ^{40}K. The nucleus of ^{40}K can therefore improve its nuclear stability by decaying either to ^{40}Ar or to ^{40}Ca. Consequently, 88.8% of ^{40}K decays via β-decay to ^{40}Ca and the remainder decays to ^{40}Ar, primarily by electron capture. For similar reasons ^{138}La ($Z = 57$, $N = 81$—an odd–odd nuclei) undergoes branched decay to both ^{138}Ce ($Z = 58$, $N = 80$) and ^{138}Ba ($Z = 56$, $N = 82$), and ^{176}Lu ($Z = 71$, $N = 105$—another odd–odd nuclei) decays to both ^{176}Hf ($Z = 72$, $N = 104$) and ^{176}Yb ($Z = 70$, $N = 106$).

2.4.6 The energy of decay

Radioactivity was first discovered because the high energy of the emitted particles led them to create noticeable reactions in either photographic film or when passing through low-vacuum chambers where they could ionize the gas present, creating electric currents in the process [e.g., *L'annunziata*, 2007]. Conservation of energy requires that the energy of nuclear decay must exactly equal the total binding energy difference between the parent and daughter nuclei. In both fission and α–decay, most of this energy is transformed into the kinetic energy of the particles produced.

During nuclear fission, the energy released by splitting the nucleus in two is mostly transformed into the kinetic energy of the two fission particles, so in a ^{238}U fission event, the two fragments fly apart with a total kinetic energy approaching 180 MeV. If the fissioning U atom is contained within a crystal lattice, the energetic fission particles impact neighboring atoms, stripping at least some of their electrons, breaking chemical bonds in the process, and displacing atoms from their sites in the crystal lattice. The result is a path of crystal damage about 5 nm wide and 10–15 μm long. These damage paths, called fission tracks, can be enlarged using chemical etching so that they can be observed in an optical microscope. Counting the density of the fission tracks in comparison to the amount of uranium in a crystal is the basis of the technique of fission track geochronology, discussed in Chapter 10.

During α–decay, the total energy of the decay also is partitioned into the two fragments according to their mass, and because momentum is conserved during the decay:

$$Vn = (M\alpha / Mn) \times V\alpha \qquad (2.4)$$

V and *M* are the velocity and mass, respectively of the α particle and nucleus (*n*) that emitted the α particle. For ^{238}U, which decays to ^{234}Th by α emission, this means that the recoil velocity of the decaying ^{234}Th nuclei is 1.7% (4/234) of that of the emitted α particle.

Neither the α particle nor the recoiling residual nucleus cause enough damage to a crystal lattice to make a track that can be seen by optical microscopy after chemical etching, nevertheless, the energetic particles do damage the crystal lattice. Even at relatively low radiation doses, the crystal damage can be sufficient to affect the diffusion rates of elements through the lattice. At high radiation doses, the damage can be so severe that the crystal structure is destroyed, leading to an amorphous state called "metamict." Alpha-particle passage through some plastics, however, does create damage paths that can be chemically etched to be visible under optical microscopes. This effect can be a useful way to map the abundance of α particle emitters in a material [*Enge*, 1980].

Beta- and positron-decay result in the emission of an electron or positron whose energy is highly variable. Because energy should be conserved during a radioactive decay, the variable energy of emitted β particles and positrons indicates that at least some of the energy of β–decay must be distributed to other emissions. The most easily detected of these are gamma rays. Gamma rays due to β–decay arise from one of two mechanisms. First, when a positron encounters an electron, the two combine and annihilate one another to produce two gamma rays of energy equivalent to the rest mass of the electron. Using equation (2.2) and an electron mass of 9.11×10^{-28} g, each of the two gamma rays derived from positron-electron annihilation will produce a gamma ray of 0.51 MeV. Second, gamma rays of different, but discrete, energies are emitted when the β–decay leads to a daughter nucleus that is not in its ground energy state. Much like electron orbitals of differing energy give rise to the emission spectra of elements, quantum mechanical models for the nucleus describe the presence of discrete energy levels within the nucleus. Figure 2.10 shows how the decay to different excited nuclear states of a daughter isotope can lead to the emission of β particles and gamma rays of differing energies.

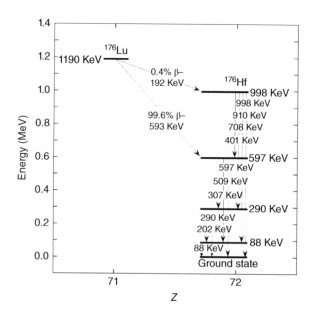

Fig. 2.10. Nuclear potential energy levels tracing one path for the β–decay of ^{176}Lu to ^{176}Hf. This decay primarily (99.6% of the time) occurs with the emission of a 593 KeV maximum energy electron, but occasionally occurs through a much lower maximum energy (192 KeV) electron emission. Both decay paths leave the ^{176}Hf daughter isotope in an excited nuclear state. To reach the ground state, the ^{176}Hf can then emit a number of different gamma rays as the nucleus moves through various excited states to its ground-state energy level. The gamma rays are of the discrete energy indicated by the difference in energy levels of the excited nuclear states. The energies listed for the β particles are maxima because a fraction of the decay energy is partitioned between electron and neutrino during β–decay.

In general, the electrons/positrons emitted by β–decay do not have discrete energies, but instead show a smooth distribution of energies with an average energy of about one-third the maximum energy (Fig. 2.11). Although apparently violating conservation of energy laws, the energy distribution of β particles instead reflects the partitioning of the decay energy between the β particle and another particle, the neutrino. With no charge and a very small mass, currently estimated at less than 0.3 eV, neutrinos interact with matter only through the weak nuclear force, allowing them to pass through matter largely unimpeded. As a result, their detection is extremely difficult, so their discovery was delayed until 1956. The type of neutrino (electron antineutrino) emitted in naturally occurring β–decay has been given the name "geoneutrino." Geoneutrinos are currently being detected using large (kiloton) liquid scintillation chambers buried underground to eliminate background sources of ionizing radiation [*Bellini et al.*, 2010]. Because a neutrino emitted by a decaying U or Th atom anywhere in Earth has a finite probability of reaching the detector, quantifying the rate of geoneutrino emission offers the potential to determine both the average abundance of U and Th in all of the Earth's interior and, at least to low spatial resolution, the distribution of these elements within Earth's interior [e.g., *Sramek et al.*, 2013].

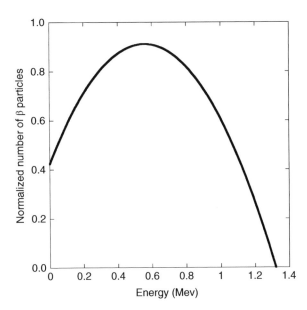

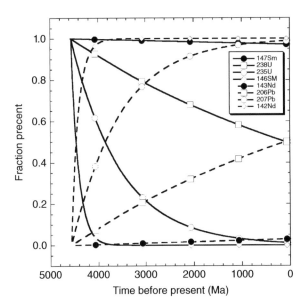

Fig. 2.11. Energy distribution of the β-particles emitted during the decay of ^{40}K. The decay has a maximum energy of 1.31 MeV, but the β-particles show a continuous distribution of energies peaking near 0.6 MeV. The energy difference is contained in the neutrino emitted by this decay. (Source: Adapted from *Cross et al.* [1983].)

Fig. 2.12. Changing relative abundance of parent (solid curves) and daughter (dashed curves) isotopes of several radioactive isotope systems used in geochronology normalized to their abundance at the time of Earth formation, 4.567 Ga (billion years ago). The lines for the daughter isotopes show only the amount of ingrowth created by the decay of the parent isotope, which is set at zero at Earth formation in this figure. Over the age of the Earth, the abundance of ^{238}U has decreased to 49% of its original abundance because the half-life of ^{238}U (4.47 Ga) is similar to the age of the Earth. In contrast, only a small fraction of radioactive isotopes with substantially longer half-lives have decayed over Earth history. For example, only about 3% of ^{147}Sm ($T_{1/2}$ = 106 Ga) has decayed over Earth history. For those isotopes with half-lives substantially shorter than the age of the Earth, most, or all, of the isotope has decayed away over Earth history. In the case of ^{235}U ($T_{1/2}$ = 0.7 Ga), only a bit over 1% of the original abundance remains, while ^{146}Sm ($T_{1/2}$ = 0.1 Ga) essentially became extinct prior to about 4.0 Ga.

2.4.7 The equations of radioactive decay

The probability that a radioactive isotope will undergo decay over any time interval is proportional to the number of atoms of the radioactive isotope present:

$$\frac{dN}{dT} = -\lambda N \tag{2.5}$$

where N is the number of atoms of the radioactive species, t is time, and λ is the decay constant for the radioactive isotope. The decay constant for a radioactive element describes its rate of decay. Many decay constants are determined by counting the number of energetic particles released by the decay of the radioactive element over some time interval, creating curves for the change in activity with time like those shown in Fig. 2.12. For radioactive isotopes with half-lives in the range of many millions or billions of years, the rate of change of radioactive decay is small over the months to years of a typical counting experiment. For these long-lived radioactive isotopes, an alternative approach for determining a decay constant is to use the radiometric system to measure an age for a rock, and then compare that age with the age determined by some other radiometric system for which the decay constant is better known. This is known as a "geologically determined" decay constant. As will be made clear in later chapters, the decay constant is a measured parameter, subject to all the usual uncertainties of any measurement. The accuracy and precision of the decay constant thus directly impacts the accuracy and precision of any age determined using radioactive decay systems, particularly when comparing ages

measured by different radiometric systems. With any single radioactive parent, however, the accuracy of the decay constant affects only the accuracy of the age determined, not the relative age difference between two rocks dated using the same radiometric system.

Integrating equation (2.5) in the form:

$$\frac{dN}{N} = -\lambda dt \tag{2.6}$$

provides the number of atoms of the radioactive species present at any time in proportion to the starting quantity of the isotope:

$$\ln(N) = -\lambda t + C \tag{2.7}$$

where C is a constant. Before any decay has occurred, N is equal to the number of atoms present when $t = 0$ (N_0), so

$$C = \ln(N_0) \tag{2.8}$$

Equation (2.6) can thus be rewritten as:

$$\ln(N/N_0) = -\lambda t \tag{2.9}$$

which is equivalent to:

$$N = N_0 e^{-\lambda t} \tag{2.10}$$

The rate of radioactive decay also is commonly expressed as "half-life", or the time needed in order for half of the starting number of atoms to experience radioactive decay. In equation form, the half-life is thus when $N/N_0 = 0.5$ so the half-life ($T_{1/2}$) is related to the decay constant by the equation:

$$T_{\frac{1}{2}} = \frac{\ln(2)}{\lambda} \tag{2.11}$$

Another useful parameter in radioactive decay is the mean-life (τ) that represents the average life expectancy of an atom of a radioactive species. The mean-life is defined as:

$$\tau = \frac{-1}{N_0} \int_{t=0}^{t=\infty} t \, dN \tag{2.12}$$

Using equation (2.6), equation (2.12) can be rewritten as:

$$\tau = \frac{1}{N_0} \int_0^\infty \lambda N t \, dt \tag{2.13}$$

and then using equation (2.10), (2.13) is rewritten as:

$$\tau = \lambda \int_0^\infty t e^{-\lambda t} \, dt \tag{2.14}$$

whose solution is:

$$\tau = - \left[\frac{\lambda t + 1}{\lambda} e^{-\lambda t} \right]_0^\infty \tag{2.15}$$

so the mean life simply is equal to the inverse of the decay constant, or:

$$\tau = \frac{1}{\lambda} \tag{2.16}$$

The amount of daughter isotope produced by radioactive decay is described simply as:

$$D = N_0 - N \tag{2.17}$$

Where the number of atoms of daughter isotope (D) produced by radioactive decay is simply the difference in the number of atoms of the parent isotope between some starting time ($t = 0$, N_0) and some later time t (N). Using equation (2.10), (2.17) can be rewritten as:

$$D = N_0 - N_0 e^{-\lambda t} \quad \text{or} \quad N_0 \left(1 - e^{-\lambda t} \right) \tag{2.18}$$

These equations assume that no atoms of the daughter isotope were present at $t = 0$, but this is not always true. If some number of daughter atoms (D_0) were present at $t = 0$, then equation (2.18) becomes:

$$D = D_0 + N_0 \left(1 - e^{-\lambda t} \right) \tag{2.19}$$

Because one often does not know the number of atoms of the radioactive species present at $t = 0$, equation (2.19) can be expressed using parameters that can be measured by using equation (2.10) to substitute N for N_0. Equation (2.19) then becomes:

$$D = D_0 + \frac{N}{e^{-\lambda t}} \left(1 - e^{-\lambda t} \right) \tag{2.20}$$

which can be rewritten as:

$$D = D_0 + N \left(e^{\lambda t} - 1 \right) \tag{2.21}$$

When the daughter element has more than one stable isotope, because isotope ratios can be measured much more precisely than the absolute abundance of any given isotope, equation (2.21) is rewritten normalizing to one of the stable isotopes of the daughter element:

$$\frac{D_r}{D_s} = \left(\frac{D_r}{D_s} \right)_0 + \frac{P}{D_s} \left(e^{\lambda t} - 1 \right) \tag{2.22}$$

where D_r is the radiogenic isotope of the daughter element, D_s is a stable isotope of that element, and P is the radioactive parent isotope. $(D_r/D_s)_0$ is the isotope ratio of the daughter element at $t = 0$. For example, in the ^{147}Sm–^{143}Nd system, D_r is ^{143}Nd, the decay product of ^{147}Sm, and D_s is ^{144}Nd, so for this system, equation (2.22) is:

$$\frac{^{143}\text{Nd}}{^{144}\text{Nd}} = \left(\frac{^{143}\text{Nd}}{^{144}\text{Nd}} \right)_0 + \frac{^{147}\text{Sm}}{^{144}\text{Nd}} \left(e^{\lambda t} - 1 \right) \tag{2.23}$$

This equation describes a straight line on a plot of the two parameters that can be measured in a sample, ^{143}Nd/^{144}Nd (y) versus ^{147}Sm/^{144}Nd (x). On such a plot (Fig. 2.13), a series of samples that formed at the same time with the same initial isotopic composition, but with different parent/daughter elemental ratios, for example different minerals that crystallized from the same magma, will define a line whose intercept with the y-axis defines the daughter isotopic composition of the magma at the time the crystals formed, and whose slope is equal to $e^{\lambda t} - 1$. Solving for "t", the time when the minerals crystallized, then gives:

$$t = \frac{\ln(\text{slope} + 1)}{\lambda} \tag{2.24}$$

This line is called an *isochron*. The isochron method is a critical approach for geochronology as it simultaneously allows determination of both the age and initial isotopic composition of the daughter element of a sample. Applications of the isochron method are described in many of the following chapters, particularly Chapters 4, 6, 7 and 14.

The equations above describe the situation where a radioactive isotope decays to a stable isotope. The U and Th decay chains, however, involve a series of radioactive isotopes that decay to other radioactive isotopes. In this case, equation (2.5) becomes:

$$\frac{dN_b}{dt} = \lambda_a N_a - \lambda_b N_b \tag{2.25}$$

where N_a is the first radioactive element in the chain that decays to N_b with a decay constant of λ_a, but N_b also is radioactive with a

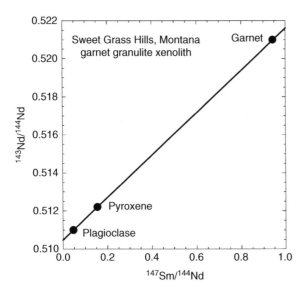

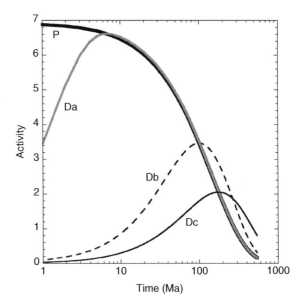

Fig. 2.13. Sm–Nd isochron diagram for minerals separated from a garnet granulite lower crustal xenolith from the Sweet Grass Hills, Montana. The three minerals in the rock formed with a relatively wide range in Sm/Nd ratios, and, as a result of radioactive decay of ^{147}Sm, now display a wide range in ^{143}Nd/^{144}Nd ratios. The three points define a line, called an isochron, that has a slope equal to 0.01115, which corresponds to an age of 1696 ± 6 Ma. The isochron intercepts the y-axis at ^{143}Nd/^{144}Nd = 0.51046 ± 0.00002, which corresponds to the isotopic composition of the Nd incorporated into these minerals when this rock formed. (Source: data from *Irving et al.* [1997].)

Fig. 2.14. Change in radioactivity with time for an initially pure radioactive parent isotope (P) with 10^8 years half-life. The parent isotope decays away at an exponential rate determined by its half-life (thick black line), but as it decays, it produces a daughter isotope that is itself radioactive. If that daughter isotope has a half-life much shorter than the parent, for example D_a (gray line) with a half-life of 10^6 years, the activity of that daughter will initially increase as the abundance of D_a increases due to production from the decay of P. Eventually, the decay rate of the daughter will equal the production rate and so its activity will stay equal to that of the parent in the condition called secular equilibrium. A similar situation is reached should the daughter isotope have a decay rate only somewhat quicker than the parent (D_b, with half-life of 10^7 years), but when the state of transient equilibrium is reached, the activity of the daughter will be higher than that of the parent as described by equation (2.30). When the daughter has a longer half-life than the parent (D_c with 1.5×10^8 years half-life), the ratio of activity of daughter to parent will continue to increase throughout the lifetime of the parent.

decay constant of λ_b. The equation accounts for the ingrowth of N_b due to production by the decay of N_a, but also the loss of N_b due to its radioactive decay to an element further down the decay chain. The solution to this differential equation is:

$$N_b = \frac{\lambda_a}{\lambda_b - \lambda_a} N_a^0 \left(e^{-\lambda_a t} - e^{-\lambda_b t} \right) + N_b^0 e^{-\lambda_b t} \qquad (2.26)$$

where N_a^0 and N_b^0 are the abundances of N_a and N_b at $t = 0$. The first term in the equation describes the production rate of N_b while the second tracks the decay of whatever N_b was present at $t = 0$.

If the half-life of the parent isotope is considerably longer than the half-life of the daughter isotope (e.g. $\lambda_a < \lambda_b$), after a sufficient time, $e^{-\lambda_b t}$ becomes small relative to $e^{-\lambda_a t}$. In this case, equation (2.26) reduces to:

$$N_b = \frac{\lambda_a}{\lambda_b - \lambda_a} N_a^0 e^{-\lambda_a t} \qquad (2.27)$$

which, because $N_a = N_a^0 e^{-\lambda_a t}$, can be further reduced to:

$$\frac{N_b}{N_a} = \frac{\lambda_a}{\lambda_b - \lambda_a} \qquad (2.28)$$

This is a state known as *transient equilibrium*. When using radioactive counting methods to determine the abundance of a radioactive isotope, the number of radioactive decays that occur

over some time interval is called the *activity* of the radioactive isotope. The activity, written as $[N_a]$, is equal to the abundance of the radioactive isotope times its decay constant, or:

$$[N_a] = \lambda_a N_a \qquad (2.29)$$

Using this definition of activity, equation (2.28) can be rewritten as:

$$[N_a] = [N_b] - N_b \lambda_a \qquad (2.30)$$

In an initially pure sample of the parent isotope (N_a), the activity of the daughter isotope (N_b) starts at zero, but grows in rapidly until it exceeds the activity of the parent, but then, when transient equilibrium is reached, the activity of N_b declines at a rate that follows the activity decline of the parent, though offset to higher activity by an amount proportional to the difference of the decay constants of elements a and b (Fig. 2.14).

In the special case where the half-life of the daughter isotope is much shorter than that of the parent isotope ($\lambda_a \ll \lambda_b$), which is true for many isotopes in the U and Th decay chains (Chapter 12), equation (2.30) reduces to:

$$[N_a] = [N_b] \qquad (2.31)$$

which describes a state known as *secular equilibrium*. In secular equilibrium, the activity of the daughter isotope is the same as that of the parent isotope. Secular equilibrium is reached after several half-lives of the daughter isotope have passed (Fig. 2.14).

2.5 NUCLEOSYNTHESIS AND ELEMENT ABUNDANCES IN THE SOLAR SYSTEM

2.5.1 Stellar nucleosynthesis

The nuclear stability rules described above provide a framework for understanding how elements are made and how our solar system ended up with the mixture of elements it contains [*Truran and Heger*, 2003]. Starting with a universe that had only hydrogen and helium, gravitational attraction of "clumps" of this gas, if in sufficient quantity, led to the formation of the first stars. If sufficiently massive, the gravitational force pulling the hydrogen in the star towards the center of mass can create pressures and temperatures sufficiently high to strip electrons from atomic nuclei, and provide enough energy to the resulting bare nuclei to overcome their Coulomb repulsion so that the nuclei can begin to interact with one another. At temperatures of about 14 million degrees kelvin, hydrogen fusion can take place. Once nuclear fusion reactions start, the energy released is sufficient to halt the gravitational collapse of the star. This creates a stable balance between gravity, which causes the star to contract, and the energy released by fusion, which drives the star to expand. At this point, the star is luminous and begins a relatively stable phase, becoming what is known as a "main sequence" star. The Sun is a main sequence star about half-way through its lifetime of stable nucleosynthesis.

The first stage of element synthesis transforms hydrogen into helium. In relatively low mass stars like the Sun, this takes place through proton–proton chain reactions that can occur in up to about the inner third of the star's radius. This sequence of reactions goes as follows:

$$^1H + {}^1H \rightarrow {}^2H + \beta^+ + \nu$$
$$^1H + {}^2H \rightarrow {}^3He + \gamma$$
$$^3He + {}^3He \rightarrow {}^4He + {}^1H + {}^1H$$

where β^+ is a positron, ν a neutrino, and γ a gamma ray, releasing about 26 MeV.

As the mass of the star increases, another sequence of reactions takes over to fuse H into He. Hans Bethe in 1938 proposed the "CNO cycle" of nucleosynthesis where ^{12}C acts as a catalyst to assemble four H atoms into one He atom [*Bethe*, 1939]. With the presence of a small amount of ^{12}C, the following sequence of reactions occurs:

$$^{12}C + {}^1H \rightarrow {}^{13}N + \gamma$$
$$^{13}N \rightarrow {}^{13}C + \beta + + \nu$$
$$^{13}C + {}^1H \rightarrow {}^{14}N + \gamma$$
$$^{14}N + {}^1H \rightarrow {}^{15}O + \gamma$$
$$^{15}O \rightarrow {}^{15}N + e^+ + \nu$$
$$^{15}N + {}^1H \rightarrow {}^{12}C + {}^4He$$

This reaction path liberates about 25 MeV of energy, but occurs only over a narrow temperature range and hence is concentrated in the cores of high mass stars, whereas the proton–proton reactions can occur in portions of the star farther from its center. A variety of other nuclear reactions also occur with H under these stellar conditions, but the CNO cycle is the most efficient at transforming hydrogen into helium. While this process is occurring, any ^{14}N present also reacts with ^4He to form ^{18}O, which can capture another ^4He to form ^{22}Ne. Elements heavier than Ne are made only in very small quantities during the hydrogen-burning phase.

Once the hydrogen is consumed in the stellar core, without the energy of fusion, the star again begins to contract due to gravity. For stars with masses less than about half that of the Sun, gravity is not sufficient to compress the stellar core to the point of initiating fusion of heavier elements. The low-mass star simply shrinks in diameter and cools to become a red dwarf. For stars of about half to ten solar masses, the gravitational contraction after hydrogen exhaustion in the stellar core brings outer, still H-rich, portions of the star to pressures and temperatures where H-burning reactions can begin in a shell surrounding the core. The energy released from these reactions causes the luminosity of the star to increase greatly and drives the outer envelope of the star to expand, increasing the star's diameter greatly, but reducing its surface temperature. The combination of these effects leads the star to be called a Red Giant [*Salaris et al.*, 2002]. In this phase, core pressures and temperatures eventually reach the point where helium fusing reactions can begin, the most prominent of which is:

$$^4He + {}^4He + {}^4He \rightarrow {}^{12}C$$

As the ^{12}C concentration builds up, it can capture a ^4He nuclei to form ^{16}O, so that by the end of the core He-burning phase the star develops a core rich in C and particularly ^{16}O. At this point, He-burning moves out into the shell surrounding the core, with H-burning reactions still taking place in an outer layer in the star. This phase of stellar evolution is known as the Asymptotic Giant Branch, or AGB star [*Lattanzio and Forestini*, 1999]. At the end of the He-burning phase, temperatures and pressures are high enough to instigate the reaction:

$$^{13}C + {}^4He \rightarrow {}^{16}O + n$$

Reactions of this type are critical because they produce free neutrons (n) that can then be captured by any heavy nuclide present to create elements of increasingly heavy mass

[*Burbidge et al.*, 1957]. When He-burning moves into the shell surrounding the core, the fusion reactions can become unstable, resulting in brief bursts of intense He-burning, called helium shell flashes, that dramatically increase the luminosity of the star for short time periods—a few hundred years—and can instigate convection within the star that dredges up the higher atomic mass material from the deep interior into shallower levels in the star. The presence of these seed heavy nuclei in neutron-rich regions in the star then leads to neutron capture reactions that can create heavier and heavier elements. At this phase of Red Giant evolution, the outer regions of the star are sufficiently cool that solids can begin to condense at the stellar surface. The dust generated by this condensation is then driven away from the star by radiation pressure, and it carries along the surrounding gas with it, creating "winds" of the newly synthesized elements that return into interstellar space eventually to be incorporated into a new star when it forms.

In stars more massive than about ten solar masses, the extremely high temperatures and pressures in their interior result in widespread nuclear fusion and hence the rapid consumption of their nuclear fuel. The lighter elements are consumed quickly in the interior of a high mass star, but there is sufficient gravity to create conditions conducive to fusion of heavier elements. High mass stars thus develop a layered structure where H-burning is occurring in an outer shell, He-burning in a shell inside that, continuing through C, O, Ne, Mg and eventually Si fusing shells. The eventual product of this heavy element fusion is iron, but as we saw in Fig. 2.2, making any element heavier than iron consumes, rather than produces, energy. Without the energy from fusion, gravity again takes over and collapses the stellar core until it reaches nuclear densities, at which point it can shrink no further. As a result, contraction is stopped, and stopped so violently that the star is disrupted in a giant explosion known as a core-collapse supernova, flinging the contents of the star back into interstellar space, eventually to be incorporated into other stars [*Smartt*, 2009].

2.5.2 Making elements heavier than iron: *s*-, *r*-, *p*-process nucleosynthesis

The main H- and He-burning phases of relatively low mass stars (<10 times the mass of the Sun) primarily turn H into He and then He into carbon. Higher mass stars continue these fusion steps up to the point where Fe and Ni are produced, but fusion reactions cannot proceed past that point as they begin to consume, rather than liberate, nuclear energy, quenching the fusion process in the interior of the star as a result. Making elements heavier than Fe generally involves neutron addition to preexisting seed nuclei. How this process occurs, and the resulting isotopes produced, depends on the rate of neutron addition, which, in turn, depends on the neutron density in various stellar environments [*Burbidge et al.*, 1957; *Meyer*, 1994].

At the neutron densities characteristic of AGB stars, free neutrons are in low enough abundance that the time between collisions of neutrons with other nuclei is long enough that unstable nuclei produced by neutron capture have time to decay back to the valley of stability on the chart of the nuclides. For example, if there are ^{43}Ca nuclei present, addition of one neutron produces stable ^{44}Ca. If ^{44}Ca captures a neutron, it produces unstable ^{45}Ca. If no additional neutron arrives before the ^{45}Ca decays, then the ^{45}Ca decays to stable ^{45}Sc, which can eventually capture a neutron to form unstable ^{46}Sc, which decays to stable ^{46}Ti, and so on, moving the atom in stair-step fashion up Z and A along the center of the valley of nuclear stability. This slow neutron capture is given the name *s*-process nucleosynthesis [*Arlandini et al.*, 1999]. At higher neutron densities, neutrons can be added to seed nuclei so rapidly that very neutron-rich, unstable, isotopes are produced. When the neutron flux is eventually reduced, the unstable nuclei will β–decay back towards the valley of stable nuclei. This rapid addition of neutrons is known as *r*-process nucleosynthesis. While originally thought to occur in core-collapse supernovae, the relative dearth of *r*-process nuclides in the galaxy suggests that the *r*-process occurs in less common events than supernovae, perhaps as a result of the collision of two neutron stars. Neutron stars are the remnants of supernovae explosions whose densities are so high that they are composed almost entirely of neutrons.

The element Sm provides a good example of the different contribution of *s*- and *r*-process nucleosynthesis to element production (Fig. 2.15). Starting, for example, with ^{140}Ce, addition of a neutron creates unstable ^{141}Ce that β-decays to ^{141}Pr. Adding a neutron to ^{141}Pr creates unstable ^{142}Pr that β-decays to stable ^{142}Nd. Sequential addition of neutrons then creates stable ^{143}Nd, ^{144}Nd, ^{145}Nd, ^{146}Nd and eventually unstable ^{147}Nd that β-decays first to unstable ^{147}Pm and again to long-lived ^{147}Sm. Sequential addition of neutrons to ^{147}Sm will produce stable ^{148}Sm, ^{149}Sm, and ^{150}Sm, ending with unstable ^{151}Sm that will β-decay to stable ^{151}Eu. This is the path of *s*-process nucleosynthesis in this mass range.

The *r*-process, in contrast, adds neutrons so fast that β-decay of unstable nuclei cannot keep up, so neutron-rich isotopes are made, moving the seed isotope out horizontally on the chart of the nuclides (Fig. 2.15). When the neutron flux is reduced, these very neutron-rich isotopes will β–decay back towards the valley of stability. For example, ^{149}Nd, ^{151}Nd, and ^{152}Nd will β-decay through their unstable Pm isobars until they reach stable ^{149}Sm, ^{151}Eu, and ^{152}Sm, respectively. In contrast, ^{144}Sm, ^{148}Sm, and ^{150}Sm are blocked from production by β-decay from the neutron-rich side of the nuclear chart by stable ^{144}Nd, ^{148}Nd, and ^{150}Nd. Of the Sm isotopes, ^{148}Sm and ^{150}Sm are thus made primarily by the *s*-process, ^{147}Sm and ^{149}Sm are made by both *s*-process and *r*-process, and ^{152}Sm and ^{154}Sm are made primarily by *r*-process nucleosynthesis.

A small number of stable isotopes exist on the proton-rich side of the valley of stability, for example ^{144}Sm (Fig. 2.15), which is stable because of its magic number (82) of neutrons. This isotope of Sm cannot be made by the *s*-process because neutron addition to Nd, starting at ^{142}Nd, continues to ^{146}Nd then to ^{147}Sm.

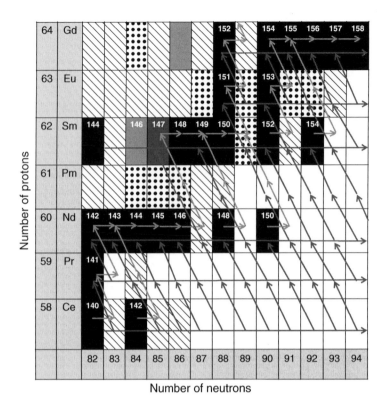

Fig. 2.15. Portion of the chart of the nuclides in the light rare earth element mass range showing the path of *s*- and *r*-process nucleosynthesis. The gray arrows show sequential neutron additions via the *s*-process followed by the β-decay that decreases *N*, and increases *Z*, by 1. The dark gray arrows show *r*-process nucleosynthetic paths that rapidly increase *N* creating a number of very neutron-rich isotopes that then β-decay back to stable isotopes. (*See insert for color representation of the figure.*)

Contributions of *r*-process nucleosynthesis to ^{144}Sm are blocked from the neutron-rich side of the nuclear chart by stable ^{144}Nd. Another synthesis mechanism thus must be sought for such proton-rich isotopes. Originally, these isotopes were believed to originate through proton capture by nuclei on the valley of stability, giving rise to the term *p*-process nucleosynthesis, but the efficiency of proton capture reactions is so low that this mechanism is unlikely to produce these isotopes in the quantity observed. Instead, proton-rich nuclei are most likely made by reactions that involve the absorption of either energetic gamma rays or neutrinos by a nuclear neutron that cause it to eject an electron and become a proton. Extremely high fluxes of both gamma rays and neutrinos are expected in the final stages of evolution of high mass stars on their way to supernova explosions, so *p*-process nucleosynthesis likely occurs in core-collapse supernovae, although the site of *p*-process nucleosynthesis remains uncertain.

2.5.3 Element abundances in the solar system

Both the stellar winds associated with AGB stars and the stellar disruption caused by supernova explosions hurl newly synthesized elements back into interstellar space. These elements eventually become incorporated into newly formed stars where the process of nucleosynthesis continues anew. While the lifetime of a low-mass star like the Sun may be as long as 10^{10} years, a high-mass star may last less than 10^7 years before it becomes a supernova. In a universe that is about 14 billion years old, the elemental composition of the galaxy and our own solar system reflects the elemental contributions from numerous stars and the likelihood that at least some fraction of the elements have been through more than one stellar nucleosynthetic cycle.

Figure 2.16 shows the element abundances in our solar system as deduced from measurements of the Sun's photosphere and from primitive meteorites [*Lodders*, 2003; *Palme and Jones*, 2003]. The shape of the abundance curve versus mass clearly reflects several aspects of nuclear stability. First, H and He are by far the most abundant elements in the solar system (note the log scale), reflecting the fact that element processing in stars still has a long way to go to reach the lowest nuclear energy state of a pure-iron universe. Elements of even-*Z* are more abundant than those with odd-*Z*, reflecting the additional stability created by nucleon paring. Carbon-12 and ^{16}O are particularly abundant, reflecting their end-product relationship in the He-burning phase of nucleosynthesis. Iron is unusually abundant amongst the

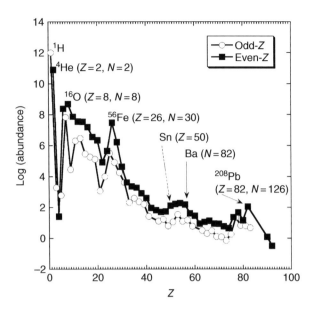

Fig. 2.16. Abundance of the elements in our solar system [*Lodders*, 2003]. Elements with an even number of protons (*Z*) are shown by filled squares, those with an odd number of protons are open circles.

heavy elements as a result of this nuclide having the highest binding energy per nucleon of any element. Element abundances decline rapidly from the iron peak with small peaks in abundance near the nuclear magic numbers of 50, 82 and neutron numbers of 126. The abundance peak at ^{208}Pb is particularly evident, as this isotope has filled nuclear shells of both protons (Z = 82) and neutrons (N = 126).

2.6 ORIGIN OF RADIOACTIVE ISOTOPES

2.6.1 Stellar contributions of naturally occurring radioactive isotopes

Besides the stable isotopes that constitute the elements in our solar system shown in Fig. 2.16, stellar nucleosynthesis also creates a vast array of radioactive isotopes. Some of these radioactive isotopes, for example ^{56}Ni (half-life of 6 days) decay so rapidly, and are so abundant, that they contribute substantially to the light output of the supernova in the days/weeks following the initial explosion. Others have long enough decay times that they survive transport through interstellar space and are incorporated into newly formed stars and planetary systems, such as our own. For those with long enough decay times, their abundance in the solar system likely reflects something of an average of all of the many stellar contributions to the region of the Milky Way from which the Sun formed. Their abundances are termed "galactic background," which nevertheless vary in different parts of the galaxy in proportion to the density of star forming regions, which, in turn reflects the mass present in a given portion of the galaxy (Fig. 2.17).

Table 2.1 lists the stellar-sourced radioactive isotopes that were present when our solar system formed. Those radioisotopes with half-lives on the shorter end of those listed in Table 2.1 may also include a contribution from the last stellar contribution to the solar nebula, enhancing their abundance over the galactic background.

2.6.2 Decay chains

The early work on radioactivity in uranium and thorium ores found that the ores contained several elements that had differing radioactive emissions, but the same chemical properties. Following this observation, T.W. Richards in 1914 reported that lead separated from uranium ores had a different atomic weight than did normal lead. With the eventual discovery of isotopes, these observations were transformed into the realization that U and Th undergo many different radioactive decays on their way to stable lead, creating a variety of elements along the way. The steps in the U and Th decay chains are summarized in Tables 2.2 through 2.4. These decay chains are unique amongst the stellar-produced radioactive isotopes where most transformations go from parent to daughter isotope in a single step. The U and Th decay chains produce a wide range of radioactive isotopes, some with half-lives long enough to be useful to track a variety of geologic processes as explored in Chapter 12.

2.6.3 Cosmogenic nuclides

As outlined in section 2.4, the extremely high pressure and temperature conditions in stellar interiors produce both nuclei and individual nucleons of sufficient energy to penetrate the Coulomb barrier of the nucleus, thus instigating nuclear reactions and creating new stable and radioactive isotopes. Outside of stars, nucleons with sufficient energy to cause nuclear reactions are found in cosmic rays, in the particles emitted by radioactive decay of other isotopes, and in man-made settings that include nuclear reactors, particle accelerators, and nuclear bombs.

Cosmic rays consist of very high energy nuclear particles, about 90% of which are single protons, 9% are α particles and 1% are nuclei of heavier elements [*Meyer*, 1969]. A small fraction of cosmic rays consist of high-energy β particles. One source of cosmic rays is the Sun. Solar cosmic rays, mostly protons (98%) and α particles, have energies generally within the range of 1–50 MeV. The flux of solar cosmic rays reaching Earth averages about 70,000 protons m^{-2} s^{-1}, but is variable throughout the 11-year solar cycle reflected in the periodic abundance of sunspots and solar flares.

Cosmic rays of much higher energy are known as galactic cosmic rays (GCR), with fluxes at 1 GeV energies of about 30,000 m^{-2} s^{-1} (Fig. 2.18) [*Amato*, 2014]. The sources of GCR are not entirely clear, but are definitely outside of the solar system, likely emitted by charged particle acceleration in the outflows of supernovae. Extremely high energy (> 10^{15} eV) GCR may originate from extragalactic sources, perhaps from active galactic nuclei

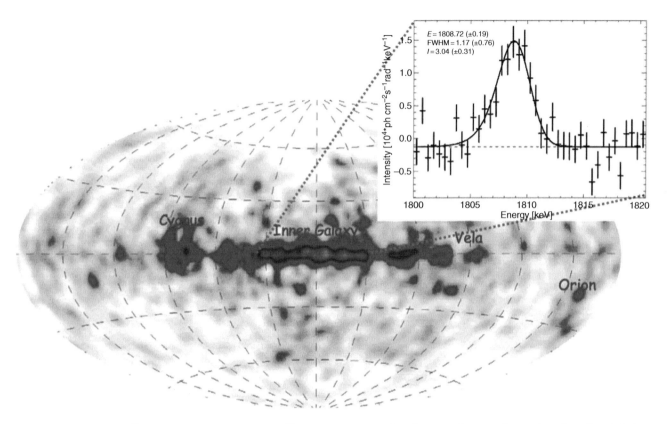

Fig. 2.17. COMPTEL map of ^{26}Al abundance and the INTEGRAL/SPI ^{26}Al gamma-ray spectrum from the inner Milky Way galaxy. Darker colors reflect the highest abundances of ^{26}Al near the galactic center and other areas of active star formation, with gray and white reflecting lower abundances at the outskirts of the galaxy. The graph in the upper left shows the distribution of gamma ray intensity versus energy. E shows the peak energy, FWHM is the "full-width, half-maximum" of the distribution of energy, and I is the standard deviation of the data. (Source: Courtesy of Roland Diehl.) (*See insert for color representation of the figure.*)

Table 2.1 Stellar-sourced radioactive isotopes

Parent isotope	Daughter isotope	Half-life (10^6 years)
^{26}Al	^{26}Mg	0.73
^{60}Fe	^{60}Ni	1.5
^{53}Mn	^{53}Cr	3.7
^{107}Pd	^{107}Ag	6.5
^{182}Hf	^{182}W	9
^{129}I	^{129}Xe	15.7
^{146}Sm	^{142}Nd	68
^{244}Pu	Fission Xe	80
^{235}U	^{207}Pb	704
^{40}K	^{40}Ar, ^{40}Ca	1270
^{238}U	^{206}Pb	4469
^{232}Th	^{208}Pb	14,010
^{176}Lu	^{176}Hf	37,100
^{187}Re	^{187}Os	41,600
^{87}Rb	^{87}Sr	49,500
^{147}Sm	^{143}Nd	106,000
^{190}Pt	^{186}Os	450,000

Table 2.2 ^{238}U decay chain

Parent	Decay mode	Half-life	Daughter
^{238}U	α	4.51×10^9 years	^{234}Th
^{234}Th	β	24.1 days	^{234}Pa
^{234}Pa	β	1.2 min, and 6.7 h	^{234}U
^{234}U	α	2.48×10^5 years	^{230}Th
^{230}Th	α	7.52×10^4 years	^{226}Ra
^{226}Ra	α	1622 years	^{222}Rn
^{222}Rn	α	3.82 days	^{218}Po
^{218}Po	α (99.98%), β	3.05 min	^{214}Pb, ^{218}At
^{218}At	α	1.3 s	^{214}Bi
^{214}Pb	β	26.8 min	^{214}Bi
^{214}Bi	β (99.96%), α	19.7 min	^{214}Po, ^{210}Tl
^{214}Po	α	10^{-4} s	^{210}Pb
^{210}Tl	β	1.32 min	^{210}Pb
^{210}Pb	β, α (10^{-6} %)	22 years	^{210}Bi, ^{206}Hg
^{210}Bi	β, α (5×10^{-5} %)	5 days	^{210}Po, ^{206}Tl
^{206}Hg	β	8.5 min	^{206}Tl
^{210}Po	α	138 days	^{206}Pb
^{206}Tl	β	4.3 min	^{206}Pb

that have central black holes with sufficient magnetic field strength to accelerate charged particles to such high energies. Their flux is more than 16 orders of magnitude lower than the peak GCR flux at ~300 MeV. The majority of solar cosmic rays are of insufficient energy to penetrate Earth's magnetic field.

Galactic cosmic rays can penetrate Earth's magnetic field, but the majority of their primary particles interact with atoms in the upper atmosphere to make a cascade of secondary particles including protons, neutrons, and muons that can reach Earth's surface.

Table 2.3 ^{235}U decay chain

Parent	Decay Mode	Half-life	Daughter
^{235}U	α	7.13×10^8 years	^{231}Th
^{231}Th	β	25.6 h	^{231}Pa
^{231}Pa	α	3.48×10^4 years	^{227}Ac
^{227}Ac	β (98.8%), α	22 years	^{227}Th, ^{223}Fr
^{227}Th	α	18.2 days	^{223}Ra
^{223}Fr	β, α (4×10^{-3} %)	22 min	^{223}Ra, ^{219}At
^{223}Ra	α	11.7 days	^{219}Rn
^{219}At	α (97%), β	0.9 min	^{215}Bi, ^{219}Rn
^{219}Rn	α	3.9 s	^{215}Po
^{215}Bi	β	8 min	^{215}Po
^{215}Po	α, β (5×10^{-4} %)	0.0018 s	^{211}Pb, ^{215}At
^{211}Pb	β	36.1 min	^{211}Bi
^{215}At	α	0.0001 s	^{211}Bi
^{211}Bi	α (99.68%), β	2.15 min	^{207}Tl, ^{211}Po
^{207}Tl	β	4.8 min	^{207}Pb
^{211}Po	α	0.5 s	^{207}Pb

Table 2.4 ^{232}Th decay chain

Parent	Decay mode	Half-life	Daughter
^{232}Th	α	1.39×10^{10} years	^{228}Ra
^{228}Ra	β	6.7 years	^{228}Ac
^{228}Ac	β	6.1 h	^{228}Th
^{228}Th	α	1.9 years	^{224}Ra
^{224}Ra	α	3.6 days	^{220}Rn
^{220}Rn	α	54.5 s	^{216}Po
^{216}Po	α	0.16 s	^{212}Pb
^{212}Pb	β	10.6 h	^{212}Bi
^{212}Bi	β (66.3%), α	60.6 min	^{212}Po, ^{208}Tl
^{212}Po	α	3×10^{-7} s	^{208}Pb
^{208}Tl	β	3.1 min	^{208}Pb

Table 2.5 Cosmic ray produced radioactive isotopes

Isotope	Decay mode[a]	Half-life	Daughter
^{3}H	β–	12.3 years	^{3}He
^{3}He		Stable	
^{7}Be	ec	53.3 days	^{7}Li
^{10}Be	β–	1.39×10^6 years	^{10}B
^{14}C	β–	5730 years	^{14}N
^{21}Ne		Stable	
^{22}Na	β+	2.6 years	^{22}Ne
^{26}Al	β+	730,000 years	^{26}Mg
^{32}Si	β–	101 years	^{32}P
^{32}P	β–	14.3 days	^{32}S
^{35}S	β–	87.5 days	^{35}Cl
^{36}Cl	β–	301,000 years	^{36}Ar
^{39}Ar	β–	269 years	^{39}K
^{41}Ca	ec	130,000 years	^{41}K
^{53}Mn	ec	3.8×10^6 years	^{53}Cr
^{60}Fe	β–	10^5 years	^{60}Ni
^{81}Kr	ec	210,000 years	^{81}Br
^{85}Kr	β–	10.7 years	^{85}Rb
^{129}I	β–	15.9×10^6 years	^{129}Xe

[a] β+ positron, β– electron, ec – electron capture.

With energies in the MeV range and beyond, cosmic rays can penetrate the Coulomb barrier of the nucleus, fragmenting the impacted nucleus in the process known as spallation. Spallation leads to production of a number of radioactive and stable nuclei that are either small fragments of the impacted nucleus (e.g., ^{3}He) or atoms with slightly lower atomic mass than the target nuclide. Another product of such violent collisions are free neutrons and muons. Muons are elementary particles similar to an electron that have a single negative charge, but a much larger mass of about 106 MeV. Both the neutrons and muons can then go on to additional reactions with the nuclei of nearby atoms. Primary target atoms for spallation by cosmic rays are N, O, Ar, and C in the atmosphere and C, O, Si, Mg, Al, and Fe in rocks. Table 2.5 provides a list of the most useful isotopes produced by cosmic ray interactions. Most of these isotopes are produced directly by spallation. The exceptions include ^{14}C that is produced by capture of a cosmic-ray-produced secondary neutron by ^{14}N, which emits a proton to become ^{14}C. Tritium (^{3}H) is produced when ^{14}N captures an energetic secondary neutron to split into ^{12}C and ^{3}H. Isotopes produced by interaction with cosmic rays are termed "cosmogenic" isotopes and can be either radioactive or stable isotopes. The many and varied applications of cosmogenically produced isotopes in geochronology are described in Chapter 13.

2.6.4 Nucleogenic isotopes

The α–decay of ^{147}Sm and particularly U, Th, and their decay-chain daughter isotopes produces α particles of sufficient energy to react with the nuclei of surrounding atoms. Isotopes produced through this route are termed "nucleogenic" [*Wetherill*, 1954]. The interaction of the α particles with surrounding atoms also can produce free neutrons, which can then go on to instigate additional nuclear transformations. Some important nucleogenic isotopes, and the reactions that produce them, include:

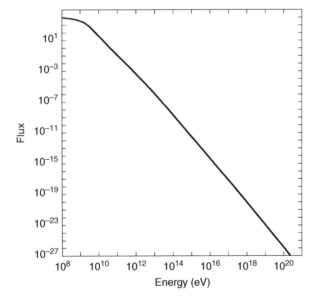

Fig. 2.18. Flux of cosmic rays versus energy. Flux is provided in units of particles per square meter per second. The lowest energy particles ($<10^{10}$ eV)) are mostly solar cosmic rays, intermediate energy particles (10^{10} to 10^{15} eV) are galactic cosmic rays and the highest energy region is attributed to extragalactic cosmic rays. (Source: Lafebre, https://commons.wikimedia.org/wiki/File: Cosmic_ray_flux_versus_particle_energy.svg#. Used under CC BY-SA 3.0 https://creativecommons.org/licenses/by-sa/3.0/deed.en)

^{6}Li $+ n \rightarrow$ ^{3}H $+ \alpha$: ^{3}H then β^{-} decays to ^{3}He with a half-life of 12.3 years

^{17}O $+ \alpha \rightarrow$ ^{20}Ne $+ n$

^{18}O $+ \alpha \rightarrow$ ^{21}Ne $+ n$

^{19}F $+ \alpha \rightarrow$ ^{22}Na $+ n$: ^{22}Na then β^{+} decays to ^{22}Ne with a half-life of 2.6 years

^{39}K $+ n \rightarrow$ ^{39}Ar $+ p$: ^{39}Ar then β^{-} decays back to ^{39}K with a half-life of 269 years

where n is a neutron, p a proton, and α an α particle. Another common way to write such reactions is, for example:

$$^{39}\text{K}\,(n, p)\,^{39}\text{Ar}$$

which signifies that neutron capture by ^{39}K will produce ^{39}Ar and a proton. Nucleogenic isotopes have not seen much use in geochronology, though recent attempts to use the (U–Th)/Ne system as a chronometer show promise [*Cox et al.*, 2015].

2.6.5 Man-made radioactive isotopes

By exploiting many of the same reaction paths described earlier in this chapter, particle accelerators, fission reactors, and nuclear bombs can produce a vast array of radioactive isotopes. For example, after the Fukushima Daiichi nuclear power plant was damaged by the tsunami generated by the 2011 Tohoku earthquake, a wide variety of fission-produced radionuclides were released into the environment. These included ^{90}Sr, ^{99}Tc, ^{129}Te, ^{131}I, ^{134}Cs, ^{136}Cs, ^{137}Cs, ^{140}Ba and ^{140}La. A general background level of ^{239}Pu and ^{240}Pu is found everywhere on Earth as a result of fallout from nuclear bomb testing in the 1950s and 1960s. Another product of atmospheric nuclear bomb testing was a big spike in the abundance of ^{14}C, in the amount of several tons, almost doubling the amount of ^{14}C in the atmosphere for a short time.

2.7 CONCLUSIONS

The properties of the nucleus allow us to address first-order questions ranging from the energy source of stars, the composition of the solar system, and the history of Earth. The production of radioactive isotopes in stars, radioactive decay chains, cosmic ray interactions, or nuclear reactors creates many isotopic clocks. The range in decay half-lives of these radioactive species provides chronological information over time periods from the age of the universe to the age of a human artifact. The availability of radioactive isotopes of elements with very diverse chemical behavior provides chronometers for a wide range of geochemical processes from igneous differentiation, volatile loss at a planetary scale, core formation, deposition of sediments, uplift of mountains or burial of sediments, mixing in the oceanic water column, the age of groundwater, as well as a vast array of applications in archeology and paleoecology. In the end, radioactive isotope geochronology provides the chronometer for Earth history. The remaining chapters in this book describe the many and varied ways in which radioactive isotopes can be used to decipher Earth history.

2.8 REFERENCES

Amato, E. (2014) The origin of galactic cosmic rays. *International Journal of Modern Physics: Conference Series* **1**. DOI: 10.1142/S0218271814300134.

Arlandini, C., Kappeler, F., Wisshak, K., *et al.* (1999) Neutron capture in low-mass asymptotic giant branch stars: cross sections and abundance signatures. *The Astrophysical Journal* **525**, 886–900.

Basdevant, J.-L., Rich, J., and Spiro, M. (2005) *Fundamentals in Nuclear Physics: From Nuclear Structure to Cosmology*. Springer, Berlin.

Bellini, G., Benziger, J., Bonetti, S., *et al.* (2010) Observation of geo-neutrinos. *Physics Letters* **B687**, 299–304.

Bethe, H. A. (1939) Energy production in stars. *Physical Review* **55**, 434–456.

Burbidge, E. M., Burbidge, G. R., Fowler, W. A., and Hoyle, F. (1957) Synthesis of the elements in stars. *Reviews of Modern Physics* **29**, 547–654.

Chadwick, J. (1932) Possible existence of a neutron. *Nature* **129**, 312.

Cox, S. E., Farley K. A., and Cherniak, D. J. (2015) Direct measurement of neon production rates by (α,n) reactions in minerals. *Geochimica et Cosmochimica Acta* **148**, 130–144.

Cross, W. G., Ing H., and Freedman, N. (1983) A short atlas of beta-ray spectra. *Physics in Medicine & Biology* **28**, 1251–1260.

Davis, E. A. and Falconer, I. J. (2005) *J.J. Thomson and the Discovery of the Electron*. Taylor & Francis, London.

Enge, W. (1980) Introduction to plastic nuclear track detectors. *Nuclear Tracks* **4**, 283–308.

Friedlander, G., Kennedy, J. W., Macias, E. S., and Miller, J. M. (1981) *Nuclear and Radiochemistry*. Wiley.

Irving, A. J., Kuehner, S., and Carlson R. W. (1997) 1.70 Ga Sm–Nd age for a garnet granulite xenolith from a minette sille, Sweetgrass Hills, Northern Montana supports Proterozoic collision of Hearne and Wyoming cratons. *EOS, Transactions of the American Geophysical Union* **78**, 786.

Krane, K. S. (1987) *Introductory Nuclear Physics*. Wiley.

L'Annunziata, M. F. (2007) *Radioactivity: Introduction and History*. Elsevier, Amsterdam.

Lattanzio, J. and Forestini, M. (1999) Nucleosynthesis in AGB stars. In *Asymptotic Giant Branch Stars*, Le Bertre, T., Lebre, A., and Waelkens, C. (eds). *Publications of the Astronomical Society of the Pacific* **111**, 125–126.

Lodders, K. (2003) Solar system abundances and condensation temperatures of the elements. *Astrophysical Journal* **591**, 1220–1247.

Mayer, M. G. (1950a) Nuclear configurations in the spin–orbit coupling model. I. Empirical evidence. *Physical Review* **78**, 16–21.

Mayer, M. G. (1950b) Nuclear configurations in the spin–orbit coupling model. II. Theoretical considerations. *Physical Review* **78**, 22–23.

Meshik, A. P. (2005) The workings of an ancient nuclear reactor. *Scientific American* **293**, 82–91.

Meyer, B. S. (1994) The *r*-, *s*- and *p*-process in nucleosynthesis. *Annual Review of Astronomy and Astrophysics* **32**, 153–190.

Meyer, P. (1969) Cosmic rays in the galaxy. *Annual Review of Astronomy and Astrophysics* **7**, 1–38.

Palme, H. and Jones, A. (2003) Solar system abundances of the elements. In *Treatise on Geochemistry*, Davis, A. M., (ed). Elsevier, Amsterdam.

Radash, L. (1968) Rutherford, Boltwood, and the age of the Earth: The origin of radioactive dating techniques. *Proceedings, American Philosophical Society* **112**, 157–169.

Reed, B. C. (2014) *The History and Science of the Manhatten Project.* Springer-Verlag, Berlin.

Rutherford, E. (1911) The scattering of alpha and beta rays by matter and the structure of the atom. *Philosophical Magazine* **6**, 21–47.

Salaris, M., Cassisi, S., and Weiss, A. (2002) Red giant branch stars: the theoretical framework. *Publications of the Astronomical Society of the Pacific* **114**, 375–402.

Smartt, S. J. (2009) Progenitors of core-collapse supernovae. *Annual Review of Astronomy and Astrophysics* **47**, 63–106.

Sramek, O., McDonough, W. F., Kite, E. S., Lekic, V., Dye, S. T., and Zhong, S. (2013) Geophysical and geochemical constraints on geoneutrino fluxes from Earth's mantle. *Earth and Planetary Science Letters* **361**, 356–366.

Stuewer, R. H. (1997) Gamow, alpha decay, and the liquid-drop model of the nucleus. *Astronomical Society of the Pacific Conference Series* **129**, 30–43.

Thomas, A. W. and Weise, W. (2001) *The Structure of the Nucleon.* Wiley.

Truran, J. W. J. and Heger, A. (2003) Origin of the elements. In: *Treatise on Geochemistry*, Davis, A. M., (ed). Elsevier, Amsterdam.

Wetherill, G. W. (1954) Variations in the isotopic abundances of neon and argon extracted from radioactive minerals. *Physical Review* **96**, 679–683.

CHAPTER 3

Analytical methods

3.1 INTRODUCTION

Geochronological techniques rely on the application of a wide variety of analytical chemistry approaches to the analysis of natural materials. Although a few techniques can proceed directly from whole-rock or thin-section samples, most procedures start with some form of sample preparation that can range from crushing a rock to powder to performing a variety of mineral separation procedures to isolate from the sample only a single, or multiple, minerals of interest. Once the sample is ready, the element to be analyzed is extracted from the sample. This can be done nonselectively through sputtering the sample surface with high-energy ions or lasers, through more selective methods that involve heating the sample to release volatile elements, or by dissolving the sample and isolating the element of interest through a variety of classical wet-chemistry techniques. Once the element to be analyzed has been liberated from its rock or mineral matrix, some form of mass spectrometer is employed to determine the isotopic composition of the element, and through the technique of isotope dilution, the concentration of the element in the sample. In this chapter, we outline some of the general analytical approaches used in geochronology, with the more technique-specific procedures described in the appropriate later chapters. We note that several radiation-damage-based techniques, including electron spin resonance (ESR), optically stimulated luminescence/thermoluminescence (OSL/TL), and fission-track dating, are important geochronologic tools useful in a wide range of applications, as discussed in Chapter 10, but involve considerably different analytical procedures and considerations than we discuss here. Some of these are highlighted in Chapter 10; here we focus primarily on techniques requiring parent and daughter nuclide quantification through elemental and isotopic analysis.

3.2 SAMPLE PREPARATION

The diversity of analytical techniques used in geochronology lead to a wide range of sample requirements, and hence sample preparation procedures. Some techniques, such as laser-ablation or secondary ion mass spectrometry (SIMS or ion probe) U–Pb dating of zircon can be applied to thin slices of rock prepared in such a way as to allow optical examination with a petrographic microscope, chemical analysis with a variety of electron-beam instruments such as scanning-electron microsopes (SEMs) and electron probes, followed by isotopic analysis by "drilling" into select minerals with either an ion-beam or laser and injection of the liberated material into a mass spectrometer. This approach allows the spatial relationships of the analyzed minerals in the rock to be preserved and observed, though it can complicate aspects of a geochronological analysis, for example, the precision of the isotope ratio measurements. Most other techniques require manual isolation of select minerals from the rock, while still other techniques are applied at the "whole rock" scale, which usually means analyzing an aliquot of finely ground powder prepared by crushing many tens or hundreds of grams of rock.

The first step of any geochronological analysis involves selecting the sample. The first guiding principle here is to choose a sample that best matches the requirement of the geochronometer to be used. If the goal is to determine the crystallization age of an igneous rock, then choosing a sample that has suffered minimal alteration is a good start. On the other hand, if the timing of the alteration is the more important question, then choosing a highly altered sample could be the best choice. The second choice is sample size. Again, the starting sample size depends on the requirements of the analysis. For whole-rock analyses, coarse-grained rocks require a larger sample to adequately average the mineral constituents in the rock compared to fine-grained rocks. The problem becomes more severe if the element of interest is strongly concentrated into a rare phase in the rock [*Clifton et al.*, 1969], which is a particular problem for Re and Os [*Reisberg and Meisel*, 2002]. If the target of the analysis is separated minerals, the initial sample size is chosen to yield enough of the mineral(s) needed for analysis, recognizing that the mineral separation procedures, to be described later, do not allow for perfect preservation and separation of all the mineral grains in the

Geochronology and Thermochronology, First Edition. Peter W. Reiners, Richard W. Carlson, Paul R. Renne, Kari M. Cooper, Darryl E. Granger, Noah M. McLean, and Blair Schoene.

rock. For most modern applications, a typical granitoid or compositionally similar crystalline rock, or an immature clastic sedimentary rock like a sandstone, a sample of roughly 2–4 kg in mass will yield thousands of zircon grains for analysis. Prior to the development of techniques capable of single zircon analysis, however, the sample sizes needed to obtain enough zircon for a single analysis sometimes ranged into the tens to hundreds of kilograms [*Silver and Deutsch*, 1963].

Once the sample is chosen, it must be prepared for analysis. As described in the introduction, sometimes this simply entails making a thin section of the rock using a rock saw and polishing equipment. Most of the time, however, the first step involves some form of crushing. In its lowest technological form, crushing involves a hammer, but automated alternatives include jaw crushers, to make millimeter- to centimeter-size chips, disk mills to break the rock to the grain size of individual crystals, and ball or ring mills to reduce the rock to submicron powder. The hardness of most rocks and minerals can result in significant erosion of the crushing apparatus, so care must be taken to choose crushing apparatus that will not contaminate the elements to be analyzed. This is mostly a problem when preparing whole-rock powders for analysis. Tungsten-carbide is a particularly effective material for crushing apparatus, but causes significant contamination of W, usually an ultra-trace element in most rocks, as well as trace elements such as Cr and Ta. Steel crushing apparatus can cause significant contamination of Re and Os. Particularly when making whole-rock powders, crushing apparatus made from agate (SiO_2) or ceramic (mostly Al_2O_3) are commonly used. Both these materials introduce minimal contamination of most trace elements, but obviously can affect Si and Al analysis. Crushing clean quartz and then comparing a trace element analysis of the quartz before and after crushing is an effective way to determine the amount of contamination introduced by crushing. Some newer methods use pulsed high-voltage to fracture minerals preferentially along grain boundaries.

Many geochronologic techniques require aliquots of single, or in some cases multiple, grains or crystals of a certain type of mineral. Examples include U/Pb dating of zircon, $^{40}Ar/^{39}Ar$ dating of micas or feldspar, construction of mineral isochrons using any of the Sr, Nd, Pb, or Hf systems, OSL/ESR, fission-track, or (U–Th)/He systems. Preparing relatively pure mineral separates from whole-rock samples can be time- and cost-intensive, sometimes dangerous, and mostly unglamorous, but because it is a first step for so many different types of geochronologic analyses we review some of the basic and most common approaches here. Once liberated from their surrounding minerals by crushing and sieving to obtain a grain-size fraction small enough to eliminate most intergrown grains, minerals can be separated from one another by hand-picking, density, or magnetic methods, and often by a combination of all three. Hand-picking involves a pair of tweezers or a needle for smaller grains, a binocular microscope, steady hands, and patience. Density separations include hydrodynamic methods that combine flowing fluids, usually water, mechanical agitation, and gravity. A commonly used tool for this is the Wilfley Table—essentially a large, semi-automated gold panning device. A more efficient density separation, though generally applicable only to smaller sample sizes, involves dense organic liquids such as methylene iodide (3.32 g/cm^3), acetylene tetrabomide (2.97 g/cm^3), and bromoform (2.89 g/cm^3). These liquids pose serious inhalation/exposure health hazards, requiring the separations to be done in fume hoods. Although previously commonly used, the densest of these liquids, thalium malonate formate, also called "Clerici solution" (4.25 g/cm^3), is highly toxic and hence is now rarely used. Increasingly, the dense organic liquids are being replaced by sodium-polytungstate (3.10 g/cm^3) or lithium-polytungstate (2.82 g/cm^3). Both are water soluble, but Na-polytungstate can be quite viscous, which inhibits the density separation of grains placed in this liquid. A common approach for this separation involves adding some of the liquid to a separatory funnel followed by some amount of the crushed, sieved, rock, which is stirred into the liquid (Fig. 3.1); minerals of higher density than the liquid will sink, while those less dense will float. The dense minerals are extracted by opening the stopcock at the bottom of the funnel and allowing the liquid to flow through a filter paper to trap the dense grains.

Separation by magnetic properties starts with a hand magnet for ferromagnetic minerals, followed by a device such as a "Frantz" isodynamic magnetic separator, which combines a strong magnetic field with gravity and mechanical agitation, to sort minerals according to their paramagnetic susceptibility [*Krogh*, 1982].

3.3 EXTRACTION OF THE ELEMENT TO BE ANALYZED

With the exception of the fission-track technique (Chapter 10), where grains are etched in acid to reveal the fission tracks that are then counted using an optical microscope, most geochronologic methods require the analysis of the isotopic composition of the daughter element and usually either the mass or concentration of both parent and daughter elements in an aliquot. In most cases, the analysis is done with some type of mass spectrometer chosen to be best suited to analysis of the element involved. When spatial resolution is particularly important, for example in the analysis of an individual grain in a thin section or of a single spot in a zoned crystal, either the energetic primary ion beam of an ion probe, or the ultraviolet photons of a laser, can be used to sputter material from the mineral surface (Fig. 3.2) for introduction into the mass spectrometer. These instruments can analyze spots as small as tens of microns to many tens of nanometers, although the precision of the analysis drops off as spot size decreases simply because the amount of material analyzed becomes smaller. Because neither the primary ion beam nor laser are selective in what they sputter from the sample surface, both the ion probe and laser ablation require a mass spectrometer capable of separating the element to be analyzed from all the other elements present in the sample. For these spot-analysis techniques, often multigrain separates are placed on a grain

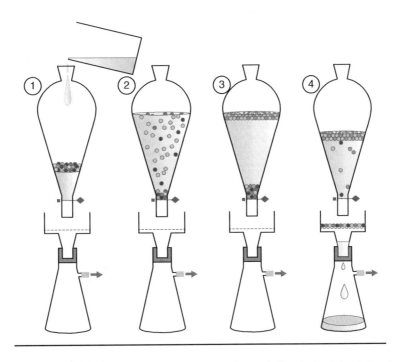

Fig. 3.1. Schematic of heavy liquid separation. Dense liquid is added to a separatory funnel (1) to which crushed and sieved minerals are added (2). The dense grains (black) sink in the liquid while the less dense grains (gray) float (3). The lower layer of the liquid is then drained through a filter (4) that captures the dense grains. The filter can then be replaced allowing the remaining low density grains to be removed from the funnel and captured by the filter paper. (Source: image from http://www.bgs.ac. uk/scienceFacilities/laboratories/mpb/media.html, reproduced by permission of the British Geological Survey.) (*See insert for color representation of the figure.*)

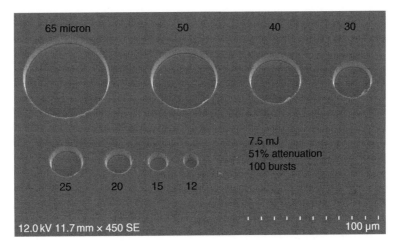

Fig. 3.2. SEM image of ablation pits created by laser ablation analysis of zircon. (Source: courtesy of George Gehrels and the Arizona LaserChron Center.)

mount where the grains are fixed to double-sided, low-vapor-pressure, conductive carbon tape and analyzed without sectioning. Some techniques, including ion probe and fission track, require exposures of polished internal sections of crystals. These may be prepared by pouring separates into restricted areas on a flat slide, embedding them in epoxy or polytetrafluoroethylene (PTFE), and grinding and polishing to expose internal surfaces.

When spatial resolution is less important, how the element to be analyzed is extracted into the mass spectrometer depends on its chemical properties. When the element to be analyzed is a gas,

for example He or Ar, the sample can be transferred to planchets that are placed into a vacuum system. Gases to be analyzed are often released by heating samples in vacuum using either a resistance or inductance furnace, or by a laser. Both furnace and laser heating, usually with closed-loop power-temperature control, are used for step-heating experiments in which gas is sequentially released in multiple steps at progressively higher temperature to derive diffusion kinetics and/or step-heating age spectra, as in $^{40}Ar/^{39}Ar$ dating and $^{4}He/^{3}He$ thermochronology. Lasers typically used for these purposes, as well as for total fusion analyses often used in ^{40}Ar/^{39}Ar dating, are usually in the infrared band and based on CO_2, Nd:YAG (yttrium, aluminum, garnet), or diode sources. Furnace and laser heating releases not only the gases of interest but also water, CO_2, hydrocarbons, and other contaminants. The gases are then purified with the use of chemical getters or cryogenic traps so that only the element of interest is eventually introduced into the gas-source mass spectrometer [*Wieler*, 2014].

Numerous geochronologic techniques, particularly those that employ thermal ionization mass spectrometers (TIMS) or inductively coupled plasma mass spectrometers (ICP-MS), require an initial digestion of solid whole rock or mineral grains in order to introduce the sample as a solution either directly into analytical instruments or, more commonly, into chromatographic columns for a preanalysis elemental separation. Digestions usually involve multiple treatments in acid-washed PTFE beakers/vials using some combination of concentrated nitric, hydrochloric, hydrofluoric, and/or perchloric acid. Many minerals dissolve in these acids at relatively low temperatures (<100 °C) and room pressure, so the procedure usually is carried out using loosely sealed Teflon vessels placed on a hot plate for hours to days. Some minerals, particularly zircon and garnet, require much more aggressive digestion techniques. The most common approach involves tightly sealed Teflon vessels contained within strong steel jackets (bombs) that can be heated to over 200 °C where they produce substantial internal pressures [*Krogh*, 1973]. Once destruction of the silica matrix of the silicate mineral is accomplished, usually through HF digestion that forms volatile SiF_4, the residual F often must be removed by evaporation in oxidizing acids (HNO_3 or $HClO_4$), because many fluoride salts are insoluble in the HCl and HNO_3 used later in the chemical separations. In contrast, many high-field-strength elements, such as Hf, Ti, Nb, and Ta, remain in solution best when there is at least some amount of fluorine present.

Because of the extremely low detection limits of most types of mass spectrometers, the amount of sample analyzed often can be in the nano- to picogram range. At these amounts, making sure that the analyzed material comes from the sample and not from the laboratory environment is critical. Minimizing background contamination, known as "blank," requires considerable care in technique and laboratory design. As a result, many geochronologic procedures are carried out in specially filtered fume hoods with positive air pressure, or entire cleanroom laboratories. Typical Sr–Nd–Hf–Pb, or U-series clean laboratories may meet class 1000 or 100 standards, which means a maximum of 1000 or 100 particles of size 0.5 mm per cubic foot of air. For comparison, ordinary indoor room air typically has a rating of 1,000,000. Geochemistry oriented cleanrooms usually try to minimize the amount of metal present, even to the extent of using plastic for applications such as cabinet hinges, drawer pulls and slides, sink faucets, and electrical and light boxes. Even stainless steel is strongly corroded by the inorganic acids, for example HCl, used in most chemical separation procedures. Some elements important in geochronology, particularly Pb (in solder or galvanized piping) and Sr (in gypsum wall board), can be very abundant

in some building materials, so minimizing the use of such materials in the laboratory, or covering them with some impenetrable paint or plastic is necessary.

The contribution of the blank to a concentration measurement is obvious. If, for example, a sample contains 10 pg of Pb, and the Pb chemical processing blank is 5 pg, if the blank is not adequately accounted for, the measured Pb concentration in the sample will be incorrect by 50%. Not correcting for blank also can have a significant effect on the measured isotopic composition, and hence on the age calculated from the data. For example, most of the Pb in industrial applications, and in the tetraethyl-lead that used to be added to gasoline to increase its octane rating, comes from lead ore sources that typically have quite low ^{206}Pb/^{204}Pb (~18.6) and ^{207}Pb/^{204}Pb (~15.4) ratios, from which a ^{207}Pb/^{206}Pb ratio of 0.828 can be calculated for a blank originating from this source of Pb. Using the example above, if the sample being analyzed is a 1 Ga zircon, and the Pb it contains was entirely produced by decay of U, its ^{207}Pb/^{206}Pb ratio should be 0.0725. Ten picograms of Pb extracted from the zircon above will consist of 4.525×10^{-14} mol of ^{206}Pb and 3.28×10^{-15} mol of ^{207}Pb. Five picograms of laboratory blank with the isotopic composition given above will consist of 5.81×10^{-15} mol of ^{206}Pb and 4.81×10^{-15} mol of ^{207}Pb. The mixture of blank and sample Pb will thus have a ^{207}Pb/^{206}Pb ratio of 0.158, which would translate to an age of 2.4 Ga if the blank correction were not applied. In practice, the best approach is to minimize the ratio of blank to sample as much as possible. When the blank is large enough compared to the sample sizes being worked with to influence the result, proper determination of both the magnitude and the reproducibility of the blank is critical so that accurate blank corrections can be applied to the measured samples.

3.4 ISOTOPE DILUTION ELEMENTAL QUANTIFICATION

Most geochronologic techniques require high-precision isotope ratios to be determined for the daughter element, but also high-precision mass, molar, or concentration ratios of the parent to daughter element. The most precise way to measure elemental concentrations, and concentration ratios, is the technique of isotope dilution [*Stracke et al.*, 2014]. This technique involves the addition of a known amount of a "spike" that consists of a strongly artificially enriched isotope of the element to be measured. An example using Sr is given below; alternative forms are given by other sources [e.g., *Faure and Mensing*, 2005]. In a common spike used for Sr concentration determination, ^{84}Sr is enriched to 83 atom percent, whereas in normal Sr, ^{84}Sr is a rare isotope at only 0.4% abundance. The ^{84}Sr/^{86}Sr ratio of a spiked sample is equal to:

$$\left(\frac{^{84}Sr}{^{86}Sr}\right)\text{mixture} = \frac{\left(^{84}Sr\right)\text{sample} + \left(^{84}Sr\right)\text{spike}}{\left(^{86}Sr\right)\text{sample} + \left(^{86}Sr\right)\text{spike}} \quad (3.1)$$

where the mixture is composed of the amount (in atoms or moles) of ^{84}Sr and ^{86}Sr contributed by the sample and spike solution. In this equation, the $^{84}Sr/^{86}Sr$ ratio of both the spike and sample are known by previous measurement, and that of the mixture by measurement of the spike–sample mixture. The amounts of ^{84}Sr and ^{86}Sr added by the spike also are known because the concentration of the spike solution is previously determined by combining aliquots of it with known amounts of normal Sr, usually involving solutions gravimetrically prepared to known concentrations from ultrapure Sr salts. The only unknown in the equation is then the amount of ^{84}Sr contributed by the sample. The equation can be rearranged to solve for the ^{84}Sr concentration in the sample as:

$$[^{84}Sr]\,Sample = \left[\frac{\left(\frac{^{84}Sr}{^{86}Sr}\right)Mix - \left(\frac{^{84}Sr}{^{86}Sr}\right)Spike}{\left(\frac{^{84}Sr}{^{86}Sr}\right)Sample - \left(\frac{^{84}Sr}{^{86}Sr}\right)Spike} \right] \times \frac{Spike\,Weight}{Sample\,Weight}$$
$$\times\,[^{84}Sr]\,Spike \qquad (3.2)$$

where the term (Spike Weight $\times\,[^{84}Sr]$Spike) provides the number of moles of ^{84}Sr added by the spike, with the spike having a known concentration of ^{84}Sr (in moles) per gram of spike solution. The concentration of ^{84}Sr in the sample also is then given in moles per gram, which can be converted to the concentration of Sr by dividing by the atom percent of ^{84}Sr and multiplying by the atomic weight of Sr in the sample. Because of radiogenic contributions to ^{87}Sr, both the atomic percentage of each isotope and the atomic weight of Sr in any given sample is not a constant, but must be calculated based on the measured Sr isotopic composition of the sample.

The accuracy of the concentrations measured by this approach will depend primarily on the accuracy with which the spike isotopic composition and concentration are known and whether perfect mixing is achieved between sample and spike. An advantage of this technique is that once sample and spike have been well mixed, any loss of the sample–spike mixture, for example through imperfect yields during chemical separation, will not affect the concentration calculation as it depends only on the isotopic composition of the spike–sample mixture. Achieving a perfect mixture of spike and sample, however, is a common concern in geochronological applications. The following two examples illustrate ways in which imperfect application of this technique can lead to inaccurate results.

(1) Most isotopically enriched spikes also contain enough of all the other isotopes of the element to require correction for the spike contribution to the isotopic abundance of all the isotopes in a spike–sample mixture. When the spike is very strongly enriched in a rare isotope of the element being analyzed, the magnitude of the correction to the isotopic composition of the sample is small enough that a sufficiently accurate correction can be made for the mixture of the spike and sample, that both the sample concentration and its isotopic composition can be obtained from a single spiked measurement. Highly enriched isotopes, however, are not available for all elements, so when the spike is made from one of these less-enriched spikes, or very high precision on the isotopic composition of the element is needed, a common analysis approach involves splitting the sample into two aliquots after dissolution, one without spike that is used for precise determination of the sample isotopic composition, and another aliquot to which spike is added for concentration determination. Critically important in this approach, however, is that the sample solution be perfectly homogeneous at the time the spike aliquot is removed. The presence of any undissolved material, or even worse, precipitates that may concentrate the element of interest, will mean that the solution will not contain all the element present in the original sample. Spiking an aliquot of this solution thus will provide a concentration, and hence parent/daughter ratio, determination that does not reflect the true concentration of the element in the original sample.

(2) Some elements, for example Os, can have different oxidation states in the sample before and after dissolution. The different oxidation states allow the formation of different Os molecules in solution that may not have equal solubilities in the different acids used in chemical separation approaches. Early dissolution techniques for Os used a combination of HF and HBr acids to keep the solution strongly reducing in order to hinder the formation of volatile OsO_4 that could be lost during sample evaporation [*Walker*, 1988; *Reisberg and Meisel*, 2002]. Under these conditions, however, the Os liberated from the sample and the Os added with the enriched spike could exist in the sample solution as different molecules, and hence never mix perfectly. The different molecules containing Os would respond differently to the Os separation chemistry, hence yielding a final Os concentrate that did not sample equal amounts of sample and spike Os, and hence yielded inaccurate concentration measurements. The solution to this problem for Os was to move to a strongly oxidizing dissolution in a sealed vessel so that both the sample and spike Os were driven to the OsO_4 molecule where perfect spike–sample mixing could occur in the gas phase present in the dissolution vessel [*Shirey and Walker*, 1995].

3.5 ION EXCHANGE CHROMATOGRAPHY

Many geochronologic techniques require purification of the element to be analyzed before it is loaded into the mass spectrometer, for example to eliminate the isobaric interference of ^{87}Rb on ^{87}Sr, or ^{187}Re on ^{187}Os. Although many classical wet-chemistry techniques can be used to achieve this elemental separation, by far the dominant choice for geochronology is liquid ion-exchange chromatography [*Schoenbaechler and Fehr*, 2014]. The overall goal of such separations is to recover essentially all of the element of interest from the rock, and to separate it from other elements that would interfere with accurate isotopic or concentration determinations for that element. Techniques that typically use chromatography include TIMS and ICP-MS-based methods for Rb–Sr, Sm–Nd, Lu–Hf, Re–Os, and U,Th/Pb, that require dissolution for some stage of the analysis and for which isobaric interferences in bulk solutions are problematic. Noble-gas-based techniques do not use chromatography, although U and Th (and Sm) may be separated and concentrated by this approach when used for (U–Th)/He dating (e.g., Chapter 11).

Liquid chromatography usually involves ion-exchange resins suspended within quartz glass or plastic columns. Liquid chromatography is based on the exchange of sample ions in a solution as they flow through a column that contains "resin" coated with a variety of organic chemicals. The resin coatings have molecular sites that can exchange with ions in solution, causing the ions to stick to the resin and be removed from the flowing sample liquid. The degree to which various elemental ions stick, or not, to the resin, defined as their "partition coefficient," determines how quickly a given element flows through the column. The partition coefficient for any element depends on the type of resin used in the column and the type and concentration of the acid used for the sample solution. An acid solution containing the dissolved sample, and all the elements it contains, is loaded to the top of the column followed by an acid used to elute the sample. Elements with low partition coefficients for the resin pass through the column quickly, while those with higher partition coefficients move more slowly through the column and hence require more acid to be flushed through the column (Fig. 3.3).

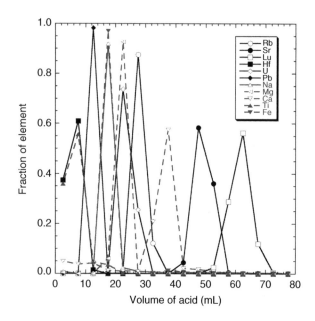

Fig. 3.3. Elution curves for a cation ion exchange column designed to separate Hf, Rb, Sr, and Lu from major elements in the sample. In this case, the sample is loaded in a solution of 1N HCl and 0.1N HF. The HF causes high-field-strength elements such as Ti and Hf to form fluoride anions that do not stick to the resin. These elements are washed off the column with the addition of 10 mL of this acid mixture. At that point, the acid is changed to 2.5N HCl. Most of the major elements (dashed lines), with the exception of Ca, elute quickly in this acid. Rb and Sr are reasonably well isolated in this step. Once Sr is eluted, the acid strength is increased to 4N in order to wash out the REE, including Lu, Sm, and Nd.

A large number of resins exist for liquid chromatography. Some selectively exchange only with cations in solution, others bind only with anions, and still others are element specific. The general characteristics that determine the quality of separation of any given element include:

(1) The difference in partition coefficients between elements and the resin in different acids and acid strengths.

(2) The length to diameter ratio of the column. In ideal chromatography theory, the elution starts with all the sample ions attached to the resin in an infinitesimally thin band at the top of the column. Resins, however, only have so many exchangeable molecular sites, so if a large sample containing many ions is loaded onto a small diameter column, the exchangeable sites in the resin beads at the top of the column fill up, causing the remaining ions to travel down the column until they find a resin site to adhere to. This will result in a broadening of the sample band on the resin, leading the elution peaks of each element to broaden, thereby causing the interelement separation to deteriorate. To compensate, the column can either be lengthened or increased in diameter. Increasing the diameter increases the volume of resin and thereby provides more exchangeable sites at the top of the column. Increasing the length provides a longer, and generally slower, flow path that enhances the ability of the column to separate elements that have different partition coefficients for the resin.

(3) The exchangeable ion on the resin. Most cation resins use H^+ as the ion on the resin that exchanges with cations in the sample solution. Other cations, however, can be used. For example, H^+ can be replaced by NH_3^+ by passing an ammonia solution through the resin prior to sample loading. Exchanging with NH_3^+ instead of H^+ enhances the separation of the rare earth elements (REE) from one another.

(4) The grain size of the resin. Smaller grain size resins tend to provide improved separation because of their larger surface area per volume, but they also tend to slow the flow rates of liquid through the column.

3.6 MASS SPECTROMETRY

The first applications of radioactive geochronology occurred before there were means to accurately measure the isotopic composition of elements. Rutherford in 1905 determined the first radioisotopic age using measurements of uranium and helium made in a mineral using classical analytical chemistry techniques. He could calculate an age on the assumptions that all the helium in the sample was produced by uranium decay, and that no helium was trapped initially in the crystal when it formed. If a further assumption was made that none was lost by outgassing from the sample at any point during its history, the age would approximate the formation age; if not, it would provide a minimum age. A similar approach was used through the 1960s in a technique called "lead-alpha" that took uranium/thorium-rich minerals or rocks, used α-particle counters to measure the amount of α-emitting radioactive elements, which would be dominated by U and Th in the sample, and then used wet-chemical techniques to determine the Pb concentration in the sample. Again, the assumption in this technique is that all the Pb present in the sample was created by radioactive decay of U and Th, with no initial Pb present in the sample, and none lost from the sample after its formation. "Chemical dating" methods that quantify element contents or concentrations, rather than isotopic compositions, are still used today for some systems and applications, including in (U–Th)/He (Chapter 11) and "total Pb" dating of monazite and other minerals that measure only the Th, U, and Pb concentrations of the minerals in order to derive an age [*Williams et al.*, 1999]. Other geochronologic techniques where isotope measurements are not needed include fission-track dating, which relies on optically observing the crystal damage tracks caused by U and Th fission as well as U concentrations, and some other radiation-damage-based techniques. These approaches are discussed in detail in Chapter 10.

When it became clear that radioactive decay transformed an isotope of the parent element into an isotope of the daughter element, a clear path to improvement in the use of radioactive decay as a chronologic tool switched the focus, for most systems, from element abundance to the relative proportion of the decay product isotope in the daughter element. The instrument of choice for isotope ratio measurement is the mass spectrometer. The basic physical principles employed in the mass spectrometer came to light through the investigations into the atomic nature of material at the turn of the 19th century. Eugen Goldstein in 1886 observed that an electric arc through a gas creates particles (electrons) that are attracted towards a positively charged plate (anode) and other particles (positive ions) that are repelled by the anode and attracted towards a negatively charged plate (cathode). The production of ions and their acceleration by electric fields is the first step in isotope separation. The second critical step was discovered by Wilhelm Wien in 1899, who showed that ions accelerated in an electric field could be deflected by magnetic fields, with the radius of curvature of the deflection proportional to the charge to mass ratio of the ion and the strength of the magnetic field. These features were combined to create the first mass spectrographs by Canadian physicist Arthur Dempster working at the University of Chicago in 1918 (Fig. 3.4), and by British chemist Francis Aston at Cambridge in 1919. These first mass spectrographs used photographic emulsions as ion detectors that, when developed, recorded the impact of beams of charged particles of different masses that had been separated by mass during their flight through the magnetic field. The photographic plates were soon replaced with electronic detection of the incoming ion beams to form the first mass spectrometers. With their mass

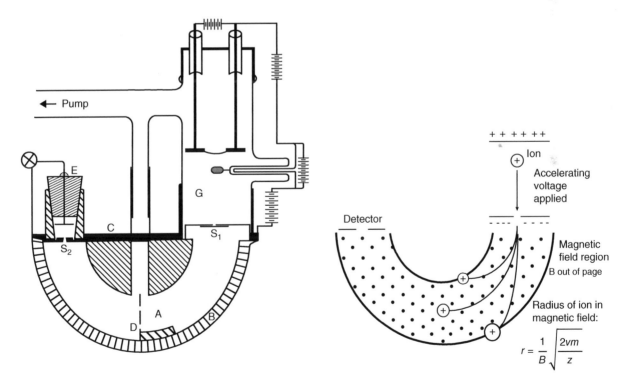

Fig. 3.4. Left: Dempster's first mass spectrometer. Ions are generated in section "G" and accelerated by an electric field through the source slit (S_1) after which they enter into a magnetic field (region A) that causes them to curve through a 180° arc and separates the ions according to their charge to mass ratio. Either the accelerating voltage or magnetic field strength can be adjusted so that an ion beam of only a single mass passes through the detector slit (S_2) to hit the detector plate (E). (Source: *Dempster* [1918]. Reproduced with permission of American Physical Society.) Right: schematic path of an accelerated ion with velocity *v*, mass *m*, and charge *z*, through an orthogonal magnetic field *B*. Single-detector mass spectrometers vary *B* and *v* to direct and measure incoming intensities of ions of different *m/z* in the detector. Multidetector systems measure multiple *m/z* ions with a single set of *B* and *v*.

spectrographs and spectrometers, Dempster discovered ^{235}U, and Aston was awarded the Nobel Prize for chemistry in 1922 for his discovery of 212 naturally occurring isotopes. Through his examination of the isotopic composition of elements, Aston formulated his "whole number rule" that stated that all isotopes were integer multiples of the mass of the hydrogen atom. This rule is basically an extension and correction of William Prout's 1815 theory that the atomic weights of all elements were integer multiples of the mass of hydrogen. The important difference is that Prout was not yet aware that many elements consist of mixtures of two or more isotopes of different mass.

Since the time of Aston and Dempster, mass spectrometry has become one of the most useful tools in analytical chemistry, and has taken on so many different forms and applications that a full description is not possible here. All mass spectrometers, however, depend on:

(1) ionization of the sample
(2) acceleration and focusing of the ions into a collimated beam by electric fields
(3) some method to separate the ions according to their mass to charge ratio
(4) a detector that quantifies the magnitude of the ion beams at each mass
(5) a vacuum that allows the unimpeded transit of ions from ion source to detector.

Various means are available to perform all of the steps above. The selection of which is best for a given geochronologic application depends on the element to be analyzed, its mass, ionization potential, vapor point, whether spatial resolution is needed, and the nature of potential interfering species present with different techniques.

3.6.1 Ionization

The first step in all mass spectrometry is to turn the element to be analyzed into an ion so that it can be accelerated by an electric field. Many methods exist to add or remove an electron from an atom to leave behind an electrically charged ion. An electrical arc was the first ionization method applied in the late 1800s. While arcs are still employed in some types of mass spectrometers, for most mass-spectrometry applications, sample ionization has moved to methods that either create a more constant ion beam or that provide high spatial resolution or elemental selectivity during ionization. Such methods include the following.

3.6.1.1 Electron bombardment

In this method, an electrically heated tungsten filament emits electrons that are accelerated by an electric field across a gap into which neutral atoms of the element to be ionized are introduced, usually as gases. For geochronology, electron bombardment ionization is mostly used in the analysis of the noble gases, He, Ar, Ne, Kr, and Xe. Most electron bombardment sources also apply a magnetic field across the electron path that causes the electrons to spiral on their path from the filament to the collecting anode (Fig. 3.5). The spiraling increases the time that the electrons spend in the area where they can encounter sample atoms, and hence improves sample ionization efficiency.

Maximum ionization efficiency to produce singly, positively charged ions occurs at electron energies in the range of 70–90 V, but the energy of the electron beam often can be adjusted to either increase ionization efficiency or minimize the creation of doubly charged ions, whichever is more important for a given measurement. Electron bombardment produces ions with a small spread in energy, which, as we will see, is important during the mass analysis step. Electron bombardment sources are used primarily in the analysis of gases that can be extracted from a solid sample either by heating or crushing (e.g., noble gases), or by conversion of a solid (e.g., carbonate, silicate) to a gas (e.g., CO_2, O_2) through chemical treatment. The sample gas is introduced into the ion source through a small orifice that controls the flow rate so that a large portion of the introduced atoms are

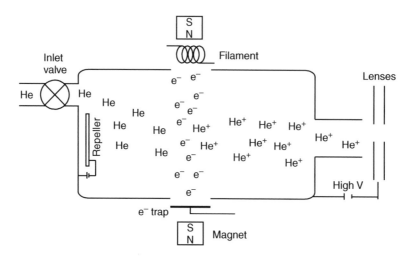

Fig. 3.5. Electron impact ion source schematic, showing He as the analyte, ionized by electrons generated from the filament and attracted by the electron trap, repelled by the slight voltage on the repeller near the inlet source, and accelerated towards the lenses and mass spectrometer on the right-hand side by a high-voltage potential.

ionized by the electron bombardment. Sample atoms that are not ionized are pumped away by a vacuum pump. As electron-impact ionization is very efficient, ions will be produced from any gas atom/molecule that intersects the electron beam with the result that any contaminating species in the sample introduced into the ionization region will be ionized along with the sample. For the usually small quantities of gas involved in geochronologic techniques using the noble gases, a technique, called static mass spectrometry, introduces all of the available sample gas into the high vacuum of the mass spectrometer and then closes off the instrument to vacuum pumps [*Wieler*, 2014]. The ion signal decreases with time as the gas sample is irreversibly adsorbed onto the inner walls of the mass spectrometer or implanted into the detector.

3.6.1.2   Thermal ionization

As its name implies, this ionization technique involves applying the element to be analyzed, usually dissolved in a small drop of acid, onto a refractory metal filament. The sample is dried into a thin solid film on the filament. The filament is then placed into the vacuum of the mass spectrometer and heated with electric current to produce ions [*Carlson*, 2013]. Ionization efficiencies, defined as the fraction of sample atoms ionized, in thermal ionization can range to as high as 50% or more, or less than 0.1%, depending on the evaporation temperature and first ionization potential of the element being analyzed. Thermal ionization works well for many of the elements employed in geochronology, for example, Cr, Rb, Sr, Sm, Nd, Re, Os, W, Pb, and U. Thermal ionization does not work well for elements with high ionization potentials (e.g., Hf) due to the low ion yields. Some improvement can be achieved in ionization efficiency through the use of multiple filaments, often either two or three, in which the sample is deposited on one filament and the other(s) heated to very high temperatures (>1800 °C). The sample filament is then heated to lower temperatures to evaporate the sample as neutral atoms. Atoms evaporated from the sample filament will be ionized if they contact the hot ionization filament. Other approaches to increase ionization efficiency include analyzing molecules, for example oxides (e.g., NdO^+ instead of Nd^+). Ionization efficiencies for NdO^+ can be up to a factor of five higher than Nd+, but require correction for oxygen isotopic composition in order to extract the Nd isotope ratios [*Carlson*, 2013]. Another example of exploiting surface chemistry for enhancing ion yields is the production of negatively charged molecular ions through the addition of materials such as $Ba(NO_3)_2$, $Ba(OH)_2$, and various rare earth element oxides to the filament. For example, with the addition of $Ba(OH)_2$ to the filament, OsO_3^- ions can be produced at ionization efficiencies of about 5% [*Creaser et al.*, 1991; *Volkening et al.*, 1991], which is quite efficient compared to other ionization techniques for this element. For a mass spectrometer to be capable of analyzing both positive and negative ions (not at the same time), the instrument must be able to change the polarity of the accelerating voltage in the ion source, and the polarity of the magnetic field produced by the magnet. Also required is an amplifier that can detect both negative and

positive ion currents. One type of detector commonly used in mass spectrometry, the Daly detector, cannot be used for negative ions for reasons discussed later in this chapter.

Thermal ionization for positive ions produces mostly singly charged atomic species, with a simple background of mostly single oxides (e.g., NdO^+, PbO^+, etc.), and few, if any, doubly charged ions, thus minimizing potential interfering species during mass analysis. Because the technique also depends on the evaporation temperature of the element being analyzed, in some cases additional chemical purification can be performed on the filament itself. For example, Rb evaporates and ionizes at a much lower temperature than Sr, so small amounts of Rb contamination in a separated sample of Sr often can be "burned off" before reaching the filament temperatures needed for ionization of Sr.

Another advantage of thermal ionization is that it produces ions with a very small spread in energy. The kinetic energy (KE) of an ion produced thermally is equal to:

$$KE = \frac{3}{2}kT \qquad (3.3)$$

where k is the Boltzman constant and T is temperature in Kelvin. At a typical Sr run temperature of 1673 K, the energy of the ion is 0.2 eV, which is small compared to the typical 10 kV energies used to accelerate ions into the mass spectrometer. Both the simple interference spectrum and the low energy spread in the ions produced by thermal ionization allow the use of relatively simple mass analyzers to accurately quantify isotope ratios. The main drawback to thermal ionization is that it is very inefficient for elements that do not thermally ionize easily, for example Hf, and Os and W as positive ions. Another drawback is that the evaporation of the sample from the filament induces a mass-dependent shift in the isotopic composition of the element that enters the mass spectrometer because lighter isotopes evaporate preferentially compared to heavier ones. This mass fraction often, but not always, can be corrected using the approaches described in section 3.6.3. Mostly as a result of the clean mass spectrum produced by thermal ionization, this technique is capable of the highest precision isotope-ratio measurements, at least for the elements that ionize efficiently by this technique.

3.6.1.3   Inductively coupled plasma ionization

This technique takes the ionization process out of the vacuum of the mass spectrometer and instead accomplishes sample evaporation and ionization in a 5000–10,000 °C argon plasma created by a strong radiofrequency field inside a quartz torch [*Olesik*, 2014] (Fig. 3.6). The use of the inductively coupled plasma (ICP) source is a relatively new addition to geochronology [*Halliday et al.*, 1995]. This technique can be applied to essentially every element in the periodic table, and provides a complimentary approach for many of the elements analyzed by thermal ionization, but also works well for elements that do not ionize efficiently (e.g., Hf and Th) by thermal ionization. Often, ICP is coupled with laser ablation to provide analyses *in situ* of both Hf isotopic composition and U–Pb ages in zircon when spatial

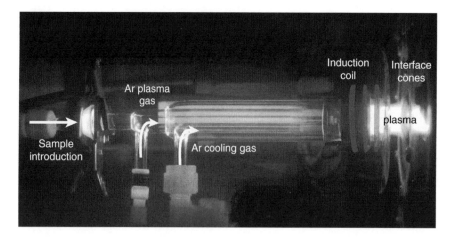

Fig. 3.6. Photograph of the ICP torch. In this image, argon flows through the three concentric quartz tubes from left to right. The small central tube carries the sample (introduced as aerosol at left) mixed with a small Ar gas flow. The Ar is turned into a plasma by a strong radiofrequency field inside the copper helical coil at the end of the torch. The sample flows through the plasma and is evaporated and ionized before it impacts the entrance cones to the mass spectrometer on the right-hand side of the photograph. (Source: Courtesy of Brad Hacker.) (*See insert for color representation of the figure.*)

resolution is more important than precision. Analytical improvements in ICP have made it an increasingly strong competitor with thermal ionization methods, although for the elements for which it works best, e.g., Sr, Nd, Os, and Pb, thermal ionization still provides some advantages in measurement precision due to its higher sample utilization efficiency, expressed as ions detected divided by the number of atoms in the sample, its less complex molecular spectrum, and lower operating costs.

In the ICP, various argon gas flows are used to establish and maintain the plasma, to cool the torch, and to carry the sample as either an aspirated solution or as very fine particles created by ablation with an ultraviolet (UV) laser. In the hot portion of the torch, the sample is first evaporated, chemically disassociated, and then ionized. The ions produced must pass from the atmospheric pressure of the torch through 0.5–1 mm diameter holes into the vacuum of the mass spectrometer. ICP ionization has the advantage of enough energy to completely ionize all elements whether introduced as gases, aspirated liquids, or submicron-size particles created by UV laser ablation of solid samples. Although the ionization efficiency of this technique is very high, the high ion and electron density in the plasma before its entry into the mass spectrometer allows numerous chemical reactions to occur that serve to neutralize some ions and enhance the formation of ionized molecules of some elements. In addition, only those ions that follow gas-flow trajectories that carry them through the tiny holes into the mass spectrometer can be analyzed. These two factors lead to sample utilization efficiencies generally of order 1% or less in most ICPs, although these are improving as the technique develops. Both the higher temperatures of ionization and the fact that ionization occurs over a range of electric field values on entry into the mass spectrometer results in a relatively wide energy spread in the ions produced. ICP ion sources can produce a quite complicated molecular spectrum, particularly of oxides, and a wide variety of molecules of Ar, such as oxides, hydroxides, and nitrides. ICP ionization also produces

a small, but non-negligible, amount of doubly charged ions. One advantage of this technique over other methods for sample ionization is that the ionization process is so efficient that it is not strongly dependent on the element involved. In ICP mass spectrometry (ICP-MS), there is a quite sizeable mass dependency to the efficiency of acceleration of ions into the mass spectrometer, with heavier ions being more successfully transmitted into the mass spectrometer. This phenomenon introduces a mass-dependent shift in the measured isotopic composition of the element being analyzed, which can be corrected using the variety of approaches described in section 3.6.3.

3.6.1.4 Ion-impact ionization

At the heart of both SIMS (or ion-probe) [*Ireland*, 2014] and accelerator mass spectrometry (AMS) [*Jull and Burr*, 2014] is the use of energetic ions to either ionize a chemically purified sample (AMS) or actively ablate a solid sample and ionize a portion of its constituent atoms in the process (SIMS). The process of ionization here is simply the energy of collision between atoms. For AMS, accelerated Cs^+ ions impacting the chemically purified sample deposited on a surface produce negative ions of the sample that are then extracted into the mass spectrometer. In SIMS, the typical primary (Cs^+ or O^-) ion beam impacts the sample surface with energies of order 10 kV. Ions of this energy are sufficient to break chemical bonds and ionize a portion of the atoms liberated by sputtering off the surface (Fig. 3.7). Because of the simultaneous breaking of sample bonds in the solid sample, and ejection of the sputtered material, ionization yields in SIMS are relatively low compared to other ionization techniques, with ion to neutral atom ratios typically in the range of 10^{-3} to 10^{-5}. SIMS-style ionization can produce either negative or positive secondary ions, depending on the target element and the element employed in the primary ion beam. Besides the low ion yields, SIMS-style ionization creates an extremely rich molecular spectrum, a large energy spread in the extracted ions, and substantial

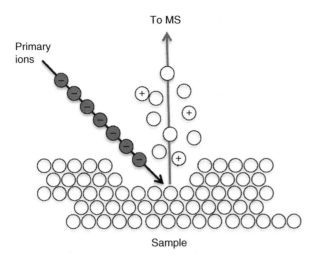

To MS

Primary ions

Sample

Fig. 3.7. Schematic example of the sputtering and ionization process involved in the SIMS ion source. A beam of energetic (~10 Kv) negative ions of oxygen is focused on the sample surface. The impact of the primary ions breaks chemical bonds on the sample surface liberating a mixture of molecules, neutral ions, and positive ions of the sample. The ions are extracted by an electric field to be introduced into the mass analyzer.

mass-dependent fractionation of the ions before they enter the mass spectrometer. These place additional requirements on the mass resolving ability of the mass spectrometer. A clear advantage of SIMS ionization is that the primary ion beam can be focused to micron to submicron size and hence provide spatial resolution in isotopic analysis that cannot be matched by other techniques. For geochronology, SIMS is a leading technique for Al–Mg dating, and for U–Pb dating of zircon and other minerals when spatial resolution is critical. An excellent comparison of thermal ionization, laser ablation ICP-MS, and SIMS for U–Pb zircon chronology is provided by [*Schaltegger et al.*, 2015].

3.6.1.5 Laser resonance ionization

This approach to ionization uses the absorption of light photons of a specific wavelength to selectively ionize elements by matching the photon energy to an element-specific electronic transition. For example, when used for the Re–Os isotope system, absorption of a photon from a laser adjusted to a wavelength of 297.15 nm will cause a ground-state electron in an Os atom to jump to an excited electronic level within the atom. If a second photon is absorbed before the excited state has returned to ground state, the electron will be ejected from the Os atom to make an Os⁺ ion. Rhenium does not absorb light of this wavelength and thus will not be ionized. If the laser is adjusted to 297.70 nm, however, two-photon absorption can ionize Re, but not Os. This element selective ionization is unique to laser resonance ionization, but the technique has a number of weaknesses that have limited its application. First, some method must be applied to get the sample into the atomic state, for example either sputtering with a primary ion beam as in SIMS, use of another laser to evaporate the sample, or evaporation of the

sample from a sample holder that can be heated electrically. The neutral atoms produced by this process must be contained within the area illuminated by the laser if they are to be ionized. Second, many of the lasers used for this technique are pulsed lasers that are needed to get the photon-density high enough to ensure the two-photon ionization. Most pulsed-lasers pulse the light on for tens of nanoseconds at frequencies of tens to perhaps 100 pulses per second. One hundred pulses in a second, of 10 ns each, means that the laser light is shining on the sample for 1 ms per second, so the time efficiency of this technique is very low. Laser resonance ionization will become more important in geochronology as laser technology improves, but at the moment, the level of complexity of these instruments restricts their use to analysis of extremely rare materials, for example, presolar grains [*Savina et al.*, 2003].

3.6.2 Extraction and focusing of ions

Once the sample atoms have been ionized, the ions must be accelerated and focused into a narrow beam so that they can be sent into some form of mass analyzer that separates them according to mass. A typical "ion source" involves a series of metal plates to which a cascading series of voltages is applied. The first goal of these voltages is to provide the accelerating force that extracts the ions from the region of their formation. The second goal is to reorient the direction of motion of the ions so that they are all moving in the same direction and can be put through a small slit, often called the source or image slit, that forms the image for the mass analyzer. An example of one commonly used ion source is shown in Fig. 3.8. Most ion sources will maintain the ionization region at high voltage, of order a few to ten kilovolts, and then gradually step the voltage down to ground potential at the image slit. Ions leaving such an ion source are thus directed into the mass spectrometer with kinetic energies equal to the accelerating voltage. The electric field between two parallel plates held at different voltages will be parallel to the plates, but when there are gaps in the plates, the electric fields will curve into the gaps (Fig. 3.8a).

When the electric field lines are perpendicular to the path of the ion, the ion experiences an acceleration in its direction of motion given by:

$$\frac{1}{2}mv^2 = qV \tag{3.4}$$

where m is mass, v is velocity, q is the charge on the ion, and V is the voltage drop experienced by the ion. If the ion is traveling in a direction that is not perpendicular to the electric field lines, the ion will experience a force that curves its trajectory following the equation:

$$r = \frac{mv^2}{qV} \tag{3.5}$$

where r is the radius of curvature of the ion trajectory. Ion sources use the curvature of the electric field between the plates in the ion source to curve the ion trajectories (Fig. 3.8b) so that

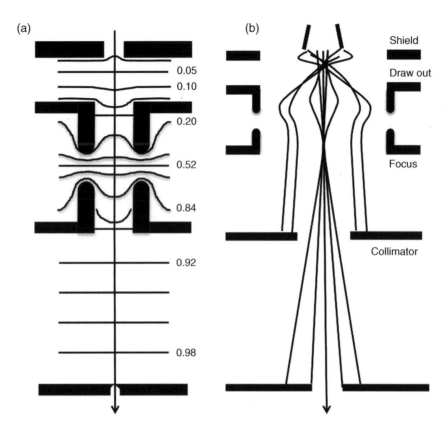

Fig. 3.8. A thick-lens ion source designed by Leonard Dietz. The thick black regions are cross-sections of metal plates to which different voltages are applied. The thin lines in (a) show the electric field lines generated between the plates with the numbers alongside showing the fractional reduction in voltage compared to the voltage of the filament. The filament is at the highest voltage whereas the slit at the bottom of the figure is normally held at ground potential. The thin lines in (b) illustrate the ion trajectories created by these electric field lines. The ions in (b) are emitted from two thermal filaments at the top of the ion source. (Source: *Dietz* [1959]. Reproduced with permission of AIP Publishing.)

they end up focused into a narrow beam such that the majority of the ions can pass through the source slit. The ions will leave the source slit and enter the mass analyzer with some amount of angular spread in their direction of motion, so one of the requirements of the mass analyzer is that it be able to refocus this diverging ion beam such that all the ions will arrive at the detector after having been separated by mass. The source slit becomes the image that the mass analyzer refocuses onto a narrow slit in front of the detector, called the detector or collector slit. As we will see, the size of the source and detector slits play an important role in defining the mass resolution of the mass spectrometer.

3.6.3 Mass fractionation

The processes of ion formation and acceleration generally have some mass dependency that can change the apparent isotopic composition of the element being analyzed. Depending on the analysis technique being employed, the mass fractionation can change the measured isotopic composition of the element by as much as ten percent per mass unit. If there were no means to correct for this instrument-induced mass fractionation, then the isotope ratio precision would be no better than the

magnitude of mass fractionation, severely limiting many, and completely eliminating some, isotope geochronology techniques. Fortunately, there are several ways to correct for instrument-induced mass fractionation. In gas source mass spectrometry, the mass spectrometer can be rapidly switched between a gas of known isotopic composition and the sample gas so that whatever mass fractionation is occurring in the instrument can be corrected for simply by comparing the difference between the isotope ratios measured for the sample and those of the standard gas. In ICP-MS, the mass fractionation is primarily, but not entirely, a function of mass, not chemical species, so it is possible to add to a sample element a nearby (in mass) element that does not interfere with the isotopes of interest in the sample, and use the offset of the isotopic composition of the added element from a known, or assumed, isotope composition to correct the sample isotopic composition for the instrument-induced mass fractionation.

Another method that can be used to correct for the isotope mass fractionation is available when the element being analyzed has enough isotopes not modified by radiogenic additions to use one of the stable isotope ratios as a measure of instrument mass fractionation. When there are not enough "constant" isotope ratios present in an element, for example for U and Pb, a more complicated approach to correct for fractionation involves the addition of two or more enriched isotopes, of a known isotope ratio, of the element. For U, this can be relatively straightforward as one can use a ^{233}U–^{236}U spike, as neither of these isotopes occur in significant abundance in nature but are available from reactors and, though radioactive, have long enough half-lives to be useful for this purpose. One then uses the known $^{233}U/^{236}U$ ratio of the spike to monitor instrument fractionation and correct the measured $^{235}U/^{238}U$. The same can be done

with Pb by using a ^{202}Pb–^{205}Pb spike [*Galer*, 1999]. Other approaches to "double spiking" can be used for most multi-isotope elements, but the approach becomes more complicated when the spike must be one of the stable isotopes of the elements, as a deconvolution of the sample and spike contribution to that isotope must be done [*Dodson*, 1970; *Galer*, 1999].

Figure 3.9 provides an example that shows how the measured Sr isotopic composition of a sample changed during the course of an analysis by TIMS. During this analysis, the measured $^{87}Sr/^{86}Sr$ ratio changed from 0.7070 to 0.7126, a total range

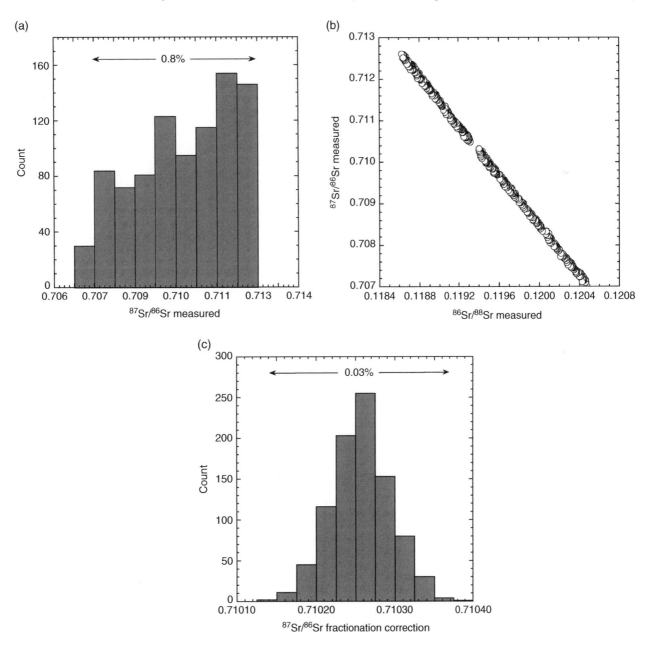

Fig. 3.9. Isotope ratio variation during the measurement of Sr by thermal ionization. (a) Histogram of the measured $^{87}Sr/^{86}Sr$ ratios. (b) The measured $^{87}Sr/^{86}Sr$ ratio is perfectly anti-correlated with the measured $^{86}Sr/^{88}Sr$ ratio. (c) Using the deviation of the $^{86}Sr/^{88}Sr$ ratio from an assumed constant ratio equal to 0.1194 allows for the correction of the measured $^{87}Sr/^{86}Sr$ ratios for the instrument-induced mass fractionation. This fractionation correction reduces the total range in $^{87}Sr/^{86}Sr$ from 0.8% to 0.03%.

of 0.8% (Fig. 3.9a). Most of this variability, however, is caused by the fact that during evaporation from a hot filament, light isotopes are preferentially evaporated first, leaving the material on the filament increasingly enriched in the heavier isotopes, so that as time goes by the Sr evaporating from the filament also becomes isotopically heavier. This effect is called Rayleigh distillation. The mass dependency of this effect is shown clearly in Fig. 3.9b where it can be seen that the measured $^{87}Sr/^{86}Sr$ ratio is perfectly correlated with the measured $^{86}Sr/^{88}Sr$ ratio. While the $^{87}Sr/^{86}Sr$ ratio can change due to ^{87}Rb decay, the $^{86}Sr/^{88}Sr$ ratio should be constant, or nearly so, in nature because both are stable isotopes with no radioactive parents. If we take the $^{86}Sr/^{88}Sr$ ratio as a constant, and a value of 0.1194 is commonly assumed for this ratio, then we can use the difference between the measured $^{86}Sr/^{88}Sr$ ratio and 0.1194 as a measure of the mass fractionation during the analysis to correct the measured $^{87}Sr/^{86}Sr$ ratio for this fractionation. Doing so results in the data shown in Fig. 3.9c where the total range in $^{87}Sr/^{86}Sr$ ratio has been reduced to only 0.03%, and the histogram of the data looks more like a Gaussian distribution that reflects the random error on this isotope ratio measurement.

For this mass-fractionation correction to be accurate, one must know the mass dependency of the fractionation occurring in the instrument in order to extrapolate the offset in $^{86}Sr/^{88}Sr$ to $^{87}Sr/^{86}Sr$. Unfortunately, this is not known from first principles, so many different mass dependencies have been proposed. The most common in use presently is the so-called exponential correction, which follows the equation:

$$Rij = (Rij)_m \times \left(\frac{Mi}{Mj}\right)^{FF} \tag{3.6}$$

where FF is the fractionation factor, defined as

$$FF = \ln\left(\frac{(Rkj)_t}{(Rkj)_m}\right) / \ln\left(\frac{Mk}{Mj}\right) \tag{3.7}$$

R is the isotope ratio of isotopes i, j, and k, with j and k being the stable isotope pair used to monitor mass fractionation, M is the atomic mass of the indicated isotope, and t and m are the true and measured ratio for the stable isotope pair, respectively. For example, if the measured $^{86}Sr/^{88}Sr$ ratio is 0.1204, $FF =$ 0.363065 (mass of $^{86}Sr = 85.909267$ Da, mass of $^{88}Sr =$ 87.905619 Da). If the measured $^{87}Sr/^{86}Sr$ ratio for this $^{86}Sr/^{88}Sr$ ratio were then 0.7073, the mass fractionation correction to the measured $^{87}Sr/^{86}Sr$ ratio would be 1.004209 for a corrected ratio of 0.710277 (mass of $^{87}Sr = 86.908884$ Da).

The exponential law, however, is not the only one employed for mass fractionation. Common alternatives include:

linear:

$$Rij = (Rij)_m \times [(FF \times (Mi - Mj)) + 1] \text{ where } FF = \left(\frac{(Rkj)_t}{(Rkj)_m} - 1\right) \div (Mk - Mj) \tag{3.8}$$

and power law:

$$Rij = (Rij)m \times FF^{(Mi-Mj)} \text{ where } FF = \frac{(Rkj)_t}{(Rkj)_m}^{\frac{1}{(Mk-Mj)}} \tag{3.9}$$

The different laws produce slightly different results that mainly become important for very high precision isotope-ratio measurements. In these cases, checking whether the mass-fractionation correction completely removes any correlation between the fractionation-corrected ratios and the measured ratios is an important step to verify the accuracy of the mass dependency law being used.

3.6.4 Mass analyzer

Separation of an ion beam according to its ratio of mass to charge can be accomplished by the travel time of the ions over a long path, or by the radius of curvature the ions follow when passing through either electric or magnetic fields. These three mass-separation methods produce the following types of mass spectrometers.

3.6.4.1 Magnetic sector mass spectrometer

The early mass spectrographs and spectrometers constructed by Aston and Dempster relied on magnetic fields to perform the mass separation of an ion beam. This type of mass analyzer remains the most commonly used for the type of isotope ratio measurements needed in geochronology. A charged particle traveling through a magnetic field aligned perpendicular to the ion flight path experiences a force that deflects the ion trajectories in a curved path of radius described by the equation:

$$r = \sqrt{\frac{2mv}{qB^2}} \tag{3.10}$$

where r is the radius of curvature, m is the ion mass, v its velocity, q is the charge on the ion, and B is the strength of the magnetic field. The key aspect of this equation is that the radius followed by the ion is proportional to its momentum (mv), and thus ions of the same energy, but different mass, will follow different trajectories through the magnetic field. Lighter ions will be curved more, and heavier ions less (Fig. 3.10). For a variety of reasons, the type of mass spectrometer that has found most application in geochronology uses only a sector of a circular flight path, often just a 60° or 90° section of a circle. This gives rise to the name "magnetic sector," for the type of mass spectrometer using this geometry pioneered by Alfred Nier in the 1950s. A key characteristic of a magnetic sector mass spectrometer is that the magnet serves as an optical element in the system, refocusing the diverging ion beam leaving the ion source and bringing it back into focus at a point that lies along the line connecting the ion-source exit slit and the center of the circle defined by the magnetic sector (Fig. 3.10). This is possible because ions diverging from the centerline of the ion beam have to traverse different distances within the magnetic field, forcing those on wider radius paths to be curved more, and those on inner paths to be curved less, so that their trajectories on leaving the magnet converge with the central flight path at a single point, the focal point of the magnet.

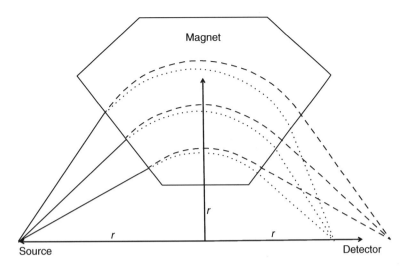

Fig. 3.10. Flight paths through a 90° sector magnet. The ions leave the ion source with a certain angular dispersion represented by the three solid black lines. In the magnet, represented by the prism, the beams are curved by the magnetic field that separates the ions by their mass to charge ratio, with the dotted line representing the lighter and the dashed the heavier of the two isotopes. The magnet also corrects for the dispersion of the beam leaving the ion source, so that each isotope beam is refocused at a point along the line connecting the source slit and the detector slit that passes through the center of the circle defined by the radius (r) of the magnet.

Because the magnetic field deflects the beam according to its mass, each isotope beam is separated and refocused at a point along a plane, separated by a distance that depends on the mass difference of the ions, the energy imparted by the ion source, and the size of the magnet.

3.6.4.2 Quadrupole mass analyzer

Although simple, and hence relatively inexpensive, the use of a quadrupole electric field to separate ions by mass was not introduced until the late 1950s by W. Paul and colleagues at the University of Bonn. The technique works by injecting an ion beam into a two-dimensional electric field in the shape of a rectangular hyperbola that is established by four circular rods (Fig. 3.11). Opposite pairs of the rods are maintained at the same voltage while the pair at 90° are held at the opposite voltage. The application of a combination of radiofrequency alternating electric field along with DC voltage to the rods causes the ions to follow oscillating paths through the quadrupole field whose wavelengths are proportional to the mass to charge ratio of the ion. The combination of frequencies and DC field can be adjusted so that the oscillating paths of ions outside of a selected mass/charge range become unstable to the point that the ions contact the rods and are neutralized (Fig. 3.11). Ions of the correct mass/charge ratio follow a stable oscillation so that they exit the quadrupole field and enter the detector. Besides its inexpensive construction, the quadrupole mass spectrometer has the advantage that the radiofrequency and DC fields can be adjusted very quickly, allowing this type of mass spectrometer to be capable of scanning a wide range of mass/charge ratios in a relatively short time. Although capable of isotope ratio measurement, most quadrupole mass spectrometers lack the mass resolution and isotope ratio precision required for some isotopic applications.

Because of its simplicity, and hence relatively low cost, the quadrupole mass spectrometer is increasingly used in some geochronology applications, for example, ^{4}He quantification by ^{3}He isotope dilution used in (U–Th)/He thermochronology (Chapter 11) and for laser ablation ICP-MS U–Pb dating of zircons (Chapter 8).

3.6.4.3 Time of flight

From equation (3.4) above, ions accelerated by a voltage potential leave the ion source with a constant kinetic energy for a given charge. Since kinetic energy is proportional to both mass and velocity, a constant kinetic energy means that light ions will be traveling faster than heavier ions. For example, rearranging equation (3.4) to solve for velocity, a singly charged atom of ^{87}Sr (mass = 86.9089 Da) accelerated in an ion source to 10 kV will have a velocity of:

$$v = \sqrt{\frac{2 \times 10,000 \, volts \times 1.60 \times 10^{-19} \, \text{Coulomb}}{86.9089 \, \text{Da} \times 1.66 \times 10^{-27} \, \frac{\text{kg}}{\text{Da}}}} \qquad (3.11)$$

With a volt having SI units of $\frac{kg \times m^2}{A \times S^3}$ and remembering that an amp (A) is equal to a Coulomb per second, equation (3.11) solves to an ion velocity of:

$$v = 148,932 \, \text{m/s}.$$

Solving the same equation for an ion of ^{86}Sr (mass = 85.9093 Da), the velocity of the ^{86}Sr ion accelerated by the same ion source is 149,796 m/s. A time-of-flight (TOF) mass spectrometer provides a long flight path, often a few meters, which allows the ions to separate by mass simply by the time they take to complete the flight path. In the example above, if the Sr ions travel a 2-m flight path from ion source to detector, they will arrive at the

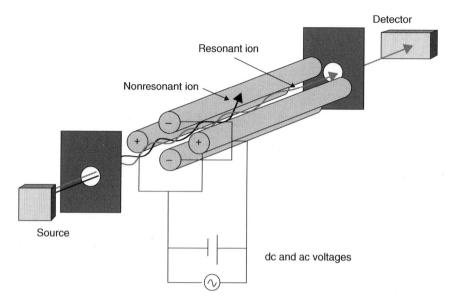

Fig. 3.11. Schematic illustration of the means of mass separation in a quadrupole mass analyzer. In this image, the gray trajectory maps the path of an ion whose mass to charge ratio is resonant with the frequency of the voltage applied to the quadrupole. The black path is for an ion of different mass to charge ratio that is forced into oscillations of increasing amplitude until it is forced out of the quadrupole. (Source: http://www.chemicool.com/definition/quadrupole_mass_spectrometry.html)

detector separated by 78 ns. This method of mass separation will only work with a signal of very short duration (a few nanoseconds) and hence requires an ion source that can be quickly turned off and on. An advantage of TOF mass spectrometers is that they are capable of sequentially detecting all elements emitted by a sample. For example, in the TOF example given above with a 2-m flight path, ^{1}H and ^{238}U will arrive at the detector 1.44 and 22.22 ms, respectively, after the ion beam enters the mass spectrometer. The mass resolution of a TOF mass spectrometer is limited by the time duration of the ion beam entering the mass spectrometer and the time of flight separation of the ions. For example, the difference in arrival time of ^{87}Sr and ^{87}Rb (86.9092 Da) in the 2-m TOF example above is 0.3 ns, which is beyond the limit that ion detection systems can distinguish one ion arrival from the next arrival. In addition, because the pulse duration of the ion source for a high-resolution TOF mass spectrometer is of order 1 ns, ^{87}Rb and ^{87}Sr would not be resolved from one another in this example. TOF mass analyzers have yet to see extensive use in geochronology, but are common in other isotopic studies, particularly in determining the mass spectrum of various organic molecules [*Henkel and Gilmour*, 2014].

3.6.4.4 Mass resolution

The goal of the mass analyzer is to separate the sample ion beam into discrete beams for each of the isotopes in the element being analyzed. To obtain accurate isotope ratios, this separation, the mass resolution of the mass spectrometer, must be sufficient so that the ion beam from one isotope is far enough removed in space from that of the next closest isotope so that no ions from the nearby mass enter the ion detector at the same time as the mass being measured. Mass resolution is usually defined as the width of an individual isotope beam divided by the distance between masses.

$$\text{resolution} = \frac{M}{\Delta M} \tag{3.12}$$

where M is the mass and ΔM the width of an individual isotope beam of that mass. In practice, the resolution is often measured from a mass scan (Fig. 3.12) using the equation:

$$\text{resolution} = \frac{\delta M \times \left[\frac{m2 + m1}{2}\right]}{\Delta M \times (m2 - m1)} \tag{3.13}$$

where $m1$ and $m2$ are the masses of adjacent peaks, δM is the distance between the midpoint of the peaks for these two masses, and ΔM is the width of the peak at either 10% or 50% of its full height. Mass scans are accomplished by varying the strength of the magnetic field in a magnetic sector, or the frequency applied to a quadrupole, so that an ion beam of a given mass moves across the entrance to an ion detector fixed in space at the focal point of the ion beam. As the beam moves across the detector, a plot of ion intensity versus mass, referred to as a mass scan, is created with peaks in signal intensity occurring sequentially as different isotopes move across the detector (Fig. 3.12). The same effect can be accomplished by scanning the high-voltage used to accelerate the ions out of the ion source, but because the efficiency with which ions are extracted from the ion source increases with increasing voltage, scanning the accelerating voltage also causes unwanted changes in signal size.

The resolution needed to just resolve an isotope from one a mass unit away thus depends on the mass. ^{88}Sr will just be resolved from ^{87}Sr at a mass resolution of 87.5, whereas just resolving ^{206}Pb from ^{207}Pb will require a resolution of 206.5. Resolution in a magnetic sector mass spectrometer depends primarily on the widths of the image slit (the slit at the end of the ion source) and the detector slit (the slit placed just in front of the detector) and the size (radius) of the magnet according to the equation:

$$\text{resolution} = \frac{r}{(S+D)} \tag{3.14}$$

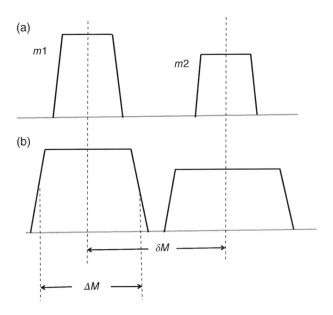

Fig. 3.12. Mass scan of two adjacent isotopes of masses $m1$ and $m2$ at two different resolutions. The width of the source slit is constant between (a) and (b), but the collector slit is twice as wide in (b) as in (a), allowing each ion beam to fully enter the detector over a wider range of apparent mass. Because only the slit widths were changed between these scans, the separation between the peaks (δM) remains the same, so the resolution is reduced in scan (b). (Source: *Carlson* [2013]. Reproduced with permission of Elsevier.)

where r is the radius of the magnetic sector and S and D are the width of the source and detector slits, respectively. Higher resolution thus is achieved by increasing the size of the magnet or by making the slits narrower, but the latter also cuts down on the ion transmission through the instrument, limiting its sensitivity. In practice, in isotope-ratio mass spectrometry, resolutions need to be substantially higher than the minimum needed to separate two neighboring isotopes because one wants the detector to measure the entire incoming beam and none of the neighboring beam. Slight variations in magnetic field strength or the accelerating voltage in the ion source will cause the exact position of the beam to wander somewhat by the time it reaches the detector. As a result, a common goal in magnetic sector mass spectrometry is a resolution low enough to allow "flat-topped peaks" while still keeping neighboring isotopes far enough apart so that only one is in the detector at a given time (Fig. 3.13). The "peak shape" of a mass spectrometer refers to the rate of change of ion beam intensity in the detector as the magnetic field is slowly varied to move the ion beam across the detector. When the focused ion-beam width is the same size as the entrance slit to the detector, scanning the beam across the entry slit will produce a conical peak shape (Fig. 3.13c). If the ion beam is smaller than the entry slit, however, the same scan will produce a peak with a flat top that marks the range of the mass scan where all of the ion beam enters the detector (Fig. 3.13a, b).

3.6.4.5 Double-focusing mass analyzers

As the ion beam travels from source to detector, any ion that impacts either a stray atom in the imperfect vacuum of the flight tube, or suffers a glancing blow off one of the surfaces of the flight tube, will experience deflection of its flight path and the loss of some amount of energy. Both processes result in a small fraction of the ion beam diverging from the flight path of the majority of ions. This is expressed in mass scans as a "tail" to the main isotope peak.

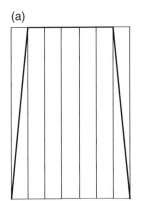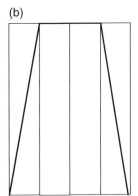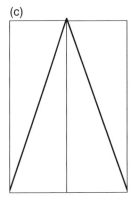

Fig. 3.13. Illustration of the dependence of peak shape on source and detector slit width. In each panel, the magnetic field is being slowly changed so that the ion beam is scanned across the detector slit producing the signal at the detector denoted by the bold black line. The boxes show the width of the ion beam in three examples for different source slit widths, but the same detector slit width. The slope of the side of the peak reflects the width of the incoming beam. The flat top is created when the incoming beam is thinner than the detector slit so that over a range in magnetic field, the whole beam passes through the detector slit into the detector. In (c), the beam width is the same as the detector width, so as soon as all of the beam enters the detector, the beam encounters the opposite edge of the detector slit causing the signal size to go down as the beam continues its scan out of the detector slit. (Source: *Carlson* [2013]. Reproduced with permission of Elsevier.)

The term "abundance sensitivity" refers to the amount of stray ions from an isotope peak that scatter under the peak of a nearby isotope (Fig. 3.14).

The ability of a magnetic sector to focus an ion beam at the detector depends critically on the energy spread of the ions. The peak broadening (tailing) that contributes to abundance sensitivity is caused by both small-angle scattering of the ions during their flight, but also the energy loss that occurs when they scatter

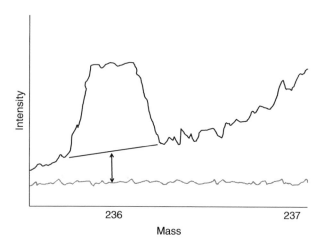

Fig. 3.14. Mass spectrometer scan of the mass range from 235.5 to 237. The lower gray trace is with no ion beam in the mass spectrometer and shows the electronic baseline of the detector. The upper trace is with a large U ion beam showing the very small peak for ^{236}U (a typical ^{238}U/^{236}U ratio is of order ten million) and the elevated baseline caused by the tailing of the very large ^{238}U ion beam off to the right of the scan. The abundance sensitivity is defined by the magnitude of the offset of the tail from ^{238}U above the electronic baseline, as shown by the arrow to the linear line extrapolation of the tail beneath the ^{236}U peak. Typical abundance sensitivities at one mass unit away from a large peak in single focusing magnetic sector instruments are of order a few times 10^{-6} of the signal intensity of the large peak.

off the remaining gas molecules in the imperfect vacuum of the mass spectrometer. Because the magnetic sector deflects the ion beam in proportion to its momentum, ions that lose energy through collision in, or prior to entering, the magnetic field will be curved more than those that do not, and hence end up creating an apparently low-mass tail to a given isotope peak. The same problem exists for ion sources that produce an ion beam with a significant spread in ion energy, for example ICP and SIMS ion sources. We saw in equations (3.4) and (3.5) that ions traveling in electric fields undergo accelerations proportional to their kinetic energy ($1/2\ mv^2$). As a result, curved metal plates can be used to create a constant electric field that will cause ions to enter into curved trajectories whose radius of curvature is determined by equation (3.5). The result is that low-energy ions will be curved more than high-energy ions with the consequence that a curved electric field can be used much as a magnetic sector to separate ions, but in the case of electric fields, the separation will be proportional to the ion energy. The combination of an electric sector and magnetic sector results in a double-focusing mass spectrometer. When the electrostatic sector is placed before the magnetic sector (Fig. 3.15), as it is in many magnetic sector ICP-MS and SIMS instruments, the electrostatic sector focuses the ion beam according to its kinetic energy, so if a metal plate with a thin slit is placed at the focus of the electrostatic sector, only an ion beam with a narrow energy dispersion will pass through to enter the magnetic sector. Because the magnetic sector focuses on the basis of the momentum (mv) of the ions, the smaller the range of velocity (v) of the ions entering the magnetic sector, the better it can refocus the ion beam on the basis of only the mass of the ion beam. Similarly, placing the electrostatic sector after the magnet (as in reverse geometry double-focusing mass spectrometers) can remove ions that have lost energy through collisions during flight, and hence remove those ions so that they do not contribute to peak tailing and abundance sensitivity.

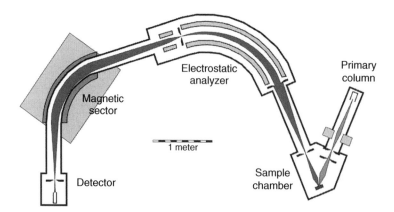

Fig. 3.15. Schematic illustration of a double-focusing mass spectrometer combining electrostatic and magnetic sectors. This instrument is a secondary ion mass spectrometer (SIMS or ion probe) that sputters the solid sample using a primary ion beam (right) to create ions from the sample (dark gray) that are then accelerated into an electrostatic analyzer to allow selection of a small energy window of the sputtered sample ions. The ions that meet the energy window of the electrostatic analyzer pass through a slit into a magnetic sector that performs the mass separation of the beam into its constituent isotopes, refocusing the isotope beam at the detector slit. (Source: *Williams* [1998]. Reproduced with permission of Society of Economic Geologists.)

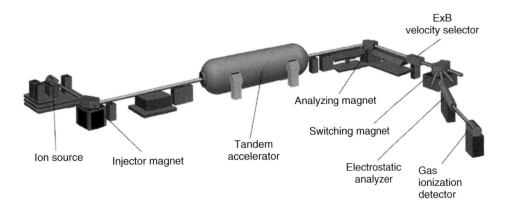

Fig. 3.16. Illustration of the Purdue accelerator mass spectrometer and the various mass and energy filters needed to achieve very high abundance sensitivities. (Source: from *Granger et al.* [2013]. Reproduced with permission of the Geological Society of America.)

A similar effect can be achieved by what is known as a retarding potential lens that, in essence, is the reverse of the ion source, slowing the main ion beam with an opposing electric field to near zero velocity so that the ions that enter the lens with lower energies than the main beam cannot penetrate the electric field of the lens to reach the detector.

The best abundance sensitivity is obtained in a very specialized type of mass spectrometer, an AMS. As illustrated by the AMS at Purdue University (Fig. 3.16), negative ions are produced by a Cs-sputter source and accelerated by a kV electric field into the mass spectrometer. An initial mass separation is performed at what is called the injector magnet that allows only ions of a narrow mass range into a tandem van de Graaff accelerator that uses static electricity to generate of order +10 million volts on a metal plate at the middle of the accelerator vessel. The negatively charged sample ions are accelerated towards this very high voltage. By doing so, the ion beam obtains sufficient energy to penetrate either a thin carbon foil or a small region of low-pressure gas at the high-voltage terminal. Passing through this "stripper" completely destroys any molecules that might have been present in the ion beam and transforms the negatively charged ions into high positive charge ions, for example C^{3+} when used for radiocarbon analysis. These positive ions are then repelled by the center plate at high voltage and hence gain even more energy. The very energetic ions are then passed through other magnetic and electrostatic sectors for additional mass and energy separation before they arrive at the ion detector. By involving so many steps of mass and energy filtering, AMS completely eliminates any molecular background and allows the detection of very low abundance isotopes separated by only one mass from very abundant isotopes. For example, abundance sensitivities in an AMS are of order 10^{-15} or smaller compared to of order 10^{-6} in a typical single-focusing TIMS. Such high-abundance sensitivities are essential to the application of AMS in cosmogenic nuclide dating (Chapter 13) given that, for example, typical $^{14}C/^{12}C$ ratios in modern living organisms are on the order of 10^{-12}.

3.6.5 Detectors

Once the ion beam leaving the source has been separated by mass, the next step in isotope ratio measurement is quantifying the magnitude of the ion beam for each isotope. Photographic film, darkened by impacting ions, was the first ion detector used in mass spectrographs, but was replaced by electronic detection methods in the transition to mass spectrometry. Ion detectors fall into two categories, Faraday cups and electron multipliers.

3.6.5.1 Faraday cups

Faraday cups are rectangular buckets most commonly made out of steel or graphite. They are connected to electrical ground through a very high-ohmage resistor, often of order 10^{10} to 10^{13} ohms (Fig. 3.17). Every positive ion that enters the Faraday cup is electrically neutralized by an electron that travels from ground through the resistor. Using Ohm's law:

$$\text{voltage} = \text{current} \times \text{resistance} \qquad (3.15)$$

so a picoamp current traveling through a 10^{11} ohm resistor will create a 0.1 V signal. This signal is amplified electronically and then converted to digital format to be read and stored by the instrument control computer. The signal size that can be detected using Faraday cups is limited by what is known as "Johnson noise," caused by the electron currents that arise within a conductor simply due to thermally induced movement of electrons [*Johnson*, 1928; *Nyquist*, 1928]. Johnson noise in high-ohmage resistors creates currents on the order of 10^{-16} amps. Other issues that can compromise the accuracy of ion signal quantification using Faraday cups are some nonideal behavior of high-ohm resistors, such as their tendency to have resistances that are slightly dependent on temperature (temperature coefficient) and on the magnitude of current traversing the resistor (voltage coefficient). The former creates the need for temperature-controlled amplifier housings in most modern mass spectrometers, and the latter leads the resistors to be placed in low vacuum enclosures in order to avoid humidity changes in the resistor that can change their behavior.

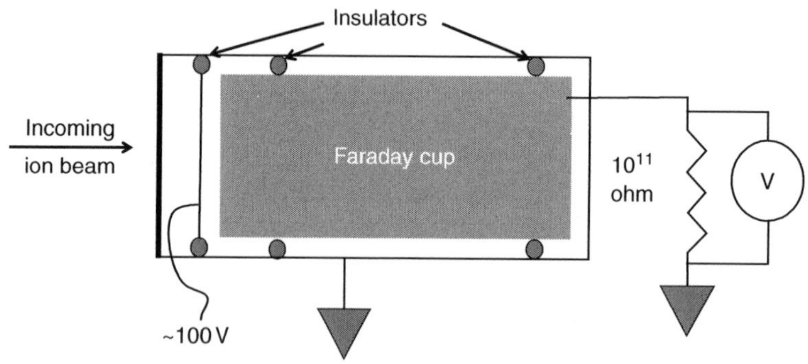

Fig. 3.17. A typical Faraday cup design and schematic signal quantification electronics. The gray rectangle shows the side view of the cup. The ion beam enters the hollow cup through a small, electrically grounded, slit placed in front of the Faraday. The Faraday cup is connected to ground only through a high-ohmage resistor so that the ion current flowing into the Faraday creates a current across the resistor that can be converted to a voltage proportional to the ion current size. Most Faraday cups will have one or more plates with a thin slit near the entrance to the cup to which a voltage is applied in order to repel secondary electrons back into the Faraday cup when they are ejected from the cup material by the energetic impact of the incoming ions. (Source: *Carlson* [2013]. Reproduced with permission of Elsevier.)

A major requirement in Faraday cup ion detection is that the cup must accurately retain the electrical charge of every ion that enters the cup. At the accelerating voltages used in many mass spectrometers, ions have enough energy to eject secondary charged particles when they impact a solid surface. The most common secondary particle is an electron, but secondary positively changed and negatively charged ions can also be emitted by energetic ion impacts. If the sample ion beam impacts anything in front of the Faraday cups, for example, the mounting hardware for the cup, then a spray of secondary charged particles can be created. If these secondary charged particles enter the Faraday cups, they will be counted as signal. If secondary charged particles created by impacts inside the Faraday cup escape the Faraday cup, they can either reduce or amplify (depending on whether they are positively or negatively charged) the amount of charge contributed by the sample ions, thus affecting the quantification of the ion signal size. Keeping secondary charged particles from affecting the signal detection by Faraday cups can be a challenge that is addressed by placing one or more slitted-plates connected to either positive or negative voltages at the entrance of the Faraday cup in order to repel electrons or secondary ions back into the Faraday cups (Fig. 3.17). Another method to keep secondary electrons under control involves placing permanent magnets along the sides of the Faraday cups. The magnetic field causes the electrons to spiral, thus enhancing the probability that they will impact the inner wall of the Faraday cup, and be electrically neutralized, before they can escape. Machining the Faraday cup out of graphite instead of steel is another approach to reducing this problem because graphite produces fewer positive secondary ions than does steel, so the focus on secondary particle control can concentrate on only negatively charged particles.

3.6.5.2 Ion multipliers
The other common detector in mass spectrometers takes advantage of the production of secondary electrons during energetic

ion collisions with solid surfaces. Three types of ion multipliers are in common use for mass spectrometry.
(1) Discrete dynode electron multipliers (DDEM, Fig. 3.18a) consist of a stacked series of curved metal plates, usually made from beryllium-copper alloy, called dynodes. The first of these, the conversion dynode, is impacted by the sample ion beam and emits a number of secondary electrons. A voltage, typically 1–2 kV, split evenly among the dynodes by the resistor string connecting them, causes these electrons to be accelerated into the second dynode where they impact to emit more electrons. The process of electron amplification continues through other dynodes, often somewhere between 12 and 20 depending on the amount of amplification desired, until the electron beam is directed into a Faraday cup to quantify its magnitude.
(2) Continuous dynode electron multipliers (CDEM, Fig. 3.18b) consist of a curved glass cone, coated on the inside with a semiconductor that allows a voltage difference to be maintained between the front of the cone and its base. Ions impacting the front of the cone emit electrons that are then accelerated down into the narrower portion of the cone, bouncing against the cone walls and emitting more electrons on the way. CDEMs can be made smaller than DDEMs, and are thus useful in multiple detector arrays where more than one ion beam can be detected simultaneously. They have significantly poorer gain stability, shorter life times, and lower maximum count rates than DDEMs, and thus tend to be used for more specialized applications where small detectors are needed.
(3) Daly detector (Fig. 3.18c). The conversion dynode of a Daly detector is a door-knob shaped piece of metal offset at right angles from the flight path of the ion beam. When the Daly is on, a high negative voltage, of order 20 kV, is applied to the Daly knob, which causes a positive ion beam to turn and impact the knob at high energy. The secondary electrons produced in this impact are than accelerated in the opposite

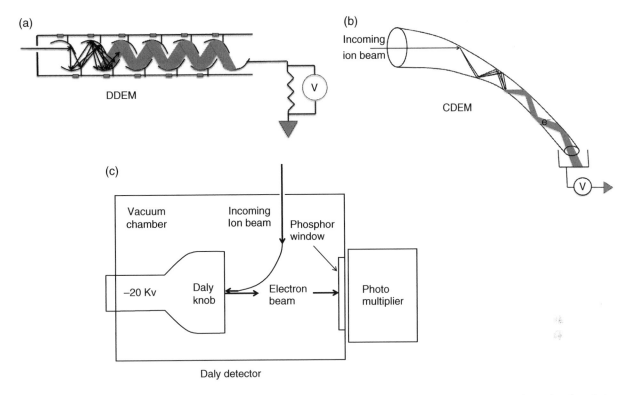

Fig. 3.18. Schematic examples of three types of electron multipliers: (a) the discrete dynode multiplier, (b) the continuous dynode multiplier, and (c) the Daly detector.

direction to impact a window coated with a material that emits light photons when struck by the electrons. The light then passes through a window into a discrete dynode electron multiplier whose first dynode emits electrons when impacted by light photons. This type of multiplier is known as a photomultiplier. The Daly detector has slightly better electron conversion efficiency, lower background, and the ability to work at higher count rates than a DDEM. The Daly, however, only works with positive ions whereas a DDEM can be set up to detect either positive or negative incoming ions.

The signal amplification provided by an electron multiplier depends on the voltage between the dynodes and the number of dynodes, but often can approach 10^5 to 10^7 electrons per incident ion at the conversion dynode. The advantage of an electron multiplier over a Faraday cup is signal amplification, thus allowing substantially smaller ion currents to be quantified before encountering the limit imposed by Johnson noise. The weakness of the electron multiplier is that the electron yield at each dynode is a probabilistic parameter, meaning that the magnitude of the signal at the end of the multiplier is not exactly proportional to the size of the ion signal impacting the conversion dynode, but fluctuates up and down about a mean value depending on exactly how many electrons are emitted at each dynode. There also is a slight mass dependency in the signal amplification of an electron multiplier because heavier ions emit more electrons than light

ions impacting the conversion dynode. Electron multipliers also have an upper limit to the size of signal they can measure because large signals result in very large electron currents at the final dynodes that serve to defocus the electron cascade because of the high charge densities. These properties of electron multipliers limit their maximum precision for isotope ratio determinations to 10^{-3} to 10^{-4} compared to precisions of 10^{-6} achievable with Faraday cups.

The magnitude of the electron beam entering the Faraday cup at the base of the electron multiplier can be quantified either in analog mode, which is basically the same as described previously for the direct Faraday cup approach, or through what is known as pulse counting. Each individual ion impacting the conversion dynode results in a short pulse of electrons into the Faraday cup at the base of the electron multiplier (Fig. 3.19). Analog measurement sums all the electrons to arrive at an average signal size. Pulse counting ignores the magnitude of each pulse, but instead counts each pulse as a single event, and sums the "events." At low ion currents, the incoming pulses are separated sufficiently in time that they can be discriminated electronically. In this case, each pulse corresponds to a single ion impacting the conversion dynode. Pulse counting thus reduces the mass dependency of electron multiplication because even if the pulse height produced by, for example, the impact of a single ion of ^{88}Sr is slightly higher than the pulse height of a single ion of ^{86}Sr, it only counts each pulse as one incoming ion. Another

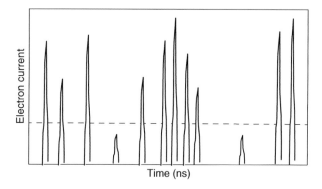

Fig. 3.19. Example of pulse counting applied to the signal supplied by an electron multiplier. Ion arrivals create bursts of electrons in the multiplier that can be of varying size. Pulse counting ignores the magnitude of the pulse and simply counts each pulse as a single ion. One can set a discrimination level (dashed line) below which the pulse is not counted. The width of the peak shows the recovery time of the electron multiplier after each ion arrival, which is the "dead time" of the electron multiplier.

advantage to pulse counting is that it can virtually eliminate the noise of the multiplier from the incoming signal. From thermal emission alone, the metal of the dynodes will occasionally emit an electron. This electron will be accelerated in the same manner as electrons emitted by ion impacts and thus create a small background current at the Faraday cup end of the electron multiplier. If the thermal electrons are emitted from any dynode other than the conversion dynode, however, the pulse of electrons that eventually reaches the Faraday cup end of the multiplier will be smaller in magnitude than the pulse created by the electrons emitted due to an impacting ion. One can thus set a "discrimination window" for the pulse height that counts only the pulses big enough to correspond to arriving ions, and not the pulses deriving from random electron emissions in the multiplier itself, thus eliminating the background noise from the multiplier (Fig. 3.19).

At larger signal sizes, pulse counting encounters the problem of "dead time." Electron pulses from an electron multiplier have durations of many nanoseconds. For large ion currents, the pulses at the end of the multiplier will arrive too close together in time to be discriminated from one another by the pulse-counting electronics, so two or more pulses will be counted as one. The window of time needed between the arrivals of two electron pulses so that both can be counted is known as the detector "dead time." Most detector dead times are on the order of tens of nanoseconds. When the incoming ion beam is big enough that detector dead time becomes important, the measured signal compared to the real signal is given by:

$$CPS_{Meas} = CPS_{Real} \times e^{-\tau CPS_{Real}} \qquad (3.16)$$

where CPS is the counts per second of the measured (Meas) and real signal, and τ is the dead time. An approximate correction for the dead time can be applied with an expansion of the above equation to:

$$CPS_{Real} \approx CPS_{Meas} \times e^{\tau \times CPS_{Meas} \times e^{\tau \times CPS_{Meas} \times e^{\tau \times CPS_{Meas}}}} \qquad (3.17)$$

which often is simplified to:

$$CPS_{Real} \approx \frac{CPS_{Meas}}{(1 - CPS_{Meas} \times \tau)} \qquad (3.18)$$

When count rates become too high, generally more than a few million counts per second, the inaccuracy of the dead time correction begins to dominate the uncertainty of the ion signal quantification. At these count rates, it is time to switch from electron multiplier to a Faraday cup detector.

3.6.5.3 Gas energy-loss detectors

A final form of ion detection takes advantage of the very high energy of the ions produced by AMS. Ions at the many million volt energies used in AMS can travel considerable distances through either solids or gases, losing energy as they go through collisions with the atoms they encounter on their path. The gas ionization detectors employed in AMS allow the ions to travel through a path filled with low-pressure gas (Fig. 3.20). When the ions impact a gas molecule, they cause it to ionize. The detector accelerates the electron produced by the ionization so that it impacts one of a series of metal plates spaced along the flight path of the ion through the detector. Through this method, the detector not only records the arrival of the ion, but can track the position where gas ionization occurs and hence the rate at which the sample ion is slowed down by collisions with the gas. The rate of energy loss is proportional not only to the energy of the incoming ion, but also to the nuclear charge of the ion. The detector thus can discriminate between the arrival of, for example, ^{10}Be and ^{10}B not because of their difference in mass, but because the ^{10}B loses energy more quickly in the detector as a result of its higher nuclear charge.

3.6.6 Vacuum systems

Eliminating chemical reactions that would neutralize sample ions, and minimizing collisions between ions and neutral gas atoms, requires that the ion beam travel in an environment of very low gas pressure, or practically speaking, a vacuum. In the days of Aston, most vacuum chambers were glass tubes either welded together or sealed with wax. These have been replaced for the most part by stainless steel tubes and chambers either welded together or attached to one another with compression fittings that prevent leakage of air into the evacuated lines. For relatively low-vacuum applications like backing lines for turbo pumps at ~0.1–1.0 Pa (~10^{-2} to 10^{-3} Torr, ~10^{-2} to 10^{-3} mbar), compressed fluoroelastomer O-rings provide seals between detachable line components. For higher vacuum applications, including ultrahigh vacuum (UHV, ~10^{-7} Pa) purposes, only low-vapor-pressure components, typically metals, can be used; UHV chambers are often connected by deformable copper gaskets compressed by stainless steel fittings, often with knife-edge flanges. After pumping to remove most gas molecules in UHV systems using pumps described below, the chambers or line are often heated, or "baked," to temperatures of 200–300 °C to remove adsorbed gases in interior parts.

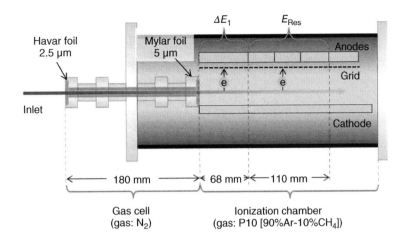

Fig. 3.20. Schematic illustration of a gas-filled energy loss detector [*Matsubara et al.*, 2014]. The energetic ion beam enters the detector from the left, passes through two thin foils and then travels through the gas-filled chamber where each collision with a gas molecule releases an electron. (Source: Matsubara, http://cdn.intechopen.com/pdfs-wm/47051.pdf. Used under CC BY 3.0.)

A large number of methods exist to pump gas molecules from inside the chamber, the choice of which depends on the vacuum desired, the amount of gas that needs to be removed, sensitivity of the measurements that depend on the vacuum to either vibration or contamination from the pumping system, and cost. Commonly used vacuum pumps include the following.

(1) Mechanical pumps: a wide variety of such pumps exist, but all are based on some method of changing the volume of a chamber, for example, by moving a piston up and down in a cylinder, or rotating a rod off center to a cylinder as in a rotary vane pump, to alternately suck gas on the down-stroke and expel the gas, usually to the room air, on the up-stroke. Many mechanical pumps achieve seals with their moving parts by immersing them in oil, but the oil vapor can result in the build-up of a hydrocarbon background in the mass spectrometer that can interfere with some measurements. As a result, many modern mechanical pumps are moving away from oil with only minor reduction in the ultimate vacuum they can achieve. Most vacuum pumps of this type can produce ultimate vacuums in the range of ~1 to 0.01 Pa. In most isotope geochemistry and geochronologic applications, mechanical pumps cannot achieve the high vacuum necessary inside mass spectrometer lines, but they are instead used to create low pressure on the exhaust side of other types of pumps capable of achieving higher vacuum, particularly turbo pumps.

(2) Sorption pumps, which are typically passive sorption devices that trap reactive gases. Some consist of a can filled with what is known as "molecular sieve," which is millimeter-sized pellets or balls of some microporous material such as zeolite, diatomaceous earth, or activated carbon. To pump, the can is submerged in liquid nitrogen, causing the gas inside the pump to freeze onto the molecular sieve. Sorption pumps can reach ultimate pressures in the range of 10^{-4} Pa, but once their sieve is saturated with the gas they pump,

their pumping capacity stops. A related passive pumping device often used in noble gas mass spectrometry is a "getter." Getters are metal alloys available as pellets or coatings on high surface-area-to-volume shapes that react with and accumulate gases. Both molecular sieves in sorption pumps and getters can be regenerated and reused by heating, but this requires isolation of the pump from the vacuum chamber so that the condensed gases can be expelled into the atmosphere. Sorption pumps contribute minimal contamination to the vacuum system, but only pump until the sieve becomes saturated. Some types do not pump any gas that cannot be frozen at liquid nitrogen temperature. They require large volumes of liquid nitrogen to cool, and need hours of heating before they can be reused. As a result, sorption pumps have mostly been replaced by oil-free mechanical vacuum pumps for what is commonly called "rough pumping," which is needed before most pumps capable of reaching higher vacuums can be employed.

(3) Diffusion pumps, which employ a fluid, either oil or mercury, that can be boiled and recondensed, so that the vapor can be expelled in jets directed down at the cooled outer wall of the pump body. The vapor jets impact the background gas molecules from the vacuum system and push the gas molecules towards the bottom of the pump, where they are pumped out by a mechanical pump. Diffusion pumps can compress the background gas by several orders of magnitude through this process so that while the base of the pump is maintained at a background pressure of some 0.1 Pa by the mechanical pump, the upper portions of the pump can reach pressures approaching 10^{-7} Pa. Diffusion pumps also can handle relatively large gas volumes and thus were the preferred pumps for imperfectly sealed vacuum systems or those supporting ion sources that produced substantial amounts of gas. While once the dominant high-vacuum pump, the need to reduce hydrocarbon background from oil diffusion pumps, the

health dangers of mercury diffusion pumps, and improvements in other forms of high-vacuum pumping are making diffusion pumps less commonly used.

(4) Turbo pumps, a style of pump that involves a rapidly rotating (up to 70–100 thousand r.p.m.) turbine interior to the pump, which pushes the gas atoms to the base of the pump where they are removed by a mechanical backing pump. Improvements in the bearings that support the spinning turbine have increased the reliability of these pumps to the point where they are becoming the primary method of obtaining high vacuums (e.g., 10^{-4} to 10^{-7} Pa). They offer the advantage of emitting minimal contaminating gases or stray ions, and can pump all gases, including relatively unreactive gases such as noble gases. Downsides to turbo pumps include their high cost and that they produce a small amount of vibration.

(5) Ion pumps, which work by imposing a strong electric field, of order a few kV, across a small area, which causes easily ionized gas atoms to be ionized. The ionized gas atom is then accelerated by the electric field into a titanium plate that chemically adsorbs the gas atom. The electron emitted by ionization of the gas atom is accelerated in the opposite direction, but because the whole pump is surrounded by a strong permanent magnet, the electron path becomes helical, increasing the probability that it will impact a gas atom and cause its ionization. This cycle of ionization and acceleration of the ions into the titanium cathode is the working principle of the ion pump. Ion pumps are relatively clean, vibration-free, and do not emit any gases of their own, but they can be a source of stray ions, especially if extensive use leads to build up of adsorbed gases. Only some types of ion pumps are capable of pumping gases with high ionization energies (e.g., noble gases), they are heavy, and they occasionally need to be heated in order to remove poorly bound gas atoms on the titanium plates. They also can be used only when the vacuum is better than about 10^{-4} Pa. Above that pressure, the ion and electron currents in the pump heat the anode and cathode to the point where they begin to emit, rather than pump, gas.

(6) Cryopumps use internal veins or plates cooled either with liquid nitrogen or helium. At liquid helium temperatures, all gases will condense on the cold plates. The flow of liquid helium is provided by an internal "refrigerator" that creates the liquid He via compression and expansion, and forces its flow into the plates in the center of the pump. Like any type of sorption pump, cryopumps occasionally need to be "regenerated," or heated, to boil off the gases that have condensed on their cold plates, which reduce the pumping efficiency.

3.7 CONCLUSIONS

The wide variety of radioisotopic systems comprising modern geochronologic methods leads to a wide variety of analytical methods. Methods focusing on radiation damage manifestations

such as luminescence, electron-spin resonance, and fission-track dating use distinct techniques that differ from most other dating methods based mostly on quantifying proportions of parent and daughter nuclides.

Although some techniques, such as whole-rock analysis for ^{40}Ar/^{39}Ar dating or laser ablation or SEM analyses of thin-sections, require little in the way of sample preparation, most techniques require mineral separation to derive a relatively pure concentrate of a specific mineral type from a bulk sample. This usually involves steps of crushing followed by separations based on density and magnetic properties, and often hand-picking. Extraction of elements to be analyzed may be achieved directly by laser ablation, or ion-impact methods (SIMS), but often involves acid digestion to prepare a solution or heating by furnace or laser to release gases. Ion exchange chromatography is often applied to solutions to purify and concentrate elements, whereas "gettering" and cryogenic purification are usually applied to gases released from samples for noble gas work. In all cases, "blanks" or background levels of isotopes or elements in routine procedures are often a primary concern that limits analytical precision and accuracy, and preparation facilities may take place in carefully controlled clean rooms with low levels of airborne contaminants.

Geochronologic analyses based on parent–daughter nuclides often involves measurement of isotopic compositions as well as elemental concentrations or mass/molar quantities. A common high-precision method for measuring an elemental concentration or abundance in a sample is isotope dilution, in which a known amount of an element with a precisely known and artificially enriched isotopic composition (spike) is mixed with a sample in either solution or gas form. Measurement of the resulting isotopic composition of the sample–spike mixture can then be used to solve for the amount of the natural (normal) element in the sample.

The analytical staple of much modern geochronology is measurement of isotopic compositions by mass spectrometry. Although this includes a wide variety of instruments and approaches, most mass spectrometry comprises five essential parts.

(1) *Ionization* of the element whose isotopic composition is to be measured: common methods include electron bombardment from a heated filament (used in noble gas applications), resistance heating of a concentrated and purified sample (e.g., TIMS), plasma heating of a solution or aerosol of the sample (ICP-MS), ion-impact (e.g., secondary ion production in SIMS and AMS applications), laser resonance (not widely used in geochronologic applications but future technological developments may change this).

(2) *Acceleration and focusing* of ions prior to detection: once formed, ions exiting the ion source are focused into a narrow beam by voltage potentials, series of electrostatic lenses, and a physical exit slit. In some systems, further focusing and reduction of energy dispersion of the ions is achieved by additional electric fields and slits, after traveling through

the mass analyzer. Mass fractionation produced throughout this and other steps of mass spectrometry needs to be corrected by comparing measured nonradiogenic isotope ratios and/or analyses of standards.

(3) *Mass separation of nuclides*: most geochronologic applications use either magnetic sector or quadrupole mass spectrometers to separate ion beams of isotopes of interest. Magnetic sectors produce the best mass resolution whereas quadrupoles are capable of scanning or jumping between masses quickly.

(4) *Ion detection*: establishing the relative abundances of isotopes requires measuring the relative strength of ion beams at the end of their flights. Commonly used detectors are Faraday cups, which use a high-ohmage resistor to convert the current of an ion beam into a voltage, and electron multipliers and Daly detectors, which measure counts of individual ions per time, generally on ion beams of lower intensity.

(5) *Vacuum*: all of the component parts described above must operate at extremely low gas pressure, to minimize collisions of ions with gas atoms. Vacuum generation inside mass spectrometers is achieved with a number of different types of pumps, but the most commonly used pumps in modern instruments are turbomolecular and ion pumps capable of achieving pressures typically $< 10^{-5}$ Pa and $< 10^{-7}$ Pa, respectively. Mechanical "rough" pumps are also commonly used to achieve pressures around 0.01–1.0 Pa, especially as backing for turbo pumps.

Future directions in geochronologic analysis may continue to emphasize high spatial resolution as exemplified in recent progress with SIMS, nanoSIMS, and atom-probe tomography [e.g., *Valley et al.*, 2015], and innovative techniques and instrument design for sample dating *in situ* on planetary surfaces.

3.8 REFERENCES

Carlson, R.W. (2013) Thermal ionization mass spectrometry. In *Treatise on Geochemistry*, McDonough, W. F. (ed.). Elsevier, Amsterdam.

Clifton, H. E., Hunter, R. E., Swanson, F. J., and Phillips, R. L. (1969) Sample size and meaningful gold analysis. *US Geological Survey Professional Paper* 625-C, C1–C17.

Creaser, R. A., Papanastassiou, D. A., and Wasserburg, G. J. (1991) Negative thermal ion mass spectrometry of osmium, rhenium, and iridium. *Geochimica et Cosmochimica Acta* 55, 397–401.

Dempster, A. J. (1918) A new method of positive ray analysis. *Physical Review* 11, 316–325.

Dietz, L. A. (1959) Ion optics of the V type surface ionization filament used in mass spectrometry. *Review of Scientific Instruments* 30, 235–241.

Dodson, M. H. (1970) Simplified equations for double-spiked isotopic analyses. *Geochimica et Cosmochimica Acta* 34, 1241–1244.

Faure, G. and Mensing, T. M. (2005) *Isotopes. Principles and Applications*, 897. John Wiley & Sons.

Galer, S. J. G. (1999) Optimal double and triple spiking for high precision lead isotopic measurement. *Chemical Geology* 157, 255–274.

Granger, D. E, Lifton, N. A., and Willenbring J. K. (2013) A cosmic trip: 25 years of cosmogenic nuclides in geology. *Bulletin, Geological Society of America* 125, 1379–1402.

Halliday, A. N., Lee, D. C., Christensen, J. C., *et al.* (1995) Recent developments in inductively coupled plasma magnetic sector multiple collector mass spectrometry. *International Journal of Mass Spectrometry and Ion Processes* 146/147, 21–33.

Henkel, T. and Gilmour, J. (2014) Time-of-flight secondary ion mass spectrometry, secondary neutral mass spectrometry, and resonance ionization mass spectrometry. In *Treatise on Geochemistry*, McDonough, W. F. (ed.), 411–424. Elsevier, Amsterdam.

Ireland, T. R. (2014) Ion microscopes and microprobes. In *Treatise on Geochemistry*, McDonough, W. F. (ed.), 385–409. Elsevier, Amsterdam.

Johnson, J. (1928) Thermal agitation of electric charge in conductors. *Physical Review* 32, 97.

Jull, A. J. T. and Burr, G. S. (2014) Accelerator mass spectrometry. In *Treatise on Geochemistry*, McDonough, W. F. (ed.), 375–383. Elsevier, Amsterdam.

Krogh, T. E. (1973) A low-contamination method for hydrothermal decomposition of zircon and extraction of U and Pb for isotopic age determinations. *Geochimica et Cosmochimica Acta* 37, 485–494.

Krogh, T. E. (1982) Improved accuracy of U–Pb zircon dating by selection of more concordant fractions using a high gradient magnetic separation technique. *Geochimica et Cosmochimica Acta* 46, 631–635.

Matsubara, A., Saito-Kokubu, Y., Nishizawa, A., Miyake, M., Ishimaru, T., and Umeda, K. (2014) Quaternary geochronology using accelerator mass spectrometry (AMS)—current status of the AMS system at the TONO Geoscience Center. In *Geochronology—Methods and Case Studies*. Morner, N.-A. (ed.). INTECH. DOI: 10.5772/58549.

Nyquist, H. (1928) Thermal agitation of electric charge in conductors. *Physical Review* 32, 110.

Olesik, J. W. (2014) Inductively coupled plasma mass spectrometers. In *Treatise on Geochemistry*, McDonough, W. F. (ed.), 309–336. Elsevier, Amsterdam.

Reisberg, L. and Meisel, T. (2002) The Re–Os isotopic system: a review of analytical techniques. *Geostandards Newsletter* 26, 249–267.

Savina, M. R., Pellin, M., Tripa, C., Veryovkin, I., Davis, A., and Calaway, W. (2003) Analysing individual presolar grains with CHARISMA. *Geochimica et Cosmochimica Acta* 67, 3215–3225.

Schaltegger, U., Schmitt, A. K., and Horstwood, M. S. A. (2015) U–Th–Pb zircon geochronology by ID-TIMS, SIMS and laser ablation ICP-MS: recipes, interpretations, and opportunities. *Chemical Geology* 402, 89–110.

Schoenbaechler, M. and Fehr, M. A. (2014) Basics of ion exchange chromatography for selected geological applications. In *Treatise on Geochemistry*, McDonough, W. F. (ed.), 124–144. Elsevier, Amsterdam.

Shirey, S. B. and Walker, R. J. (1995) Carius tube digestions for low-blank rhenium–osmium analysis. *Analytical Chemistry* **67**, 2136–2141.

Silver, L. T. and Deutsch, S. (1963) Uranium–lead isotopic variation in zircon—a case study. *Journal of Geology* **71**, 721–758.

Stracke, A., Scherer, E.E., and Reynolds, B.C. (2014) Application of isotope dilution in geochemistry. In *Treatise on Geochemistry*, McDonough, W.F. (ed.), 71–86. Elsevier, Amsterdam.

Valley, J. W., Reinhard, D. A., Cavosie, A. J., *et al.* (2015) Presidential Address. Nano-and micro-geochronology in Hadean and Archean zircons by atom-probe tomography and SIMS: new tools for old minerals. *American Mineralogist* **100**(7), 1355–1377.

Volkening, J., Walczyk, T., and Heumann, K. G. (1991) Osmium isotope ratio determinations by negative thermal ionization mass spectrometry. *International Journal of Mass Spectrometry and Ion Processes* **105**, 147–159.

Walker, R. J. (1988) Low-blank chemical separation of rhenium and osmium from gram quantities of silicate rock for measurements by resonance ionization mass spectrometry. *Analytical Chemistry* **58**, 2923–2927.

Wieler, R. (2014) Noble gas mass spectrometry. In *Treatise on Geochemistry*, McDonough, W.F. (ed.), 355–373. Elsevier, Amsterdam.

Williams, I. S. (1998) U–Th–Pb geochronology by ion microprobe. In *Applications of Microanalytical Techniques to Understanding Mineralizing Processes*, McKibben, M. A., Shanks, W. C., III, and Ridley, W. I. (eds), 1–35. Special Publication, *Reviews in Economic Geology* 7. Society of Economic Geologists, Littleton, CO.

Williams, M. L., Jercinovic, M. J., and Terry, M. P. (1999) Age mapping and dating of monazite on the electron microprobe: deconvoluting multistage tectonic histories. *Geology* **27**, 1023–1026.

CHAPTER 4

Interpretational approaches: making sense of data

The uncertainty of a date is as important as the date itself. –
Ken Ludwig
All models are wrong, but some are useful. – George Box

4.1 INTRODUCTION

At the time of publication, 105 years have passed since Arthur Holmes published *The Age of the Earth*, the first attempt at an absolute (numerical) geologic timescale. In the intervening century, improvements in laboratory techniques and mass spectrometer technology have revolutionized the precision and accuracy of published radio-isotopic dates, as well as the sheer quantity of data produced by a growing number of laboratories. The results have significantly refined the geologic timescale, making it possible to address new scientific questions. Indeed, the scope of the scientific questions that geochronology and thermochronology can address is a function of the achievable precision and accuracy, which govern the level at which data can be interpreted. As our ability to make more and better measurements improves, our ability to interpret the data must improve as well.

Important questions that often arise in interpreting measured geochronological and thermochronological data include:

- How do I combine multiple measurements of a single quantity?
- How do I calculate and express the uncertainty in my measurements?
- What assumptions are involved when modeling geochronological data, for instance with a weighted mean or a linear regression?
- What should I do when I measure more scatter in my system than my model predicts?

This chapter seeks to provide answers to these questions and a statistical context for interpreting geochronological and thermochronological data.

4.2 TERMINOLOGY AND BASICS

4.2.1 Accuracy, precision, and trueness

The terms *precision* and *accuracy* are used to characterize the differences among the measured data points themselves, and the difference between the average of those measured values and the true value, respectively [*Bevington and Robinson*, 2003; *Schoene et al.*, 2013]. While these terms are in common scientific usage, the term *trueness* is relatively new, introduced by the metrology community, a branch of statistics that deals specifically with quantifying what we know (and do not know) about measured data.

High-precision data have relatively small scatter about the mean, while lower precision data have more point-to-point variability (see Fig. 4.1). The distinction between "high" and "low" precision data is entirely relative and contextual, though: papers published with the words "high precision" in their titles a decade or two ago might now be considered "low precision." And a "high precision" date for one chronometer (e.g., Re–Os) might be considered "low precision" for another (e.g., U–Pb) simply because it is easier to analyze the latter. A clearer way to express the precision of analyses is simply with its relative uncertainty, e.g., 0.1%, and the confidence level of the uncertainty (see Section 4.3).

Accuracy, on the other hand, has traditionally been understood to characterize how well the mean of a data set agrees with the true value of the parameter being measured (Fig. 4.1). Instead of using "high" or "low" to describe accuracy, as for the precision, most authors simply refer to a result as accurate or inaccurate. An accurate result agrees with the true value at a stated or implied confidence level: for instance, within its 95% confidence interval (again, see Section 4.3). Because this understanding takes the precision of a measurement into account, the two concepts are linked. Imprecise data could have a mean

Geochronology and Thermochronology, First Edition. Peter W. Reiners, Richard W. Carlson, Paul R. Renne, Kari M. Cooper, Darryl E. Granger, Noah M. McLean, and Blair Schoene.

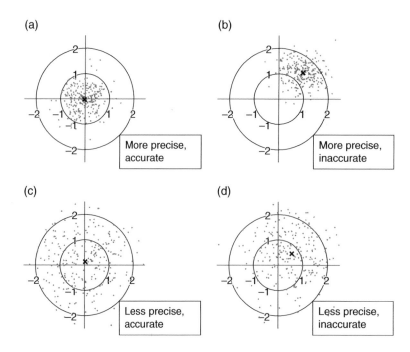

Fig. 4.1. The conventional understanding of accuracy and precision. Measured data, plotted as 250 points, have a true mean at the origin of each plot, and the mean of the measured data is plotted with an "x". Circles centered on the true value are drawn for reference. If the data sets are unbiased as in (a) and (c), the mean of a data set with lower precision is likely to be farther from the true value than the mean of a more precise data set. Because the accuracy depends on the difference between the measured and true values, in the language of the VIM this means a less precise data set is likely to be less accurate. (*See insert for color representation of the figure.*)

relatively far from the true value and still be accurate (Fig. 4.1c), whereas more precise data could have a mean closer to the same true value but be inaccurate (Fig. 4.1b).

To better distinguish the ideas surrounding accuracy and precision, the international metrology community has added an additional conceptual term, the "*trueness.*" Metrology is the science of measurement, often closely related to the field of statistics, and its Joint Committee for Guides in Metrology publishes an International Vocabulary of Basic and General Terms in Metrology known as the VIM [*JCGM*, 2012]. In the VIM, the trueness is the hypothetical difference that would be observed between a measurement carried out an infinite number of times, until its precision was infinitesimal, and the true value. By removing the effect of precision, the trueness becomes the total of any systematic measurement effects that might bias a measurement from its true value.

4.2.2 Random versus systematic, uncertainties versus errors

The metrology community also draws a distinction between two terms that are often used synonymously in the geo- and thermochronology communities, "error" and "uncertainty." An *error* is defined to be the difference between the measured value and the true value. As such, an error is a single number that has the same units as the measurement itself: a measurement of an age standard of 101 Ma that has a true age of 100 Ma has an error

of 1 Ma. An *uncertainty* as defined by the VIM is a parameter that characterizes the dispersion of the values attributed to the quantity being measured [*JCGM*, 2012]. For instance, the same age standard measurement might be reported as 101 ± 2 Ma (2σ), where the interval "± 2 Ma" is the uncertainty, and the "(2σ)" expresses the confidence level attached to that interval. By using the "σ" notation, a normal (Gaussian) distribution is implied, and in a normal distribution about 95% of $\pm 2\sigma$ confidence intervals contain the true mean.

A measurement's error can come from several different effects, and their contributions are called components of the error (Fig. 4.2). One component comes from *random effects*. For instance, there is thermal noise (also known as Johnson noise) in any electrical system with finite resistance, which has a normal distribution with a mean of zero. Any measurement, like the voltage across two points of that circuit, contains thermal noise and no two successive measurements will be exactly the same. As a random component of the error, thermal noise will not bias the measured value from the true value over many measurements averaged together.

Systematic components of the error, however, by definition bias the measured value from the true value. While making multiple measurements decreases the random component of error in the measurement (see Section 4.3.2), it cannot reduce a systematic component. Systematic components of error are shared across multiple analytical results. For instance, all radio-isotopic dates for a given isotopic system rely on the same decay constant or constants. Because decay constants are not perfectly known, the difference between their measured and true values will bias every calculated date by about the same amount, and the magnitude of this bias is a systematic component of error in the date. The true value of the decay constant is not known, and since an error by definition is the difference between the estimated and true value, the systematic component of error cannot be directly calculated. One can, however, estimate a range of reasonable values that the error might take, from for instance the scatter in repeat measurements of the decay constant. This uncertainty becomes a systematic uncertainty in every calculated date for that isotope system.

This distinction between error and uncertainty becomes useful when discussing issues of accuracy, precision, and trueness. The uncertainty is related to the precision alone, and has no bearing on how close the outcome of the measurement is to its true value.

If the error of the measurement is well within its uncertainty, as in the example above, then the result is considered accurate. Evaluating the trueness attempts to remove consideration of the random components of error and isolate only systematic components.

4.2.3 Probability density functions

A *probability density function* describes the relative likelihood that a measurement or variable takes on any of its possible values. This concept is easier to visualize than to define: the familiar "bell curve" is the probability distribution function for the *normal* or *Gaussian distribution* (Fig. 4.3). The normal distribution is the default assumption for most geo- and thermochronological variables, for good reason: most uncertainties in

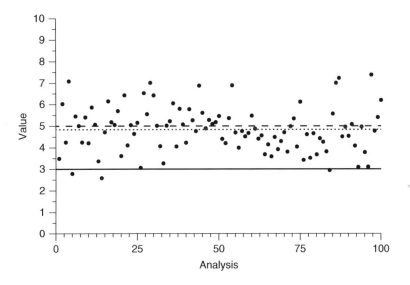

Fig. 4.2. A simulation with 100 replications of a measurement with a mean of 5 (dashed line) and a standard deviation of 1. The mean of the measurements (dotted line) is 4.9, so that the random component of the error is $4.9 - 5 = -0.1$. If the measurement has been biased by a large systematic component of error, and the true value is 3 (solid line), then the systematic component of the error is $5 - 3 = 2$. The total error is therefore $-0.1 + 2 = 1.9$.

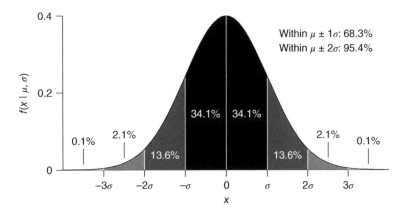

Fig. 4.3. The normal (Gaussian) distribution. On the x-axis are possible values of the variable of interest (e.g., an isotopic date). The mean value of the normal distribution is labeled μ, and its standard deviation σ.

the measurements for these fields are really the sum of many small effects.

For instance, an isotope ratio measurement made on a mass spectrometer might be the result of many ion arrivals at two detectors, each an independent random event, as well as thermal noise in the amplifier circuits used to measure the current induced by the ion beam, and an uncertainty in correcting for the mass-dependent bias in the measurement. The *central limit theorem* says that, even if the uncertainties caused by these effects do not have a normal distribution themselves (for instance, the distribution of ion arrivals follows a Poisson distribution), the sum of many such effects will approach a normal distribution. This chapter will therefore devote the most time to interpreting properties of normal distributions (see below).

The two parameters that technically define a normal distribution are its *mean*, usually symbolized with the Greek letter mu (μ) and its *variance*, usually symbolized as the Greek letter sigma, squared (σ^2). If the variable of interest (e.g., a date) has units of years, then the variance will have units of years^2. This is difficult to relate to a reported uncertainty, so the *standard deviation* ($\sigma = \sqrt{\sigma^2}$) is almost always reported instead. The variance, though, generally plays the most important role calculating and quantifying uncertainties, as demonstrated in the rest of this chapter.

4.2.4 Univariate (one-variable) distributions

For a single variable, for instance a reported date, the normal distribution takes the shape of the familiar "bell curve" (Fig. 4.3). The formula for the curve is written $f(x|\mu, \sigma)$, where the probability density is a function f of the variable of interest x, if you know (the vertical bar symbolizes "given") both the *mean* μ of the distribution and its *standard deviation* σ (equation 4.1).

$$f(x|\mu, \sigma) = \frac{1}{\sqrt{2\pi\sigma^2}} e^{-\frac{1}{2}\frac{(x-\mu)^2}{\sigma^2}} \qquad (4.1)$$

The actual values returned by the probability density function f are less important than the area under the curve it creates (Fig. 4.3). The area under the entire curve is unity, and the probability that μ lies between two prospective values x_1 and x_2 is simply the area under the curve between those two values. For instance, 68.3% of the area underneath the normal distribution lies within the interval $\pm 1\sigma$ of the mean, and 95.4% within $\pm 2\sigma$ of the mean. If a reported uncertainty is normally distributed, then there should be a $\sim$68.3% chance that the true value lies within $\pm 1\sigma$ of the mean, and $\sim$95.4% within $\pm 2\sigma$ of the mean.

When reporting an uncertainty, the interval concerned (e.g., $\pm 2\sigma$) is called the *confidence interval*, the estimated probability that the true value occurs in this interval is called the *confidence level*, and both are required to make sense of a stated uncertainty. A measurement reported in text, a data table, or a figure as "100 ± 2", the "2" could represent a 68% or 95% confidence interval – or something else – without further clarification, e.g., "100 ± 2 (2σ)." Confidence intervals for data from a single variable (known as *univariate data*) are often called "error bars," but more precisely and informatively described as, for instance, "95% confidence intervals" using the definitions above.

4.2.5 Multivariate normal distributions

In geo- and thermochronology, we usually measure multiple variables at once, or are interested in more than one related result. Considering the probability distributions of these variables together results in *multivariate probability density functions*. Examples include the $^{206}Pb/^{238}U$ and $^{207}Pb/^{235}U$ ratios in the U–Pb system, the $^{36}Ar/^{40}Ar$ and $^{39}Ar/^{40}Ar$ ratios in the Ar–Ar system, and the initial $\delta^{234}U$ and U–Th date in the U–Th system. When examining multiple variables, we must not only consider the probability density of each variable alone, but also the relationships between variables to correctly quantify uncertainty.

When two or more variables share a common source of uncertainty or variability, those variables have *correlated uncertainties*. One example would be the simultaneous measurement of three isotopes, a, b, and c, to calculate two isotope ratios, a/c and b/c, where random noise causes variability in repeated measurements of all three isotopes. When c randomly has values that are higher than average, the ratios a/c and b/c are more likely to be lower than average. Likewise, when c randomly is lower than average, a/c and b/c are more likely to be higher than average. This is an example of *positive correlation*: both a/c and b/c tend to be either higher than average or lower than average at the same time. For *negatively correlated* variables, when one variable is higher than average, the other tends to be lower than average.

The degree of correlation between two variables can be expressed as a *correlation coefficient*. Correlation coefficients, denoted ρ, range from -1 to 1, with -1 being perfect negative and 1 being perfect positive correlation. *Statistically independent* variables have zero correlation.

The two-dimensional analog of an uncertainty bar is an uncertainty ellipse (Fig. 4.4), popularized in the isotope geochemistry community by Ken Ludwig and his data visualization software Isoplot [e.g., *Ludwig*, 1983, 1991, 2012]. The ellipse itself represents a contour of equal relative likelihood in a bivariate (two-variable) normal distribution. For variables x and y with uncorrelated uncertainties, the ellipse becomes a circle, and for variables with correlated uncertainties, the ellipse "tilts." Figure 4.4 illustrates a positive correlation between the uncertainties in x and y.

The *multivariate normal distribution* generalizes these concepts to measuring many variables at the same time. All variables taken one-by-one have univariate normal distributions, and because considering these separately ignores other variables, these are examples of *marginal distributions*. Each unique pair of variables can have a different correlation coefficient. The probability distribution function for a multivariate normal distribution is written

$$f(\boldsymbol{x}|\boldsymbol{\mu}, \boldsymbol{\Sigma}) = \frac{1}{\sqrt{|2\pi\Sigma|}} e^{-\frac{1}{2}(x-\mu)^T \Sigma^{-1}(x-\mu)} \qquad (4.2)$$

Here, the variables of interest are arranged in a vector, *x*, as are their means, *μ*. For the multivariate analog of the variance (σ^2 in equation 4.1), we consider not only the uncertainties of all the variables, but also their uncertainty correlations. These appear in the equation above as components of the covariance matrix **Σ**. This matrix maps each variable to the others, and contains information about the uncertainty correlation between each of the variables. For a full explanation of the relationship between statistical covariance and correlation, see Section 4.3.4.

Multivariate normal distributions do not have the same familiar confidence intervals that univariate normal distributions do.

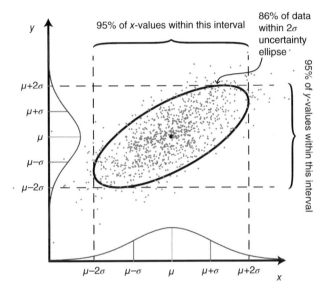

Fig. 4.4. Anatomy of an uncertainty ellipse. One thousand bivariate normally distributed random numbers are plotted, along with the 2σ ellipse that illustrates the bivariate analog of their mean and standard deviation. On the x-axis and y-axis, the probability distributions for the x and y variables are plotted. They are univariate normal distributions, and known as marginal distributions. Note that the horizontal and vertical tangents of the ellipse correspond to the 2σ uncertainty envelopes for the x and y variables. The ellipse encloses a smaller number of points than the 95% confidence intervals, including only 86% of the simulated data. Therefore, a 2σ uncertainty ellipse is not a 95% uncertainty envelope. (*See insert for color representation of the figure.*)

For univariate distributions, recall that only 68% of ±1σ confidence intervals overlap the true mean, and only about 95% of ±2σ confidence intervals overlap the true mean. Figure 4.4 shows how the two-dimensional uncertainty ellipse is related to the one-dimensional normal probability density function for each variable. Some of the data points are within the 95% (±2σ) confidence for one variable but are outside the same interval for the other variable, and the ellipse does not include all the data points in the confidence interval for each variable. For that reason, a 2σ uncertainly ellipse contains less than 95% of the data, and represents only an 86% confidence interval.

As more variables are measured, more dimensions are added to the uncertainty envelope. For instance, if three ratios are considered together (e.g., for the total Pb/U isochron of *Ludwig* [1998]), the multivariate uncertainty envelope takes the shape of an *ellipsoid* in three dimensions; higher dimensional envelopes are analogous in shape but difficult to visualize. The probability that the true value lies inside, for instance, 1σ or 2σ uncertainty envelope decreases as the dimension of the data increases, as shown in Table 4.1.

To understand why this is important, imagine that a date has been very precisely determined (for instance, by astrochronology) to be 100 Ma. This result is tested by measuring two U–Pb dates (e.g., $^{206}Pb/^{238}U$ and $^{207}Pb/^{235}U$) with correlated uncertainties, and the results plotted as a 2σ uncertainty ellipse on a concordia diagram. Should one expect the ellipse to overlap the point on the plot that represents 100 Ma if all three dates are actually 100 Ma? Not necessarily. If you repeated this experiment many times over, you would expect the point that represents 100 Ma (for instance, the 100 Ma point on the concordia curve) to fall outside the ellipse for about 14% of your measurements.

4.3 ESTIMATING A MEAN AND ITS UNCERTAINTY

In order to further explore interpretational approaches for real data, it is helpful to go back to the basics and understand a few fundamental calculations and assumptions.

Table 4.1 Confidence levels for univariate and multivariate normally distributed data sets. The dimension of the data determines the confidence level for a particular uncertainty envelope. For instance, there is only a ~39% chance that a 1σ uncertainty envelope in two dimensions (a 1σ uncertainty ellipse) contains the true value of a measurement

Dimension of data	Uncertainty envelope			Confidence level		
	≤1σ	≤2σ	≤3σ	90%	95%	99%
1	0.6827	0.9545	0.9973	≤1.645σ	≤1.960σ	≤2.576σ
2	0.3935	0.8647	0.9889	≤2.146σ	≤2.448σ	≤3.035σ
3	0.1988	0.7385	0.9707	≤2.500σ	≤2.795σ	≤3.368σ
4	0.0902	0.5940	0.9389	≤2.789σ	≤3.080σ	≤3.644σ
5	0.0374	0.4506	0.8909	≤3.039σ	≤3.327σ	≤3.884σ

4.3.1 Average values: the sample mean, sample variance, and sample standard deviation

The natural measure of central tendency for approximately normally distributed data is the *arithmetic mean* $\bar{x}$, and the natural measure of the scatter between the data points is the *sample standard deviation s*, the square root of the *sample variance s^2*.

$$\bar{x} = \frac{1}{n} \sum_{i=1}^{n} x_i$$

$$s^2 = \frac{1}{n-1} \sum_{i=1}^{n} (x_i - \bar{x})^2 \qquad (4.3)$$

$$s = \sqrt{s^2}$$

Equations (4.3) are probably the most commonly used equations for interpreting data, so they deserve some explanation and analysis. The equations describe n successive measurements of a variable x, the focus of our study. Here, x could stand for anything that can be measured quantitatively: the diameter of a mineral, the mass of a sample, or an isotope ratio. The i in each of the summations is an index: for instance, the first measurement ($i = 1$) is x_1, the second ($i = 2$) is x_2, through the last measurement ($i = n$), which is x_n. The statistics $\bar{x}$, s, and s^2 can be calculated for data with any type of statistical distribution, but the $\pm 1s$ and $\pm 2s$ intervals are only 68% and 95% confidence intervals for the normal distribution. Box 4.1 explains the difference between μ versus $\bar{x}$ and σ versus s. Both the variance and the standard deviation describe the data-point-to-data-point variability of the data—increasing the sample size, n, does not change their expected value. The reason there is an $(n-1)$ in the denominator of the sample variance is explored in detail in Section 4.3.5.

Box 4.1   What's the difference between μ versus $\bar{x}$ and σ versus s?

Equations (4.1) and (4.2) use the Greek letter variables μ and σ (mu and sigma), while equations (4.3) use $\bar{x}$ and the Roman letter s. The difference is that the Greek letter variables are most often used in the statistics literature to describe the "true" population statistics. For instance, if you were measuring the uranium concentration of zircon crystals in a granite, then μ and σ might describe the true mean and standard deviation of all of the zircons in that unit. This information is generally unfeasible or impossible to obtain: instead, you might collect a hand sample, separate out its zircons, and measure the uranium concentration in just 100 grains. Because this is just a small sample of the very large number of actual zircon crystals, $\bar{x}$ and s are termed the sample mean and sample standard deviation, signifying that they are only estimates of the true mean and variability of the zircon uranium concentrations. Almost all scientific work involves sampling, so the "sample" is most often dropped from the beginning of terms like the "sample mean" and "sample variance," a convention this chapter will follow.

Geochronology and thermochronology literature often uses the symbol σ in the place of s for the sample standard deviation, like reporting uncertainties in text or a table as "$\pm 2\sigma$." While this nomenclature does not match statistical conventions, it is widely understood that, for instance in the example above, every zircon from a granite has not been analyzed. The Greek letter σ is also advantageous in that it is most often used to represent the standard deviation, whereas s could also represent the SI unit for time, the second.

When calculating the mean, variance, and standard deviation in this way, two important assumptions are made—that the data are *independent and identically distributed* (iid). Violating either assumption results in either overestimation or underestimation of uncertainties, and can lead to erroneous data interpretation. Figure 4.5 illustrates an iid data set (a), and two data sets that violate this assumption (b and c).

The statistical parameters defined in equations (4.3) are all appropriate for describing the independent, identically distributed data illustrated in Fig. 4.5a. However, in Fig. 4.5b, the data show some random variation superimposed on a linear trend, violating the assumption that the data are independent of one another. If there is any trend in the data, whether it is linear, sinusoidal, or a different function, then the variance and standard deviation do not correspond to predictable confidence intervals about the mean value. At the least, this implies that an unaccounted for and perhaps previously unrecognized affect is influencing the data. For this reason, Equations (4.3) should not be applied, and a different model of the data behavior is needed before it can be interpreted statistically. For Fig. 4.5b, the linear regression algorithm detailed in Section 4.4.1 would be more appropriate, and both the slope and y-intercept of the data should be estimated and interpreted.

In Fig. 4.5c, the point-to-point scatter in the data increases during the course of the 100 successive measurements. The data seem approximately normally distributed over short time periods (say, 20 measurements), but the standard deviation of the first 20 measurements is significantly lower than the standard deviation of the last 20 measurements. This violates the assumption that the data be *identically distributed*: different values of standard deviation, and therefore different normal distributions, describe different portions of the data. Evaluating the mean and standard deviation of the data in Fig. 4.5c using equations (4.3), which give each data point equal weight in the result, causes two related problems. First, no single standard deviation can describe the data—the standard deviation changes dramatically during the analysis. Second, giving the same weight to the less precise data at the end of the analysis as the more precise data at the beginning of the analysis will likely result in a mean that is farther away from the true value.

Instead of evaluating the mean and standard deviation of the data set altogether, a better approach here would be to bin or divide the data into groups that share the same or nearly the same variability and calculate the mean and standard deviation of each group separately. The best average for the entire data set would then be estimated using a weighted mean of the binned data (see Section 4.3.7). Data sets like Fig. 4.5c, where the point-to-point scatter is variable, are known as *heteroscedastic*, whereas the data in Fig. 4.5a is *homoscedastic*.

4.3.2 Average values: the standard error of the mean

When the iid assumptions for calculating a mean, variance, and standard deviation have been met, then it is possible to ask how well the mean is known. The more measurements n of a variable that are made, the more information is available and the

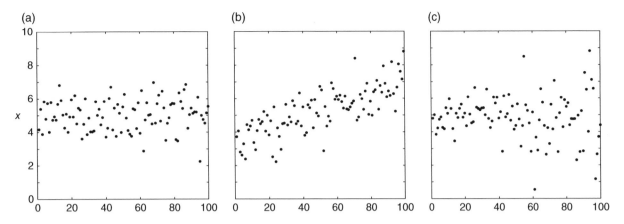

Fig. 4.5. Example data sets with $n = 100$ successive measurements, plotted with the measured variable, x, on the vertical axis. These plots approximate the visual output from many measurement devices, such as mass spectrometers. (a) Independent, identically distributed data. (b) Data appear to show some random variability around a clear linear trend. (c) Data show increasing variability with time during the analysis.

better the mean of that variable is known. Intuitively, this means we should have a more precise estimate of the mean and its uncertainty. This is not reflected in the variance or standard deviation of the data set, whose expected values do not change with increasing n. Instead, the variance and standard deviation in equations (4.3) can be used to calculate the *variance of the mean*, $s_{\bar{x}}^2$, and the *standard deviation of the mean*, $s_{\bar{x}}$.

$$s_{\bar{x}}^2 = \frac{1}{n(n-1)} \sum_{i=1}^{n} (x_i - \bar{x})^2 = \frac{\sigma^2}{n}$$

$$s_{\bar{x}} = \sqrt{\frac{s^2}{n}} = \sqrt{s_{\bar{x}}^2} = \frac{s}{\sqrt{n}}$$

(4.4)

The standard deviation of the mean is also called the *standard error of the mean*, or simply the *standard error*, and these three terms are often used synonymously in statistics, geochronology, and thermochronology literature (see Box 4.2). The standard error is used to calculate and report the uncertainty in many replicate measurements of the same mean value, $\bar{x}$, and for normally distributed data, the uncertainty in the mean is normally distributed. After n measurements, the true value of the mean would be found within the interval $\bar{x} \pm s_{\bar{x}}$ about 68% of the time and about 95% of the time for the interval $\bar{x} \pm 2 s_{\bar{x}}$.

The word "error" in the term "standard error" is confusing, since it is not an *error* as described in Section 4.2.2. Like many other historical terms in science and mathematics, the term "standard error" is a misnomer, but it is fixed in our widely used and accepted terminology and unlikely to change. The standard error of a measurement can be used to estimate the *uncertainty* of that measurement, and an *error* is better defined as the difference between the measured value and the true value.

4.3.3 Application: accurate standard errors for mass spectrometry

In mass spectrometry and many other measurement systems, the scatter between data points depends on an operator-chosen

Box 4.2 Common questions

Should I use the standard deviation or standard error?

Even experienced scientists find themselves asking whether to report the standard deviation or standard error for a measurement. Use the standard deviation when you're interested in characterizing the variability of a population and your measurements can resolve those differences. Use the standard error when you've made repeated measurements of the same true value, and the variability of those measurements is simply the result of some (e.g., instrumental) noise in the measurement. Two examples will help illustrate these differences: measuring the average grain size in a poorly sorted sedimentary rock and making repeat measurements of an isotope ratio that does not change over time. The following questions can guide your choice of standard deviation versus standard error.

Will making repeated measurements continuously improve my knowledge of the mean?

If so, then the standard error is the statistic to report: as n increases, the standard error decreases by the square root of n. Measuring the size of just a few grains in a sedimentary rock would give a rough idea of the average grain size, but measuring many more would help you pin down that average very precisely.

Are successive measurements resolvable by your measurement process?

If they are not, then the scatter you observe is due to your measurement process, and does not reflect any variability in what you are measuring. This means that the standard deviation describes the variability in your measurements, but the standard error best describes your uncertainty in the measurement. For the example of measuring isotope ratios, the standard deviation is often determined by the mass spectrometer setup (see Section 4.3.3 below), and the standard error should be used to estimate the uncertainty in the analysis.

If successive measurements are readily resolvable by your measurement process, then the standard deviation and standard error will have different meanings. Consider the example of measuring the grain-size distribution in a poorly sorted sedimentary rock. One parameter of interest might be the variability in size from grain to grain—Just how poorly sorted is this sample?—and the standard deviation of the measured grain sizes would quantify this variability. On the other hand, you might be most interested in the mean grain size. In this case, the standard error will describe how well you know the mean itself. At large n, you will have precisely constrained the mean, even though there is considerable variation in size from one grain to the next.

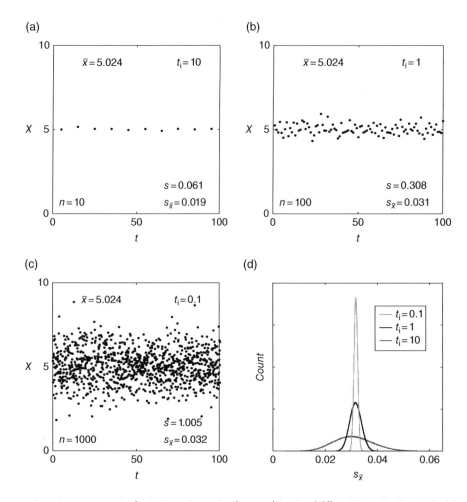

Fig. 4.6. Simulated mass spectrometer measurements of an isotope ratio, x, using the same data set and different integration times t_i. (a–c) Data generated with a true mean of $\bar{x} = 5$ and a true standard error of $s_{\bar{x}} = 0.032$. The estimated mean is exactly the same for each scenario because the underlying data are the same. Decreasing the integration time increases the standard deviation, but the standard error is approximately the same for each scenario. (d) Histograms showing variability of the calculated standard error for each integration time over 10^6 random simulations. Note that higher values of n yield better uncertainty estimates. (*See insert for color representation of the figure.*)

"integration time" over which a signal is measured and integrated before an output value is calculated. For instance, an instrument might offer the option of outputting an isotope ratio every 10 s, 1 s, or 0.1 s for a 100-s analysis (Fig. 4.6a-c). There is a trade-off between the scatter among the data points and the number of data points measured for a given analysis time: the 10 data points measured with 10-s integrations will have a lower standard deviation than the 100 data points measured with 1-s integrations, which have a still lower standard deviation than the 1000 data points measured with 0.1-s integration times.

If the "integration time" is left for the instrument operator to choose, then which choice is best? All three scenarios integrate the same information, the 100 s of intensities for the two isotopes that make up the isotope ratio, and therefore the mean value in all the three scenarios will be exactly the same. The standard errors should be similar, but not necessarily equal.

To better understand the behavior of the system, set up a simulation of the process using a random number generator:

Generate 100 s of synthetic mass spectrometer data, sample it at integration times of 10, 1, and 0.1 s, and then calculate the standard error for each case using equations (4.4). The results for one such simulation are illustrated in the plots of Fig. 4.6a–c. Although one would expect the value of the standard error the be the same no matter what integration time is used, Fig. 4.6 shows that calculated standard errors for real data can be different from the expected value. This is because measured data contain real variability: repeating the same experiment twice under ideal conditions will produce similar but not identical results due to random variation in the measurements.

In Fig. 4.6a–c, the standard error of each data set gets closer to its true value as n increases. To see how much better our standard error estimate becomes, we could repeat the same simulation many times over and compile the results. Figure 4.6d contains three histograms, each with the results of 10^6 simulations of the 100-s mass spectrometer run with a true standard error of 0.032. The histogram for the $t_i = 10$-s integrations (so that

$n = 10$ for each calculated standard error) shows that 95% of the calculated standard errors are between 0.017 and 0.046, a span of about 0.03 that is about the same size as the true standard error. However, the histogram for the $t_i = 0.1$-s integrations with $n = 1000$ shows that 95% of the estimated standard errors fall within an interval of about 0.003, a significantly tighter distribution. Therefore, while the standard error is expected to be about the same for each choice of integration time, the higher the number of measurements, n, that are used to calculate the standard error, the closer the calculated standard error is likely to be to its true value. Another way of looking at this phenomenon is that both the mean and the standard deviation (and all other statistics) have an uncertainty. The uncertainty in the mean depends on the standard error of the measurements, and the uncertainty in the standard deviation also depends on the number of measurements made. In detail, the uncertainty in the mean also depends on n when n is low (fewer than about 100), which is the subject of Section 4.3.6.

4.3.4   Correlation, covariance, and the covariance matrix

Multivariate normal distributions are defined not only by their means or best estimate of the average values, and their standard deviations, used to estimate the uncertainties in the mean, but also by the degree of statistical correlation between these uncertainties. Statistical correlation is introduced in Section 4.2.5, and often results when two variables share a common source of uncertainty or scatter. The correlation coefficient, ρ, can be used to describe the degree of correlation. Another useful parameter for describing statistical correlation is the *covariance*, the two-variable analog of the variance calculated in equations (4.3).

$$\sigma_{xy}^2 = \frac{1}{n-1} \sum_{i=1}^{n} (x_i - \bar{x})(y_i - \bar{y}) \qquad (4.5)$$

The covariance and the correlation coefficient are related by the equation

$$\rho_{xy} = \frac{\sigma_{xy}^2}{\sigma_x \sigma_y} \qquad (4.6)$$

The correlation coefficient ρ_{xy} in equation (4.6) can be thought of as a covariance that has been scaled by the uncertainties in x and y, normalizing its values to be between –1 and 1 inclusive. Subscripts like the "xy" in equations (4.5) and (4.6) are used when discussing the covariance or correlation between two particular variables, but these are often dropped from the ρ term when the identities of these two variables are understood. Using equations (4.5) and (4.6) to calculate the correlation coefficient for the $n = 1000$ measurements plotted in Fig. 4.4 yields a correlation coefficient of $\rho_{xy} = 0.7$.

Covariance terms are the building blocks of *covariance matrices*, like the one that appears in the probability distribution function for a multivariate normal distribution (equation 4.2), or later in weighted least squares regression (Sections 4.4.2 and 4.4.3). Covariance matrices are helpful when organizing variance and covariance terms. The number of pairs of variables increases even

faster than exponentially with each new variable. In a covariance matrix, variance terms are placed on the diagonal, at the position corresponding to the first row and first column, the second row and second column, etc. The covariance terms are placed in the off-diagonal positions—the covariance between the first and second variables goes in the first column and second row, and also the second column and first row, creating a matrix that is symmetric across the diagonal. For n variables, a covariance matrix takes the form

$$\Sigma = \begin{bmatrix} \sigma_1^2 & \sigma_{12}^2 & \cdots & \sigma_{1n}^2 \\ \sigma_{12}^2 & \sigma_2^2 & \cdots & \sigma_{2n}^2 \\ \vdots & \vdots & \ddots & \vdots \\ \sigma_{1n}^2 & \sigma_{2n}^2 & \cdots & \sigma_n^2 \end{bmatrix}$$

4.3.5   Degrees of freedom, part 1: the variance

In equations (4.3)–(4.5), a factor of $1/(n-1)$ is used to calculate the variance and covariance, rather than simply $1/n$. The $n-1$ can be understood to represent the *degrees of freedom*, or the number of values in a statistic that are "free to vary." An example calculating the mean and variance of three values, x_1, x_2, and x_3 helps to illustrate this concept. Following equations (4.3):

$$\bar{x} = \frac{1}{3}(x_1 + x_2 + x_3) \qquad (4.7)$$

$$s^2 = \frac{1}{3-1}\left[(x_1 - \bar{x})^2 + (x_2 - \bar{x})^2 + (x_3 - \bar{x})^2\right] \qquad (4.8)$$

For the calculation of the variance, a substitution can be made by solving equation (4.7) for x_3.

$$x_3 = 3\bar{x} - x_1 - x_2 \qquad (4.9)$$

and this result can be substituted into equation (4.8) to give a formula for the variance in terms of only x_1, x_2, and the mean.

$$s^2 = \frac{1}{3-1}\left[(x_1 - \bar{x})^2 + (x_2 - \bar{x})^2 + (3\bar{x} - x_1 - x_2 - \bar{x})^2\right] \qquad (4.10)$$

Since the mean is fixed, only two values in equation (4.10) are free to vary. In other words, the variance has only $3 - 1 = 2$ degrees of freedom, and more generally

$$\nu = n - 1 \qquad (4.11)$$

where the Greek letter ν (nu) is commonly used for the degrees of freedom. The same argument, substituting the final measured value as a function of the mean and previous measured values, can be scaled up to any value of n. Using equation (4.11) to calculate the degrees of freedom for the variance makes the assumption that all n data points are independent and identically distributed, the same assumption that applies to calculating the sample variance.

4.3.6   Degrees of freedom, part 2: Student's t distribution

The sample statistics presented in equations (4.3) and (4.4) are used in the majority of real-world statistical situations, where the true mean and standard deviation of a normally distributed variable are not known and it is not feasible to measure an entire population. Sometimes it is also infeasible to extensively sample the population, and values of n can fall well below 100. While equations (4.3) and (4.4) for the mean, variance, standard deviation, and standard error of the mean are all accurate at low n, the confidence levels given in Table 4.1 are too narrow.

The problem is that at low n (in practice, fewer than about 100 measurements), the uncertainty in the estimated sample mean is no longer normally distributed, but rather follows a *Student's t distribution*. Unlike the normal distribution in equation (4.1), the Student's t distribution depends on n, which is used to calculate its degrees of freedom $\nu = n - 1$. It has a similar shape to, but "heavier tails" than, the normal distribution, assigning larger probabilities to values farther from the mean (Fig. 4.7). The smaller the degrees of freedom, more pronounced this effect is; the larger the degrees of freedom the more a Student's t distribution resembles a normal distribution. At infinite degrees of freedom, the normal and Student's t distributions are identical, though for most practical purposes they are approximately the same for $\nu \geq 100$.

To illustrate how the uncertainty in the sample mean is different for the normal and Student's t distributions, find the difference between the measured sample mean $\bar{x}$ and the true mean μ, and then divide this difference by the sample standard error, $s/\sqrt{n}$. Although the true mean μ is unknown, the resulting one-size-fits-all variable, called a *Student's t statistic*, is simply the difference between the measured and true means, scaled by the standard error. The shifting and scaling is more generally referred to as *statistical normalization*.

$$t = \frac{\bar{x} - \mu}{s/\sqrt{n}} \qquad (4.12)$$

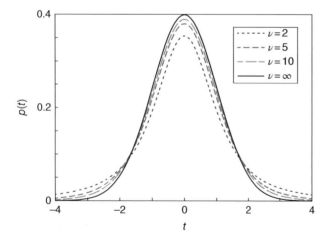

Fig. 4.7. The probability distribution function for a Student's t distributed variable. For small degrees of freedom, the Student's t distribution has a larger probability density farther from the mean. The greater the degrees of freedom, the more the Student's t distribution resembles the normal (Gaussian) distribution, which is effectively a Student's t distribution with infinite degrees of freedom.

The probability distribution for the Student's t statistic, illustrated in Fig. 4.7, depends only on the degrees of freedom. For a measured low-n data set, this distribution can then be rescaled and shifted to fit the measured mean and standard error. For instance, for $n = 6$ measurements with a mean of 10 and a standard error of 2, the $\nu = 5$ probability distribution function can be stretched horizontally by a factor of two so that it is twice as wide (and half as tall), then shifted to the right so that it is centered around a mean value of 10.

Compared to a normal distribution, the heavier tails of a Student's t distribution indicate that confidence intervals are generally wider for a given confidence level. For instance, 95% of the area under the $\nu = 5$ curve lies within the interval $t \leq 2.671$, so 95% of $\pm 2.671\ s_{\bar{x}}$ confidence intervals produced by repeating this experiment many times will overlap the true value of the measurement. This is in contrast to a normal distribution (or Student's t with infinite degrees of freedom), where the 95% of calculated $\pm 1.960\ \sigma_{\bar{x}}$ confidence intervals contain the true mean (see Table 4.1). We can apply the statistical normalization by using a multiplication factor for the standard deviation derived from the Gaussian case. When we do this, symmetric confidence intervals around the mean take the form

$$\bar{x} \pm t_{\alpha,\nu}\sigma_{\bar{x}} \qquad (4.13)$$

where $t_{\alpha,\nu}$ is the value of the Student's t multiplier with ν degrees of freedom corresponding to the desired confidence level α. Table 4.2 contains selected values of $t_{\alpha,\nu}$ for two-sided confidence intervals—with minimum and maximum values, the type of interval used most often in geochronology and thermochronology. Expanded tables can be found in statistics textbooks and reputable online sources.

Using a Student's t multiplier from Table 4.2 requires two important sets of assumptions. First, like the assumptions for calculating sample statistics such as the variance, standard deviation, and standard error, the data points must be independent and identically distributed (iid).

The second assumption causes significantly more confusion— each of the n data points must be *discrete observations*. If one or more of the n data points incorporates an average of multiple measured values, then ν will be no longer be equal to $n - 1$,

Table 4.2 Values of $t_{\alpha,\nu}$, also known as a Student's t multiplier, for two-sided confidence intervals, provided for selected confidence levels α and degrees of freedom, $\nu = n - 1$

Degrees of freedom (ν)	Confidence level (α)		
	0.90	0.95	0.99
1	6.314	12.706	63.657
2	2.920	4.303	9.925
5	2.015	2.571	4.032
10	1.812	2.228	3.169
50	1.676	2.009	2.678
100	1.660	1.984	2.626
∞	1.645	1.960	2.576

and will in general be larger. The Welch–Satterthwaite formula, used to calculate the effective degrees of freedom in more complicated scenarios [*Satterthwaite*, 1946; *Welch*, 1947], should be employed to determine the degrees of freedom for a weighted mean. This is appropriate, in that weighted means are mostly averages that themselves comprise one or more averages. The $n-1$ term in the calculation of the sample variance and standard error (equations 4.3), known as the Bessel correction, however, always stays the same.

As an example, every 10 successive observations in Fig. 4.6c are averaged to create a data point in Fig. 4.6b, so that $n=1000$ becomes $n=100$. The measured mean of the 100 observations is exactly the same as the measured mean of the 1000 observations, which has been calculated in two steps: calculating the intermediate averages in Fig. 4.6b, then calculating the overall average, $\bar{x}=5.024$. Therefore, the degrees of freedom in both cases is the same, $\nu = n - 1 = 999$.

It is common practice in geochronology and thermochronology disciplines to report uncertainties at the $\sim$95% confidence level. For this reason, it is important to use the Student's t multiplier for data sets with small degrees of freedom, instead of simply reporting $\bar{x} \pm 2s_{\bar{x}}$.

4.3.7 The weighted mean

The assumptions required to calculate a conventional mean, namely that the data are iid, can be overly restrictive for some data sets. In the simplest instance a data set might be comprised of several repeated measurements of the same value, but each measurement might have a different uncertainty. This violates the assumption that the data are identically distributed, as illustrated in Fig. 4.8. Intuitively, it makes sense that the most representative mean would be closer to the more precise analyses, with less precise analyses playing a smaller role in the calculation. The

best average, *a weighted mean*, would weight each analysis by a factor related to its uncertainty.

There are several ways that one might weight each data point, and the only constraint is that the sum of these factors, or weights, should be unity. For an arithmetic mean, where the data are identically distributed, the weights should be the same: $1/n$. Expressing the weighted mean, $\bar{x}$, as the sum of the weights, a_i multiplied by the n data points x_i, the formula for the weighted mean (equation 4.14)

$$\bar{x} = \sum_{i=1}^{n} a_i x_i \tag{4.14}$$

$$\bar{x} = \sum_{i=1}^{n} \frac{1}{n} x_i = \frac{1}{n} \sum_{i=1}^{n} x_i \tag{4.15}$$

is the same as the equation for an arithmetic mean (equations 4.3 and 4.15).

If the data x_i have different uncertainties, represented by their variances σ_i^2, then the weights in geochronology and thermochronology are almost always chosen to be inversely proportional to those variances. This produces weighted means with the smallest propagated uncertainties. Because the weights must sum to unity and be inversely proportional to the variances, the formula for the weighted mean is

$$\bar{x} = \sum_{i=1}^{n} \frac{x_i}{\sigma_i^2} \bigg/ \sum_{i=1}^{n} \frac{1}{\sigma_i^2} \tag{4.16}$$

where the sum in the denominator normalizes the $1/\sigma_i^2$ relative weights.

These weights make intuitive sense: the larger the variance, the smaller the $1/\sigma_i^2$ weight in the numerator of equation (4.16). If the uncertainty in x_1 is one half the uncertainty in x_2, then x_1 will get four times as much weight in the weighed mean.

The variance of the weighted mean, $\sigma_{\bar{x}}^2$, is calculated using the formula

$$\sigma_{\bar{x}}^2 = 1 \bigg/ \sum_{i=1}^{n} \frac{1}{\sigma_i^2} \tag{4.17}$$

and the (1σ) uncertainty in the weighted mean is the square root of its variance. This uncertainty is the weighted mean analog of the standard error of the arithmetic mean, and for equal weights reduces to equation (4.4). A step-by-step derivation of equation (4.15) can be found in section 5.1 of *McLean et al.* [2011].

Again, equation (4.15) makes intuitive sense for simple test cases. Each new data point will have an associated a_i^2, which increases the sum in the denominator of equation (4.15) and therefore decreases the weighted mean's uncertainty. This is another way of saying that every new piece of information, no matter what its uncertainty, helps constrain the weighted mean and improve its precision. More precise data will have a smaller a_i^2, which will make for a larger contribution to the sum in the denominator of equation (4.15), which will in turn decrease the weighted mean uncertainty more than a data point with a

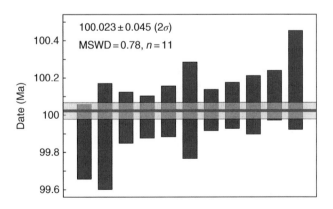

Fig. 4.8. Example of a weighted mean plot. Each vertical rectangle represents a separate analysis, centered vertically at the mean and spanning a $\pm 2\sigma$ ($\sim$95%) confidence interval. The horizontal line is plotted at the weighted mean of the analyses, and its $\pm 2\sigma$ ($\sim$95%) confidence interval shown as a shaded region above and below. The weighted mean, uncertainty, its confidence level, the MSWD (mean square weighted deviation), and n are all reported together with the data. The meaning of the MSWD is further explored later in the chapter.

larger uncertainty. Compared to equations (4.4) for the standard error, equation (4.15) seems to be missing a term in the denominator that depends on n. This factor is already in the denominator of the right-hand side of equation (4.15), as the n terms in the summation.

A weighted mean value and its uncertainty cannot be clearly interpreted, however, if the assumptions that attend its calculation are not met. Overestimated and underestimated uncertainties in the data can lead to an overestimated and underestimated uncertainty in the weighted mean. If the data do not represent repeat measurements of the same "true" value, then the weighted mean represents one, but not necessarily the best, estimate of their average, and the weighted mean uncertainty is meaningless. Testing the assumptions involved in evaluating a weighted mean is discussed in detail in Section 4.5.1.

4.4 REGRESSING A LINE

The mean is the simplest interpretation of data that consists of a single variable, like a date. When two or more variables are involved, their relationships can be more complicated. One of the most commonly explored is the linear relationship, which for two variables can be written $y = ax + b$, where x and y are the variables of interest and a and b are constants. Since x and y have been measured, the task is to estimate a and b, the slope and y-intercept of the line.

There are two broad measurement scenarios addressed by linear regression. In the first, one is interested in how y-values, which are subject to noise, depend on x-values that have been determined with zero or negligible uncertainty. In this case, the x variable is known as the *independent variable* and the y-value as the *dependent variable*. Examples include experiments where the x-values can be controlled by the experimenter, who measures an effect y, or when the y-variable of interest is a function of an easily and precisely measurable quantity like time, elevation, or depth. In this case, the values of a and b are usually found using *ordinary least-squares* or *weighted least-squares* algorithms. These are discussed in Sections 4.4.1 and 4.4.2, and are directly analogous to the mean and weighed mean discussed in Section 4.3.

There is another measurement scenario that is commonly encountered in geochronology and thermochronology: measured data with variability or estimated uncertainties in both the x-variables and y-variables. This is true for many systems where measured isotope ratios are plotted on the x-axis and y-axis, like isochrons or isotopic fractionation lines. Section 4.4.3 describes the algorithm developed by *York* [1966, 1968] to fit a straight line to these data, and Section 4.5 further explores the assumptions that underpin isochron calculations.

4.4.1 Ordinary least-squares linear regression

The ordinary least-squares algorithm (OLS) fits data from an independent variable, x, and a dependent variable, y, where both represent discrete data. The OLS algorithms that follow are the best choice for calculating a slope and intercept and their uncertainties, providing several critical assumptions about the data are met. First, the relationship between the x- and y-variables must actually be linear. Although equations 4.22 and 4.23 will return estimates and uncertainties for a and b regardless of the shape of the data, these cannot be easily interpreted unless y is a linear function of x. Second, the dependent variable data should be homoscedastically distributed (see Section 4.3.1) around the best fit line, so that its scatter does not increase or decrease as the value of x increases or decreases. Finally, interpreting the estimates and uncertainties of the slope and y-intercept using the confidence intervals introduced in Section 4.3.2 requires that the y-variable uncertainties must be normally distributed (see Section 4.2.5) about the best fit line, and the data points themselves must be independent measurements, sharing no common source of uncertainty.

The ordinary and weighted least squares problems for two variables are most easily represented using vector and matrix notation. The measured y-values are assembled in a column vector (a matrix with one column), y. The x-values make up the first column of a two-column matrix, X, and the second column of X is filled with values of 1. If the slope and y-intercept of the linear relationship, a and b, are the first and second rows in the column vector β, when the measurements of y have zero uncertainty, the system can be expressed in the equation

$$y = X\beta \tag{4.18}$$

Taking apart equation (4.18), the first value y_1 in the vector y is calculated by multiplying the first row of X, $[x_1\ 1]$ by the column vector with elements a and b, yielding

$$y_1 = \begin{bmatrix} x_1 & 1 \end{bmatrix} \begin{bmatrix} a \\ b \end{bmatrix} \quad \text{or} \quad y_1 = ax_1 + b \tag{4.19}$$

The same product can be assembled for any row i of the matrix X, $[x_i\ 1]$, such that each measurement could be written as $y_i = ax_i + b$. This equation is the linear relationship hypothesized for the data, expressed for each data point.

Unfortunately, measurements of the dependent variable never have zero uncertainty, so that the equalities in equations (4.18) and (4.19) do not hold. Instead, the y-values contain some noise, represented by a column vector r with n rows. The discussion below assumes that the residuals are normally distributed, which is an excellent a priori assumption. Since the data are noisy, they will not all follow the hypothesized linear relationship perfectly, and will instead scatter about the line

$$y = X\beta + r \tag{4.20}$$

The differences between the measured data and the best fit line are known as *residuals*. The OLS *best fit line* minimizes *the sum of the squared residuals*, and the best fit slope and y-intercept that accomplish this are written as $\hat{\beta}$. The "hat" on top of a variable name usually indicates that it is an estimate of that parameter that

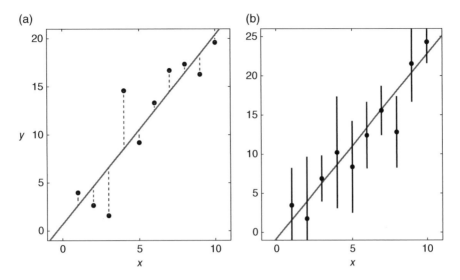

Fig. 4.9. Ordinary least squares and weighted least squares fits to $n = 10$ random, linearly related data points. Following convention, the independent variable is plotted on the horizontal (x-)axis and the dependent variable on the vertical (y-)axis. The best fit line is plotted for each. The dashed vertical lines in (a) are the normally distributed *residuals,* or differences between the measurements and the best fit model. The black vertical bars in (b) represent 2σ uncertainties in the measurements of y. (*See insert for color representation of the figure.*)

is based on measured values, as opposed to its true value (which we usually never know). Likewise, the predicted y-values of the measurements, based on the best-fit slope and y-intercept, are denoted $\hat{y}$, and the equation of the best fit line can be written

$$\hat{y} = \mathbf{X}\hat{\boldsymbol{\beta}} \tag{4.21}$$

Figure 4.9a illustrates an OLS best fit line, with the residuals shown as dashed lines. The residuals $r_i = y_i - \hat{y}$ are normally distributed around the regression line.

To solve equation (4.16) for $\hat{\boldsymbol{\beta}}$, the best fit line parameters, it would be appealing to multiply each side on the left by $\mathbf{X}^{-1}$, so that $\hat{\boldsymbol{\beta}} = \mathbf{X}^{-1}\boldsymbol{y}$. Unfortunately, $\mathbf{X}$ is not necessarily a square matrix (if it was, there would be only two measurements, and the best fit line would simply connect the two points), so it does not have a true inverse. Instead, multiplying each side of equation (4.19) by $\mathbf{X}^T$, the transpose of $\mathbf{X}$, yields a square $(\mathbf{X}^T\mathbf{X})$ term in front of $\boldsymbol{\beta}$, which is invertible if all the data are not perfectly collinear. Multiplying through by the inverse of the $(\mathbf{X}^T\mathbf{X})$ term yields

$$\mathbf{X}^T\boldsymbol{y} = \mathbf{X}^T\hat{\boldsymbol{\beta}}$$
$$(\mathbf{X}^T\mathbf{X})^{-1}\mathbf{X}^T\boldsymbol{y} = (\mathbf{X}^T\mathbf{X})^{-1}\mathbf{X}^T\hat{\boldsymbol{\beta}} \tag{4.22}$$
$$\hat{\boldsymbol{\beta}} = (\mathbf{X}^T\mathbf{X})^{-1}\mathbf{X}^T\boldsymbol{y}$$

This solution calculates the best fit line parameters, $\hat{\boldsymbol{\beta}}$, as a function of the independent variable x, inside the matrix $\mathbf{X}$, and the measured data of the dependent variable y.

4.4.2 Weighted least-squares regression

If the independent x-variable data have zero or negligible uncertainties, but the y-variable data all have different uncertainties, then the best fit to the data should give more weight to the more

precise analyses and less weight to the less precise analyses, just as with a weighted mean. As for the univariate weighted mean, the optimal weighting scheme uses the inverse of the variance of each data point. It can be shown that this weighting scheme produces the smallest possible uncertainties in the slope and y-intercept of the data.

The easiest way to incorporate these weights into the diagonal of a square n-by-n matrix, $\mathbf{W}$, so that the inverse variance for the first measurement goes into the first row and first column of $\mathbf{W}$, the inverse variance for the second measurement goes into the second row and the second column of $\mathbf{W}$, and so forth. More technically, the matrix $\mathbf{W}$ is the inverse covariance matrix (Section 4.3.4), usually denoted $\boldsymbol{\Sigma}^{-1}$. The weighted least-squares fit parameters, $\hat{\boldsymbol{\beta}}$, can now be calculated as

$$\hat{\boldsymbol{\beta}} = (\mathbf{X}^T\mathbf{W}\mathbf{X})^{-1}\mathbf{X}^T\mathbf{W}\boldsymbol{y} \tag{4.23}$$

An example weighted best fit linear regression line is plotted in Fig. 4.9b, which shows the different uncertainties for each of the measurements plotted on the y-axis as ±2σ uncertainty bars.

4.4.3 Linear regression with uncertainties in two or more variables (York regression)

The weighted least squares algorithm presented in Sections 4.4.1 and 4.4.2 assumes that there is zero or negligible uncertainty in the independent variable plotted on the x-axis. However, the distinction between independent and dependent variables does not exist for all data sets. For instance, when plotting one measured isotope ratio against another, like on an isochron or concordia plot, then neither ratio counts as an "independent" variable, and both isotope ratios have measurement uncertainties.

Additionally, the uncertainties between the x- and y-variables may be correlated with one another. For instance, consider measuring three isotopes, a, b, and c, together, and plotting the isotope ratio a/c on the x-axis and b/c on the y-axis. Measurements of all three isotopes are inherently noisy, so that the intensity of each beam is variable. When the intensity of isotope c is slightly lower than average, the ratios a/c and b/c will have lower-than-average denominators, and both will be higher than average. Likewise, when the intensity of isotope c is slightly higher than average, the ratios a/c and b/c will both be slightly lower than average. Because both a/c and b/c tend to move together, their uncertainties are said to be correlated. A general linear regression algorithm should account for both the uncertainties in the two variables, and any uncertainty correlation they have.

This statistical problem has been addressed by several pioneering papers, elegant cross-disciplinary works that made

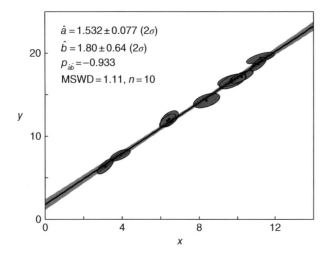

$\hat{a} = 1.532 \pm 0.077 \ (2\sigma)$
$\hat{b} = 1.80 \pm 0.64 \ (2\sigma)$
$\rho_{\hat{a}\hat{b}} = -0.933$
MSWD $= 1.11, n = 10$

Fig. 4.10. Weighted least-squares linear regression (a "York fit") with correlated uncertainties in the x-axis and y-axis variables. The best fit line is shown in black and its 2σ uncertainty envelope is shaded. The best fit slope $\hat{a}$ and y-intercept $\hat{b}$ have uncertainties, and these uncertainties are correlated, with correlation coefficient $\rho_{\hat{a}\hat{b}}$. The correlation coefficient is needed to propagate uncertainties in any function of $\hat{a}$ and $\hat{b}$, such as the uncertainty envelope for the best fit line. (*See insert for color representation of the figure.*)

contributions to both isotope geochemistry and statistics [*York*, 1966, 1968; *Titterington and Halliday*, 1979]. A more recent summary of this work is presented in *York et al.* [2004], and the same concepts have been extended to straight-line regression through more than two variables by *McLean* [2014]. These papers all share a common approach.

Figure 4.10 illustrates a best fit line to data with correlated uncertainties in both the x-variable and y-variable. The 2σ uncertainty (86% confidence interval) for each data point is represented with an ellipse (see Section 4.2.5 and Fig. 4.4). Strictly, the shortest distance between each of the individual data points, marked with small black dots, and the line would be the perpendicular distance between the two. However, the perpendicular distance might not be the closest in terms of the point's uncertainty.

Figure 4.11a shows a point whose perpendicular projection from the point to the best fit line lies just outside its uncertainty ellipse. Elsewhere, the same best fit line lies inside the uncertainty envelope, demonstrating that the perpendicular distance is not the closest when taking into account the uncertainty. Instead, it is more useful to calculate the distance from the point to the line in units of uncertainty, so that a 1σ ellipse represents all points at a "distance" 1 from the point, a 2σ ellipse represents a distance of 2, and so on. A distance calculated this way is called a *Mahalanobis distance*, and can be estimated with the formula

$$d^2 = r^T \Sigma^{-1} r \qquad (4.24)$$

where d is the Mahalanobis distance, r is the residual vector, and Σ^{-1} is the inverse of the covariance matrix for the data point. The residual vector is the shortest line segment between the point and the line, whose first component is the difference between the measured x value and the x-value of the nearest point (by Mahalanobis distance) on the line. The covariance matrix had two rows and two columns. The value in the first row and first column is the variance in x, the value in the second row and second column is the uncertainty in y, and the entries in the first row and second column and the second row and first column are the same, the covariance between x and y.

Because the Mahalanobis distance takes the uncertainty into account, the sum of the squared Mahalanobis distances from each of the data points to the line is analogous to the weighted

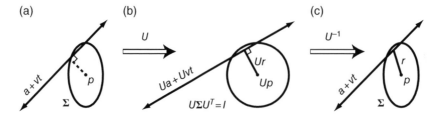

Fig. 4.11. The shortest distance between a point, labelled p, and a best fit line (with arrows) is technically the perpendicular distance, illustrated with the dashed line in (a). However, the intersection between the dashed line and the best fit lines lies just outside the uncertainty ellipse for the data, which is determined by its covariance matrix Σ. The regression line passes inside the uncertainty ellipse elsewhere. (b,c) illustrations of methods for finding the closest point on the line in terms of the point's uncertainty. A linear transform, U, is applied to the system, which transforms the data, ellipse, and line, turning the ellipse into a circle. The perpendicular distance from the point to the line is now the point farthest inside the uncertainty envelope. The inverse of the linear transform is then applied to produce (c), which includes the best fit residual r. (Source: *McLean* [2014]. Reproduced with permission of Elsevier.)

sum of the squared differences between the data and the mean for a weighted mean or between the points and a line for a weighted (ordinary) least squares fit for linear regression. The best fit line is therefore the one that minimizes the sum of the squared Mahalanobis distances from the data to the best fit line. Figure 4.11 illustrates one way to determine the smallest Mahalanobis distance, using a linear transform to turn the uncertainty ellipse into a circle.

Unlike the weighted mean and weighted least-squares algorithms, there is no deterministic equation that solves for the best fit line for data with correlated uncertainties in both dimensions. Instead, a numerical solver [e.g., *Marquardt*, 1963] is used to iteratively find the slope and *y*-intercept that minimize the sum of the squared Mahalanobis distances, a form of numerical optimization. Derivation of the best fit line and its uncertainty is outside the scope of this chapter, but is explained in detail in *York et al.* [2004] and *McLean* [2014].

4.5 INTERPRETING MEASURED DATA USING THE MEAN SQUARE WEIGHTED DEVIATION

Now that we have explored the most common statistical applications to measured data, fitting data to a mean or a line, we can begin to test the assumptions that went into calculating each one. For normally distributed data whose uncertainties have been estimated during measurement, which constitutes the broad majority of data in geochronology and thermochronology, the most powerful tool used for this data exploration is the *reduced chi-squared statistic*, also known in geochronology and thermochronology as the mean square weighted deviation (MSWD).

4.5.1 Testing a weighted mean's assumptions using its MSWD

While the weighted mean and its uncertainty given in equations (4.16) and (4.17) are used specifically for data that are not identically distributed, they do assume that the data are independent, the uncertainties have been accurately estimated, and that each analysis represents a repeat measurement of the same true mean value. If the data are not independent—if they share a common source of uncertainty, for instance from sample-standard bracketing, or if a systematic uncertainty like a decay constant is to be included in the weighted mean uncertainty—then a modified weighted mean algorithm [*McLean et al.*, 2011] must be used. However, if the data do not represent repeat measurements of the same true value, then the results of equations (4.16) and (4.17) are no longer meaningful, and other statistics should likely be used to describe not only the mean but also the scatter in the data (see the overdispersion parameter in *Vermeesch* [2010]).

We would like to test the fundamental assumption that the observed variability in a data set is consistent with (noisy) measurements of the same true value. One way is to compare the measured differences between the data points and the weighted mean to their expected differences, as represented by their uncertainties, is by calculating the *reduced chi-square statistic* or *MSWD*,

$$\chi^2_{\text{red}} = \text{MSWD} = \frac{1}{n-1} \sum_{i=1}^{n} \frac{(x_i - \bar{x})^2}{\sigma_i^2} \qquad (4.25)$$

The numerators in the summed ratios are the squared difference between the data points x_i and the mean $\bar{x}$, and the denominator is the expected value of this squared difference, the squared (1σ) uncertainties in the data points. If the scatter in the data points is commensurate with the estimated uncertainties, this ratio should be approximately 1, and after dividing the sum by the $n-1$ degrees of freedom, the average of these ratios should also be 1.

Note that this statistic is called the reduced chi-square statistic throughout science, but has been called the MSWD (mean square weighted deviation) in the geochronology and thermochronology communities following its use in the popular software package Isoplot [e.g., *Ludwig*, 1983, 1991, 2012] and a paper exploring its properties, *Wendt and Carl* [1991]. This chapter hereafter mostly refers to the statistic as the MSWD, consistent with geochronology and thermochronology literature.

The ratio of the numerator to the denominator terms in equation (4.25) is never exactly equal to 1. By chance alone, some measurements will be closer to the weighted mean than others, and the value of each ratio will vary. The higher the value of n the more ratios are averaged together, and the closer the mean (i.e., the MSWD) should be to 1, just as the more times you flip a fair coin, the closer the heads-to-tails ratio should be to 1. The distribution of the expected MSWD depends on the degrees of freedom, like Student's t discussed above, and values for several different $\nu = n - 1$ are illustrated in Fig. 4.12. The mean or center

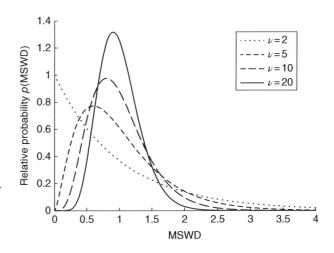

Fig. 4.12. Probability distribution functions for the reduced chi-squared distribution, illustrated for several degrees of freedom, ν. Each distribution is asymmetric, with a mode < 1 and mean of 1. The greater the degrees of freedom, the more the distribution resembles a normal distribution, with mean 1 and standard deviation $\sqrt{2/\nu}$. (*See insert for color representation of the figure.*)

Table 4.3 Acceptable ranges for the MSWD. The confidence level (α) represents the proportion of MSWD values that should fall in this range if all uncertainties are estimated correctly and all assumptions that attend the calculation have been met. Common practice in the geochronology and thermochronology communities is to reject estimated MSWDs that fall outside of the $\alpha = 0.95$ range for a given number of degrees of freedom.

ν	Confidence level (α)					
	0.90		0.95		0.99	
1	0	2.71	0	3.84	0	6.63
2	0	2.30	0	3.00	0	4.61
3	0.00	2.09	0.00	2.61	0.00	3.78
4	0.04	1.97	0.02	2.38	0.00	3.32
5	0.10	1.89	0.06	2.24	0.02	3.03
6	0.15	1.83	0.10	2.13	0.04	2.82
7	0.19	1.78	0.14	2.05	0.07	2.66
8	0.23	1.74	0.18	1.99	0.10	2.54
9	0.27	1.70	0.21	1.93	0.12	2.44
10	0.30	1.67	0.24	1.89	0.15	2.35
11	0.33	1.64	0.27	1.85	0.17	2.28
12	0.35	1.62	0.29	1.81	0.20	2.22
13	0.38	1.60	0.32	1.78	0.22	2.17
14	0.40	1.58	0.34	1.75	0.24	2.12
15	0.42	1.56	0.35	1.73	0.25	2.08
20	0.49	1.49	0.43	1.63	0.33	1.92
25	0.54	1.45	0.48	1.56	0.38	1.81
30	0.58	1.41	0.52	1.52	0.43	1.74
35	0.61	1.38	0.56	1.48	0.46	1.68
40	0.63	1.36	0.58	1.45	0.49	1.63
45	0.65	1.34	0.61	1.42	0.52	1.59
50	0.67	1.32	0.62	1.40	0.54	1.56

of the area under the curve of each distribution in Fig. 4.12 is 1, but the distribution is right-skewed. As the degrees of freedom ν increases, the distribution more resembles a normal distribution, with a mean of 1 and a standard deviation of $\sqrt{2/\nu}$. Table 4.3 provides a list of acceptable ranges for MSWDs.

The MSWD statistic can provide a sensitive test of the assumptions underpinning a weighted mean. If the data are repeat measurements of the same value, and the uncertainties have been estimated correctly, then a weighted mean's MSWD value should be consistent with the probability distribution functions in Fig. 4.12. For example, the MSWD = 0.78 calculated for the $n = 11$ data points in Fig. 4.8 is entirely consistent with the $\nu = n - 1 = 10$ probability distribution function illustrated in Fig. 4.12 and the $\alpha = 0.95$ interval for $\nu = 10$ in Table 4.3.

However, if the MSWD is much less than 1, then the measurement uncertainties are likely overestimated. For the example illustrated in Fig. 4.8 with 10 degrees of freedom, Fig. 4.12 shows that there is a near-zero likelihood that MSWD < 0.1, assuming the uncertainties used in its calculation have been estimated correctly.

Alternatively, if the data represent measurements of two or more different values instead of a single "true" value, then the resulting MSWD would be greater than expected. How much greater depends on the relative size of the scatter in the true values being measured and the measurement uncertainties. Variability in measured data that is, for instance, ten times greater

than the uncertainties results in a MSWD $\gg 1$, strong evidence against the assumption that the data are repeat measurements of a single value. Uncertainties that are ten times more than the variation in true values, however, can mask that variation, producing subtly greater MSWD than expected values that can reasonably be attributed to chance.

The ability to resolve "excess scatter" in measured data, or even "excess cluster," where the scatter in the data is significantly smaller than the estimated uncertainties, also depends on the number of measurements made. The measured MSWD should fall in a smaller and smaller range as n increases, for instance falling in the range $\pm 2\sqrt{2/\nu}$ about 95% of the time for large n data sets. The same principle applies to locating underestimated uncertainties.

One assumption built into equation (4.25) is that the data point uncertainties (σ_i^2) are all analytical uncertainties alone, and do not incorporate systematic uncertainties. Incorporating systematic uncertainties in equation (4.25) will result in an underestimate of the MSWD. This is because systematic uncertainties do not increase the random point-to-point scatter in the data, in the numerator of the ratio in the summation, but erroneously including them will increase the denominator term. The resulting ratios will no longer average to a value of 1. Formulas for a weighted mean and its MSWD that include systematic uncertainties are derived in *McLean et al.* [2011].

4.5.2 Testing a linear regression's assumptions using its MSWD

The principles outlined above for analyzing the estimated MSWD for a weighted mean apply to linear regression as well. In both cases, there is a difference between the measured data and the model fit, and this difference can be compared to the average value it is expected to be based on its uncertainty. For a weighted ordinary least-squares line fit, the equation to calculate the reduced chi-squared statistic is

$$\text{MSWD} = \frac{1}{n-2} \sum_{i=1}^{n} \frac{(y_i - \hat{y})^2}{\sigma_i^2} \tag{4.26}$$

The $n - 2$ in the denominator is the degrees of freedom for a linear regression with two variables. For a mean, the number of degrees of freedom was $n - 1$, where the 1 represented the number of estimated parameters, $\bar{x}$. For the linear regression, there are two estimated parameters, the slope and the y-intercept. As with the mean in equation (4.7), the equation for the variance of the fit parameters can be rearranged as a function of the measured data set and the two best fit line parameters, so that the variance can be written as a function of only $n - 2$ measured values.

When there are uncertainties in both the measured x and y-variables, then the distance to the best fit line in both directions is part of the MSWD calculation. For each data point, this is the residual vector r_i, and the uncertainty in both variables and any uncertainty correlation between them is captured in the covariance matrix $\boldsymbol{\Sigma}$. The more general form for a χ^2_{red} for any model fit with ν degrees of freedom is

$$\chi^2_{red} = \frac{1}{\nu}\sum_{i=1}^{n} \boldsymbol{r}_i^T \Sigma_i^{-1} \boldsymbol{r}_i \tag{4.27}$$

Note the similarity with equation (4.24), the squared Mahalanobis distance.

4.5.3 My data set has a high MSWD—what now?

There is no unique solution to having an MSWD that is higher than the expected random variation depicted in Fig. 4.12. *This means that one or more of the assumptions required to perform the calculation has not been met.* Unfortunately, there is no mathematical way to determine which assumption or assumptions are incorrect without further geological or analytical context. What is certain is that the best fit value has been estimated incorrectly, and that the values and uncertainties of the model fit parameters, like the weighted mean or slope and y-intercept, do not have the same meaning they would if the fit were valid. To see why this is the case, consider an example where the uncertainty in one data point has been drastically underestimated. This means that the uncertainty is wrong, and that the point will be weighted too much in the weighted mean or linear regression. Because of this, the calculated weighted mean would represent a biased estimate for the data set and be different from the value obtained if the uncertainty of the point was estimated correctly. For this reason, publishing and interpreting data sets with high MSWD should be approached cautiously, and the reader should be alerted to the impact on the results.

There are a number of avenues to explore in order to diagnose the cause and perhaps fix a too-high estimated χ^2_{red}, each often unsatisfying without further research or analyses.

(1) Perhaps the uncertainties have been underestimated. This could be tested analytically using a set of homogeneous reference materials that closely mimic the material being measured, or by closely scrutinizing the analysis and accounting for any overlooked uncertainty contributions. Inflating the uncertainties in the measured data points artificially can remedy the high χ^2_{red}, and generally acts to decrease the relative differences between the weights. For data sets with uncertainties in two or more variables, like that in Section 4.4.3, it is unclear a priori which variable's uncertainties should be inflated. Inflating, for instance, the x- versus the y-variable uncertainties will change the best fit line parameters, yielding two different answers.

(2) Perhaps the data set contains one or more *outliers*, or measurements that have been erroneously biased from the rest of the data set. Unfortunately, there is no foolproof test to determine whether or not a measurement is an outlier, even for normally distributed data. Most measured data points should cluster near the estimated mean or the best fit line. But the tails of the normal distribution, illustrated in Fig. 4.3,

assign finite probability to results far from the mean. This means that making an accurate measurement far from the mean is always possible.

Simple 2σ "filtering", where any results farther than $\pm 2\sigma$ from the mean or best fit model are rejected from consideration in the calculation, will reject outliers but also about 5% of the valid data. In the absence of outliers, this artificially reduces the standard deviation and standard error of the analyses, while at the same time decreasing the precision of the measurement, making this a losing proposition for routine use. Other algorithms for outlier detection, including Chauvenet's criterion, are available, but suffer from the same drawbacks. Since there is no quantitative definition of an outlier, there is no foolproof way to identify them.

(3) Perhaps the data do not fit the model being used. For instance, if the χ^2_{red} for a weighted mean is too high, the data might not represent repeat measurements of the same true value. After rejecting this simpler model to explain the data, one could fall back on a more complicated model, introducing a new parameter. One such model includes an *overdispersion* term, as in *Vermeesch* [2010], which assumes that the true data are not repeat measurements of the same true value, but vary in true value from one to the next, following a normal distribution with a finite standard deviation. Instead of solving for just the mean, the measured data and uncertainties can be used to solve for the mean and standard deviation of the underlying, true distribution.

(4) Finally, perhaps the geological model being used to interpret the data is incorrect. This could be due to any number of factors, including open-system behavior or analyses biased by contamination. In this case, the numerical results of any statistical algorithm that does not take these factors into account will yield inaccurate results. While it may be personally difficult to abandon a data set, sometimes it is the best option, avoiding misinterpreting geochronologically or thermochronologically important data.

There are a number of avenues to explore in order to diagnose the cause and perhaps fix a too-high of χ^2_{red} but the best require additional geological context or analyses. This is often impractical after leaving the field or having consumed most of a sample in analysis.

4.5.4 My data set has a really low MSWD—what now?

This case essentially means that the data are far less variable than their uncertainties would suggest. Unlike the case for high MSWD, this mostly results from overestimating uncertainties, or very low ν. Figure 4.12 shows that for $\nu = 2$, the probability distribution for the MSWD does not have a mode, meaning that very low values are quite likely. This makes sense intuitively, since the most likely result of an accurate measurement is close to the true value, and very few measurements might not accurately reflect the true scatter.

Overestimating uncertainties also gives low χ^2_{red}, and can result from scientists taking too "conservative" an approach to estimating variance. It is common for some fields to include the entire range of possible values in an uncertainty estimate, such as the absolute maximum and minimum estimate for slip on a fault. While this quantity is meaningful for a single measurement, it is not proper to use a range to calculate a MSWD. The researcher must instead incorporate a more reasonable estimate of the expected variance or standard deviation of the quantities being measured. Another common mistake is incorporating systematic uncertainties into each data point. For equations (4.25) and (4.26), only analytical uncertainties should be included in each σ_i term.

4.6 CONCLUSIONS

A broad variety of statistical tools are available for making sense of measured geochronologic and thermochronologic data. Correctly applying these tools and interpreting the results requires understanding how the data were acquired and for what purpose, and correctly estimating not just the variables being measured but their uncertainties and uncertainty correlations. The best statistical model to use usually depends on the number of assumptions one can make about the data set, and these assumptions can be tested using the reduced chi-square statistic. When these assumptions fail to explain the observed variability in the data, options include reevaluating the data set and uncertainties, searching for outliers, adjusting the model used to fit the data, or abandoning geologic interpretation of the data altogether. These are realistic, but often worst-case scenarios. On the whole, well-informed interpretations of precise and accurate geochronological and thermochronological data have played an important role in the most important Earth Science advances over the last century, and will certainly continue to do so well into the future.

4.7 BIBLIOGRAPHY AND SUGGESTED READINGS

Bevington, P. R. and Robinson, D. K. (2003) *Data Reduction and Error Analysis for the Physical Sciences*, 3rd edn. McGraw Hill.

JCGM (2012) *International Vocabulary of Metrology – Basic and General Concepts and Associated Terms (VIM)*. Technical Report, Joint Committee for Guides in Metrology.

Ludwig, K. R. (1983) *Plotting and Regression Programs for Isotope Geochemists, for Use with HP- 86/87 Microcomputers.* Technical Report, US Geological Survey, Open-File Report 83-849. http://pubs.er.usgs.gov/publication/ofr83849

Ludwig, K. R. (1991) *ISOPLOT: A Plotting And Regression Program For Radiogenic-Isotope Data*, Version 2.53. Technical Report, US Geological Survey, Open-File Report 91–445. http://pubs.er.usgs.gov/publication/ofr91445

Ludwig, K. R. (1998) On the treatment of concordant uranium-lead ages. *Geochimica et Cosmochimica Acta* **62**(4), 665–676.

Ludwig, K. R. (2012) *Isoplot/Ex Version 3.75: A Geochronological Toolkit for Microsoft Excel.* Special Publication 4, Berkeley Geochronology Center, 75 pp.

Marquardt, D. W. (1963) An algorithm for least-squares estimation of nonlinear parameters. *SIAM Journal of Applied Mathmatics* **11**(2), 431–441.

McLean, N. M. (2014) Straight line regression through data with correlated uncertainties in two or more dimensions, with an application to kinetic isotope fractionation. *Geochimica et Cosmochimica Acta* **124**, 237–249.

McLean, N. M., Bowring J. F., and Bowring S. A. (2011) An algorithm for U–Pb isotope dilution data reduction and uncertainty propagation. *Geochemistry, Geophysics, Geosystems* **12**. DOI: 10.1029/2010GC003478.

Satterthwaite, F. E. (1946) An approximate distribution of estimates of variance components. *Biometrics Bulletin* **2**, 110–114. DOI: 10.2307/3002019

Schoene, B., Condon, D. J., Morgan, L., and McLean, N. (2013) Precision and accuracy in geochronology. *Elements* **9**(1), 19–24.

Titterington, D. M. and Halliday, A. N. (1979) On the fitting of parallel isochrons and the method of maximum likelihood. *Chemical Geology* **26**(3–4), 183–195.

Vermeesch, P. (2010) Helioplot, and the treatment of overdispersed (U–Th–Sm)/He data. *Chemical Geology* **271**(3–4), 108–111.

Welch, B. L. (1947) The generalization of "Student's" problem when several different population variances are involved. *Biometrika* **34**, 28–35. DOI: 10.2307/2332510

Wendt, I. and Carl, C. (1991) The statistical distribution of the mean squared weighted deviation. *Chemical Geology: Isotope Geoscience Section* **86**(4), 275–285.

York, D. (1966) Least-squares fitting of a straight line. *Canadian Journal of Physics* **44**, 1079.

York, D. (1968) Least squares fitting of a straight line with correlated errors. *Earth and Planetary Science Letters* **5**, 320–324.

York, D., Evensen, N. M., Martinez, M. L., and Delgado, J. D. (2004) Unified equations for the slope, intercept, and standard errors of the best straight line. *American Journal of Physics* **72**(3), 367–375.

CHAPTER 5

Diffusion and thermochronologic interpretations

5.1 FUNDAMENTALS OF HEAT AND CHEMICAL DIFFUSION

5.1.1 Thermochronologic context

Familiarity with fundamental quantitative theory of diffusion is necessary for useful thermochronologic interpretation both because the daughter product loss that characterizes thermochronometric behavior is usually modeled as a diffusive process, and because interpreting geologic histories from the thermal history constraints that thermochronometers provide requires understanding heat flow and the thermal field in the Earth. This chapter aims to provide users with fundamentals of diffusion for both contexts.

While conventional geochronology relies on a fundamental assumption of closed-system behavior for both parent and daughter nuclides, thermochronology relies on the principle that daughter product loss may occur after mineral formation, and that the rate of this loss is primarily sensitive to temperature. In theory all parent-daughter systems in minerals are subject to some daughter-product loss at sufficiently high temperatures, but this does not mean that all geochronology is thermochronology, because (i) for some systems (e.g., U/Pb in monazite, zircon) the temperatures and durations sufficient to mobilize significant fractions of daughter products within or out of a crystal would result in wholesale melting or recrystallization, and (ii) some minerals form at temperatures well below those at which significant daughter-product mobility occurs. But thermochronometric systems are available and have been applied to deduce thermal histories of rocks ranging from temperatures lower than ~20–50 °C (e.g., optically stimulated luminescence or ^{4}He/^{3}He methods) to higher than 400–800 °C (e.g., U/Pb in apatite, titanite, rutile), and "intermediate-temperature" thermochronology, in the ~150–400 °C range, using ^{40}Ar/^{39}Ar methods, has a long and well established history. Although in this chapter we focus attention primarily on low-temperature thermochronology, the principles and interpretive contexts are similar with higher temperature methods.

The kinetics describing the temperature sensitivity to loss of daughter products often assume a relatively simple process like annealing of crystal defects or interstitial migration and escape of atoms. But in some cases more complex diffusional processes need to be considered, including isotopic or element exchange between phases, concurrent volume and grain-boundary diffusion, interphase partitioning, multicomponent diffusion, anisotropic diffusion, local sinks, and finite reservoirs [e.g., *Giletti*, 1991; *Jenkin*, 1995; *Ganguly*, 2002; *Baxter*, 2003]. Indeed some of the most interesting and useful thermochronologic results require more than the simplest interstitial-impurity assumptions and infinite external-reservoir assumptions to make sense [e.g., *Kelley and Wartho*, 2000]. Some cases may also require consideration of the role of fluids, deformation, and recrystallization on thermochronologic systems [e.g., *Mulch and Cosca*, 2004; *Camacho et al.*, 2005].

5.1.2 Heat and chemical diffusion equation

Although complex models may be required for some systems, useful thermochronometric results can often be interpreted using basic laws following first-order kinetics, volume diffusion, or some more empirical combination of the two. Even in cases where the actual mechanism(s) of daughter product mobility and loss, or annealing, are not well understood at the atomic scale, it may be argued that most thermochronologic systems are governed by some type of diffusive processes, and the behavior of many systems seems to be adequately described, for most purposes, by conventional thermally activated volume-diffusion laws.

From a mechanistic perspective, volume diffusion can be considered an emergent phenomenon of Brownian motion of particles and the second law of thermodynamics. Random walks of particles that have a spatially variable chemical potential (e.g., as arises from concentration) result in an expanding probability distribution of particle locations. The rate of diffusion, or the diffusion coefficient in a random walk problem can be derived from

Geochronology and Thermochronology, First Edition. Peter W. Reiners, Richard W. Carlson, Paul R. Renne, Kari M. Cooper, Darryl E. Granger, Noah M. McLean, and Blair Schoene.
© 2018 John Wiley & Sons Ltd. Published 2018 by John Wiley & Sons Ltd.

a fundamental description using the Einstein relation and the mean square distance of migrating particles. A recent description of this relationship in a thermochronologic context is that of *Gautheron and Tassan-Got* [2010]. Although a rigorous understanding of the fundamental physics governing diffusion is not required for a strong functional understanding of thermochronologic systems, it should be considered, as frequently pointed out by one famous thermochronologist, that escape of a radiogenic atom impurity from the middle of a typical crystal is not an efficient one-way journey over a distance on order the size of the crystal. It is closer to an epic random trek, modulated by concentration gradient, over a distance many orders of magnitude larger than the direct-path distance between its starting and exit points.

The practical theory of volume diffusion derives in many ways from the theory of heat conduction (transfer of energy due to a temperature gradient), and because the theory of heat conduction predates that of mass diffusion, we begin with heat.

Newton's Law of cooling specifies that the rate of cooling of an object is proportional to the temperature difference between it and its surroundings. This is true regardless of whether heat is transferred via conduction, radiation, or convection. Following on this, *Fourier* [1822] more specifically stated that "the heat flux resulting from thermal conduction is proportional to the magnitude of the temperature gradient and opposite to its sign." If the temperature gradient is defined as the difference in temperatures T_2 and T_1 over an infinitesimal distance within a representative volume bounded by x to $x + dx$ (a width of dx), this can be written as (Fig. 5.1):

$$q \propto -\frac{dT}{dx} \tag{5.1}$$

where q is the heat flux (W/m^2) and dT/dx is the temperature gradient in the positive x direction. Replacing the proportionality sign with an equality and introducing k (W/mK) as thermal conductivity leads to

$$q = -k\frac{dT}{dx} \tag{5.2}$$

If the incoming (q_1) or outgoing (q_2) fluxes are different and assuming energy differences can only be expressed as temperature differences, then the temperature within the representative volume must change with time as

$$\frac{\partial T}{\partial t} \propto -\frac{\partial q}{\partial x} \tag{5.3}$$

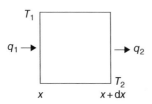

Fig. 5.1. Setup of a one-dimensional unit for understanding contrasting temperatures and fluxes of heat on either side of an infinitesimally small region.

Replacing the proportionality sign with an equality and the assumption that heat differences are manifest only as temperature differences according to the density ρ, and heat capacity c, of the material composing the volume (e.g., no volume or chemical reaction changes), then

$$\rho c\frac{\partial T}{\partial t} = -k\frac{\partial q}{\partial x} \tag{5.4}$$

From equation (5.2) and assuming constant k, we find that

$$\frac{\partial q}{\partial x} = -k\frac{\partial^2 T}{\partial x^2} \tag{5.5}$$

Combining and rearranging equations (5.4) and (5.5), we obtain

$$\frac{\partial T}{\partial t} = \frac{k}{\rho c}\frac{\partial^2 T}{\partial x^2} \tag{5.6}$$

which could be called a simple form of the heat equation. Thanks to the superposition of solutions allowed by systems of this kind, we could also account for heat production (e.g., through radioactive decay) in, as well as heat advection through, our representative volume by adding terms to obtain

$$\frac{\partial T}{\partial t} = \frac{k}{\rho c}\frac{\partial^2 T}{\partial x^2} + \frac{A}{\rho c} + \frac{\nu}{\rho c} \tag{5.7}$$

where A is the volumetric heat production (W/mK) and ν is the velocity of the material in the positive x direction. This equation can also be simply extended to a second or third dimension, as we will see later.

Returning to equation (5.6), we note that the units of $k/\rho c$ reduce to length-squared per time (e.g., m^2/s), which, because we are still dealing with heat, we call thermal diffusivity K (normally kappa), and we restate equation (5.6) as,

$$\frac{\partial T}{\partial t} = K\frac{\partial^2 T}{\partial x^2} \tag{5.8}$$

Whereas this derivation described transfer of energy driven by a temperature gradient (heat), exactly analogous steps lead to an equivalent derivation for the diffusion of species such as atoms, ions, electrons, etc., in a system where their transfer can be thought of as governed by their concentration gradient (which is related to their chemical potential gradient, at least in a simple system) across the representative volume. Thus equation (5.8) can be written as

$$\frac{\partial T}{\partial t} = D\frac{\partial^2 C}{\partial x^2} \tag{5.9}$$

where C is a concentration (molar, mass-wise, etc.) of the diffusing species, and D is its diffusivity (m^2/s), which we have implicitly assumed is spatially invariant. Equation (5.9) is sometimes called Fick's second law. (Equation (5.2) is sometimes called Fick's first law, but Fourier actually proposed essentially the same relationship for heat a few decades earlier, and Newton preceded it all with a similar viscosity law). Analogous equations also exist for hydraulic flow (Darcy's Law) and current (Ohm's Law).

Generalizing to a three-dimensional system in which D varies spatially would yield

$$\frac{\partial C}{\partial t} = \frac{\partial}{\partial x} D \frac{\partial C}{\partial x} + \frac{\partial}{\partial y} D \frac{\partial C}{\partial y} + \frac{\partial}{\partial z} D \frac{\partial C}{\partial z} \quad (5.10)$$

Additional terms accounting for advection (here assumed to be unidirectional in z) and production of the diffusing species, could be added to yield

$$\frac{\partial C}{\partial t} = \frac{\partial}{\partial x} D \frac{\partial C}{\partial x} + \frac{\partial}{\partial y} D \frac{\partial C}{\partial y} + \frac{\partial}{\partial z} D \frac{\partial C}{\partial z} + P + v \frac{\partial C}{\partial z} \quad (5.11)$$

(here v is assumed to be positive downwards). Many applications related to diffusive loss from mineral grains assume a spherical geometry for a diffusion domain's spherical coordinates. The loss of accuracy associated with conversion of more complex shapes to spheres is not large in most cases, provided the surface-area to volume ratio of actual diffusion domain and its spherical representation are the same [*Meesters and Dunai*, 2005; *Gautheron and Tassan-Got*, 2010]. The diffusion equation for spherical coordinates is

$$\frac{\partial C(r,t)}{\partial t} = \frac{D(t)}{a^2} \frac{\partial}{\partial r} \left[r^2 \frac{\partial C(r,t)}{\partial r} \right] \quad (5.12)$$

where a is the radius of the sphere. This length scale, or its equivalent in plane-sheet or other domain shapes, is a fundamental aspect of all diffusion problems involving a domain of finite size.

5.1.3 Temperature dependence of diffusion

Most applications involving migration and loss of radioisotopic daughter products from minerals assume a fundamental control by thermally activated volume diffusion, rather than another type of kinetic law. Although there is strong evidence that many problems of interest do indeed show behavior consistent with diffusion, other types of atomic migration and loss laws may explain the behavior of daughter products. For example, some work in materials science assumes that diffusion of noble gases from metals and many other solids is too rapid to apply to observations at temperatures above room conditions. In these cases, noble gases are lost from metals and other solids via a first-order single-jump mechanism with fundamentally different kinetic laws.

Nonetheless, a great deal of evidence from laboratory experiments and natural observations is consistent with a diffusion rate D following a fundamental Arrhenius law:

$$D = D_0 \exp \left[-\frac{E_a + PV_a}{RT} \right] \quad (5.13)$$

The E_a and V_a are activation energy and activation volume, describing the temperature and pressure dependencies, respectively, of D. D_0 is unimaginatively termed the preexponential factor, and can be defined as the diffusivity at infinite temperature, thereby defining a speed limit for diffusion. The empirical agreement of experimental data to this functional form is typically envisioned as a manifestation of a mechanistic situation of something like that depicted in Fig. 5.2.

Here, an impurity such as a neutral noble gas atom migrates from one interstitial position to the next by overcoming energy barriers associated with lattice distortion, which are represented by E_a and V_a. Qualitatively, the effect of increasing P is to decrease diffusion rates because of the increased energy required to displace lattice ions. However, this effect is thought to be very small compared with the effect of temperature, which provides energy to the migrating atom. At constant P, the law of mass action for the migration "reaction," adapted to diffusion, suggests that the ΔG represented in Fig. 5.2 is related to rate of migration by

$$D = A \exp \left[-\frac{\Delta G}{RT} \right] \quad (5.14)$$

where the preexponential factor must have the same units as the diffusion rate (length squared per time) and therefore contains information about the average distance covered by an interstitial jump as well as the frequency of jumps. Casting the ΔG in terms of enthalpy and entropy yields

$$D = A \exp \left[-\frac{\Delta H - T\Delta S}{RT} \right] \quad (5.15)$$

so

$$D = A \exp \left[\frac{\Delta S}{R} \right] \exp \left[-\frac{\Delta H}{RT} \right] \quad (5.16)$$

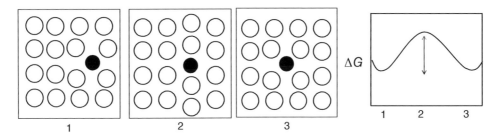

Fig. 5.2. Interstitial diffusion. Black circle is interstitial impurity that causes slight lattice strain on surrounding ions. Movement from position 1 to 3 requires additional lattice strain, associated with an energy barrier ΔG.

Combining the preexponential factor and the exponential that does not depend on T into a new preexponential D_0, and renaming the enthalpy of migration E_a yields the familiar temperature-dependent Arrhenius relation for diffusion.

$$D = D_0 \exp\left[-\frac{E_a}{RT}\right] \tag{5.17}$$

This analysis associates some significance to thermodynamic variables associated with lattice jumps of impurity atoms. It should be recognized that this is restricted to interstitial diffusion of impurity atoms (at constant pressure). Crystal defects are expected to significantly modify this formulation, and more sophisticated forms of diffusion laws take into account the abundance and migration energies of the defects themselves.

Taking the natural log of both sides of equation (5.17) yields a linear relationship between $\ln(D)$ and inverse temperature, in which $-E_a/R$ is the slope and D_0 is the intercept. Linear relationships between between $\ln(D)$ and inverse temperature, at least over some range of temperature, are often observed in diffusion data for natural samples and form the basis of the Arrhenius trend seen in multiple examples later in this and other chapters.

5.1.4   Some analytical solutions

Although many "realistic" problems involving diffusion require numerical solutions, several analytical solutions are available for simple scenarios. These are often used for interpreting diffusion kinetics from experiments whose conditions mimic the initial and boundary conditions of the problems. These solutions also serve as useful guides or approximate solutions to more complicated scenarios involving more realistic conditions. Many of these, as well as many others, can be found in the authoritative resources *Crank* [1956], *The Mathematics of Diffusion*, and *Carslaw and Jaeger* [1959], *Conduction of Heat in Solids*. Several recent references on geochemical kinetics also provide valuable derivations, analytical solutions, and examples, including *Lasaga* [1998] and *Zhang* [1998].

5.1.5   Anisotropic diffusion

Before proceeding it needs to be recognized that this chapter focuses on crystallographically isotropic diffusion. Although daughter products appear to display isotropic or only weakly anisotropic diffusion for several important thermochronologic systems, for example in crystallographically isotropic minerals, and also He in apatite and titanite, important examples exist of strongly contrasting diffusion kinetics parallel and perpendicular to the c-axis in some minerals including Ar in micas, He in zircon, and He in rutile. He in zircon displays strong anisotropy, though not as strong as and different in form from predictions of molecular dynamics simulations. Implications of this behavior for thermochronology are discussed in detail in Chapter 11, and this treatment can be extended to other minerals and systems.

5.1.6   Initial infinite concentration (spike)

One of the simplest examples of a diffusion problem is that of migration of a finite mass M of a species, initially distributed in an infinitely thin layer, into surrounding media in one dimension symmetric to its initial position. This may be envisioned as something like diffusion of a contaminant into surrounding sediment (on its side), or diffusion of heat from a very narrow dike. Here the concentration of the species as a function of time t and position x is

$$C(x,t) = \frac{M}{(4\pi Dt)^{1/2}} \exp\left[-x^2/4Dt\right] \tag{5.18}$$

where M has units of mass or moles per length-scale squared. The solution to this problem may be likened to that of diffusion of a contaminant from a thin layer of sediment into surrounding sediment in the vertical directions (Fig. 5.3). The same problem in three dimensions amounts to diffusion of a point-source in the radial dimension r,

$$C(r,t) = \frac{M}{(4\pi Dt)^{3/2}} \exp\left[-r^2/4Dt\right] \tag{5.19}$$

which is only really different from the one-dimensional case in that the decay of the infinite concentration spike decays faster (Fig. 5.3b).

5.1.7   Characteristic length and time scales

The form of equation (5.18) is strikingly similar to the formula for the normal (Gaussian) distribution

$$f(x) = \frac{1}{\sigma(2\pi)^{1/2}} \exp\left[\frac{-(x-\mu^2)}{2\sigma^2}\right] \tag{5.20}$$

expanded around a mean μ of zero. Further comparison shows that variance σ^2 in probability theory is analogous to $2Dt$ in the one-dimensional problem in equation (5.18) and Fig. 5.3a. Each curve of increasing t can be thought of as a normal distribution with increasing σ^2. This underscores the fact that diffusion theory is rooted in the probability of spatial–temporal distributions (of species, energy, charge, etc.) generated by random motions. It also shows the fundamental importance of the product Dt as a scaling variable in diffusion problems. Although the exact analogy of σ^2 and Dt depends on the problem, in this problem the relationship is simply $\sigma^2 = 2Dt$. By analogy with standard deviation (the square root of the variance in a normal distribution) we can define the characteristic diffusion distance as $\sqrt{(2Dt)}$ (Fig. 5.4). For this particular problem, the diffusion distance defines the position where the concentration of the diffusant is $1/e$ of the concentration at $x = 0$. As Fig. 5.4 shows, the characteristic diffusion distance is proportional to the square root of time, and more specifically the square root of Dt.

This proportionality holds true for any diffusion problem. Although the constant of proportionality depends on the particular problem, one can nevertheless define a characteristic length scale of diffusion L, such that

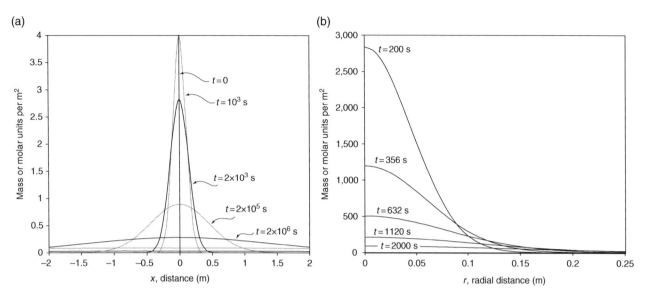

Fig. 5.3. The initial spike problem. A species of mass M and initially infinite concentration exists at time zero at position $x = 0$ (a delta function). With time, the species diffuses to either side into the positive and negative x domains, following equation (5.18).

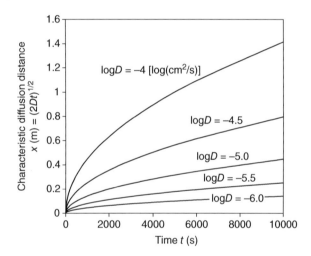

Fig. 5.4. Characteristic diffusion distance (the point at which concentration is $1/e$ times the concentration at zero distance) as a function of time for various values of the diffusion coefficient D.

$$L \propto \sqrt{Dt} \qquad (5.21)$$

or, bearing in mind that L can take different values depending on the problem,

$$L \sim \sqrt{Dt} \qquad (5.22)$$

Similarly, one can define a characteristic timescale of diffusion τ

$$\tau \sim \frac{L^2}{D} \qquad (5.23)$$

The scaling variables provide useful approximations for describing the progress of diffusion and the approach to equilibrium in space and time in diffusion problems.

5.1.8 Semi-infinite media

Analytical solutions for a few simple one-dimensional problems involving semi-infinite media are often used for a wide variety of geologic problems (Fig. 5.5). We review three here. Figure 5.5a shows an infinite half-space bounded at $x = 0$ by another infinite half-space with a different, and in this case fixed, concentration of diffusant, simulating a well-mixed reservoir with a very high or otherwise unchanging concentration, juxtaposed with one with a different concentration. Figure 5.5b shows a similar scenario except neither region of the couple has a fixed concentration, analogous to a coupling of two media with different species or heat concentrations but no internal advection. Figure 5c shows the case of a slab of half width h, centered at $x = 0$, analogous to a cooling dike or diffusing contaminant distributed across a finite width rather than the infinitely thin layer as above.

In all these cases the concentration $c(x,t)$ can be generalized to be a function of only the concentrations each region c_1 and c_2, and another function involving the error function (erf) of $x/(Dt)^{1/2}$ or some combination of h and $x/(Dt)^{1/2}$. Specifically,

$$c(x,t) = c_2 + (c_1 - c_2)\zeta \qquad (5.24)$$

For the case shown in Figs 5.5a and 5.5b,

$$\zeta = 1 - \mathrm{erf}\left(\frac{x}{2\sqrt{Dt}}\right) \qquad (5.25)$$

and

$$\zeta = \left(\frac{1}{2}\right)\left[1 - \mathrm{erf}\left(\frac{x}{2\sqrt{Dt}}\right)\right] \qquad (5.26)$$

respectively. For the case shown in Fig. 5.5c,

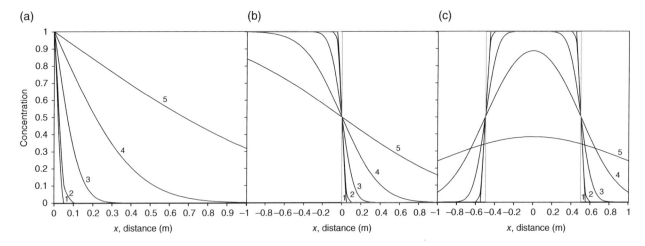

Fig. 5.5. Simple diffusion problem solutions. (a) Fixed concentration at $x = 0$ that diffuses into the positive x domain with time, following equation (5.25). (b) Initial normalized concentrations of unity (negative x domain) and zero (positive x domain) with step-function separation at $x = 0$. Concentrations approach equilibrium by rotating about a fixed point of the mean concentration at $x = 0$. (c) Initial normalized concentration of unity between $-0.5 < x < 0.5$, and zero concentration outside this range.

$$\zeta = \left(\frac{1}{2}\right)\left[\mathrm{erf}\left(\frac{h-x}{2\sqrt{Dt}}\right) + \mathrm{erf}\left(\frac{h+x}{2\sqrt{Dt}}\right)\right] \qquad (5.27)$$

5.1.9 Plane sheet, cylinder, and sphere

Even though the geometry of diffusion domains may be complex in reality, many important and common diffusion problems in thermochronology can be well represented by diffusional loss from one of three morphologies. Two of these, the plane sheet and the cylinder, are effectively infinite in one dimension, but diffusive loss in this dimension is considered negligible considered to the orthogonal direction. The third is the sphere. Although these morphologies are convenient and often used, exact analytical solutions to the diffusion equation for them involve infinite series, which makes them somewhat unwieldy for some applications. Because of this, useful approximations involving fractional losses as a function of the controlling Dt/a^2 (or equivalent) parameters are often used. Derivations of the analytical solutions can be found in *Crank* [1956] and *McDougall and Harrison* [1999]. More sophisticated numerical methods including Monte-Carlo [*Gautheron and Tassan-Got*, 2010] and lattice–Boltzmann methods [*Huber et al.*, 2011] allow numerical simulation of diffusion with arbitrary domain shapes and anisotropic diffusivities.

The plane sheet or infinite slab solution is based on a separation of variables approach to the one-dimensional diffusion equation (equation 5.9). For a plane sheet with half-thickness l and an initial concentration C_0, the concentration of a diffusant as a function of distance x away from the mid-plane position is

$$C(x,t) = \frac{4C_0}{\pi}\sum_{n=0}^{\infty}\frac{(-1)^n}{(2n+1)}$$
$$\times \exp\left[-Dt(2n+1)^2\pi^2/(4l^2)\right]\cos\left[\frac{(2n+1)\pi x}{2l}\right] \qquad (5.28)$$

The solutions for both infinite cylinder and sphere come from coordinate transformations. For an infinite circular cylinder with radius a, the concentration of a diffusant as a function of distance r away from the axis of the cylinder is

$$C(r,t) = \frac{2C_0}{a}\sum_{n=0}^{\infty}\exp\left[\frac{\left(-Dt\,\alpha_n^2\right)J_0(r\alpha_n)}{\alpha_n J_1(a\alpha_n)}\right] \qquad (5.29)$$

In this equation $J_0(r\alpha_n)$ represents the Bessel function (of the first kind, order zero) of r times the nth root of the Bessel function, and $J_1(r\alpha_n)$ is the Bessel function (of the first kind, order one) of α times the nth root of the Bessel function. Subroutines for both Bessel functions and for root generation are available in common mathematical software programs (although only the former is available in a well-known spreadsheet program).

For a sphere the analogous solution is

$$C(r,t) = \frac{2aC_0}{\pi r}\sum_{n=1}^{\infty}\frac{(-1)^n}{n}\sin\left[\frac{n\pi r}{a}\right] \times \exp\left[-n^2\pi^2\left(\frac{Dt}{a^2}\right)\right] \qquad (5.30)$$

5.2 FRACTIONAL LOSS

Equations (5.28)–(5.30) above can be integrated across the diffusion domain dimensions and combined with equivalent expressions for initial homogeneous concentrations to give fractional retention F and loss f. Although these analytical solutions are the most accurate, their convergence for small extents of fractional loss can require a huge number of terms, preventing their easy use in many thermochronologic intepretations such as age calculations and step-heating experiments. Fortunately they can be approximated with simpler and finite equations that are valid for certain overlapping fractional loss ranges, even at low loss ranges. The most commonly used approximations involving

Table 5.1 Analytical solutions and approximations for fractional loss for various domain morphologies as a function of Dt/a^2

Geometry	Equation	Applicable f range
Plane sheet (half-width = l)	$f = 1 - \dfrac{8}{\pi^2} \displaystyle\sum_{n=0}^{\infty} \dfrac{1}{(2n+1)^2} \exp\left[-(2n+1)^2 \pi^2 Dt/(4l^2)\right]$	all f
	$f \approx \dfrac{2}{\sqrt{\pi}} \left(\dfrac{Dt}{l^2}\right)^{1/2}$	$f \le 0.60$
	$f \approx 1 - \dfrac{8}{\pi^2} \exp\left[-\pi^2 \dfrac{Dt}{4l^2}\right]$	$f \ge 0.45$
Cylinder (radius = a)	$f = 1 - 4 \displaystyle\sum_{n=1}^{\infty} \dfrac{1}{\alpha_n^2 \exp\left[-\alpha_n^2 \frac{Dt}{a^2}\right]}$	all f
	$f = \dfrac{4}{\sqrt{\pi}} \left(\dfrac{Dt}{a^2}\right)^{1/2} - \dfrac{Dt}{a^2}$	$f \le 0.60$
	$f \approx 1 - \dfrac{9}{13} \exp\left[-5.78 \dfrac{Dt}{a^2}\right]$	$f \ge 0.60$
Sphere (radius = a)	$f = 1 - \dfrac{6}{\pi^2} \displaystyle\sum_{n=1}^{\infty} \dfrac{1}{n^2} \exp\left[-n^2 \pi^2 \dfrac{Dt}{a^2}\right]$	all f
	$f \approx 6\left(\dfrac{Dt}{a^2}\right)^{1/2} - 3\dfrac{Dt}{a^2}$	$f \le 0.85$
	$f \approx 1 - \dfrac{6}{\pi^2} \exp\left[-\pi^2 \dfrac{Dt}{a^2}\right]$	$f \ge 0.85$

fractional losses come from *Crank* [1975] and other sources cited in *McDougall and Harrison* [1999] (Table 5.1). *Fechtig and Kalbitzer* [1966] also present and discuss approximations with more and slightly different ranges of loss conditions.

5.3 ANALYTICAL METHODS FOR MEASURING DIFFUSION

Here we take a brief look at some of the principal experimental methods for measuring diffusion rates and their dependences on temperature and other factors. More detail is provided on specific diffusants in chapters covering specific systems. Most analytical methods for estimating diffusion rates measure either fractional loss during step-heating experiments and convert these to apparent diffusivities, or they directly measure concentration profiles [e.g., by nuclear reaction analysis (NRA), Rutherford backscatter spectrometry (RBS), laser ablation inductively coupled plasma mass spectrometry (LA-ICP-MS), and potentially other methods including secondary ion mass spectrometry (SIMS)] and fit these curves to analytical expressions from which diffusion rate can be extracted. More rarely, studies use sorption or inward diffusion (essentially the opposite of step-heating fractional loss) to estimate diffusion rates. All these methods require the assumption or validation of an initial concentration profile that has been perturbed, usually by heating, as well as a boundary condition assumption of the concentration of the diffusing species at the edge of the medium of interest. It is also worth noting that the length scale over which diffusion is characterized varies with the method. For the most part, step-heating experiments elucidate the behavior of the scale of individual grains or domains if domains are smaller than the physical grains, and NRA and RBS methods typically characterize diffusion over tens to a few hundred nanometers.

5.3.1 Step-heating fractional loss experiments

Step-heating diffusion experiments have a long history of use in noble gas thermochronology [e.g., *Fechtig and Kalbitzer*, 1966]. The protocol typically involves placing grain(s), preferably of known, crystallographically oriented dimensions, in vacuum, holding the sample at a known temperature for a duration of time and measuring the quantity of gas released. The heating is usually done with a conventional resistance heating or radiofrequency induction furnace, or a radiant-heat "microfurnace" delivering either incandescent or laser generated radiation [*Farley et al.*, 1999]. Gases released during heating are prepared and analyzed using conventional methods, and the fractional release for each heating step is calculated following the final degassing step. Using the characteristic relationship between fractional loss and the dimensionless parameter Dt/a^2, combined with the duration of each heating step t, the fractional release is converted to an apparent D/a^2 at the temperature of the step. If the dimension of the diffusion domain is known, this can then be converted to D.

Necessary assumptions of this method include uniform distribution of the diffusing species throughout the sample, a single domain or series of domains with uniform, or at least known distribution of sizes, negligible ramp-up, cool-down, and overshoot in attainment of temperature in each step, and no compositional or mineralogic change in the sample during the experiment.

Section 5.2 provides both the exact and approximate equations relating fractional loss to D/a^2 for a single step experiment. However, because only the first step can conform to the assumption of uniform concentration of the diffusant, converting f to D/a^2 for every successive step requires correcting for the effects of changes in the concentration profile. This can be done by essentially lumping all fractional releases from previous steps into a virtual

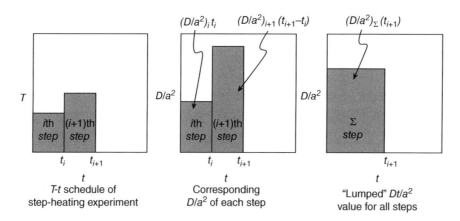

Fig. 5.6. Schematic of approach to determine D/a^2 of any step following the first one (for which expressions in Table 5.1 can be used) in a step-heating diffusion experiment, using equations (5.31)–(5.39). This example shows two steps of duration t_i and $t_{i+1} - t_i$. These correspond to Dt/a^2 values as shown in the middle panel. A summed Dt/a^2 for all steps up to and including the final one represents the equivalent square pulse D/a^2 for duration t_{i+1}(right panel). We seek the value of $(D/a^2)_{(t_{i+1}-t_i)}$, and it is the difference between the area in the block in the right panel and the area of the first step block in the middle panel (equation 5.32). Dt/a^2 values for the two terms on the right-hand side of equation (5.32) can be associated with specific fractional losses using the approximation expressions in Table 5.1, leading to the equations (5.33)–(5.40).

single release, associating it with a single effective D/a^2, and subtracting these effects from the fractional loss equations [e.g., *Fechtig and Kalbizter*, 1966; *Lovera et al.*, 1989; *McDougall and Harrison*, 1999]. If we want to calculate D/a^2 corresponding to the $(i+1)$th step in an experiment, and the duration of the isothermal holding time for this step was $(t_{i+1}) - t_i$, as shown in Fig. 5.6, we can recast the sum of the integral t values for all steps up to and including the $(i+1)$th step as

$$\left(\frac{D}{a^2}\right)_S t_{i+1} = \left(\frac{D}{a^2}\right)_i t_i + \left(\frac{D}{a^2}\right)_{i+1} (t_{i+1} - t_i) \quad (5.31)$$

and solve for the Dt/a^2 value for just the last step

$$\left(\frac{D}{a^2}\right)_{i+1} (t_{i+1} - t_i) = \left(\frac{D}{a^2}\right)_S t_{i+1} - \left(\frac{D}{a^2}\right)_i t_i \quad (5.32)$$

We then rewrite equation (5.32) with the two terms on the right-hand side replaced by inverted expressions in Table 5.1 so that they are cast in terms of fractional loss f instead of Dt/a^2. This procedure is shown graphically in Fig. 5.6.

For spherical diffusion domains, using the approximate loss expressions in Table 5.1, this leads to

$$\left(\frac{D}{a^2}\right)_{i+1} = \frac{-\frac{\pi^2}{3}(f_{i+1} - f_i) - 2\pi\left(\sqrt{1 - \frac{\pi}{3}f_{i+1}} - \sqrt{1 - \frac{\pi}{3}f_i}\right)}{\pi^2(t_{i+1} - t_i)} \quad (5.33)$$

for $f \leq 0.85$, and

$$\left(\frac{D}{a^2}\right)_{i+1} = \frac{\ln\left[\frac{1-f_i}{1-f_{i+1}}\right]}{\pi^2(t_{i+1} - t_i)} \quad (5.34)$$

for $f \geq 0.85$. *Fechtig and Kalbitzer* [1966] used slightly different limits on the approximate solutions, and added another expression for fractional losses less than 10%:

$$\left(\frac{D}{a^2}\right)_{i+1} = \frac{\pi\left(f_{i+1}^2 - f_i^2\right)}{36(t_{i+1} - t_i)} \quad (5.35)$$

For plane sheet geometry, the expressions are

$$\left(\frac{D}{a^2}\right)_{i+1} = \frac{\pi\left(f_{i+1}^2 - f_i^2\right)}{4(t_{i+1} - t_i)} \quad (5.36)$$

for $f \leq 0.60$, and

$$\left(\frac{D}{a^2}\right)_{i+1} = \frac{4\ln\left[\frac{1-f_i}{1-f_{i+1}}\right]}{\pi^2(t_{i+1} - t_i)} \quad (5.37)$$

for $0.45 \leq f \leq 1.0$.

For infinite cylinder geometry, the expressions are

$$\left(\frac{D}{a^2}\right)_{i+1} = \frac{(f_i - f_{i+1}) - \frac{4}{\sqrt{\pi}}\left(\sqrt{\frac{4}{\pi} - f_i} - \sqrt{\frac{4}{\pi} - f_{i+1}}\right)}{(t_{i+1} - t_i)} \quad (5.38)$$

for $f \leq 0.60$, and

$$\left(\frac{D}{a^2}\right)_{i+1} = \frac{\ln\left[\frac{1-f_{i+1}}{1-f_i}\right]}{-5.78(t_{i+1} - t_i)} \quad (5.39)$$

for $0.60 \leq f \leq 1.0$.

Once D/a^2 is known for several temperatures, the simple linear relationship

$$\ln\left(\frac{D}{a^2}\right) = \ln\left(\frac{D_0}{a^2}\right) + \left(\frac{-E_a}{R}\right)\left(\frac{1}{T}\right) \quad (5.40)$$

provides the ability to extract E_a and D_0/a^2 from the slope and intercept, respectively, of the array of step-heating data plotted as $\ln(D/a^2)$ as a function of inverse temperature.

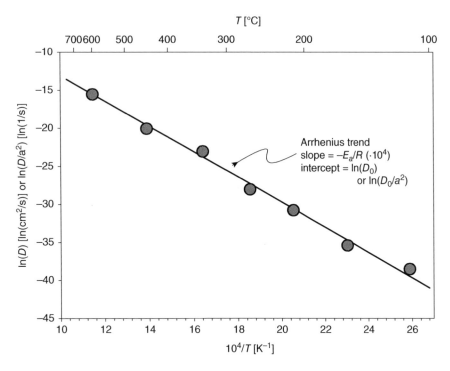

Fig. 5.7. Example Arrhenius trend (black line) fit to experimental data measured at different temperatures (gray circles).

The trend representing the data shown in Fig. 5.7 is often called an Arrhenius trend, after the Arrhenius Law that describes the first-order kinetic loss of the diffusant as a function of temperature. Because of the dependence of fractional losses on diffusion domain size a, Arrhenius trends derived from step-heating experiments can only constrain the bulk term D_0/a^2. If the dimension of the diffusion domain is known, D_0/a^2 can be converted to D_0 by multiplying by the square of the domain size. For some minerals the domain size a corresponds to a dimension of the physical specimen itself (e.g., radius of a spherical grain or half-thickness of a sheet), and this can be shown by a one-to-one correlation between measuring D_0/a^2 (or D/a^2 at a fixed temperature) and the reciprocal scale of the physical grain dimension [*Reiners and Farley*, 1999; *Farley*, 2000, 2007; *Evenson et al.*, 2014]. But diffusion domains need not correspond to easily observable physical grain dimensions, and domains may be defined by subgrain features such as distances between fractures or defects. Approaches that estimate diffusion rates from concentration profiles derive D_0 directly, without the need for understanding diffusion domain dimensions, provided the concentration profiles are observed over length scales smaller than the domains.

Although in practice many studies propagate only analytical error into uncertainties for each step on D/a^2, and then into E_a and D_0/a^2, in reality the uncertainties on any step depend on the uncertainties of all previous steps, because of the "correction" to the initial concentration profile for each step [*Lovera et al.*, 1989].

The applicability of diffusion kinetics derived from step-heating diffusion experiments to reality is subject to a number of assumptions. One is that of uniform concentration of the diffusant across the diffusion domain. Up until this point most of our discussion has implicitly assumed that the initial condition of a diffusion domain, whether in a model or natural sample that we subject to step-heating experiments, bears a uniform concentration profile, when in reality, natural samples that have cooled slowly or been reheated, have rounded profiles. Sample preparation may avoid this problem by creating "fresh" internal surfaces of larger domains.

Also, it is worth remembering that in most cases, experiments are performed at temperatures much higher and over timescales many orders of magnitude smaller than those most pertinent to loss of daughters in geologic conditions of interest. This is also true for most other experimental approaches as well, and means that one must assume that the primary mechanisms and characteristics of diffusion are the same and can be extrapolated over orders of magnitude in temperature and time. It must be assumed that experimental charges experience no significant compositional or mineralogic changes in vacuum. This is a particularly important concern for experiments on phases whose stability is restricted to temperatures or oxygen or water fugacities very different from the conditions in the experiment. This presents difficulties for hydrous phases such as micas and metaloxyhydroxides.

Another concern is the possibility of anisotropic diffusion. Fractional-loss interpretations as described above assume the presence of only one set of kinetic parameters, so if diffusion occurs with different kinetic in different crystallographic orientations, the interpretations will reflect some convolution of these directions. This can be accounted for by performing experiments on crystallographically oriented slabs with large aspect ratios, so that most diffusive loss occurs in a particular crystallographic direction [e.g., *Farley*, 2000, 2007; *Guenthner et al.*, 2013]. Chapter 11 discusses the implications of anisotropic diffusion in more detail.

5.3.2 Multidomain diffusion

Another concern for step-heating diffusion experiments is the domain-size distribution. Because fractional loss depends on the inverse square of the length scale of the diffusion domain, a, samples comprising small domains will yield much higher apparent D/a^2 and D_0/a^2 (and lower closure temperatures) than those with large domains. In most minerals, domain size appears to scale with the physical grain size of the mineral. But when samples comprise polycrystalline mixtures of diverse crystal sizes (e.g., some types of goethite, hematite, and biogenic apatite), or include complex internal structures (e.g., slowly cooled alkali feldspars), the presence of multiple diffusion domains may be evident in Arrhenius trends. If the domain distribution comprises a continuum of sizes with subequal concentrations of diffusants, identification of the multidomain properties may be difficult to discern, and discrete apparent kinetic parameters of the distribution may be difficult to interpret, though this does not preclude a multidomain diffusion interpretation of the data or a thermal history derived from it. But if only a few domains with large size differences are present, their manifestation in an Arrhenius trend is one of discrete steps of decreasing apparent $\ln(D/a^2)$ (Fig. 5.8). It should be said that some studies have attributed these types of trends to experimental artifacts, with no significance to diffusion in nature [*Cassata and Renne*, 2013]. In the $^{40}Ar/^{39}Ar$ system wherein ^{39}Ar is used as the diffusant, inference of very small domains (i.e., < 5 μm) may be confounded by recoil artifacts [*Onstott et al.*, 1995; *Villa*, 1997] during irradiation (see Chapter 9).

Both cycled step-heating schedules shown in Fig. 5.8 suggest relatively sharp steps to lower apparent diffusivity that reflect exhaustion of the diffusant from a small domain. In this example, these steps are fairly obvious, due to the discrete and large differences between the model domains. Samples comprising continua of domain sizes or larger numbers of discrete domains (relative to heating steps) with smaller sizes would be harder to distinguish. In some step-heating experiments, Arrhenius trends from such domain spectra might be misinterpreted as having low slope and intercept, and therefore artificially low E_a and D_0/a^2. Because of this, cycled step-heating schedules involving both prograde and retrograde cycles are often used to elucidate more realistic activation energies in retrograde portions (Fig. 5.8b).

The appearance of Arrhenius trends from samples with identical diffusion kinetics and domain distributions is highly dependent on heating schedule. Because of this, a useful tool for interpreting multiple domain sizes (or other kinetic differences) in step-heating experiments is one that removes the effect of temperature from differences between observed diffusivity (D/a^2). If one assumes a single E_a among all domains and throughout an experiment, changes in the apparent intercept or D_0/a^2 value during an experiment can be described by comparing D/a^2 of any step to a reference D/a^2 assuming the only change is

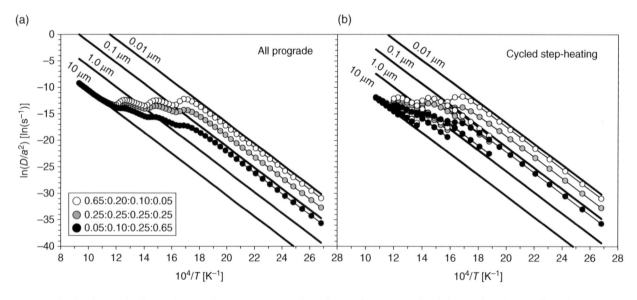

Fig. 5.8. Simulated Arrhenius plots for step-heating release experiments involving all prograde steps (a) and cycled prograde and retrograde steps (b) all of 3600 s durations for samples with E_a of 165 kJ/mol, D_0 of 0.01 cm^2/s, and comprising mixtures of four discrete domain sizes of varying size (labeled diagonal lines). Gray fill represents a mixture with equal proportions of the domains, and the white and black filled symbols represent higher proportions of smaller and larger domains, respectively.

D_0/a^2, which, if D_0 is also assumed constant, reflects differences in a, or domain size. In some applications, the reference D_0/a^2 and E_a are established from the last steps [e.g., *Reiners and Farley*, 1999], but a more common and precedent approach [*Lovera et al.*, 1991; *McDougall and Harrison*, 1999] uses the first few steps to characterize a reference E_a and D_0/a^2. Then, the apparent D/a^2 of each successive step can be compared to the $(D/a^2)_0$ at the same temperature,

$$\ln\left(\frac{D}{a^2}\right)_0 - \ln\left(\frac{D}{a^2}\right) \tag{5.41}$$

Assuming identical E_a for the steps/domains yields

$$\ln\left[\left(\frac{D_0}{a^2}\right)_0 \left(\frac{a^2}{D_0}\right)\right] \tag{5.42}$$

Assuming identical D_0 and dividing by a factor of two then yields the simple index $\ln(a/a_0)$, which represents the difference log ratio of size difference between the domains, or, in some interpretations, the effects of domain size reduction (e.g., through fracturing) or expansion (e.g., through annealing) during the experiment. Figure 5.9 shows the $\ln(a/a_0)$ plot for both experiments in Fig. 5.8 (the $\ln(a/a_0)$ trend is not sensitive to heating schedule). It also shows three inflection points suggesting the presence of four distinct domain sizes.

5.3.3 Profile characterization

Other approaches to measuring diffusion kinetics characterize concentration profiles within minerals more directly. One such approach is the use of laser ablation to target He or Ar concentrations in profiles over tens of microns in minerals whose

presumably uniform concentrations have been perturbed by heating and loss from a polished surface [*Wartho et al.*, 1999; *van Soest et al.*, 2011] (see Chapter 11 for more on this).

Another approach measures concentration profiles (over much smaller length scales) representing perturbations of thin layers of implanted ions, created by heating. These avoid assumptions about the uniformity of the initial distribution of diffusants, and may also avoid complications associated with crystal defects and features that modify bulk transport over larger length scales. The initial profile itself is created by implanting ions of the diffusant itself in a thin layer (nm thick) and then heating the sample for a precisely known duration and temperature. The concentration of the diffusant in the initial layer is modeled as having a distribution of

$$C(x,0) = \frac{N_{imp}}{\sqrt{2\pi\Delta R}} \exp\left[\frac{-(x-R)^2}{2\Delta R}\right] \tag{5.43}$$

where N_{imp} is the dose of implanted ions, x is the distance away from the surface of the mineral, R is the distance between the concentration peak maximum and the surface, and ΔR is the "range straggle" or "Bohr straggle" that describes the deviation from a peak of infinite concentration, due to varying energies of the initially implanted ions [*Cherniak et al.*, 1991]. After heating, the concentration distribution relaxes to one predicted by one-dimensional diffusion at a rate $D(T)$ symmetric about the peak maximum. Incorporating the range straggle, the post-heating distribution of number of nuclides as a function of depth x and time t takes the form below [*Ryssell and Ruge*, 1986]:

$$N(x,t) = \frac{N_m/2}{\sqrt{\left(1 + \frac{2Dt}{\Delta R^2}\right)}}$$

$$\left(\exp\left[-\frac{(x-R)^2}{2\Delta R^2 + 4Dt}\right] \times \left[1 + \operatorname{erf}\left(\frac{\frac{R\sqrt{4Dt}}{\sqrt{2\Delta R}} + \frac{x\sqrt{2\Delta R}}{\sqrt{4Dt}}}{\sqrt{2\Delta R^2 + 4Dt}}\right)\right]\right.$$

$$\left. + \exp\left[-\frac{(x-R)^2}{2\Delta R^2 + 4Dt}\right] \times \left[1 + \operatorname{erf}\left(\frac{\frac{R\sqrt{4Dt}}{\sqrt{2\Delta R}} + \frac{x\sqrt{2\Delta R}}{\sqrt{4Dt}}}{\sqrt{2\Delta R^2 + 4Dt}}\right)\right]\right) \tag{5.44}$$

where N_m is the maximum concentration of the implanted species prior to heating, R is the depth range in the crystal, and ΔR is the range straggle (full-width-at-half-maximum of the initial implanted distribution) [*Cherniak et al.*, 2009]. The post-heating distributions are measured using RBS [e.g., *Cherniak et al.*, 1991], elastic recoil detection analysis (ERDA) [e.g., *Ouchani et al.*, 1998], or NRA [e.g., *Cherniak et al.*, 2009]. Interestingly, diffusion kinetics inferred from ion implantation and relaxation in this way often differ significantly from those inferred from step-heating degassing experiments (see examples in Chapter 11 and *Cassata et al.* [2011]).

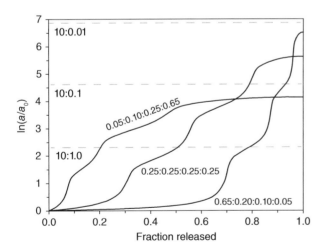

Fig. 5.9. The ln(a/a₀) (also called log(r/r₀) plot in many applications). This index is one half the difference in the vertical displacement (apparent difference in ln(D/a²)) of successive steps in the step-heating experiments shown in Fig. 5.8. This is used to constrain the distribution and proportions of domain sizes in step-heating results measured on multidomain samples. These ln(a/a₀) trends are not sensitive to heating schedule, and are the same for both examples shown in Fig. 5.8. Dashed horizontal lines represent tenfold differences in domain sizes.

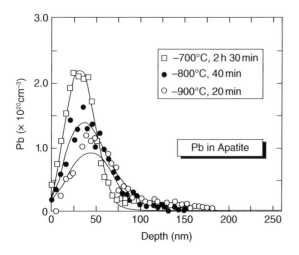

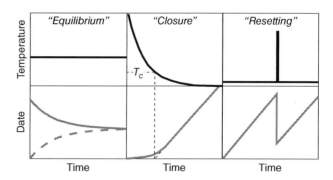

Fig. 5.10. Pb concentrations in apatite, measured by Rutherford backscatter spectroscopy, after heating at different temperatures and durations and perturbation of an initial peak created by Pb ion implantation. (Source: *Cherniak et al.* [1991]. Reproduced with permission of Elsevier.)

Fig. 5.11. "End-member" interpretations for thermochronometric cooling ages. Left: equilibrium or steady-state age, reflecting a balance of radiogenic production and diffusive loss. The equilibrium age may be approached from either old or young ages [see *Wolf et al.*, 1998] for more on equilibrium ages. Middle: closure ages, in which the age primarily reflects the duration of time since cooling below a low temperature [e.g., *Dodson*, 1973]. Right: resetting age, in which the age reflects an extent and date of partial resetting (fractional loss), as well as the original date.

5.4 INTERPRETING THERMAL HISTORIES FROM THERMOCHRONOLOGIC DATA

5.4.1 "End-members" of thermochronometric date interpretations

Because daughter products may be lost via diffusion through time, with no external context or additional geologic insight the significance of a thermochronometric date in isolation is fundamentally ambiguous. Thus it is instructive to consider several possible types of what might be considered end-member interpretations of bulk grain thermochronometric dates, each of which corresponds to a characteristic way of viewing the thermal history that led to the resulting date (Fig. 5.11). In the first, the concentration of daughter product in a grain, and therefore its apparent date, reflects a balance between radiogenic production and loss, or the approach to it, an equilibrium or steady-state date. The second and probably most common end-member interpretation is as a closure date, whereby the apparent date simply reflects

the duration of time elapsed since an episode of rapid cooling to low temperatures, or slightly more precisely, a part of the thermal history in which the rate of diffusive loss of the daughter product decreased much more rapidly than its production (i.e., $\lambda\tau \ll 1$; *Lovera et al.* [1989]). In detail, the accumulation of daughter in the closure context is actually the early part of an approach towards equilibrium from the point of zero daughter concentration. Finally, either increasing or steady (or in between) daughter concentration with time may be interrupted by a resetting event that causes partial resetting.

These end-member contexts can be appreciated more quantitatively with a generalized approximate analytical solution (equation 5.45) to the spherical diffusion-production equation, subject to the simplification that ingrowth of the daughter product N progresses linearly with time [*Wolf et al.*, 1998]. This assumption is valid when the half-life of the parent P is sufficiently large compared with the duration t over which ingrowth is considered to occur while the system is held at constant temperature, and is therefore subject to constant diffusion rate D of the daughter product:

$$\frac{N}{P} = t' = \frac{a^2}{D}\left[\frac{1}{15} - \sum_{n=1}^{\infty}\frac{6}{\pi^4 n^4}\exp\left(-n^2\pi^2\frac{D}{a^2}t\right)\right] + \frac{N^*}{P}\sum_{n=1}^{\infty}\frac{6}{\pi^2 n^2}\exp\left(-n^2\pi^2\frac{D}{a^2}t\right) \quad (5.45)$$

where a is the radius of the spherical diffusion domain, which for many but not all systems scales with the size of the physical grain itself, t' is the apparent thermochronometric date of the system after time t at constant D, and N^*/P is the preexisting daughter/parent ratio and therefore date, prior to the period of interest at constant D.

5.4.2 Equilibrium dates

The usefulness of equation (5.45) comes from evaluating the approach of N/P (or t') to various values for various combinations of t, the duration of time over which the grain with diffusion domain radius a is held at constant T and therefore D. For example when either D or t is very large in equation (5.45), the exponential terms converge to zero, and the solution simplifies to

$$t' = t_{eq} = \frac{a^2}{D}\left(\frac{1}{15}\right) \qquad (5.46)$$

where t_{eq} is the equilibrium date, where daughter production and diffusive loss are balanced, and the thermochronometric date is essentially saturated. This t_{eq} depends solely on the square of the diffusion domain radius a and D, which for a given thermochronometric system with fixed kinetic properties depends solely on T. Thus for isothermal holding, all (ideal) thermochronometric systems can be thought of as approaching

t_{eq}, which is older for lower temperatures and larger domain size (Fig. 5.12).

As can be seen in Fig. 5.12, thermochronometers held at higher T (therefore with younger t_{eq}) approach their equilibrium date much more quickly than those held at lower T. As shown by *Wolf et al.* [1998], the e-folding time t^*, of the approach to t_{eq} is approximately

$$t^* \approx \frac{a^2}{D}\left(\frac{1}{\pi^2}\right) \qquad (5.47)$$

and the t_{eq} is essentially reached within $5\,t^*$.

5.4.3 Partial retention zone

Thermochronometric partial retention zones (PRZs) or "fossil" PRZs are often identified qualitatively in geologic studies as a characteristic distribution of apparent dates with depth or elevation (e.g., Chapter 11). A more robust quantitative

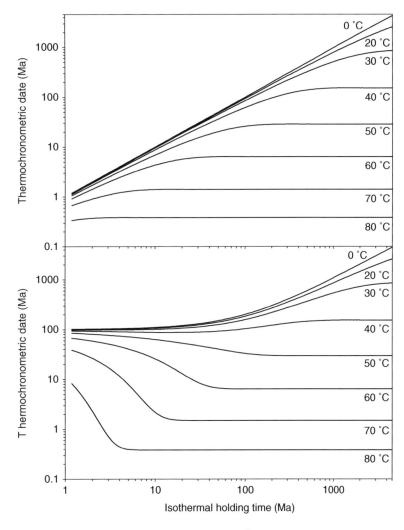

Fig. 5.12. Apparent date evolution for a thermochronometer with $E_a = 138\,kJ/mol$ and $\ln(D_0/a^2) = 14.05$ (corresponding to Durango apatite kinetics with a radial domain size of 50 μm), for isothermal holding at various temperatures. Upper panel shows the approach to t_{eq} for grains with an initial date of zero; lower panel shows date evolution starting from an initial date of 100 Ma.

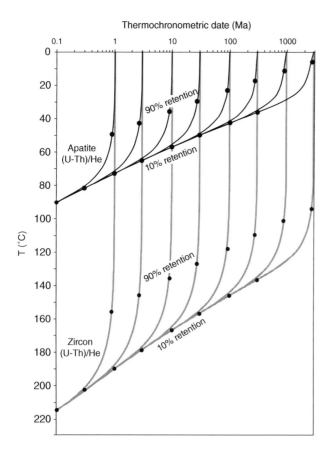

Thermochronometric date (Ma)

Fig. 5.13. Partial retention zones and apparent dates for the apatite and zircon (U–Th)/He systems for isothermal holding at temperatures shown on y-axis, for durations of 1, 3, 10, 30, 100, 300, 1000, and 3000 Ma. The equilibrium date (t_{eq}, equation 5.46) is reached quickly for lower temperature samples, resulting in a steep depth-date trend at shallow depths, but the t_{eq} requires much longer ingrowth times for higher temperatures and older dates, at greater depth, resulting in a shallow temperature-date slope. Temperature is plotted increasing downwards to correspond with the typical situation of increasing temperature with depth in the crust. The partial retention zone (PRZ) can be defined as the range of temperatures where $0.1 \leq t'/t_{eq} \leq 0.9$, and these limits for each isothermal holding duration are shown as filled circles. Note that both the upper and lower temperature limits of the PRZ decrease with time. The apatite system assumes Durango apatite kinetics ($E_a = 138$ kJ/mol; $D_0 = 31.6$ cm²/s; *Farley* [2000]) and the zircon system assumes typical kinetics ($E_a = 168$ kJ/mol; $D_0 = 0.46$ cm²/s; *Reiners et al.* [2004]); both assume a spherical domain with radius of 50 μm.

definition of the PRZ is temperature (or depth, assuming a single geothermal gradient) range where isothermal holding leads to dates that are 10% to 90% of the holding time [e.g., *Reiners and Brandon*, 2006] or, more appropriately, $0.1 \leq t'/t_{eq} \leq 0.9$ [*Wolf et al.*, 1998].

Figure 5.13 shows the development of a thermochronometric PRZ using the kinetics of Durango apatite and typical zircon as examples. The temperatures of the 10% and 90% limits of the PRZ decrease with increasing isothermal holding times, showing that the temperature (and corresponding depths) of PRZs depend strongly on the "age" of the thermochronometric date profile.

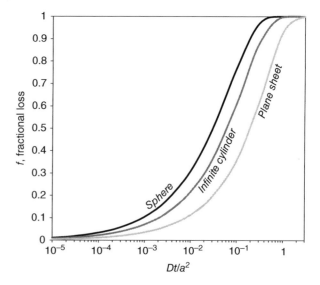

Fig. 5.14. Fractional daughter loss from a diffusion domain with radius a as a function of dimensionless parameter Dt/a^2.

5.4.4 Resetting dates

Equation (5.45) is also useful for evaluating how preexisting daughter/parent ratios and their corresponding apparent dates can be partially or completely reset by a thermal event of a given duration and temperature and therefore diffusivity. If the resetting event is short compared to the half-life of the parent nuclide(s), equation (5.45) simplifies to

$$\frac{N}{P} = t' = \frac{N^*}{P} \sum_{n=1}^{\infty} \frac{6}{\pi^2 n^2} \exp\left(-n^2 \pi^2 \frac{D}{a^2} t \right) \tag{5.48}$$

which shows that the preexisting daughter/parent ratio and corresponding date decreases with an exponential dependence on the dimensionless parameter Dt/a^2, where t in this case is the duration of isothermal holding of the resetting event. Because D/a^2 is an exponential with temperature, this means that the corresponding date decrease for an event of Dt/a^2 has a "double-exponential dependence" on T. Figure 5.14 shows the fractional loss (approximately equal to fractional resetting in most cases) resulting from different Dt/a^2. Note that Dt/a^2 could also be cast as an integral form over varying T, as discussed below.

In some treatments Dt/a^2 is called the Fourier number and denoted by *Fo* or, as used here, ω. This ω can be generalized to any arbitrary thermal history, rather than a square-pulse event of constant temperature for a discrete duration, as

$$\omega(T, t) = \frac{D_0}{a^2} \int_0^t \exp\left[\frac{-E_a}{R} \frac{T}{T(t')} \right] dt' \tag{5.49}$$

If two thermochronometers, i and j, with different kinetic properties experience partial resetting in the same event, the ratio of their Dt/a^2 (for a square-pulse event) or ω (for an arbitrary thermal history) can be defined as

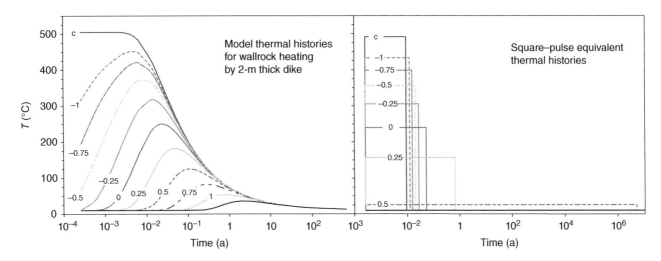

Fig. 5.15. Left: model thermal histories for wallrock samples (initially at 10 °C) at various distances from the contact with a 2-m thick dike with an initial temperature of 1000 °C. In this example, the magma and wallrock are assumed to have identical thermal properties aside from initial temperature, and latent heat is ignored. Distances on each curve are log of nondimensional distance ζ, where $\zeta = 2d/T_m$, where d is the real distance (m) and T_m is the dike half-thickness. Right: square-pulse equivalent thermal histories calculated using equations (5.50)–(5.52).

$$\frac{\omega_i}{\omega_j} \equiv \chi_{i-j} = \frac{D_{0i}}{D_{0j}} \frac{a_j^2}{a_i^2} \frac{\int_0^t \exp\left[\frac{-E_{ai}}{R} \frac{1}{T(t')}\right] dt'}{\int_0^t \exp\left[\frac{-E_{aj}}{R} \frac{1}{T(t')}\right] dt'} \quad (5.50)$$

The temperature of the square-pulse resetting event, or its integrated equivalent, is then

$$T = \frac{E_{aj} - E_{ai}}{R} \left[\ln\left(\chi_{i-j} \frac{D_{0j}}{D_{0i}} \frac{a_i^2}{a_j^2}\right)\right]^{-1} \quad (5.51)$$

and the duration of the event is

$$t = \frac{a_i^2}{D_{0i}} \exp\left[\frac{E_{ai}}{R} \frac{1}{T}\right] \omega_i \quad (5.52)$$

or the corresponding equation for the jth thermochronometer.

The utility of these expressions comes through the fact that in some types of nonmonotonic thermal histories, fractional resetting extents of two or more thermochronometers can often be estimated from thermochronometric dates, yielding ω estimates from equation (5.48) or similar expressions in Table 5.1. The ratio of these ω's can then be used as above to estimate temperatures and durations of square-pulse equivalent thermal histories of partial resetting events, or as appropriate, more complex thermal history forms. This approach requires that each of the thermochronometers is only partially reset, so that a finite ω for each may be calculated, and that the thermochronometer pairs have distinct and sufficiently well known activation energies. The fission-track and (U–Th)/He systems in apatite are one example of a thermochronometer pair with these characteristics.

Figure 5.15 shows an example of square-pulse equivalent thermal histories calculated for a series of model thermal histories for heating of wallrock adjacent to a dike. This shows that the derivative square-pulse histories always have lower maximum temperatures and longer durations than the "real" histories. For higher temperature perturbations that cause larger extents of resetting, this difference is minimal, but for lower temperature perturbations, which result in less extensive resetting, retrodicted thermal histories imply unrealistically long duration and low-temperature thermal perturbations.

In some cases partial resetting events caused by nonmonotonic thermal histories may have counterintuitive thermochronometric results. For some thermal histories and thermochronometer pairs, activation energy differences may be large enough to cause thermochronometric date "inversions." That is, systems with higher closure temperatures may experience larger fractional daughter loss and date resetting than systems with lower closure temperatures (Fig. 5.16). This has been used to understand thermal histories of short-duration and high-temperature events in several settings including heating from overlying lava flows, wildfire, and meteorite impact histories [*Stockli et al.*, 2000; *Reiners*, 2009; *Cassata et al.*, 2010].

5.4.5 Closure

5.4.5.1 Closure temperature

Steady accumulation of daughter product following an episode of rapid cooling to temperatures where little diffusive loss occurs (i.e., a $\lambda\tau << 1$ episode), is the most commonly assumed scenario for interpreting thermochronometric dates in most tectonic and geomorphic applications. The assumption of monotonic cooling in this context provides the opportunity to define closure temperature. By definition, closure temperature T_c, is the temperature of a sample when its daughter-product concentration was zero, which is in turn defined by the one-to-one extrapolation of date and time (Fig. 5.17). In other words, and as originally defined by *Dodson* [1973], T_c is the temperature of a sample at the time

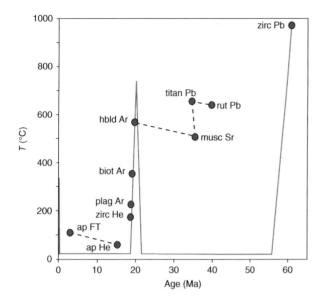

Fig. 5.16. Nominal closure temperatures (see section below) of several thermochronometers plotted as a function of dates that result from the thermal history shown by the solid gray line. This history involves rapid cooling from high temperatures (<900 °C) and ending at 55 Ma, followed by two reheating episodes from an ambient T of 25 °C: (i) to 750 °C at 20 Ma with symmetric monotonic 1-Myr prograde and retrograde paths; (ii) recent (almost indistinguishable from zero age) heating to 350 °C for 30 min (e.g., from near-surface heating by wildfire, localized shear, volcanism, impact-related metamorphism, etc.). This thermal history produces apparent "date inversions" among four different pairs of thermochronometers (dashed lines), whereby systems with higher T_c values have younger ages than those with lower T_c values. If interpreted in the context of monotonic cooling, such data would probably be considered unreliable for analytical or petrologic reasons. Allowing for nonmonotonic histories, these data require at least two different reheating events to produce the date inversions observed among the high-T_c and low-T_c samples.

corresponding to its apparent date. Although straightforward in concept, the full mathematical derivation of closure temperature is far from simple [*Dodson*, 1973] and several variants and simplified derivations exist [e.g., *McDougall and Harrison*, 1999; *Ganguly and Tirone*, 1999; *Harrison and Zeitler*, 2005]. The form presented here largely follows the abridged version presented in *Dodson* [1973], but without presentation of the supporting appendices.

The analytical formulation of closure temperature is based on the assumption of a monotonic cooling path with $1/T \propto t$, at least over a temperature interval where fractional daughter-product retention changes between chronometrically discernible fractions. For nearly all practical purposes, however, the precise function form of a cooling path is not critical; indeed a later part of the derivation for volume diffusion makes the assumption of linear cooling in time. However, the above form of the T–t relation simplifies the derivation of an analytical formulation.

The assumption of $1/T \propto t$ leads to the concept of a time constant t, which is the time required for diffusivity in the mineral of interest to decrease by a factor of $1/e$, given by

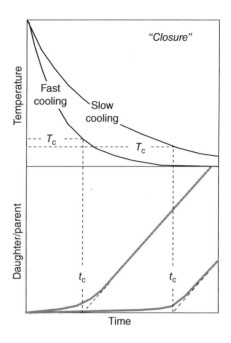

Fig. 5.17. The graphical concept of closure temperature, shown for two different cooling paths. Cooling (with the form $1/T$ proportional to t) leads to decreasing daughter loss, leading to initially partial daughter retention (curved part of gray lines), and eventually to essentially complete daughter retention (linear portion of gray line). T_c is the temperature corresponding to the extrapolation of the linear part of the daughter accumulation trend with time (the closure time, t_c).

$$D = D_0 \exp[-E_a/(RT_0) - t/\tau] \tag{5.53}$$

and

$$D = D(0) \exp[-t/\tau] \tag{5.54}$$

where T_0 is some high temperature at which diffusivity is $D(0)$. From equations (5.53) and (5.54) and the assumption that t is significantly smaller than the decay constant of the parent nuclide(s), one can solve for t as

$$\tau = \frac{R}{E_a dT^{-1}/dt} \tag{5.55}$$

which leads to

$$\tau = \frac{-RT^2}{(E_a dT/dt)} \tag{5.56}$$

The variable τ will be important for understanding several aspects of thermochronology beyond closure temperature, including multidomain diffusion.

The original *Dodson* [1973] treatment of closure temperature first derived T_c for a system whose daughter product loss is governed not by volume diffusion but by simple first-order loss [e.g., *Hanson and Gast*, 1967]. The behaviors of most thermochronometers are considered to generally follow volume diffusion, but there are some whose kinetics may be similar to first-order loss behavior. In any case, derivation of the relationship between T_c, τ, cooling rate, and first-order kinetics provides the foundation for understanding T_c for systems governed by volume diffusion. The rigorous derivation, however, is complex (e.g., see appendix A.5.4 of *McDougall and Harrison* [1999]), and is simply summarized here.

For first-order kinetics the rate of daughter-product loss follows the general Arrhenius relationship

$$k = k_0 \exp\left(-\frac{E_a}{RT}\right) \tag{5.57}$$

The concentration of daughter product x as a function of time is then given by the difference between production and loss, as

$$\frac{dx}{dt} = \lambda C_p - k(t)x \tag{5.58}$$

where λ and C_p are the decay constant and concentration of the parent nuclide, respectively. Adapting equation (5.54) to equations (5.57) and (5.58) yields

$$\frac{dx}{dt} = \lambda C_p - k(0)\exp(-t/\tau)x \tag{5.59}$$

If the initial temperature is high enough such that $k(0)$ is sufficiently rapid so that the system effectively begins cooling with zero daughter product, then equation (5.59) can be used to derive (see *Dodson et al.* [1973] or *McDougall and Harrison* [1999, appendix A.5.4]) the fundamental relationship between the T_c, the kinetics of loss, and the cooling rate (here represented by τ):

$$\frac{E_a}{RT_c} = \ln(\gamma\tau k_0) \tag{5.60}$$

where γ is $\exp(c)$ and c is Euler's constant, 0.5772. Rearranging, using equation (5.57), and designating $T = T_c$ because of our interest in cooling rate near T_c, we find

$$T_c = \frac{E_a / R}{\ln\left(\dfrac{-R\, T_c^2 \gamma k_0}{E_a dT/dt}\right)} \tag{5.61}$$

Equation (5.61) is transcendental for T_c and must be solved numerically. But it converges rapidly as an iterative solution using an initial estimate of T_c on the right-hand side.

Equations (5.60) and (5.61) provide not only the equation for T_c for first-order kinetic loss behavior, but also serve as a guide for deriving T_c for a system following thermally activated volume diffusion. According to *McDougall and Harrison* [1999], the approach taken by *Dodson* [1973] is not mathematically rigorous but appropriate. Alternative interpretations of the derivation, with varying degrees of rigor, have also been presented [e.g., *Ganguly and Tirone*, 1999; *Harrison and Zeitler*, 2005]. The approach summarizes the description of the original Dodson formulation as outlined by *McDougall and Harrison* [1999], starting with the postulation that the rate of loss k_0 can be replaced by a loss parameter corresponding to a combination of terms in the infinite series solution for fractional retention of radiogenic daughter product experiencing loss by volume diffusion

$$F = \sum_{n=1}^{\infty} \left(\frac{B}{\alpha_n^2}\right) \exp\left(-\alpha_n^2 Dt/a^2\right) \tag{5.62}$$

As shown by *Crank* [1956], B and α_n take values depending on the geometry of the diffusion domain. For a sphere of radius a, $B = 6$ and $\alpha_n = n\pi$; for a cylinder with radius a, $B = 4$ and $\alpha_n = $ the nth root of $J_0(x)$ (the Bessel function of the first kind order zero), and for an infinite plane sheet with half thickness a, $B = 2$ and $\alpha_n = (n - 1/2)\pi$.

Equation (5.62) effectively expresses diffusive loss, which is analogous to the first-order rate coefficient k_0, as a weighted sum of Dt/a^2 terms. This suggests that equation (5.62) can be rewritten as a sum of weighted diffusive loss terms

$$\frac{E_a}{RT_c} = \sum_{n=1}^{\infty} \frac{B}{\alpha_n^2} \ln\left(\gamma\tau\alpha_n^2 D_0/a^2\right) \tag{5.63}$$

which can be simplified to

$$\frac{E_a}{RT_c} = \ln\left(\gamma\tau D_0/a^2\right)\sum_{n=1}^{\infty} \frac{B}{\alpha_n^2} + 2B\sum_{n=1}^{\infty} \frac{\ln(\alpha_n)}{\alpha_n^2} \tag{5.64}$$

Using the relationship that

$$\sum_{n=1}^{\infty} \frac{B}{\alpha_n^2} = 1 \tag{5.65}$$

means that

$$\frac{E_a}{RT_c} = \ln\left(\gamma\tau D_0/a^2\right) + 2B\sum_{n=1}^{\infty} \frac{\ln(\alpha_n)}{\alpha_n^2} \tag{5.66}$$

Calling

$$g = \exp\left[\sum_{n=1}^{\infty} \frac{\ln(\alpha_n)}{\alpha_n^2}\right] \tag{5.67}$$

allows us to write

$$\frac{E_a}{RT_c} = \ln\left(\gamma\tau D_0/a^2\right) + \ln(g) \tag{5.68}$$

and

$$\frac{E_a}{RT_c} = \ln\left(g\gamma\tau D_0/a^2\right) \tag{5.69}$$

and combining $g\gamma$ into the "geometric term" A, yields

$$\frac{E_a}{RT_c} = \ln\left(A\tau D_0/a^2\right) \tag{5.70}$$

Values of A (55 for sphere, 27 for cylinder, and 8.7 for plane sheet) can be determined from

$$A = \exp\left[c + 2B\sum_{n=1}^{\infty} \frac{\ln(\alpha_n)}{\alpha_n^2}\right] \tag{5.71}$$

Using the definition of τ (equation 5.56),

$$T_c = \cfrac{E_a / R}{\ln\left[\cfrac{-AR\,T_c^2\left(D_0 / a^2\right)}{E_a\,dT/dt}\right]} \qquad (5.72)$$

As with equation (5.61), equation (5.72) for volume-diffusion T_c requires a numerical solution, but converges rapidly with a simple iteration procedure. It should be noted that some formulations for T_c leave out the negative sign in the logarithmic term. Doing so implies that the term dT/dt should actually be regarded as a rate of cooling, not strictly change in temperature with time, as written; thus we consider the above expression to be more self-consistent.

It may be noted that although a fundamental assumption of the derivation of T_c is a cooling path such that $1/T$ (not T) progresses linearly with t, the only information about the form of the cooling path in equation (5.72) seems to suggest that dT/dt is constant. This apparent paradox arises through the definition of τ, and the conversion of $d(T^{-1})/dt$ to $(-1/T^2)(dT/dt)$ between equations (5.55) and (5.56). This step essentially introduces an approximation to the closure temperature formula as the T in the T^2 is assigned a single value of T_c. The same relationship also holds for the first-order loss T_c as derived in equation (5.61). Although the magnitude of error associated with this approximation is negligible, it is worth noting, as other applications such as the derivation of closure depth, below, also assume constant dT/dt.

The analytical formulation, and to some degree the concept, of closure temperature as defined by *Dodson* [1973] also relies on a number of other assumptions besides the condition of a monotonic cooling path with $1/T \propto t$, and $\tau \ll 1$. These include: (i) constant kinetic properties (including domain size and shape) of daughter-product loss during cooling; (ii) infinite partitioning or external removability of the daughter product in/through a fast-diffusing matrix outside the diffusion domain; (iii) full retention of daughter product following closure and no significant loss of daughter as might be caused by approach to an equilibrium date; (iv) homogeneous distribution and zero loss (or gain) of parent nuclides; (v) isotropic diffusion of the daughter product; and (vi) and an initial temperature prior to cooling that was sufficiently high so that the original daughter-concentration profile is sufficiently different from the post-cooling profile (i.e., full loss of any memory prior to the cooling history; see subsequent sections).

Figure 5.18 shows approximate closure temperatures for a wide range of systems used as thermochronometers, assuming diffusion parameters and domain sizes as cited in the compilation in *Reiners* [2009]. Closure temperatures generally increase through systems based on loss of He, Ar, Sr, Nd, and Pb. The anomalously steep curves for the Rb/Sr systems in micas are associated with the combination of low E_a and D_0 inferred for these systems relative to most thermochronometers.

As shown in Fig. 5.19, variations in cooling rate have a secondary effect on closure temperature compared with variations in

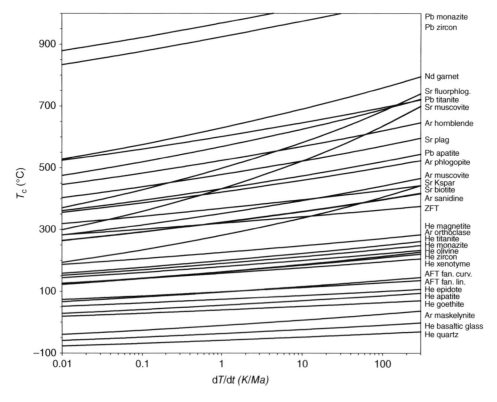

Fig. 5.18. Closure temperatures for various thermochronometers as a function of cooling rate. Kinetic parameters are as compiled in *Reiners* [2009].

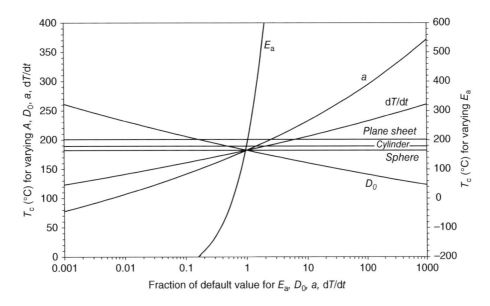

Fig. 5.19. Variations in closure temperature (T_c) as a function of E_a, a, dT/dt, D_0, and domain geometry. Default values (fraction of default values = 1) for example are $E_a = 169$ kJ/mol, $D_0 = 0.5$ cm^2/s, $a = 60$ mm, and $dT/dt = -10$ K/Ma, which yield a T_c of 182 °C. T_c for variations in E_a is shown on right y-axis; all other variations are shown on left y-axis. Fractional variations in E_a have by far the strongest control on T_c.

domain size a (or the lumped parameter D_0/a^2), and all systems have a much lower effect on closure temperature than E_a. For typical values of other parameters and a cooling rate of 10 K/Ma, variations of only 10% in E_a produce on order 25% differences in T_c; this leverage is stronger at higher cooling rates. This underscores the importance of accurate experimental constraints on activation energies. In practice, correlations between E_a and D_0 are often observed in experiments both among and within different minerals. Whether this is because of experimental artifacts relating slope and intercept, or through a conceivably real effect like a diffusion compensation relationship, it will counteract wide variations in bulk retention properties including closure temperature. Regardless of their origin, correlations between E_a and D_0 should not be seen as relieving experimentalists (or dynamics simulators) of the responsibility for accurately determining kinetic properties independently from one another, for at least the reason that there is no single compensation law, and differences in these relationships among minerals may bear important clues about how diffusion actually works at the atomic scale.

For systems whose diffusion properties change as a function of radiation damage, cooling rate has an additional effect that is tied to the rate of radiation damage accumulation. In the (U–Th)/He system in apatite and zircon, for example, closure temperature may depend on parent-nuclide concentration, as well as cooling rate, because slowly cooled samples accumulate damage that changes bulk He diffusion kinetic properties (see Chapter 11).

5.4.5.2 The Ganguly extensions
The Dodson closure temperature formulation is valid for systems in which the grain/domain has a uniform concentration of the daughter product at T_0 and the final daughter concentration profile is sufficiently changed so that T_c is independent of T_0. For many high-temperature thermochronologic systems such as garnet Sm/Nd, Lu/Hf, and the U/Pb system in several phases, diffusive flux out of a grain may be sufficiently small, e.g., as a result of slow diffusivity, large domain size, rapid cooling, or some combination of these factors, so that these conditions are not met.

Ganguly and Tirone [1999] extended the treatment of closure temperature to these systems with this complication by recognizing that Dodson effectively restricted his T_c formulation for situations in which $M >> 1$, where

$$M = \frac{D(T_0)\tau}{a^2} \tag{5.73}$$

and $D(T_0)$ is the diffusivity at T_0. M is thus similar to the dimensionless parameter Fo (or Dt/a^2), which here is associated with the magnitude of daughter-product loss at T_0 over duration τ. When $M >> 1$, a more general form of equation (5.66) above, which also takes account of closure temperature at different positions x, within the diffusion domain (e.g., for a sphere, x is radial distance from the core out to a), is

$$\frac{E_a}{RT_c(x)} - \frac{E_a}{RT_0} = 2\sum_{n=1}^{\infty}\frac{\beta_n(x)}{\alpha_n}\left[C + \ln\left(\alpha_n^2\right) + \ln(M)\right] \tag{5.74}$$

where C is Euler's constant (0.5772). *Ganguly and Tirone* [1999] showed that the minimum values of M so that daughter-product concentration even in the core of a diffusion domain can be considered negligible are ~0.3 for a sphere, ~0.5 for a cylinder, and ~1.1 for a plane sheet (Fig. 5.20). Under these conditions, equation (5.74) can be used to derive

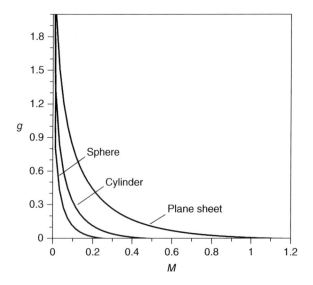

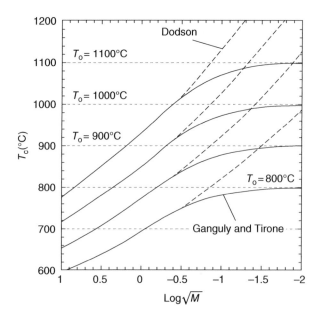

Fig. 5.20. Spatially-weighted $g(x)$ (equation 5.79) as a function of M (equation 5.73) for three diffusion domain geometries. The contribution of g to the closure temperature determination becomes negligible for M values greater than about ~0.3 for a sphere, ~0.5 for a cylinder, and ~1.1 for a plane sheet. (Source: *Ganguly and Tirone* [1999]. Reproduced with permission of Elsevier.)

Fig. 5.21. Closure temperatures calculated by the *Dodson* [1973] formula (dashed lines) and the *Ganguly and Tirone* [1999] model as a function of dimensionless parameter. The point of divergence between the closure temperatures of these two models represents the point where M becomes sufficiently large that the memory function requires designation of a lower closure temperature. (Source: *Ganguly and Tirone* [1999]. Reproduced with permission of Elsevier.)

$$\frac{E_a}{RT_c(x)} = \ln\left(\frac{\tau D_0}{a^2}\right) + \left[C + 4\sum_{n=1}^{\infty}\frac{\beta_n(x)}{\alpha_n}\ln(\alpha_n)\right] \quad (5.75)$$

Dodson [1986] referred to the quantity in the brackets as the "closure function," or $G(x)$. Taking a volume average of all values of $G(x)$ leads to G, and

$$\frac{E_a}{RT_c} = \ln\left(\frac{\tau D_0}{a^2}\right) + G \quad (5.76)$$

Then defining $A \equiv \exp(G)$ and substituting the definition of τ yields the classic Dodsonian bulk T_c (equation 5.72), where G takes on values of 4.0066 for a sphere, 3.29506 for a cylinder, and 2.15821 for a plane sheet.

When M is not sufficiently large for full memory loss, however, *Ganguly and Tirone* [1999] showed that equation (5.75) becomes

$$\frac{E_a}{RT_c(x)} = \ln\left(\frac{\tau D_0}{a^2}\right) + G(x) + d(x) \quad (5.77)$$

which is also

$$= \ln(M) + \frac{E_a}{RT_0} + G(x) + g(x) \quad (5.78)$$

where, as before, $G(x)$ can be replaced with $A \equiv \exp(G)$, but here $g(x)$ is

$$g(x) = -2\sum_{n=1}^{\infty}\frac{\beta_n(x)}{\alpha_n}\left[C + \ln(b_n) + \sum_{m=1}^{\infty}\frac{(-b_n)^m}{m(m!)}\right] \quad (5.79)$$

where $b_n = \alpha_n^2 M$. *Ganguly and Tirone* [1999] showed that the series involving b_n can be truncated at $n = 10$, allowing one to

evaluate the spatially weighted $g(x)$ as a function of domain morphology and M as shown in Fig. 5.20. Tables of $g(x)$ values for spherical, cylindrical, and plane sheet geometries are available in *Ganguly and Tirone* [2001].

The upshot of the Ganguly and Tirone approach for bulk closure temperature essentially results in replacing A with A' in equation (5.72) (the classic Dodsonian T_c equation for volume diffusion), where $A' = \exp(G+g)$, and g is the spatially weighted $g(x)$ determined as in equation (5.79) and shown in Fig. 5.20. For significantly small values of M (equation 5.73), this leads to lower T_c values (Fig. 5.21).

5.4.5.3 Closure profiles

The leap between equations (5.74) and (5.75) essentially involves deriving a weighted average value of T_c from a closure profile. In essence, the time of zero-daughter concentration, and therefore the post-cooling daughter–parent concentration varies continuously from core to rim of a grain. This leap to derive bulk T_c reflects the general practical ease of measuring daughter to parent ratios (i.e., apparent ages) of whole bulk grains, relative to their measurement at particular locations within grains. However, several analytical approaches including ^{40}Ar/^{39}Ar, ^{4}He/^{3}He, and microprobe measurements *in situ* are able to resolve intragrain age variations, which makes the closure profile more useful (Fig. 5.22).

Rewriting (5.76) with the one-dimensional spatial resolution as

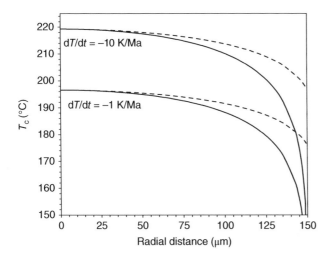

Fig. 5.22. Theoretical T_c profiles (solid lines) and integrated T_c profiles (dotted lines) for the titanite (U–Th)/He system, calculated by the method of *Dodson* [1986], assuming spherical grains with 150 μm radius. Integrated T_c profiles are He T_c values of a portion of the crystal from core to rim at that radial distance. These profiles do not include the effects of alpha ejection on the diffusive profiles.

$$\frac{E_a}{RT_c(x)} = \ln\left(\frac{\tau D_0}{a^2}\right) + G(x) \qquad (5.80)$$

Dodson (1986) summarized the summations of the $G(x)$ term for various positions within the grain, as well as the volume integrations for the bulk grain case, as shown in Table 5.2. Combined with equation (5.56), this leads to a T_c for each radially or axisymmetric position in a domain. $G(x)$ depends only on the domain shape, but the actual profile of closure temperatures within the domain [e.g., solve equation (5.80) for $T_c(x)$] depend on this as well as the other parameters involved in the bulk closure temperature.

Alternative formulations to equation (5.80) are presented in other sources [e.g., *McDougall and Harrison*, 1999], which differ only in whether or not Euler's constant is lumped into closure function $G(x)$; i.e., Dodson's form of $G(x)$ contains $\ln(c)$, whereas in *McDougall and Harrison* [1999] the $\ln(c)$ appears in another term, and the closure function is instead called $4S_2(x)$.

5.4.5.4 Multidomain diffusion revisited

In many cases, the physical grain size of a mineral does not correspond to the diffusion domain size. Internal domains with smaller dimensions and potentially variable kinetic properties may actually control the release of daughter products as a function of time and temperature, both in the laboratory and in nature over geologic time. The potential for multiple kinetic properties of subgrain domains and the potential for interaction of diffusing species between these domains presents complexities and, as usual for thermochronology, opportunities for interpretations. One type of intragranular multikinetic behavior is multidomain diffusion, and is particularly important in $^{40}Ar/^{39}Ar$ thermochronology.

Table 5.2 Spatially resolved values for $G(x)$, the closure function, where x is the fractional spatial coordinate from the core of a diffusion domain, for the three primary domain geometries

x	Plane sheet	Cylinder	Sphere
0	0.98916	1.60161	1.96351
0.05	0.99375	1.61553	1.96702
0.10	1.00758	1.61712	1.97761
0.15	1.03089	1.63666	1.99548
0.20	1.06407	1.66450	2.02093
0.25	1.10769	1.70115	2.05445
0.30	1.16257	1.74733	2.09668
0.35	1.22976	1.80398	2.14853
0.40	1.31069	1.87237	2.21114
0.45	1.40721	1.95417	2.28609
0.50	1.52180	2.05161	2.37546
0.55	1.65779	2.16773	2.48206
0.60	1.81976	2.30671	2.60984
0.65	2.01421	2.47454	2.76443
0.70	2.25074	2.68013	2.95426
0.75	2.54432	2.93748	3.19265
0.80	2.92013	3.27039	3.50233
0.85	3.42548	3.72396	3.92672
0.90	4.16673	4.40072	4.56518
0.95	5.48358	5.63193	5.74177
0.96	5.91600	6.04299	6.13842
0.97	6.47749	6.58088	6.65997
0.98	7.27456	7.35123	7.41124
0.99	8.64699	8.69205	8.73850
0.995	10.02635	10.05229	10.07383
Mean	2.15821	3.29506	4.00660

Source: *Dodson* [1986]. Reproduced with permission of Trans Tech Publications, Ltd.

One of the first studies to recognize and interpret a simple version of multidomain diffusion was *Turner et al.*'s [1966] study of the Bruderheim L-chondrite meteorite. This study also presented the first step-heating age spectrum (Fig. 5.23). Apparent $^{40}Ar/^{39}Ar$ ages from the meteorite systematically increased with increasing fractional release, from about 500 Ma at about 0–10% release to a maximum close to 2.5 Ga from 90–100%. Recognizing that samples of this meteorite were not simple single-phase domains of a uniform size, *Turner et al.* modeled the release of Ar as occurring from spherical diffusion domains. This ignores the potential for large differences in activation energy, frequency factor, and K concentration among the phases comprising the domains. But their simple model of Ar release showed that domains with a lognormal size distribution reproduced the age spectrum well if each domain contained an amount and internal distribution of radiogenic ^{40}Ar that would have been produced by a geologically reasonable history involving formation at 4.5 Ga, followed by reheating at ~0.5 Ga that produced fractional losses inversely proportional to the size of each domain (~90% for the whole sample). This example points out that apparent ages of samples may not yield apparent ages that correspond to discrete formation or cooling (or reheating) events [*Huneke*, 1976] or that form plateaux in age spectra. In this example, the maximum age of the meteorite samples is only ~2.5 Ga, despite the fact that the model that best explained the data involved formation at 4.5 Ga followed by complete Ar retention until 0.5 Ga.

Turner et al.'s study was important not only for pointing out that more information than is explicitly represented by apparent ages can be gleaned from age spectra. It also set the stage for interpreting complex variation in ages and daughter-product

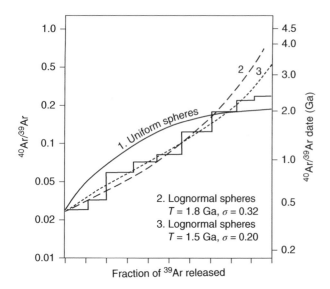

Fig. 5.23. *Turner et al.*'s [1966] ^{40}Ar/^{39}Ar study of the Bruderheim chondrite. This was one of, if not the, first age spectra, and the interpretation of differing radiogenic Ar contents in spheres of varying sizes (*T* is mean age of sample, σ is scale factor of lognormal size distribution of spheres) was one of, if not the, first multidiffusion domain models. (Source: *Turner* [1968]. Reproduced with permission of Elsevier.)

release kinetics from multiple domains in general. Later, *Gillespie et al.* [1984] derived an approach for interpreting the age spectra and kinetics of partially degassed granitic xenoliths erupted in Neogene basaltic lavas that also led the way for multidomain theory.

Much attention in the context of multidomain diffusion models has been paid to the ^{40}Ar/^{39}Ar system in slowly cooled alkali feldspars (K-spar). The potential for complex Ar retention properties in K-spar due to complex internal structures was recognized for a long time [e.g., *Foland*, 1974; *Zeitler and Fitz Gerald*, 1986; *Parsons et al.*, 1999]. *Zeitler* [1987] showed that K-spar step-heating age data as well as apparent kinetic changes could be explained by the presence of a range of domain sizes. He also showed how this domain-size distribution could be revealed by the Arrhenius trends based on the step-heating, but that these trends depend on the step-heating schedule. *Lovera et al.* (1989) published a seminal paper quantifying multidomain theory, particularly as applied to ^{40}Ar/^{39}Ar studies of K-spar.

Lovera et al. [1989] introduced K-spar multiple diffusion domain (MDD) theory in the context of several specimens from the granitic Chain of Ponds Pluton in Maine studied by *Heizler and Harrison* [1988]. Heizler and Harrison attributed the variable ages and complex Arrhenius trends of these samples (Fig. 5.24) to modification during heating *in vacuo*, using only the low-temperature release patterns to derive closure temperatures for the samples. They then associated each sample's age with the closure temperature inferred from the low-*T* part of its Arrhenius trend to derive an apparent cooling history.

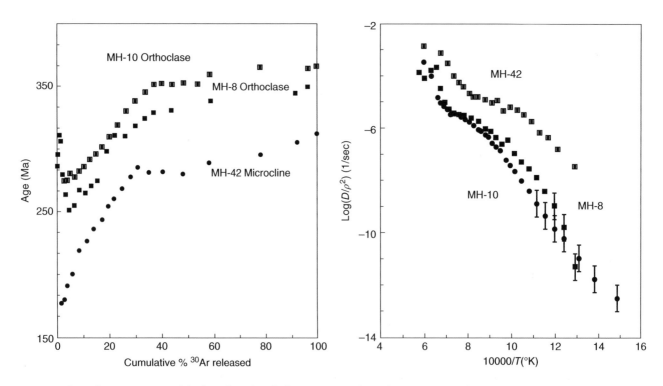

Fig. 5.24. Left: step-heating age spectra of the three Chain of Ponds pluton K-spars [*Heizler et al.*, 1988] as presented in *Lovera et al.* [1989]. Right: Arrhenius trends derived from the same experiments. (Source: *Lovera et al.* [1989]. Reproduced with permission of Elsevier.)

Later experiments with more complex heating schedules involving both prograde and retrograde heating-step progressions [*Lovera et al.*, 1991], as well as multiple irradiation experiments [*Lovera et al.*, 1993] showed that most K-spars do not actually breakdown or become modified with respect to their Ar release kinetics until temperatures near their melting temperatures. Thus, if *Heizler and Harrison*'s [1988] cooling history were correct, the cooling history interpreted from the low-T portions of the Arrhenius trends should predict an ^{40}Ar concentration profile across a single domain that in turn produces the ^{40}Ar/^{39}Ar age spectrum seen in Fig. 5.24. *Lovera et al.* showed that *Heizler and Harrison*'s [1988] thermal history was in fact not consistent with the observed age spectrum, and in fact any thermal history for a single domain was incapable of predicting age spectra with the multiple slope changes seen in these, and many other, K-spar specimens.

Complex age spectra and Arrhenius trends such as those in Fig. 5.24 could conceivably be explained by a range of phenomena, including multiple domains with varying activation energies, or nested diffusion domains that release Ar through several linked pathways. But, as suggested by the similarity of the results of Fig. 5.24 to Turner's Bruderheim meteorite results and the multidomain Arrhenius trend examples above, *Lovera et al.* [1989] suggested that the structure of both the age spectrum and the Arrhenius trend could be reasonably explained (by an arguably parsimonious model) in which multiple, noninteracting domains release Ar as controlled only by differing D_0/a^2. Each of the Chain of Ponds pluton K-spar results in Fig. 5.24 can be reasonably explained by mixtures of three to four domains with size differences (assuming a single D_0) of about a factor of 100 and varying proportions from about 10% to 50%. The mathematics of multidomain diffusion theory are nontrivial, but are outlined in detail (at least in the context of ^{40}Ar/^{39}Ar dating of K-spar) in *Lovera et al.* [1989, 1991] and *McDougall and Harrison* [1999].

Key to the MDD model as developed by Lovera and coworkers is the correspondence between changes in the Ar age spectrum and changes in Ar release kinetics, when compared as a function of the total fraction of Ar released. This is important because it means that distinct reservoirs with distinct radiogenic ^{40}Ar contents release their Ar inventories in kinetically distinct ways. The only way for this to happen is if the kinetic differences observed in the laboratory correspond to the kinetic differences present during accumulation and retention of Ar in geologic history. If crystals and their domains were modified by processes such as hydrothermal recrystallization, deformation, or annealing at any time after cooling and retention of the Ar inventories (including during step-heating experiments), there would be little or no correspondence between kinetic and age spectrum changes.

Lovera et al. [2002] established a quantitative measure of the correspondence between kinetic properties and age spectra, C_{fg}. This index uses the $\log(r/r_0)$ parameter (Fig. 5.25) as the measure of kinetic property changes, because of its insensitivity to the specific heating schedule used in the step-heating experiment (many ^{40}Ar/^{39}Ar studies use r or ρ to represent a). Among

K-spar samples lacking obvious evidence of low-temperature recrystallization or debilitating excess Ar, *Lovera et al.* [2002] found roughly two-thirds showed good correlations between kinetic and age changes ($C_{fg} > 0.8$), and about 40% showed excellent correlations (>90%).

Figure 5.25 shows an example of K-spar ^{40}Ar/^{39}Ar data measured and interpreted in the context of the MDD model. The Arrhenius trend in Fig. 5.25b can be reproduced with a distribution of domains with a single E_a and varying D_0/r^2; the Arrhenius trend for each domain by itself would produce the light gray lines of varying width (proportional to relative abundances). The reference Arrhenius trend taken from the first few steps (E_a and D_0/r^2 shown) is then combined with all successive steps to derive the $\log(r/r_0)$ plot in Fig. 5.25a, which shows a high degree of correlation with the age spectrum.

The MDD model is most frequently used for the ^{40}Ar/^{39}Ar system in K-spar, but it has also been applied to ^{40}Ar/^{39}Ar in muscovite [*Harrison and Lovera*, 2014] and some meteoritic material [*Cassata et al.*, 2010; *Shuster et al.*, 2010], the ^{40}K/^{40}Ca system in K-spar [*Harrison and Lovera*, 2014], and the (U–Th)/He system in hematite [*Farley and Flowers*, 2012].

The principal practical utility of the MDD approach is its ability to provide much greater constraint on thermal histories than simple bulk-grain dating can provide. In a similar way that the closure profile approach or the bulk-grain dating of multiple specimens with distinct kinetics essentially provide a spectrum of thermochronometers with distinct thermal sensitivities, the MDD approach exploits the variation in domain sensitivities to reconstruct continuous time–temperature paths. This is typically done through a Monte-Carlo approach that seeks to predict histories that minimize misfits to the age spectra using the kinetics inferred from the step-heating experiment. In general, restricting the range of potential inverse models to monotonic cooling histories provides t–T solutions with a sufficiently limited range of variation to be practically useful (e.g., Fig. 5.25). When transient heating (i.e., nonmontonic cooling) is allowed, a wider range of potential histories becomes possible. In most cases allowing nonmonotonic cooling, however, contouring of the density of best fit paths in the time–temperature domain usefully points to the highest probability rapid-cooling episodes [*Quidelleur et al.*, 1997; *Harrison et al.*, 2000].

5.5 FROM THERMAL TO GEOLOGIC HISTORIES IN LOW-TEMPERATURE THERMOCHRONOLOGY: DIFFUSION AND ADVECTION OF HEAT IN THE EARTH'S CRUST

The thermal histories, or the ranges of possible thermal histories, of rocks that the approaches discussed here provide is sometimes the goal of thermochronology. Depending on the application and objective, these thermal histories may be simple, for example a "cooling age" in the closure context, or more complex, as

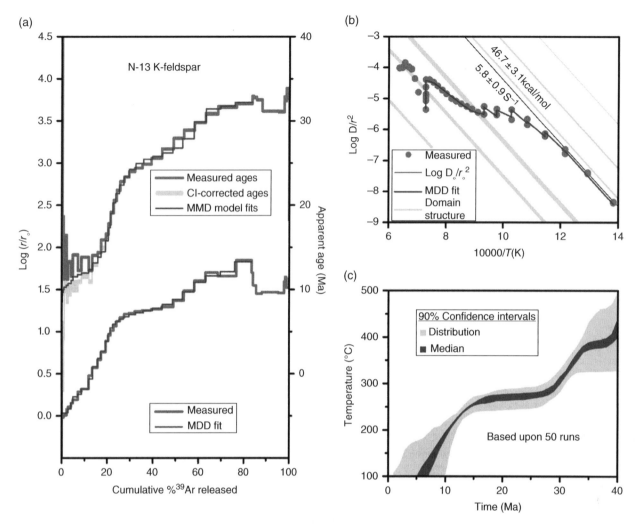

Fig. 5.25. (a) Example of K-spar step-heating experiment, showing Cl-corrected age spectra (upper curves; right vertical axis) and log(r/r_0) trend derived from Arrhenius trend in (b). (b) Arrhenius trend of Ar release from experiment, showing E_a and D_0/r^2 of reference trend, and multiple diffusion domain (MDD) model trend for the step-heating schedule based on a combination of domains with the same E_a as the reference trend but varying D_0/r^2 as shown in the light gray lines. (c) Model time–temperature history derived from inverse modeling of combined kinetic data and age spectra, assuming monotonic cooling of the sample. (Source: *Harrison and Lovera* [2014]. The Geological Society of London.)

might be constrained from multikinetic or multidomain diffusion behavior, to provide a range of permissible time–temperature paths. Solutions for more complex histories may come from forward or inverse modeling, and several useful software packages exist for this purpose, including HeFTy and QTqt [*Ketcham*, 2005; *Gallagher*, 2012].

More often, thermal histories are not the final objective of thermochronology, and instead they are used to elucidate geologic histories of one sort or another. A wide range of phenomema cause temperature changes in rocks. Extraterrestrial materials cooled from initial accretion and heating from short-lived isotopes may experience transients from impact events, volcanism, and solar radiation. On Earth, surface heating (variation in solar radiation and climate, wildfires), frictional heating in faults, magmatism, fluid flow, changes in basal heat flow, and burial and exhumation are typically responsible for temperature

changes in rocks. Although thermochronology has been used to understand all these phenomena, by far the most attention to thermochronologic data is oriented around understanding thermal histories that are interpreted in terms of exhumation processes.

Exhumation, the motion (or sometimes history or path) of a rock towards the surface is one of the simplest geologic processes leading to cooling. Exhumation is either tectonic or erosional, depending on the mechanism of denudation that brings a rock and the surface closer [*Ring et al.*, 1999]. Tectonic exhumation includes ductile thinning of lithosphere and, more commonly, normal faulting or tectonic denudation, which moves footwall rocks towards the surface by removing overlying rock in the hanging wall. The other type of exhumation is erosional, in which rock, regolith, or soil at a specified point at or under Earth's surface is removed by physical or chemical denudation.

In most cases, exhumation causes cooling as rocks pass through progressively lower temperature isotherms en route to the surface. A large fraction of the intermediate to low-temperature ($\sim < 250\,°C$) cooling histories of most rocks exposed at or near the surface today is due to either erosional or tectonic exhumation. Fluid flow, magmatism, frictional heating, and other effects undoubtedly contribute, and in some cases dominate temperature changes, but in most cases these are (correctly or not) regarded as much less important than history of a rock's distance from the interface between rock and air, water, or ice, where the surface temperature boundary condition over most of the planet has a relatively small range (e.g., ~ 0 to $20\,°C$). However, the details of the time–temperature path, or thermal history, of exhumation may be complicated by a number of factors. As one of the principal goals of thermochronology is to use thermal histories to infer exhumation histories to infer geologic processes, it is therefore important to understand the relationships between time, depth, and temperature in exhuming terrains. This requires understanding the thermal field at and beneath the Earth's surface.

Some of the same principles that are used for understanding diffusion of species in crystals are also used for understanding the diffusion of heat in the Earth. A wide variety of approaches may be used for modeling crustal thermal fields and the thermal histories of rocks passing through them. These vary widely in complexity and applicability [e.g., *Ehlers et al.*, 2005; *Braun et al.*, 2006]. Depending on the objectives of a study and the confidence in pertinent constraints on many parameters (including their variation in the past, deformation, fluid flow, etc.) complex three-dimensional models for crustal thermal fields may be constructed that may be used to test hypotheses for interpreting thermochronologic data. These typically require numerical solutions, which may be amenable to solving using programs like Pecube [*Braun*, 2003; *Braun et al.*, 2012]. However, in many cases, a great deal of insight may be gained by simply considering the significance of thermochronologic ages in the context of relatively simple one- and two-dimensional models of crustal thermal fields, if straightforward consideration of a few key phenomena are also considered. Here we summarize some of the more generalized approaches for simple crustal thermal field modeling.

5.5.1 Simple solutions for one- and two-dimensional crustal thermal fields

A general equation describing our situation can be written as spatial–temporal variation in temperature in a homogeneous isotropic solid (what we are assuming to be the crust), whose thermal diffusivity (and conductivity) are independent of temperature:

$$\frac{k}{\rho C}\nabla^2 T - \bar{u}\nabla T + \frac{A'}{\rho C} = \frac{\partial T}{\partial t} \tag{5.81}$$

where T and t represent the usual variables temperature and time, k is thermal conductivity (W/m/K), $\boldsymbol{u}$ is velocity vector (m/s), ρ is density (kg/m^3), C is specific heat (J/kg/K), and A' is

volumetric heat production (W/m^3). To simplify things we will collapse the thermal conductivity term into thermal diffusivity κ (m^2/s), and the heat production term into a heat generation term A with units K/s, and we will assume heat generation is spatially uniform. For present purposes we will also concern ourselves with the two-dimensional problem, and will consider only vertical displacements, so equation (5.81) simplifies to:

$$\kappa\left(\frac{\partial^2 T}{\partial x^2} + \frac{\partial^2 T}{\partial z^2}\right) + u\frac{\partial T}{\partial z} + \frac{A'}{\rho C} = \frac{\partial T}{\partial t} \tag{5.82}$$

where x is horizontal distance across the surface, z is vertical distance, which is positive downwards, and u is vertical velocity *relative to the surface* (m/s), positive upwards (i.e., erosion positive). For this set of solutions we will be dealing with steady state, so the time-transient term $dT/dt = 0$. We will assume that topography is periodic in x, with

$$h = h_0 \cos\left(\frac{2\pi x}{\lambda}\right) \tag{5.83}$$

where h_0 is the amplitude (distance from mean elevation to top of the ridge of bottom of the valley), and λ is the wavelength of the topography. Other variables are the temperature at the surface T_s (K), the temperature at the base of the layer (e.g., base of an accretionary complex) T_L (K), and the surface lapse rate φ (K/m). One last thing to bear in mind is that the way we are dealing with topography involves an approximation to the depth dependence of the perturbation of isotherms by topography [*Turcotte and Schubert*, 2014].

For the case of no exhumation, no heat production, and no topography, equation (5.82) can be solved to yield the simple linear dependence of temperature with depth

$$T(z) = T_S + (T_L - T_S)(z/L) \tag{5.84}$$

Another way to express this is by lumping $(T_L - T_S)/L$ together and calling it the geothermal gradient (dT/dz). Zeroeth-order thermochronologic interpretations can be made using this equation, with some big caveats. For example, an exhumation rate may be inferred by asserting a closure temperature T_c, surface temperature, and geothermal gradient, to derive a closure depth, Z_c. Assuming a steady-state exhumation rate, Z_c represents the vertical distance traveled in the time represented by the thermochronologic age, and the ratio of the two is an estimate of exhumation rate. Aside from the question of steady state, and the assertion of a closure temperature (which of course varies with cooling rate and therefore exhumation rate), this makes no allowance for change in the geothermal gradient due to the exhumation (vertical advection), which we will treat below.

Equation (5.82) can be solved for various combinations of heat production, exhumation, and topography. For heat production alone,

$$T(z) = T_S + (T_L - T_S)\frac{z}{L} - \frac{A}{2\kappa}z^2 + \frac{A}{2\kappa}Lz \tag{5.85}$$

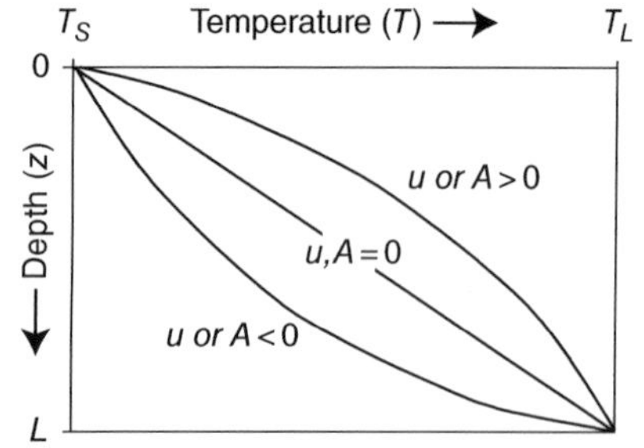

Fig. 5.26. Schematic effects of heat advection and heat production on a one-dimensional temperature profile.

and for exhumation alone:

$$T(z) = T_S + (T_L - T_S)\frac{1 - \exp\left[\frac{-uz}{\kappa}\right]}{1 - \exp\left[\frac{-uL}{\kappa}\right]} \qquad (5.86)$$

which is perhaps the most generally important of these solutions for most thermochronologic applications.

Setting aside topography for the moment, Fig. 5.26 shows qualitatively the effects of either exhumation or heat production on the one-dimensional crustal thermal field. Both advection of heat towards the surface through exhumation and heat production produce a concave down temperature profile (higher geothermal gradient near the surface). Negative advection (e.g., deposition) or negative heat production (e.g., endothermic metamorphic reactions) produce the opposite profile.

As Fig. 5.26 shows, advection of heat by exhumation (or deposition) raises (or lowers) the geothermal gradient near the surface (base), which can dramatically affect the depth of a thermochronometer's closure temperature. This is why the zeroeth-order interpretation described above is not straightforward, especially in situations of rapid exhumation and young cooling ages. A convenient way to estimate the importance of advection in the problem is to ratio the heat transported to the surface by advection to that by diffusion, or the Peclet number,

$$Pe = \frac{uL}{\kappa} \qquad (5.87)$$

When Pe is significantly less than 1, most heat is lost by diffusion, and the simple linear solution may be appropriate. When Pe is significantly greater than 1, most heat is lost by advection, and the exhumation rate needs to be considered in the thermal field solution. For reference, terrestrial subaerial erosion rates span a range of roughly 0.01 (e.g., Kansas) to 10 km/Ma (e.g., Taiwan). Combined with typical thermal diffusivities of about 32 km^2/Ma and a crustal length scale of roughly 40 km, this yields Pe of about 0.01 to 12. Thus advection of heat by exhumation is important to consider for rapidly exhuming regions, as we show later.

Topography causes several different effects important for practical interpretations [*Stüwe et al.*, 1994; *Mancktelow and Graseman*, 1997; *Brandon et al.*, 1998; *Braun*, 2002; *Willett and Brandon*, 2013]. One of these is bending isotherms in the

subsurface, which is shown using the solution to equation (5.82) ignoring exhumation and heat production and accounting for topography alone:

$$T(z,x) = T_S + (T_L - T_S)(z/L) + \left[\frac{(T_L - T_S)}{L} - \phi\right]h_0$$

$$\cos\left[\frac{2\pi x}{\lambda}\right]\exp\left[\frac{-2\pi z}{\lambda}\right] \qquad (5.88)$$

The shape of subsurface isotherms follows that of overlying topography, but with a damped amplitude, the dampening increasing for shorter wavelength topography and higher temperature isotherms (Fig. 5.27). This also means that, at least at a given (shallow) depth, temperatures are higher beneath valleys than beneath ridges.

For the case of heat production and exhumation by no topography:

$$T(z) = \left[T_S + T_L - T_S + \frac{AL}{u}\right]\left[\frac{1 - \exp(-uz/\kappa)}{1 - \exp(-uL/\kappa)}\right] - \frac{Az}{u} \qquad (5.89)$$

For the case of heat production, topography, but no exhumation:

$$T(z) = T_S + (T_L - T_S)(z/L) - \frac{Ay^2}{2\kappa} + \frac{ALz}{2\kappa}$$

$$+ \left[\frac{(T_L - T_S)}{L} + \frac{AL}{2\kappa} - \phi\right]h_0\cos\left[\frac{2\pi x}{\lambda}\right]\exp\left[\frac{-2\pi z}{\lambda}\right] \qquad (5.90)$$

And finally, for the case of heat production, erosion, and topography

$$T(y) = T_S + \left[T_L - T_S + \frac{AL}{u}\right]\left[\frac{1 - \exp(-uy/\kappa)}{1 - \exp(-uL/\kappa)}\right]$$

$$- \frac{Ay}{u} + \left[\frac{\left(T_L - T_S + \frac{AL}{u}\right)\left(\frac{u}{\kappa}\right)}{1 - \exp(-uL/\kappa)} - \frac{A}{u} - \phi\right]h_0\cos\left[\frac{2\pi x}{\lambda}\right]\exp\left[\frac{-2\pi y}{\lambda}\right] \qquad (5.91)$$

5.5.2 Erosional exhumation

Erosion is responsible for far more exhumation than tectonic mechanisms alone, and is the primary driving force for cooling of rocks in both tectonically active and inactive regions. Thermochronology is uniquely well suited to providing constraints on erosional histories that in turn provide understanding of a range of phenomena including tectonic and climatic histories. Thermochronologic approaches to constraining timing and rates of erosion typically take one of four approaches:

(1) dating multiple samples collected over a large area to map out a spatial pattern of erosion rates (e.g., to elucidate variations caused by tectonic or climatic (or both) variation across a region);

(2) using cooling ages from samples collected over a large vertical distance but small horizontal distance to interpret temporal variations in erosion rates in a single location (e.g., the "vertical transect" or age–elevation relationship (AER) approach);

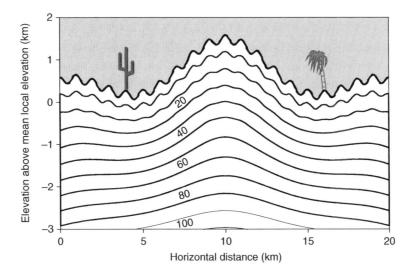

Fig. 5.27. Isotherms beneath model topography with three different waveforms: h_0 of 0.1, 0.5, and 1.0 km, expressed over wavelengths of 1, 10, and 40 km, respectively. The shortest wavelength topography ($\lambda = 1$ km) is essentially imperceptible for isotherms hotter than ~40 °C, and more than 2 km beneath the surface. The longest wavelength topography ($\lambda = 40$ km) is still well expressed in perturbations of the depth of the ~180 °C isotherm (not shown). (Source: *Reiners and Shuster* [2009]. Reproduced with permission of AIP Publishing.)

(3) measuring continuous $t–T$ histories or cooling ages from multiple systems with different closure temperatures in the same sample or location to interpret temporal variations in erosion rates in a single location;

(4) documenting the spectrum of cooling ages observed in a large number of detrital grains collected in one or a few samples of modern detritus from a catchment of known dimensions in order to characterize spatial or temporal variations in erosion rates.

5.5.3 Interpreting spatial patterns of erosion rates

A bulk-grain cooling age observed at the surface of a landscape can be divided into its closure depth to provide an estimate of erosion rate at that location, assuming a steady erosion rate through time (Fig. 5.28). The crux of this approach is estimating the closure depth, which depends on both closure temperature and geothermal gradient, which in turn depend on the exhumation rate as well as those aspects of the thermal field that are independent of exhumation. Unless the topography is perfectly flat, it also requires some correction for the effects of topography on closure depth.

Brandon et al. [1998] created a useful framework for deriving steady erosion rates ($\dot{\varepsilon}$) from single cooling ages based on a thermal field model depicted by either equation (5.86) or (5.89). The objective is first to find a closure depth Z_c that simultaneously satisfies the equation for the thermal field and the closure temperature. This will involve a solution with two equations that are solved iteratively or numerically. Starting with equation (5.86), we rewrite it with $\dot{\varepsilon}$ in place of u, and G_0 (the geothermal gradient in the absence of exhumation) times L in place of $(T_L - T_S)$, as

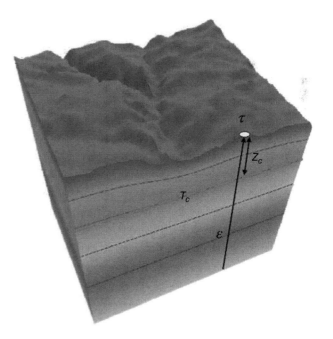

Fig. 5.28. Schematic landscape and underlying crust depicting a sample at surface today with age τ, which passed through its closure temperature T_c at its closure depth Z_c. The rate of erosion is simply the closure depth divided by the age, but closure depth depends on several parameters including erosion rate, and its determination may involve varying degrees of sophistication and precision depending on the application. (Source: *Braun et al.* [2006]. Reproduced with permission of Cambridge University Press.) (*See insert for color representation of the figure.*)

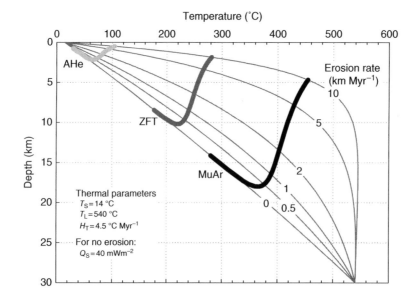

Fig. 5.29. Relationship between depth and temperature, and closure depth for three thermochronometers (apatite (U–Th)/He, zircon fission-track, and muscovite K/Ar). The thermal field is similar to that presented in equation (5.96), but a small radiogenic heat contribution is also included [see *Reiners and Brandon*, 2006]. As erosion rate increases, the temperature–depth profile is increasingly perturbed from the linear solution. As erosion rate increases, the closure depth for any thermochronometer initially increases (due to faster cooling rate increasing the closure temperature) then decreases (due to the faster advection decreasing the depth of the closure isotherm). (Source: *Reiners and Brandon* [2006]. Annual Reviews.)

$$T(z,\dot{e}) = T_S + G_0 L \frac{1 - \exp\left(-\dfrac{\dot{e}z}{\kappa}\right)}{1 - \exp\left(-\dfrac{\dot{e}L}{\kappa}\right)} \quad (5.92)$$

This equation can be solved for $z = Z_c$, and the geothermal gradient $\partial T/\partial z$ at the T_c and Z_c as

$$\left.\frac{\partial T}{\partial z}\right]_{T_c} = \frac{\dot{e}}{\kappa}\left[\frac{G_0 L}{1 - \exp\left(-\dot{e}L/_\kappa\right)} - (T_c - T_S)\right] \quad (5.93)$$

This can be combined with the definition of erosion rate in this case $\partial z/\partial t = \dot{e}$ to yield the cooling rate at the closure temperature

$$\left.\frac{\partial T}{\partial t}\right]_{T_c} = \left.\frac{\partial T}{\partial z}\right]_{T_c} \frac{\partial z}{\partial t} = \frac{\dot{e}^2}{\kappa}\left[\frac{G_0 L}{1 - \exp\left(-\dot{e}L/\kappa\right)} - (T_c - T_S)\right] \quad (5.94)$$

The RHS of equation (5.94) can be equated with the equation for cooling rate at T_c from the closure temperature equation,

$$\left.\frac{\partial T}{\partial t}\right]_{T_c} = \frac{A\left(D_0/_{a^2}\right)RT_c}{E_a}\exp[-E_a/(RT)] \quad (5.95)$$

which yields the first of two equations in two unknowns, T_c and $\dot{e}$. The second equation comes from the solution of equation (5.92) for $z = Z_c$,

$$z_c = \frac{-\kappa}{\dot{e}}\left[1 - \frac{T_c - T_S}{G_0 L}\left(1 - \exp\left(-\dot{e}L/_\kappa\right)\right)\right] \quad (5.96)$$

Solving these two equations simultaneously, numerically, and/or iteratively, yields mutually consistent T_c, Z_c, and $\dot{e}$ (Fig. 5.29).

Figure 5.30 shows an example of the relationship between observed cooling age and the steady-state erosion rate for a variety of thermochronometers using this approach, for a typical crustal thermal model as described in *Reiners and Brandon* [2006]. The solutions shown in Fig. 5.30 are actually strictly relevant to a sample collected at the mean elevation of the topography (average of an appropriate horizontal length scale). If a sample is collected at an elevation that is significantly different from the mean elevation of the topography, a correction to the closure depth needs to be made to derive an erosion rate. The elevation of the mean topography is defined by the elevation averaged over a length scale commensurate with the critical length scale of the thermochronometer ($\lambda_c = T_c/G$, where G is the geothermal gradient; (*Braun* [2002, 2005]). Assuming no change in topography since closure, samples collected at high or low elevations in short-wavelength topography must have effective closure depths (Z_c') that are larger or smaller than the mean elevation Z_c. So the final Z_c' used to calculate erosion rate is

$$Z_c' = Z_c + (h - h_m) \quad (5.97)$$

where h is the elevation of the sample and h_m is the elevation of the mean topography, averaged over a length scale similar to the critical wavelength of the thermochronometer. This Z_c' is then used to estimate the erosion rate, $\dot{e} = Z_c'/t$ (Fig. 5.31).

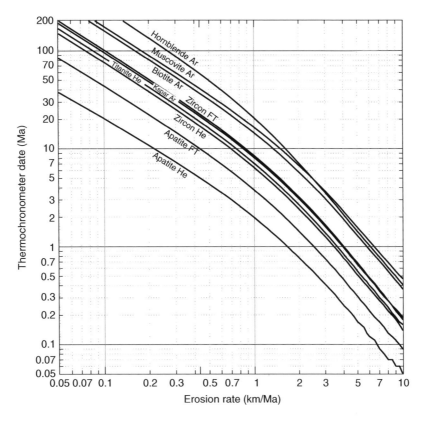

Fig. 5.30. Relationship between cooling age (for multiple thermochronometers) and inferred steady-state erosion rate for a simple thermal model as described in *Reiners and Brandon* [2006]. The hornblende Ar/Ar curve differs from that of the other systems because a different crustal thermal model was used for that system to accommodate the higher temperature behavior of the system. (Source: *Reiners and Brandon* [2006]. Annual Reviews.)

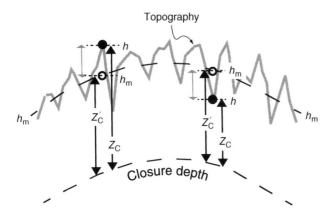

Fig. 5.31. The conversion of Z_c to Z_c', the effective closure depth accounting for the difference in sample elevation from mean elevation of topography averaged over the critical wavelength. (Source: *Reiners et al.* [2003]. Reproduced with permission of American Journal of Science.)

Another approach to estimating steady erosion rates from cooling ages in a landscape uses a thermal field without a fixed lower boundary temperature [*Willett and Brandon*, 2013]. The thinking here is that only the unusual cases will lead to

a constant basal temperature if rock is advecting upward by erosion; underplating at the base of an accretionary wedge or thickening that leads to a middle crustal layer or approximately constant temperature might be examples. The downside of a lack of basal temperature condition is that the geothermal gradient increases continuously, which might be unrealistic in situations of prolonged exhumation. This debate raises questions about the geodynamic processes causing exhumation in the first place, and also raises questions about transience in the thermal field, which we discuss below. In any case, in most cases involving slow to moderate erosion rates, the question of the basal temperature condition does not have a major impact on the interpretations.

The *Willett and Brandon* [2013] model overcomes the problem of coupling between erosion rate, closure depth, and closure temperature by truncating logarithmic terms involving erosion rate with Taylor series expansions, and making a slight approximation to the closure temperature function that leads to negligible inaccuracy. They derived an explicit expression for erosion rate as a function of surface temperature, cooling age, the closure temperature as defined for a cooling rate of 10 K/Ma, the kinetics of the thermochronometer, and the modern geothermal gradient (in contrast to the *Brandon et al.* [1998]

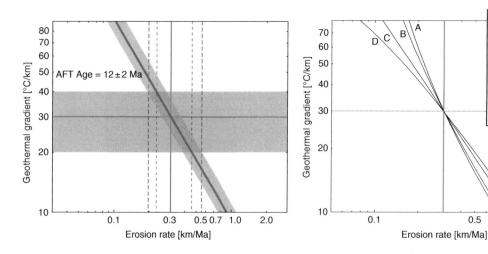

Fig. 5.32. Estimating steady erosion rate from the *Willett and Brandon* [2013] approach. Left: parametric solution of geothermal gradient and erosion rate for an apatite fission-track (AFT) thermochronometer age of 12 Ma, illustrating the propagation of uncertainties into the erosion rate. Right: the same approach but for multiple samples collected at different elevations. For this "vertical transect" approach, no change in erosion rate or geothermal gradient with time should result in intersecting of each solution at a common point. (Source: *Willett and Brandon* [2013]. Reproduced with permission of American Geophysical Union.)

model that specified the geothermal gradient in the absence of erosion):

$$\dot{\varepsilon} = \frac{E_a + T_{c10} - T_0 \left[\ln(\tau) + \ln\left(\dfrac{A\,T_{c10}^2}{T_{c10} - T_0} \right) + 2 \right]}{\tau G \left[\ln(\tau) + \ln\left(\dfrac{A\,T_{c10}^2}{T_{c10} - T_0} \right) - \dfrac{T_0}{T_{c10} - T_0} + 1 \right]} \qquad (5.98)$$

Although equation (5.98) explicitly depends on the T_c for $dT/dt = 10\,\text{K/Ma}$, it actually accounts for the dependence of T_c on cooling rate, and the T_c appropriate for the solution can be approximated (with a high degree of accuracy) as

$$T_c = \frac{E_a + 2\,T_{c10}}{2 + \ln\!\left(A\,T_{c10}^2\right) - \ln(dT/dt)} \qquad (5.99)$$

As with the previous approach, this solution is only valid for samples collected at mean elevation of topography. As Willet and Brandon treated the problem, the elevation of mean topography is best expressed as the average topography over a circle with radius of πZ_c^{est}, where

$$Z_c^{est} = \frac{(T_{c10} - T_0)}{G} \qquad (5.100)$$

The difference between the mean topography and elevation of the sample is then h (which is defined differently than in the *Brandon et al.* [1998] version), and equation (5.98) becomes

$$\dot{\varepsilon} = \frac{E_a + T_{c10} - (T_0 - hG) \left[\ln(\tau) + \ln\left(\dfrac{A\,T_{c10}^2}{T_{c10} - T_0 + hG} \right) + 2 \right]}{\tau G \left[\ln(\tau) + \ln\left(\dfrac{A\,T_{c10}^2}{T_{c10} - T_0 + hG} \right) - \dfrac{T_0 - hG}{T_{c10} - T_0 + hG} + 1 \right]} \qquad (5.101)$$

This approach allows a convenient parametric estimate of erosion rate, in which an observed age for a given system leads to a coupled solution for the geothermal gradient and erosion rate (Fig. 5.32), showing how uncertainty in geothermal gradient (typically the largest source of uncertainty) propagates into erosion rate uncertainty. A similar approach can be taken for multiple samples collected at different elevations. Assuming no change in erosion rate or geothermal gradient through time, these solutions should yield concordant solutions (Fig. 5.32).

The approaches described above assume steady topographic form as well as steady erosion in time. If erosion rates change through time, isotherms may move relative to rocks. For example, the onset of rapid erosion can lead to advection of a given isotherm towards the surface at the same rate as a rock, so that rocks may experience no cooling at all during much of their path of even (and especially) rapid exhumation. Conversely, slowing of exhumation results in lower advection and decreasing geothermal gradients with time, leading to one cause of "thermal relaxation," which may cause rocks to cool even in the absence of exhumation.

Transience in the erosion rate can have large effects on apparent erosion rates, with the magnitude obviously depending on that of the time-derivative of the erosion rate. Accounting for transience in exhumation simultaneously with the cooling-rate dependence on the closure temperature and with exhumation on the depth of the closure isotherm is not as simple a solution as the steady erosion case, but it is possible [e.g., *Brown and Summerfield*, 1997; *Brandon et al.*, 1998; *Moore and England*, 2001; *Rahl et al.*, 2007; *Willett and Brandon*, 2013]. In most cases, accounting for transience requires an estimate for the onset of exhumation, or more generally, the change in its rate. This usually comes from geologic observations, but is often not well constrained or potentially logically circular. Fortunately, it is often possible to constrain the approximate magnitude of the effect of transience on an otherwise steady rate estimate for various thermochronometers (e.g., Fig. 5.33; *Brandon et al.* [1998]). And in some cases, a vertical transect of cooling ages, combined with the parametric solution for geothermal gradient and erosion rate, can effectively solve for the time of onset (Fig. 5.34; *Willett and Brandon* [2013]).

Interpreting spatial variations in erosional exhumation rates is complicated by other factors as well, many of which are discussed in *Braun et al.* [2005]. One of the most important additional caveats, which applies particularly to tectonically active regions undergoing contraction, is the potential for large components of horizontal advection, in addition to erosional exhumation, which is by definition vertical. In such cases, rocks may cross isotherms that are at high angles to overlying topography, and ages observed in any location may contain information about erosion rates in other locations relative to the various fixed points in a reference frame such as the topographic form of a range, major fault, etc. [*Batt et al.*, 2001; *Thomson et al.*, 2010; *Reiners et al.*, 2015].

5.5.4 Interpreting temporal patterns of erosion rates

In cases of high topographic relief over short horizontal distances, differences in thermochronometer cooling ages at different elevations have the potential to constrain erosion rates and their variations over time. This is the idea behind the vertical transect, or AER. In its simplest form, topographic relief with a high amplitude but small wavelength is assumed to have been steady in time, and thus differences in cooling ages of samples at different elevations reflect different durations of time since the samples crossed the (ostensibly flat) closure isotherm beneath the topography. The slope of the AER thus provides an estimate of the exhumation rate (Fig. 5.35). Often, changes in slope within the AER are interpreted to represent the timing of, and magnitude of, changes in erosion rates [e.g., *Fitzgerald et al.*, 1995].

At face value the AER constrains the erosion rate only over the time interval represented by the cooling ages composing it; the AER can also be placed in the context of the mean local elevation (at the appropriate wavelength for the thermochronometer) and a regional geothermal gradient, and the apparent zero-age intercept formed by the trend of the youngest (presumably deepest or lowest-elevation) samples can be compared with the expected depth of the closure isotherm for the system. If the intercept is deeper than the predicted closure isotherm depth, this implies an increase in erosion rates more recently than the youngest cooling age; if the intercept is shallower, a decrease. This is analogous to comparing the (topographically corrected) ages of samples at low elevations to those that would be predicted for a steady erosion rate, or also to the parametric intersections approach of *Willett and Brandon* [2013] above.

Assumptions of the vertical transect or AER approach include that: (i) the topography does not bend isotherms significantly; (ii) the isotherms have not moved relative to the surface; (iii) no structures have moved samples relative to one another after crossing the closure isotherm; and (iv) the topography has remained steady in time.

The effect of bending of isotherms can be significant when estimating erosion rate from AERs. In some cases assessing this effect may require numerical modeling of specific and three-dimensional topography, but magnitudes can also be easily estimated for some common situations. As shown in Fig. 5.36, samples collected over wavelengths that approach or are longer than the critical wavelength of a thermochronometer will yield age differences that are smaller than would be exhibited for samples collected over shorter wavelengths, because of topographic bending of isotherms; i.e., samples collected at wavelengths much longer than the critical wavelength will all have the same age (assuming spatially uniform erosion), so their AER will appear to be vertical, implying infinitely fast erosional exhumation. More generally, assuming a steady and spatially uniform erosion rate, an AER will be rotated from its "true erosion-rate-indicative" slope in proportion to the ratio of the amplitudes of the closure isotherm to the topography [*Braun*, 2002, 2003].

This amplitude ratio for isotherm to topography is termed the admittance ratio and is higher for longer wavelengths and lower temperatures (Fig. 5.37). A simple practical use of the admittance ratio is to correct an AER for the effects of topographic bending of isotherms in order to derive a more realistic estimate of exhumation rate. This is done by "unrotating" the AER based on an estimate of the wavelength of the principal topography over which the vertical transect was collected, and an estimate of the closure temperature of the thermochronometer. For example, collection of a vertical transect over a horizontal distance of 6 km (corresponding to a symmetric 12 km wavelength), for a thermochronometer with a closure temperature of about $60\,^\circ C$, would imply an admittance ratio, α, of roughly 0.4. As shown by *Braun* [2002)] the AER of this vertical transect will yield a slope approximating the apparent erosion rate if multiplied by a factor of $1 - \alpha$. Thus the actual erosion rate is only about 60% of the apparent slope [*Reiners et al.*, 2003].

5.5.5 Interpreting paleotopography

In an earlier section we discussed how spatial variations in erosion rate may be interpreted from spatial variations in cooling ages, but this was done without consideration of how this variation may affect topography itself. Many studies have also used cooling age and thermal history differences of samples collected at different locations to infer changes in topographic relief [e.g., *Shuster et al.*, 2005, 2011; *Ehlers et al.*, 2006; *Schildgen et al.*, 2009] by assuming that isostatic rebound or lithospheric deformation does not produce compensatory differences in rock uplift rates, or that these differences can be constrained [e.g., *Braun and Robert*, 2005; *Braun et al.*, 2006]. Interpretations of the position, form, and scale of paleotopography can also be made from more sophisticated comparisons of cooling ages and elevation across landscapes. By comparing cooling ages of thermochronometers at different positions relative to the long-wavelength, high-relief river (formerly glacier) valleys on the west slope of the Sierra Nevada, *House et al.* [1998, 2002] showed that these large

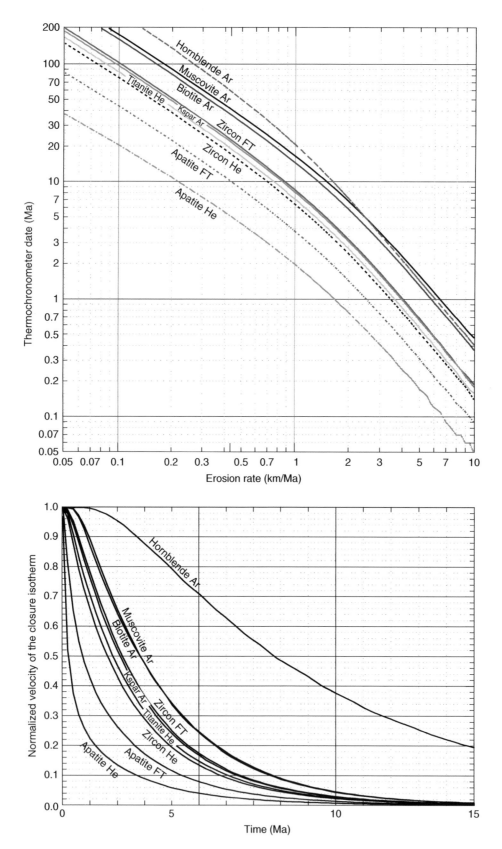

Fig. 5.33. Ratio of vertical rock velocity relative to surface (erosional exhumation rate) to velocity of the closure isotherm of a given thermochronometer, for a crustal column with thermal parameters as described in Figs 5.29 and 5.30, and *Reiners and Brandon* [2006] that experienced an instantaneous exhumation rate increase from zero to 1 km/Ma. The curves show the approximate amount of time required for a given closure isotherm to stop moving to various fractions of the velocity of the rock carrying it. This provides a first-order constraint on the magnitude of inaccuracy in the erosion rate estimate. For example, for the apatite He system, the closure isotherm slows to about 20% of the rock exhumation velocity within 1 Ma, and to about 5% within 4 Ma. Broadly speaking this means that once erosion at this rate has occurred for ~4 Ma, the erosion rate estimate will be accurate to better than about 5%. Curves for the other systems show that thermochronometers with higher closure temperatures take longer to achieve steady state. This is because lower-temperature isotherms cannot move as far in the crust before getting bunched up close to the surface and slowing. (Source: *Reiners and Brandon* [2006]. Annual Reviews.)

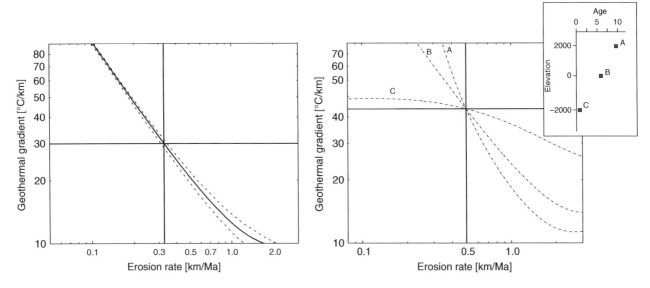

Fig. 5.34. Solutions for coupled closure temperature, erosion rate, and closure depth for time-transient erosional exhumation, from *Willett and Brandon* [2013]. Left: parametric estimate of coupled geothermal gradient and erosion rate for an apatite fission-track (AFT) age of 12 Ma similar to that shown for the steady case in Fig. 5.35a, except here the onset of erosion is assumed to be 30 Ma. The effect of transience is reflected in the fact that if the geothermal gradient is currently very low, then an apparent 12 Ma AFT age must require very rapid exhumation rates that have not yet equilibrated the thermal field. Right: a vertical transect of three cooling ages characterized by onset from zero to 0.5 km/Ma at 15 Ma. In this case the parametric solution pinpoints the geothermal gradient, erosion rate, and onset of exhumation. This is analogous to identifying a break-in-slope in a vertical transect by more informal methods. (Source: *Willett and Brandon* [2013]. Reproduced with permission of American Geophysical Union.)

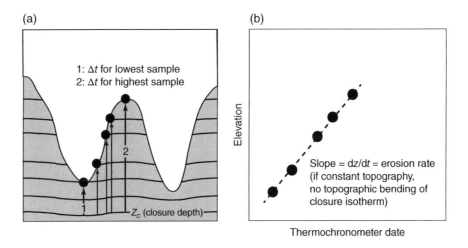

Fig. 5.35. Concept of the vertical transect, or age–elevation relationship.

valleys, and therefore the topographic relief of the range itself, were present as early, and in fact even larger than today, 60–80 million years ago. This interpretation is supported by spectral analyses of the relationship between age and elevation.

As shown in Fig. 5.38, the general rule is that AERs over short wavelengths yield slopes approximating erosion rate, and AERS over wavelengths much longer than the critical wavelength ($\lambda_c = T_c \cdot dz/dt$) yield age-invariant (apparently vertical) AERs. However, if topographic relief is increasing at long wavelengths, then ages at low elevation will be younger than

predicted by the AER at short wavelengths. And if topographic relief is decreasing, ages at low elevation will be older than predicted by the AERs at short wavelengths (Fig. 5.38).

This general relationship between cooling ages and elevation over different wavelengths of topography has been formalized by Braun into a spectral approach for deducing relief changes. At short wavelengths, the gain of the age/elevation ratio (units of Ma/km) will be the inverse of the erosion rate. At long wavelengths, the gain will be zero (e.g., vertical AER) if relief is not changing, negative if relief is decreasing, and positive if relief is increasing. Specifically, at long wavelengths, the change in relief, β, over the timescale of the average cooling ages is

$$\beta = \frac{1}{1 - G_L/G_S} \tag{5.102}$$

where G_L and G_S are the gain at long and short wavelengths. The Braun approach applied to the Sierra Nevada suggests a background erosion rate (measured at short wavelengths) of about

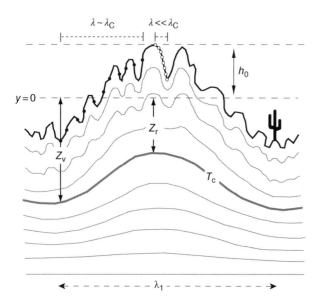

Fig. 5.36. Bending of isotherms between topography of varying wavelengths (λ). Topography at the longest wavelength in this example has amplitude h_0, and the depths to the closure isotherm (as measured from mean elevation) between the valleys and ridges are Z_v and Z_r, respectively. (Source: *Reiners* [2007]. Reproduced with permission of Mineralogical Society of America.)

0.04 km/Ma, and an overall decrease in relief during the Cenozoic of about 50% (Fig. 5.39).

5.6 DETRITAL THERMOCHRONOLOGY APPROACHES FOR UNDERSTANDING LANDSCAPE EVOLUTION AND TECTONICS

Thermochronologic analysis of multiple grains from detrital samples in modern and ancient sediment has a long history in tectonic and geomorphic studies [e.g., *Cerveny et al.*, 1988; *Copeland and Harrison*, 1990; *Brandon and Vance*, 1992]. The advantage of detrital approaches is the ability to characterize ages of grains from a wide region from a single sample, which provides information on the provenance of sediment as well as the exhumation rates that produce it over the landscape. Large-scale provenance questions typically focus on paleogeography and major paleodrainage patterns as the relative contributions of distinct regions to a basin over time [e.g., *Dickinson and Gehrels*, 2003; *Rahl et al.*, 2003], or changes in the durations of time over which minerals are exhumed through closure depths and then deposited ("lag time"), which can elucidate evolutionary stages of growth, steadiness, and decay of orogens [e.g., *Spotila*, 2005]. But provenance questions may also be more focused, for example on the spatial patterns of erosion within individual basins [e.g., *Stock et al.*, 2006] and how these rates and the distribution of topography across regions may have changed over time [*Stock and Montgomery*, 1996; *Brewer et al.*, 2003; *Ruhl and Hodges*, 2005].

In other parts of this chapter we have discussed approaches to interpreting erosional exhumation rates using thermochronologic ages of samples taken directly from bedrock. A single point-wise cooling age from a single sample links a time with a

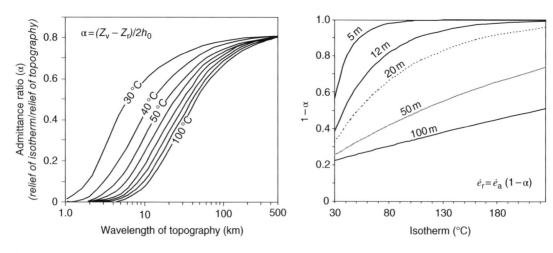

Fig. 5.37. Left: the admittance ratio as applied to the effect of topographic bending of isotherms. Lower temperature isotherms and longer wavelength topography result in higher thermal perturbations at depth. In this case, the admittance ratio cannot reach unity, because of a surface-elevation lapse rate that sets the surface temperature lower at high elevations [see *Reiners et al.*, 2003; *Reiners*, 2007]. Right: the factor by which the apparent slope of the AER must be multiplied to correct for topographic bending of isotherms. $\dot{e}_a$ and $\dot{e}_r$ are the apparent and "real" erosion rates inferred from the slope of the AER. (Source: *Reiners* [2003]. Reproduced with permission of American Journal of Science.)

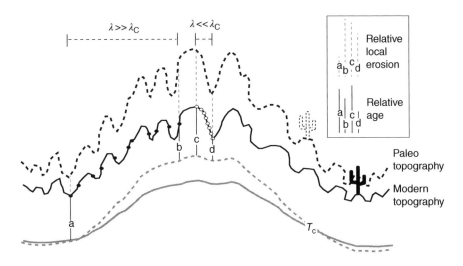

Fig. 5.38. Differences in amounts of erosion (gray dashed vertical lines) and relative cooling ages (gray vertical lines) implied by a scenario of decreasing topographic relief (Source: *Reiners* [2007]. Reproduced with permission of Mineralogical Society of America.)

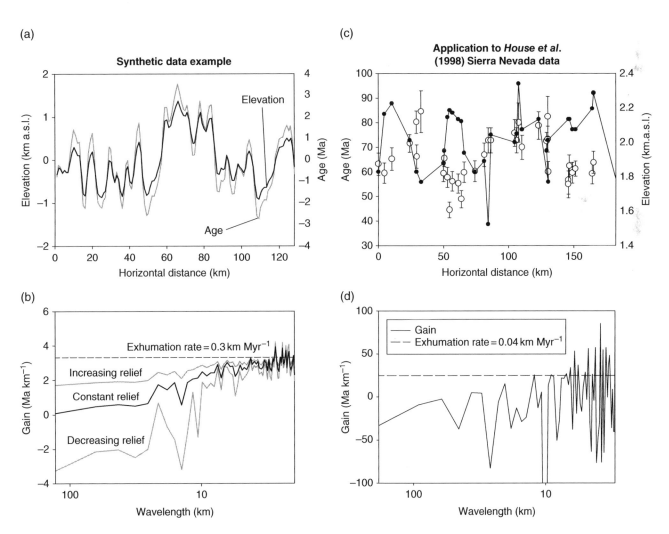

Fig. 5.39. The Braun spectral method for deducing topographic relief changes. (a,b) Synthetic data examples showing AERs over different wavelengths. The gain is the inverse of erosion rate at short wavelengths. At long wavelengths, the gain is zero for constant relief, negative for decreasing relief, and positive for increasing relief. (c,d) Application to the Sierra Nevada data of *House et al.* [1998], suggesting a ~50% decrease in relief since the average age of the samples (~60 Ma). Data from *Braun et al.* [2002] and *Braun* [2005]. (Source: *Reiners* [2007]. Reproduced with permission of Mineralogical Society of America.)

closure temperature, and with some assumptions described above, a closure depth. This yields a time-averaged erosion rate (averaged over the duration since closure), and has been called the "time-of-flight" approach. Dating multiple samples over a range of elevation and short horizontal distance yields the AER, which provides an estimate of the erosion rate near the time of exhumation through the closure depth—the "AE" or "AER" approach. Dating multiple samples over a range of elevation and longer horizontal length scales (greater than the closure depths) can yield estimates of spatial patterns in erosion rates over time-scales on order of the cooling ages of the samples. This elucidates either spatial variations in isostatic or tectonic rock-uplift rates, or in their absence, changes in topographic relief [*Braun et al.*, 2002].

Detrital themochronologic approaches differ from bedrock ones in several ways, starting with the fact that the basic unit of data is a distribution of ages, often depicted as either a histogram or, more usefully for quantitative interpretations, a probability distribution. We will proceed with the assumption that this distribution represents cooling ages of the original bedrock sources of the detrital grains, and that the grains or bulk multigrain sample have not been buried or otherwise heated sufficiently to reset or partially reset thermochronologic systems either during transport or after deposition.

When a distribution of cooling ages is determined for a detrital sample collected from a drainage basin with a known hypsometry, and reasonable assumptions can be made that grains were not derived from outside the basin (e.g., by glacial transport or significant basin configuration changes), then this distribution reflects three principle properties of its source region.

Assuming that the bedrock within a basin contains a uniform modal distribution of the mineral of interest, and that delivery of this mineral is not spatially fractionated during erosion and transport, then the first property is the spatial pattern of ages of the mineral of interest within the bedrock in the basin. This spatial variation may arise in several ways. A simple one is topographic relief combined with a simple AER such as a constant age increase with elevation. In this case the ages vary linearly with elevation, producing a suite of flat "isochrones," contours of constant age across the landscape [*McPhillips and Brandon*, 2010], with uniform vertical spacing (Fig. 5.40). In detail, even in the case of

spatially uniform and temporally constant erosion rates, low-temperature thermochronometric isochrones will not be flat, because of topographic bending of isotherms near the closure depths (Fig. 5.40b). Isochrones may also be deformed by tilting or more complex deformation occurring after exhumation through closure depths (Fig. 5.40c). Finally, even if isochrones are flat or nearly so with respect to elevation across a basin, their vertical spacing may not vary linearly with depth, due to changes in erosional exhumation rates with time (Fig. 5.41).

Taking the above considerations together, the first property of a basin reflected in a detrital age distribution or spectrum is the horizontal and vertical patterns of ages, which reflect erosion rates and their changes through time, post-closure deformation, and to a lesser degree for most systems, topographic relief.

The second basin property reflected in the detrital age spectrum is the hypsometry, or distribution of elevation across the basin. Again assuming uniform modal abundance and delivery of the mineral of interest across the landscape, this combines with the AER, or isochrones distribution, to influence the detrital age spectrum (Fig. 5.42). For example, a basin resembling a plateau with a narrow canyon, where every elevation erodes at the same rate, would be expected to yield a sharp probability peak at the age represented by the plateau elevation, and only a small fraction of (presumably younger) grains derived from the canyon.

Of course the example of the narrowly incised canyon suggests, intuitively, the possibility that erosion rates may not be

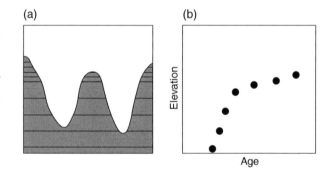

Fig. 5.41. (a) Incised topographic profile with schematic flat and unevenly spaced isochrones. (b) Schematic age–elevation relationship (AER) for this scenario, showing break in slope of AER typically interpreted as an abrupt change in erosion rate.

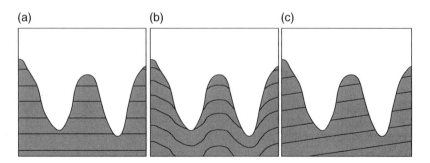

Fig. 5.40. Schematic incised topographic profile with (a) flat and evenly spaced isochrones, (b) topographically bent isochrones, and (c) deformed (tilted) isochrones.

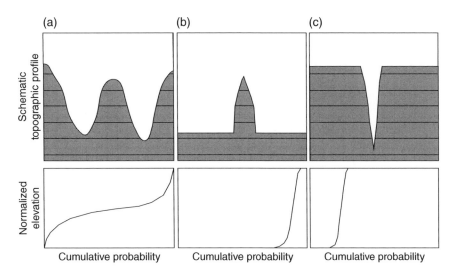

Fig. 5.42. Three simple topographic profiles representing distinct schematic probability distributions of elevation: (a) "normal" distribution of elevation; (b) low-relief plain with a narrow high mountain; (c) plateau with narrow gorge.

spatially uniform. It is the distribution of erosion rates across the basin that is the third principal property reflected in the detrital age spectrum. If the canyon is in fact incising faster than the plateau is eroding, then the detrital age spectrum will contain an abundance of dates characteristic of the low elevations—typically younger dates—that is higher than predicted from combining the AER and hypsometry of the basin and assuming uniform erosion rates (Fig. 5.42).

Assuming for the moment that isochrones are not perturbed by topography nor tilted by postclosure deformation, and all exhumation and cooling occurred due to erosion that did not involve large-magnitude changes in the Peclet number that would complicate shallow crustal thermal fields, then a probability distribution of detrital thermochronologic ages contains information on (i) the variation in erosion rate through time, (ii) the hypsometry of the basin (with respect to production of the dated mineral of interest), and (iii) the spatial distribution of erosion. Several studies have combined observed detrital age spectra with information or assumptions on two out of three of these in order to derive constraints on the third.

Stock and Montgomery [1996] outlined an approach for using the range of ages in a detrital population of ancient sediment (e.g., derived from a now modified drainage basin or landscape), combined with estimates of the isochrones spacing in that landscape (e.g., by converting paired cooling ages to cooling rates to erosion rates, or observing preserved isochrones in the relict landscape), and assuming erosion from all elevations in the paleobasin, to estimate the paleotopographic relief in the landscape. To date this approach has never been applied to paleodetrital samples, but the way it works can be demonstrated with modern detritus from a drainage basin of known relief (Fig. 5.43).

The Icicle Creek drainage basin in the Washington Cascades contains approximately 2.4 km of topographic relief, and 57 apatite grains from a sample of modern river sediment range in age

from about 10 to 46 Ma. In detail the distribution of ages is somewhat unusual because of a notable gap between about 10 and 17 Ma. The difference between the oldest and youngest ages in the population is approximately 35 Ma. If we multiply this Dt by an estimated erosion rate (Dz/Dt) over this duration, we will have an estimate of the topographic relief in the basin, which we can compare to the observed relief. A few samples of bedrock at intermediate elevation in the basin have apatite fission track ages of about 45 Ma and (U–Th)/He ages of 20 Ma. Dividing this age difference into the approximate difference in the closure temperatures of these two systems (~45 °C) yields an estimated cooling rate of 1.8 °C/Ma. Multiplying this by the inverse of a reasonable estimate for the geothermal gradient of 25 °C/km, we obtain an estimated erosional exhumation rate for the basin for the time period of interest of 0.07 km/Ma. Multiplying this by the Dt observed in the detrital sample yields 2.4 km, which is the same as the observed relief. Obviously there are many assumptions involved in this calculation, but the fact that reasonable estimates for parameters along the way lead to a result that matches observations is consistent with a spatial pattern of erosion rates that is reasonably uniform. Note that a slightly more sophisticated analysis, preferably supported by a larger data set of grains and more robust cooling rates from more intermethod age comparisons, could yield not just total relief but an estimate of the probability distribution of elevation (i.e. the hypsometry) of the basin.

If erosion (and the mode and erosional productivity of the mineral of interest) is really spatially uniform throughout a basin, then the elevations and ages of bedrock throughout the basin are related to one another through their probability distributions, such that each elevation is assigned an age of equal probability (Fig. 5.44). This allows creation of a synthetic AER (e.g., a model "vertical" transect). A simple interpretation of the results of this approach for Icicle Creek would suggest rapid exhumation

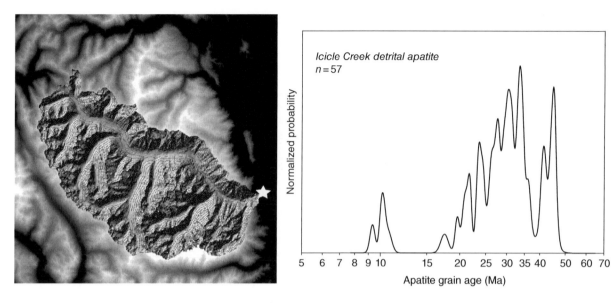

Fig. 5.43. Left: topographic contour map of the approximately 600-km² Icicle Creek drainage basin in the Washington Cascades. White star denotes sample location southwest of the town of Leavenworth. Right: probability distribution of apatite (U–Th)/He ages from Icicle Creek sediment (note logarithmic scale to reduce overemphasis of the youngest and relatively precise ages). (Source: *Reiners* [2007]. Reproduced with permission of Mineralogical Society of America.)

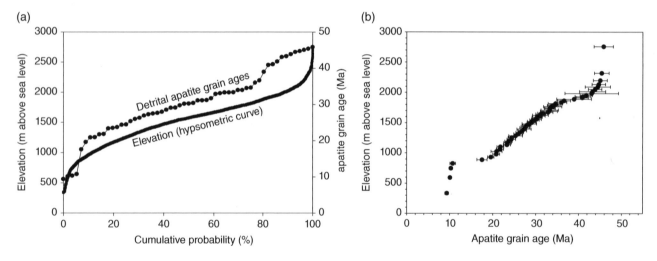

Fig. 5.44. (a) Cumulative probability distributions of elevation (continuous line) and detrital apatite (U–Th)/He ages (filled circles) in the Icicle Creek drainage basin. The elevation distributions shown here are actually modified by removing elevations (about 20% of the basin) occupied by extremely apatite-poor bedrock. If the erosion is spatially uniform, then equivalent age and elevation probabilities lead to a synthetic age–elevation relationship (b) that can be interpreted as a bedrock "vertical transect." (Source: *Reiners* [2007]. Reproduced with permission of Mineralogical Society of America.)

around 45 Ma, followed by very slow erosion until about 35 Ma, erosion rates of about 0.05 km/Ma from ~35–20 Ma, very slow erosion from roughly 20–11 Ma, and rapid erosion at rates of 0.3–0.4 km/Ma until about 9 Ma. Extension of this lowest-elevation rate would intercept zero ages at depths far greater than the approximate closure depth (beneath mean regional elevation), suggesting that erosion slowed dramatically sometime since ~9 Ma.

The assumption of spatially uniform erosion could be tested by comparing the synthetic AER in Fig. 5.44b to ages measured in bedrock samples collected across a large range of elevation over a short horizontal distance. Conversely, if the actual bedrock AER

is known a priori, then it can be combined with the probability distribution of elevation in the basin (the hypsometry) to predict what the detrital age probability distribution would look like if erosion were uniform throughout the basin. Differences between the predicted and observed age distribution can then be used to constrain the spatial pattern of erosion.

Stock et al. [2006] compared observed detrital age spectra with those predicted from bedrock AERs and basin hypsometry in two small (~3 and 30 km²) drainage basins on the eastern flank of the Sierra Nevada (Fig. 5.45). In one, they found that the observed probability distribution of detrital ages closely matched the spatially uniform erosion prediction, but the other basin contained a

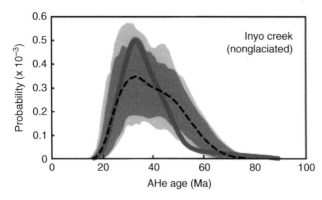
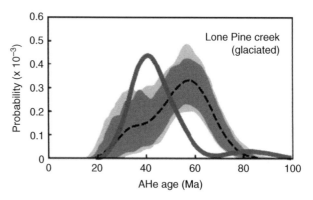

Fig. 5.45. Probability distributions of detrital apatite (U–Th)/He ages from 3 to 30 km² drainage basins on the east flank of the Sierra Nevada, California. Dark gray lines are smoothed (using estimated 11% uncertainty on ages) observed age spectra. Dashed lines are the predicted age spectra combing the observed bedrock AER from nearby locations, the hypsometry (elevation probability distribution) of each basin, and an assumption of spatially uniform erosion. The light and dark gray fields represent Monte Carlo simulations of detrital age spectra predicted from picking only 52 (light) or 100 (dark) detrital grains. The observed and predicted distributions for Inyo Creek are not sufficiently different to disprove the uniform erosion hypothesis, whereas the observed distribution from Lone Pine Creek contains a much higher (and statistically significant) proportion of younger grains, consistent with more rapid recent erosion, possibly by glacial action, at low elevations. (Source: *Stock et al.* [2006]. Reproduced with permission of Geological Society of America.)

significantly higher proportion of young ages than predicted for the uniform erosion model. A major difference between these two basins is the recent glacial erosion in Lone Pine, leading the researchers to suggest that glacial erosion preferentially eroded lower elevations, liberating detritus with younger ages that were recently transported by fluvial erosion.

In another detrital study of the Sierra Nevada, in this case from much larger (>3000 km²) drainage basins on the west flank, *McPhillips and Brandon* [2010] combined bedrock AERs, hypsometry, and observed probability distributions of detrital ages to infer isochrones tilted about 3.4° to the west, consistent with geologic observations, and opposite spatial patterns of erosion in each basin. The San Joaquin basin to the north contains an overabundance of old ages compared to bedrock and spatially uniform predictions, whereas the Kings basin to the south contains an overabundance of young ages. Reconciling the observations with predictions in the Kings drainage requires incision of low-elevation regions in the western part of the range at a rate approximately 10 times faster than nearby higher elevations, but also allows for erosion of higher elevations in the eastern portion, where tilted isochrones expose younger ages at higher elevation. In the San Joaquin basin the results suggest the opposite, with erosion rates several orders of magnitude faster in the low-elevation interfluves in the western part of the basin. Taken together the results are consistent with decreasing topographic relief in the San Joaquin, and increasing relief in the Kings drainage, which may be attributable to larger scale tectonic forcing such as the inferred lithospheric dip in the southern Sierra Nevada.

5.7 CONCLUSIONS

Quantitative understanding of diffusion is important in thermochronology because diffusion (or a similar phenomenon) describes the migration and/or loss of daughter products in

minerals, as well as the flow of heat and the thermal field in rocks, understanding of which is necessary to interpret geologic processes from thermal histories. Diffusion can be thought of as the emergent macroscopic phenomenon resulting from random walk in a concentration gradient, and species and heat share similar forms of the diffusion equation. The Arrhenius law for thermally activated volume diffusion for interstitial impurities in a crystal can be derived conceptually from thermodynamic considerations of migration through a crystal lattice, with complications due to anisotropy, tortuosity, and energy-minima defects.

Useful analytical solutions for concentrations of a diffusant as a function of time and distance can be derived for simple scenarios and initial and boundary conditions, and useful characteristic time and length scales describe the approximate durations and distances over which critical extents of diffusion or flux occur. Although these characteristic scales depend on specific characteristics of the problem, in general the diffusion length scale L, timescale t, and diffusion rate D relate as $L \sim (Dt)^{1/2}$. Analytical solutions to some of the most important diffusion problems, such as sphere, plane sheet, and cylinder, involve infinite sums that can be unwieldy, but approximate analytical solutions are commonly used to describe fractional losses of integrated concentration profiles across these geometric forms.

A common analytical method for measuring the kinetics (activation energy E_a and frequency factor D_0, or frequency factor divided by the square of the diffusion domain size D/a^2) of diffusion of gases like He or Ar in minerals include step-heating degassing experiments in which fractional losses are converted to D/a^2 values across a range of temperatures. A simple system of a single domain (size) and set of kinetics yields an Arrhenius trend in which the slope of a line in $\ln(D/a^2)$ versus inverse temperature space is proportional to E_a, and the intercept is $\ln(D_0/a^2)$. Assumptions for this approach include isotropic diffusion, uniform initial distribution of the diffusant, no mineralogic changes during heating, single (or uniform) domain structure, and, usually, extrapolation of the diffusion kinetics to temperatures and timescales of geologic interest. A step-heating experiment on a sample comprising multiple domains with distinct sizes will show a distinctive form that may be most evident in cycled step heating, including both prograde and retrograde heating sequences. Other analytical methods for measuring diffusion include characterization of concentration profiles by laser-ablation or ion-probe measurements, or, at shorter length scales, RBS or ERDA.

Measured dates, concentration profiles, track length distributions, and other thermochronometric observables provide non-unique constraints on time–temperature histories that should be combined with other lines of evidence for geologic interpretation. For example, bulk-grain dates could represent equilibrium, partial resetting, or closure ages. Equilibrium ages are those in which diffusion and production are balanced and the date is a function of (besides the thermochronometer kinetics) temperature and domain size. This leads to the conception of the partial-retention zone in the crust, which is the idealized depth region between the 10 and 90% daughter-product retention isopleths in a section of static crust where each depth is held isothermally. In contrast, a partial resetting date reflects fractional loss of daughter product in a past short-lived event that heated the sample from temperatures at which no significant daughter product is lost. In this case, fractional resetting extents of thermochronometers with contrasting kinetics may be combined to estimate the duration and temperature of the event. Finally, many thermochronologic dates are interpreted as closure ages representing the duration of steady accumulation of daughter products following an episode of rapid cooling from high to low temperature where little diffusive loss occurs. The closure temperature is that which corresponds to the extrapolation of the 1:1 time–date correlation to zero date in the sample's thermal history, and depends on the cooling rate as well as kinetics, geometry, and diffusion domain size of the thermochronometer. Closure temperatures can also be derived for systems without an initial daughter concentration at high temperature, and for individual positions within a diffusion domain, yielding closure profiles.

The first application of multidomain diffusion theory was also one of the earliest $^{40}Ar/^{39}Ar$ experiments, in which a continuous step-heating age spectrum of a meteorite was interpreted as reflecting a distribution of domain sizes that all formed ~4.5 Ga and experienced 90% resetting at ~0.5 Ga. Multidomain diffusion is commonly interpreted for $^{40}Ar/^{39}Ar$ results on slowly cooled K-feldspars, particularly when changes in apparent diffusion kinetics are correlated with changes in $^{40}Ar/^{39}Ar$ ages during the step-heating experiment.

Most often, determination of a thermochronologic date or thermal history is an intermediate step towards a more valuable interpretation of geologic process. One of the primary causes of cooling of interest is exhumation, or advection of a rock towards the Earth's surface. Users are typically interested in the timing, rate, and magnitude of exhumation, which may be either erosional or tectonic. Studies of erosional exhumation commonly focus on spatial patterns of dates across a landscape, temporal patterns using "vertical" transects or depth profiles, temporal patterns using multiple thermochronometers with different closure temperatures, or detrital populations.

Interpretations of exhumation can use sophisticated numerical modeling or simpler approximations, but especially in cases involving rapid and/or transient exhumation, any interpretation should consider parameters affecting the thermal field of the crust and the depth of the closure temperature (closure depth), which may depend on topography, thermal diffusivity, heat production, and exhumation rate itself. Due to competing effects of closure temperature and isotherm advection, closure depths are usually maximum at intermediate exhumation rates. A useful dimensionless variable in this arena is the Peclet number, which is the product of the rate and length scale of exhumation divided by thermal diffusivity. Relatively simple models of the crustal thermal field can be combined with observed thermochronologic ages to deduce time-averaged erosion rates. Time-transience in erosional exhumation can have large effects on interpretations, with higher temperature thermochronometers taking longer to reach steady state and having a greater range of potential closure depths.

Age–elevation relationships, typically in "vertical" transects, are commonly used to interpret erosion-rate changes in time. Simple applications of this approach assume unchanging topography, no structures that perturb the thermal field or position of crustal sections, and that topography has a minimal effect on the depth of the closure isotherm. Topography can have a significant effect on the depth of closure isotherms of low temperature thermochronometers, particularly when it is long wavelength. The ratio between the amplitude of surface topography and that of an isotherm is the admittance ratio. Estimating admittance allows one to correct for topographic effects on AERs. More generally, the relationship between thermochronologic ages and elevation over different wavelengths can be used to estimate both the background erosion rate as well as changes in topographic relief over time.

Thermochronologic ages of large populations of detrital grains contain a variety of information about erosion in the source area of the grains. If isochrones are not perturbed by topography nor tilted by post-closure deformation, and all exhumation and cooling occurred due to erosion that did not involve large-magnitude changes in the Peclet number that would complicate shallow crustal thermal fields, then a probability distribution of detrital thermochronologic ages contains information on (i) the variation in erosion rate through time, (ii) the hypsometry of the basin (with respect to abundance/production of the dated mineral of interest), and (iii) the spatial distribution of erosion. Studies combine observed detrital age populations with information or assumptions on two out of the three of these in order to derive constraints on the third.

5.8 REFERENCES

Batt, G. E., Brandon, M. T., Farley, K. A., and Roden-Tice, M. (2001) Tectonic synthesis of the Olympic Mountains segment of the Cascadia wedge, using two-dimensional thermal and kinematic modeling of thermochronological ages. *Journal of Geophysical Research: Solid Earth (1978–2012)* **106**(B11), 26731–26746.

Baxter, E. F. (2003) Quantification of the factors controlling the presence of excess ^{40}Ar or ^{4}He. *Earth and Planetary Science Letters* **216**(4) 619–634.

Brandon, M. T. and Vance, J. A. (1992) Tectonic evolution of the Cenozoic Olympic subduction complex, Washington State, as deduced from fission track ages for detrital zircons. *American Journal of Science* **292**, 565–565.

Brandon, M. T., Roden-Tice, M. K., and Garver, J. I.. (1998) Late Cenozoic exhumation of the Cascadia accretionary wedge in the Olympic Mountains, northwest Washington State. *Geological Society of America Bulletin* **110**(8), 985–1009.

Braun, J. (2002) Estimating exhumation rate and relief evolution by spectral analysis of age–elevation datasets. *Terra Nova* **14**(3), 210–214.

Braun, J. (2003) PECUBE: a new finite-element code to solve the 3D heat transport equation including the effects of a time-varying, finite amplitude surface topography. *Computers & Geosciences* **29**(6) 787–794.

Braun, J. (2005) Quantitative constraints on the rate of landform evolution derived from low-temperature thermochronology. *Reviews in Mineralogy and Geochemistry* **58**(1), 351–374.

Braun, J. and Robert, X. (2005) Constraints on the rate of post-orogenic erosional decay from low-temperature thermochronological data: application to the Dabie Shan, China. *Earth Surface Processes and Landforms* **30**(9), 1203–1225.

Braun, J., Van Der Beek, P., and Batt, G. (2006) *Quantitative Thermochronology: Numerical Methods for the Interpretation of Thermochronological Data*. Cambridge University Press.

Braun, J., Van Der Beek, P., Valla, P., *et al.* (2012) Quantifying rates of landscape evolution and tectonic processes by thermochronology and numerical modeling of crustal heat transport using PECUBE. *Tectonophysics* **524**, 1–28.

Brewer, I. D., Burbank, D. W., and Hodges, K. V. (2003) Modelling detrital cooling-age populations: Insights from two Himalayan catchments. *Basin Research* **15**(3), 305–320.

Brown, R. W. and Summerfield, M. A. (1997) Some uncertainties in the derivation of rates of denudation from thermochronologic data. *Earth Surface Processes and Landforms* **22**(3), 239–248.

Camacho, A., Lee, J. K. W., Hensen, B. J., and Braun, J. (2005) Short-lived orogenic cycles and the eclogitization of cold crust by spasmodic hot fluids. *Nature* **435**(7046), 1191–1196.

Carslaw, H. S. and Jaeger, J. C. (1959) *Conduction of Heat in Solids*, 2nd edn. Clarendon Press, Oxford.

Cassata, W. S. and Renne, P. R. (2013) Systematic variations of argon diffusion in feldspars and implications for thermochronometry. *Geochimica et Cosmochimica Acta* **112**, 251–287.

Cassata, W. S., Shuster, D. L., Renne, P. R., and Weiss, B. J. (2010) Evidence for shock heating and constraints on Martian surface temperatures revealed by ^{40}Ar/^{39}Ar thermochronometry of Martian meteorites. *Geochimica et Cosmochimica Acta* **74**(23), 6900–6920.

Cassata, W. S., Renne, P. R., and Shuster, D. L. (2011) Argon diffusion in pyroxenes: implications for thermochronometry and mantle degassing. *Earth and Planetary Science Letters* **304**(3), 407–416.

Cerveny, P. F., Naeser, N. D., Zeitler, P. K., Naeser, C. W., and Johnson, N. M. (1988) History of uplift and relief of the Himalaya during the past 18 million years: Evidence from fission-track ages of detrital zircons from sandstones of the Siwalik Group. In *New Perspectives in Basin Analysis*, Kleinspehn, K. L. and Paola, C. (eds), 43–61. Springer, New York

Cherniak, D. J., Lanford, W. A., and Ryerson. F. J. (1991) Lead diffusion in apatite and zircon using ion implantation and Rutherford backscattering techniques. *Geochimica et Cosmochimica Acta* **55**(6), 1663–1673.

Cherniak, D. J., Watson, E. B., and Thomas, J. B. (2009) Diffusion of helium in zircon and apatite. *Chemical Geology* **268**(1), 155–166.

Copeland, P. and Harrison, T. M. (1990) Episodic rapid uplift in the Himalaya revealed by ^{40}Ar/^{39}Ar analysis of detrital K-feldspar and muscovite, Bengal fan. *Geology* **18**(4), 354–357.

Crank, J. (1956) *The Mathematics of Diffusion*. Oxford University Press, 347 pp.

Crank, J. (1975) *The Mathematics of Diffusion*, 2nd edn. Clarendon Press, Oxford

Dickinson, W. R. and Gehrels, G. E. (2003) U–Pb ages of detrital zircons from Permian and Jurassic eolian sandstones of the Colorado Plateau, USA: paleogeographic implications. *Sedimentary Geology* **163**(1), 29–66.

Dodson, M. H. (1973) Closure temperature in cooling geochronological and petrological systems. *Contributions to Mineralogy and Petrology* **40**(3), 259–274.

Dodson, M. H. (1986) Closure profiles in cooling systems. *Materials Science Forum* **7**, 145–154.

Ehlers, T. A., Chaudhri, T., Kumar, S., *et al.* (2005) Computational tools for low-temperature thermochronometer interpretation. *Reviews in Mineralogy and Geochemistry* **58**(1), 589–622.

Ehlers, T. A., Farley, K. A., Rusmore, M. E., and Woodsworth, G. J. (2006) Apatite (U–Th)/He signal of large-magnitude accelerated glacial erosion, southwest British Columbia. *Geology* **34**(9), 765–768.

Evenson, N. S., Reiners, P. W., Spencer, J., and Shuster, D. L. (2014) Hematite and Mn-oxide (U-Th)/He dates from the Buckskin-Rawhide detachment system western Arizona: gaining insights into hematite (U–Th)/He systematics. *American Journal of Science* **314**(10), 1373–1435.

Farley, K. A. (2000) Helium diffusion from apatite: general behavior as illustrated by Durango fluorapatite. *Journal of Geophysical Research: Solid Earth (1978–2012)* **105**(B2), 2903–2914.

Farley, K. A. (2007) He diffusion systematics in minerals: evidence from synthetic monazite and zircon structure phosphates. *Geochimica et Cosmochimica Acta* **71**(16), 4015–4024.

Farley, K. A. and Flowers. R. M. (2012) (U–Th)/Ne and multidomain (U–Th)/He systematics of a hydrothermal hematite from eastern Grand Canyon. *Earth and Planetary Science Letters* **359**, 131–140.

Farley, D. R., Estabrook, K. G., Glendinning, S. G., *et al.* (1999) Radiative jet experiments of astrophysical interest using intense lasers. *Physical Review Letters* **83**(10), 1982.

Fechtig, H. and Kalbitzer, S. (1966) The diffusion of argon in potassium-bearing solids. In *Potassium Argon Dating*, Schaeffer, O. A. and Zähringer, J. (compilers), 68–107. Springer, Berlin.

Fitzgerald, P. G., Sorkhabi, R. B., Redfield, T. F., and Stump, E. (1995) Uplift and denudation of the central Alaska Range: a case study in the use of apatite fission track thermochronology to determine absolute uplift parameters. *Journal of Geophysical Research* **100** 20–175.

Foland, K. A. (1974) Ar 40 diffusion in homogeneous orthoclase and an interpretation of Ar diffusion in K-feldspars. *Geochimica et Cosmochimica Acta* **38**(1), 151–166.

Fourier. J. (1822) *Theorie analytique de la chaleur, par M. Fourier.* Chez Firmin Didot, père et fils.

Gallagher, K.(2012) Transdimensional inverse thermal history modelling for quantitative thermochronology. *Journal of Geophysical Research* **117**, B02408.

Ganguly, J. (2002) Diffusion kinetics in minerals: principles and applications to tectono-metamophic processes. *EMU Notes in Mineralogy* **4**, 271–309.

Ganguly, J. and Tirone, T. (1999) Diffusion closure temperature and age of a mineral with arbitrary extent of diffusion: theoretical formulation and applications. *Earth and Planetary Science Letters* **170**(1), 131–140.

Ganguly, J. and Tirone, M. (2001) Relationship between cooling rate and cooling age of a mineral: theory and applications to meteorites. *Meteoritics & Planetary Science* **36**(1), 167–175.

Gautheron, C. and Tassan-Got, L. (2010) A Monte Carlo approach to diffusion applied to noble gas/helium thermochronology. *Chemical Geology* **273**(3), 212–224.

Giletti, B. J. (1991) Rb and Sr diffusion in alkali feldspars, with implications for cooling histories of rocks. *Geochimica et Cosmochimica Acta* **55**(5), 1331–1343.

Gillespie, A. R., Huneke, J. C., and Wasserburg, G. J. (1984) Eruption age of a 100,000-year-old basalt from $^{40}Ar-^{39}Ar$ analysis of partially degassed xenoliths. *Journal of Geophysical Research: Solid Earth (1978–2012)* **89**(B2), 1033–1048.

Guenthner, W. R., Reiners, P. W., Ketcham, R. A., Nasdala, L., and Giester, G. (2013) Helium diffusion in natural zircon: Radiation damage, anisotropy, and the interpretation of zircon (U–Th)/He thermochronology. *American Journal of Science* **313**(3), 145–198.

Hanson, G. N. and Gast, P. W. (1967) Kinetic studies in contact metamorphic zones. *Geochimica et Cosmochimica Acta* **31**(7), 1119–1153.

Harrison, T. M. and Lovera, O. A. (2014) The multi-diffusion domain model: past, present and future. *Geological Society, London, Special Publications* **378**(1), 91–106.

Harrison, T. M. and Zeitler, P. K. (2005) Fundamentals of noble gas thermochronometry. *Reviews in Mineralogy and Geochemistry* **58**(1), 123–149.

Harrison, T. M., Yin, A., Grove, M., Lovera, O. M., Ryerson, F. J., and Zhou, X. (2000) The Zedong Window: A record of superposed Tertiary convergence in southeastern Tibet. *Journal of Geophysical Research: Solid Earth (1978–2012)* **105**(B8), 19211–19230.

Heizler, M. T. and Harrison, T. M. (1988) Multiple trapped argon isotope components revealed by $^{40}Ar^{39}Ar$ isochron analysis. *Geochimica et Cosmochimica Acta* **52**(5), 1295–1303.

House, M. A., Wernicke, B. P., and Farley, K. A. (1998) Dating topography of the Sierra Nevada, California, using apatite (U–Th)/He ages. *Nature* **396**(6706), 66–69.

House, M. A., Kohn, B. P., Farley, K. A., and Raza, A. (2002) Evaluating thermal history models for the Otway Basin, southeastern Australia, using (U–Th)/He and fission-track data from borehole apatites. *Tectonophysics* **349**(1), 277–295.

Huber, C., Cassata, W. S., and Renne, P. R. (2011) A lattice Boltzmann model for noble gas diffusion in solids: the importance of domain shape and diffusive anisotropy and implications for thermochronometry. *Geochimica et Cosmochimica Acta* **75**, 2170–2186.

Huneke, J. C. (1976) Diffusion artifacts in dating by stepwise thermal release of rare gases. *Earth and Planetary Science Letters* **28**(3), 407–417.

Jenkin, G. R. T., Rogers, G., Fallick, A. E., Farrow, C. M. (1995) Rb–Sr closure temperatures in bi-mineralic rocks: a mode effect and test for different diffusion models. *Chemical Geology* **122**(1), 227–240.

Kelley, S. P. and Wartho, J. A. (2000) Rapid kimberlite ascent and the significance of Ar–Ar ages in xenolith phlogopites. *Science* **289**(5479), 609–611.

Ketcham, R. A. (2005) Forward and inverse modeling of low-temperature thermochronometry data. *Reviews in Mineralogy and Geochemistry* **58**(1), 275–314.

Lasaga, A. C. (1998) *Kinetic Theory in the Earth Sciences.* Princeton University Press, 822 pp.

Lovera, O. M., Richter, F. M., and Harrison, T. M. (1989) The $^{40}Ar/^{39}Ar$ thermochronometry for slowly cooled samples having a distribution of diffusion domain sizes. *Journal of Geophysical Research: Solid Earth (1978–2012)* **94**(B12), 17917–17935.

Lovera, O. M., Richter, F. M., and Harrison, T. M.. (1991) Diffusion domains determined by ^{39}Ar released during step heating. *Journal of Geophysical Research: Solid Earth (1978–2012)* **96**(B2), 2057–2069.

Lovera, O. M., Heizler, M. T., and Harrison, T. M. (1993) Argon diffusion domains in K-feldspar II: Kinetic properties of MH-10. *Contributions to Mineralogy and Petrology* **113**(3), 381–393.

Lovera, O. M., Grove, M., and Harrison. T. M. (2002) Systematic analysis of K-feldspar $^{40}Ar/^{39}Ar$ step heating results. II: relevance of laboratory argon diffusion properties to nature. *Geochimica et Cosmochimica Acta* **66**(7), 1237–1255.

Mancktelow, N. S. and Grasemann, B. (1997) Time-dependent effects of heat advection and topography on cooling histories during erosion. *Tectonophysics* **270**(3), 167–195.

McDougall, I. and Harrison, T. M. (1999) *Geochronology and Thermochronology by the $^{40}Ar/^{39}Ar$ Method.* Oxford University Press.

McPhillips, D. and Brandon, M. T. (2010) Using tracer thermochronology to measure modern relief change in the Sierra Nevada, California. *Earth and Planetary Science Letters* **296**(3), 373–383.

Meesters, A. G. C. A. and Dunai, T. J. (2005) A noniterative solution of the (U–Th)/He age equation. *Geochemistry, Geophysics, Geosystems* **6**(4).

Moore, M. A. and England, P. C. (2001) On the inference of denudation rates from cooling ages of minerals. *Earth and Planetary Science Letters* **185**(3), 265–284.

Mulch, A. and Cosca, M. A. (2004) Recrystallization or cooling ages: in situ UV-laser 40Ar/39Ar geochronology of muscovite in mylonitic rocks. *Journal of the Geological Society* **161**(4), 573–582.

Onstott, T. C., Miller, M. L., Ewing, R. C., Arnold, G. W., and Walsh. D. S. (1995) Recoil refinements: implications for the ^{40}Ar/^{39}Ar dating technique. *Geochimica et Cosmochimica Acta* **59**(9), 1821–1834.

Ouchani, S., Dran, J-C., and Chaumont, J. (1998) Exfoliation and diffusion following helium ion implantation in fluorapatite: implications for radiochronology and radioactive waste disposal. *Applied geochemistry* **13**(6), 707–714.

Parsons, I., Brown, W. L., and Smith, J. V. (1999) ^{40}Ar/^{39}Ar thermochronology using alkali feldspars: real thermal history or mathematical mirage of microtexture?. *Contributions to Mineralogy and Petrology* **136**(1–2), 92–110.

Quidelleur, X., Grove, M., Lovera, O. M., Harrison, T. M., Yin, A., and Ryerson, F. J. (1997) Thermal evolution and slip history of the Renbu Zedong Thrust, southeastern Tibet. *Journal of Geophysical Research: Solid Earth (1978–2012)* **102**(B2), 2659–2679.

Rahl, J. M., Reiners, P. W., Campbell, I. H., Nicolescu, S., and Alle,. C.M. (2003) Combined single-grain (U–Th)/He and U/Pb dating of detrital zircons from the Navajo Sandstone, Utah. *Geology* **31**(9), 761–764.

Rahl, J. M., Ehlers, T. A., and van der Pluijm, B. A. (2007) Quantifying transient erosion of orogens with detrital thermochronology from syntectonic basin deposits. *Earth and Planetary Science Letters* **256**(1), 147–161.

Reiners, P. W. (2007) Thermochronologic approaches to paleotopography. *Reviews in Mineralogy and Geochemistry* **66**(1), 243–267.

Reiners, P. W. (2009) Nonmonotonic thermal histories and contrasting kinetics of multiple thermochronometers. *Geochimica et Cosmochimica Acta* **73**(12), 3612–3629.

Reiners, P. W. and Brandon. M. T. (2006) Using thermochronology to understand orogenic erosion. *Annual Reviews in Earth and Planetary Science* **34**, 419–466.

Reiners, P. W. and Farley, K. A. (1999) Helium diffusion and (U–Th)/He thermochronometry of titanite. *Geochimica et Cosmochimica Acta* **63**(22), 3845–3859.

Reiners, P. W., and Shuster, D. L. (2009) Thermochronology and landscape evolution. *Physics Today* **62**(9), 31–36.

Reiners, P. W., Zhou, Z., Ehlers, T. A., *et al.* (2003) Post-orogenic evolution of the Dabie Shan, eastern China, from (U–Th)/He and fission-track thermochronology. *American Journal of Science* **303**, 489–518.

Reiners, P. W., Spell, T. L., Nicolescu, S., and Zanetti, K. A. (2004) Zircon (U–Th)/He thermochronometry: He diffusion and comparisons with ^{40}Ar/^{39}Ar dating. *Geochimica et Cosmochimica Acta* **68**(8), 1857–1887.

Reiners, P.W., Thomson, S.N., Vernon, A., *et a.* (2015) Low-temperature thermochronologic trends across the central Andes, 21°S–28°S. *Geological Society of America Memoir*, **212**, 215–249.

Ring, U., Brandon, M. T., Willett, S. D., and Lister, G. S. (1999) Exhumation processes. *Geological Society, London, Special Publications* **154**(1), 1–27.

Ruhl, K. W. and Hodges, K. V. (2005) The use of detrital mineral cooling ages to evaluate steady state assumptions in active orogens: An example from the central Nepalese Himalaya. *Tectonics* **24**(4).

Ryssel, H., and Ruge, I. (1986) *Ion Implantation*. John Wiley & Sons, New York.

Schildgen, T. F., Ehlers, T. A., Whipp, D. M., van Soest, M. C., Whipple, K. X., and Hodges, K. V. (2009) Quantifying canyon incision and Andean Plateau surface uplift, southwest Peru: a thermochronometer and numerical modeling approach. *Journal of Geophysical Research: Earth Surface (2003–2012)* **114**(F4).

Shuster, D. L., Ehlers, T. A., Rusmoren, M. E., and Farley, K. A. (2005) Rapid glacial erosion at 1.8 Ma revealed by ^{4}He/^{3}He thermochronometry. *Science* **310**(5754), 1668–1670.

Shuster, D. L., Balco, G., Cassata, W. S., Fernandes, V. A., Garrick-Bethell, I., and Weiss, B. J. (2010) A record of impacts preserved in the lunar regolith. *Earth and Planetary Science Letters* **290**(1), 155–165.

Shuster, D. L., Cuffey, K. M., Johnny W. Sanders, J. W., and Balco, G. (2011) Thermochronometry reveals headward propagation of erosion in an alpine landscape. *Science* **332**(6025), 84–88.

Spotila, J. A. (2005) Applications of low-temperature thermochronometry to quantification of recent exhumation in mountain belts. *Reviews in Mineralogy and Geochemistry* **58**(1), 449–466.

Stock, G. M., Ehlers, T. A., and Farley, K. A. (2006) Where does sediment come from? Quantifying catchment erosion with detrital apatite (U–Th)/He thermochronometry. *Geology* **34**(9), 725–728.

Stock, J. D. and Montgomery, D. R. (1996) Estimating palaeorelief from detrital mineral age ranges. *Basin Research* **8**(3), 317–327.

Stockli, D. F., Farley, K. A., and Dumitru, T. A. (2000) Calibration of the apatite (U–Th)/He thermochronometer on an exhumed fault block, White Mountains, California. *Geology* **28**(11), 983–986.

Stüwe, K., White, L., and Brown, R. (1994) The influence of eroding topography on steady-state isotherms. Application to fission track analysis. *Earth and Planetary Science Letters* **124**(1), 63–74.

Thomson, S. N., Brandon, M. T., Reiners, P. W., Zattin, M., Isaacson, P. J., and Balestrieri, M. L. (2010) Thermochronologic evidence for orogen-parallel variability in wedge kinematics during extending convergent orogenesis of the northern Apennines, Italy. *Geological Society of America Bulletin* **122**(7–8), 1160–1179.

Turcotte, D. L. and Schubert, G. (2014) *Geodynamics*. Cambridge University Press.

Turner, G., Miller, J. A., and Grasty. R. L. (1966) The thermal history of the Bruderheim meteorite. *Earth and Planetary Science Letters* **1**(4), 155–157.

Turner, G. (1968) The distribution of potassium and argon in chondrites. In *Origin and Distribution of the Elements*, Ahrens, L. H. (eds), 387–398. International Series of Monographs in Earth Sciences, Elsevier.

Van Soest, M. C., Hodges, K. V., Wartho, J-A., *et al.* (2011) (U–Th)/He dating of terrestrial impact structures: the Manicouagan example. *Geochemistry, Geophysics, Geosystems* **12**(5).

Villa, I. M. (1997) Direct determination of ^{39}Ar recoil distance. *Geochimica et Cosmochimica Acta* **61**(3), 689–691.

Wartho, J-A., Kelley, S. P., Brooker, R. A., Carroll, M. R., Villa, I. M., and Lee, M. R. (1999) Direct measurement of Ar diffusion profiles in a gem-quality Madagascar K-feldspar using the ultra-violet laser ablation microprobe (UVLAMP). *Earth and Planetary Science Letters* **170**(1), 141–153.

Willett, S. D. and Brandon. M. T. (2013) Some analytical methods for converting thermochronometric age to erosion rate. *Geochemistry, Geophysics, Geosystems* **14**(1), 209–222.

Wolf, R. A., Farley, K. A., and Kass, D. M. (1998) Modeling of the temperature sensitivity of the apatite (U–Th)/He thermochronometer. *Chemical Geology* **148**(1), 105–114.

Zeitler, P. K. and Fitz Gerald, J. D. (1986) Saddle-shaped ^{40}Ar/^{39}Ar age spectra from young, microstructurally complex potassium feldspars. *Geochimica et Cosmochimica Acta* **50**(6), 1185–1199.

Zeitler, P. K. (1987) Argon diffusion in partially outgassed alkali feldspars: insights from ^{40}Ar^{39}Ar analysis. *Chemical Geology: Isotope Geoscience section* **65**(2), 167–181.

Zhang, Y. (1998) *Geochemical Kinetics*. Princeton University Press, 664 pp.

CHAPTER 6

Rb–Sr, Sm–Nd, and Lu–Hf

6.1 INTRODUCTION

Rb–Sr was a leading geochronological tool from the 1950s through the 1970s. When Sm–Nd began to be actively used in the 1970s, and Lu–Hf in the 1980s, their chronological applications led the way. Though still extremely useful geochronometers, all three systems now see more use in isotope geology than geochronology as they are particularly useful for tracking the history of a diverse suite of natural geochemical processes ranging from the history of mantle differentiation, the formation history of continents, and the history of continental erosion as recorded in oceanic sediments and seawater. Their continued use as geochronometers can be attributed to the distinct geochemical behavior of these elements, and hence the number of geological processes that they can be used to date.

As the methodology of application of these three radioactive systems as geochronometers is similar, their discussion is combined into a single chapter. For all three systems, finding natural materials that initially contain none of the daughter elements is rare, so chronological applications of these systems require an approach that can simultaneously solve for both the age and initial isotopic composition of the daughter element incorporated into the sample at the time of its formation. Although they have overlapping use as geochronometers, the diverse geochemical behavior of the elements involved, and the variable closure temperature of the systems in different materials, allow these systems to be used to determine the ages of a variety of events that may not be able to be dated by any other system, ensuring that their contribution to the field of geochronology will continue well into the future.

6.2 HISTORY

Although [87]Rb was first identified as a radioactive isotope in 1937 [*Hemmendinger and Smythe*, 1937; *Mattauch*, 1937],

the modern approach to Rb–Sr dating began in the 1950s [*Aldrich et al.*, 1956]. The first minerals targeted for Rb–Sr analysis were those with extremely high Rb/Sr ratios, for example, lepidolite, the lithium end-member mica. In such minerals, if they are old enough, the Sr they contain is sufficiently radiogenic that correction for the isotopic composition of the Sr that the mineral incorporated at the time of its crystallization is not critical for the age determination. As measurement techniques improved, the Rb–Sr system was applied to more common minerals, most with lower Rb/Sr ratios than mica. This led to the need to simultaneously solve for both the radiogenic and inherited component of [87]Sr in the mineral. L.O. Nicolaysen accomplished this feat in 1961 with the invention of the isochron approach [*Nicolaysen*, 1961] as described in Chapter 2, which is essential to all three systems discussed in this chapter. While adequate precision on [87]Sr/[86]Sr ratios to enable Rb–Sr dating could be obtained with the manually operated mass spectrometers used through the mid-1960s, the Sm–Nd system did not begin to be used until the 1970s [*Notsu et al.*, 1973; *Lugmair et al.*, 1975] when computer-controlled mass spectrometers provided factors of ten to a hundred improvement in isotope-ratio measurement precision. Although pursued since the 1950s, the modern era of Lu–Hf dating also was delayed until 1980 [*Patchett and Tatsumoto*, 1980a] by the need for computer-controlled mass spectrometers. Full use of this system, however, had to wait until the advent of inductively coupled plasma (ICP) multicollector mass spectrometers [*Blichert-Toft et al.*, 1997]. Although Hf can be analyzed by thermal ionization mass spectrometry, the ionization efficiency of Hf by this technique is very low. The orders of magnitude improvement in ionization efficiency of Hf by ICP allowed the necessary improvement in isotope ratio precision, and reduction in sample size, that drove a wide diversity of applications of this system beginning in the late 1990s.

Geochronology and Thermochronology, First Edition. Peter W. Reiners, Richard W. Carlson, Paul R. Renne, Kari M. Cooper, Darryl E. Granger, Noah M. McLean, and Blair Schoene.
© 2018 John Wiley & Sons Ltd. Published 2018 by John Wiley & Sons Ltd.

6.3 THEORY, FUNDAMENTALS, AND SYSTEMATICS

6.3.1 Decay modes and isotopic abundances

The decay schemes for these three systems are:

$^{87}Rb \rightarrow {}^{87}Sr + \beta^- + \nu$, where ν is a neutrino and β^- is an electron;

$^{147}Sm \rightarrow {}^{143}Nd + \alpha$, where α is an alpha particle;

$^{176}Lu \rightarrow {}^{176}Hf + \beta^- + \nu$.

The approximate isotopic compositions of all the elements involved in these systems are given in Tables 6.1 through 6.3.

6.3.2 Decay constants

The half-life of ^{147}Sm is based on α-decay counting measurements mostly conducted during the 1960s and 1970s [*Begemann et al.*, 2001]. The resulting average value from these measurements is $(1.06 \pm 0.01) \times 10^{11}$ years, which corresponds to a decay constant of $(6.54 \pm 0.06) \times 10^{-12}$/year. Ages determined using this decay constant agree well with U–Pb ages determined on the same rocks, so this decay constant has been in use, and seen little controversy, since the introduction of the Sm–Nd system in the early 1970s.

The values of the decay constants for both ^{87}Rb and ^{176}Lu have been subject to much more controversy. The long half-lives of both ^{87}Rb and ^{176}Lu coupled with the fact that they emit relatively low energy β particles when they decay, which can be difficult to detect, contribute to the difficulty of determining accurate decay constants for these parent isotopes by radioactive counting. The first value for the ^{87}Rb decay constant that became widely used in geochronology was $(1.39 \pm 0.06) \times 10^{-11}$/year (half-life = 49.9 Ga) [*Aldrich et al.*, 1956], which was determined by comparing the ages for a number of Rb-rich minerals, such as mica, with U–Pb ages for U-rich minerals, such as monazite and uraninite, from the same igneous rocks. The various determinations of the ^{87}Rb decay constant were reviewed again in the late 1970s to come up with a new value of 1.42×10^{-11}/year (48.8 Ga half-life) [*Steiger and Jager*, 1977]. Although this value began to be used by most workers in the field, those who studied old rocks, for example meteorites and lunar rocks, noted that this value for the half-life resulted in disagreement of Rb–Sr and U–Pb ages. To reconcile the ages, they recommended a ^{87}Rb decay constant of $(1.402 \pm 0.008) \times 10^{-11}$/year (half-life = 49.4 Ga) [*Minster et al.*, 1982]. The most recent evaluation of the ^{87}Rb half-life [*Villa et al.*, 2015] examines a combination of recent radioactive counting results, geological comparisons, and also a measure of the build-up of ^{87}Sr in a solution containing pure Rb. The value recommended in this comparison is $(1.397 \pm 0.004) \times 10^{-11}$/year (half-life = 49.61 ± 0.16 Ga). When comparing Rb–Sr ages from different studies, the range of Rb decay constants in use can produce a considerable range in the calculated age from the same isochron. For example, an isochron slope that yields an age of 4.00 Ga using the 1.39×10^{-11}/year decay constant, translates to an age of 3.92 Ga using the 1.42×10^{-11}/year decay constant.

Like the branching decay of ^{40}K, ^{176}Lu potentially can decay either by β^- emission to ^{176}Hf, or by electron capture to ^{176}Yb. Unlike ^{40}K, however, the branching ratio of ^{176}Lu decay is not well known and likely is quite small. The percentage of ^{176}Lu decays that go to ^{176}Yb instead of ^{176}Hf is important in light of the difficulty in measuring the ^{176}Lu decay constant by radioactive counting. Both radioactive counting and geologic determinations of the ^{176}Lu decay constant have provided a range of values, with counting results from 1.7 to over 1.9×10^{-11}/year, and geologic

Table 6.1 Rb–Sr isotopic composition in atom%

Element	Mass				
	84	85	86	87	88
Rb		72.17		27.83	
Sr[a]	0.5573		9.865	6.955	82.62

[a] Sr atomic abundances assume a $^{87}Sr/^{86}Sr$ ratio of 0.705.

Table 6.2 Sm–Nd isotopic composition in atom%

Element	Mass										
	142	143	144	145	146	147	148	149	150	152	154
Sm			3.075			14.99	11.24	13.82	7.38	26.74	22.75
Nd[a]	27.168	12.197	23.794	8.290	17.177		5.748		5.626		

[a] Nd atomic abundances assume a $^{143}Nd/^{144}Nd$ ratio of 0.51263.

Table 6.3 Lu–Hf isotopic composition in atom%

Element	Mass					
	174	175	176	177	178	179	180
Lu		97.42	2.58				
Hf[a]	0.161		5.258	18.595	27.282	13.621	35.083

[a] Hf atomic abundances assume a $^{176}Hf/^{177}Hf$ ratio of 0.282785.

determinations from 1.82 to 1.98×10^{-11}/year. This range in values, and the apparent difference between values determined by counting and by geologic cross-comparison, could reflect any or all of:

(1) a larger than expected percentage of decays that create ^{176}Yb instead of ^{176}Hf;

(2) the experimental difficulty of counting low-energy β-particles;

(3) use of geological comparisons that do not provide the same age for Lu–Hf and the reference dating system, because one or the other system has been disturbed by metamorphism;

(4) some process in nature, but not the laboratory, that accelerates the decay of ^{176}Lu.

As of this date, no one has detected the decay of ^{176}Lu to ^{176}Yb. This null result places an upper limit of < 0.5% on the percentage of ^{176}Lu decays that create ^{176}Yb instead of ^{176}Hf. The ^{176}Lu to ^{176}Hf route thus is the only Lu decay of importance at the current level of precision of Lu–Hf systematics.

With the advent of modern mass spectrometric analysis of Hf isotope variation, the first geologic determination of the ^{176}Lu half-life was made by comparison of the slope of the Lu–Hf isochron measured for a group of igneous meteorites known as eucrites. This work produced an isochron slope = 0.0934 ± 0.0040 [*Patchett and Tatsumoto*, 1980b]. Normally, one uses the isochron slope and the equation

$$Slope = e^{\lambda t} - 1 \tag{6.1}$$

to solve for "t," the age of the sample. When attempting to determine the decay constant, however, one instead uses a "t" for the isochron derived by some other dating method and solves for λ. In this example, several eucrites were previously dated at 4.55 Ga by U–Pb. On the assumption that the Lu–Hf whole-rock eucrite isochron should also record this age, the slope of the isochron translates to a decay constant of $(1.96 \pm 0.08) \times 10^{-11}$/year. As with any geologic determination of a half-life, however, one must be sure that the system of interest is recording the same age as the reference age. In the case of whole-rock meteorites, later metamorphic events, particularly impact-related shocks, can disturb isochron systematics, so additional support for the assumption that the U–Pb age of individual eucrites, and the Lu–Hf whole-rock eucrite isochron, should provide the same age is needed. In the early 2000s, two studies used terrestrial Lu-rich minerals, precisely dated by U–Pb, for another attempt at geologic calibration of the ^{176}Lu decay constant. These studies returned decay constants of $(1.865 \pm 0.015) \times 10^{-11}$/year and $(1.867 \pm 0.008) \times 10^{-11}$/year [*Scherer et al.*, 2001; *Soderlund et al.*, 2004]. Similar values were determined by [*Amelin*, 2005] in a comparison of Lu–Hf isochron and U–Pb ages measured for phosphate grains in meteorites. These studies combine to provide a best estimate of the ^{176}Lu decay constant of 1.867×10^{-11}/year, which translates to a half-life of 37.13 Ga.

Although there are many arguments in favor of this half-life for ^{176}Lu, using this value causes the eucrite whole-rock isochron to correspond to an age of 4.78 Ga, which is older than the accepted age of the solar system. A suggestion to explain this observation is that ^{176}Lu has an excited nuclear state (an isomer) that can be induced by capture of a very energetic (838 keV) gamma-ray by ground-state ^{176}Lu [*Albarède et al.*, 2006]. The excited isomer of ^{176}Lu has a very short half-life of 3.7 h and decays to ^{176}Hf. Gamma-ray irradiation thus would speed-up the decay of ^{176}Lu, reducing its half-life, which would lead to inaccurately old ages for samples that had suffered such irradiation compared to those that had not. While an interesting idea, such irradiation should induce other effects, such as changing the isotopic composition of Lu, that have not been observed. In addition, other radiometric systems, for example Rb–Sr, applied at the whole-rock scale to meteorites often produce very scattered isochrons, so the expectation that the Lu–Hf whole-rock eucrite isochron necessarily should give the same age as U–Pb dates for individual eucrites is questionable. Although the debate about the true half-life of ^{176}Lu likely will continue for at least a while, the good agreement of the three geologic determinations that examined cases where there is a reasonable expectation that the U–Pb and Lu–Hf systems should provide the same age, argue strongly for a ^{176}Lu decay constant of 1.867×10^{-11}/year.

If nothing else, this debate should alert the reader that the value of the decay constant for a radioactive isotope must be determined experimentally, and that a variety of issues can produce uncertain, and sometimes inaccurate, results. As age precisions in geochronology continue to improve, keeping in mind the uncertainty in the decay constant and its effect on both the precision and accuracy of the ages determined by geochronological techniques is essential.

6.3.3 Data representation

As with many isotopic systems, the variations in Sr, Nd, and Hf isotopic composition can be quite small. Consequently, the isotopic compositions are often reported relative to some standard value, usually the isotopic composition the undifferentiated mantle (bulk-silicate Earth or BSE) would have at the same age of the sample. For all three systems, these relative values are given the epsilon (ε) notation denoting their difference, in parts in 10,000, compared to the standard value, using the equations:

$$\varepsilon Sr(t) = \left[\frac{\left[\frac{^{87}Sr}{^{86}Sr}\right]_{Sample}}{\left[\frac{^{87}Sr}{^{86}Sr}\right]_{BSE}} - 1 \right] \times 10,000 \tag{6.2}$$

$$\varepsilon Nd(t) = \left[\frac{\left[\frac{^{143}Nd}{^{144}Nd}\right]_{Sample}}{\left[\frac{^{143}Nd}{^{144}Nd}\right]_{BSE}} - 1 \right] \times 10,000 \tag{6.3}$$

$$\varepsilon Hf(t) = \left[\frac{\left[\frac{^{176}Hf}{^{177}Hf}\right]_{Sample}}{\left[\frac{^{176}Hf}{^{177}Hf}\right]_{BSE}} - 1 \right] \times 10,000 \tag{6.4}$$

Table 6.4 provides the parent/daughter ratios and daughter isotopic compositions of a number of materials that have been used to represent the composition of the Earth. A common starting point for estimating bulk-Earth composition is the Sun, because the Sun contains over 99.8% the mass of the solar system and hence dominates the average composition of the solar system. One type of primitive meteorite (CI chondrite) has chemical composition close to the composition of the Sun in all but the most volatile elements, and hence often is used as a starting point to estimate the composition of the bulk-Earth [*McDonough and Sun*, 1995; *Palme and O'Neill*, 2014]. The bulk-Earth includes core, mantle, crust, hydrosphere, and atmosphere. Because Rb, Sr, Sm, Nd, Lu, and Hf are all lithophile elements, meaning that they are soluble primarily in silicates, their concentration in the core should be negligible, so their concentration in the silicate portion of Earth (BSE) is simply their concentration in the whole Earth times 1.48; the mass ratio of the silicate Earth (67.5%) to core (32.5%). Table 6.4 includes estimates of the composition of both BSE and the incompatible element-depleted mantle that is the source of basaltic volcanism along all the world's ocean ridges.

In the literature, there has been some variability in the estimates of BSE composition. For Sm–Nd and Lu–Hf, both parent and daughter elements have high condensation temperatures from the gas in an initially hot solar nebula [*Lodders*, 2003], and both are much more soluble in silicates than in metals. These properties lead to their classification as refractory (high condensation temperature) lithophile (soluble in silicates) elements. Volatile loss and metal-silicate separation are the most important chemical fractionation mechanisms that occur during planet formation. Neither process fractionates refractory lithophile elements from one another. As a result, the assumed BSE values for Sm/Nd and Lu/Hf ratios are based on the values measured in primitive meteorites [*Bouvier et al.*, 2008] on the assumption that these ratios in the bulk-Earth should be the same as they are in the average solar system, as represented by the Sun. Strontium also is a refractory lithophile element, but Rb is a moderately volatile lithophile element with a boiling point of 688 °C. Earth is quite depleted in volatile elements compared to the Sun or primitive meteorites [e.g., *McDonough and Sun*, 1995]. This is reflected quite clearly in the Sr isotopic composition of Earth compared to chondrites. CI chondrites have present-day $^{87}Sr/^{86}Sr$ ratios of approximately 0.76. Measurements of the first

high-temperature condensates of the solar system, the so-called calcium–aluminum rich (CAIs) found in some primitive meteorites, show that the solar system began with a $^{87}Sr/^{86}Sr$ near 0.698975 [*Hans et al.*, 2013]. Rearranging the Rb–Sr isochron equation as

$$\frac{^{87}Rb}{^{86}Sr} = \frac{\frac{^{87}Sr}{^{86}Sr}_P - \frac{^{87}Sr}{^{86}Sr}_0}{e^{\lambda t} - 1} \tag{6.5}$$

and putting in the CI chondrite Sr isotopic compositions above along with a 4.567 Ga age for the solar system, returns a CI chondrite $^{87}Rb/^{86}Sr = 0.926$. In contrast, most modern mid-ocean ridge basalts (MORB), that globally sample Earth's upper mantle, have $^{87}Sr/^{86}Sr$ ratios between 0.702 and 0.704 [*Hofmann*, 2003]. Continental crust displays a much higher, and wider range, of Sr isotopic compositions, with an average estimated to be near 0.715, still much lower than that of CI chondrites. Chemical differentiation of the Earth has moved a good portion of Earth's Rb into the continental crust, creating the chemically complimentary reservoirs of Rb-rich continental crust and the Rb-poor mantle source of MORB [*Hofmann*, 1988]. Although the continental crust has high Rb concentrations and Rb/Sr ratios, it constitutes less than 0.5% of the mass of Earth, so adding it back into the whole mantle affects the mantle's Rb/Sr and $^{87}Sr/^{86}Sr$ ratios only by a small amount. Because of the volatile-depletion of Earth, along with the extraction of the continental crust from the mantle, the BSE value for Rb/Sr ratio cannot be deduced from solar or meteoritic compositions. Instead, the ratio is estimated from a comparison of $^{87}Sr/^{86}Sr$ and $^{143}Nd/^{144}Nd$ in modern mantle-derived volcanic rocks [*O'Nions et al.*, 1977]. In these rocks, the Sr and Nd isotope ratios correlate reasonably well, so the BSE value for $^{87}Sr/^{86}Sr$ is selected where the correlation goes through $\varepsilon_{Nd} = 0$; the Nd isotopic composition equal to the average of primitive meteorites. The Sr–Nd isotope correlation in modern mantle-derived rocks, however, is not perfect; so various values of the modern day $^{87}Sr/^{86}Sr$ ratio of the BSE have been used in the literature. These mostly fall within the range 0.7045 to 0.7050; a range that likely is smaller than the uncertainty with which this ratio can be estimated. Assuming a BSE $^{87}Sr/^{86}Sr$ of 0.705, using the equation above, this equates to a BSE $^{87}Rb/^{86}Sr$ ratio of approximately 0.09, about a factor of ten lower than that of CI chondrites as a result of Earth's depletion in volatile Rb.

Table 6.4 Reference reservoir parameters for Rb–Sr, Sm–Nd, and Lu–Hf

Ratio	Solar system initial (4.567 Ga)	CI chondrite	Bulk silicate earth	MORB mantle
$^{87}Rb/^{86}Sr$		0.926	0.09	0.0535
$^{87}Sr/^{86}Sr$	0.698975	0.75989	0.7049	0.7025
$^{147}Sm/^{144}Nd$		0.1980	0.1960	0.2148
$^{143}Nd/^{144}Nd$	0.506686	0.51269	0.51263	0.51320
$^{176}Lu/^{177}Hf$		0.0338	0.0336	0.0395
$^{176}Hf/^{177}Hf$	0.279784	0.282793	0.282785	0.28330

MORB, mid-ocean ridge basalt.

6.3.4   Geochemistry

All six elements involved in these systems are incompatible lithophile elements, which means that they are concentrated in silicate phases, not metal or sulfide, and that when melting occurs in the mantle, they preferentially partition into the melt. The mineral-melt distribution coefficients of various major minerals in the mantle are listed in Table 6.5. These distribution coefficients reflect the ratio of the abundance of the element in the mineral compared to a melt in chemical equilibrium with the mineral. Values less than 1 imply that the element prefers the melt, i.e. is incompatible, whereas distribution coefficients greater than 1 show that the element prefers the mineral to the melt, i.e. is compatible. A perfectly incompatible element, meaning that all the element enters the melt during even small degrees of melting in the mantle, would have a distribution coefficient of zero.

Because of its large ionic size, Rb is strongly incompatible in most mantle minerals, with the exception of some relatively rare micas stable in the mantle, such as phlogopite. Sr, Nd, and Sm also are incompatible in most mantle phases, with the exception of the high-pressure phase Ca-perovskite, where all three elements are compatible. Given the bulk composition of the mantle, Ca-perovskite will be a minor phase in the lower mantle, hence the bulk distribution coefficient (the sum of the distribution coefficients of each mineral present times their modal abundance in the whole rock) of the lower mantle will be less than 1 for these three elements, meaning that they will preferentially partition into the melt should melting, or crystal fractionation, occur at lower mantle pressures. Lu and Hf are incompatible in most upper mantle minerals with the exception that Lu is compatible in garnet. Given the modal percentages of garnet in most upper mantle rocks, however, the bulk partition coefficient for Lu in the mantle is less than 1. The dominant phase in the lower mantle, bridgemanite, which is the high-pressure analog of Mg-rich olivine, excludes most of the same elements as do upper mantle phases, with the exception of its ability to accommodate ions of high charge, such as Hf. Much like garnet, bridgemanite strongly prefers the heavy rare-earth elements (REE), like Lu, over lighter REE, like Sm and Nd.

Magmatic fractionation (see Box 6.1) can result in huge changes in Rb/Sr ratio, but generally not until the latter stages of fractionation, for example during the formation of extreme compositions such as granites and high-silica rhyolites when mineral phases like plagioclase and apatite crystallize and rapidly extract Sr from the remaining melt. Magmatic and metamorphic processes also fractionate Sm from Nd and Lu from Hf, but the magnitude of fractionation is generally small because in most minerals the partition coefficients for these elements are not dramatically different from one another. The primary exceptions are garnet that strongly concentrates both Sm and Lu over their daughters Nd and Hf, and zircon that can have weight percent of Hf, but ppm levels of Lu, resulting in extremely low Lu/Hf ratios. Table 6.6 provides parent–daughter concentrations for a small selection of rock types to indicate the magnitude of fractionation present

Box 6.1   Parent–daughter fractionation via partial melting

Simple batch partial melting assumes that all the mineral phases melt in the same proportion as their abundance in the whole rock. Under this condition, the concentration of any element in the melt is given by:

$$C_m = \frac{C_0}{F + D \times (1 - F)}$$

and in the complementary residue of melting:

$$C_r = \frac{C_0 \times D}{F + D \times (1 - F)}$$

where C is the concentration in the starting solid (0), melt (m), and residual solid (r), D is the distribution coefficient and F the percentage of partial melting [*Shaw*, 1970]. When the distribution coefficients of parent and daughter element are similar, melting or crystallization cause very little fractionation of the parent/daughter ratio, but when the distribution coefficients are different, partial melting can result in large changes in parent/daughter ratio as Fig. 6.B1.1. illustrates in particular for Rb–Sr and Lu–Hf.

The curves in Fig. 6.B1.1 show that during mantle melting, melts will have higher Rb/Sr, but lower Sm/Nd and Lu/Hf ratios than the starting solid. In contrast, the residual solid left after extraction of melt will have low Rb/Sr and high Sm/Nd and Lu/Hf ratios compared to the starting solid. The partition coefficients in Table 6.5 show the strong effect garnet has on the partitioning of Sm, Nd, Lu, and Hf, with garnet strongly retaining Lu and much more Sm than Nd. As a result, garnet can have very high Lu/Hf and Sm/Nd ratios, which makes it a very useful mineral for Sm–Nd and Lu–Hf chronology. Rb and Sr are strongly fractionated from one another by minerals like the potassium feldspars, micas, clays, and amphiboles that strongly concentrate Rb over Sr, and minerals such as plagioclase, calcite, and apatite that strongly concentrate Sr over Rb.

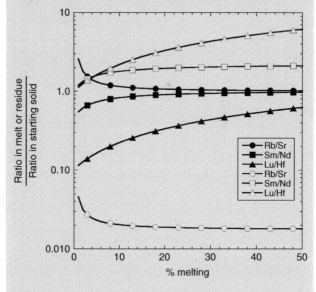

Fig. 6.B1.1. Relative change in parent/daughter ratio in the melt (solid symbols) and residue of melting (open symbols) as a result of modal batch partial melting in the mantle.

in these systems. Figure 6.1 provides a richer sampling of the compositional variation of over 6000 samples of igneous rocks from western North America contained in the NAVDAT database [*Walker et al.*, 2006]. Perhaps the most obvious feature of the data in Fig. 6.1 is the strong increase in Rb and

Table 6.5 Mineral-melt distribution coefficients for minerals important during melting of the mantle

Mineral[a]	Rb	Sr	Sm	Nd	Lu	Hf
Olivine	0.0003	0.00004	0.0011	0.00042	0.02	0.0011
Orthopyroxe	0.0002	0.0007	0.02	0.012	0.12	0.024
Clinopyroxene	0.0004	0.091	0.15	0.088	0.276	0.14
Garnet	0.0002	0.0007	0.23	0.064	7	0.4
Bridgemanite	0.023	0.009	0.07	0.023	1.0	1.8
Ca-Perovskite	0.4	1.7	20	16	16	2.1
Garnet-peridotite	0.0003	0.017	0.052	0.024	0.71	0.072

[a] Distribution coefficients at 3 GPa from *Salters and Stracke* [2004] for bridgemanite and Ca-perovskite from *Corgne et al.* [2005]. The bulk distribution coefficient for garnet peridotite assumes a modal composition of 60% olivine, 11% orthopyroxene, 19% clinopyroxene, and 9% garnet. The bulk distribution coefficient is defined as the sum of the abundance of each mineral in the whole rock times its distribution coefficient.

Table 6.6 Rb, Sr, Sm, Nd, Lu, and Hf concentrations (in ppm) in various meteorites, igneous rocks, and the estimated composition of the bulk-silicate Earth (BSE)

Sample	Rb	Sr	Sm	Nd	Lu	Hf
CI chondrite[a]	2.30	7.25	0.148	0.457	0.0246	0.103
BSE[b]	0.60	19.9	0.406	1.25	0.0675	0.283
MORB[c]	2.88	129	3.82	12.03	0.53	2.79
E-MORB[c]	10.56	207	3.72	14.86	0.38	2.54
Average OIB[d]	31	660	10.0	38.5	0.30	7.80
Parana CFB[e]	19	163	3.97	14.5	0.47	2.71
Alkaline basalt[e]	69.7	2223	9.7	58.9	0.19	7.15
Kimberlite[f]	75	601	13.7	110	0.08	5.71
Continental crust[g]	49	320	3.9	20	0.30	3.7
Arc andesite[h]	46	587	3.92	20.89	0.23	3.56
Average TTG[i]	56	493	3.03	18.16	0.12	4.0
High-silica rhyolite[j]	433	2.5	7.5	31	1.7	4.9

[a] *Anders and Grevesse* [1989]
[b] *McDonough and Sun* [1995]
[c] average normal (MORB) and enriched (E-MORB) mid-ocean ridge basalt [*Gale et al.*, 2013]
[d] average ocean island basalt [*Sun and McDonough*, 1989]
[e] Parana (South America) continental flood basalt and continental alkalic basalt [*Farmer*, 2003]
[f] *Le Roex et al.* [2003]
[g] average continental crust [*Rudnick and Gao*, 2003]
[h] average arc andesite [*Kelemen et al.*, 2003]
[i] average Archean TTG (tonalite, trondhjemite, granodiorite) [*Moyen and Martin*, 2012]
[j] *Christiansen et al.* [1984].

decrease of Sr in Si-rich rocks such as granites and rhyolites. This results in extreme Rb/Sr ratios in such rocks. In contrast, while the more Si-rich rocks tend towards lower Sm/Nd and Lu/Hf ratios compared to Si-poor rocks, the distinction is nowhere near as great as for Rb/Sr ratio.

Rubidium is the only moderately volatile element in this group, as all the other elements are very refractory. As will be discussed later, this characteristic of Rb makes the Rb–Sr system a very powerful tool to understand the timing of volatile depletion of planets and planetesimals. Both rubidium and strontium are soluble in water and so are often strongly fractionated during rock alteration and metamorphism, as either, or both, elements are removed by whatever fluids might be generated or involved. The high solubility of Sr in water results in a very long residence time in seawater, allowing it to be well mixed throughout the world's oceans. As a result, the isotopic composition of Sr in the ocean at any given time is nearly constant, so its variability in oceanic sediments through time provides a sensitive measure of the ratio through geologic history of continental weathering (contributing high $^{87}Sr/^{86}Sr$ because of the high Rb/Sr ratio

of the average continental crust) to hydrothermal Sr addition to the ocean along the volcanic mid-ocean ridges, which contributes low $^{87}Sr/^{86}Sr$ due to the low Rb/Sr ratio of the oceanic crust and mantle [*Veizer and Mackenzie*, 2003]. In contrast, Nd and Hf have very low solubility in seawater, and hence are removed rapidly from the water column following their delivery to the ocean either by rivers or by hydrothermal activity on the ocean floor. As a result, their isotopic composition in the ocean, as measured primarily in materials that form by precipitation from seawater, for example Mn-nodules or ferromanganese crusts on the ocean floor, are good tracers of local variations in the source of Nd and Hf to the oceans. For example, Nd and Hf isotopic variations in seawater precipitates have been used to track the migration of waters from the North Atlantic southward along the floor of the Atlantic basin [*Rutberg et al.*, 2000]. The distinct geochemical behavior of these elements provides the foundation for the use of the Rb–Sr, Sm–Nd, and Lu–Hf systems to trace geochemical processes operating over all of Earth history, thus establishing their dominant role in the field of isotope geochemistry.

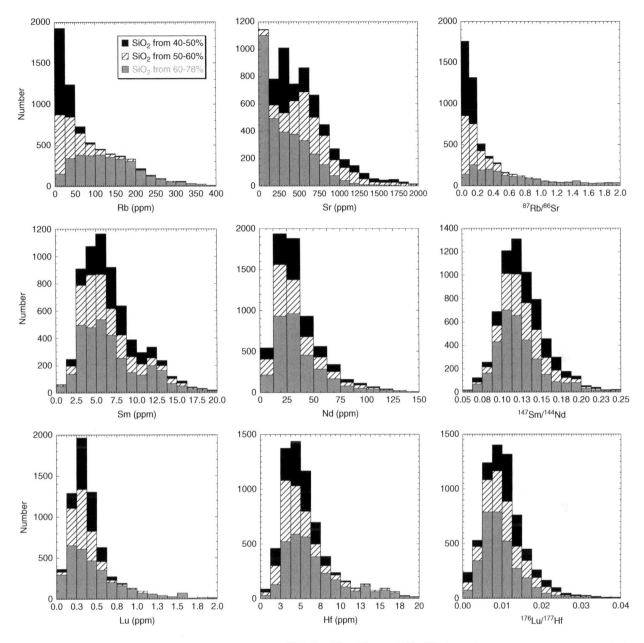

Fig. 6.1. Histograms of Rb, Sr, Sm, Nd, Lu, and Hf concentrations and ^{87}Rb/^{86}Sr, ^{147}Sm/^{144}Nd, and ^{176}Lu/^{177}Hf ratios for over 6000 Cenozoic igneous rocks from western North America. The data in each plot are shown separately for samples with SiO$_2$ contents between 40 and 50 wt% (basalt), 50 and 60 wt% (broadly andesite), and 60 and 78 wt% (dacite to rhyolite). (Source: Data from NAVDAT.)

6.4   ISOCHRON SYSTEMATICS

Only a very small number of minerals have parent/daughter ratios in Rb–Sr, Sm–Nd and Lu–Hf that are high enough to make the initial isotopic composition of the daughter element insignificant compared to radiogenic ingrowth after mineral formation. The most obvious exception is mica, which can have exceptionally high Rb/Sr ratios that enable the calculation of ages directly from their measured Rb/Sr and ^{87}Sr/^{86}Sr ratios [*Patel et al.*, 1999; *Blackburn et al.*, 2008]. For most other minerals, the primary approach to age determinations using these systems is the isochron method. Isochrons provide not only the age of the sample, but also the isotopic composition of the daughter element it incorporated when the sample formed. As will be seen in later sections of this chapter, the initial isotopic composition can provide much information on the source of a given sample, including the nature and timing of the source formation from some model reservoir, for example the BSE.

The standard isochron equations for these systems are:

$$\left(\frac{^{87}\mathrm{Sr}}{^{86}\mathrm{Sr}}\right)_{\mathrm{p}} = \left(\frac{^{87}\mathrm{Sr}}{^{86}\mathrm{Sr}}\right)_{0} + \frac{^{87}\mathrm{Rb}}{^{86}\mathrm{Sr}} \times \left(e^{\lambda_{87}t} - 1\right) \qquad (6.6)$$

$$\left(\frac{^{143}\text{Nd}}{^{144}\text{Nd}}\right)_p = \left(\frac{^{143}\text{Nd}}{^{144}\text{Nd}}\right)_0 + \frac{^{147}\text{Sm}}{^{144}\text{Nd}} \times \left(e^{\lambda 147 t} - 1\right) \qquad (6.7)$$

$$\left(\frac{^{176}\text{Hf}}{^{177}\text{Hf}}\right)_p = \left(\frac{^{176}\text{Hf}}{^{177}\text{Hf}}\right)_0 + \frac{^{176}\text{Lu}}{^{177}\text{Hf}} \times \left(e^{\lambda 176 t} - 1\right) \qquad (6.8)$$

Where "p" reflects the present-day, or measured, isotope ratios, "0" is the isotope ratio of the daughter element that was incorporated into the sample when it formed, and λ is the decay constant.

Isochrons require measurement of at least two, but hopefully many more, samples in order to define a line on a plot of daughter isotopic composition (ordinate) versus parent/daughter ratio (abscissa). Chapter 4 explores the statistical methods used to calculate a best fit line, the isochron, for a set of data points that have uncertainties in both the measured isotopic composition and parent/daughter ratio. Once a best-fit line is calculated it will be an isochron, and will provide an accurate age only if:

(1) all the samples used to define the isochron formed at the same time and with the same initial daughter isotopic composition;

(2) no changes to parent/daughter ratio within the samples occurred since the time of their formation;

(3) the samples have a sufficient range in parent/daughter ratios, and enough time has passed since their formation, to evolve a significant difference in daughter isotope ratios so that the individual sample points on an isochron diagram define a line whose slope can be determined precisely.

Not all linear arrays on an isochron diagram are necessarily isochrons. As will be discussed later, a variety of mixing processes also can produce linear data arrays on isochron plots that may, or may not, have any chronological significance.

Table 6.7 provides Rb–Sr, Sm–Nd, and Lu–Hf data for minerals separated from Apollo norite 77215; a rock from the ancient lunar crust. This rock consists of only two major minerals, plagioclase and pyroxene. This simple mineralogy, coupled with constraints on the amount of material available for study, limit the number of analyses that can be used to produce an isochron to just separates of these two minerals and to a whole-rock analysis.

Table 6.7 Isotopic data for lunar norite 77215

Sample	Whole rock	Plagioclase	Pyroxene
Rb (ppm)	2.67	6.59	1.09
Sr (ppm)	92.6	216	7.80
^{87}Rb/^{86}Sr	0.0834	0.0884	0.4041
^{87}Sr/^{86}Sr	0.704124	0.704723	0.724811
Sm (ppm)	3.52	8.35	2.224
Nd (ppm)	12.13	47.78	6.082
^{147}Sm/^{144}Nd	0.1754	0.1566	0.221
^{143}Nd/^{144}Nd:	0.512016	0.511491	0.513320
error	0.000004	0.000004	0.000005
Lu (ppm)	0.642	0.444	0.897
Hf (ppm)	2.70	3.501	3.988
^{176}Lu/^{177}Hf	0.0338	0.01801	0.03193
^{176}Hf/^{177}Hf	0.282778	0.281421	0.282601

Source: Data from *Carlson et al.* [2014].

Isochron diagrams for these data are shown in Fig. 6.2 . All three systems define lines with minimal scatter, although in some cases, such as Rb–Sr, the plagioclase and whole-rock data are so close to one another that the line is close to being a tie-line that simply connects two points. All three systems define ages that overlap within the age uncertainty that arises from the precision by which the slopes of the isochrons are defined by the data. The isotopic compositions of the Sr, Nd and Hf that this rock incorporated at the time of its formation are indicated by the intersection of the isochrons with the *y*-axis where the parent/daughter ratio is zero.

In the isochron diagrams shown in Fig. 6.2, the measurement uncertainties on the data are much smaller than the size of the points used in the figure, so the amount of scatter about the best fit line to the data is difficult to see. To better observe the amount of data scatter, one can construct what are known as δY or εY diagrams. These diagrams are essentially isochron diagrams, but they replace the isotopic compositions shown on the *y*-axis with the relative deviation of the isotope ratios of each sample from the best fit line. Take for example the Sm–Nd data given in Table 6.7. The slope of the best fit line is 0.02844 ± 0.0003 and the initial ^{143}Nd/^{144}Nd is 0.50703 ± 0.00006. One can calculate the ^{143}Nd/^{144}Nd ratio of the isochron at any ^{147}Sm/^{144}Nd ratio using the equation:

$$\frac{^{143}\text{Nd}}{^{144}\text{Nd}} = \left(\frac{^{143}\text{Nd}}{^{144}\text{Nd}}\right)_0 + \left(\frac{^{147}\text{Sm}}{^{144}\text{Nd}} \times \text{slope}\right) \qquad (6.9)$$

where $(^{143}\text{Nd}/^{144}\text{Nd})_0$ is the initial Nd isotope composition provided by the isochron. The εY diagram is then constructed by comparing the measured Nd isotopic composition for a given sample against the Nd isotopic composition calculated from the isochron at the sample's ^{147}Sm/^{144}Nd ratio (Table 6.8). εY is calculated using the equation:

$$\varepsilon Y = \left(\left[\frac{\left(\frac{^{143}\text{Nd}}{^{144}\text{Nd}}\right)_\text{Sample}}{\left(\frac{^{143}\text{Nd}}{^{144}\text{Nd}}\right)_\text{Isochron}}\right] - 1\right) \times 10,000 \qquad (6.10)$$

δY is calculated in the same way, but with a multiplier of 1000 instead of 10,000.

The εY diagram for the Sm–Nd data for 77215 is shown in Fig. 6.3 . In this representation of the same data shown in Fig. 6.2, the offset of the individual points from the best fit isochron, now represented by the horizontal line in Fig. 6.3 at εY = 0, are shown at a resolution that makes it possible to see that the datum for plagioclase lies off the isochron by an amount larger than its measurement uncertainty. The sloped lines in Fig. 6.3 reflect the slope and initial isotopic composition uncertainties calculated for the isochron. The positively sloped line is the comparison of the isochron with the line calculated from the isochron slope plus its uncertainty, using the isochron initial Nd isotopic composition minus its uncertainty. The negatively sloped line uses the isochron slope minus its uncertainty, and the isochron initial Nd isotopic composition plus its uncertainty. All the data

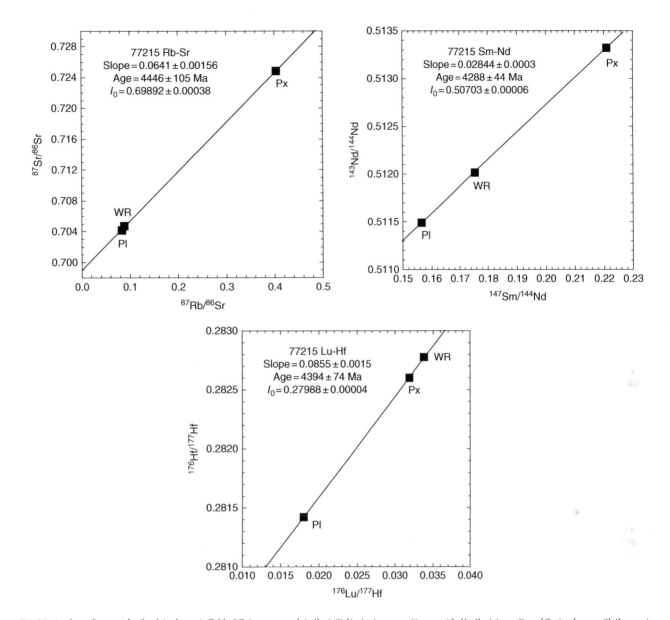

Fig. 6.2. Isochron diagrams for the data shown in Table 6.7. I_0 corresponds to the initial isotopic composition provided by the intersection of the isochrons with the y-axis. Abbreviations are pyroxene (Px), plagioclase (Pl), and whole rock (WR).

Table 6.8 εY calculation for the Sm–Nd data for lunar norite 77215

Sample	Whole rock	Plagioclase	Pyroxene
$^{147}Sm/^{144}Nd$	0.1754	0.1754	0.221
Measured $^{143}Nd/^{144}Nd$	0.512016	0.512016	0.513320
Isochron $^{143}Nd/^{144}Nd$	0.512018	0.512018	0.513315
εY	−0.04	−0.04	0.10

points lie within the error envelope created by the sloped lines that correspond to the ±44 Ma age uncertainty provided by the isochron.

Another important observation to make about isochron data is the distribution of points along the isochron. The Rb–Sr data for

the plagioclase and the whole rock point from 77215 are so similar because a very large fraction of the whole-rock's Sr is contained in the plagioclase. For example, if this rock were composed only of plagioclase and pyroxene, the measured whole-rock Sr concentration would suggest that the rock consists of 41% plagioclase and 59% pyroxene, which is not an unreasonable modal distribution for the two minerals in this rock. Using this modal mineralogy and the concentrations of Rb measured in pyroxene and plagioclase leads to a calculated whole-rock Rb concentration of 3.3 ppm, which is somewhat, but not greatly, higher than the measured value. These results suggest that, at least for Rb–Sr, the whole rock is primarily just a mixture of plagioclase and pyroxene. Doing the same exercise with the Lu–Hf data, however, leads to a

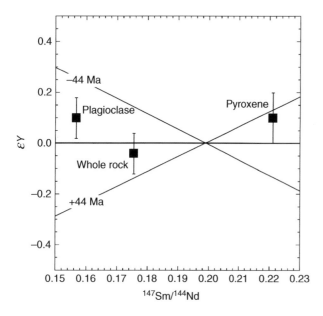

Fig. 6.3. εY diagram for the Sm–Nd data for 77215 shown in Table 6.8.

quite different conclusion. In this system, the measured whole rock has a higher Lu/Hf ratio than either the measured plagioclase or pyroxene separate. This obviously is not possible if these are the only two mineral phases in the rock. A likely explanation is that the whole rock also contains apatite as a trace phase. Apatite has an extremely high Lu/Hf ratio and is a phase known to exist in this rock, but in quantities too low to allow its separation and analysis as a discrete data point. Apatite likely will have a low Rb/Sr ratio and Sr concentration similar to plagioclase, so its low modal abundance in the rock will limit its influence on the whole-rock Rb–Sr data.

Inspection of the element concentrations of the individual data points thus can provide supporting information to evaluate whether the isochron fulfills all the criteria needed for it to provide a meaningful age. If all the mineral separates have elemental concentrations consistent with their known partition coefficients, this is evidence that the minerals formed by crystallization from the same magma and their parent–daughter inventories have remained closed since then. If the concentrations and modal percentages of the minerals sum up to the concentrations measured in the whole rock, this is evidence in support of both the quality of the analytical results and that no important trace phases were missed in the mineral separation (as they were in the Lu–Hf analyses for 77215). If the rock being analyzed is a mixture of unrelated components, for example a breccia composed of fragments from many different rocks, then the likelihood of it meeting the requirements for the isochron to provide a meaningful age is low.

6.4.1 Distinguishing mixing lines from isochrons

Isochrons constructed from minerals separated from a single igneous rock have a good chance of fulfilling the requirements

for an isochron to provide an accurate age because, in most circumstances, the minerals will have crystallized over a short time interval from a magma that likely had a constant isotopic composition for the daughter element. Isochrons also can be constructed from whole-rock analyses of groups of igneous rocks that are presumed to have been erupted or emplaced over a short time interval from a common parent magma. Many examples of such whole-rock isochrons exist in the literature. If the whole rocks used to construct the isochron indeed formed from an isotopically homogeneous parental magma(s) and over a time interval that was short compared to the temporal resolution of the dating system being employed, then the whole-rock isochron can provide a valid estimate of the age of rock formation. On the other hand, given the sometimes large range in initial isotopic variability seen in modern lavas erupted in close spatial association, and the considerable time periods over which some magmatic provinces are constructed, whether whole-rock isochrons provide accurate ages must be examined on a case by case basis.

Variability in the initial isotopic composition of the daughter usually translates to scatter of the data points about an isochron, but some common petrogenetic processes can lead to a series of igneous rocks whose initial isotopic compositions correlate with parent/daughter ratios. Essentially, the isochron for the rocks starts with a nonzero slope that further on in time will result in an age determination that is inaccurately old compared to the true age of rock formation. A good example is seen by comparing the data and discussion in [*McCulloch and Compston*, 1981] with those of [*Chauvel et al.*, 1985] concerning whether the Sm–Nd whole-rock isochron for komatiites from Australia provides the correct age. The most common cause of such inaccuracy is mixing between two isotopically distinct materials. Mixing between two, or more, isotopically distinct, unrelated, components can lead a set of whole-rock samples to produce a linear array on an isochron diagram, but one that has limited, or no, chronological significance. The problem is particularly acute for systems like Sm–Nd or Lu–Hf where the degree of fractionation between parent and daughter during igneous differentiation is small. When constructing whole-rock isochrons, the desire is strong to include samples that will expand the range in parent/daughter ratio in order to improve on the precision with which the isochron slope can be determined. This goal can push in the direction of including samples that may not all derive from simple igneous differentiation of a common, isotopically homogeneous, parent magma, and hence result in a chronologically meaningless mixing line, not an isochron [*Wilson and Carlson*, 1989].

For example, a common process that occurs during magmatic evolution is contamination of a primitive magma by the wall rock through which the magma travels on its way to eruption. Assimilation of the compositionally distinct wall rock can create mixtures that have elemental and isotopic compositions intermediate between the primitive magma and the wall rock being assimilated. The elemental concentrations in a simple two-component mixture are given by the equation:

$$[E]_{Mix} = (X_a \times [E]_a) + (X_b \times [E]_b) \qquad (6.11)$$

where $[E]$ denotes the concentration of element "E" in the two end-member components "a" and "b," and "X" denotes the fraction of the end member in the mixture. In the simple two-component case:

$$X_a + X_b = 1 \qquad (6.12)$$

For the same mixing scenario, the isotopic composition of the mixture [*DePaolo*, 1981], using Nd as an example, can be approximated as:

$$\left(\frac{^{143}Nd}{^{144}Nd}\right)_{Mix} = \left(\frac{^{143}Nd}{^{144}Nd}\right)_b + \frac{X_a \times \left[\left(\frac{^{143}Nd}{^{144}Nd}\right)_a - \left(\frac{^{143}Nd}{^{144}Nd}\right)_b\right]}{\left(X_b \times \frac{[Nd]_b}{[Nd]_a}\right) + X_a} \qquad (6.13)$$

As this equation uses the concentration of the element in the end members, the equation is accurate as long as the isotopic composition of the two end members is not so different as to significantly affect the atom percent of the radiogenic isotope in the element, and hence the atomic mass of the element. This condition is usually met by Nd and Hf, and will remain valid in Sr unless the Sr isotopic composition of the end members is extremely different. For example, if component "a" has a Sr concentration of 300 ppm and a $^{87}Sr/^{86}Sr = 0.703$, while component "b" has a Sr concentration of 50 ppm and $^{87}Sr/^{86}Sr = 1.00$, the $^{87}Sr/^{86}Sr$ of a 50:50 mixture of a:b using the equation above will be 0.1% higher than if the mixture is calculated by first changing the concentrations of each end member to atomic abundance of ^{86}Sr and ^{87}Sr in each component, and then doing the mixture of the isotopes individually.

Consider the case of a zero age mantle-derived basaltic magma that mixes with a 4.4 Ga granite to create a series of rocks that are mixtures between these two end members. For this example, both the granite and the basalt are assumed to derive originally from a mantle source that is evolving with chondritic Sm/Nd ratio. When the granite is created at 4.4 Ga, its Sm–Nd evolution diverges from that of the mantle as a result of its much lower Sm/Nd ratio. Table 6.9 shows the present-day composition of the end members

and mixed rocks produced in this model. The left-hand column shows the ratio of basalt:granite in the mixed magma.

Figure 6.4 shows the isochron diagram constructed using the data in Table 6.9. In this example, the ^{147}Sm–^{143}Nd correlation (Fig. 6.4a) displays perfect linearity and could easily be misinterpreted as an isochron. The slope of the correlation corresponds to an age of 4.4 Ga with an initial $\varepsilon^{143}Nd$ of 0. The zero-age mixed rocks in this example have thus inherited an apparent age from the old granite involved in the mixing relationship.

The point to be learned from this example is that mixing can create linear relationships on isochron diagrams that are not isochrons and hence do not provide meaningful ages. Distinguishing a mixing line from an isochron is not always easy. In the mixing model outlined in Table 6.9, the mixing will produce a perfect linear relationship between the Nd isotopic composition and 1/Nd concentration of the mixtures (Fig. 6.4b), so plots of daughter isotopic composition versus the inverse concentration of that element are a common test for a mixing relationship. If the Nd concentration is then further modified by additional fractionation, for example through fractional crystallization, the perfect 1/Nd versus $^{143}Nd/^{144}Nd$ correlation will be degraded, so the lack of a 1/Nd versus $^{143}Nd/^{144}Nd$ correlation does not always convincingly demonstrate that the isochron is providing a meaningful age, but it helps. When minerals crystallize from a single magma, all the minerals initially will have the same Nd isotopic composition that they inherit from the magma, but they will have Nd concentrations that depend on the partitioning of Nd between the magma and the growing crystals. At the time of formation of such a rock, there thus will be no correlation between Nd isotopic composition and 1/Nd concentration of the minerals. Ingrowth of radiogenic Nd will eventually change the Nd isotopic composition of the minerals, but the ingrowth will correlate with the Sm/Nd ratio, not necessarily 1/Nd concentration. Consequently, the lack of such a correlation can be used to support the interpretation of the isochron as providing a chronologically meaningful age.

6.5 DIVERSE CHRONOLOGICAL APPLICATIONS

The distinct geochemical behavior of the six elements involved in these radioactive geochronometers allow them to be applied to a wide range of rock types and hence decipher the ages of many different geologic processes. The following sections give just a limited example of the diversity of problems to which these systems can be applied. The key is choosing situations that meet the criteria for a suite of samples to form a chronologically significant isochron.

6.5.1 Dating diagenetic minerals in clay-rich sediments

Detrital sediments, in general, are poor dating targets because they consist of a mixture of minerals from different sources formed at different times. As such, they violate the isochron

Table 6.9 Sm–Nd systematics of mixtures between 4.4 Ga granite and modern basalt

Sample	[Sm] ppm	[Nd] ppm	$^{147}Sm/^{144}Nd$	$^{143}Nd/^{144}Nd$
Basalt	3.25	10	0.1964	0.512630
80:20	3.60	14	0.1554	0.511438
60:40	3.95	18	0.1326	0.510775
50:50	4.125	20	0.1247	0.510544
40:60	4.3	22	0.1181	0.510354
20:80	4.65	26	0.1081	0.510062
Granite	5.00	30	0.1007	0.509848

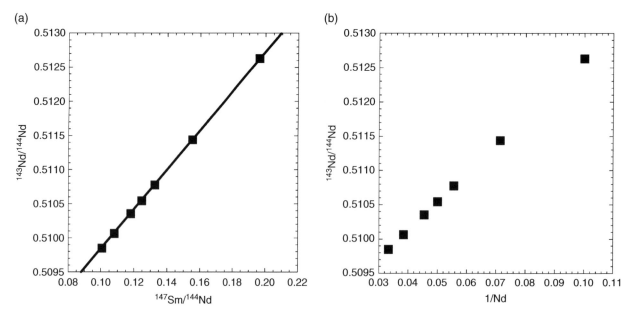

Fig. 6.4. (a) Sm–Nd isochron and a plot of Nd isotopic composition versus (b) 1/Nd concentration for the mixing model outlined in Table 6.9.

requirements of a short formation interval and a constant daughter isotopic composition at the time of formation. One way to get around this problem is to look at very fine-grained fractions of such sediments on the assumption that the fine-grained, clay-rich, component of the sediment forms within the sediment during weathering and diagenesis of the sediment [*Clauer*, 1979]. The formation of these clays involves the growth of a new mineral whose Rb–Sr systematics are reset through interaction with the pore water in the sediment, working in the direction of creating an isotopically homogeneous starting point for the diagenetic minerals. Along with clays, another commonly used diagenetic mineral targeted for both Rb–Sr and K–Ar dating is glauconite, an iron potassium phyllosilicate that can have very high Rb/Sr ratios and thus produce substantial ingrowth of radiogenic [87]Sr that can help to overprint whatever variation might exist in the isotopic composition of the Sr incorporated into the mineral at the time of its growth. Various leaching techniques have also been developed to remove Rb and Sr from loosely bound crystallographic sites within the mineral, leaving behind only the Rb and Sr that was incorporated into strongly bound atomic sites in the diagenetic mineral when the mineral formed [*Morton and Long*, 1980].

Table 6.10 provides an example of this approach applied to clay-rich sediments from Estonia [*Clauer et al.*, 1993]. This study explored the consequences of using different reagents (HCl, NH_4Cl, humic acid, acetone, NH_4-EDTA and equilibration with ion-exchange resin) to treat two different size fractions of the sediment in order to remove components of the sediments that may have formed at different times than the diagenetic minerals. As seen in the concentration data in Table 6.10, the leaches removed only a very small portion of the Rb and Sr from the samples, but in all cases, the leaches removed mostly Sr, not Rb.

To go along with the low Rb/Sr ratios in the material removed by leaching, all the leaches have [87]Sr/[86]Sr ratios much lower than any of the residue. The [87]Sr/[86]Sr ratio of Cambrian seawater was near 0.709, so the leaches likely removed primarily a small amount of carbonate that formed by precipitation from the pore water that was dominated by Cambrian seawater. Lines fit to the residues of the various leach steps (Fig. 6.5) provide ages of 694 Ma for the coarse fraction and 588 Ma for the fine fraction. The age provided by the coarse-grained fraction is older than the deposition age of the sediment, likely because this size fraction includes detrital grains that contribute not the age of sediment deposition, but the age when the detrital grains formed in their igneous protolith. The fine-grained fraction provides an age much closer to the age of sediment deposition likely because the primary constituents of this size fraction are illite clays that grew during the diagenesis of the sediment. If so, the fine-grained clays likely meet the requirement of an isochron to provide an accurate age because they grew over a short time interval and all incorporated the same [87]Sr/[86]Sr from the pore fluid at the time of their growth. This cannot be confirmed by the data, however, so one must always question whether the requirements for an isochron are being met by data sets of this nature. Nevertheless, creative selection of samples, or manipulation of the samples, e.g., through leaching, grain-size separation, or mineral separation can extract useful geochronological information out of samples that at first glance would not be obvious candidates to provide meaningful isochrons.

6.5.2 Direct dating of ore minerals

With the exception of Re–Os dating of molybdenite (covered in Chapter 7) and the obvious case of U and Th ores, very few ore

Table 6.10 Rb–Sr isotope data for the Cambrian Lontova Formation, Estonia

Fraction		[Rb] ppm	[Sr] ppm	$^{87}Rb/^{86}Sr$	$^{87}Sr/^{86}Sr$
Coarse (0.8–2 μm)	Untreated	189.4	128.8	4.264	0.74883
	Residue, HCl leach	172.3	110.8	4.508	0.75213
	HCl leach	0.03	0.13	0.607	0.71689
	Residue, NH$_4$Cl leach	185.4	113.0	4.756	0.75335
	Residue, humic acid leach	174.2	125.1	4.036	0.74669
	Residue, resin extract	196.7	84.0	6.787	0.77300
	Resin extract	0.01	0.26	0.125	0.71201
	Residue, NH4-EDTA leach	179.0	68.8	7.546	0.78142
	NH4-EDTA leach	0.004	0.07	0.162	0.71201
Fine (<0.4 μm)	Untreated	180.5	87.8	5.954	0.75590
	Residue, HCl leach	181.9	42.5	12.41	0.80558
	HCl leach	0.04	0.72	0.169	0.71340
	Residue, acetone leach	197.7	89.4	6.408	0.75622
	Residue, resin extract	178.0	57.3	9.010	0.78160
	Resin extract	0.002	0.05	0.087	0.71307
	Residue, NH4-EDTA leach	189.6	52.8	10.42	0.79535

Source: Adapted from *Clauer et al.* [1993].

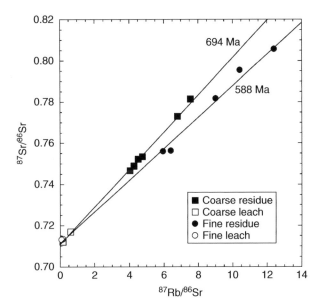

Fig. 6.5. Rb–Sr isochron diagram for the data given in Table 6.10. The lines in the figure are fit only to the data for the residues of leaching. The extrapolation of these lines to the abscissa, however, passes just slightly below the data for the leaches.

Table 6.11 Rb–Sr data for sphalerites and their fluid inclusions from the Polaris deposit

Sample	$^{87}Rb/^{86}Sr$	$^{87}Sr/^{86}Sr$	Error	Age[a] (Ma)
01646 Sphalerite	0.654	0.711973	0.000028	326
01646 Inclusion	0.044	0.709190	0.000034	
01648 Sphalerite	0.282	0.710061	0.000027	309
01648 Inclusion	0.040	0.709015	0.000025	
01649 Sphalerite	0.217	0.709735	0.000024	317
01649 Inclusion	0.032	0.708913	0.000028	
01650 Sphalerite	0.836	0.712996	0.000029	
01651 Sphalerite	0.394	0.710754	0.000024	358
01651 Inclusion	0.023	0.708893	0.000028	
01652 Sphalerite	0.495	0.711299	0.000027	362
01652 Inclusion	0.027	0.708925	0.000026	
01653 Sphalerite	0.581	0.711668	0.000026	355
01653 Inclusion	0.050	0.709025	0.000031	

Source: *Christensen et al.* [1995]. Reproduced with permission of Elsevier.
[a] The age calculated for the tie-line connecting the sphalerite-inclusion data for each sample.

minerals contain enough of the long-lived radioactive species to be dated directly. Nevertheless, there are clearly strong reasons to want to decipher the timing and sequence of events that produce economic ore deposits. One ore mineral that has been used in this quest is sphalerite (ZnS), where the Rb–Sr system has been shown to be an effective dating tool. The question with sphalerite dating, however, is whether the Rb and Sr reside in the mineral itself, or in fluid inclusions trapped during the growth of the sphalerite. The example shown in Table 6.11 and Fig. 6.6 is an attempt to address this question. This study examined sphalerites from the Polaris Zn–Pb ore deposit in the Canadian Arctic [*Christensen et al.*, 1995]. Sphalerite grains were separated from

various areas within the ore deposit. They were cleaned of any surface coating of carbonate by leaching in acetic acid. The cleaned sphalerite grains were then crushed to powder under water in order to separate the mineral from its fluid inclusions on the assumption that the fluid inclusions liberated by crushing ended up in the water. For this experiment, each combination of sphalerite and its fluid inclusions can be used to construct a two-point isochron, which range in age from 309 to 362 Ma. The isochron fit to just the sphalerite analyses (Fig. 6.6a) provides an age of 374 ± 21 Ma and initial $^{87}Sr/^{86}Sr$ of 0.70863 ± 0.00016. The isochron diagram of Fig. 6.6b shows the inclusion data to scatter slightly above the line defined just by the sphalerites, which explains why the two-point tie lines between the sphalerites and their inclusions provide slightly younger ages than the line fit to all the sphalerite data. Although the inclusion data are characterized by low Rb/Sr ratios, and a small range in Rb/Sr ratios, with the exception of the one inclusion with the highest measured $^{87}Sr/^{86}Sr$, the inclusion data on their own provide an

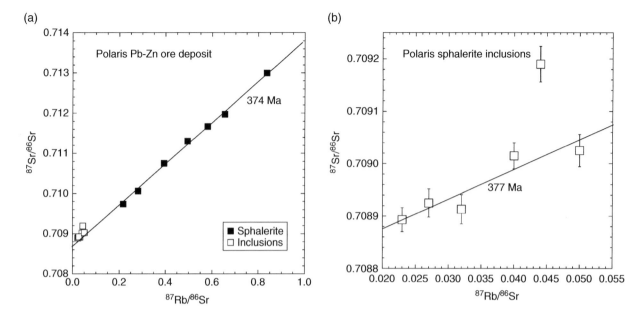

Fig. 6.6. (a) Rb–Sr data for sphalerites (filled symbols) and their fluid inclusions (open symbols) from the Polaris Pb-Zn deposit, Canada. (b) An expanded scale to better illustrate the data distribution for the fluid inclusion data.

isochron whose slope corresponds to an age of 377 ± 160 Ma with an initial $^{87}Sr/^{86}Sr = 0.70877 \pm 0.00008$. Given the uncertainties on the ages and initial isotopic compositions of the two isochrons shown in Fig. 6.6, the data support the interpretation that the sphalerites and their inclusions formed at the same time, from a fluid with the same Sr isotopic composition. The scatter of the two-point sphalerite-inclusion tie lines to younger ages suggests that the sphalerites, or their inclusions, continued to exchange Sr with their surroundings for a period of up to many tens of million years after the 374 Ma formation age of the ore deposit.

6.5.3 Dating of mineral growth in magma chambers

Silica-rich magmas, such as rhyolites, can develop very high Rb/Sr ratios, and can crystallize minerals with both high (e.g., mica) and low (e.g., feldspars) Rb/Sr ratios. The extreme Rb/Sr ratios found in some evolved lavas potentially can allow very high temporal resolution. Isochrons based on minerals crystallizing from a magma were mentioned earlier in the chapter as one of the likely approaches to ensure that the minerals formed at the same time, and with the same initial daughter isotopic composition—two of the requirements for an isochron to provide a meaningful age. At very high temporal resolution, however, one may be able to resolve the time it takes for crystals to grow in a magma chamber. This approach is now common in high-resolution studies of zircon U–Pb ages in granites [*Schoene et al.*, 2012], but started with Rb–Sr applied to silicic magmas, an example of which is provided in Table 6.12. In this study [*Christensen and DePaolo*, 1993], glass and sanidine separates were made from samples of the 0.74 Ma old Bishop Tuff eruption in southern California. The data can be seen to scatter considerably on the Rb–Sr isochron

Table 6.12 Rb–Sr data for glass–sanidine pairs separated from the Bishop Tuff

Sample	Material	$^{87}Rb/^{86}Sr$	$^{87}Sr/^{86}Sr$	Glass–sanidine age (Ma)
BT6	Glass	127.8	0.70976	1.84
	Sanidine	14.53	0.70684	
BT39	Glass	275.0	0.70968	0.82
	Sanidine	16.40	0.70670	
BT27	Glass	16.60	0.70627	1.44
	Sanidine	0.97	0.70596	
BT24	Glass	6.28	0.70662	7.21
	Sanidine	0.42	0.70603	

Source: *Christensen and DePaolo* [1993]. Reproduced with permission of Springer.

diagram of Fig. 6.7. Two-point tie lines between the data for the glass and sanidine separates from the same sample provide ages that range from 0.82 to 7.21 Ma. Fitting just the sanidine data in Fig. 6.7 provides a line of slope corresponding to an age of 3.7 Ma. The data for the glass separates scatter widely on the isochron diagram of Fig. 6.7, but a best fit line through the scatter corresponds to an age of 0.94 Ma.

All of these ages are older than the known 0.74 Ma eruption age of the Bishop tuff. Why? Remember that the requirements for an isochron to provide an accurate age include instantaneous (at least relative to the chronological resolution of the system) formation of all the samples used to define the isochron and that all the samples initially had the same daughter isotopic composition. The dispersion in apparent ages for this data set thus allow a number of explanations.

(1) The observation that all the sanidine-glass tie lines for each sample correspond to ages older than the eruption age of the tuff might indicate that the sanidines crystalized in the magma chamber long before the actual eruption.

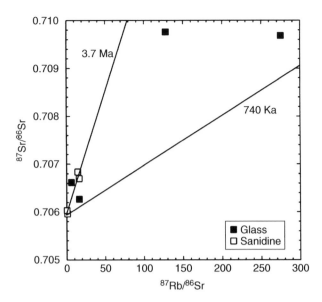

Fig. 6.7. Rb–Sr isochron diagram for the data shown in Table 6.12.

(2) The 3.7 Ma isochron fit to the sanidine data alone might suggest that the sanidines are not related to the magma (glass), but instead crystallized from an older magma and were simply picked up as xenocrysts (unrelated crystals) by the 0.74 Ma eruption.

(3) The scatter, and pre-eruption, age provided by the glass separates alone may suggest that the magma itself was not isotopically homogeneous at the time of eruption. A process that could lead to such initial isotopic heterogeneity would be mixing with older magmas, or their crystallization products, as might be reflected by the sanidines.

This case provides a good example of the complications that can enter into geochronological studies, particularly at high temporal resolution, in a rock that has experienced a complex history of pre-eruption crystallization, mixing with older wall-rock, and incorporation of an unrelated cargo of crystals derived from older magmatic products.

6.5.4 Garnet Sm–Nd and Lu–Hf dating

While the Rb–Sr system often experiences considerable fractionation of parent from daughter element during a variety of natural processes, few processes are as effective at separating Sm from Nd and Lu from Hf. A clear exception is fractionation involving the mineral garnet. The crystal structure of garnet can more easily accommodate the heavy REEs than lighter REEs and Hf (e.g., Table 6.5). Consequently, garnet grains often have very high Sm/Nd and Lu/Hf ratios. While the high Sm/Nd and Lu/Hf ratios of garnet allow it to provide very high temporal resolution in these systems, there are at least three complications in using garnet crystals for precise Sm–Nd and Lu–Hf chronology. These include:

(1) Garnet crystals often contain numerous mineral inclusions, some of which can have high Hf contents and low Lu/Hf

ratios (e.g., rutile, zircon), and others with high Nd concentrations and moderate Sm/Nd ratios (e.g., apatite, monazite). These inclusions often can be removed by leaching crushed garnet grains in various acids [*Pollington and Baxter*, 2011], although there is some concern that harsh acid treatments may modify the parent/daughter ratios in the garnet crystals remaining after such severe leaching.

(2) Garnet commonly is a metamorphic mineral that grows within a rock by subsolidus recrystallization of existing minerals. As with the previous example of the growth of diagenetic clays in shales, the initial isotopic composition in the garnet crystals formed through such a process is thus a function of the other minerals and fluids that contribute to their growth. If the parent/daughter ratios in the garnet grains are sufficiently large, and the garnet crystals are old enough, the amount of radiogenic ingrowth in the garnet crystals may be sufficiently large to overwhelm the variability in the isotopic composition of the daughter element incorporated into the garnet crystal at the time of its growth. In younger garnet grains, and those garnet grains with only moderate Sm/Nd and Lu/Hf ratios, initial isotope ratio variation may compromise the chronological accuracy.

(3) Garnet often forms at high enough metamorphic temperatures that its Sm–Nd and Lu–Hf systems are close to, or above, the closure temperature of the radioactive systems for at least a portion of the history of the rock. The ages obtained under such circumstances thus may not reflect the initial garnet growth, but some age between the time of crystal growth and the time when the rock cooled below the closer temperature of the radiometric system in use. Interpretation of the ages in such circumstances is further complicated by the fact that diffusion of the daughter Hf may be slower than the parent Lu, and the fact that the concept of a closure temperature is not simple for a radiometric system that starts with a lot of the daughter element [*Ganguly*, 2010].

Table 6.13 provides an example of garnet Lu–Hf and Sm–Nd dating applied to a high-grade metamorphic rock from the Archean Pikwitonei Domain in Manitoba, Canada [*Smit et al.*, 2013]. This study investigated the effect of grain size on Sm–Nd and Lu–Hf age dating of garnet crystals, as one would expect smaller grains to have lower closure temperatures for their isotope chronometers because of the shorter diffusion distances needed to reach the borders of the garnet grain. There are a number of ways to obtain ages from data sets like that shown in Table 6.13. One can fit an isochron to all of the garnet mineral separates. Doing so provides a Lu–Hf age of 2704 ± 12 Ma and a Sm–Nd age of 2601 ± 15 Ma.

One can also calculate ages from the two-point tie line that connects the garnet grain data with the whole-rock data on the isochron diagram. Using the whole rock data allows correction for the isotopic composition of the Hf or Nd incorporated into the garnet crystal at the time of its growth. Figure 6.8 shows that the Sm–Nd garnet grain ages calculated in this manner are

Table 6.13 Lu–Hf and Sm–Nd data for whole rocks and garnets from the Pikwitonei Domain, Canada

Sample	^{176}Lu/^{177}Hf	^{176}Hf/^{177}Hf	Tie-line age[a] (Ma)	^{147}Sm/^{144}Nd	^{143}Nd/^{144}Nd	Tie-line age[a] (Ma)
Whole Rock	0.002486	0.281197		0.09932	0.510873	
0.35–0.55 mm:						
A-2	0.0276	0.282475	2659	0.2276	0.513085	2614
A-3	0.03912	0.283085	2692	0.4521	0.516934	2604
0.75–0.90 mm:						
C-1	0.09026	0.285723	2693	0.1468	0.511702	2647
C-2	0.11224	0.286881	2704	0.2026	0.512649	2607
2.0–2.3 mm						
G-1	0.05281	0.283812	2713	0.1305	0.511433	2722
G-2	0.02314	0.282248	2658	0.1215	0.511257	2625

Source: *Smit et al.* [2013]. Reproduced with permission of Elsevier.
[a] Ages calculated for the two-point line connecting the garnet and bulk-rock data.

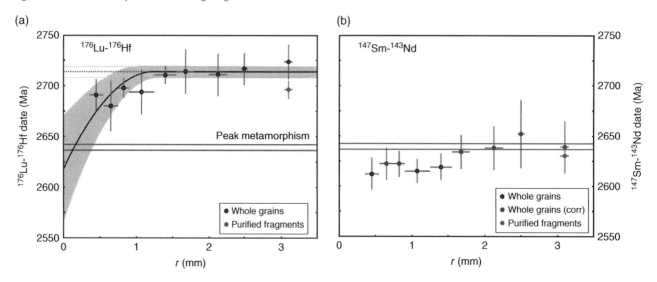

Fig. 6.8. Two-point garnet–whole-rock tie line ages plotted against grain size for the data provided in Table 6.13. (Source: *Smit et al.* [2013]. Reproduced with permission of Elsevier.)

younger than the Lu–Hf garnet-WR ages, and that the ages in both systems show a slight, barely resolved, younging in the ages as the garnet grains get smaller. The younger ages for smaller grain sizes indicate some loss of the radiogenic daughter product, as diffusive loss of the parent element would translate into older, not younger, model ages. This is somewhat unexpected as the diffusion of Lu in garnet grains is much faster than that of Hf [*Ganguly*, 2010], but the concentration of Lu in the garnet grains is maintained by the distribution coefficient of Lu between the garnet grains and their surrounding minerals. Consequently, although Lu likely is diffusively exchanged between the garnet grains and their surroundings, as much Lu diffuses into the garnet grains as diffuses out of them. Since the isotopic composition of the Lu is the same in both the garnet grains and the minerals that surround them, this diffusion is of no consequence for the Lu–Hf ages. In contrast, the Hf in the minerals surrounding the garnet grains is much less radiogenic than the Hf in the garnet grain, so any diffusive exchange of Hf between the garnet grains and their surrounding minerals will result in a reduction in ^{176}Hf/^{177}Hf compared to that which would have evolved in the garnet grain in the absence of diffusion. Even if the Lu/Hf

ratio of the garnet grain is not affected by diffusion at temperatures above the Lu–Hf closure temperature, such diffusion will result in lower garnet grain ages because of the diffusive lowering of the ^{176}Hf/^{177}Hf ratio in the garnet grain.

What is more obvious in Fig. 6.8 than the age variation with grain size is that the Lu–Hf data provide older ages than do the Sm–Nd data. The interpretation provided by these authors is that the closure temperature of the Sm–Nd system in garnet grains is some 150 °C lower than the >900 °C closure temperature for the Lu–Hf system in garnet (Fig. 6.9). The peak metamorphic temperature determined for these rocks is about 760 °C [*Mezger et al.*, 1990] leading to the interpretation that the Lu–Hf ages for these garnet grains date their growth, with only the smallest grains showing some lowering of their age due to diffusional exchange of Hf in the garnet grains with surrounding minerals. In contrast, the Sm–Nd ages date the time of cooling of the rock below the closure temperature of the Sm–Nd system. This explanation also would explain the minimal variation in the individual garnet–whole-rock Lu–Hf ages compared to the larger range in the Sm–Nd ages. For Lu–Hf, the rock was always below the closure temperature of the Lu–Hf system in garnet, so one

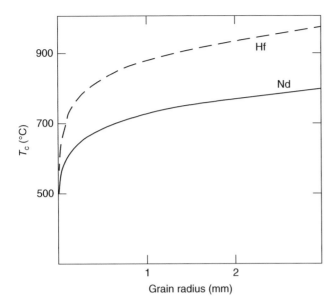

Fig. 6.9. Closure temperatures for the Lu–Hf and Sm–Nd systems in garnet grains versus grain size calculated for a cooling rate of 2 °C/Ma and an activation energy of 250 kJ/mole. (Source: *Smit et al.* [2013]. Reproduced with permission of Elsevier.)

would expect very little variability in ages dependent on grain size. Basically, the grains begin keeping time in the Lu–Hf system as soon as metamorphic conditions reach the point of allowing garnet grains to form. For Sm–Nd, however, the peak metamorphic temperature is close to the closure temperature of the system, so one would expect to see some open-system behavior as the radiogenic Nd being produced in the garnet grains diffuses out, to exchange with less radiogenic Nd from the surrounding minerals in the whole rock.

6.6   MODEL AGES

Besides the isochron method of age calculation, model age approaches have found considerable use in most radiometric systems. Model age approaches calculate the time when the isotopic composition of a sample began to diverge from the isotopic evolution of the model system because some event changed the parent/daughter ratio from that of the model system. We have already used this concept in the garnet grain–whole-rock tie line ages from the example in Table 6.13 by calculating the time at which the isotopic evolution of the garnet diverged from that of the host whole rock.

Model ages for the Rb–Sr, Sm–Nd and Lu–Hf systems are calculated using the equations:

$$T_{Model} = \frac{1}{\lambda_{Rb}} \times \ln\left[\left[\frac{\left(\frac{^{87}Sr}{^{86}Sr}\right)_{Model} - \left(\frac{^{87}Sr}{^{86}Sr}\right)_{Sample}}{\left(\frac{^{87}Rb}{^{86}Sr}\right)_{Model} - \left(\frac{^{87}Rb}{^{86}Sr}\right)_{Sample}}\right] + 1\right]$$

(6.14)

$$T_{Model} = \frac{1}{\lambda_{Sm}} \times \ln\left[\left[\frac{\left(\frac{^{143}Nd}{^{144}Nd}\right)_{Model} - \left(\frac{^{143}Nd}{^{144}Nd}\right)_{Sample}}{\left(\frac{^{147}Sm}{^{144}Nd}\right)_{Model} - \left(\frac{^{147}Sm}{^{144}Nd}\right)_{Sample}}\right] + 1\right]$$

(6.15)

$$T_{Model} = \frac{1}{\lambda_{Lu}} \times \ln\left[\left[\frac{\left(\frac{^{176}Hf}{^{177}Hf}\right)_{Model} - \left(\frac{^{176}Hf}{^{177}Hf}\right)_{Sample}}{\left(\frac{^{176}Lu}{^{177}Hf}\right)_{Model} - \left(\frac{^{176}Lu}{^{177}Hf}\right)_{Sample}}\right] + 1\right]$$

(6.16)

where λ_{Rb}, λ_{Sm}, and λ_{Lu} are the decay constants of ^{87}Rb, ^{147}Sm, and ^{176}Lu, respectively. The model parameters used in the equations above depend on the assumption of which isotopic evolution the sample, or its source region, was following prior to its formation. In the example provided in Table 6.13, the "model" reservoir was the whole rock and the "sample" was the individual garnet grain. The model age approach is flexible because you can adjust the "model" and "sample" parameters to address a wide variety of questions of potential interest. For example, if you want to calculate when the incompatible-element-depleted source of MORBs became incompatible-element depleted, you could use the composition of a modern MORB (sample) in comparison to the expected isotopic evolution of the undifferentiated Earth's mantle (model), commonly called BSE, or alternatively the "chondritic uniform reservoir" or CHUR reservoir, which is the same as BSE for Sm–Nd and Lu–Hf, but not Rb–Sr. For example, if you want to know when a continental granite was produced by melting of the modern, incompatible-element-depleted mantle, you would compare the granite isotopic composition with a model for the isotopic evolution of the depleted mantle (see Box 6.2).

This latter model evolution is depicted in Fig. 6.10 in both isotope ratio (left panel) and ε notation (right panel). Earth started with some value of $^{143}Nd/^{144}Nd$ and $^{176}Hf/^{177}Hf$ when the planet formed at 4.567 Ga. These ratios increase through time in proportion to the Sm/Nd and Lu/Hf ratio of the Earth. If the parent/daughter ratios are equal to those of the BSE, then the isotopic evolution follows the BSE lines in Fig. 6.10. In a hypothetical reservoir with higher Sm/Nd and Lu/Hf ratios, for example, the incompatible-element-depleted mantle that melts to produce modern MORBs, the isotopic composition will increase more rapidly, following the depleted mantle, "DM," lines in Fig. 6.10. In the models depicted in Fig. 6.10, the present-day isotopic composition of the granite is measured and then its daughter isotopic composition is extrapolated back in time, as shown by the dashed lines, until it intersects the evolution of either the BSE or DM evolution lines. In this example, the T_{BSE} of the granite is 1.8 Ga and the T_{DM} model age is about 2.3 Ga. Which is more accurate? That cannot be determined from these data alone, but requires some other argument that supports the

Box 6.2   Choosing the right model

A critical feature of model ages is that the accuracy of the age depends on the accuracy of the isotopic evolution assumed for the model reservoir. For example, the line shown in Fig. 6.11 for the Nd and Hf isotopic evolution of the depleted mantle is the simple connector line between the average value of the Nd and Hf isotopic composition of modern mid-ocean-ridge basalts (MORB) and the starting isotopic composition of Earth. Figure 6. B2.1 shows that mafic rocks, presumably derived by melting of the mantle, show a wide range in initial Nd isotopic composition throughout Earth history. Many of these rocks have positive initial ε_{Nd}, which means that their mantle source regions must have had a time-averaged Sm/Nd ratio higher than the chondritic ratio that defines the ε_{Nd} evolution of the BSE or CHUR reservoir, where $\varepsilon_{Nd} = 0$ by definition. How the depleted mantle obtained its superchondritic Sm/Nd ratio is a long-running question with major implications for the chemical differentiation of the planet. The continental crust is strongly enriched in incompatible elements compared to any estimate of the composition of the mantle. Even though the continental crust is small in volume, constituting only about 0.5% the mass of the mantle, its strong enrichment in incompatible elements means that its extraction from the mantle must have created a chemically complimentary portion of the mantle that is depleted in

incompatible elements [*Hofmann*, 1988]. This mantle is the depleted mantle. Its depletion in incompatible elements means that it is characterized by superchondritic Sm/Nd and Lu/Hf ratios, and hence positive ε_{Nd} and ε_{Hf} (Fig. 6.B2.1).

In this model for the genesis of the depleted mantle, the parent/daughter ratios, and hence isotopic evolution, of the depleted mantle through time depend on the rate of continent formation, the rate of continental recycling back into the mantle, and the volume of mantle affected by continent extraction. The straight isotopic evolution line for the depleted mantle shown in Fig. 6.11 makes the implicit assumption that the depleted mantle formed in a single step that accompanied Earth formation. More complicated, and likely more accurate, models for the compositional evolution of the depleted mantle assume various rates of continent extraction from some volume of mantle to predict the isotopic evolution of the depleted mantle [*DePaolo*, 1980]. These models predict isotope evolution curves for the mantle that are not straight lines because the Sm/Nd ratio for the depleted mantle changes with time. Figure B6.2 shows examples of depleted mantle evolution curves that would be expected for a constant rate of continent formation, or one that assumes continent formation slows exponentially over Earth history. While these models provide geologically reasonable explanations for the creation of depleted mantle, the comparison of these evolution curves with the actual initial isotopic data for mantle-derived rocks (Fig. 6. B2.1) reveals so much scatter in the rock data that no particular model is strongly supported. Which of these models best represents the true evolution of Earth's mantle will remain a research topic for quite some time. Until this question is resolved, however, the accuracy of mantle model ages will be impacted by the uncertainty in the isotopic evolution of the model depleted mantle. When using model ages, one must always keep in mind the possibility that the model employed may, or may not, accurately reflect the history of the sample or the source materials from which it formed.

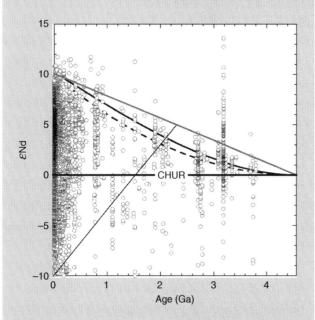

Fig. 6.B2.1. Initial Nd isotopic composition of mafic rocks (MgO > 7 wt%) versus time compared to models for the isotopic evolution of the mantle. The line labeled "CHUR" shows the Nd isotope evolution of a mantle source with chondritic Sm/Nd ratio. The dashed curve shows a volume of mantle from which the continental crust is extracted at a rate of 0.35% of its current volume every 100 Ma. The solid curve assumes that the rate of continent extraction from the mantle falls off exponentially with time, whereas the gray line assumes a constant, superchondritic, Sm/Nd for the depleted mantle since Earth formation. The thin, positively sloped line is a schematic representation of the hypothetical isotopic evolution of a zero age granite. This line provides model ages for the source of the granite that range from 1.5 Ga (CHUR) to 2 Ga (constant rate crust formation) to 2.1 Ga (exponentially declining rate of continent formation) to 2.3 Ga (instantaneous continent formation at 4.567 Ga) to illustrate the importance of the assumed model evolution to the calculated model age. (Source: Data from EarthChem.)

derivation of the granite by melting of a source with either BSE or DM isotope characteristics.

6.6.1   Model ages for volatile depletion

The model age approach depicted in Fig. 6.10 involved estimating when a given sample was produced by melting of the mantle, but similar model age approaches can be used to track a wide range of potential processes involved in creating a given sample. Take, for example, estimating the time when Earth acquired its depletion in volatile elements compared to the Sun. Figure 6.11 shows the initial Sr isotopic compositions of a variety of materials. The initial $^{87}Sr/^{86}Sr$ of the solar system (0.698975 [*Hans et al.*, 2013]) is estimated from the initial $^{87}Sr/^{86}Sr$ determined for the calcium–aluminum-rich inclusions in primitive meteorites that are believed to be the first solids formed in the solar system. Because of the high Rb/Sr ratio in the Sun, as represented by Cl chondrites, the $^{87}Sr/^{86}Sr$ of the volatile-rich solar system rises rapidly (Fig. 6.11). One type of volatile-depleted igneous meteorite, the angrites, has extremely low initial $^{87}Sr/^{86}Sr$ (0.698978 [*Sanborn et al.*, 2015]). The lowest initial $^{87}Sr/^{86}Sr$ measured in a lunar sample is 0.698985 [*Borg et al.*, 2011] and the lowest $^{87}Sr/^{86}Sr$ for a terrestrial rock is the calculated initial of 0.70048 for pyroxenes separated from 3.45 Ga komatiites from the Barberton greenstone belt in South Africa [*Jahn and Shih*, 1974].

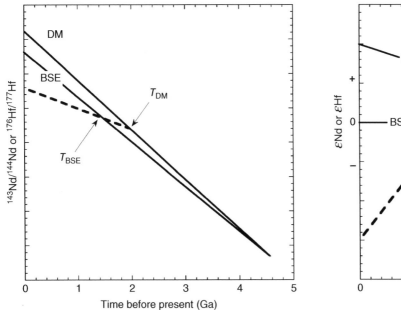

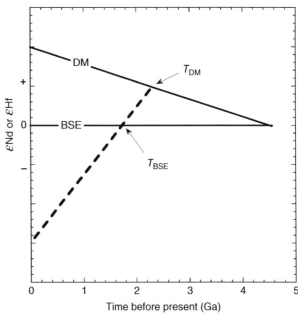

Fig. 6.10. Schematic isotope evolution diagrams illustrating the model age approach. The solid lines show the isotopic evolution of a bulk-silicate-Earth (BSE) and depleted-mantle (DM) model reservoir. The dotted line starts at the measured isotopic composition of a granite and is then extrapolated back in time using the measured parent/daughter ratio of the granite to determine the slope. The point of intersection of the dotted line with the evolution of the model reservoirs gives the model age.

One can use these data to calculate the minimum time when the Rb/Sr ratio of the source of these rocks became lower than solar by assuming that the Rb/Sr ratio of the source was zero, which allows the initial $^{87}Sr/^{86}Sr$ of the sample to be extrapolated horizontally in Fig. 6.11 until it intersects the $^{87}Sr/^{86}Sr$ growth line of the CI chondrite. For the Barberton komatiite, this point of intersection is 4.46 Ga (Fig. 6.11). For the angrite and lunar sample, their $^{87}Sr/^{86}Sr$ ratios are reached in less than 4 Ma after the 4.567 Ga age of the solar system in a material evolving with Solar Rb/Sr ratio (Fig. 6.11). If the source materials of the Barberton komatiite had nonzero Rb/Sr ratios between 4.567 Ga and 3.45 Ga, as is likely, then the initial $^{87}Sr/^{86}Sr$ of the komatiite would extrapolate back in time with a negative slope, intersecting the CI chondrite evolution curve at some age older than 4.46 Ga. These model ages are thus minimum estimates of the time when the source of the samples obtained their low Rb/Sr ratios. For example, to reach the $^{87}Sr/^{86}Sr$ of the Barberton komatiite from the solar system initial $^{87}Sr/^{86}Sr$ in the time period from 4.567 Ga to 3.45 Ga would require a $^{87}Rb/^{86}Sr$ ratio of 0.0956. If this ratio were maintained over all of Earth history, the Earth today would have a $^{87}Sr/^{86}Sr$ ratio equal to 0.70508, not far from the value estimated for that of the modern BSE. Thus, Earth, and the materials that accumulated to make up Earth, must have become volatile depleted compared to solar very early in solar system history. The model age does not precisely define the time of volatile depletion of Earth, because it depends on the assumed Rb/Sr ratio of the source of the

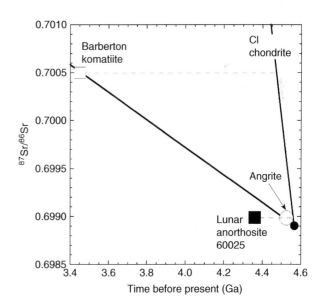

Fig. 6.11. Sr isotope evolution of the Sun, as represented by measured values in a type of primitive meteorite (CI chondrite) whose composition is similar to that of the Sun, at least for these elements. The data points show the initial $^{87}Sr/^{86}Sr$ measured in igneous pyroxenes separated from a 3.45 Ga Barberton komatiite flow, and calculated from isochrons for a lunar anorthosite and a volatile-depleted angrite meteorite. The dotted lines show the extrapolation of the initial Sr isotopic compositions horizontally until they intersect the $^{87}Sr/^{86}Sr$ growth line of the CI chondrite. This point of intersection gives the Rb–Sr model age of the samples.

komatiites. Nevertheless, this model age approach shows conclusively that the Rb/Sr ratio of Earth must have become much lower than solar by at least 4.46 Ga, and likely much earlier, which represents a first-order conclusion about the timing of Earth's volatile depletion.

6.6.2 Model ages for multistage source evolution

As an example of a common use of model ages in these systems, consider the case of granites from China where the data for one sample measured by [*Liu et al.*, 2009] are provided in Table 6.14. These granites contain zircons that provide U–Pb ages of 500 Ma that likely date the formation age of the granite. Figure 6.12 , however, shows that the data provide very different model ages from the different systems, with T_{BSE} model ages of 601 Ma, 1077 Ma and 2240 Ma for Rb–Sr, Lu–Hf, and Sm–Nd, respectively. All are older than the 500 Ma crystallization age of the granite, which shows that this granite was not made by the direct partial melting of a BSE mantle, but the discrepancy between the ages points to a more complicated evolution than considered by the model age approach.

The model age calculation assumes only two stages to the compositional evolution of the sample. The first stage follows the isotopic evolution of the model reservoir, and the second stage follows an isotopic evolution dictated by the measured parent/daughter ratio and isotopic composition of the sample. In the simple case where a given sample forms directly by partial melting of the mantle reservoir, the model age will accurately date the time of formation of the sample, but also requires that the isotopic composition of the sample be the same as that of the model reservoir at the time of its formation. The fact that the isotopic compositions of the granite at the time of its formation in the example here are not the same as any of the model reservoirs shows that the two-stage model does not accurately describe the history of this sample and its source materials. The history of the granite almost certainly involved a 500 Ma interval from the time of its crystallization to the present day when it was evolving with the measured parent/daughter ratios to end up with

the measured isotopic compositions. The fact that at 500 Ma the sample did not have BSE or DM isotopic composition implies that the granite was not generated by partial melting of a source material that had followed either the BSE or DM isotopic evolution path. This indicates the need to add at least one more stage to the evolution of this sample—an intermediate stage where the isotopic evolution of the source material of the granite diverged from that of the model mantle reservoir. An example of such an evolution would be generation of the granite by the melting of crustal rock that may have been extracted from the BSE or DM reservoir long before the event that melted the crustal rock to make the granite.

Introducing this third-stage to the evolution of the sample adds the unknowns of the parent/daughter ratio during this intermediate stage, and the duration of that stage. Without additional information about the sample, there is no way to obtain the necessary parameters to definitely model the isotopic evolution of this intermediate stage, but there are ways to approximate that evolution. For example, because Sm and Nd are neighboring REEs and thus share similar chemical behaviors, the amount that Sm and Nd are fractionated from one another by partial melting is much less than, for example, Rb from Sr. The Rb–Sr model age for the granite in the example above is only somewhat older than the 500 Ma formation age of the granite. The granite, however, has an extremely high Rb/Sr ratio that quite likely reflects the fact that Rb was strongly incompatible during the partial melting event that made the granite, but Sr may have been moderately compatible. The more the measured parent/daughter ratio of the sample is different from that of its source material, the farther the model age will be from the true age of source formation. Given the more limited Sm–Nd fractionation expected during granite formation, we can start with the assumption that the Sm–Nd model age for the sample is the better approximation for the true age of the source. The Rb/Sr ratio of the source materials of the granite can then be calculated by using the difference in Sr isotopic composition of the sample at 500 Ma and that of the BSE mantle reservoir at the 2240 Ma Sm–Nd BSE model age. Plugging these values into the Rb–Sr model age equation, we obtain:

$$2240 - 500 \, \text{Ma} = \frac{1}{\lambda_{Rb}} \times \ln \left[\frac{0.70427 - 0.71490}{0.0906 - \left(\frac{^{87}Rb}{^{86}Sr} \right)_{Sample}} + 1 \right]$$

(6.17)

Solving this equation suggests that the source materials of the granite may have had a $^{87}Rb/^{86}Sr$ ratio equal to 0.523, a value typical of many basaltic rocks that might have served as the source material that when melted produced the PH5-4 granite. The three-stage history of the granite is illustrated in Fig. 6.12a.

This study also involved determination of the Hf isotopic composition of zircons contained in the granites. This approach is becoming more common because of the relative ease of measuring the Hf isotopic composition of the high Hf content zircons

Table 6.14 Chemical and isotopic results for granites from southwest China

	PH5-4	PDX01-1 zircon	T_{BSE} Ma	T_{DM} Ma
$^{87}Rb/^{86}Sr$	7.57			
$^{87}Sr/^{86}Sr$	0.76796		601	962
$^{87}Sr/^{86}Sr$ (500 Ma)	0.71490			
$^{147}Sm/^{144}Nd$	0.1532			
$^{143}Nd/^{144}Nd$	0.511998		2241	2955
ε_{Nd}	−12.3			
$^{143}Nd/^{144}Nd$ (500 Ma)	0.511496			
ε_{Nd}(500 Ma)	−9.6			
$^{176}Lu/^{177}Hf$	0.0323	0.001391		
$^{176}Hf/^{177}Hf$		0.282131	1077	1618
ε_{Hf}		−23.1		
$^{176}Hf/^{177}Hf$ (500 Ma)		0.282118		
ε_{Hf} (500 Ma)		−12.5		

Source: *Liu et al.* [2009]. Reproduced with permission of Elsevier.

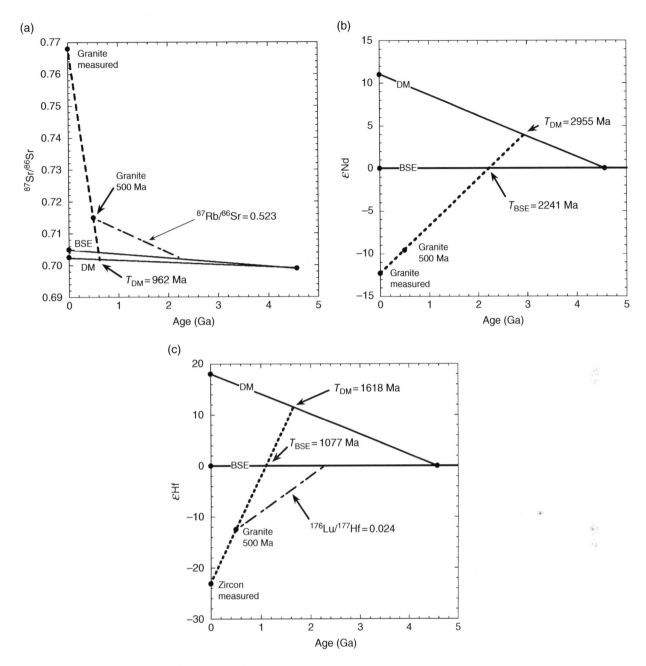

Fig. 6.12. Sr, Nd, and Hf isotope evolution diagrams for the data from Table 6.14. The solid lines show the isotopic evolution of the BSE and DM mantle reservoirs. The thick dashed line shows the isotopic evolution of the sample based on the measured parent/daughter ratio. The dot-dash lines in (a) and (c) show the evolution needed to bring the Rb–Sr and Lu–Hf model ages into agreement with the Sm–Nd model age.

using laser-ablation ICP-MS, rather than going to the effort of making the difficult Lu–Hf measurements in whole rocks. Because of the very low Lu/Hf ratio in the zircons, the measured Hf isotopic composition needs only limited correction for the ingrowth of radiogenic Hf that occurred after the formation of the zircon. Using the measured Lu/Hf ratio and Hf isotopic composition of the zircon in the model age equation results in model ages between 1.0 and 1.6 Ga (Fig. 6.12c). These model ages almost certainly underestimate the true age of source

formation because they use the very low Lu/Hf ratio of the zircon, not the higher Lu/Hf ratio of the whole rock. While the zircon data can provide a good estimate of the Hf isotopic composition of the granite at the time of its formation, calculating the model age would use the whole-rock Lu/Hf ratio, not the ratio in the zircon. In this particular example, however, and somewhat surprisingly, the Lu/Hf ratio of the granite is quite close to that of the model mantle reservoir. Consequently, using the whole-rock Lu/Hf ratio creates Hf isotope evolution lines that are nearly parallel to the mantle evolution lines, with the consequence that the model ages are substantially older than the age of the Earth. As in the case for Rb–Sr, the likely explanation is that the whole-rock Lu/Hf ratio is different from that of the source of the granite. As was done with Rb–Sr, we can use the Sm–Nd model age to estimate the Lu/Hf ratio of the source of the granite.

$$2,240 - 500\,\text{Ma} = \frac{1}{\lambda_{Lu}} \times \ln\left[\frac{0.28247 - 0.282118}{0.0347 - \left(\frac{^{176}\text{Lu}}{^{177}\text{Hf}}\right)_{Sample}} + 1\right]$$

(6.18)

This approach suggests that the $^{176}\text{Lu}/^{177}\text{Hf}$ ratio of the source material of the granite between the 2240 Ma Sm–Nd model age and the 500 Ma crystallization age was 0.024 (Fig. 6.12c); a value that is typical of mafic crustal rocks that might have served as the source material that when melted produced this granite.

The approach above produces a reasonable story for the whole history of the granite and its source materials, but obviously is dependent on the accuracy of the Sm–Nd model age. In all likelihood, the Sm/Nd ratio measured in the sample is lower than in its source rocks, which would mean that the calculated Nd model ages underestimate the true formation age of the source material of the granite. There is also the question of whether the source material of the granite was derived from melting of a BSE or DM type mantle, again with consequences for the true age of formation of the granite's source material.

Perhaps the key observation to extract from the above example is that the chronological accuracy of model ages should be viewed with considerable suspicion because the ages depend on how well the actual history of the sample matches the model being used to calculate the model age. Nevertheless, model age approaches are extremely useful in extracting fundamental information regarding the history of many samples. In the example above, the model age data show that the granite was not produced by direct melting of the mantle, but instead by remelting of some source material that had separated from the mantle well over a billion years prior to the 500 Ma formation age of the granite. The important conclusion derived from this analysis is that the granite reflects the reworking of older crustal material and is not a new addition to the continental crust.

6.7 CONCLUSION AND FUTURE DIRECTIONS

While the chronological applications of Rb–Sr, Sm–Nd, and Lu–Hf are increasingly taking a back-seat to their applications in isotope geology, there remain many geological processes for which they are well suited, and where the more popular dating methods of U–Pb and K-Ar cannot be applied as effectively. The key to their use involves the distinct geochemical behavior of the parent and daughter elements in each system and applying them to geological processes that cause parent–daughter fractionation. The examples given in this chapter were selected to highlight some of the chronological applications of these systems that cannot be as successfully pursued with other techniques discussed in the book. The combination of the diverse problems to which these systems can be applied, as well as their increasing application as thermochronometers for higher temperature processes, likely will see their continued application in geochronology for quite some time.

6.8 REFERENCES

Albarède, F., Scherer, E. E., Blichert-Toft, J., Rosing, M., Simionovici, A., and Bizzarro, M. (2006) γ-ray irradiation in the early solar system and the conundrum of the ^{176}Lu decay constant. *Geochimica et Cosmochimica Acta* **70**, 1261–1270.

Aldrich, L. T., Wetherill, G. W., Tilton, G. R., and Davis, G. L. (1956) Half-life of ^{87}Rb. *Physics Reviews* **103**, 1045–1047.

Amelin, Y. (2005) Meteorite phosphates show constant ^{176}Lu decay rate since 4557 million years ago. *Science* **310**, 839–841.

Anders, E. and Grevesse, N. (1989) Abundances of the elements: meteoritic and solar. *Geochimica et Cosmochimica Acta* **53**, 197–214.

Begemann, F., Ludwig, K. R., Lugmair, G. W., *et al.* (2001) Call for an improved set of decay constants for geochronological use. *Geochimica et Cosmochimica Acta* **65**, 111–121.

Blackburn, T. J., Stockli, D. F., Carlson, R. W., and Berendsen, P. (2008) (U–Th)/He dating of kimberlites - A case study from north-eastern Kansas. *Earth and Planetary Science Letters* **275**, 111–120.

Blichert-Toft, J., Chauvel, C., and Albarede, F. (1997) Separation of Hf and Lu for high-precision isotope analysis of rock samples by magnetic sector-multiple collector ICP-MS. *Contributions to Mineralogy and Petrology* **127**, 248–260.

Borg, L. E., Connelly, J. N., Boyet, M., and Carlson, R. W. (2011) Chronological evidence that the Moon is either young or did not have a global magma ocean. *Nature* **477**, 70–73.

Bouvier, A., Vervoort, J. D., and Patchett, P. J. (2008) The Lu–Hf and Sm–Nd isotopic composition of CHUR: constraints from unequilibrated chondrites and implications for the bulk composition of terrestrial planets. *Earth and Planetary Science Letters* **273**, 48–57.

Carlson, R. W., Borg, L. E., Gaffney, A. M., and Boyet, M. (2014) Rb–Sr, Sm–Nd and Lu–Hf isotope systematics of the lunar Mg-suite: the age of the lunar crust and its relation to the time of Moon formation. *Philosophical Transactions Royal Society of London Series A* **372**. DOI: 10.1098/rsta.2013.0246.

Chauvel, C., Dupre, B., and Jenner, G. A. (1985) The Sm–Nd age of Kambalda volcanics is 500 Ma too old! *Earth and Planetary Science Letters* **74**, 315–324.

Christensen, J. N. and DePaolo, D. J. (1993) Time scales of large volume silicic magma systems: Sr isotopic systematics of phenocrysts and glass from the Bishop Tuff, Long Valley, California. *Contributions to Mineralogy and Petrology* **113**, 100–114.

Christensen, J. N., Halliday, A. N., Leigh, K. E., Randell, R. N., and Kesler, S. E. (1995) Direct dating of sulfides by Rb–Sr: a critical test using the Polaris Mississippi Valley type Zn–Pb deposit. *Geochimica et Cosmochimica Acta* **59**, 5191–5197.

Christiansen, E. H., Bikun, J. V., Sheridan, M. F., and Burt, D. M. (1984) Geochemical evolution of topaz rhyolites from the Thomas Range and Spor Mountain, Utah. *American Mineralogist* **69**, 223–236.

Clauer, N. (1979) A new approach to Rb–Sr dating of sedimentary rocks. In *Lectures in Isotope Geology*, Jager, E. and Hunziker, J. C. (eds). Springer, Heidelberg.

Clauer, N., Chaudhuri, S., Kralik, M., and Bonnot-Courtois, C. (1993) Effects of experimental leaching on Rb–Sr and K–Ar isotopic systems and REE contents of diagenetic illite. *Chemical Geology* **103**, 1–16.

Corgne, A., Liebske, C., Wood, B. J., Rubie, D. C., and Frost, D. J. (2005) Silicate perovskite-melt partitioning of trace elements and geochemical signature of a deep perovskitic reservoir. *Geochimica et Cosmochimica Acta* **69**, 485–496.

DePaolo, D. J. (1980) Crustal growth and mantle evolution: inferences from models of element transport. *Geochimica et Cosmochimica Acta* **44**, 1185–1196.

DePaolo, D. J. (1981) Trace element and isotopic effects of combined wallrock assimilition and fractional crystallization. *Earth and Planetary Science Letters* **53**, 189–202.

Farmer, G. L. (2003) Continental basaltic rocks. In *Treatise on Geochemistry*. Rudnick, R. L. (ed.). Elsevier, Amsterdam.

Gale, A., Dalton, C. A., Langmuir, C. H., Sun, Y., and Schilling, J.-G. (2013) The mean composition of ocean ridge basalts. *Geochemistry, Geophysics, Geosystems* **14**, 489–518.

Ganguly, J. (2010) Cation diffusion kinetics in aluminosilicate garnet and geologic applications. *Reviews in Mineralogy and Geochemistry* **72**, 559–601.

Hans, U., Kleine, T., and Bourdon, B. (2013) Rb–Sr chronology of volatile depletion in differentiated protoplanets: BABI, ADOR and ALL revisited. *Earth and Planetary Science Letters* **374**, 204–214.

Hemmendinger, A. and Smythe, W. R. (1937) The radioactive isotope of rubidium. *Physical Review* **51**, 1052–1053.

Hofmann, A. W. (1988) Chemical differentiation of the Earth: the relationship between mantle, continental crust, and oceanic crust. *Earth and Planetary Science Letters* **90**, 297–314.

Hofmann, A. W. (2003) Sampling mantle heterogeneity through oceanic basalts: isotopes and trace elements. In The Mantle and Core, *Vol. 2,* Treatise on Geochemistry, Carlson, R. W. (ed). Elsevier, Amsterdam.

Jahn, B.-M. and Shih, C. (1974) On the age of the Overwacht Group, Swaziland Sequence, South Africa. *Geochimica et Cosmochimica Acta* **38**, 873–885.

Kelemen, P. B., Hanghoj, K., and Greene, A. R. (2003) One view of the geochemistry of subduction-related magmatic arcs with an emphasis on primitive andesite and lower crust. In *Treatise on Geochemistry*, Rudnick, R. L. (ed.). Elsevier, Amsterdam.

Le Roex, A. P., Bell, D. R., and Davis, P. (2003) Petrogenesis of Group 1 kimberlites from Kimberley, South Africa: evidence from bulk-rock geochemistry. *Journal of Petrology* **44**, 2261–2286.

Liu, S., Hu, R., Gao, S., *et al.* (2009) U–Pb zircon, geochemical and Sr-Nd-Hf isotopic constraints on the age and origin of early Paleozoic I-type granite from the Tengchong-Baoshan Block, Western Yunnan Province, SW China. *Journal of Asian Earth Sciences* **36**, 168–182.

Lodders, K. (2003) Solar system abundances and condensation temperatures of the elements. *Astrophysical Journal* **591**, 1220–1247.

Lugmair, G. W., Scheinin, N. B., and Marti, K. (1975) Sm–Nd age and history of Apollo 17 basalt 75075: evidence for early differentiation of the lunar interior. *Proceedings of the 6th Lunar Science Conference*; 1419–1429.

Mattauch, J. (1937) Das Paar ^{87}Rb–^{87}Sr und die isobarenregel. *Naturwissenschaften* **25**, 189.

McCulloch, M. T. and Compston, W. (1981) Sm–Nd age of Kambalda and Kanownwa greenstones and heterogeneity in the Archean mantle. *Nature* **294**, 322–327.

McDonough, W. F. and Sun, S.-S. (1995) The composition of the Earth. *Chemical Geology* **120**, 223–253.

Mezger, K., Bohlen, S. R., and Hanson, G. N. (1990) Metamorphic history of the Archean Pikwitonei Granulite Domain and the Cross Lake Subprovince, Superior Province, Manitoba, Canada. *Journal of Petrology* **31**, 483–517.

Minster, J.-F., Birck, J.-L., and Allegre, C. J. (1982) Absolute age of formation of chondrites studied by the ^{87}Rb–^{87}Sr method. *Nature* **300**, 414–419.

Morton, J. P. and Long, L. E. (1980) Rb–Sr dating of Paleozoic glauconite from the Llano region, central Texas. *Geochimica et Cosmochimica Acta* **44**, 663–672.

Moyen, J.-F. and Martin, H. (2012) Forty years of TTG research. *Lithos* **148**, 312–336.

Nicolaysen, L. O. (1961) Graphic interpretation of discordant age measurements on metamorphic rocks. *Annals New York Academy Sciences* **91**, 198–206.

Notsu, K., Mabuchi, H., Yoshioka, O., and Ozima, M. (1973) ^{147}Sm–^{143}Nd dating of a Roberts Victor eclogite. *Geochemical Journal* **7**, 51–54.

O'Nions, R. K., Hamilton, P. J., and Evensen, N. M. (1977) Variations in ^{143}Nd/^{144}Nd and ^{87}Sr/^{86}Sr ratios in oceanic basalts. *Earth and Planetary Science Letters* **34**, 13–22.

Palme, H. and O'Neill, H. S. C. (2014) Cosmochemical estimates of mantle composition. In *Treatise on Geochemistry*, Carlson, R. W. (ed.). Elsevier, Amsterdam.

Patchett, P. J. and Tatsumoto, M. (1980a) Hafnium isotope variations in oceanic basalts. *Geophysical Research Letters* **7**, 1077–1080.

Patchett, P. J. and Tatsumoto, M. (1980b) Lu–Hf total-rock isochron for the eucrite meteorites. *Nature* **288**, 571–574.

Patel, S. C., Frost, C. D., and Frost, B. R. (1999) Contrasting responses of Rb–Sr systematics to regional and contact metamorphism, Laramie Mountains, Wyoming, USA. *Journal of Metamorphic Geology* **17**, 259–269.

Pollington A.D. and Baxter, E. F. (2011) High precision microsampling and preparation of zoned garnet porphyroblasts for Sm–Nd geochronology. *Chemical Geology* **281**, 270–282.

Rudnick, R. L. and Gao, S. (2003) Composition of the Continental Crust. In *The Crust*, Rudnick R.L. (ed.), Vol. 3, *Treatise of Geochemistry*, Holland, H. D. (ed),. Elsevier-Pergamon, Oxford.

Rutberg, R. L., Hemming, S. R., and Goldstein, S. L. (2000) Reduced North Atlantic deep water flux to the glacial Southern Ocean inferred from neodymium isotope ratios. *Nature* **405**, 935–938.

Sanborn, M. E., Carlson, R. W., and Wadhwa, M. (2015) ^{147,146}Sm–^{143,142}Nd, ^{176}Lu–^{176}Hf, and ^{87}Rb–^{87}Sr systematics in the angrites: implications for chronology and processes on the angrite parent body. *Geochimica et Cosmochimica Acta* **171**, 80–99.

Scherer, E., Muenker, C., and Mezger, K. (2001) Calibration of the lutetium-hafnium clock. *Science* **293**, 683–687.

Schoene, B., Schaltegger, U., Brack, P., Latkoczy, C., Stracke, A., and Gunther, D. (2012) Rates of magma differentiation and emplacement in a ballooning pluton recorde by U–Pb TIMS-TEA, Adamello batholith, Italy. *Earth and Planetary Science Letters* **355–356**, 162–173.

Shaw, D. M. (1970) Trace element fractionation during anatexis. *Geochimica et Cosmochimica Acta* **34**, 237–243.

Smit, M. A., Scherer, E. E., and Mezger, K. (2013) Lu–Hf and Sm–Nd garnet geochronology: Chronometric closure and implications for dating petrological processes. *Earth and Planetary Science Letters* **381**, 222–233.

Soderlund, U., Patchett, P. J., Vervoort, J. D., and Isachen, C. E. (2004) The ^{176}Lu decay constant determined by Lu–Hf and U–Pb isotope systematics of Precambrian mafic intrusions. *Earth and Planetary Science Letters* **219**, 311–324.

Steiger, R. H. and Jager, E. (1977) Subcommission on geochronology: convention on the use of decay constants in geo- and cosmochronology. *Earth and Planetary Science Letters* **36**, 359–362.

Sun, S.-S. and McDonough, W. F. (1989) Chemical and isotopic systematics of oceanic basalts: implications for mantle composition and processes. In *Magmatism in the Ocean Basins*, Saunders, A. D. and Norry, M. J. (eds),. Geological Society of America.

Veizer, J. and Mackenzie, F. T. (2003) Evolution of sedimentary rocks. In *Treatise on Geochemistry*, Mackenzie, F. T. (ed.). Elsevier, Amsterdam.

Villa, I. M., Bievre, P. D., Holden, N. E., and Renne, P. R. (2015) IUPAC-IUGS recommendation on the half life of ^{87}Rb. *Geochimica et Cosmochimica Acta* **164**, 382–385.

Walker, J. D., Bowers, T. D., Black, R. A., Glazner, A. F., Farmer, G. L., and Carlson, R. W. (2006) A geochemical database for western North American volcanic and intrusive rocks (NAVDAT). *Geological Society of America Special Paper* **397**, 61–71.

Wilson, A. H. and Carlson, R. W. (1989) A Sm–Nd and Pb isotope study of Archaean greenstone belts in the southern Kaapvaal Craton, South Africa. *Earth and Planetary Science Letters* **96**, 89–105.

CHAPTER 7

Re–Os and Pt–Os

7.1 INTRODUCTION

Osmium (Os) receives radioactive decay contributions from both rhenium (Re) and platinum (Pt). The utility of this system derives from the distinct geochemical behavior of the elements involved compared to most of the elements that make up other radioactive chronometers.

(1) Rhenium, Pt, and Os belong to the group of elements known as "highly siderophile elements" (HSE). Siderophile means "iron-loving" and conveys the fact that in a rock with coexisting iron metal and silicate, these elements will be strongly enriched in the metal over their abundance in the silicates. As a result, most of Earth's Re, Pt, and Os (and gold, another HSE) are in the core, leaving the mantle and crust with very low abundances of these elements (see Box 7.1). All three elements also are much more soluble in sulfides then in silicates, but the degree to which they partition into sulfides depends on the composition of the sulfide.

(2) Like all the elements that make up other long-lived radioactive chronometer systems, Re and Pt behave as incompatible elements during partial melting of the mantle. In contrast, Os is strongly compatible. By definition, when partial melting occurs in a rock, incompatible elements are concentrated in the melt, while compatible elements remain in the residual solids. This means that partial melting of the mantle produces melts that have high Re,Pt/Os ratios and leaves the residue of melting with low Re/Os ratios. This property eventually results in very distinct Os isotopic compositions between crust and mantle.

(3) Complexing with organic molecules, along with the presence of reducing conditions that promote the formation of sulfides, leads the HSE to be strongly enriched in organic-rich sediments including black shales and petroleum. This property, nearly unique to Re and Os among the radioactive dating systems, allows the Re–Os system to provide ages and track the formation of organic-rich materials. Re–Os dating thus has been used effectively in pinpointing the age of boundaries within the Phanerozoic geologic timescale.

(4) Molybdenite (MoS_2) strongly concentrates Re, but nearly completely excludes Os. Consequently, the Re–Os system can provide remarkably precise age information on the formation of molybdenite ores. Although the fractionation of Re from Os is not as extreme in other sulfides, the concentration of Re and Os into sulfides allows this system to be widely applicable for dating a variety of ore-forming processes.

These properties of the Re–Os system have driven the development of analytical techniques that largely overcome the very low abundances (10^{-9} to 10^{-12} g/g) of these elements in terrestrial materials to exploit the nearly unique information provided by Pt–Re–Os systematics. Detailed reviews of this system, and the analytical procedures that allow its use, include [*Shirey and Walker*, 1998; *Reisberg and Meisel*, 2002; *Carlson*, 2005].

7.2 RADIOACTIVE SYSTEMATICS AND BASIC EQUATIONS

Osmium contains two isotopes produced by radioactive decay of Re and Pt.

$$^{187}Re \rightarrow {}^{187}Os + \beta \quad \text{half-life} = 4.16 \times 10^{10} \text{ years}$$
$$^{190}Pt \rightarrow {}^{186}Os + \alpha \quad \text{half-life} = 4.69 \times 10^{11} \text{ years}$$

Rhenium consists of 37.4 atom% stable ^{185}Re and 62.6 atom% radioactive ^{187}Re (Table 7.1). Platinum-190, however, is a rare (0.01 atom%) isotope of Pt. Both ^{186}Os and ^{187}Os are relatively low abundance isotopes of Os. Given the isotopic make up of the elements, the decay of ^{187}Re can substantially affect the abundance of ^{187}Os, but both the low abundance of ^{190}Pt and its very long half-life combine to make its contribution to ^{186}Os quite small, thus limiting its utility as a chronometer.

In the 1950s and 1960s, the decay constant of ^{187}Re was determined to be $(1.61 \pm 0.17) \times 10^{-11}$/year by comparison of

Geochronology and Thermochronology, First Edition. Peter W. Reiners, Richard W. Carlson, Paul R. Renne, Kari M. Cooper, Darryl E. Granger, Noah M. McLean, and Blair Schoene.
© 2018 John Wiley & Sons Ltd. Published 2018 by John Wiley & Sons Ltd.

The strongly siderophile nature of these elements is reflected in their relatively high absolute abundances in various types of iron meteorites (Table 7.2). Given the strong preference these elements have for metal over silicate, core formation on Earth should have left the mantle with Re, Os, and Pt (and other HSE like Ir and Pd) concentrations several orders of magnitude lower than those observed in mantle rocks (Fig. 7.B1.1). Because such a large proportion of these elements would fractionate into the metal, the core-forming metal would start with Re/Os and Pt/Os ratios similar to the bulk composition of the Earth. The same is not true of the silicate Earth (mantle and crust) where the small amount of HSE left behind in the mantle should strongly reflect the differences in metal–silicate partition coefficients of these elements, resulting in a chondrite-normalized HSE abundance pattern very different from chondritic, with high Pt/Os and low Re/Os ratios compared to the modern mantle (Fig. 7.B1.1). Nevertheless, mantle rocks believed to approximate the bulk composition of the mantle, called "fertile peridotites," have Os isotopic compositions that overlap those of chondrites [*Meisel et al.*, 2001]. The observation that the Re/Os, Pt/Os, and the $^{187}Os/^{188}Os$ ratios in fertile peridotites are close to chondritic has been used to suggest that a small amount of material with chondritic HSE abundances was accreted to Earth as a "late veneer" after core formation was complete [*Chou et al.*, 1983; *Morgan*, 1986]. For example, the present-day Os concentration of the fertile mantle is close to 4 ppb [*Morgan*, 1986]. The mass of the mantle is about 4×10^{27} g, which means there currently are 1.64×10^{19} g of Os in the mantle. Primitive chondrites have an average Os concentration of about 460×10^{-9} g/g, so 3.6×10^{25} g of this chondrite could supply all the Os found today in Earth's mantle. The total mass of the Earth is 5.98×10^{27} g, so adding an amount of chondrite equivalent to only 0.6% of Earth's mass to the mantle after core–mantle exchange stopped could explain both the HSE abundances and the fact that the chondrite-normalized HSE abundance pattern of the primitive mantle is relatively flat (Fig. 7.B1.1).

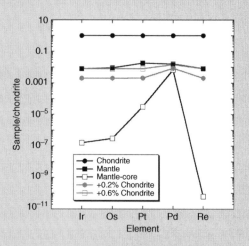

Fig. 7.B1.1. Abundances of several highly siderophile elements in CI chondrites (filled circles [*Horan et al.*, 2009]) and the modern primitive mantle (filled squares [*Becker et al.*, 2006]). The open squares show the calculated concentrations in the mantle if it were in chemical equilibrium with the core. The very low HSE concentrations are reflective of the very high partition coefficients of these elements for iron metal. The gray symbols reflect the HSE patterns that result from additions to this low HSE mantle of small amounts of chondrite corresponding to 0.2% and 0.6% of Earth's mass. The HSE concentrations of the mantle after core formation are so low that the fractionated HSE pattern is very quickly flattened by even small additions of material with chondrite-like HSE absolute and relative abundances.

Re–Os isochrons for molybdenites whose ages had been measured by other methods, particularly U–Pb. A different approach to determining the ^{187}Re half-life involved measuring the rate of ingrowth of ^{187}Os in a solution containing pure Re [*Lindner et al.*, 1986]. This method yielded a ^{187}Re decay constant of $(1.64 \pm 0.05) \times 10^{-11}$/year. The most commonly used Re decay constant currently is based on a Re–Os isochron for one chemically related group of iron meteorites (the IIIA group) that show excellent linearity on a Re–Os isochron diagram [*Smoliar et al.*, 1996]. A ^{187}Re decay constant of 1.666×10^{-11}/year is calculated from the IIIA Re–Os isochron slope.

As with any geologic determination of a radioactive half-life, the accuracy of the resulting half-life depends on whether the age calculated by some other means dates the same event recorded by the system of interest. Because Re–Os is uniquely suited to dating iron meteorites, no other direct age determination on the IIIA iron meteorites was possible, forcing a comparison of the iron meteorite isochron with the ages determined for angrites, a type of igneous meteorite consisting mostly of olivine, pyroxene, and plagioclase. When the IIIA isochron work was done, only two angrites had precise U–Pb ages, and both agreed at 4558 Ma [*Lugmair and Galer*, 1992]. More recent work on angrites has shown various members of this meteorite group to have ages ranging from 4557.8 to 4563.4 Ma [*Amelin*, 2008; *Brennecka and Wadhwa*, 2012]. The IIIA Re–Os isochron defined a slope of 0.07887. Under the difficult to prove assumption that angrites and IIIA iron meteorites formed at the same time, an angrite age of 4563.4 Ma coupled with the slope of the IIIA Re–Os isochron provides a ^{187}Re decay constant of 1.664×10^{-11}/year. Given that the uncertainty on the Re–Os isochron slope for the IIIA iron meteorites translates to a decay constant uncertainty of $\pm 0.2\%$, and the difference between the decay constant calculated assuming a 4557.8 instead of 4563.4 Ma age for angrites is only 0.15%, this geological determination of the ^{187}Re decay constant results in a value uncertain to at least 0.2%. In reality, uncertainty in the calibration of the spikes used to determine Re and Os concentrations for the IIIA limits the accuracy of the ^{187}Re decay constant to about $\pm 1\%$. At the current time, the ^{187}Re decay constant in most common use is thus $(1.666 \pm 0.017) \times 10^{-11}$/year.

The half-life of ^{190}Pt was determined by counting radioactive decays in a sample of Pt to provide a decay constant of $(1.07 \pm 0.04) \times 10^{-12}$/year [*Begemann et al.*, 2001]. In contrast, a geologic determination of the ^{190}Pt decay constant made by measuring a Pt–Os isochron in Pt ores from the Noril'sk (Siberia) igneous intrusions, dated by U–Pb at 251.2 ± 0.3 Ma, provided a Pt–Os slope of 0.0003875 ± 0.0000032, from which a decay constant of $(1.542 \pm 0.015) \times 10^{-12}$/year can be calculated [*Walker et al.*, 1997]. Why the two measurements differ by this much is not clear. One contributing factor is the uncertainty of the relative abundance of the rare isotope ^{190}Pt. Recent determinations of its abundance have resulted in an adjustment of the Noril'sk determined decay constant to 1.477×10^{-12}/year [*Begemann et al.*, 2001], which is the value currently in use.

Table 7.1 Portion of the nuclear chart around Pt, Os and Re showing the relative abundance of each isotope (in atom percent)

Z	A	184	185	186	187	188	189	190	191	192	193	194	195	196	198
78	Pt							0.01		0.79		32.9	33.8	25.3	7.2
77	Ir								37.3		62.7				
76	Os	0.02		1.58	1.6	13.3	16.1	26.4		41.0					
75	Re		37.4		62.6										
	N	108	109	110	111	112	113	114	115	116	117	118	119	110	112

Table 7.2 Median and range[a] of Re and Os concentrations[b] in various natural materials

	[Re] ppb	[Os] ppb	$^{187}Re/^{188}Os$
C-chondrites	51	633	0.39
	35–71	429–886	0.31-0.43
O-chondrites	61	720	0.42
	26–89	289–1072	0.36–0.47
E-chondrites	54	626	0.42
	24–76	277–861	0.41-0.44
Iron meteorites	386	3800	0.51
	35–4840	244–102000	0.32-0.99
Peridotite	0.13	2.83	0.31
	0.009–1.77	0.003 – 9.2	0.03–27
Komatiite	0.58	1.66	1.8
	0.37–3.65	0.75–2.47	0.9–14
Mid-ocean ridge basalt	0.65	0.08	10.8
	0.16–1.2	<0.001-0.35	7–235
Ocean island basalt	0.32	0.092	18.9
	0.02–1.6	<0.001–1.2	0.6–2510
Continental basalt	0.45	0.13	39
	0.05–3.6	0.003–2.1	0.26–3150
Metaliferous Sediment	0.62	0.075	6.8
	0.04–11	0.002–1.5	1.2–112
Black shale	43.9	0.3	623
	15–517	0.095–3.69	292–1300
Minerals:			
sulfide	52–2550	4.7–122	
chromite	0.22-0.64	13–67	
molybdenite	700–160000	0	

Source: *Shirey and Walker* [1998] and *Carlson* [2005].
[a] The first line in each row gives the median value whereas the second line gives the observed range.
[b] Concentrations are expressed in parts per billion (ppb), which corresponds to 10^{-9} g/g.

The standard decay equations for the Re–Os and Pt–Os systems are:

$$\left(\frac{^{187}Os}{^{188}Os}\right)_m = \left(\frac{^{187}Os}{^{188}Os}\right)_0 + \left(\frac{^{187}Re}{^{188}Os}\right)_m \times \left(e^{\lambda t} - 1\right) \qquad (7.1)$$

and

$$\left(\frac{^{186}Os}{^{188}Os}\right)_m = \left(\frac{^{186}Os}{^{188}Os}\right)_0 + \left(\frac{^{190}Pt}{^{188}Os}\right)_m \times \left(e^{\lambda t} - 1\right) \qquad (7.2)$$

where m is measured, 0 corresponds to $t = 0$, and λ is the decay constant of ^{187}Re or ^{190}Pt. Early studies using the Re–Os system placed ^{186}Os in the denominator instead of ^{188}Os. With the realization that ^{186}Os abundances are variable in nature due to the decay of ^{190}Pt, and that more precise isotope ratios can be measured when normalizing to the more abundant isotope ^{188}Os,

most modern discussions of the Pt–Re–Os system use ^{188}Os in the denominator of these equations.

As in many isotopic measurements, variations in Os isotopic composition are small enough where convention sometimes references the measured, or calculated initial, Os isotopic composition relative to a standard value for the bulk Earth. For Os, the isotopic variations are large enough so that the reference is expressed in percent deviation using the equation:

$$\gamma Os = \left(\frac{\left(\frac{^{187}Os}{^{188}Os}\right)_{SA}}{\left(\frac{^{187}Os}{^{188}Os}\right)_{BSE}} - 1\right) \times 100 \qquad (7.3)$$

SA stands for sample and BSE is the bulk silicate Earth. BSE is assumed to represent the composition of the whole mantle prior to formation of continental and oceanic crust. In order to calculate the initial γ_{Os}, the sample Os isotopic composition is back-calculated to the age of the sample using its measured Re/Os ratio, and the BSE Os isotopic composition is similarly calculated at the sample age, so:

$$\gamma Os(t) = \left(\frac{\left[^m_{SA}\left(\frac{^{187}Os}{^{188}Os}\right) - ^m_{SA}\left(\frac{^{187}Re}{^{188}Os}\right) \times \left(e^{\lambda t}-1\right)\right]}{^P_{BSE}\left(\frac{^{187}Os}{^{188}Os}\right) - ^P_{BSE}\left(\frac{^{187}Re}{^{188}Os}\right) \times \left(e^{\lambda t}-1\right)} - 1\right) \times 100 \qquad (7.4)$$

where m is the measured ratio in the sample and p is the present-day value in the BSE.

Because the composition of the BSE in terms of Re–Os parameters is not known precisely, the literature contains some variability in the isotopic evolution parameters for the BSE reference used by different studies. A common approach uses an estimate of the modern day mantle $^{187}Os/^{188}Os = 0.1296$ that derives from measurement of the Os isotopic composition of modern mantle peridotites from various localities, and corrected for the consequences of partial melt removal back to an estimate of the undifferentiated mantle [*Meisel et al.*, 2001]. The corresponding BSE $^{187}Re/^{188}Os$ ratio is then calculated using this modern day mantle isotopic composition and the solar system initial $^{187}Os/^{188}Os$ (0.09524) that was determined from the isochron for the IIIA group of iron meteorites discussed previously. The corresponding $^{187}Re/^{188}Os$ is 0.4353. A less commonly used estimate for BSE Os isotopic evolution uses the modern day average Os isotopic composition determined for a particularly primitive group of meteorites, the carbonaceous chondrites, that have an average

^{187}Os/^{188}Os = 0.1263 and resulting ^{187}Re/^{188}Os = 0.3939. At the present day, these two normalizing values for the BSE are different by 2.6% (γ_{Os} = 2.6), but the difference decreases going back in time as both values converge on the solar system initial ^{187}Os/^{188}Os determined for iron meteorites at 4.567 Ga. Care is thus required in comparing Os isotope data sets produced by different groups to make sure that the normalizing values used to calculate γ_{Os} are the same.

7.3 GEOCHEMICAL PROPERTIES AND ABUNDANCE IN NATURAL MATERIALS

In terms of their cosmochemical properties, Re, Os, and Pt are refractory elements with more than half of each element condensing from a hot gaseous hydrogen-rich solar nebula into metal alloys at temperatures between 1500 °C (Re, Os) and 1100 °C (Pt) [*Lodders*, 2003]. As a result of their refractory nature, the abundance of Re, Os, and Pt does not vary much within undifferentiated meteorites such as the chondrites [*Horan et al.*, 2003]. Table 7.2 shows the various groups of chondrites to have of order 50 ppb (parts per billion or 10^{-9} g/g) Re and 0.6–0.7 ppm Os. Similarly, Pt concentrations in chondrites show a relatively narrow range between 1.1 and 1.3 ppm. The chondrites also show a very limited range in Re/Os ratio, with both ordinary and enstatite chondrites overlapping at ^{187}Re/^{188}Os ~0.42. Carbonaceous chondrites, for reasons that are not yet understood, have resolvably lower average ^{187}Re/^{188}Os ratios than other chondrite groups that also are reflected in their slightly lower modern day average ^{187}Os/^{188}Os ratios [*Walker et al.*, 2002].

Re and Pt are both moderately incompatible during partial melting with Re partition coefficients being similar to those of aluminum or ytterbium. Because Os is a compatible element during partial melting in the mantle, mantle peridotite that has been "depleted" by removal of partial melt will have slightly higher Os concentration and a lower Re/Os ratio than peridotite that has not experienced melting. As a result, ingrowth of ^{187}Os is retarded in depleted mantle. Figure 7.1 shows that "depleted" peridotites overlap with fertile mantle Os isotopic compositions, but extend to lower ^{187}Os/^{188}Os. The lower ^{187}Os/^{188}Os ratios indicate that at least some portions of the mantle were affected by partial melt removal long enough ago to result in observable isotopic differences in Os. The compatibility of Os in the solids during partial melting is unique amongst all the elements used in radioactive dating schemes. These distinct geochemical properties of the parent and daughter element allow the Re–Os system to provide important information on the history of partial melting in the mantles of the terrestrial planets and rocky planetesimals.

The compatibility of Os, but incompatibility of Re, during partial melting also results in melt compositions that have high to very high Re/Os ratios, and hence rapid ingrowth of radiogenic Os. While offering the potential for precise chronometry of even relatively young igneous rocks, the compatibility of Os can lead to exceedingly low (<10^{-12} g/g) Os concentrations in evolved melts, which complicates the Os analyses that would take advantage of the high Re/Os ratios for geochronology.

The limited solubility of the HSE in silicate minerals means that their concentrations in most silicate-dominated rocks are controlled by trace phases, generally iron–nickel–copper sulfides, HSE-rich sulfides such as laurite (RuS_2) or erlichmanite (OsS_2), or alloy phases such as osmiridium or platinum. The degree to which these elements can be concentrated into ultratrace phases can provide a daunting analytical challenge. For example, studies that have tried to determine the distribution of Os between the various mineral phases that make up a mantle peridotite find that only about 15% of the Os is contained in the main mineral phases [*Hart and Ravizza*, 1995; *Luguet et al.*, 2007]. Where is the rest of the Os? Only a single 80 μm diameter grain of osmiridium (20 g/cm^3, 50 wt% Os) contained in a kilogram of peridotite would raise the Os concentration of that kilogram to the 4 ppb Os content typical of mantle peridotite. The strong concentration of Os, and other HSE, into ultratrace phases gives rise to what is known as the "nugget" effect, where different analyses of the same rock powder may give a significantly higher or lower Os concentration because the different sample aliquots either did, or did not, contain an extra grain of the trace phase. Consequently, even though techniques exist to analyze Os concentrations in rocks to better than a percent precision, these analyses often do not provide comparable reproducibility for whole-rock Os concentrations because of the influence of the nugget effect.

The tendency of Re and Os to concentrate into trace phases can result in extreme fractionation of parent from daughter element. A number of phases including osmiridium and chromite can have Re/Os ratios near zero. Many igneous rocks, and the sulfides that form from them, can have very high Re/Os ratios, in the range of hundreds to thousands, culminating in molybdenite that can have hundreds of ppm Re, but less than 10^{-9} g/g Os, resulting in Re/Os ratios above 10^5. Another rock type where high Re/Os ratios can be found is organic-rich sediments where Os concentrations can range into the 10^{-9} g/g range with Re concentrations often hundreds of times higher. The high Re/Os ratios of both molybdenites and organic-rich sediments enable Re–Os to provide important chronological information on the origin of these materials.

7.4 ANALYTICAL CHALLENGES

The Re–Os and Pt–Os systems were slow to be added to the radioactive geochronology toolbox, in part because of the very low abundance of these elements in most natural materials (Table 7.2) and in part because their chemical behavior makes their separation from rocks a challenge. In the mantle, Os is present at the 10^{-9} g/g range whereas in crustal rocks, the concentration can drop to the 10^{-12} g/g range or lower. Just to make the point of how rare Os is in crustal rocks, 10^{-12} g/g is a part-per-million of a part-per-million, or in other words, Os is of order a million times less abundant in crustal rocks than are elements

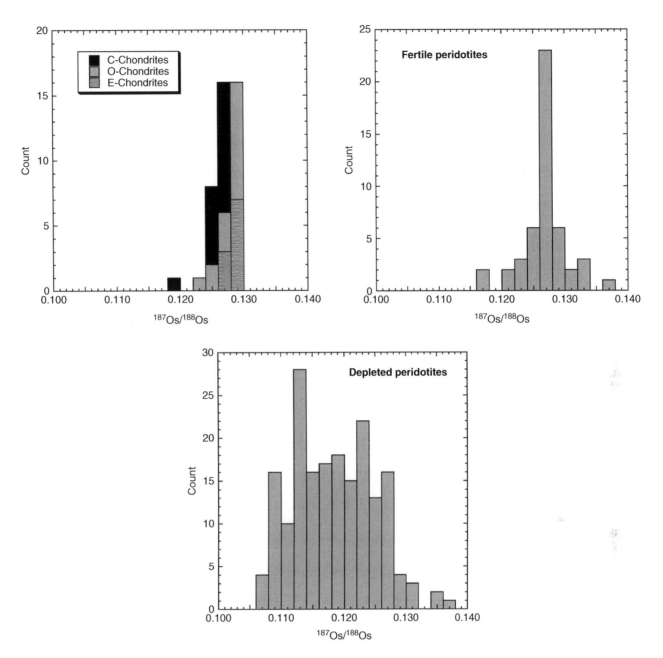

Fig. 7.1. Histogram of modern Os isotopic compositions for various types of chondritic meteorites [*Walker et al.*, 2002] compared to both fertile terrestrial mantle peridotite (peridotite that has not experience partial melt removal as indicated by Al_2O_3 concentrations near 4 wt%) and terrestrial mantle peridotite that has been "depleted" by partial melt removal as indicated by Al_2O_3 concentrations near 1 wt% [*Carlson et al.*, 2005].

such as Sr, Nd, Hf, and Pb, that fall under the title "trace elements."

The difficulty of separating Re and Os from rocks was avoided in early Re–Os work by concentrating on analysis of rare Os-rich alloys using an ion-microprobe to directly sputter Os from the sample, or an Os concentrate of the sample, for isotopic analysis [*Allegre and Luck*, 1980; *Hart and Kinloch*, 1989]. The desire to extend Re–Os analyses beyond rare ore minerals required separation procedures that could cleanly concentrate nanograms or picograms of Os from grams of rock. Beyond the simple rarity

of these elements, Os has many stable oxidation states and hence forms many molecular compounds that can have quite different chemical properties. For example, as a metal, Os is highly refractory with a boiling point of just over 5000 °C. In its most oxidized state, as OsO_4, Os is volatile and can sublime from the solid even at room temperature. These diverse chemical properties both aid and complicate the separation of Os from rocks. Good summaries of chemical separation approaches for Re and Os are given in *Shirey and Walker* [1995] and *Reisberg and Meisel* [2002].

Standard rock-dissolution approaches use a combination of hydrofluoric acid to break apart silicate minerals, combined with oxidizing acids such as nitric or perchloric acid. The strongly oxidizing conditions in this type of dissolution will turn Os into OsO_4 that will be lost by volatilization unless the dissolution vessels are tightly sealed. One way around this problem is to use dissolution methods [*Birck et al.*, 1997] that use reducing acids, e.g., hydrofluoric and hydrobromic, but these often result in poor chemical equilibrium between Os extracted from the dissolved rock and the isotopically enriched Os spike used to determine Os concentration by isotope dilution. An alternative uses the classic analytical technique for analysis of the HSE, known as nickel-sulfide fire assay [*Hoffman et al.*, 1978], that involves mixing rock powder with alkaline fluxes such as sodium-carbonate or sodium-borate, nickel metal and sulfur powder. The powder mixture is then melted at temperatures over 1000 °C to form a melt of the rock plus flux along with an immiscible nickel-sulfide melt. Most HSE, including Pt and Os, are strongly concentrated in the nickel-sulfide melt that sinks to the bottom of the molten-rock–flux mixture. The Os and Pt are contained in alloy phases within the nickel-sulfide and can be separated by dissolving away the nickel-sulfide, leaving the insoluble alloys for further analysis. Although effective for Os and Pt separation from large samples, this approach works poorly for Re, as Re does not quantitatively partition into the NiS bead.

Overcoming these difficulties led to the adaptation of an old technique, Carius tube dissolution, developed originally by chemist Georg Carius in the mid-1800s. In this approach, the finely ground sample powder is mixed with isotopic spikes for Re, Pt, and Os, and a strongly oxidizing mixture of hydrochloric and nitric acids in a thick-walled glass or quartz tube [*Shirey and Walker*, 1995]. After adding sample, spike, and dissolution acids, the opening of the tube is welded shut so that it can withstand the high pressures that develop as the Carius tube is heated to over 240 °C, within a steel explosion shield. In this approach, all Os is converted to OsO_4 so that spike and sample Os can effectively mix in the gas phase that, in the sealed tube, cannot escape. Before the tube is opened, the OsO_4 is frozen by immersing the bottom of the tube in a mixture of solid-CO_2 and methanol. Once open, the solution is unfrozen, and mixed with either liquid bromine or carbon-tetrachloride, which effectively extracts the Os from the acid mixture. The Os is then back-extracted into hydrobromic acid, which also reduces the OsO_4 so that it is no longer volatile.

The problem with the Carius tube approach is that this mixture of acids is not particularly effective at dissolving most silicate minerals, the primary exception being olivine. The use of a glass or quartz tube precludes the addition of hydrofluoric acid that would help break down the crystal structure of silicates, but also would dissolve the tube walls in the process. The Carius tube technique thus provides accurate results only when the majority of the Re, Os, and Pt, are contained in metal or sulfide phases, which are readily dissolved in the combination of hydrochloric and nitric acids. Fortunately, in most, but not all cases, the majority of the Re, Os, and Pt in a rock can be dissolved using this approach. Improved dissolution can be achieved by pushing the Carius tube approach to higher temperatures, but the higher pressures produced require more specialized apparatus, such as high-pressure ashers, to avoid explosions.

An alternative analytical approach, called sparging, allows the dissolution step to occur while the Carius tube is connected to an ICP-MS. In this case, as OsO_4 is formed, a steady flow of argon gas through the dissolution vessel carries the OsO_4 into the torch of the ICP-MS where it is disassociated and ionized for mass analysis in the mass spectrometer [*Hassler et al.*, 2000]. The difficulty of controlling the release rate of OsO_4 during dissolution, however, makes high-precision analysis using this approach difficult.

The refractory nature of Os and its high ionization potential lead to exceedingly low efficiencies for the production of Os^+ ions by the type of thermal ionization mass spectrometry used for Sr and Nd. This led to early Os analyses being accomplished by using an ion-probe to sputter-ionize chemically separated Os off a small plate onto which separated Os had been deposited [*Allegre and Luck*, 1980]. An alternative used a technique called resonance ionization mass spectrometry where separated Os in the source of a mass spectrometer is illuminated with two lasers whose wavelengths combine to selectively ionize neutral atomic Os atoms [*Walker and Fassett*, 1986]. When it was found that Os very efficiently produces negatively charged oxide (OsO_3^-) ions when loaded with Ba or REE salts on a platinum filament, negative thermal ionization mass spectrometry (NTIMS) analysis took over as the technique of choice for Os isotope analysis [*Creaser et al.*, 1991; *Volkening et al.*, 1991]. With this technique, ionization efficiencies are sufficient to easily analyze picograms of Os, with the primary limit on sample size being the Os blank added during sample preparation.

7.5 GEOCHRONOLOGIC APPLICATIONS

7.5.1 Meteorites

The main application of Re–Os for chronological measurements applied to meteorites is for iron-rich meteorites. Because Re and Os are both refractory elements, undifferentiated stony meteorites that have not experienced metal–silicate separation display quite a small range in Re/Os ratios (Fig. 7.2a). Re–Os data for stony meteorites do show a correlation between Re/Os and $^{187}Os/^{188}Os$ consistent with a 4.56 Ga age for these rocks, but as Fig. 7.2a illustrates, the available data show considerable scatter, and on their own define rather poor isochrons. The scatter likely reflects disturbance of the Re–Os systematics due to the shock metamorphism experienced when the meteorite parent body collided with another object in space. Another potential explanation for the scatter is terrestrial weathering that will rapidly degrade the main carrier of Re and Os—iron metal and iron sulfide. Alteration could lead to selective loss of Re or Os, but if done recently, the isotopic composition of Os would not have time to record the change in Re/Os ratio, hence moving the

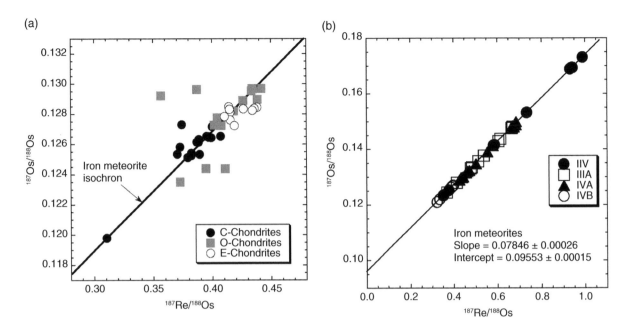

Fig. 7.2. Re–Os isotope data for various types of (a) stony chondritic meteorites and (b) iron meteorites. Data from *Walker and Morgan* [1989], *Shen et al.* [1996], *Smoliar et al.* [1996] and *Walker et al.* [2002].

point for the altered sample to the right or left away from the 4.56 Ga isochron.

Given the very strong preference of Pt, Re, and Os for iron metal over silicate, the iron metal fraction in meteorites contains the majority of the budget of these elements in the bulk rock. If the meteorite parent body were to experience melting, the density difference between iron metal and silicate will lead to gravitational separation of iron from silicate melt and the formation of an iron metal core. The preference of Re, Os, and Pt for metal over silicate is so strong that the metal ends up with such a large proportion of these elements during core formation that there is very limited fractionation between these three elements. As a result, many iron meteorites, which represent the fragmented cores of small planetesimals, have Re/Os ratios similar to those in chondrites (Fig. 7.2b). When the core of the planetesimal begins to cool and crystallize, both Re and Os will concentrate in the solid metal, but Os has a much stronger preference for the solid over liquid metal than does Re. Consequently, while both the Re and Os concentration will decrease in the remaining liquid present during core crystallization, the Re/Os ratio of the liquid fraction will increase as crystallization of the core proceeds. The process of core crystallization on planetesimals, as sampled by different iron meteorites, created a range in Re/Os ratios and the potential to construct an isochron from the Re–Os system. Figure 7.2b shows that the Os isotopic composition of many different chemical types of iron meteorites indeed shows a very strong correlation with Re/Os ratio. For this correlation to be an isochron, all the samples used to define the isochron must have formed at the same time and started with the same initial $^{187}Os/^{188}Os$. Geochemical studies of the iron meteorites indicate that not all derive from the same parent body, so both their

synchronicity of formation and whether they started with identical Os isotopic compositions can be questioned. As discussed in Chapter 6, these two requirements are constant concerns for "whole rock" isochrons—isochrons constructed from analyses of different rocks rather than minerals separated from one rock.

One approach to test whether this correlation indeed is an isochron is to consider the individual chemical groups of meteorites separately, as presented in Table 7.3.

Both the slopes and intercepts calculated for all groups of iron meteorites separately overlap within uncertainty suggesting that these different meteorite groups indeed formed at roughly the same time and with the same initial Os isotopic composition. When comparing isochron results, however, keeping in mind the significance of the uncertainties on the slopes of the lines is critical in the interpretation of the meaning of the correlation. For example, the most precise line in Fig. 7.2b is defined by the IIA iron meteorites, but even the small uncertainty on this slope corresponds to an age uncertainty of ±14 Ma, a time interval that we now know from other dating approaches covers a very large fraction of the formation and differentiation interval of planetesimals. The relatively high precision of the IIA isochron slope derives, in part, because this group of meteorites shows the largest range in Re/Os ratios, and hence $^{187}Os/^{188}Os$. While this

Table 7.3 Line fitting results for the data shown in Fig. 7.3b

Meteorite Group	Slope	Intercept	MSWD
IIA	0.07844 ± 0.00025	0.09557 ± 0.00017	8
IIIA	0.07889 ± 0.00041	0.09525 ± 0.00022	2.7
IVA	0.07890 ± 0.00170	0.09526 ± 0.00091	49

MSWD, mean square weighted deviation.

Table 7.4 Re–Os concentration data for molybdenites

Sample	Re (ppm)	^{187}Re (10^{-6} mol)	^{187}Os (ppb)	^{187}Os (10^{-10} mol)	Common Os[a] (ppb)	^{188}Os (10^{-13} mol)	^{187}Re/^{188}Os	^{187}Os/^{188}Os	Age (Ma)
LP-1	268.4	0.9023	29.67	1.587	0.4	2.784	3,241,000	570.0	10.56
LP-3b	373.0	1.254	41.43	2.216	0.4	2.784	4,504,000	796.0	10.61
LP-5	450.7	1.515	51.20	2.739	1.4	9.744	1,555,000	281.1	10.85

Source: *Stein* [2014]. Reproduced with permission of Elsevier.
[a] Common Os is the sum of the abundance of all the Os isotopes using some assumed, low, value for the ^{187}Os/^{188}Os ratio.

contributes to a precisely determined slope for the best fit line, the IIA data also scatter about the best fit line more than expected given analytical uncertainty, as indicated by the relatively large MSWD (mean square weighted deviation—a measure of the magnitude of scatter of the data about the best fit line—see Chapter 4). A MSWD equal to 1 means the scatter of the points about the isochron is exactly the amount expected from the uncertainty of the individual points. A MSWD < 1 means less scatter than the data uncertainty would predict, likely as a result of overestimation of the individual errors, and a MSWD > 1 indicates that the data scatter away from the isochron outside of the individual uncertainties of each data point. The scatter of some points away from a single line could reflect postformation disturbance of the Re–Os system in some of these samples. Alternatively, displacement from a single line could be an indication of a sample that either formed at a different time, or with a different initial Os isotopic composition, from the other samples. These complications should always be kept in mind when interpreting the significance of the age indicated by any whole-rock isochron. Often the best way to resolve these issues is with additional data. For example, examination of the Hf–W isotope system (Chapter 14) in IIA irons shows them to have formed synchronously, within about 1 Ma of the beginning of solar system formation. This result suggests that the cause of the scatter of the IIA data about the Re–Os isochron is postformation disturbance of the Re–Os system in the samples used to construct the isochron.

7.5.2 Molybdenite

The use of the Re–Os system to date molybdenite formation was one of the first applications of Re–Os, and with improvements in analytical technique, has become an important chronometer for Mo ore formation [*Stein et al.*, 2001; *Stein*, 2014]. Molybdenite presents an interesting problem for Re–Os dating, however, in that it can have original Os concentrations so low that an analytical blank can impede the determination of the abundances of the nonradiogenic Os isotopes. In this case, one can calculate an age directly from the measured abundance of ^{187}Re and ^{187}Os with the equation:

$$^{187}\text{Os (moles)} = {}^{187}\text{Re (moles)} \times \left(e^{\lambda t} - 1\right) \quad (7.5)$$

An example of this approach is provided by data from the Los Pelambres copper–molybdenum porphyry deposit in Chile (Table 7.4). Ages can be calculated for individual samples using

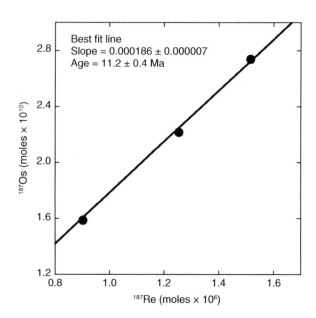

Fig. 7.3. Re–Os isochron plot for the molybdenite samples in Table 7.4.

the equation above or from a combination of samples using the isochron approach (Fig. 7.3). Ages also could be calculated using the conventional isochron plots of ^{187}Re/^{188}Os versus ^{187}Os/^{188}Os, but the concentration of ^{188}Os is so low in the analyzed samples that the data for ^{188}Os abundance are the most uncertain quantity measured in this data set. That uncertainty would then propagate into the ratios normalized to ^{188}Os. For example, for the ~20 mg sample sizes used in the measurements reported in Table 7.4, the total amount of all Os isotopes, excluding the radiogenic contribution to ^{187}Os, extracted from the samples was 8 pg. The reported blanks in these analyses were 2–3 pg, which is close to the limit possible for samples of this size. With this range in blanks and sample sizes, the blank correction contributes an uncertainty of 12% to the abundance of ^{188}Os. In contrast, the reported uncertainties in the abundance of ^{187}Re and ^{187}Os are about 0.1% and 0.1–0.2%, respectively. Treating these data in the conventional approach, e.g. normalizing the data to ^{188}Os, thus would greatly increase the uncertainty of the measured data.

In order to calculate accurate ages, however, one must still correct for the amount of ^{187}Os incorporated into the molybdenite at the time of its formation. The abundance data for ^{187}Os in Table 7.4 are calculated assuming that the initial Os in the

Table 7.5 Ni, Cu, Re, and Os concentration and Os isotope data for sulfide inclusions

Sample	Ni wt%	Cu wt%	Re (ppb)	Os (ppb)	^{187}Re/^{188}Os	^{187}Os/^{188}Os
DP4	5.8	2.3	196	6.56	146.4	7.67
DP6	5.2	2.6	378	13.8	133.8	6.722
DP9a	7.5	5.2	215	38.46	27.15	1.306
DP9b	8.4	2.2	189	44.38	20.69	1.303
DP9 average	7.9	3.7	201	41.66	23.43	1.304
DP11	10.3	4.4	598	1760	1.64	0.2376

Source: *Richardson et al.* [2001]. Reproduced with permission of Elsevier.

molybdenite had a typical crustal Os isotopic composition with ^{187}Os/^{188}Os ~0.2. If the Os incorporated into the molybdenite at the time of its formation instead was sourced from some more radiogenic crustal material, for example one with ^{187}Os/^{188}Os = 1.0, then the ^{187}Os concentration in sample LP1 would be reduced to 1.585×10^{-10} mol, and the calculated age consequently reduced to 10.54 Ma, a difference of only 0.2%. In samples with this high a Re/Os ratio, the correction for initial Os thus is not very significant, but for samples with lower Re/Os ratios, choosing the right value for initial Os isotopic composition is of more concern. Assuming a range of initial ^{187}Os/^{188}Os ratios in the age calculations for each sample will reveal the sensitivity of the age to the assumed initial ^{187}Os/^{188}Os. If the lower Re/Os ratio is caused by a larger amount of initially incorporated Os, for example in the iron meteorites shown in Fig. 7.2, then the traditional approach involving the normalization to ^{188}Os is preferred because, in this approach, the initial ^{187}Os/^{188}Os does not have to be assumed, it is the *y*-intercept of the isochron defined by the data.

7.5.3   Other sulfides, ores, and diamonds

Molybdenite provides an extreme example of the concentration of Re over Os in a mineral, but sulfides, in general, can strongly concentrate all of the HSE including Pt, Re, and Os. Most sulfides, however, do not discriminate between Re and Os to the same degree as does molybdenite. The ability of sulfides to strongly concentrate these elements allows the Re–Os system to be particularly useful in chronological studies of the diverse suite of rocks that contain sulfides, from mantle peridotite, to sulfide-ores, to organic-rich sediments. The direct application of Re–Os for dating ore minerals, rather than more traditional radiometric techniques that often date silicate alteration minerals assumed to form at the same time as the ore, provides a critical check on the time sequence, and hence process, of ore formation in situations where several events may be superimposed in the production of a given ore deposit.

Concentrations of Re and Os in sulfides often are sufficient, and analysis techniques so sensitive, that individual sulfide grains can be used for age determinations. A good example is provided by the sulfides contained within diamonds [*Pearson and Shirey*, 1999]. Diamonds form deep in the upper mantle and often contain a wide variety of mineral inclusions, with iron-nickel sulfide

being the most abundant. Diamond is such a pure mineral that it contains insufficient quantities of any radioactive trace element that would allow the diamond to be dated directly. This is not so for the mineral inclusions captured during diamond growth. The first diamond dating results came from Sm–Nd measurements in garnet inclusions, but given the concentrations of Sm and Nd in those garnets, and the amounts needed for accurate analysis, hundreds of inclusion-bearing diamonds had to be sampled in order to provide enough garnet for analysis [*Richardson et al.*, 1984]. In contrast, Re and Os concentrations in sulfide inclusions in diamonds can be high enough for analysis of single sulfide grains as small tens of microns in size [*Pearson et al.*, 1998]. An example is provided in Table 7.5 that contains data for four sulfide grains separated from four diamonds from kimberlites from the Kimberley vicinity, South Africa.

All of the data together define a best fit line on a Re–Os isochron diagram (Fig. 7.4a) whose slope corresponds to an age of 2.97 ± 0.27 Ga. The data do not all lie within uncertainty of the best fit line, as indicated by the large MSWD of 46. Most of this scatter is due to sample DP9. The "a" and "b" portions of this sample represent two fragments of the same sulfide, broken when it was extracted from the diamond. The expanded scale of the same isochron diagram (Fig. 7.4b) shows that samples DP9a and 9b have overlapping Os isotopic compositions, but very distinct Re/Os ratios. The two fragments thus define a very shallow slope on the isochron diagram that corresponds to an "age" of 28 ± 92 Ma that is a hundred times lower than the age suggested by the line fit to all the data. Why this discrepancy?

This example exposes many of the requirements that must be met in order for an isochron to provide an accurate age. One such requirement is that all the samples used to define the isochron must have formed at the same time with exactly the same initial Os isotopic composition. In this case, the individual sulfides came from separate diamonds that may have come from different kimberlites, so there is no guarantee that they meet this requirement. The one exception is sample DP9 where "a" and "b" are simply two fragments of the same sulfide. From other age dating information, we know that the Kimberley kimberlites were erupted at roughly 85 Ma, and a line of 85 Ma slope does indeed pass through DP9a and DP9b within their measurement error (Fig. 7.4b). Thus, a possible interpretation of the data is that the sulfides formed at the same time as the kimberlite eruption. In this case, the older 2.89 Ga age may have no chronological

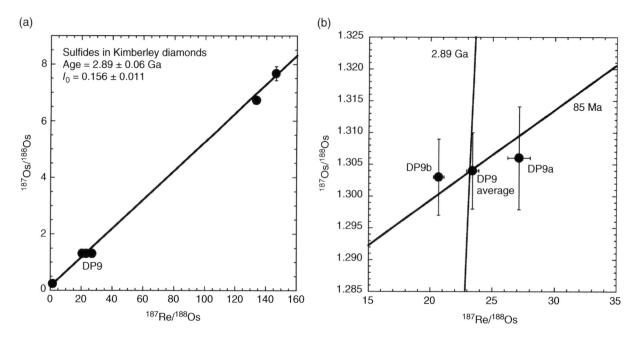

Fig. 7.4. (a) Re–Os isochron diagrams for the sulfide data listed in Table 7.5; (b) expanded scale of the region around the DP9 point in (a). (Source: adapted from *Richardson et al.* [2001]. Reproduced with permission of Elsevier.)

significance. Alternatively, the individual sulfides that define the 2.89 Ga line might all have formed at 85 Ma, but with distinct Os isotopic compositions representing their derivation from some much older source. In this case, for the 2.89 Ga age to have any chronological significance would require that the steep correlation of Re/Os versus $^{187}Os/^{188}Os$ was inherited from the source material of the individual sulfides, implying no fractionation of Re from Os during formation of the sulfides. While this is not impossible, neither is it likely.

Another requirement for an isochron to have chronological significance is that none of the samples on the isochron can have experienced a change to their Re/Os ratios, or Os isotopic composition, other than simple ingrowth of ^{187}Os, after their initial formation. This requirement is always hard to verify. In this case, however, because the individual sulfide grains were contained within diamond, which has neither Re nor Os, and the tight atomic structure of diamond leads to extremely slow diffusion of elements through diamond, the included sulfide grains likely were protected from chemical exchange with their surroundings and thus experienced no modification of their Re–Os system at the whole-grain scale since their entrapment in the diamond. Sample DP9, on the other hand, is two fragments of the same sulfide, with the two fragments showing quite different Ni and Cu contents (Table 7.5). Before their capture by the kimberlite, these diamonds were residing in the shallow mantle beneath Kimberley, likely at temperatures of 1000 °C or more. At this temperature, Ni-rich and Cu-rich iron sulfides dissolve into one another to form a phase known as monosulfide solid solution, so DP9 likely was a chemically homogeneous grain of monosulfide solid solution when it was in the mantle.

At surface pressures and temperatures, however, monosulfide solid solution exsolves into two different sulfides, Ni-rich pyrrhotite and Cu-rich chalcopyrite. Re follows Cu into the chalcopyrite while Os prefers pyrrhotite. The limited variation in Os isotopic composition, but quite different Re/Os ratios, of the two fragments of DP9 suggest that this separation into zones with different Re/Os ratios occurred within the last tens of million years, most likely at the time the diamond was brought to the surface by the kimberlite and cooled to surface temperatures. If so, combining the data for the two fragments of DP9 provides the composition of the whole sulfide when it was in the mantle. The combination of DP9a and DP9b results in a point that falls on the 2.89 Ga isochron defined by the sulfides from the other diamonds. In this case, a possible explanation for the data is that the diamonds and their included sulfides formed in the sub-Kimberley mantle at 2.89 Ga and that the transformation from monosulfide solid solution to mixed chalcopyrite-pyrrhotite occurred when the diamonds and their included sulfides were brought to the surface by the kimberlite. The fact that this 2.89 Ga age overlaps both the peak in crust formation age near Kimberley and the ages of other mantle samples from this area provides additional support for the interpretation that these diamonds, and their inclusions, formed almost 3 billion years ago.

This case provides a good example of the need for background knowledge of the samples used to construct isochrons, in Re–Os or any other radiometric system. Not all correlations between Re/Os and $^{187}Os/^{188}Os$ need reflect accurate ages for geologic events, so it is always wise to question whether the samples meet the requirements for that correlation to be an isochron.

7.5.4 Organic-rich sediments

Another nearly unique application of the Re–Os system is to the dating of organic-rich sediments and oil. Besides being important for the understanding of the history of petroleum generation, organic-rich black shales are a common marker of the anoxic conditions associated with extinction events, so having the ability to directly determine ages for these rocks can provide important constraints on the geologic timescale. Organic-rich shales both concentrate Re and Os and are characterized by high Re/Os ratios, so they provide good targets for Re–Os geochronology. Table 7.6 provides the data for one example of measurements made on shales from the Exshaw Formation in Alberta, Canada that marks the Devonian–Mississippian boundary. Unlike the molybdenite example described in section 7.5.2, the black shale data in Table 7.6 have enough Os for an accurate measurement of ^{188}Os abundance, and the shales incorporate enough Os during their formation to require the more conventional approach to data analysis using the isochron approach. These data define an excellent Re–Os isochron (Fig. 7.5).

Once again, understanding of the geologic history of the dated samples is critical in order to properly interpret the isochron age obtained. Shales of this nature comprise two components: (i) detrital grains (mostly clays), quartz, and feldspar; (ii) Re and Os extracted from seawater, the so-called hydrogenous component, by the organic matter in the sediments. As was the case in the sediment Rb–Sr dating example given in Chapter 6, the detrital grains will carry Re–Os systematics established during the formation of these grains either by igneous events in some crustal section or through weathering and alteration of that crust associated with erosion and sediment transport. Unless all the detrital grains derive from exactly the same igneous source, which is unlikely, one would not expect them to meet the criteria to form a whole-rock isochron because they derive from different sources of different age and different initial Os isotopic composition. The hydrogeneous component, on the other hand, consists of the Os extracted from seawater, which can be assumed to be reasonably homogeneous, at least over short time intervals. Thus, if the sediments were deposited over a short time interval, in a well-mixed body of water, the hydrogeneous component could provide a Re–Os isochron that dates the sediment deposition.

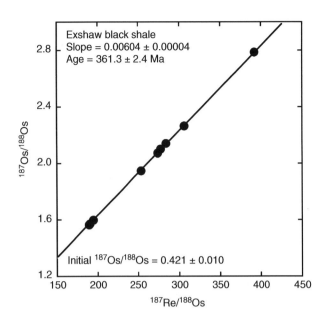

Fig. 7.5. Re–Os isochron diagram for the data listed in Table 7.6.

In samples with very high total organic carbon content, the hydrogenous component will dominate the Re–Os abundance in the sediment and thus "drown out" the signal from the detrital component. The authors of this study further enhanced the ratio of hydrogenous to detrital component in their analysis by using a dissolution method that is more effective on the hydrogenous component then on silicate minerals. Using this approach, the excellent isochron of Fig. 7.5 was produced. The age provided by the isochron was interpreted as reflecting the time of deposition of this shale, and the initial Os isotopic composition that of the Os in the seawater into which this sediment was deposited. Further support for the conclusion that the isochron provides the depositional age of this sediment is that zircons extracted from volcanic tuff layers above and below the Devonian–Mississippian boundary, along with estimates of the sedimentation rate between the tuff layers, provides an age of 360.7 ± 0.7 Ma for this boundary, overlapping the age determined from Re–Os when using a dissolution technique that primarily samples the hydrogeneous component. Being able to isolate the hydrogenous from detrital component is the key factor in allowing the Re–Os system to accurately date the deposition of these shales as it allows the results to fulfill two main criteria for an isochron— starting at the same time, from the same initial Os isotopic composition. These developments have allowed the Re–Os system to become one of the most useful approaches for determining precise deposition ages for organic-rich sediments.

7.5.5 Komatiites

Because of the compatibility of Os, but incompatibility of Re, during partial melting of mantle rocks, the Re/Os ratio of mantle-derived magmas can range widely and extend to very high

Table 7.6 Re–Os data from the Exshaw Formation Black Shales

Sample	Re (ppb)	Os (ppb)	^{187}Re/^{188}Os	^{187}Os/^{188}Os
DS53	15.63	0.3678	253.49	1.9475
DS54	17.54	0.3759	283.84	2.1405
DS55A	15.47	0.3414	273.73	2.0725
DS55B	16.36	0.3635	277.30	2.1001
DS55C	21.18	0.4263	306.14	2.2650
DS56	16.4	0.4933	190.35	1.5704
DS57	35.97	0.5958	391.83	2.7866
DS58A	15.18	0.4587	189.33	1.5632
DS58B	16.61	0.4906	194.46	1.5971

Source: *Selby et al.* 2005. Reproduced with permission of Elsevier.

Table 7.7 Re–Pt–Os data for komatiite samples from the Komati Formation, South Africa

Sample	Re (ppb)	Pt (ppb)	Os (ppb)	$^{187}Re/^{188}Os$	$^{190}Pt/^{188}Os$	$^{187}Os/^{188}Os$	$^{186}Os/^{188}Os$
BV03 whole rock	0.0374	3.49	0.9443	0.1802		0.11430	
BV03 olivine	0.145	0.268	0.6612	0.1055		0.10948	
BV03 chromite	0.2414	6.21	17.87	0.0649		0.10714	
BV10 whole rock	0.0410	3.33	1.263	0.1560		0.11296	
BV10 olivine	0.00335	0.08	0.3557	0.0452		0.10569	
BV10 chromite	0.2376	6.69	51.33	0.0222		0.10484	
BV15a	0.0525	3.4	3.109	0.0754	0.001316	0.1080382	0.1198336
BV15b	0.0461		3.422	0.0832	0.001310	0.1082836	0.1198344
BV16	0.1790	3.84	2.425	0.3779	0.001307	0.1258510	0.1198343

Source: *Puchtel et al.* [2014]. Reproduced with permission of Elsevier.

values, which potentially can be used to provide precise forma-tion ages for these magmas. A dating example of komatiitic lavas from the type section of the komatiite, the Komati Formation of the Barberton greenstone belt in South Africa, is given in Table 7.7. Komatiites are large degree (20–30%) partial melts of the mantle. Because of the high-degrees of melting involved, the Re and Os concentrations in komatiites are higher than in lower degree melts such as basalt. While the higher concentra-tions make analyses easier, as a result of their higher extents of melting, the range in Re/Os ratios in komatiites is much smaller than in basalts, which then limits the precision with which Re–Os isochrons can be defined. In the example in Table 7.7, two sam-ples were analyzed for whole rock, olivine, and chromite mineral separates. Using the data for the two minerals and whole rock, the isochrons for each sample are 3702 ± 150 Ma (BV03) and 3621 ± 451 Ma (BV10). The large uncertainties on these ages primarily reflect the limited range in Re/Os and $^{187}Os/^{188}Os$ in the samples, as shown in Fig. 7.6, which limits the precision with which the slope of the isochron can be defined. Increasing the range in Re/Os ratios by including data for three additional whole rocks improves the isochron and increases the precision of the age to 3483 ± 91 Ma (Fig. 7.6).

The Komati data set provides a good example of the advan-tages and disadvantages of isochrons determined from minerals separated from a single igneous rock as opposed to the use of a group of whole-rock samples from what is assumed to be a related sequence of rocks. Minerals that crystallize from a high-temperature magma like a komatiite have a high probability of forming with the same initial Os isotopic composition, that of the host magma, and over a time period that is short compared to the age of the lavas. Internal isochrons of igneous rocks, by definition composed of data for the whole rock and the minerals they contain, thus most obviously match the isochron require-ment of all data starting at the same time, from the same initial isotopic composition. Unfortunately, however, the minerals in the komatiite simply do not strongly fractionate Re from Os and hence have Re/Os ratios not much different from the host magma or the whole rock. The limited spread in Re/Os, and hence $^{187}Os/^{188}Os$, ratios between minerals and whole rocks leads to an isochron whose slope is not defined to high precision,

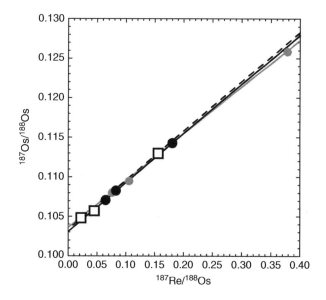

Fig. 7.6. Isochron plot for the data of Table 7.7. Filled black circles are the data for sample BV03, open squares for sample BV10, and gray circles are the data for the additional whole rocks. The isochrons for the individual samples have large slope, and hence age, uncertainties as a result of the limited spread in Re/Os ratios and consequently $^{187}Os/^{188}Os$ ratios. Adding the one sample with high Re/Os ratio dramatically improves the precision of the slope of the best fit line.

resulting in a relatively imprecise age. The 219 Ma difference between the BV03 mineral isochron and the whole-rock iso-chron is a long time, even by geologic standards, but the whole-rock and BV03 internal isochron in fact overlap within their respective errors. Adding in the three additional whole-rock analyses expands the range in Re/Os and $^{187}Os/^{188}Os$ ratios in this data set, which results in a more precisely defined slope for the isochron. For this whole-rock isochron to provide an accurate age for these rocks, however, all the different lava flows used to define the isochron must have formed over a short time period relative to the resolution of the Re–Os dating scheme, and they all must have formed with the same initial Os isotopic composi-tion. While the former requirement is likely in a sequence of lava flows like those in the Komati Formation, there is no guarantee that the latter requirement is fulfilled as the different lavas may have been derived from different mantle sources, or experienced

Table 7.8 Re–Os analyses of a basalt from the FAMOUS area of the Atlantic mid-ocean ridge

Sample	Re (10⁻¹² g/g)	Os (10⁻¹² g/g)	^{187}Re/^{188}Os	^{187}Os/^{188}Os	Uncertainty
Sulfide 1	713,000	158,000	21.6	0.1310	0.0005
Sulfide 2	323,0000	679,000	22.7	0.1305	0.0009
Spinel	126,000	1200	514	0.1874	0.0045
Olivine 1	687	11.39	288	0.1399	0.0023
Olivine 2	1153	2.242	2492	0.2411	0.0074
Plagioclase	277.9	0.925	1456	0.1892	0.0028
Glass 1	3518	13.16	1281	0.1816	0.0041
Glass 2	3163	11.53	1310	0.1846	0.0024
Matrix 1	1489	295.7	24.0	0.1359	0.0006
Matrix 2	1264	248.6	24.3	0.1374	0.0005
Matrix 3	1151	234.6	23.8	0.1359	0.0009

Source: adapted from *Gannoun et al.* [2004].

different amounts of contamination from the crust they penetrated to erupt, both of which could result in variable initial ^{187}Os/^{188}Os. In this case, additional assurance on the accuracy of the Re–Os age is derived from the observation that the whole-rock Re–Os isochron age agrees well with the 3482 ± 5 Ma U–Pb zircon age obtained from an evolved gabbro within the Komati sequence.

As the data in Table 7.7 show, the ^{186}Os/^{188}Os ratio in three komatiites shows a range of only 7 ppm. Using current techniques, this ratio can only be measured to a precision of 5–10 ppm, so the data in Table 7.7 show too little range in ^{186}Os/^{188}Os to define a Pt–Os isochron. Because of the small contribution of ^{190}Pt decay to ^{186}Os, the Pt–Os chronometer has found only limited application.

7.5.6 Basalts

Lower degree melts, such as basalt, start with lower Os concentrations (of order 10^{-10} g/g) compared to komatiites. Fractionation of the magmas, particularly when the fractionation involves minerals such as sulfides and chromite, can lead to rapid reduction of Os concentration in the magma, driving Os concentrations down to the point where measurement becomes difficult. At such low Os concentrations, contamination by surrounding rocks, or simply through alteration, also can be a significant problem. These factors complicate the application of Re–Os to dating basalts, but are compensated somewhat by the fact that basalts can have both extremely high, and a wide range of, Re/Os ratios, as the example in Table 7.8 illustrates.

These data well illustrate a number of aspects of Re and Os geochemistry. The most obvious is the very strong concentration of Re and Os into sulfide and their exclusion from silicate minerals. If the glass is assumed to be the best representative of the magmatic Re and Os concentrations, then one can deduce that Os is more compatible than Re in the sulfides in this rock because the Re/Os ratio of the sulfide is significantly lower than that of the glass. Another interesting feature of these data is the very different Re/Os ratios of the two olivine separates. Equilibrium partitioning of Re and Os between melt and olivine should result in a limited range in Re/Os ratio in the olivine. The large difference

in Re/Os ratios between the two olivine separates suggests that these separates were not pure olivine, but were mixtures of olivine with another phase. In a mid-ocean ridge basalt, olivine crystallizes with chromite and sulfide, both of which would be expected to have a much lower Re/Os ratio than olivine. A possible explanation of the olivine data is thus that the olivine in the two separates included small amounts of chromite or sulfide inclusions. If the mineral inclusions in the olivine formed at the same time, and from material with the same Os isotopic composition, as the rest of the minerals in the rock, as is likely given that they all crystallized from the same magma, then the presence of the mineral inclusions does not compromise the use of these olivines to help define the isochron for the basalt.

Plotting all these data on an isochron diagram (Fig. 7.7) shows considerable scatter about any single line. The most deviant point is the measurement for the spinel. In this study, the spinel separate weighed 0.51 mg and hence provided a total amount of Os

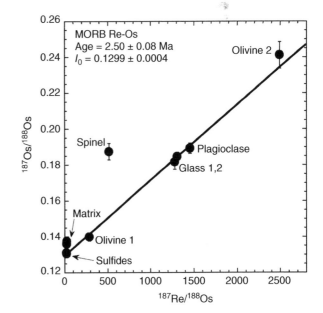

Fig. 7.7. Re–Os isochron diagram for the samples from the mid-ocean ridge basalt data shown in Table 7.8.

of 0.6 pg. As a result of the small sample size, blank correction is critical as is the chance that the spinel was slightly contaminated with seawater Os through alteration. The other aberrant points are the three measurements of "matrix" that consisted of the fine-grained crystalline material in between the phenocrysts of plagioclase and olivine. Because of its small grain size and porosity, such matrix material is the first to be affected by seafloor weathering. The Os dissolved in modern seawater has a $^{187}Os/^{188}Os$ ratio of just less than 1, compared to the values of ~0.13 measured in the sulfides. Even a small addition of this Os to the matrix through weathering would elevate the $^{187}Os/^{188}Os$ ratio of the matrix. For example, 121 mg was used in the analysis Matrix-1, providing a total of 36 pg of Os. If only 0.7% (0.25 pg) of the Os measured in this sample derives from seawater, this would raise the measured $^{187}Os/^{188}Os$ from 0.13 to 0.136. Excluding the spinel and matrix points, the remaining data define a precise line of slope = 0.0000416 that corresponds to an age of 2.50 ± 0.08 Ma, with an initial $^{187}Os/^{188}Os = 0.1299 \pm 0.0004$. The initial Os isotopic composition of this isochron overlaps the value of 0.1296 assumed for the bulk mantle. These data illustrate both the potential of Re–Os to provide precise ages for young basaltic rocks, and the difficulty in realizing this potential due to the low Re and Os contents of silicate minerals.

7.5.7 Dating melt extraction from the mantle—Re–Os model ages

The removal of partial melt from a portion of the mantle will leave the residue of melting with a much reduced Re/Os ratio. This geochemical property of the Re–Os system allows for one of its most commonly used applications, determination of the timing of melt extraction events from the mantle [*Walker et al.*, 1989; *Carlson et al.*, 2005]. The peridotites that make up the main volume of the mantle generally have Os concentrations in the 10^{-9} g/g range; concentrations that are high enough to both make Os isotope measurements relatively easy, and leave the Os isotopic composition of mantle rocks relatively immune to changes caused by alteration, contamination, or metasomatism by infiltrating melts that in most cases will have much lower Os concentrations.

Low degrees of melting, such as involved in the production of basaltic magmas, will leave some Re in the residue, but higher degrees of melting, such as involved in komatiite production, can completely remove Re from the residue. In the ideal case where a section of mantle is subjected to different amounts of melt extraction during the same melting event, the resulting range in Re/Os ratio in the mantle residue could be used to construct a Re–Os isochron that would give the time of the melt extraction event. Unfortunately, well defined Re–Os isochrons for mantle rocks are rare. One reason is that while Os in the residue is in reasonably high concentration, the Re concentration in a melt residue can become so low that it can be very sensitive to Re introduction via alteration or from addition by melts that may

have passed through the sample at some time after the original melt extraction event [*Walker et al.*, 1989]. This latter problem is particularly acute in the study of mantle xenoliths (fragments of mantle carried to the surface by an explosive volcanic eruption). By definition, the xenolith is unrelated to the magma that serves only as its transport agent. If the xenolith is a mantle residue of some past melting event, its low Re concentration makes it susceptible to significant Re addition should it be infiltrated by even a very small amount of the host magma, as such magmas tend to have high Re contents. One potentially could avoid this problem by separating clean minerals from the xenolith, but as described in previous examples, the low concentrations of Re and Os in the dominant silicate minerals of a peridotite make such analyses very difficult.

To avoid the problem of Re contamination of mantle-derived rocks, two approaches unique to the Re–Os system of mantle rocks have been devised. Table 7.9 and Fig. 7.8 provide an example of the problem and one approach to a solution. Fertile mantle, defined as mantle that has not experienced the extraction of partial melts, has Al_2O_3 concentrations near 4 wt%. Like Re, aluminum is an incompatible element during melting, so residues of melting will have lower Al_2O_3 contents that depend on how much melt has been extracted. Figure 7.8a shows that Al_2O_3 and Re concentrations display a very rough correlation in these samples, indicating that these two elements are removed in roughly equal proportions during the melt extraction.

The Re–Os isochron for these samples (Fig. 7.8c) shows considerable scatter with a best fit line corresponding to an age of 880 Ma, but with an uncertainty of about the same magnitude as the age. In contrast, the samples define a much better correlation between their Os isotopic composition and Al concentration (Fig. 7.7b), a feature seen in many studies of ultramafic rocks. This commonly observed correlation between Al content and Os isotopic composition in peridotites has been given the name "alumichron" [*Reisberg and Lorand*, 1995]. An interpretation of the improved correlation when using Al_2O_3 concentration instead of Re/Os ratio is that Al is less severely affected by secondary events, such as alteration or metasomatism by passing magmas, than is Re. Because ^{187}Os is not generated by the decay of Al, however, there is no direct way in which the alumichron can be turned into an age. The solution is the Re–Os model age.

As with other isotope systems like Rb–Sr, Sm–Nd, and Lu–Hf, discussed in Chapter 6, a model age in the Re–Os system reflects the time when the Os isotopic composition of a given sample diverged from some model evolution path. Re–Os model ages are generally referenced to the Os isotope evolution expected for undifferentiated mantle. Study of the Re–Os system of mantle peridotites suggests that the best approximation for the $^{187}Os/^{188}Os$ of the modern, undifferentiated, mantle is 0.1296 [*Meisel et al.*, 2001]. The initial $^{187}Os/^{188}Os$ for the solar system is 0.09524, as determined from the isochron for the IIIA group of iron meteorites discussed previously [*Smoliar et al.*, 1996]. Connecting these two points for a 4.567 Ga age for the solar system indicates that the mantle has had, on average, a

Table 7.9 Al, Re, and Os concentrations and Os isotopic composition of peridotite samples from the Pyrenees

Sample	Al$_2$O$_3$ (wt%)	Re (ppb)	Os (ppb)	^{187}Re/^{188}Os	^{187}Os/^{188}Os
71-321	3.10	0.267	4.53	0.281	0.1276
71-324	2.96	0.248	3.92	0.302	0.1259
71-322	1.53	0.055	3.91	0.067	0.1173
72-442	1.18	0.035	2.89	0.058	0.1178
73-104	1.04	0.013	5.95	0.010	0.1153
71-325	0.62	0.044	3.86	0.054	0.1151
71-336	3.47	0.164	3.43	0.228	0.1284
71-339	3.41	0.218	3.47	0.300	0.1268
71-335	2.86	0.233	4.12	0.270	0.1226
72-425	2.77	0.146	4.15	0.168	0.1225
FON-4	3.77	0.283	3.81	0.354	0.1288
FON-3	0.72	0.037	2.77	0.064	0.1211
70-354	3.80	0.165	3.35	0.235	0.1295
Sem 2	3.63	0.364	3.93	0.442	0.1271
70-116	3.55	0.167	3.52	0.226	0.1280
70-5	3.40	0.082	3.05	0.128	0.1259
71-264	2.13	0.118	4.12	0.136	0.1221
DES 7	4.05	0.202	3.66	0.263	0.1284
TUR 7	4.18	0.370	3.87	0.456	0.1284

Source: *Reisberg and Lorand* [1995]. Reproduced with permission of Nature Publishing Group.

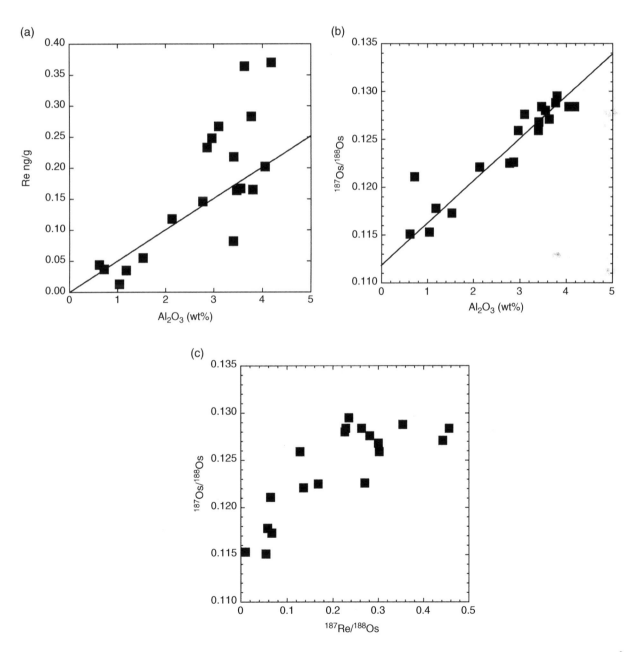

Fig. 7.8. Re, Os, and Al data for an obducted section of mantle peridotite in the Pyrenees. The line in part (a) delineates a constant Re/Al$_2$O$_3$ ratio of 5×10^{-9}.

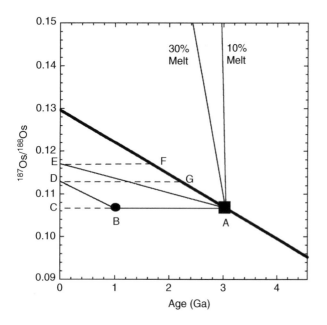

Fig. 7.9. Model evolution lines appropriate for the calculation of T_{MA} and T_{RD} Re–Os model ages.

$^{187}Re/^{188}Os = 0.4345$ over the age of the Earth. These parameters define the mantle Os isotope evolution curve shown in Fig. 7.9. Given the long half-life of ^{187}Re, the Os isotope evolution "curve" for the mantle is very close to a straight line over the history of the Earth. Model ages can be determined with respect to this curve by calculating the time when a sample's Os isotopic composition diverged from the mantle value because of some event that caused the Re/Os ratio of the sample to be different from that of the mantle. The evolution of the Os isotopic composition in any material is given by the equation:

$$\left(\frac{^{187}Os}{^{188}Os}\right)_p = \left(\frac{^{187}Os}{^{188}Os}\right)_t + \left(\frac{^{187}Re}{^{188}Os}\right)_p \times \left(e^{\lambda t} - 1\right) \quad (7.6)$$

where p is present day, t is some time in the past, and λ is the decay constant of ^{187}Re. The model age of a sample is the time at which its Os isotopic composition is the same as that of the mantle, in other words, the t when $(^{187}Os/^{188}Os)_{Sample} = (^{187}Os/^{188}Os)_{Mantle}$. Using this requirement in the equation above, and solving for t, gives the model age as:

$$T_{MA} = \frac{1}{\lambda} \times \left[\ln \left[\frac{\left(\frac{^{187}Os}{^{188}Os}\right)_{Mantle} - \left(\frac{^{187}Os}{^{188}Os}\right)_{Sample}}{\left(\frac{^{187}Re}{^{188}Os}\right)_{Mantle} - \left(\frac{^{187}Re}{^{188}Os}\right)_{Sample}} \right] + 1 \right] \quad (7.7)$$

This type of model age is commonly referred to as the Re–Os mantle model age, or T_{MA}, and is identical in its approach to the model ages discussed in Chapter 6. There is another model age approach unique to the Re–Os system. This model age, called "time of Re–depletion," or T_{RD} [*Walker et al.*, 1989], calculates the time when a sample had its Re/Os ratio lowered to zero as

the result of complete extraction of Re during a partial melting event. The T_{RD} approach modifies the T_{MA} equation (7.7) by assuming that $(^{187}Re/^{188}Os)_{Sample} = 0$, so:

$$T_{RD} = \frac{1}{\lambda} \times \left[\ln \left[\frac{\left(\frac{^{187}Os}{^{188}Os}\right)_{Mantle} - \left(\frac{^{187}Os}{^{188}Os}\right)_{Sample}}{\left(\frac{^{187}Re}{^{188}Os}\right)_{Mantle}} \right] + 1 \right] \quad (7.8)$$

Re–depletion model ages are particularly useful in the study of mantle xenoliths because they help avoid the consequences of Re contamination from the host magmas that transported the xenolith to Earth's surface, particularly for xenoliths transported by young eruptions. For older eruptions, one uses the measured Re/Os and $^{187}Os/^{188}Os$ ratio in the xenolith to back calculate the Os isotopic composition to the time of eruption, which is then used for $(^{187}Os/^{188}Os)_{Sample}$ in the equation above [*Walker et al.*, 1989].

Figure 7.9 illustrates the workings of Re–Os model ages as well as some interpretive concerns with Re–Os model age approaches. The data for this figure are listed in Table 7.10. In the figure, the Os isotope evolution of the mantle is shown by the solid black line. In the model shown in the figure, at 3 Ga (point A), two portions of the mantle experience a partial melting event. One portion is melted to 10%, the other to 30%, and all the melt is removed. Both melts have high Re/Os ratios and hence evolve ^{187}Os rapidly with $^{187}Os/^{188}Os$ rising off the top of the figure. The residue of 30% partial melt removal loses all of its Re to the extracted melt, leaving the residue with a Re/Os ratio equal to zero. Ingrowth of ^{187}Os in this residue thus stops, and as a result, if nothing else happens to this material, its $^{187}Os/^{188}Os$ ratio measured today (point C) will be the same as it was when the residue was formed at 3 Ga. In contrast, the residue of 10% partial melt removal has a nonzero Re/Os ratio, and hence continues to evolve ^{187}Os to point E on Fig. 7.9. Using the data in Table 7.10, both the melts and residues will have T_{MA} ages of 3 Ga. For the one example of the residue of 30% partial melting, the T_{RD} and T_{MA} ages are equal because this sample's Re/Os ratio is zero. The T_{RD} for the residue of 10% melting, however, is 1.7 Ga (point F), which grossly underestimates the time of melt extraction in this model. The T_{RD} values of the two melts will be some time way into the future as the mantle will require a long time to catch up to

Table 7.10 Re–Os parameters for melts and melt-residues created during a partial melting event at 3 Ga

Material	Re (ppb)	Os (ppb)	$^{187}Re/^{188}Os$	$(^{187}Os/^{188}Os)_p$
BSE	0.27	3.0	0.4345	0.1296
10% melt	2.0	0.10	96.4	5.048
30% melt	0.9	1.0	4.34	0.3298
Residue (10%)	0.08	3.3	0.1168	0.1133
Residue (30%)	0	3.8	0.0	0.1073

the high ^{187}Os/^{188}Os in these samples. The T_{RD} approach is not appropriate for materials with Re/Os ratios higher than that of the model reservoir.

Where the T_{RD} approach can be useful is in samples that have had their Re/Os ratios modified at some time after the initial melt extraction event. In the model illustrated in Fig. 7.9, the residue of 30% melting is captured by an explosive volcanic event occurring at 1 Ga (point B in Fig. 7.9) that brings the residue to the surface as a mantle xenolith. During transport, the xenolith is infiltrated by a small amount of the host melt. In this case, the addition of Re will reinitiate ingrowth of ^{187}Os, taking the sample from point B to D in Fig. 7.9. The T_{RD} for such a sample requires that the measured ^{187}Os/^{188}Os in the sample be corrected back to the time of the eruption of its host magma using its measured Re/Os ratio, after which the T_{RD} can be calculated by inserting the ^{187}Os/^{188}Os ratio this sample would have had at 1 Ga into the T_{RD} equation.

The T_{RD} approach also can be used with the alumichrons described earlier on the premise that the correlation between ^{187}Os/^{188}Os and aluminum content can be extrapolated to an Al_2O_3 content where the residue would have had a Re/Os ratio of zero. The exact value of Al_2O_3 where the Re/Os goes to zero is not known, so different applications of model ages using the alumichron approach use the ^{187}Os/^{188}Os at either $Al_2O_3 = 0$ or 1 wt%. For the correlation shown in Fig. 7.8, the Al_2O_3 versus ^{187}Os/^{188}Os correlation passes through ^{187}Os/^{188}Os = 0.1119 at $Al_2O_3 = 0$ or 0.1161 at $Al_2O_3 = 1$ wt%. Using these ^{187}Os/^{188}Os ratios in the T_{RD} equation provides T_{RD} model ages for the Pyrenean peridotites of 1.83–2.39 Ga, substantially older than the age provided by the highly scattered Re–Os isochron for these samples (Fig. 7.8c).

As discussed here and in Chapter 6, model age approaches greatly expand the use of isotope systems to extract the timing of geologically important events. The accuracy of the ages obtained, however, depends on how well any given sample's evolution matches that predicted by the model.

(1) For the T_{MA} approach to provide an accurate age, the sample's Re/Os ratio must have changed only once in a single event that caused it to deviate from the mantle Re–Os isotope evolution curve. Changing the Re/Os ratio a second time will cause the T_{MA} age to either overestimate or underestimate the age of the initial differentiation event depending on whether the second event takes the sample's Re/Os ratio further from, or brings it closer to, the undifferentiated mantle's Re/Os ratio.

(2) In the T_{RD} approach, an accurate age is obtained only if the sample deviated from mantle evolution during a single event that left the sample with Re/Os = 0. The T_{RD} approach allows for a second event, of independently known age, that added back Re to the sample analyzed and created the Re/Os ratio measured in the sample today. If not all the Re was removed from the sample during differentiation, its T_{RD} will provide only a minimum estimate of the true time of the event that caused it to leave the mantle evolution curve.

Samples with ^{187}Os/^{188}Os ratios higher than that of undifferentiated mantle will produce meaningless, future, T_{RD} model ages.

(3) Both model ages are sensitive to the model mantle evolution assumed. For example, if the undifferentiated mantle evolved with carbonaceous chondrite Re–Os parameters (^{187}Re/^{188}Os = 0.3939, present day ^{187}Os/^{188}Os = 0.1263) instead of those shown in Fig. 7.9, the T_{MA} model age of the 10% residue would be 2.75 Ga and that of the 30% residue would be 2.83 Ga instead of 3.0 Ga. The difference increases as the model age of the sample decreases. For example, if the differentiation event shown in Fig. 7.9 occurred at 1 Ga instead of 3 Ga, the T_{MA} of the 10% and 30% residues would be 431 and 606 Ma, respectively.

7.6 CONCLUSIONS

The Re–Os system has become a commonly used tool in geochronology and isotope geology in part because it can be used to date materials that cannot be easily dated by other methods, for example, organic-rich sediments, iron meteorites, molybenite, and other sulfides such as sulfide inclusions in diamond. One of its most common applications is to determine the timing of melt extraction events from the mantle. This question cannot be addressed by internal isochrons of mantle materials because their storage temperatures, even in the uppermost mantle, are above the closure temperatures of most chronometers. Closure temperature is less important at the whole-rock scale because of the large diffusion distances involved. Nevertheless, whole-rock age dating of mantle samples has proven difficult using Rb–Sr, Sm–Nd, or U–Pb because the extremely low concentrations of all of these elements in the mantle lead these systems to be extremely sensitive to overprinting by later events, such as the migration of small volume melts in the mantle and contamination by the host magma of mantle xenoliths. Re–Os model ages have shown great promise in providing an age dating technique that extends below the crust into the upper mantle. The low concentrations of Re and Os in most crustal igneous rocks has slowed the development of its chronological applications to these rocks, but the very high Re/Os ratios in crustal rocks carry the promise of precise ages even for young rocks once the analytical limitations are overcome. The Pt–Os system has seen only limited application as the radioactive contributions to ^{186}Os are, at best, very small. For the full utility of this system to be realized, better isotope ratio precisions than currently possible are needed.

7.7 REFERENCES

Allegre, C. J. and Luck, J.-M. (1980) Osmium isotopes as petrogenetic and geological tracers. *Earth and Planetary Science Letters* **48**, 148–154.

Amelin, Y. (2008) U–Pb ages of angrites. *Geochimica et Cosmochimica Acta* **72**, 221–232.

Becker, H., Horan, M. F., Walker, R. J., Gao, S., Lorand, J.-P., and Rudnick, R. L. (2006) Highly siderophile element composition of the Earth's primitive upper mantle: constraints from new data on peridotite massifs and xenoliths. *Geochimica et Cosmochimica Acta* **70**, 4528–4550.

Begemann, F., Ludwig, K. R., Lugmair, G. W., *et al.* (2001) Call for an improved set of decay constants for geochronological use. *Geochimica et Cosmochimica Acta* **65**, 111–121.

Birck, J.-L., Barman, M. R., and Capmas, F. (1997) Re–Os isotopic measurements at the femtomole level in natural samples. *Geostandards Newsletter* **21**, 19–27.

Brennecka, G. A. and Wadhwa, M. (2012) Uranium isotope compositions of the basaltic angrite meteorites and the chronological implications of the early solar system. *Proceedings of the National Academy of Sciences* **109**, 9299–9303.

Carlson, R. W. (2005) Application of the Pt–Re–Os isotopic systems to mantle geochemistry and geochronology. *Lithos* **82**, 249–272.

Carlson, R. W., Pearson, D. G., and James, D. E. (2005) Physical, chemical, and chronological characteristics of continental mantle. *Reviews of Geophysics* **43**. DOI:2004RG00156.

Chou, C.-L., Shaw, D. M., and Crocket, J. H. (1983) Siderophile trace elements in the Earth's oceanic crust and upper mantle. *Journal of Geophysical Research* **88**, A507–A518.

Creaser, R. A., Papanastassiou, D. A., and Wasserburg, G. J. (1991) Negative thermal ion mass spectrometry of osmium, rhenium, and iridium. *Geochimica et Cosmochimica Acta* **55**, 397–401.

Gannoun, A., Burton, K. W., Thomas, L. E., Parkinson, I. J., Calsteren, P. von, and Schiano, P. (2004) Osmium isotope heterogeneity in the constituent phases of mid-ocean ridge basalts. *Science* **303**, 70–72.

Hart, S. R. and Kinloch, E. D. (1989) Osmium isotope systematics in Witwatersrand and Bushveld ore deposits. *Economic Geology* **84**, 1651–1655.

Hart, S. R. and Ravizza, G. E. (1995) Os partitioning between phases in lherzolite and basalt. In *Reading the Isotopic Code*, Basu, A. and Hart, S. R. (eds). American Geophysical Union, Washington, DC.

Hassler, D. R., Peuker-Ehrenbrink, B., and Ravizza, G. E. (2000) Rapid determination of Os isotopic composition by sparging OsO$_4$ into a magnetic sector ICP-MS. *Chemical Geology* **166**, 1–14.

Hoffman, E. L., Naldrett, A. J., Loon, J. C. v., Hancock, R. G., and Manson, A. (1978) The determination of all the platinum-group elements and gold in rocks and ore by neutron activation analysis after preconcentration by a nickel sulphide fire-assay technique on large samples. *Analytica Chimica Acta* **102**, 157–166.

Horan, M. F., Walker, R. J., Morgan, J. W., Grossman, J. N., and Rubin, A. E. (2003) Highly siderophile elements in chondrites. *Chemical Geology* **196**, 27–42.

Horan, M. F., Alexander, C. M. O. D., and Walker, R. J. (2009) Highly siderophile element evidence for early solar system processes in components from ordinary chondrites. *Geochimica et Cosmochimica Acta* **73**, 6984–6997.

Lindner, M., Leich, D. A., Borg, R. J., Russ, G. P., Bazan, J. M., Simons, D., and Date, A. R. (1986) Direct laboratory determination of the ^{187}Re half-life. *Nature* **320**, 246–248.

Lodders, K. (2003) Solar system abundances and condensation temperatures of the elements. *Astrophysical Journal* **591**, 1220–1247.

Lugmair, G. W. and Galer, S. J. G. (1992) Age and isotopic relationships among angrites Lewis Cliff 86010 and Angra dos Reis. *Geochimica et Cosmochimica Acta* **56**, 1673–1694.

Luguet, A., Shirey, S. B., Lorand, J.-P., Horan, M. F., and Carlson, R. W. (2007) Residual platinum-group minerals from highly depleted harzburgites of the Lherz massif (France) and their role in HSE fractionatino in the mantle. *Geochimica et Cosmochimica Acta* **71**, 3082–3097.

Meisel, T., Walker, R. J., Irving, A. J., and Lorand, J. P. (2001) Osmium isotopic compositions of mantle xenoliths: a global perspective. *Geochimica et Cosmochimica Acta* **65**, 1311–1323.

Morgan, J. W. (1986) Ultramafic xenoliths: clues to Earth's late accretionary history. *Journal of Geophysical Research* **91**, 12,375–12,387.

Pearson, D. G. and Shirey, S. B. (1999) Isotopic dating of diamonds. In *Application of Radiogenic Isotopes to Ore Deposit Research and Exploration*. Lambert, D. and Ruiz, J. (eds). Society of Economic Geologists, Boulder, CO.

Pearson, D. G., Shirey, S. B., Harris, J. W., and Carlson, R. W. (1998) Sulphide inclusions in diamonds from the Koffiefontein kimberlite, S Africa; constraints on diamond ages and mantle Re–Os systematics. *Earth and Planetary Science Letters* **160**, 311–326.

Puchtel, I. S., Walker, R. J., Touboul, M., Nisbet, E. G., and Byerly, G. R. (2014) Insights into early Earth from the Pt–Re–Os isotope and highly siderophile element abundance systematics of Barberton komatiites. *Geochimica et Cosmochimica Acta* **125**, 394–413.

Reisberg, L. C. and Lorand, J.-P. (1995) Longevity of subcontinental mantle lithosphere from osmium isotope systematics in orogenic peridotite massifs. *Nature* **376**, 159–162.

Reisberg, L. and Meisel, T. (2002) The Re–Os isotopic system: a review of analytical techniques. *Geostandards Newsletter* **26**, 249–267.

Richardson, S. H., Gurney, J. J., Erlank, A. J., and Harris, J. W. (1984) Origin of diamonds in old enriched mantle. *Nature* **310**, 198–202.

Richardson, S. H., Shirey, S. B., Harris, J. W., and Carlson, R. W. (2001) Archean subduction recorded by Re–Os isotopes in eclogitic sulfide inclusions in Kimberley diamonds. *Earth and Planetary Science Letters* **191**, 257–266.

Selby, D., Creaser, R., Dewing, K., and Fowler, M. (2005) Evaluation of bitumen as a Re–Os geochronometer for hydrocarbon maturation and migration: a test case from the Polaris MVT deposit, Canada. *Earth and Planetary Science Letters* **235**, 1–15.

Shen, J. J., Papanastassiou, D. A., and Wasserburg, G. J. (1996) Precise Re–Os determinations and systematics of iron meteorites. *Geochimica et Cosmochimica Acta* **60**, 2887–2900.

Shirey, S. B. and Walker, R. J. (1995) Carius tube digestions for low-blank rhenium-osmium analysis. *Analytical Chemistry* **67**, 2136–2141.

Shirey, S. B. and Walker, R. J. (1998) The Re–Os isotope system in cosmochemistry and high-temperature geochemistry. *Annual Reviews in Earth and Planetary Science* **26**, 423–500.

Smoliar, M. I., Walker, R. J., and Morgan, J. W. (1996) Re–Os ages of group IIA, IIIA, IVA, and IVB iron meteorites. *Science* **271**, 1099–1102.

Stein, H. J. (2014) Dating and tracing the history of ore formation. In *Treatise on Geochemistry*, Scott, S. D. (ed.). Elsevier, Amsterdam.

Stein, H. J., Markey, R. J., Morgan, J. W., Hannah, J. L., and Schersten, A. (2001) The remarkable Re–Os chronometer in molybdenite; how and why it works. *Terra Nova* **13**, 479–486.

Volkening, J., Walczyk, T., and Heumann, K. G. (1991) Osmium isotope ratio determinations by negative thermal ionization mass spectrometry. *International Journal of Mass Spectrometry and Ion Processes* **105**, 147–159.

Walker, R. J. and Fassett, J. D. (1986) Isotopic measurement of subnanogram quantities of rhenium and osmium by resonance ionization mass spectrometry. *Analytical Chemistry* **58**, 2923–2927.

Walker, R. J. and Morgan, J. W. (1989) Rhenium–osmium systematics of carbonaceous chondrites. *Science* **243**, 519–522.

Walker, R. J., Morgan, J. W., Beary, E. S., Smoliar, M. I., Czamanske, G. K., and Horan, M. F. (1997) Applications of the ^{190}Pt–^{186}Os isotope system to geochemistry and cosmochemistry. *Geochimica et Cosmochimica Acta* **61**, 4799–4807.

Walker, R. J., Carlson, R. W., Shirey, S. B., and Boyd, F. R. (1989) Os, Sr, Nd, and Pb isotope systematics of southern African peridotite xenoliths: Implications for the chemical evolution of subcontinental mantle. *Geochimica et Cosmochimica Acta* **53**, 1583–1595.

Walker, R. J., Horan, M. F., Morgan, J. W., Becker, H., Grossman, J. N., and Rubin, A. E. (2002) Comparative ^{187}Re–^{187}Os systematics of chondrites: implications regarding early solar system processes. *Geochimica et Cosmochimica Acta* **66**, 4187–4201.

CHAPTER 8

U–Th–Pb geochronology and thermochronology

8.1 INTRODUCTION AND BACKGROUND

8.1.1 Decay of U and Th to Pb

The decays of U and Th to Pb were fundamental in the discovery of radioactivity and the subsequent exploitation of this process for dating geologic materials and determining the age of the Earth and solar system [*Dalrymple*, 1994; *Lewis*, 2002; *Davis et al.*, 2003; *Mattinson*, 2013]. Work by A. Henri Bequerel and Marie and Pierre Curie in the late 1800s described what was to be called radioactivity by studying U, Th, and their daughter isotopes [*Becquerel*, 1896a, b; *Curie and Sklodowska-Curie*, 1898]. A series of papers by *Rutherford and Soddy* [1902a, b, c] first formulated a mathematical description of exponential radioactive decay, based on radiation measurements for what was later determined to be the intermediate daughter product of ^{232}Th decay, ^{224}Ra. Pioneering work on U and Th decay over the subsequent decade led to a series of papers that may be considered the first papers published on geochronology [*Boltwood*, 1907; *Holmes*, 1911], which were based on U–Pb dating approaches that used only the elemental abundance of Pb and U, prior to the coining of the term "isotope" by Soddy in 1913 [*Soddy*, 1913a]. Holmes' paper also served as an early attempt to place dates onto the geologic timescale, an effort he would continue by using U–Pb dating for decades [e.g, *Holmes*, 1947].

While the ages reported for key period boundaries in the Phanerozoic were remarkably accurate given the infancy of the technique, the drawbacks of U–Pb elemental dating soon became apparent. The ubiquitous presence of nonradiogenic Pb, especially in the whole-rock samples dated by Holmes resulted in overestimated ages. Similarly, the realization that two parent isotopes of U and one of Th each decay to a different isotope of Pb required isotopic analysis of Pb in order to generate accurate U–Pb, or Th–Pb dates. Already by the 1920s—long before measuring isotopic ratios was possible—a good understanding of the *decay chains* of U and Th and the mathematics to describe them had been outlined [e.g., *Bateman*, 1910; *Soddy*, 1913b; *Johnstone and Boltwood*, 1920; *Russell*, 1923]. Thus, by the time it

was possible to measure isotopic ratios, much of the theoretical framework for U–Th–Pb dating was simply waiting to be put into action. Following the development of early mass spectrometers in the 1930s, the first isotopic composition of lead was published in 1941 [*Nier et al.*, 1941], unleashing new approaches to calibrating the age of the Earth, the geologic timescale, and a host of other geologic processes through all of Earth history.

A fundamental benefit of the U–Th–Pb decay schemes is that all three parent isotopes (^{238}U, ^{235}U, and ^{232}Th) each decay to the stable daughter isotopes of Pb (^{206}Pb, ^{207}Pb, and ^{208}Pb, respectively) via completely independent decay chains involving numerous intermediate isotopes (Figs 8.1 and 8.2). The intermediate daughter isotope half-lives are much shorter than their parent isotopes, which can lead to complications in calculating dates if elemental fractionation occurs during mineral growth (section 8.4.3), but also leads to opportunities to measure shorter-duration processes utilizing isotopic ratios of intermediate daughter products (Chapter 12). The long decay chains appear to require a complicated set of decay equations to calculate a U–Pb or Th–Pb date. But it can be shown mathematically that, in the case where the activities of intermediate daughter products are equal to each other and that of the parent nuclide—a condition called *secular equilibrium* [*Bateman*, 1910; *Bourdon et al.*, 2003; *Soddy*, 1913b]—these equations simplify to include only terms for the parent and final stable daughter product. Secular equilibrium is defined as:

$$N_1 \lambda_1 = N_2 \lambda_2 = N_3 \lambda_3 = \dots \tag{8.1}$$

$$[N_1] = [N_2] = [N_3] = \dots \tag{8.2}$$

where N_1 is the moles of parent isotope 1 and λ_1 is its decay constant. The second identical equation uses the nomenclature for the activity of each isotope, denoted by square brackets. A system out of secular equilibrium will return to secular equilibrium in a time proportional to the half-life of the longest-lived intermediate daughter product [*Bateman*, 1910; *Bourdon et al.*, 2003].

Geochronology and Thermochronology, First Edition. Peter W. Reiners, Richard W. Carlson, Paul R. Renne, Kari M. Cooper, Darryl E. Granger, Noah M. McLean, and Blair Schoene.
© 2018 John Wiley & Sons Ltd. Published 2018 by John Wiley & Sons Ltd.

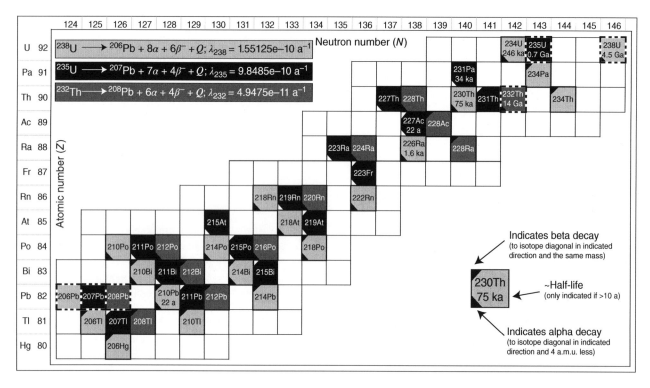

Fig. 8.1. An illustration of the U–Th–Pb decay chains. Each isotope occurring in a given decay chain is shaded to match its parent isotope, which are outlined in dashed bold boxes, as are the stable daughter isotopes of Pb. See inset for description of symbols used in each box. α is an alpha particle, β⁻ is a beta particle, and Q is energy released during the decay. (Source: *Schoene* [2014]. Reproduced with permission of Elsevier.)

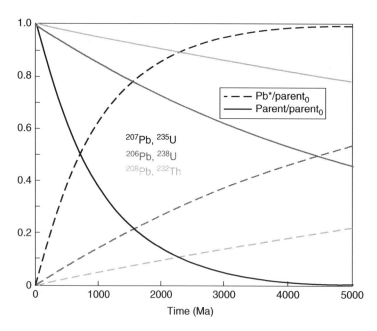

Fig. 8.2. Illustration of the decay and ingrowth rates of the parent and daughter isotopes involved in the decay of ^{238}U, ^{235}U, and ^{232}Th. Note that where the curves of parent decay and daughter ingrowth cross is the half-life. See Fig. 8.1 and text for actual decay-constant values.

Departures from and the return to secular equilibrium via radio-isotopic decay form the foundation of U-series geochronology, which has been key in determining the rates of geologic processes in the last ~800 ka (Chapter 12). For U–Th–Pb geochronology, secular equilibrium at the time the system closed is either assumed or evaluated based on alternative means (section 8.4.3). Below, the simplified age equations are outlined assuming secular equilibrium and later some detail is given on how isotope geochemists detect and correct for cases when secular equilibrium is jeopardized.

8.1.2 Dating equations

The power of U–Th–Pb geochronology derives in part from the three independent decay systems, which lead to three parent–daughter age equations shown below:

$$\left(\frac{^{206}\text{Pb}}{^{204}\text{Pb}}\right) = \left(\frac{^{206}\text{Pb}}{^{204}\text{Pb}}\right)_0 + \left(\frac{^{238}\text{U}}{^{204}}\text{Pb}\right)\left(e^{\lambda_{238}t} - 1\right) \tag{8.3}$$

$$\left(\frac{^{207}\text{Pb}}{^{204}\text{Pb}}\right) = \left(\frac{^{207}\text{Pb}}{^{204}\text{Pb}}\right)_0 + \left(\frac{^{235}\text{U}}{^{204}\text{Pb}}\right)\left(e^{\lambda_{235}t} - 1\right) \tag{8.4}$$

$$\left(\frac{^{208}\text{Pb}}{^{204}\text{Pb}}\right) = \left(\frac{^{208}\text{Pb}}{^{204}\text{Pb}}\right)_0 + \left(\frac{^{232}\text{U}}{^{204}\text{Pb}}\right)\left(e^{\lambda_{232}t} - 1\right) \tag{8.5}$$

^{204}Pb is used as the normalizing isotope as it is the only nonradiogenic isotope of Pb. A fourth commonly used equation derives from the assumption that ^{238}U/^{235}U is constant in nature (though see section 8.1.4), permitting division of equation (8.4) by equation (8.3):

$$\frac{\left(\frac{^{207}\text{Pb}}{^{204}\text{Pb}}\right) - \left(\frac{^{207}\text{Pb}}{^{204}\text{Pb}}\right)_0}{\left(\frac{^{206}\text{Pb}}{^{204}\text{Pb}}\right) - \left(\frac{^{206}\text{Pb}}{^{204}\text{Pb}}\right)_0} = \left(\frac{^{235}\text{U}}{^{238}\text{U}}\right)\frac{\left(e^{\lambda_{235}t} - 1\right)}{\left(e^{\lambda_{238}t} - 1\right)} = \left(\frac{^{207}\text{Pb}}{^{206}\text{Pb}}\right)^{*}$$

$$\tag{8.6}$$

where the $\left(^{207}\text{Pb}/^{206}\text{Pb}\right)^{*}$ refers to the radiogenic ^{207}Pb/^{206}Pb isotope ratio. A date determined this way is often called a Pb–Pb date [*Gerling*, 1942; *Holmes*, 1946; *Houtermans*, 1946]. Note that this equation is transcendental and must be solved through iteration. The ^{238}U/^{235}U has now been shown to vary as a result of fractionation by natural processes [*Brennecka et al.*, 2010, 2011; *Asael et al.*, 2013; *Andersen et al.*, 2014; *Dahl et al.*, 2014]. This equation works when the ^{238}U/^{235}U for a dated sample can be measured, and also when the mean and variability of that ratio can be measured in other similar samples, with the variability propagated as an uncertainty in the resulting date [*McLean et al.*, 2011; *Hiess et al.*, 2012; *Condon et al.*, 2015].

A Pb–Pb date determined using equation (8.6) is useful for several reasons. For example, one can date a sample without measuring uranium [*Holmes*, 1946; *Houtermans*, 1946]. For samples older than about 1–2 Ga, a Pb–Pb date is more precise than a U–Pb date as a result of how the variable magnitudes of the decay constants propagate through the uncertainty budgets (*Mattinson* [1987]); Fig. 8.3). Pb–Pb dates are also more accurate than U–Pb dates in the specific case where zero-aged (recent) Pb loss has affected a sample, assuming that the Pb-loss mechanism does not cause Pb isotope fractionation (either due to mass dependent processes or otherwise). This will be shown graphically below when discussing the common ways of plotting U–Th–Pb data.

8.1.3 Decay constants

The decay constants of ^{238}U and ^{235}U are considered to be the most accurate and most precise used in geochronology [*Renne*

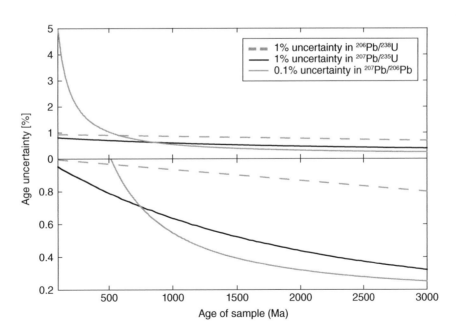

Fig. 8.3. Uncertainties in U–Th–Pb dates as a function of age. The uncertainties in Pb/U and Pb–Pb dates decrease as a percentage of the age back in time, but at different rates. Therefore, the most precise date depends the age of the sample, but also the analytical uncertainties. For example, although a ^{207}Pb/^{235}U date is more precise for the same analytical uncertainty, it is almost always less precise compared to the ^{206}Pb/^{238}U date in samples < 1 Ga because of the small amount of ^{207}Pb present in the sample relative to ^{206}Pb.

et al., 1998; *Begemann et al.*, 2001; *Mattinson*, 2010]. The commonly used values result from a single study over 40 years ago, carried out by alpha counting on samples of very pure ^{238}U and ^{235}U [*Jaffey et al.*, 1971]. Despite much scrutiny [*Begemann et al.*, 2001; *Schön et al.*, 2004; *Schoene et al.*, 2006; *Mattinson*, 2010; *Boehnke and Harrison*, 2014], the experimental approach and resulting values are widely regarded as robust. They are: 1.55125×10^{-10}/year and 9.8485×10^{-10}/year for ^{238}U and ^{235}U, respectively. Uncertainties on these values were given as $\pm 0.11\%$ and $\pm 0.14\%$, respectively [*Jaffey et al.*, 1971]. These values were adopted and recommended by the IUGS Subcommission on Geochronology [*Steiger and Jäger*, 1977] and have been used almost exclusively since. The veracity of these values has been supported by U–Pb dates of minerals argued to represent closed systems, in that if the ratio of the decay constants is accurate then dates determined by equations (8.3) and (8.4) above should yield the same answer. This was initially tested by *Mattinson* [2000], who showed that ^{238}U–^{206}Pb and ^{235}U–^{207}Pb dates agreed within the uncertainties in decay constants reported by *Jaffey et al.* [1971]. This conclusion was confirmed by a subsequent generation of large data sets deriving from dates of minerals ranging over billions of years in age [*Schoene et al.*, 2006]. Another recent study combined existing data sets to recalculate a value for λ_{235} of 9.8571×10^{-10}/year ($\pm 0.012\%$) relative to the more precisely measured λ_{238} [*Mattinson*, 2010].

Importantly, each of these studies point out that the precision of the recalculated value of λ_{235} is a good estimate of the reproducibility of the U–Pb measurements used in the recalculation. However, it is overly precise because ultimately the precision is limited by that of λ_{238} used for recalibration, namely 0.11%, as determined by the alpha counting experiments of *Jaffey et al.* [1971]. The studies cited above also note that the accuracy of the uranium decay constants needs to be confirmed by additional first-principle studies. This is important in part because decay constants from other dating methods such as Rb–Sr, K–Ar, Lu–Hf and Re–Os, as well as the U-series daughter nuclides, have been recalculated in part by comparison with U–Pb dates on samples argued to represent the same age [*Min et al.*, 2000; *Scherer et al.*, 2001; *Selby et al.*, 2007; *Renne et al.*, 2010; *Nebel et al.*, 2011; *Cheng et al.*, 2013].

The decay constant of ^{232}Th is less precisely determined compared to the U decay constants. The value recommended by *Steiger and Jäger* [1977] is 4.9475×10^{-11}/year with an uncertainty of ~1%, which derives from a poorly documented study [*Le Roux and Glendenin*, 1963]. *Amelin and Zaitsev* [2002] report a revised value for λ_{232} of 4.934×10^{-11}/year ($\pm 0.3\%$), determined by comparing ^{208}Pb/^{232}Th apatite dates to zircon and baddeleyite ^{206}Pb/^{238}U dates and using λ_{238} to recalculate λ_{232}. There is no general consensus in the literature as to what value should be used.

8.1.4 Isotopic composition of U

It was long assumed that, for the purposes of U–Th–Pb geochronology, the isotopic composition of natural U is fixed at ^{238}U/^{235}U = 137.88 (Steiger and Jäger, 1977). The accuracy of this assumption becomes important directly when using equation (8.6), such that any deviation from 137.88 in the actual sample U composition will result in inaccurate dates [*Hiess et al.*, 2012; *Condon et al.*, 2015; *Tissot and Dauphas*, 2015]. Because ^{235}U is over a hundred times less abundant than ^{238}U, many mass spectrometer protocols analyze only the ^{238}U and assume a ^{238}U/^{235}U to calculate the amount of ^{235}U. In the case of isotope dilution mass spectrometry (Chapter 3), samples are often spiked with a mixed tracer containing ^{235}U and a synthetic uranium isotope (e.g., ^{233}U or ^{236}U) in order to correct for mass fractionation during analysis, which is made possible by assuming a composition of the sample U [*Roddick et al.*, 1987; *Condon et al.*, 2015]. In both of these cases, deviation from ^{238}U/^{235}U = 137.88 would result in inaccurate dates proportional to the magnitude of the deviation if the uncertainties in the assumed value are not properly treated [*Hiess et al.*, 2012].

Numerous recent studies have now documented variability in ^{238}U/^{235}U in natural samples [*Stirling et al.*, 2007; *Weyer et al.*, 2008; *Brennecka et al.*, 2010, 2011 *Asael et al.*, 2013; *Andersen et al.*, 2014; *Dahl et al.*, 2014] (Fig. 8.4). The U isotopic composition of water and carbonate samples has been shown to vary by 1–2‰, which has been attributed to both redox-related mass fractionation of U isotopes deriving from processes related to the the multiple redox states of U and complexation to UO_2^- in oxidizing environments such as surface waters [*Langmuir*, 1978; *Schauble*, 2007; *Stirling et al.*, 2007; *Abe et al.*, 2008; *Weyer et al.*, 2008; *Bopp et al.*, 2009; *Brennecka et al.*, 2011]. Isotopic variability of uranium in extraterrestrial samples has been measured and, in addition to chemically induced mass fractionation, has been attributed to radiogenic ingrowth of ^{235}U resulting from decay of the short-lived nuclide ^{247}Cm ($t_{1/2}$ = 15.6 Ma) in the early solar system [*Chen and Wasserburg*, 1981; *Amelin et al.*, 2010; *Brennecka et al.*, 2010; *Connelly et al.*, 2012; *Tissot et al.*, 2016]. Systematic study of minerals most often used in high-temperature igneous and metamorphic systems, such as zircon and titanite, have also revealed variability of ^{238}U/^{235}U, with a mean of 137.818 ± 0.022 (2 standard deviations; *Hiess et al.* [2012]) for zircon, which can be propagated into age calculations if ^{235}U is used as a spike isotope and/or is not abundant enough to measure with adequate precision [*Condon et al.*, 2015].

The implications for these recent studies on U–Pb geochronology depend on the samples being analyzed, the analytical techniques being used and the questions being asked [*Condon et al.*, 2015; *Tissot and Dauphas*, 2015; *Tissot et al.*, 2016]. On the one hand, one can in theory measure the isotopic composition directly using a ^{233}U–^{236}U double spike, thereby avoiding assumption of the ^{238}U/^{235}U. This approach can work for isotope-dilution thermal ionization or inductively coupled plasma mass spectrometry (ID-TIMS/ID-ICPMS) geochronology where sample size is adequate to measure the ratio with sufficient precision, such as in some meteorite U–Pb geochronology studies [*Connelly et al.*, 2012]. For many ID-TIMS U–Pb geochronological studies, however, the focus is on dating small

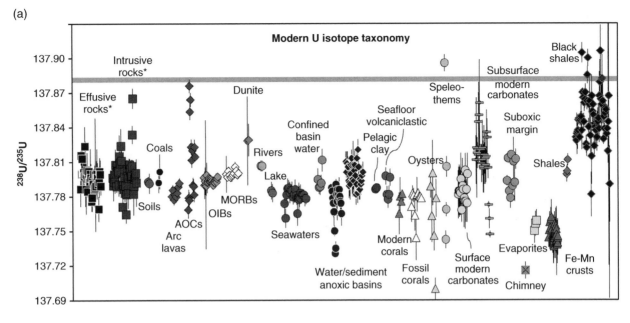

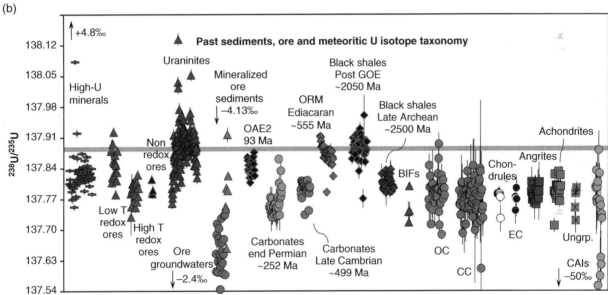

Fig. 8.4. Measured values of $^{238}U/^{235}U$ for natural samples. Modified from *Tissot and Dauphas* [2015], which also has all the original references for the data. AOC, altered oceanic crust; OIB, ocean island basalts; MORB, mid-ocean ridge basalts; OAE, ocean anoxic event; ORM, organic rich mudrocks; GOE, great oxidation event; BIF, banded iron formation; OC, oceanic crust; CC, continental crust; EC, enstatite chondrites; CAI, calcium aluminum inclusion; effusive rocks are volcanic rocks of a range of SiO_2. Horizontal gray line is equal to 137.88, the value recommended by *Steiger and Jager* [1977]. (Source: *Tissot and Dauphas* [2015]. Reproduced with permission of Elsevier.)

samples such as single minerals or mineral fragments [*Mundil et al.*, 2004; *Crowley et al.*, 2009; *Gordon et al.*, 2010; *Wotzlaw et al.*, 2013; *Rivera et al.*, 2014; *DesOrmeau et al.*, 2015; *Samperton et al.*, 2015; *Schoene et al.*, 2015], and the precision with which one can measure the U isotopic composition is low enough given modern TIMS and ICPMS instrumentation that simply propagating the natural variability in that ratio as an additional source of uncertainty is justified [*Condon et al.*, 2015].

For *in situ* dating techniques such as laser ablation (LA) ICPMS and secondary ion mass spectrometry (SIMS), uranium isotopes may or may not be able to be measured directly depending on the instrumentation, but given the low precision of those techniques, this is probably not a large source of uncertainty in the resulting date. An exception may be in Pb–Pb dates of Archean or older samples, where relatively high precision can be achieved with *in situ* techniques.

8.2 CHEMISTRY OF U, Th, AND Pb

As with other radioisotopic systems, the approaches taken to exploit the U–Th–Pb decay scheme to measure geologic time are highly dependent on how U, Th, and Pb behave chemically in natural environments. The multiple decay chains add complexity, such that a basic understanding of the geochemical behavior of the intermediate daughter products must be understood to use U–Pb and Th–Pb dating accurately, and to take full advantage of U-series geochronology as well (Chapter 12). Here, the focus is on the geochemical behavior of U, Th, and Pb, with some caveats outlined related to several longer lived intermediate daughter products in section 8.4.3. The discussion assumes that loss of other intermediate daughter products in dated samples is negligible. For more detailed descriptions of how U, Th, and Pb are used in trace element geochemistry in a host of geologic environments, the reader is referred to other excellent texts [e.g., *Albarède*, 2003; *White*, 2013; *Rollinson*, 2014].

Uranium and thorium are part of the actinide series of elements in the periodic chart. In silicate liquids at most relevant oxygen fugacities, U and Th are quadravalent cations, similar to Zr, Hf, and Ti whereas Ta and Nb are typically pentavalent (Fig. 8.5). U and Th have larger ionic radii than Zr, Hf, and Ti

and thus are commonly fractionated from the high-field-strength elements during crystallization, partial melting, and weathering [*Wood and Blundy*, 2003; *Hofmann*, 2007]. In strongly oxidizing conditions typical for low-temperature weathering environments and ocean waters, uranium is hexavalent and forms the water-soluble U^{6+} as UO_2^{2+}, which is important for U-series geochronology and using uranium as a tracer of redox conditions at the surface of the earth in the present day and in deep time. For most geochronological applications, uranium is considered quadravalent when attempting to understand fractionation of U and Th for dating rocks and minerals. The important point is that Pb, by contrast, is found as the divalent cation Pb^{2+} and is grouped with the large ion lithophile elements, similar in ionic radius to Ba, Sr, and Eu. Lead can also be found in its quadravalent state in highly oxidizing conditions, but at most oxygen fugacities in high-temperature crustal systems studied by U–Th–Pb geochronology it is assumed to be divalent. This is consistent with the observation that Pb substitutes most readily into minerals with large divalent cations as stoichiometric constituents and is not found associated with the high-field-strength elements except as radiogenic Pb.

The differences in valence and ionic radii of Pb compared to U and Th benefits the U–Th–Pb system for geochronology because a wide range of minerals fractionate parent and daughter elements during crystallization, leading to both (i) opportunities to generate data spreads on isochron diagrams and (ii) minerals with negligible daughter product at the time of crystallization, permitting single mineral geochronology. Table 8.1 shows typical U, Th, and Pb abundances in minerals used for geochronology, which are determined by the mineral–liquid partition coefficients at the temperatures, pressures, and oxygen fugacities at which those minerals are typically found.

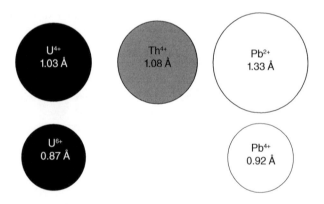

Fig. 8.5. Illustration of the most common valence states of U, Th, and Pb. Size of circles represents approximate size of cations. Note that U^{6+} is most commonly found as UO_2^{2+} and therefore should be treated as such when considering substitution into mineral structures.

8.3 DATA VISUALIZATION, ISOCHRONS, AND CONCORDIA PLOTS

8.3.1 Isochron diagrams

As outlined above, U, Th, and Pb can be fractionated significantly during geologic processes, such that rocks often contain

Table 8.1 List of minerals, simplified mineral structures, and the approximate abundances of U, Th, and Pb

Mineral	Formula	U content (ppm)	Pb_c content (ppm)	Th/U
Zircon	$ZrSiO_4$	1–10,000	< 1	0.01–3
Titanite	$CaTiOSiO_4$	1–500	5–40	< 1–20
Monazite	$(LREE)PO_4$	200–> 50,000	< 5	> 10
Apatite	$Ca_5(PO_4)_3(OH,F,Cl)$	5–200	< 5–30	< 1–20
Rutile	TiO_2	< 1–400	< 1–50	0–1
Baddeleyite	ZrO_2	50–4000	< 1–10	< 1
Xenotime	YPO_4	100–30,000	< 5	1–5
Thorite	$ThSiO_4$	> 100,000	< 2	> 2
Allanite	$Ca_2(Fe,Al)_3(SiO_4)_3(OH)$	100–1000	5–30	> 10
Perovskite	$CaTiO_3$	20–350	< 2–90	5–30

LREE, light rare earth elements.

numerous U-bearing and Th-bearing minerals with varying parent–daughter ratios. The result is that isochron correlation diagrams and isochron dates can be calculated with relatively high precision using equations (8.3–8.5). Though not commonly used anymore in the U–Pb system, these diagrams can be used identically to isochrons in other systems and must meet the same set of assumptions to yield accurate ages—namely that the minerals or rocks used in the calculation must in fact be the same age, must have the same initial lead isotopic composition, and must have remained closed-system since the time of formation. As with other geochronological techniques outlined in Chapter 6, these assumptions can be addressed in part based on geologic knowledge of the samples in question, and can also be addressed using the statistical techniques outlined in Chapter 4.

An advantage of the U–Pb method is that, due to the dual decay of ^{238}U and ^{235}U, several other types of isochron diagrams can be constructed. Equation (8.6) can also be used to formulate an isochron diagram when the initial $^{207}Pb/^{204}Pb$ and $^{206}Pb/^{204}Pb$ are unknown. As with other isochrons, if the samples are in fact isochronous they fall on a line, where the slope is proportional to the age. In this case, the initial isotopic composition of lead, $(^{238}U/^{206}Pb)_0$ and $(^{204}Pb/^{206}Pb)_0$, plots as a single point (see Fig. 8.14 later). As outlined above, the accuracy of a date calculated in such a way is dependent on knowledge of the sample's $^{238}U/^{235}U$ value. A case study of this isochron diagram is given in section 8.6.

The U–Pb system also permits isochrons with more than two dimensions, and a popular one, the total Pb–U isochron, combines the measured $^{207}Pb/^{206}Pb$, $^{238}U/^{206}Pb$, and $^{204}Pb/^{206}Pb$ ratios for more than two data points (Fig. 8.6) [*Wendt*, 1984; *Zheng*, 1992; *Ludwig*, 1998). Sample suites that meet the assumptions of an isochron fall on a plane in this diagram; the intercept with the $^{204}Pb/^{206}Pb$ axis is the $^{204}Pb/^{206}Pb$ isotopic composition of initial lead, the intercept with the $^{207}Pb/^{206}Pb$ axis is the $^{207}Pb/^{206}Pb$ composition of initial lead, and the point where the $^{204}Pb/^{206}Pb = 0$ and the calculated $^{207}Pb/^{206}Pb$ and $^{206}Pb/^{238}U$ dates are the same is the age of the sample or suite of samples. The benefit of three-dimensional isochron diagrams is that it more rigorously tests for colinearity of the sample suite and also the added constraint of the additional dimension can result in more precise age estimates compared to traditional two-dimensional isochrons [*Ludwig*, 1998]. Note that the $^{207}Pb/^{206}Pb$ and $^{238}U/^{206}Pb$ axes in this diagram form what is described below as the Tera–Wasserburg concordia diagram (Fig. 8.6).

8.3.2 Concordia diagrams

In the U–Th–Pb system, the presence of minerals with very high parent to daughter ratio present the opportunity to perform somewhat ad hoc corrections for common Pb while introducing minimal age uncertainty. The various approaches to the common Pb correction are outlined in section 8.4.4. Once the correction is done and uncertainties in this correction are accounted for, it is common to plot data on what are commonly called concordia diagrams. Concordia diagrams are the most common way to present U–Pb data in the modern literature, so understanding them is key to understanding the method.

One of the isochron approaches outlined in the previous section, the three-dimensional total Pb–U isochron, has two axes ($^{207}Pb/^{206}Pb$ and $^{238}U/^{206}Pb$) that plot ratios for which ages can be directly calculated if initial Pb has been corrected for. Using the age equations, modified to be corrected for initial Pb (in this case equations 8.3 and 8.6), as a set of parametric equations, one can plot a curve in this space that corresponds to ratios yielding the same date. This curve is the concordia curve, and similar diagrams can be constructed using any of equations (8.3)–(8.6). The two most widely used concordia diagrams are for $^{207}Pb/^{206}Pb$ and $^{238}U/^{206}Pb$, called the Tera–Wasserburg concordia diagram [*Tera and Wasserburg*, 1972a, b], and for $^{207}Pb/^{235}U$ and $^{206}Pb/^{238}U$, called the Wetherill concordia diagram [*Wetherill*, 1956a]. These diagrams contain the same information, but are preferred for different reasons by different users. A benefit of the Tera–Wasserburg diagram is that when plotting samples with high initial Pb contents of unknown isotopic composition, these data can fall on an isochron where the intersection with the $^{207}Pb/^{206}Pb$ axis is the composition of initial lead and the intersection with the concordia curve is the age—much like the three-dimensional isochron diagram outlined above. Some users also prefer the Tera–Wasserburg diagram because some of the the largest sources of uncertainty, resulting from elemental and isotopic fractionation and low abundance of ^{207}Pb, are separated onto different axes. A benefit of the Wetherill diagram is that no isotope is found on both axes, permitting evaluation of the independent decay schemes. The Wetherill diagram also visually creates a larger spread between points of different ages that scales more predictably with age.

All types of concordia diagrams in U–Th–Pb geochronology have the benefit of being able to evaluate closed-system behavior in dated samples. This is because, once corrected for initial Pb, samples that represent a single age and have not lost or gained Pb or U other than through radioactive decay should plot on the concordia curve. This internal check for closed-system behavior is one of the benefits of the U–Pb system, and improves accuracy by facilitating better data interpretations. Samples that plot off the concordia curve (ignoring analytical effects) do so for many reasons, which are discussed in the next section.

Additional insight can be gained by interpreting the distribution of data off the concordia curve (Fig. 8.7). For example, if a suite of results are colinear (as determined by similar statistical methods as applied to isochrons) and fall on the concave side of the concordia curve, the intercepts of the regressed line with the concordia curve can still yield age information. In a system where analyses fall off the concordia due to either (i) a single episode of Pb loss following initial system closure or (ii) mixing between two discrete events of mineral growth, then the upper intercept with concordia gives the time of initial mineral growth and the lower intercept gives the time of the second event,

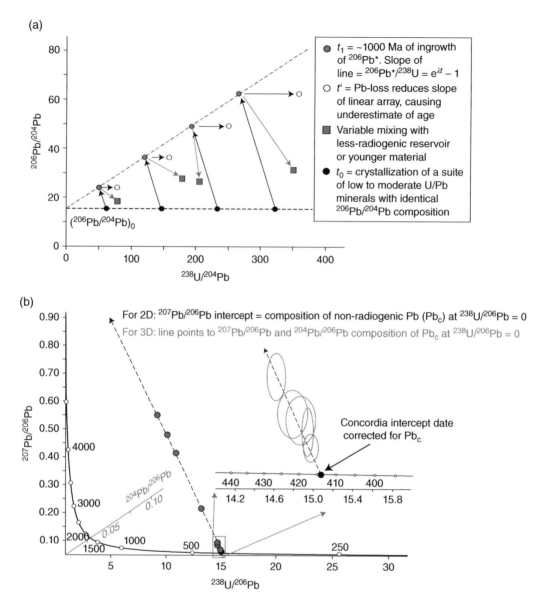

Fig. 8.6. Two-dimensional and three-dimensional isochron diagrams in the U–Pb system. (a) An example of one of three possible parent–daughter isochron diagrams in the U–Th–Pb system. The times, t_0, t_1, and t', refer to different times in the system's evolution and the data points represent possible positions of that system for both closed-system and open-system behavior. Possible explanations for that behavior are listed in the figure. (b) Tera–Wasserburg concordia diagram shown in three-dimensions. Emphasized here is the linear array in U–Pb isotopic data that is caused by a mixture of radiogenic and common Pb, and thus the three-dimensional plane is an isochron. See text in section 8.3.2 for for discussion of how this diagram also serves as the Tera–Wasserburg concordia plot.

whether it be Pb loss or secondary mineral growth. Figure 8.7 illustrates a qualitative way of arriving at these different dates. In reality, a suite of discordant analyses are often not colinear, indicating that either multiple events of Pb loss or mineral growth occurred within the sample suite or that these processes were semicontinuous. In such cases, age information can be still extracted, but the accuracy and precision of interpreted events diminishes and geologic meaning from intercept dates should be treated with skepticism [*Mezger and Krogstad*, 1997; *Corfu*, 2013; *Schoene*, 2014].

8.4 CAUSES OF DISCORDANCE IN THE U–Th–Pb SYSTEM

The causes of discordance in U–Th–Pb (particularly U–Pb) geochronology have been discussed within the context of countless studies. Reasons for discordance include: (i) mixing between two or more sample populations of different age; (ii) Pb loss; (iii) intermediate daughter product disequilibrium; and (iv) initial Pb. The first two are entirely geologic in nature, and the second two, while reflecting geologic processes, can

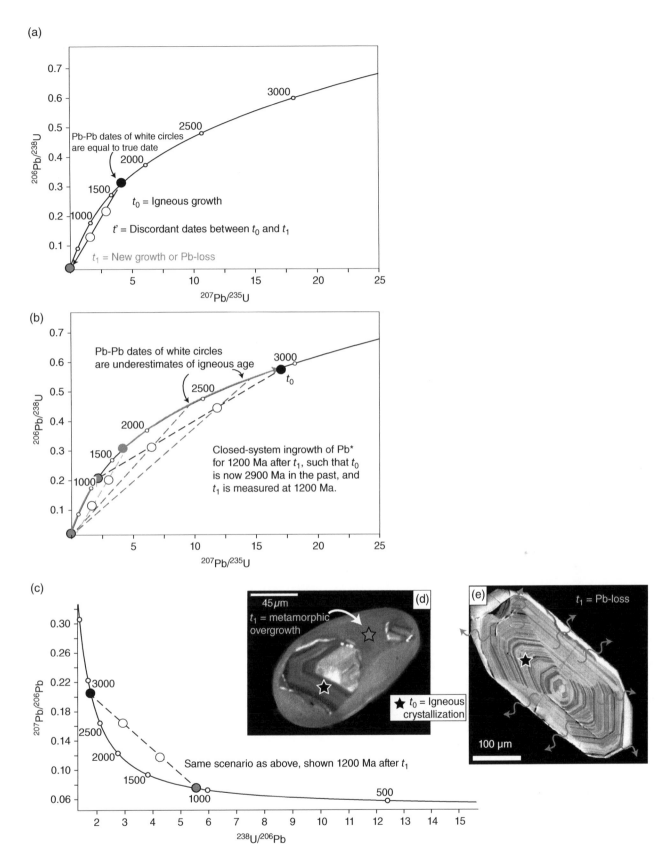

Fig. 8.7. Graphical representation of zircon growth history in (a,b) the Wetherill concordia diagram and (c) the Tera–Wasserburg diagram. (a) Example of a 1700 Ma zircon losing Pb or mixing with metamorphic overgrowth. Time t_0 is the crystallization age of the zircons; after 1700 Ma of closed-system ingrowth of Pb (t_1), the zircon suffers Pb loss or growth of new zircon around old core; t' represents zircons that are discordant following partial Pb loss or mineral overgrowth at t_1. (b) shows the same data after the system has closed again and continues to evolve up the concordia curve for 1200 Ma. The discordia line defined by open symbols now has an upper intercept with concordia representing the original igneous crystallization event at t_0, and a lower intercept age representing t_1, the time before present at which Pb loss or overgrowth occurred. (c) The same scenario as (b) but in a Tera–Wasserburg diagram. (d) An illustration of how the scenario would possibly be recorded in the event of metamorphism at t_1. (e) The case where Pb loss happens at t_1. (Source: (a)–(c) adapted from *Schoene* [2014]. (d) *Schmitz and Bowring* [2004]. Reproduced with permission of The Geological Society of South Africa.)

potentially be corrected for during data reduction. Purely analytical causes of discordance arising during measurement are also important, but beyond the scope of this text. Inaccurate U isotopic composition of the sample and inaccurate decay constants, which were discussed above, are also potential causes of slight discordance but not discussed further here.

8.4.1 Mixing of different age domains

Observing the external morphology of dated minerals and their petrographic relationship to other minerals in a rock is important for interpreting geochronologic data [*Krogh and Davis*, 1975; *Pupin*, 1980; *Vavra et al.*, 1996; *Belousova et al.*, 2006]. Additionally, internal growth histories within single datable minerals can be complicated and necessary to resolve in order to interpret geochronologic data. Mixing between two different age domains during an analysis can result in data that plots discordantly, depending on the difference in age between one domain and another (or potentially more than two domains). In some minerals, different domains in grains can be seen petrographically or under an optical microscope in mineral separates or grain mount [*Krogh and Davis*, 1975; *Bickford et al.*, 1981; *Corfu and Ayres*, 1984]. Such domains become even more pronounced if imaged by backscatter electron (BSE) [*Wayne and Sinha*, 1988; *Wayne et al.*, 1992] or cathodoluminescence (CL) [*Schenk*, 1980; *Hanchar and Miller*, 1993; *Hanchar and Rudnick*, 1995], typically done in a scanning electron microscope or electron microprobe (EMP).

In addition to identifying the presence of domains in grains that potentially record different ages, BSE or CL imaging can also reveal internal zonation within different aged domains that give insight into the growth history of the mineral. For example, concentric oscillatory zoning (Fig. 8.8) is typical of igneous growth whereas metamorphic growth may appear "splotchy," "patchy," or simply nondescript. Such images may also reveal truncations in zoning resulting from mineral resorption (dissolution resulting from disequilibrium) that can be used to evaluate mineral undersaturation or original morphology of inherited cores of older xenocrysts (phenocrysts inherited from older unrelated material). This simple yet powerful tool is used routinely prior to U–Th–Pb geochronology *in situ* to identify, target, and date growth domains in zircon, monazite, titanite, allanite, and apatite. In cases where different growth domains are visible and the ages are resolvably different (or not), the combination of grain imaging and subsequent geochronology by all analytical techniques has been important for pulling apart complex igneous and metamorphic histories of rocks [*Bowring et al.*, 1989b; *Vavra et al.*, 1996; *Hawkins and Bowring*, 1999; *Schaltegger et al.*, 1999; *Rubatto*, 2002; *Kelly and Harley*, 2005; *Schoene and Bowring*, 2007; *Corrie and Kohn*, 2007; *Crowley et al.*, 2007; *Dumond et al.*, 2008; *Gordon et al.*, 2010; *Harley and Kelly*, 2007; *Cottle et al.*, 2009b]. Even in cases where domains are too small to analyze individually, identifying them prior to analysis can help explain discordant data and still constrain the timing of the different events using intercept ages.

8.4.2 Pb loss

Although loss or gain of U, Th, or Pb are forms of open-system behavior that can result in discordant data arrays, Pb loss is most widely discussed in the literature. One reason is that Pb is generally incompatible in zircon to begin with, whereas U is compatible. In support of the importance of Pb loss is that most discordant data plot in the domain of concordia diagrams that is explained by either Pb loss or U gain (i.e., the Pb/U ratio is lower than it should be), and U gain is a difficult process to envision in the absence of new mineral growth (discussed above). As such, the mechanisms of Pb loss in U–Th–Pb geochronology have received a lot of attention in empirical, experimental, and theoretical studies. A well-understood process that can result in Pb loss is that of volume diffusion through the crystal lattice. This process forms the basis of U–Pb thermochronology in titanite, apatite, and rutile, and is discussed in detail in section 8.12. U–Pb thermochronology is analogous to other forms of thermochronology that take advantage of volume diffusion, which are discussed throughout this book.

Lead loss in zircon, the bane of many U–Pb geochronologists, has received a lot attention since the 1950s, yet still remains an unsolved problem (see *Corfu* [2013] for a recent review). These early studies often focused on Proterozoic and Archean rocks (perhaps because of how the Pb–Pb isotope systematics of these rocks tied into the determination of the age of Earth), where Pb loss was obvious and data often formed discordia arrays pointing towards the origin. Debate about whether Pb loss in monazite and uraninite was due to a protracted process or a single event [*Ahrens*, 1955a, b; *Wetherill*, 1956b; *Catanzaro*, 1963] was based on whether the discordia arrays were in fact linear. *Tilton* [1960] recognized that many zircon discordia arrays have lower intercepts of ca. 600 Ma but did not follow linear arrays consistent with a single Pb-loss event. He applied formulas for loss of Pb by volume diffusion and even calculated kinetic data for Pb diffusion in zircon. Following the detailed empirical study of *Silver and Deutsch* [1963], *Wasserburg* [1963] derived equations that also described Pb loss by diffusion, but with a diffusion coefficient that was a function of U-induced and Th-induced radiation damage to the zircon lattice. These descriptions of Pb loss were perhaps more realistic given that modern experimental data show that Pb diffusion in zircon is negligible at temperatures < 800–900 °C in nonmetamict crystals [*Lee*, 1997; *Mezger and Krogstad*, 1997; *Cherniak and Watson*, 2001). Work at least in part inspired by the fission-track community has shown that radiation damage in zircon is a result of both alpha recoil and fission-track accumulation [*Silver and Deutsch*, 1963; *Pidgeon et al.*, 1966; *Deliens et al.*, 1977; *Nasdala et al.*, 1996; *Meldrum et al.*, 1998], and at temperatures above ~250 °C, radiation damage in zircon is quickly annealed [*Ketcham et al.*, 1999]. Furthermore, radiation damage has been shown to correlate roughly with degree of discordance in some zircon suites [*Nasdala et al.*, 1998], but attempts at revisiting equations for radiation-induced Pb diffusion, for example by adopting short-circuit diffusion models

Fig. 8.8. Cathodoluminescence images of zircon. (a) Examples where igneous or metamorphic cores are overgrown by younger metamorphic or igneous zircon rims. (b) Complex zoning, with igneous or metamorphic rims overgrown by one or more zircon growth domains, representing a later igneous or metamorphic event. (c) Typical sector zoning in igneous zircon. (d, e) Examples of relatively simple igneous zircon, though some have cores that may represent older growth periods. Scale bars are 100 μm. Images with no scale bars are between 100 and 200 μm in diameter. Some images were compiled and contributed by F. Corfu; see *Corfu et al.* [2003] for references.

[*Lee*, 1995] still do not quantitatively capture the process. Additional mechanisms that may contribute to Pb loss are crystal-plastic deformation as a means of generating fast-diffusion pathways [*Reddy et al.*, 2006] and low-temperature hydrothermal dissolution–reprecipitation mechanisms [*Krogh and Davis*, 1974; *Pidgeon et al.*, 1966; *Geisler et al.*, 2002, 2003), perhaps correlated with the timing of uplift and exposure to weathering processes [*Goldich and Mudrey*, 1972; *Allègre et al.*, 1974].

In the end a combination of all these processes probably result in Pb loss in zircon. While we continue to learn more about this process, the greatest advances in the accuracy of U–Pb geochronology of zircon have not come through understanding Pb loss but instead by learning ways to avoid it. These can be summarized in three advances:

(1) the air-abrasion technique [*Krogh*, 1982], which mechanically removes the outer, often higher U and more metamict, domains of grains prior to analysis of whole grains;

(2) reduction in sample size through either the use of *in situ* dating techniques, which have sufficient spatial resolution to avoid domains that have undergone Pb loss, or by analyzing smaller and smaller amounts of material in TIMS U–Pb geochronology;

(3) the chemical abrasion technique [*Mattinson*, 2005] which partially anneals the crystal structure of zircons and then chemically dissolves discordant domains, leaving a closed-system residue amenable to analysis.

The first technique changed and dominated the field for two decades but is not widely used now because of proliferation of the second and third techniques, which will be discussed in section 8.5.1.

8.4.3 Intermediate daughter product disequilibrium

Equations (8.3)–(8.6) can be used to accurately calculate an age only if the reservoir from which a mineral crystallizes is in secular equilibrium and if there is no elemental fractionation of the intermediate daughter products during crystallization. Both of these assumptions are probably not strictly met in any system [*Mattinson*, 1973; *Schärer*, 1984]. Preferential partitioning of the intermediate daughter product into a mineral over its parent, will result in an excess amount of radiogenic Pb while the system approaches secular equilibrium. The result will be a calculated U–Pb age that is too old. An age will be underestimated if an intermediate daughter product is preferentially excluded during crystallization (Fig. 8.9).

On the brighter side, equations (8.1) and (8.2) show that only intermediate daughter products with long enough half-lives will be present or absent in enough quantity to affect a resulting date, namely ^{230}Th ($t_{1/2} = 75.4$ ka), ^{234}U ($t_{1/2} = 245$ ka), from the ^{238}U decay chain and ^{231}Pa ($t_{1/2} = 32.8$ ka) from the ^{235}U decay chain. ^{234}U is assumed to not be significantly fractionated from ^{238}U at high temperatures (though it definitely is in low-T systems), so this is generally ignored as a significant source of disequilibrium. ^{230}Th disequilibrium is important, in part because in very high Th/U minerals such as monazite, the resulting excess ^{206}Pb is easily recognizable on a conventional concordia plot [*Mattinson*, 1973]. *Schärer* [1984] and *Parrish* [1990] quantified these effects by relating the amount of intermediate daughter product lost or gained relative to secular equilibrium during mineral crystallization using versions of the following equation:

$$t_{\text{excess}} = \left(\frac{1}{\lambda_{238}}\right) \ln\left[1 + (f-1)\left(\frac{\lambda_{238}}{\lambda_{230}}\right)\right] \qquad (8.7)$$

where

$$f = \left| \frac{(\text{Th}/\text{U})_{\text{mineral}}}{(\text{Th}/\text{U})_{\text{liquid}}} \right| \qquad (8.8)$$

where f is the mineral/melt partition coefficients for the ratio of Th and U in the phase of interest ($D^{\text{Th}/\text{U}}_{\text{mineral/melt}}$). For a full derivation of these equations, see (*McLean et al.*, 2011). If partition coefficients for minerals and various melts were invariant, then f would always be the same and one could easily correct for intermediate daughter product disequilibrium. This is unfortunately not the case. For example, published zircon/melt partition coefficients are variable and probably dependent on temperature, pressure, oxygen fugacity, and magma composition [*Hanchar and van Westrenen*, 2007; *Rubatto and Hermann*, 2007a; *Burnham and Berry*, 2012]. Nonetheless, one approach that has been used to correct for intermediate daughter product disequilibria is to assume a value for f using knowledge of the magma composition and temperature and available partitioning data (*Lissenberg et al.*, 2009; *Barboni and Schoene*, 2014; *Barboni et al.*, 2015; *Samperton et al.*, 2015). Partitioning of Th and U between melt and titanite and apatite have been determined experimentally [*Prowatke and Klemme*, 2005, 2006a, b], but the range in D_U/D_{Th} observed precludes a bulk correction for intermediate daughter product disequilibrium even in magmatic minerals.

Alternatively, correcting for ^{230}Th disequilibrium can be done using estimates for the Th/U$_{\text{mineral}}$ and the Th/U$_{\text{liquid}}$, where the liquid might be a silicate melt or fluid from which the mineral crystallizes. Th/U$_{\text{mineral}}$ can be measured directly during mass spectrometry (typical for SIMS and LA-ICPMS) or estimated by assuming concordance between the U–Pb and Th–Pb dates, measuring ^{208}Pb*, and then calculating ^{232}Th (typical for ID-TIMS). The Th/U$_{\text{liquid}}$ can be estimated by using the Th/U of the rock from which the mineral is extracted [*Schärer et al.*, 1990], but this requires that a rock accurately represents a liquid composition. Another approach is by measuring Th/U of the host glass for volcanic minerals [*Schmitz and Bowring*, 2001; *Bachmann et al.*, 2010] or in melt inclusions [*Crowley et al.*, 2007; *Ickert et al.*, 2015]; these approaches come with inherent uncertainty given the range of Th/U$_{\text{melt}}$ measured within glass and melt inclusions from a single magmatic system.

For minerals where $f < 1$ (see equation 8.8), such as zircon and xenotime, the correction for ^{230}Th disequilibrium has a lower limit at $f = 0$ of -110 ka—meaning that the calculated age is 110 ka too young [*Schärer*, 1984] (Fig. 8.9). In such cases, Th/U$_{\text{liquid}}$ is commonly assumed, given that this ratio usually falls between 2 and 6 in magmas. Propagating the uncertainty of the correction into final ages results in minimal added age uncertainty, except for very young samples [*Crowley et al.*, 2007; *Bachmann et al.*, 2010; *Ickert et al.*, 2015]. For minerals where $f >> 1$, such as allanite and monazite (and sometimes titanite and apatite), the excess ^{206}Pb resulting from initial ^{230}Th disequilibrium causes calculated dates to appear too old, and can be debilitating and many geochronologists simply avoid using the

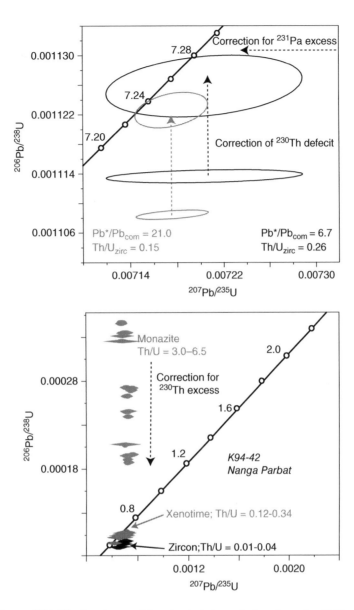

Fig. 8.9. The effect of ^{230}Th disequilibrium in ^{206}Pb/^{238}U dates. Top panel shows two typical examples from zircon where Th/U$_{zircon}$ is less than the liquid from which it crystallizes, assuming a Th/U$_{liquid}$ of ~3 (modified from *Schaltegger et al.* [2015]). This results in a deficiency of ^{206}Pb and therefore an underestimate of the age, and is therefore corrected toward an older date using equations (8.7) and (8.8). In both cases, the correction is almost the maximum of 110 ka. Bottom panel (from *Crowley et al.* [2009]): an example where excess ^{206}Pb was generated through incorporation of ^{230}Th above that of secular equilibrium, as is typical of monazite, which incorporates much more Th than U relative to the liquid from which it crystallizes. The result is ^{206}Pb/^{238}U dates that are much older than the more accurate ^{207}Pb/^{235}U dates, and the analyses plot way above the concordia curve. Also plotted are zircon and xenotime analyses, which plot below concordia, similar to in the top panel. (Sources: *Schaltegger et al.* [2015] and *Crowley et al.* [2009]. Reproduced with permission of Elsevier.)

^{206}Pb/^{238}U date in such cases [*Villeneuve et al.*, 2000; *Schoene and Bowring*, 2006; *Cottle et al.*, 2009b; *Crowley et al.*, 2009].

The effect of ^{231}Pa disequilibrium is poorly understood because (i) there are no other isotopes of Pa that can be used to estimate Pa and U partitioning during crystallization and (ii) because experimental data for Pa partitioning is not available. *Schmitt* [2007] measured [^{231}Pa]/[^{235}U] in young volcanic zircons and found values near unity, suggesting only ~15 ka age excess for older samples. However, extreme ^{231}Pa disequilibrium

has been observed in zircon [*Anczkiewicz et al.*, 2001], as well as more subtle evidence for excess ^{231}Pa consistent with < 20 ka influence on the ^{207}Pb/^{235}U date [*Crowley et al.*, 2007; *Rioux et al.*, 2015].

8.4.4   Correction for initial Pb

As with other radioisotopic systems, the correction for the initial daughter product present in a system is crucial for accurate age determinations. In U–Pb geochronology, both two-dimensional and three-dimensional isochrons can be used to solve for the isotopic composition of initial Pb (Pb$_0$; as used here, Pb$_0$ differs from Pb$_c$ in that Pb$_c$ includes laboratory blank Pb) if a data set meets the required assumptions for isochron calculations. On the Terra–Wasserburg concordia and three-dimensional isochron diagrams described above, the ^{207}Pb/^{206}Pb of Pb$_0$ is the y-intercept. Unaccounted for Pb$_c$ on a Wetherill concordia diagram creates a discordia with slope equal to the ^{206}Pb/^{207}Pb $\times$ ^{238}U/^{235}U$_{sample}$; though data are usually corrected for Pb$_c$ by some method before being plotted (Fig. 8.10). A data set in either concordia diagram can form a linear array because of mixing or Pb loss as well. Whether Pb$_c$ or mixing/Pb loss is responsible for the spread in data for young samples can be determined because the upper intercept on a Tera–Wasserburg diagram for linear arrays affected by only Pb$_c$ is > 4.5 Ga (Fig. 8.6). For Paleoproterozoic or Archean samples, however, the Pb$_c$ array can be nearly parallel to discordia arrays created by mixing/Pb loss, and it is thus dangerous to assume the source of the discordance. An example can be extracted from the Tera–Wasserburg diagram in Fig. 8.6, where an upper intercept > 4.5 Ga is illustrated for a young sample set. Interpreting a discordia array for a Proterozoic sample set would be more ambiguous because the array could both parallel concordia and also be misinterpreted as mixing with older domains.

Other methods for Pb$_c$ correction involve an assumption about, or an attempt to independently measure, its composition. For example, by measuring the moles of ^{204}Pb in a sample and

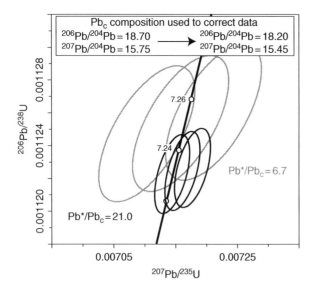

Fig. 8.10. Effect of common Pb correction in relatively high Pb*/Pb$_c$ minerals. Two examples of zircons are plotted with variable Pb*/Pb$_c$. The same range in isotopic composition of Pb$_c$ is used to correct the data. The analysis with the lower Pb*/Pb$_c$ is more heavily affected by the correction. The end-member values of Pb isotopic composition of Pb$_c$ are shown above; the third ellipse uses an intermediate Pb$_c$ composition. (Source: *Schaltegger et al.* [2015]. Reproduced with permission of Elsevier.)

estimating a mass and ^{206}Pb/^{204}Pb of the Pb$_c$ from a single or multiple known sources (e.g., blanks measured in a laboratory), one can then calculate the amount of ^{206}Pb* [*Ludwig*, 1980; *Williams*, 1998; *Schmitz and Schoene*, 2007; *McLean et al.*, 2011]. In the case of zircon, there may be no Pb$_0$, and all Pb$_c$ is introduced through one or more source of laboratory contamination. The isotopic composition of Pb$_0$ adopted for other minerals is often assumed based on a bulk-Pb evolution model [e.g., *Cumming and Richards*, 1975; *Stacey and Kramers*, 1975] given an estimated crystallization age for the mineral. Alternatively, it can be estimated by dissolving or leaching Pb from coexisting low-U phases such as K-feldspar [*Catanzaro and Hanson*, 1971; *Housh and Bowring*, 1991; *Chamberlain and Bowring*, 2001]; Several studies make direct comparisons of these techniques [*Chamberlain and Bowring*, 2001; *Schmitz and Bowring*, 2001; *Schoene and Bowring*, 2006, 2007], with the general conclusion that analyzing cogenetic phases is more robust. However, *Schoene and Bowring* [2006] argue that apatite and titanite Pb$_0$ from a syenite was derived from an evolving source that is not defined by feldspar Pb and instead prefer the three-dimensional isochron-derived Pb$_0$ composition.

Since ^{204}Pb is always the least abundant Pb isotope present (^{206}Pb/^{204}Pb >> 1000 is typical for zircon), this ratio is difficult to measure precisely and is also subject to isobaric interferences during mass spectrometry. Furthermore, in many analytical set-ups typical for LA-ICPMS U–Pb dating, ^{204}Pb is impossible to measure due to unresolvable isobaric interference ^{204}Hg, requiring different methods of Pb$_c$ correction [*Horstwood et al.*, 2003]. Most of these are similar to the ^{204}Pb correction but instead involve assuming an initial ^{207}Pb/^{206}Pb or ^{208}Pb/^{206}Pb and concordance between the U–Th systems [*Williams*, 1998]. The former, "207-correction," is essentially the same as fixing a ^{207}Pb$_0$/^{206}Pb$_0$ intercept on a Tera–Wasserburg concordia plot and regressing a line anchored there through the data, and thus assumes concordance. If Pb loss or mixing was important in a data set, then both the 208 and 207 corrections are inaccurate. *Andersen* [2002] presents a method of 204-absent Pb$_c$ correction utilizing all three decay schemes that does not assume concordance, but instead must assume a time of Pb loss.

8.5 ANALYTICAL APPROACHES TO U–Th–Pb GEOCHRONOLOGY

There are a diversity of approaches used to measuring isotope ratios of U, Th and Pb (see recent reviews in [*Corfu*, 2013; *Schoene*, 2014; *Schaltegger et al.*, 2015]. Modern techniques all involve methods that generate ions of sample ± tracer isotopes, which are then accelerated into a mass spectrometer. Chapter 3 summarizes different types of mass spectrometers and how they transform sample into isotope ratios, so this is not covered in detail here. Instead, the benefits and drawbacks of each popular method used for U–Th–Pb geochronology are outlined briefly, and then specific applications that benefit from the strengths of each method are covered in detail. In general, the methods that result in the highest precision dates require the largest sample sizes, while those that have better spatial resolution have lower precision in resulting dates. This is demonstrated in Fig. 8.11, which plots the sample size and analytical precision versus time of publication of various studies. These data show that each technique has improved in analytical precision since

its inception and that workers continue to strive for higher and higher spatial resolution as well. One important characteristic of each technique that is not plotted is the analysis time, which also varies significantly, and these are noted in the following sections (Fig. 8.12).

8.5.1 Thermal ionization mass spectrometry

Application of thermal ionization mass spectrometry to U–Th–Pb geochronology usually involves isotope dilution by spiking samples prior to acid digestion with a mixed Pb and U or Th tracer [*Parrish and Noble*, 2003; *Stracke et al.*, 2014]. Commonly used isotopes of Pb for the spike include ^{202}Pb and ^{205}Pb for geochronological applications. Studies measuring only Pb isotopic compositions on large samples amenable to sample aliquoting also employ double-spike techniques using ^{204}Pb, ^{207}Pb and/or ^{208}Pb [*Todt et al.*, 1996; *Galer and Abouchami*, 1998; *Thirlwall*, 2000; *Baker et al.*, 2004]. The benefit of having two spike isotopes is that it permits calculation of mass fractionation during mass spectrometry by measuring the deviation of the measured value from the known value for each measured sample. This is otherwise a large source of uncertainty when sample size is large, such that common Pb isotopic composition, blank corrections, and analytical uncertainties do not dominate the total uncertainty budget.

Tracer isotopes of uranium commonly used for spiking include ^{233}U, ^{235}U, and ^{236}U [*Roddick et al.*, 1987; *Richter et al.*, 2008; *Brennecka et al.*, 2010; *Connelly et al.*, 2012; *Hiess et al.*, 2012]. While ^{233}U is always used for U double spikes, the choice between ^{235}U and ^{236}U depends on sample size and resulting measurement uncertainties, in that the abundance of ^{235}U is so low in many samples that assuming a ^{238}U/^{235}U for the sample (with an uncertainty) can be more precise than measuring the ratio itself [*Condon et al.*, 2015; *Tissot and Dauphas*, 2015]. Uranium is commonly measured by TIMS as the oxide UO_2. The uncertainty contribution resulting from correcting for the isotopic composition of oxygen varies depending on which spike isotopes are chosen [*Condon et al.*, 2015].

Th–Pb geochronology is less common by TIMS than with the complementary techniques outlined below. Spike isotopes of Th include ^{229}Th and ^{230}Th, the latter of which occurs naturally in samples [*Parrish and Noble*, 2003]. It is thus necessary to assume the sample is in secular equilibrium prior to spiking. Chiefly, ionization of Th by TIMS is poor compared to U and Pb, limiting the precision of Th–Pb geochronology by TIMS on small samples; Th can be analyzed by ICPMS and Pb by TIMS, however, which can still result in high-precision data.

As with other applications of TIMS, U and Pb isotope ratio measurements take a long time—generally on the order of 2–6 h for a single mineral, depending on whether an analysis is performed on ion counters or in Faraday cups (the latter being faster because of multicollection capabilities) [*Parrish and Noble*, 2003; *Schoene*, 2014; *Schaltegger et al.*, 2015]. The main benefit

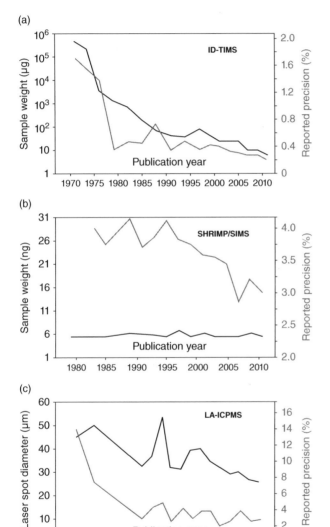

Fig. 8.11. Comparison of spatial resolution and precision of different U–Th–Pb analytical methods. Data were obtained by searching for the 10 most cited papers for a given 1-year or 2-year period and extracting the average 2-sigma uncertainty in ^{206}Pb/^{238}U dates or ratios for single analyses (not weighted means), then taking the average of the 10 papers. Conspicuous outliers were excluded from the analysis if the data were anomalously imprecise or if the sample size was anomalously large, but not for the opposite cases. Thus, data are biased towards higher precision and smaller sample size (to emphasize state of the art). Sample weight is given for ID-TIMS and SHRIMP/SIMS, but not LA-ICPMS because spot depths are not consistently reported. All uncertainties are 2-sigma. (Source: *Schoene* [2014]. Reproduced with permission of Elsevier.)

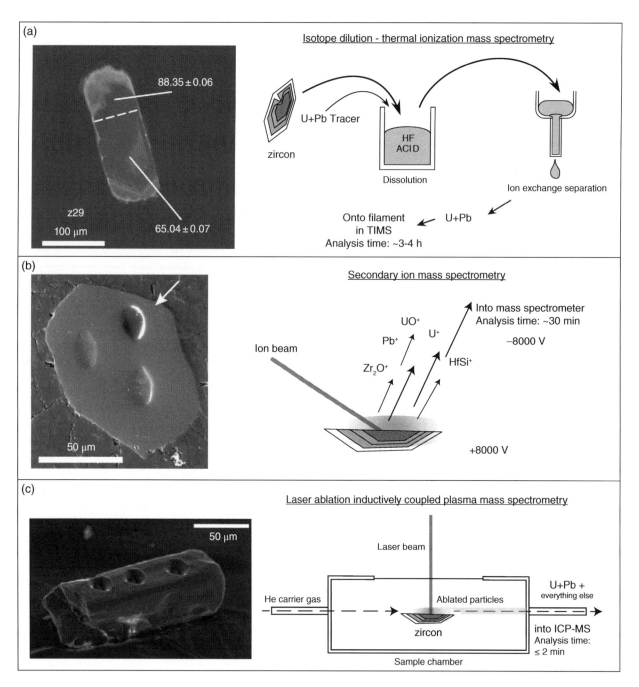

Fig. 8.12. Cartoons depicting the three most common types of U–Th–Pb analysis. Each method shows an image of a zircon that was analyzed to illustrate typical sample size. Diagrams to the right illustrate how the method works. See section 8.4 of the text for more discussion. (a) Isotope dilution thermal ionization mass spectrometry (ID-TIMS). Image of microsampled zircon that was subsequently analyzed comes from *Gordon et al*. [2010]. (b) Secondary ion mass spectrometry (SIMS). Image of zircon is from a lunar sample analyzed by *Grange et al*. (2011). (c) Laser ablation inductively coupled plasma mass spectrometry (LA-ICPMS). Image is from *Cottle et al*. [2009a]. (Sources: *Gordon et al*. [2010]. Reproduced with permission of Geological Society of America. *Grange et al*. [2011]. Reproduced with permission of Elsevier. *Cottle et al*. [2009a]. Reprodcued with permission Royal Society of Chemistry.)

is that, because stable ion beams can be generated for this long, counting statistics permit very high-precision data. Poor ion yield of a few percent and the amount of laboratory contamination of common Pb (blank) are current limitations to the precision of this technique for young, low Pb* samples [*Ludwig*, 1980;

Mattinson, 1987; *Roddick*, 1987; *Schmitz and Schoene*, 2007; *McLean et al*., 2011]. However, the ^{230}Th disequilibrium correction can dominate uncertainties for very young samples, while Pb mass fractionation in the TIMS can dominate the uncertainty for older and/or radiogenic samples. Nonetheless, dates on single high-U minerals or mineral fragments can now be measured with precisions of better than 0.1%, which is one to two orders of magnitude better than other U–Pb techniques [*Davydov et al.*, 2010; *Sageman et al.*, 2014; *Schoene et al.*, 2015]. The trade-off is that samples need to be physically manipulated, transferred into small beakers, and dissolved, thereby limiting sample size to minerals or fragments of minerals tens of microns in size. A growing knowledge of the complexity of growth histories of dated minerals, most notably zircon, pushes users of TIMS to smaller and smaller sample sizes that are accompanied by more contextual information [*Schoene et al.*, 2010b; *Wotzlaw et al.*, 2013; *Rivera et al.*, 2014; *Samperton et al.*, 2015]. This includes imaging of internal textures revealed in polished grain mounts prior to analysis, which can reveal different growth zones that may correspond to different periods of metamorphism, and coupling TIMS U–Pb geochronology with dating and geochemical techniques *in situ* [*Rivera et al.*, 2013; *DesOrmeau et al.*, 2015; *Samperton et al.*, 2015].

As discussed in section 8.4, one of the most difficult problems to overcome in U–Pb geochronology is that of Pb loss and mixing of different age domains. One breakthrough in this regard was the advent of microbeam techniques for analyzing small spots of grain interiors to isolate single growth domains and hopefully avoid zones that experienced open-system behavior. Because TIMS measurements are restricted in terms of spatial resolution, overcoming Pb loss and mixing requires physically or chemically removing such domains. In terms of Pb loss, a major breakthrough came with mechanical abrasion of zircon [*Krogh*, 1982], whereby aliquots of crystals are placed into a cylinder and blown around in circles, thereby mimicking sediment transport and abrading the outer rims of crystals. Because many zircons have high-U rims that are more susceptible to Pb loss through radiation damage, or have younger overgrowths, removing these rims can result in more concordant results. This blunt but effective approach dominated sample preparation for two decades. Because U zoning in zircon is not spherical nor always increaseing in U towards the rim, air abrasion is an imperfect solution for U–Pb geochronology.

Various workers experimented with acid leaching techniques for removing discordant domains of zircon (see reviews in *Davis and Krogh* [2000] and *Mattinson* [2005]). These techniques originally had the side effect of fractionating Pb from U and ^{206}Pb from ^{207}Pb, which was counterproductive. *Mattinson* [2005] presented a breakthrough in this regard, and showed that by annealing the zircons at 900 °C for 48–60 h prior to HF acid leaching prevented elemental and isotopic fractionation while systematically removing domains of discordant zircon. He dubbed this technique chemical abrasion TIMS (CA-TIMS). Using step leaching on large aliquots of zircon, he showed that through progressive leaching, resulting ages of leachates approached a plateau that corresponded with the true closed-system age of the zircon population. The ID-TIMS community quickly adopted this technique and revised it to be amenable to single-crystal studies [*Mundil et al.*, 2004; *Schoene et al.*, 2006],

and virtually every ID-TIMS U–Pb geochronology laboratory now employs CA-TIMS routinely. Attempts to adapt this technique to other minerals [*Peterman et al.*, 2012; *Rioux et al.*, 2010] and to further understand the details of the leaching process are ongoing. Some success has also been shown in using chemical abrasion as a pretreatment for *in situ* U–Pb [*Allen and Campbell*, 2012; *Crowley et al.*, 2014; *Kryza et al.*, 2012; *von Quadt et al.*, 2014].

8.5.2 Secondary ion mass spectrometry

Early application of SIMS to isotope measurements [*Hinthorne et al.*, 1979], and particularly in the early 1980s the development and application of the sensitive high resolution ion microprobe (SHRIMP), to U–Pb geochronology changed the landscape and potential of the U–Pb dating technique [*Compston et al.*, 1984]. As outlined in Chapter 3, the SIMS has superior spatial resolution compared to other dating methods (Fig. 8.12), which has been instrumental in characterizing age heterogeneity in single minerals either in grain mount [*Bowring et al.*, 1989b] or in depth-profiling applications [*Grove and Harrison*, 1999]. This capability has been used to constrain the timing of different metamorphic events, the ages of xenocrystic cores in magmatic minerals, isotopic variation related to diffusion of Pb, and also as a relatively rapid way of characterizing ages in a large sample population (see section 8.6). The precision of SIMS techniques for U–Pb geochronology is largely controlled by variable mass-dependent and element-dependent fractionation during sample sputtering by the ion beam, which is unpredictable and also sensitive to sample preparation techniques [*Williams*, 1998; *Stern and Amelin*, 2003; *Stern et al.*, 2009; *Schmitt and Zack*, 2012].

SIMS dates are obtained by sample-standard bracketing such that the age of a standard, commonly a naturally occurring mineral with an ID-TIMS-measured age, is assumed and deviation in the measured isotopic (^{207}Pb/^{206}Pb) and elemental (^{206}Pb/^{238}U) ratios from the known value is applied to unknown samples measured within the same analytical session. Inherent is the assumption that the standard and sample experience the same fractionation, which is mediated by matching the standard and sample as closely as possible in terms of mineralogy and even chemistry of the mineral being dated (so-called "matrix-matching" [*Williams*, 1998; *Black et al.*, 2004; *Fletcher et al.*, 2010; *White and Ireland*, 2012]. Further monitoring of sputtering behavior and isobaric interferences, for example in the abundance of oxide (e.g., U$^+$ vs. UO$^+$) or hydride (^{207}Pb vs. ^{206}PbH) species, is required to correct for elemental and isotopic ratios [*Hinthorne et al.*, 1979; *Jeon and Whitehouse*, 2015].

Both the precision and spatial resolution of SIMS U–Th–Pb geochronology have increased only slightly since the technique's inception (Fig. 8.11), and analysis time remains at about 30 min per date. Improved analytical protocol and more homogeneous and matrix-matched standards have likely increased the accuracy of dates as well [*Ireland and Williams*, 2003; *Black et al.*, 2004; *White and Ireland*, 2012]. Limitations in the precision of SIMS geochronology likely stem from uncertainties and variability in elemental and isotopic fractionation during sputtering, oxide and hydride formation, and other instrumental drift [*Ireland and Williams*, 2003; *Jeon and Whitehouse*, 2015]. However, new applications of SIMS to geologic problems are constantly being explored and given the unparalleled spatial resolution, continue to open up new avenues of research.

8.5.3 Laser ablation inductively coupled plasma mass spectrometry

Another means of measuring U–Pb dates *in situ* is by LA-ICPMS. The basics of LA-ICPMS are outlined in Chapter 3, and only the particulars relevant to U–Th–Pb geochronology are highlighted here. The first application of LA-ICPMS to measuring U, Th and Pb isotopes was in the mid-1990s [*Feng et al.*, 1993; *Fryer et al.*, 1993; *Hirata and Nesbitt*, 1995], presenting an alternative to SIMS that was cheaper and faster, while maintaining the advantage of measuring age domains within single grains either in thin-section or grain mount. Ease of access and room for method development resulted in LA-ICPMS becoming one of the most rapidly advancing geochronological techniques in the past 20 years [*Košler and Sylvester*, 2003; *Schaltegger et al.*, 2015]. While most workers have focused on geochronology of zircon, this technique is also now commonly applied to monazite, titanite, apatite, thorite, and other minerals. As a result of the rapid initiation of many laboratories globally, there is large variability in (i) the amount of material consumed during analysis, (ii) the data reduction algorithms and reported precision and accuracy, and (iii) instrumentation and analytical protocols. However, recent efforts to standardize analytical approaches and data reduction techniques in combination with interlaboratory calibrations continue to push this technique towards better precision and accuracy [*Gehrels*, 2011; *Košler et al.*, 2013; *Horstwood et al.*, 2016].

Similar to SIMS, U–Th–Pb dates are measured relative to a standard material of known age, and the offset between the measured and known date of the standard is applied to unknowns [*Horstwood*, 2008; *Košler and Sylvester*, 2003; *Slama et al.*, 2008]. Also similar to SIMS, matrix matching has been found to be critical to minimize differences in fractionation of U, Th, and Pb in standard and unknown both at the site of ablation and in the plasma during ionization. Similar to TIMS and SIMS techniques, precision on low-U and/or young samples is limited by counting statistics during analysis. Uncertainty in the Pb_c correction is also important for high Pb_c minerals. However, uncertainty in elemental fractionation during the course of an analytical session and during the course of each ablation (e.g., so-called "downhole" fractionation) is the largest contributor to the uncertainty of LA-ICPMS dates of all but the youngest zircons. Differences in composition and degree of radiation damage between samples and standards hinder the ability to reproduce the TIMS age of multiple secondary standards against a single

primary standard. Recent laboratory comparisons conclude that as a result, precision is currently limited to a few percent [*Gehrels*, 2011; *Košler et al.*, 2013]. Increasing precision and accuracy of this technique will likely come from understanding and minimizing variability in elemental fractionation during the laser ablation process [*Horn and von Blanckenburg*, 2007; *Kimura et al.*, 2011], as well as decreasing sample volume [*Cottle et al.*, 2009a]. Despite current limitations, rapid data acquisition has made LA-ICPMS the tool of choice for problems requiring large data sets, such as detrital mineral analysis, and has had a tremendous impact on increasing the accessibility of U–Th–Pb geochronology to a wide range of users [*Košler and Sylvester*, 2003; *Simonetti et al.*, 2005; *Gehrels et al.*, 2008; *Gehrels*, 2011; *Pullen et al.*, 2014; *Horstwood et al.*, 2016].

8.5.4 Elemental U–Th–Pb geochronology by EMP

One of the limitations of U–Th–Pb geochronology at the beginning of the 20th century was that mass spectrometry had not been invented, thereby limiting methods to just measuring concentrations of U, Th, and Pb. This approach neglects the contribution of nonradiogenic Pb to the sample and lacks the check for closed-system behavior afforded by dual decay of the uranium isotopes. However, in certain situations in which it is safe to assume that initial Pb is negligible and the system has remained closed, elemental dating can be accurate and done *in situ* on an EMP if concentrations of U, Th, and Pb are high-enough [*Cocherie et al.*, 1998; *Montel et al.*, 1996]. This approach has been applied to Proterozoic or older monazite, in that this mineral contains very high abundances of U and Th and therefore radiogenic Pb in old samples [*Williams and Jercinovic*, 2002; *Mahan et al.*, 2006; *Williams et al.*, 2007; *Dumond et al.*, 2015]. The limitations of the accuracy of this method include the assumptions noted above; the limitations to the precision lie in the counting statistics of an EMP and resolving spectral interferences during analysis. Reported precision using this technique is on the order of a few percent on weighted means of populations of analyses [*Williams et al.*, 2006]. The benefit of this method is that it can provide unprecedented spatial resolution of about 1 µm and therefore is well suited to characterizing (i) complexly zoned monazites that may record numerous periods of metamorphic growth and (ii) tiny inclusions within other minerals that can record critical petrogenetic information, especially when petrographic context is maintained by using an *in situ* approach [*Dumond et al.*, 2008; *Williams and Jercinovic*, 2012; *Budzyń et al.*, 2015; *Lo Pò et al.*, 2016].

8.6 APPLICATIONS AND APPROACHES

8.6.1 The age of meteorites and of Earth

8.6.1.1 Pioneering work

Perhaps the best-known isochron in geochronology is the one constructed by *Patterson* [1956] in his paper entitled "Age of meteorites and the earth." A solution to the problem of the age of the Earth had been forthcoming, following the realization 20 years earlier that $^{207}Pb/^{206}Pb$ model ages could be used to place ever-increasing minimum age constraints on the Earth. *Gerling* [1942] used the newly produced Pb isotope measurements by *Nier* [1941] to argue that the Pb isotopic composition of Pb-rich Earth materials such as galena could be used to model their age if these materials were separated at some time from a constantly evolving terrestrial reservoir. The assumptions that go into this process are:

(1) that there was and still is a homogeneous reservoir in the Earth whose isotopic composition is evolving only through radioactive decay;

(2) that periodically this reservoir is tapped to form galena deposits whose Pb isotopic composition is frozen in time (because galena has no U).

The equations used to quantify this process are modifications of equation (8.6), but can be derived here by simple mass balance of one Pb isotope. For ^{206}Pb, we can state that the total $^{206}Pb_g$ in a sample of galena extracted from the homogeneous reservoir at time t is given, relative to ^{204}Pb, by:

$$\left(\frac{^{206}Pb}{^{204}Pb}\right)_g = \left(\frac{^{206}Pb}{^{204}Pb}\right)_0 + \left(\frac{^{206}Pb}{^{204}Pb}\right)_T - \left(\frac{^{206}Pb}{^{204}Pb}\right)_t \qquad (8.9)$$

where $\left(^{206}Pb/^{204}Pb\right)_0$ is the initial isotopic composition of the uniform reservoir, $\left(^{206}Pb/^{204}Pb\right)_T$ is the radiogenic ^{206}Pb ingrown since that reservoir formed at time T, and $\left(^{206}Pb/^{204}Pb\right)_t$ is the ^{206}Pb that *would have* ingrown had the galena not been extracted from the reservoir, thereby ceasing its Pb isotopic evolution. We can substitute the radiogenic portion of the decay equation (8.3) in for the last two terms and arrive at:

$$\left(\frac{^{206}Pb}{^{204}Pb}\right)_g = \left(\frac{^{206}Pb}{^{204}Pb}\right)_0 + \left(\frac{^{238}U}{^{204}Pb}\right)\left(e^{\lambda_{238}T}-1\right) \\ -\left(\frac{^{238}U}{^{204}Pb}\right)\left(e^{\lambda_{238}t}-1\right), \qquad (8.10)$$

which simplifies into:

$$\left(\frac{^{206}Pb}{^{204}Pb}\right)_g = \left(\frac{^{206}Pb}{^{204}Pb}\right)_0 + \left(\frac{^{238}U}{^{204}Pb}\right)\left(e^{\lambda_{238}T}-e^{\lambda_{238}t}\right). \qquad (8.11)$$

Combining equation (8.11) with a similar equation for ^{207}Pb gives a version of the Pb–Pb dating equation:

$$\frac{\left(\frac{^{207}Pb}{^{204}Pb}\right)_g - \left(\frac{^{207}Pb}{^{204}Pb}\right)_0}{\left(\frac{^{206}Pb}{^{204}Pb}\right)_g - \left(\frac{^{206}Pb}{^{204}Pb}\right)_0} = \left(\frac{^{235}U}{^{238}U}\right)\frac{\left(e^{\lambda_{235}T}-e^{\lambda_{235}t}\right)}{\left(e^{\lambda_{238}T}-e^{\lambda_{238}t}\right)} \qquad (8.12)$$

Data can then be plotted on a $^{207}Pb/^{204}Pb-^{206}Pb/^{204}Pb$ diagram (section 8.3.1; Fig. 8.13). The U isotopic composition was assumed to be uniform in Earth, and therefore did not have to be measured. Figure 8.14 shows the evolution of this system in Pb–Pb space, using Canyon Diablo troilite (inferred to be primordial solar system Pb, see below) for the primordial value. In the early 1940s, the Pb isotopic compositions of meteorites had not yet been measured, but the approach described above can be used to estimate a *minimum* age for the Earth by using the most primitive galenas for $\left(^{206}Pb/^{204}Pb\right)_0$ and $\left(^{207}Pb/^{204}Pb\right)_0$, estimating t for the most evolved galenas, and therefore permitting calculation of T. Applying a variation of this approach, Gerling used Mesozoic galenas as a more recent end-member and compared them to Archean and Proterozoic galenas from Greenland and Canada, respectively, to arrive at a minimum age for the Earth of 3–4 Ga. The same type of approach was developed independently by both *Holmes* [1946] and *Houtermans* [1946] as a means of constraining the minimum age of Earth (presumably language and political barriers associated with World War II prevented these workers from becoming aware of each others' similar pursuits). Figure 8.13 shows a modified version of this approach, which involves a two-stage Pb evolution model for Earth from *Stacey and Kramers* [1975], which was motivated in part by the observation that many terrestrial Pb samples plot with higher $^{206}Pb/^{204}Pb$ than the single stage model would predict (which is identical to the meteorite isochron plotted in the figure).

Soon after these attempts, mineral separates of troilite from the Canyon Diablo meteorite (derived from Meteor Crater, AZ) were measured and shown to contain negligible uranium and thus their Pb isotopes provided a new estimate for the primordial composition of Pb in the solar system. Using these isotopic values as $\left(^{206}Pb/^{204}Pb\right)_0$ and $\left(^{207}Pb/^{204}Pb\right)_0$ pushes the age of the Earth to as old as 4.5 Ga, though it could not yet be argued that meteorites were a good representation of the hypothesized uniform reservoir from which the galenas were extracted [*Patterson et al.*, 1955]. Finally, *Patterson* [1956] argued that a sample of whole Earth Pb isotopic composition could be shown to fall on an isochron with the meteorites, this would argue that the Earth and meteorites were the same age, and this age could be calculated simply using the Pb–Pb dating equation (8.6). Such a sample of terrestrial material surely does not exist, but Patterson considered that continent-sourced marine sediments could be a candidate for homogenizing an otherwise complex crust. Numerous sediment samples fell on the same isochron as three stone and two iron meteorites, including the Canyon Diablo troilite, and yielded an isochron age of 4.55 ± 0.07 Ga. The conclusion was that this was the best estimate for the age of meteorites and Earth. Although a modern view of plate tectonics, elemental fractionation during crust formation, and subsequent weathering and erosion would discourage the assumption that marine sediments could represent a closed U–Pb system, the conclusion that Earth and meteorites derive from the same primordial material and thus have a similar age remains largely intact.

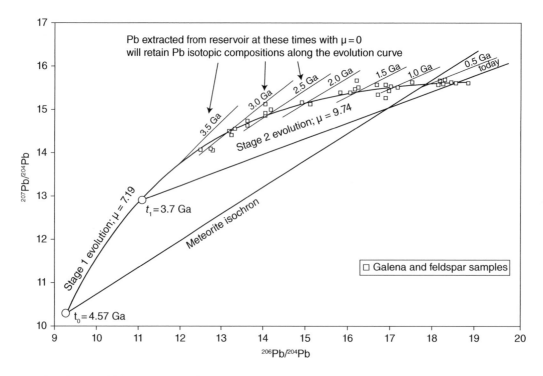

Fig. 8.13. Two-stage Pb evolution model for Earth. Diagram depicts a hypothetical reservoir in Earth that evolves from t_0 given a μ value of 7.19, where μ is the $^{238}U/^{204}Pb$ of the reservoir. At t_1, the reservoir differentiates, resulting in another reservoir with a μ of 9.74. The resulting reservoir evolves only due to the decay of U, and this reservoir is periodically tapped to crystallize U-free galena and feldspar, which are actual data and shown as open squares. The ages of those minerals correspond to the closest intersection of the labeled isochrons with the reservoir evolution path. Note this diagram is a modification of earlier single-stage Pb–Pb Earth evolution diagrams by Holmes, Houtermans, and Gerling, as discussed in text. (Source: *Stacey and Kramers* [1975]. Reproduced with permission of Elsevier.)

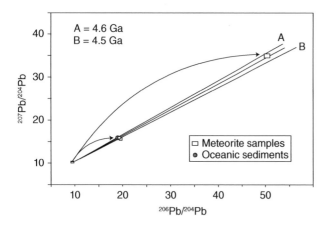

Fig. 8.14. The isochron published by *Patterson* [1956], which plots Pb–Pb isotope systematics of five meteorites and oceanic sediments. This diagram was used to argue that formation of meteorites and Earth was synchronous, thereby yielding an age of 4.55 ± 0.07 Ga for Earth. See text for discussion. (Source: *Patterson* [1956]. Reproduced with permission of Elsevier.)

Attempts to recreate Patterson's meteorite isochron using modern analytical techniques would likely fail. The reasons are that:

(1) better analytical precision and a modern statistical treatment of the data would show that in fact the materials analyzed are not co-linear;

(2) the terrestrial materials shown by Patterson to lie on the isochron did so in part coincidentally;

(3) applying the more accurate decay constants for ^{238}U and ^{235}U in use today (and were unavailable at the time), Patterson's isochron is too young by several hundred million years. In addition, as described in part below, modern attempts at resolving the processes responsible for the formation of the Earth and solar system have shown that this was not a single "event" but a drawn out and complex process, including a moon-forming impact, that a few colinear terrestrial and meteorite samples would surely obscure. Nonetheless, the original reported age for meteorites and Earth overlaps with modern age estimates for these bodies and so Patterson and other workers of that era get the much-deserved credit for determining the first accurate age for the solar system. This is arguably one of the greatest discoveries of the 20th century.

Variations on the Gerling–Holmes–Houtermans $^{207}Pb/^{204}Pb$–$^{206}Pb/^{204}Pb$ isotopic method are still used to understand bulk Earth reservoirs and geochemical cycling [*Stacey and Kramers*, 1975; *Zartman and Doe*, 1981; *Zartman and Haines*, 1988; *Kramers and Tolstikhin*, 1997; *Tornos and Chiaradia*, 2004; *Hofmann*, 2007; *Jackson et al.*, 2014]. Despite the efficiency of plate tectonics at rehomogenizing and/or mixing previously isolated reservoirs, several recent studies argue, based on isotopic evidence from young mantle-derived lavas, that primitive reservoirs may still exist in the mantle [e.g., *Jackson et al.*, 2014]. Thus, the vision of the early pioneers of Pb isotope geochemistry carries on and continues to yield new discoveries.

8.6.1.2 Modern meteorite chronology

Advances in sample preparation and mass spectrometry have enabled analyses of much smaller samples with much higher precision, such that it is now possible to resolve the ages of different early solar system materials with uncertainties on reported ages of less than a million years [*Connelly et al.*, 2008, 2012; *Amelin et al.*, 2009, 2010; *Bouvier and Wadhwa*, 2010]. A major concern in meteorites is still open-system behavior, particularly that associated with meteorite impact and residence on Earth's surface. This process can lead to U and Pb loss and also introduction of terrestrial Pb as a contaminant. The former two are overcome by using $^{207}Pb/^{206}Pb$ isochrons, while the latter is often addressed by cleaning samples through acid leaching, which can also access more and less radiogenic domains of the sample and provide a larger spread on an isochron.

As an example, Fig. 8.15 shows a Pb–Pb isochron from a calcium aluminum inclusion (CAI) from the CV chondrite Efremovka [*Connelly et al.*, 2012]. This material was leached in acid in several steps to both remove terrestrial contamination and to access different reservoirs of Pb that would create spread on the isochrons. The type of Pb–Pb isochron used in this case, and often by modern meteorite chronologists, differs from a traditional $^{207}Pb/^{204}Pb$–$^{206}Pb/^{204}Pb$ diagram by using $^{207}Pb/^{206}Pb$–$^{204}Pb/^{206}Pb$ as the axes (Fig. 8.15). On such a diagram, the y-intercept is the $^{207}Pb/^{206}Pb^*$, from which an age and uncertainty can be calculated directly. In the case where

terrestrial contamination is absent, closed-system data should plot on a line; terrestrial contamination will draw the points off the line towards the composition of the contaminant [*Amelin et al.*, 2002; *Connelly and Bizzarro*, 2009]. For these types of meteorite samples, measuring the U isotopic composition is critical due to nucleosynthetic variability in meteorite U isotopic composition. In the study shown in Fig. 8.15, the uranium isotopic composition was measured for each sample using a ^{233}U–^{236}U tracer (section 8.5.1).

A focal point of meteorite geochronology has been in synchronizing dates provided by long-lived radionuclide systems such as U–Pb and short-lived isotopes, some of which decayed to near completion within ~10^6-10^8 years years after the start of the solar system and therefore are used to calculate relative ages of various materials over timescales comparable to uncertainties of high-precision U–Pb dates (see Chapter 14 for a thorough discussion of these methods). On one hand, numerous U–Pb studies have argued that CAIs are up to several million years older than other meteorite material such as chondrules [*Amelin et al.*, 2005; *Connelly et al.*, 2008], suggesting that they are the earliest condensates of the solar nebula and were subsequently included within younger meteorites during reorganization of the inner solar system. This offset between CAI and chondrule dates is consistent with model ages based on the short-lived ^{26}Al–^{26}Mg system, leading to a model in which chondrules and other rocky and iron meteorites formed from accretion, differentiation, and destruction of protoplanetary bodies several millions of years following CAI formation. Alternatively, other Pb–Pb isochron data sets have been used to argue that chondrule formation occurred

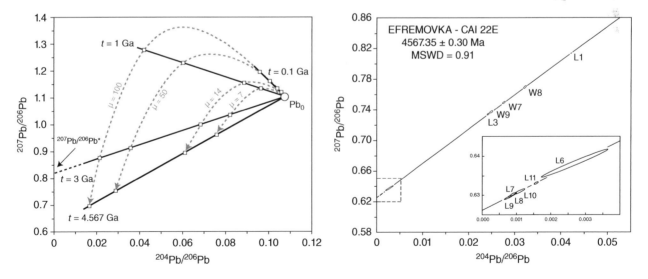

Fig. 8.15. Modern meteorite Pb–Pb geochronology. Left panel shows a hypothetical evolution diagram for minerals with variable μ, where $\mu = {}^{238}U/{}^{204}Pb$. Pb_0 is the isotopic composition of Pb_c at the beginning of closed system behavior, in this case the beginning of the solar system. Dotted lines show paths followed by individual mineral aliquots with variable μ. Isochrons are shown at various times, where t is equal to the time after the clock started 4.567 Ga. Note the intersection of those isochrons with the y-axis can be used as the $({}^{207}Pb/{}^{206}Pb)^*$ to calculate a Pb–Pb date. Also note that a system crystallized more recently (even with the same Pb_0, which is unlikely) would follow a different path due to the evolving $^{238}U/^{235}U$ in the solar system over time. Right panel shows an example of Pb–Pb geochronology from a calcium aluminum inclusion (CAI) from the Efremovka meteorite, from *Connelly et al.* [2012]. Each data point corresponds to either the leachate or residue of aliquots of the CAI. (Source: *Connelly et al.* [2012]. Reproduced with permission of The American Association for the Advancement of Science.)

over a span in time, with the oldest chondrules overlapping with CAIs [*Connelly et al.*, 2012]. Reconciling these dates with $^{26}Al–^{26}Mg$ model ages can be done if the distribution of ^{26}Al relative to Mg was not uniform in the solar nebula.

As noted in the previous section, the age of the Earth and meteorites calculated by *Patterson* [1956] would have been considerably younger if modern estimates of the U decay constants were used. The issue of decay-constant accuracy is still important for determining the absolute age of CAIs and chondrules, and therefore the oldest ages in the solar system, given that even small errors in those values can lead to inaccurate absolute $^{207}Pb/^{206}Pb$ dates of several million years in rocks ca. 4.57 Ga. Decay-constant uncertainties become especially important in early solar system studies when comparing dates from different decay systems because, e.g., 0.1% and 1% uncertainties in decay constants translates to between roughly ±4 and 40 Ma age uncertainty. For example, when comparing U–Pb to Ar-Ar dates from the early solar system, uncertainties of this magnitude can completely change the interpretation of the thermal history of early solar system material [*Renne*, 2000]. One potential solution to this problem is to calibrate other decay constants against that of ^{238}U, but this approach is hard to perfect given the difficulty of finding material amenable to different radioisotopic systems that should in fact be recording the same process (e.g., crystallization of different meteorite constituents, differing closure temperatures, etc.). Existing uncertainties in some decay constants are far larger than the length of many processes being scrutinized 4 billion years ago, and thus knowledge of decay constants remains a limitation for early solar system chronology. Regardless, ongoing advances in approaches to solar system chronology have shown that it is no longer appropriate to discuss the "age of meteorites and Earth" because we can now delineate the durations and mechanisms of processes occurring within the early solar system in ways not possible several decades ago.

8.6.2 The Hadean

8.6.2.1 The Acasta Gneiss
Evidence of what happened in the first 500 Ma of Earth history is sparse, and rocks that provide Hadean dates have been subjected to multiple periods of metamorphism and deformation such as to obscure their early history. Only two geologic terranes on Earth, both in Canada, are thought to contain rocks of Hadean age: the Acasta Gneiss in the Slave craton, and the Nuvvuagittuq supracrustal gneiss terrane in the Superior Province. The latter's age has been determined by Sm–Nd geochronology and not discussed here. The former's antiquity was determined by U–Pb geochronology of zircon with an age of ca. 4.03 Ga [*Bowring and Williams*, 1999a]. In most respects, the Acasta Gneiss is typical of other Archean gneiss complexes, consisting of mafic to felsic banded orthogneisses [*Bowring et al.*, 1989a, 1990; *Bowring and Williams*, 1999b] so the geologic history will not be discussed in detail here. Single and multigrain ID-TIMS dates from zircons from the Acasta Gneiss produced discordant arrays with

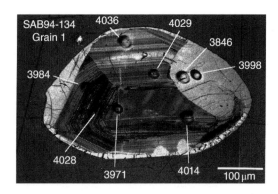

Fig. 8.16. Reflected light photomicrograph of a zircon from the Acasta Gneiss, analyzed by *Bowring and Williams* [1999]. SHRIMP U–Pb analysis pits are labeled with resulting $^{207}Pb/^{206}Pb$ dates. Uncertainties are about ± 6–40 Ma (2-sigma). (Source: *Bowring and Williams* [1999]. Reproduced with permission of Springer.)

upper intercept dates of > 3.8 Ga [*Bowring et al.*, 1989a]. *Bowring et al.* [1989b] used a SHRIMP to date domains within single zircons to illustrate a minimum igneous age of ca. 3.96 Ga, thereby reaffirming the power of is *situ* techniques to deconvolve the igneous and metamorphic histories of complexly deformed rocks.

Continuing the quest for a yet older history recorded in the rocks, *Bowring and Williams* [1999b] dated igneous cores of Acasta Gneiss zircons as old as ca. 4.03 Ga (Fig. 8.16). Whole-rock Nd isotopic data from the same sample suite produced both negative and positive εNd at 4.0 Ga, arguing for the existence of enriched and depleted reservoirs within the Earth well into the Hadean [*Bowring and Housh*, 1995; *Bowring et al.*, 1989a]. This revelation fueled the debate about when, and by what processes, the construction and recycling of continental crust began and importantly whether or not a depleted mantle reservoir was generated at the same time, which if so, places constraints on the volume of continental crust present at the time. Arguments against the Hadean antiquity for the source of the Acasta orthgneiss point to the potential for resetting whole-rock Nd systematics during Mesoarchean high-grade metamorphism, thereby decoupling the zircon dates from the Nd isotope systematics and precluding the necessity for Hadean crust–mantle differentiation [*Moorbath et al.*, 1997; *Whitehouse et al.*, 2001]. More recent whole-rock Sm–Nd isotopic data have reproduced a Mesoarchean errorchron date of ca. 3.3 Ga, though the same rocks also record ^{142}Nd anomalies, requiring a crustal source that formed within the Hadean [*Roth et al.*, 2014].

An emphasis in the last decade on using Hf-isotope data from individual closed-system zircons from the Acasta Gneiss has also contributed to this debate. When U–Pb and Pb–Pb dates from individual zircon spots are combined with Lu–Hf data (Chapter 6) from the same zircons, these data can address whether or not both enriched and depleted reservoirs existed within the Earth in the Paleoarchean. A lack of Hadean superchondritic εHf values can be used to argue that the volume of continental crust extracted was insufficient to generate a depleted mantle reservoir, whereas finding such high values opens the door for the presence of large volumes of enriched crust (i.e., low Lu/Hf, presumably high SiO_2) at the time [*Amelin et al.*, 2000; *Iizuka et al.*, 2009]. The answer to this question to whether a depleted mantle reservoir existed at the time, which hinges on analytical and geologic details of both Sm–Nd and Lu–Hf data [*Vervoort and Kemp*, 2016], remains debated, but regardless, Lu–Hf isotope systematics can be used to evaluate whether some volume of older crust was present when the Acasta Gneiss formed. Whole-rock Lu–Hf systematics have shown that some rocks (but not all), despite having undergone multiple episodes of Archean metamorphism, retain Hf isotope signatures that agree well with those of individual zircons and argue for derivation of the Acasta Gneiss, at least in part, from older crust [*Guitreau et al.*, 2014]. Subsequent identification of more [*Mojzsis et al.*, 2014], and even older, Hadean zircons of ca. 4.2 Ga [*Iizuka et al.*, 2006] from within the Acasta Gneiss also support the interaction of Acasta Gneiss with Hadean crust. These data therefore have broad implications for the geochemical cycling of early Earth and the potential for a Hadean ocean–atmosphere–biosphere system, and have therefore driven workers to look in more detail at even older, albeit detrital, zircons.

8.6.2.2 Zircon and the Hadean Earth

Given the poor preservation of Hadean rocks, much of our very limited knowledge of the Hadean Earth comes from zircon > 4.0 Ga, found exclusively by U–Pb geochronology of detrital or xenocrystic zircons in younger rocks. So far, Hadean zircons have been found in the Acasta Gneiss [*Bowring and Williams*, 1999b; *Iizuka et al.*, 2006] (see above); the Eoarchean Itsaq Gneiss, Greenland [*Mojzsis and Harrison*, 2002]; Ordovician volcanics of the Caotangou Group in the North Qinling Orogenic Belt, China [*Wang et al.*, 2007; *Diwu et al.*, 2013); metapelites within the Changdu Block of North Qiangtang, Tibetan Plateau, with a preliminary age of Meso- to Neoproterozoic [*He et al.*, 2011]; Neoproterozoic to Paleozoic metasediments from the Cathaysia Block of Southern China [*Xing et al.*, 2014]; Mesoarchean paragneisses and quartzites in the Beartooth Mountains, Wyoming Craton, USA [*Mueller et al.*, 1992; *Maier et al.*, 2012); ortho- and paragneisses of the Napier Complex, Antarctica [*Black et al.*, 1986; *Belyatsky et al.*, 2011]; felsic magmatic rocks of the Iwokrama Formation, Guyana shield [*Nadeau et al.*, 2013]; and most recently from orthogneisses in the Sao Francisco craton in northeast Brazil [*Paquette et al.*, 2015]. Typically only one or two grains of hundreds analyzed yielded such old ages; these discoveries are largely due to the ability of LA-ICPMS and SIMS U–Pb geochronology to screen a huge amount of zircon from a variety of rock types.

8.6.2.3 Jack Hills zircon

By far the most frequently exploited Hadean zircons hail from the Jack Hills Conglomerate, Yilgarn craton, Western Australia, which was deposited ca. 3.0 Ga and subsequently metamorphosed at least once [*Harrison*, 2009]. Hadean zircons from the Narryer Gneiss complex, also in the Yilgarn craton, were discovered over 30 years ago [*Froude et al.*, 1983], but shortly after samples from a limited number of outcrops of the nearby Jack Hills were found to contain up to 15% Hadean zircon [*Compston and Pidgeon*, 1986], which go back to a world-record 4.4 Ga. Since their discovery, rapid *in situ* U–Pb geochronology has been used to screen > 100,000 zircons for Hadean ages [*Maas et al.*, 1992; *Holden et al.*, 2009], permitting focused research on those old grains. These zircons have become a nearly exclusive archive of information about the Hadean Earth.

So what can be determined about the early Earth from single minerals, other than the fact that zircons crystallized? Research on these zircons has focused on:

(1) Hf isotopic evidence for early continental crust [*Amelin et al.*, 1999; *Harrison et al.*, 2005, 2008; *Blichert-Toft and Albarède*, 2008; *Bell et al.*, 2011, 2014];

(2) zircon geochemistry and oxygen isotopic signature as a recorder of magma geochemistry and redox state [*Peck et al.*, 2001; *Wilde et al.*, 2001; *Cavosie et al.*, 2004; *Watson and Harrison*, 2005; *Trail et al.*, 2007; *Harrison et al.*, 2008];

(3) inclusion suites in zircons as a means of obtaining other Hadean material and fingerprinting the rocks from which the zircons were eroded [*Maas et al.*, 1992; *Cavosie et al.*, 2004; *Hopkins et al.*, 2008; *Bell et al.*, 2015].

These different data sets have been used to argue that the early Earth was capable of producing zircon in water-rich high-silica magmatic systems in continental crust—not unlike today. While an in-depth analysis of these data is beyond the scope of this chapter, they are reviewed here because these questions can only be asked within the framework of U–Pb geochronology.

Because zircon forms a solid solution between $ZrSiO_4$ and $HfSiO_4$, hafnium contents typically range between 0.5 and 2.0 wt%, which is sufficient to measure Hf isotopes *in situ* by SIMS or LA-ICPMS [*Harrison et al.*, 2005, 2008; *Kemp et al.*, 2010; *Bell et al.*, 2011, 2014], or by solution ICPMS on the same material analyzed for age by either ICP-MS, SIMS, or TIMS [*Amelin et al.*, 1999; *Blichert-Toft and Albarède*, 2008]. Hafnium isotopes in the Jack Hills zircons, when age-corrected, show scatter at 4.0 Ga both above and below chondrites (also called CHUR, chondritic uniform reservoir). Enriched values (lower εHf, generally higher SiO_2) are consistent with the existence of a yet older enriched reservoir that partially melted and that melt saturated igneous zircon > 4.0 Ga. The most obvious such reservoir is continental crust, which needed to be depleted in Lu relative to Hf in CHUR hundreds of Ma prior to the formation of the Jack Hills zircons. An obvious question is how much continental crust was required to form these zircons, which could in part be answered by discovering the complementary depleted-mantle reservoir (following classic isotopic arguments for evolution of the crust–mantle system—Chapter 6). Although depleted Hf isotopes in Jack Hills zircons have been measured [*Harrison et al.*, 2005; *Blichert-Toft and Albarède*, 2008), these grains are rare compared to enriched compositions and nonetheless the volume of the depleted reservoir from which they were extracted is difficult to assess.

Arguments against existence of continental crust in the early Hadean include that Hadean Jack Hills zircons could have been derived from a predominantly mafic source [*Coogan and Hinton*, 2006; *Shirey et al.*, 2008], given that zircon can be found both in oceanic crust and ocean island basalts [*Grimes et al.*, 2007; *Lissenberg et al.*, 2009; *Carley et al.*, 2014]. The geochemistry and oxygen isotopes of Hadean zircons have thus been measured to constrain the composition and temperature of the magma from which they crystallized. Rare earth element (REE) patterns, for example, have been used to fingerprint magma composition by comparing REE in zircon of known provenance to those of the Jack Hills zircons [*Peck et al.*, 2001]. Although zircon REEs alone have been argued to be nonunique [*Hoskin and Ireland*, 2000], a more complete data set including U, Th, Y, and Hf can help distinguish zircon provenance (i.e. arc vs. oceanic crust, alkali vs. metaluminous) [*Belousova et al.*, 2002; *Grimes et al.*, 2007]. Using these tools, it has been argued that the Jack Hills and Mount Narryer zircon chemistry is more consistent with derivation from a continental setting rather than an oceanic setting [*Crowley et al.*, 2005; *Peck et al.*, 2001], which opens the door to, but does not require in itself, a water-rich subduction-related origin [*Trail et al.*, 2007, 2011; *Carley et al.*, 2014].

Further evidence for the origins of the Jack Hills zircons comes from oxygen isotopes and estimates for crystallization temperatures from the oxygen isotope data and Ti-in-zircon thermometry. Oxygen isotopic compositions of Jack Hills zircons are variable, but deviate strongly from mantle values and also from lunar zircons thought to have formed from differentiated basaltic magma [*Mojzsis et al.*, 2001; *Peck et al.*, 2001; *Trail et al.*, 2007; *Wilde et al.*, 2001]; see summary in *Carley et al.* [2014] (Fig. 8.17). Particularly high $\delta^{18}O$ observed in many Jack Hills zircons could be indicative of derivation from a source that has interacted with water at the surface of the Earth, which has been interpreted as evidence that sediment was weathered in a hydrosphere and taken to depth by tectonic processes (see summary in *Harrison* [2009]). Consistent with these data are proxy records for zircon crystallization temperature, derived using the Ti-in-zircon thermometer, which uses the concentration of Ti in zircon, combined with knowledge or assumptions about the Si and Ti activity in the coexisting liquid, to determine the temperature of crystallization [*Watson and Harrison*, 2005; *Watson et al.*, 2006]. A histogram of Ti-in-zircon temperatures from > 4 Ga Jack Hills zircons shows peaks at ~650–700 °C, consistent with a water-saturated minimum melt, and inconsistent with zircons from oceanic crust or the moon, which have a range of, but generally higher, temperatures [*Watson and Harrison*, 2005; *Trail et al.*, 2007; *Fu et al.*, 2008; *Carley et al.*, 2014].

Mineral inclusions have been identified in thousands of Jack Hills zircons, which can be used as an indicator of both the mineralogy of material that predated zircon crystallization and also the equilibrium mineralogy. Inclusions suites in the Jack Hills zircons typically include quartz, feldspar, and muscovite, among others [*Maas et al.*, 1992; *Cavosie et al.*, 2004; *Hopkins et al.*, 2008, 2010; *Bell et al.*, 2015]. Muscovite in particular is important because it implies that the magma from which these zircons crystallized was muscovite-saturated, implicating a two-mica peraluminous granitic magma that could have formed from melting a sedimentary protolith [*Hopkins et al.*, 2010]. In modern tectonic environments, such melts are formed exclusively through burial by thrust faulting in convergent margins. While it is not well understood what the effects of higher mantle temperature in the Archean would have had on plate tectonic processes, these data are consistent with the existence of continental crust, surface water, and plate tectonics into the Hadean period. While still a work in progress, these revelations have had a profound effect on our view of the early Earth, which has been facilitated by U–Pb geochronology and the ability to combine this technique with a myriad other microanalytical geochemical techniques.

8.6.3 *P–T–t* paths of metamorphic belts

The application of SIMS U–Pb geochronology to metamorphic minerals revolutionized our ability to resolve metamorphic events in crustal rocks. Though microscope imaging of zircons in the 1960s and 1970s showed these minerals to contain complex zonation, not until the 1980s were scientists able to resolve age differences between, for example, different metamorphic zones within single minerals [*Bowring et al.*, 1989b; *Friend and Kinney*, 1995; *Vavra et al.*, 1996]. These different periods of mineral growth could be obscured by multigrain aliquots required by ID-TIMS at the time, resulting in discordant arrays of U–Pb data that represented mixing between multiple growth periods, which could be superimposed upon Pb-loss arrays. As single grain to subgrain analysis became common in ID-TIMS, coupled with chemical abrasion this technique proves useful when high-precision dates are required to resolve different periods of growth or durations of metamorphic processes. But both SIMS and now LA-ICPMS U–Pb geochronology remain the most popular tools for dating metamorphic events because of their obvious advantages in spatial resolution and speed of analysis.

Though minerals such as zircon, monazite, allanite, xenotime, and titanite have all been observed to contain zonation associated with metamorphic growth [*Corfu et al.*, 2003], accurately placing constraints on *P–T–t* paths in metamorphic terranes requires associating those growth domains with pressures and temperatures of formation [*Harley et al.*, 2007]. This is difficult because high-U minerals amenable to geochronology are not generally participants in well-calibrated phase equilibria (though the contrasting stability fields of titanite and rutile have been used to qualitatively address pressures of growth [e.g., *Kylander-Clark et al.*, 2008]). While tempting to associate mineral growth with peak metamorphic conditions, datable minerals can (re)crystallize at many points on a *P–T* path [*Fraser et al.*, 1997; *Roberts*

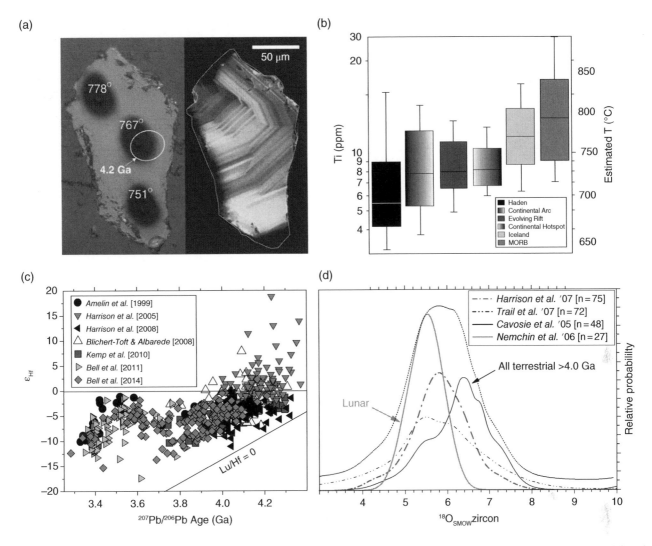

Fig. 8.17. Summary of various data sets from Jack Hills zircons. (a) SIMS measurements of Ti-in-zircon temperatures and date, and backscattered electron (left) and cathodoluminescence images (right). (Source: *Watson and Harrison* [2005]. Reproduced with permission of The American Association for the Advancement of Science.) (b) Compilation of Ti-in-zircon temperatures from various geologic environments. (Source: *Carley et al.* [2014]. Reproduced with permission of Elsevier.) (c) Compilation of εHf measurements from Jack Hills zircons. (Source: *Bell et al.* (2014). Reproduced with permission of Elsevier.) (d) Oxygen isotopic compositions of lunar zircons and Jack Hills zircons.

and Finger, 1997; *Liati and Gebauer*, 1999; *Rubatto and Hermann*, 2007b]. Attaching pressure and temperature significance to dates generated by U–Pb geochronology has been done in three main ways:

(1) dating minerals hosted as inclusions by minerals whose pressure and temperature of growth can be constrained through thermobarometry;
(2) using the geochemistry of dated minerals to infer growth during the growth or breakdown of minerals observed petrographically;
(3) applying mineral thermometers or barometers for dated minerals, which are typically based on temperature- or pressure-sensitive trace element partitioning.

An additional benefit of *in situ* techniques is the ability to retain petrographic context by analyzing grains in thin-

section. Therefore, it is possible to place maximum age constraints on the growth of minerals used in thermobarometric calculations by dating the mineral inclusions they host. An example is dating monazite or zircon inclusions within garnet as a means of constraining garnet growth [*Foster et al.*, 2004] (Fig. 8.18).

Similarly, identifying *P–T* sensitive inclusions hosted in zircon can constrain minimum ages of when those inclusions crystallized [*Hermann et al.*, 2001; *Katayama et al.*, 2001]. A limitation to this approach is that many inclusions are too small to date even by SIMS or LA-ICPMS. In such cases, monazite EMP dating (section 8.5.4) has been successful at isolating domains whose growth are interpreted to be approximately synchronous with peak metamorphism [*Baldwin et al.*, 2006]. Integrating a mineral's petrographic context with high-precision ID-TIMS analysis,

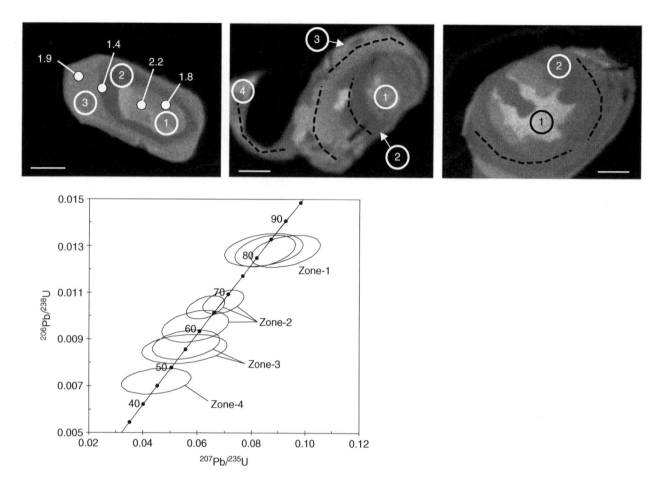

Fig. 8.18. An example of *in situ* monazite U–Pb geochronology. Cathodoluminescence images of monazite are shown in upper three panels. Intensity of grayscale corresponds to Y content in the monazite, and dots with numbers pointing to them (1.9, 2.2, etc.) are concentrations of Y determined by electron microprobe. Circles with numbers in them correspond to LA-ICPMS U–Pb geochronology spots, where numbers 1–4 correspond to zones defined by Y content. The growth zones and Y content are linked chronologically to garnet and xenotime stability during protracted metamorphism. Lower concordia diagram shows the U–Pb data from those spots. All data from sillimanite bearing metapelite from the Himalaya (sample K98-6). (Source: *Foster et al.* [2004]. Reproduced with permission of Elsevier.)

though logistically more tedious, has been carried out by removing minerals from thin-section to be dissolved and analyzed [*Corrie and Kohn*, 2007].

Because of the versatility of *in situ* instrumentation and the possibility of nondestructive analysis, U–Pb geochronology is often coupled with trace element analyses of the same grains. The goal is usually to connect changes in trace-element concentrations of the grain targeted for geochronology with growth or breakdown of minerals that can be used in pressure and/or temperature estimates. One example is using yttrium concentration in monazite as an indicator of whether garnet is in equilibrium: monazites in equilibrium with garnet have been argued to grow with low Y contents whereas in the absence of garnet, monazite can be a more important repository for Y. By making chemical maps of monazite and dating the different domains *in situ*, time constraints can be placed on metamorphic reactions involving garnet [*Gibson et al.*, 2004; *Kohn and Malloy*, 2004; *Mahan et al.*, 2006]. Similarly, zircon in equilibrium with garnet has been shown to have depleted heavy REE signatures and therefore

can also be used as an indicator of garnet growth and breakdown [*Rubatto*, 2002; *Harley and Kelly*, 2007; *DesOrmeau et al.*, 2015].

Coupling geochronology and trace-element geochemistry typically involves measuring the trace elements *in situ* by SIMS, LA-ICPMS, or EMP. In the case of EMP and SIMS elemental analysis, microbeam spots analyzed for isotopes can often be placed directly on top of the domain analyzed for trace elements. The larger volume needed for LA-ICPMS is often limiting enough that geochronology must be carried out on spots next to those analyzed for trace elements, forcing the assumption that the material analyzed represents the same period of growth (which can be defended, e.g., by using grain imagery). However, the stream of ablated material can also be split and sent into two separate ICPMS instruments—one to analyze trace elements and one to analyze U–Pb isotopes—thereby providing the chemical analysis and date from the same volume of ablated material [*Holder et al.*, 2015; *Kylander-Clark et al.*, 2013; *Yuan et al.*, 2008] (Fig. 8.19).

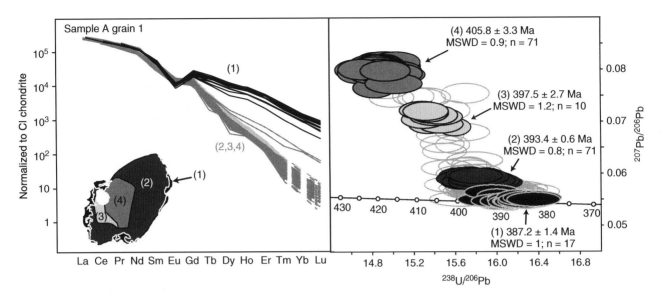

Fig. 8.19. Split stream LA-ICPMS on monazite. Trace elements and U–Pb isotopic analysis are carried out simultaneously by sending the ablated material into two separate mass spectrometers, one tuned to measure trace elements, and one tuned to measure U and Pb isotopes. The result is geochemical data coupled directly with age and petrographic/textural information. Left panel shows rare earth element plots, shaded and numbered according to zone within the monazite grain. Right panel shows corresponding U–Pb data plotted on a Tera–Wasserburg concordia diagram. Analyses are again shaded and numbered according to both the geochemical data and the monazite zone. Groups of analyses are chosen based on geochemical and age data, from which weighted means are calculated on Pb_c-corrected analyses by estimating a discordia y-intercept. Data come from a single monazite grain taken from the Western Gneiss Region in Norway. (Source: *Holder et al.* [2015]. Reproduced with permission of Elsevier.)

When high-precision time constraints provided by ID-TIMS are necessary, the solution from anion exchange chemistry not analyzed for U–Pb isotopes can be analyzed by ICP-MS for trace elements. This approach provides geochemical information for the exact same volume of mineral analyzed for age control, allowing one to correlate different age domains with the corresponding geochemical domains (called TIMS-TEA, for trace element analysis) [*DesOrmeau et al.*, 2015; *Samperton et al.*, 2015; *Schoene et al.*, 2010b]. Whereas both age and geochemistry are therefore an average signal over the volume of mineral being analyzed, this approach avoids complications arising through attempting to correlate *in situ* spot analyses with a volume of mineral analyzed for age; in other words this method compares apples to apples.

An alternative way to connect geochronology to temperature is through mineral thermometry applied directly to high-U minerals. This technique typically involves exploiting experimentally determined temperature sensitive mineral–element partition coefficients. In other words, for elements whose partitioning into high-U minerals is temperature sensitive and can be measured quantitatively, and the activity of those elements is either known or can be estimated in the system from which they grew, then the temperature of growth can be calculated. Though the equations used have both temperature and pressure terms, partition coefficients are typically more sensitive to temperature rather than pressure within the crust. Examples of widely used thermometers in high-U minerals include Ti-in-zircon, Zr-in-rutile, and Zr-in-titanite (where the nomenclature refers to the concentration of the mentioned element in the mentioned mineral) [*Watson*

and Harrison, 2005; *Baldwin et al.*, 2007; *Ferry and Watson*, 2007; *Tomkins et al.*, 2007; *Blackburn et al.*, 2012a]. Once temperature is determined, it may be linked with traditional thermobarometry to place time (*t*) points on a *P–T–t* path. Further temperature–time estimates can be determined with U–Pb thermochronology (or of course other thermochronometric systems), which is discussed in section 8.6.7.

8.6.4 Rates of crustal magmatism from U–Pb geochronology

Magmatic systems play an important role in generating oceanic and continental crust and for the thermal evolution of the crust and Earth. U–Pb geochronology from magmatic systems, particularly of the mineral zircon, has in turn been instrumental in calibrating the rates of crustal growth and the longevity and tempo of crustal magmatism [*Paterson and Ducea*, 2015; *Lundstrom and Glazner*, 2016]. Tectonic, metamorphic, and magmatic events related to crustal growth and modification can be dated with igneous zircon in conjunction with geologic mapping and field-based or laboratory-based petrology and structural geology [*Mahan et al.*, 2003; *Oberli et al.*, 2004; *Schoene et al.*, 2012]. Zircon is ideal for this kind of study because slow volume diffusion of both U and Pb means that it dates the timing of its growth within magmatic systems at high temperatures [*Cherniak and Watson*, 2001; *Lee*, 1997]. Furthermore, high uranium contents and negligible initial lead allow robust dates to be generated on single analyses, alleviating the need for isochrons and the assumptions that go with them (see previous sections).

The propensity of zircon to retain growth ages despite storage at high temperatures can also result in a range in zircon ages from a single magmatic pulse and/or hand-sample of igneous rock [*Bachmann et al.*, 2007; *Miller et al.*, 2007; *Simon et al.*, 2008; *Charlier and Wilson*, 2010; *Claiborne et al.*, 2010; *Schoene et al.*, 2012]. This has become increasingly clear as the precision of ID-TIMS geochronology has improved and the required sample size has decreased [*Rivera et al.*, 2014; *Samperton et al.*, 2015]. U-series geochronology of zircon from young volcanic systems has also played a key role in understanding zircon growth in upper crustal magmatic systems [*Charlier et al.*, 2005; *Schmitt et al.*, 2010b; *Coombs and Vazquez*, 2014; *Cooper and Kent*, 2014] (Chapter 12). The past decade has seen an increased focus on understanding zircon growth in magmatic systems because of the potential to understand the rates of magma transfer, pluton growth, and volcanic hazards.

8.6.4.1 Magma transfer and emplacement as plutons

The elliptical shape of many plutons drove early models of magma transfer in the crust to focus on vertical motion of large ellipsoid shaped magmatic bodies through the crust in a process called diapirism [*Marsh*, 1982; *Miller and Paterson*, 1999]. Noting the mechanical difficulties of diapiric ascent within a cold crust, alternative models for pluton and batholith construction involve the transfer of small batches of magma within dikes into the upper crust where they cool quickly and accumulate to form batholiths [*Petford et al.*, 1993, 2000; *Bartley et al.*, 2008]. Testing these models involves knowing where liquid existed within a system as a function of time—which is an ideal task for zircon geochronology [*Samperton et al.*, 2015].

A centerpiece for this debate has been in the Tuolumne intrusive suite in the Sierra Nevada batholith, CA (Fig. 8.20). Here, U–Pb geochronology of zircon was used to show that apparently homogeneous plutonic bodies (i.e., plutons with no obvious internal contacts to suggest multiphase intrusion) gave zircon dates that span over millions of years [*Coleman et al.*, 2004; *Glazner et al.*, 2004]. Because this period of time is longer than can be explained by simple cooling of a single intrusion of that size [*Glazner et al.*, 2004; *Annen*, 2011], these data were used to suggest that each pluton that makes up the batholith grew by incrementally emplacing batches of magma far smaller than the volume of each pluton. These results have been used to question whether magma chambers, as traditionally viewed, are long-lived features in the upper crust or whether upper crustal melt residence is ephemeral [*Coleman et al.*, 2004; *Glazner et al.*, 2004]. Further application of zircon geochronology to the Tuolumne suite [*Memeti et al.*, 2010] and other plutonic systems has seen similar results and generally confirmed that construction of plutons piecewise over hundreds of thousands to millions of years is the norm [*Matzel et al.*, 2006; *Schaltegger et al.*, 2009; *Schoene et al.*, 2012; *Mills and Coleman*, 2013; *Barboni and Schoene*, 2014; *Barboni et al.*, 2015; *Samperton et al.*, 2015], though counterexamples may also exist [*Eddy et al.*, 2016]. More detailed sampling and examination of the zircon record have also

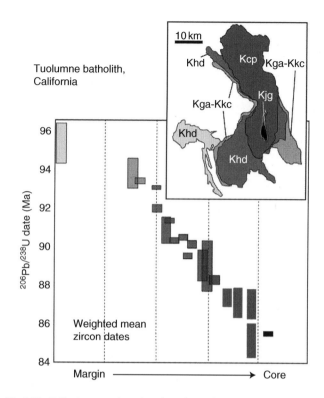

Fig. 8.20. U–Pb zircon geochronology from the Tuolumne batholith. Upper box shows simplified geologic map of the Tuolumne batholith, with individual mapped plutonic units shown in different shades of gray. Unit shades correspond to U–Pb zircon data shown in rank-order plot below [*Coleman and Glazner*, 1997; *Glazner et al.*, 2004; *Burgess and Miller*, 2008; *Memeti et al.*, 2010]. These data show that the batholith was assembled over ca. 10 Ma, which is too long for the existence of an upper crustal magma chamber, thereby arguing for pulsed emplacement of the batholith.

begun to shed light on the complexities and opportunities of interpreting zircon dates in terms of magma transfer into the upper crust.

Determining rates of magma transfer, storage, and solidification (or eruption) require accurate interpretation of the U–Pb zircon record given that zircon can grow and be transported within the crust by mobilized magma [*Miller et al.*, 2007]. In a closed system, zircon may saturate in a magma during cooling at a temperature that depends on the major element chemistry and zirconium content of the melt [*Watson and Harrison*, 1983; *Boehnke et al.*, 2013;] and then crystallize as a liquidus phase until the solidus is reached. The result is that if a magma is saturated in zircon, it may entrain xenocrystic and liquidus zircon and transport and recycle it within the crust. When combined with prolonged crystallization of zircon within a slowly cooling magma, there is opportunity to record a spectrum of zircon dates within a single hand-sample (see Fig. 8.22 below). Zircon dates that span hundreds of thousands or millions of years within single rocks are now typical. Interpreting these data in terms of magmatic processes remains a challenge [*Charlier et al.*, 2005; *Miller et al.*, 2007; *Simon et al.*, 2008; *Charlier and Wilson*, 2010; *Claiborne et al.*, 2010; *Schmitt et al.*, 2010b; *Schoene et al.*, 2012;

Rivera et al., 2014; *Samperton et al.*, 2015]. A host of analytical and numerical techniques are being applied to zircons in such systems in order to understand how to link zircon growth to processes of interest such as magma transfer, emplacement, and cooling. For example, measuring trace-element geochemistry, Hf isotope composition, and/or magmatic growth temperatures using the Ti-in-zircon thermometer (see section 8.6.3) and interpreting them within the context of petrographic relationships can both reveal the origins of igneous zircon and build time series on processes such as fractional crystallization and assimilation during magma ascent and emplacement [*Schaltegger et al.*, 2009; *Schoene et al.*, 2010b; *Schoene et al.*, 2012; *Rivera et al.*, 2013, 2014; *Barboni and Schoene*, 2014; *Samperton et al.*, 2015; *Tapster et al.*, 2016]. Both ID-TIMS and *in situ* dating techniques have been used in conjunction with complementary analytical techniques focused on linking geochemistry and isotopic signatures of dated minerals to petrologic processes. When combined with field observation and numerical modeling, geochronology has proven to be an essential and expanding tool to understanding crustal magmatism.

8.6.4.2 Magma transfer, volcanic systems, and super eruptions

Models for incremental emplacement of magma into the upper crust are also viable for producing "normal" volcanic systems that spend most their dormant life at low melt fractions, and are then periodically rejuvenated by small batches of melt that subsequently erupt [e.g, *Huber et al.*, 2012; *Cooper and Kent*, 2014]. Despite the increased focus on incremental addition and rapid cooling of magma to the upper crust as a means of building batholiths, the existence of large volumes (>500 km^3) of melt in the upper crust is required because high-SiO$_2$ volcanic eruptions of such magnitude are observed in the geologic record [*Miller and Wark*, 2008]. These so-called "super eruptions" form one end-member of upper crustal magma reservoirs that pose significant volcanic hazards [*Reid*, 2008]. Geochronology of zircon and other minerals such as allanite [*Vazquez and Reid*, 2004], have played an important role in understanding how these magma bodies form and in contrasting them with more typical volcanic eruptions that characterize most arc volcanoes.

Some early geochronologic estimates of the residence time of magmas feeding large eruptions argued that reservoirs with significant melt existed for period of over a million years, for example in Long Valley, CA [*Christensen and DePaolo*, 1993]. Application of U–Pb geochronology to the ~600 km^3 Bishop Tuff, erupted from the Long Valley caldera ca. 770 ka, shows that estimates of millions of years of pre-eruptive magma residence are unlikely within that system using both lower precision, higher spatial resolution SIMS analysis and higher precision, lower spatial resolution ID-TIMS analysis [*Reid and Coath*, 2000; *Simon and Reid*, 2005; *Crowley et al.*, 2007; *Ickert et al.*, 2015]. These data show that zircon growth within the upper crustal magmatic system was likely less than tens of thousands of years, and melt residence was likely not much longer. In another example, the

zircons from the Fish Canyon Tuff, southwest Colorado, grew over ~400 ka prior to eruption of ~1000 km^3 of crystal-rich dacitic tuff [*Wotzlaw et al.*, 2013]. The resulting model, which is corroborated by petrologic data, argues for zircon growth within a low-melt fraction reservoir for hundreds of thousands of years before injection of mafic material into the crystal mush reinvigorated the system, leading to eruption [*Bachmann et al.*, 2002; *Bachmann and Bergantz*, 2003]. In contrast, zircons extracted from the ~1800 km^3 Kilgore Tuff, erupted in eastern Idaho ca. 4 Ma, gave indistinguishable ID-TIMS U–Pb zircon dates at the thousand-year scale, suggesting a rapid transition from zircon saturation to eruption [*Wotzlaw et al.*, 2014]. A complementary study on multiple Yellowstone Caldera systems argues that injection of felsic magma into the upper crust resulted in melting of isotopically heterogeneous crust, partial homogenization and accumulation of a melt-rich reservoir, and eruption, all occurring within the uncertainties of individual zircon dates of tens of thousands of years [*Wotzlaw et al.*, 2015].

These chronological constraints on zircon growth in upper crustal magmatic systems, when combined with numerical modeling and petrologic data, illustrate that large bodies of magma beneath volcanic edifices are ephemeral features, and long-lived magmatic systems spend most of their lifetimes in a low-melt-fraction state interrupted by periodic injection and eruption of magmas. The potential for eruption of an upper crustal magma body may thus depend on the rate of magma input and the resulting heat and volatile budget. Perhaps a critical threshold of magma input is required to grow magma chambers large enough to feed super eruptions [*Annen*, 2011; *Gelman et al.*, 2013; *Coleman et al.*, 2016]. Attempts to reconstruct magma input rates using U–Pb geochronology broadly show that plutons are generally emplaced more slowly than magmas that result in super eruptions [*Barboni et al.*, 2015; *Coleman et al.*, 2016] (Fig. 8.21). However, discovery of large, rapidly emplaced plutons [*Eddy et al.*, 2016], may provide further insight into the physics and chemistry of magma transport and eruption.

Uncertainties remain in applying U–Pb geochronology to young volcanic systems. As mentioned above, data combining analytical approaches to obtain better interpretations of U–Pb data will lead to more insight into magmatic processes (Fig. 8.22). Also, the correction for ^{230}Th-disequilibrium on zircon < 1 Ma is ~10% of the age, and while the uncertainty in that correction is < 20 ka, this is a limitation in the accuracy and precision of geochronology of young zircons analyzed by ID-TIMS (see section 8.4.3). Overcoming these uncertainties requires better knowledge of the Th/U and isotopic equilibrium of melts from which zircon crystallized, and/or of zircon/melt partition coefficients as a function of pressure, temperature, and magma composition [*Crowley et al.*, 2007; *Rubatto and Hermann*, 2007a; *Rioux et al.*, 2010; *Ickert et al.*, 2015; *Boehnke et al.*, 2016]. This issue becomes more complicated in that all dating techniques average the ages of different domains in zircon that may have grown over tens to hundreds of thousands of years. There is a trade-off between very precise concentration-weighted

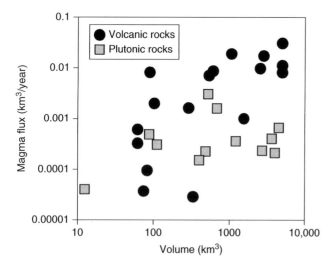

Fig. 8.21. Flux of magma for volcanic and plutonic systems, as derived from U–Pb dates of zircons. Values are calculated by estimating total volume of either a volcanic eruption or a plutonic body, making assumptions about eroded and covered volumes, and dividing by the duration of zircon crystallization, making assumptions of zircon saturation time in a liquid. (Source: *Coleman et al.* [2016]. Reproduced with permission of Mineralogical Society of America.)

and volume-weighted average ages determined by ID-TIMS and imprecise ages of much smaller domains dated by SIMS or LA-ICPMS. Application of nonradiometric techniques to investigate the timescales of magmatic processes have shown that many igneous processes can occur on timescales much shorter than is resolvable by any absolute dating technique [*Costa et al.*, 2008; *Gualda et al.*, 2012; *Chamberlain et al.*, 2014; *Till et al.*, 2015], and so the application of a wide array of petrologic and geochronologic tools becomes even more important for understanding the timescales of mechanisms that lead to volcanic eruptions and the associated hazards [*Cooper and Kent*, 2014; *Druitt et al.*, 2012).

8.6.5 U–Pb geochronology and the stratigraphic record

The geologic timescale is an essential part of sequencing events in Earth history [*Holmes*, 1947, 1959; *Harland*, 1990; *Gradstein et al.*, 2004, 2012). While every introductory geology student learns the eons, eras, and periods that break up the past 4.5 billions years, many do not appreciate the methods and uncertainties that underpin the calibration of the absolute geologic timescale. Events and processes recorded in sedimentary rocks

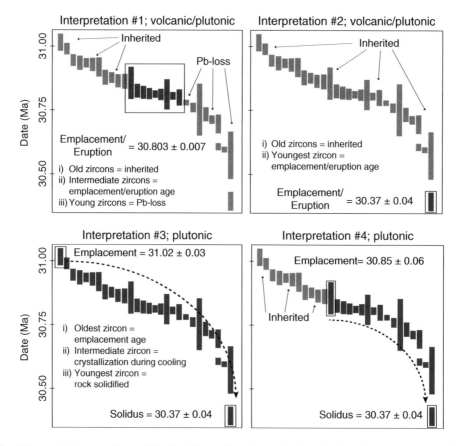

Fig. 8.22. Summary of multiple ways to interpret a zircon crystallization data set where dates scatter beyond analytical uncertainty. Data are shown as rank-order plots, where the *y*-axis is date in Ma, and each vertical bar corresponds to the 2-sigma uncertainty for a $^{206}Pb/^{238}U$ date on a single zircon fragment. Dark gray analyses are those used in the age interpretation, whereas light gray are excluded for reason indicated. Analyses used in age calculation, either as a single analysis or weighted mean, are outlined by a gray box next to quoted date. First two panels from left are applicable to volcanic and plutonic rocks, whereas the next two two panels apply only to plutonic rocks, because zircon cannot crystallize after eruption.

are the best records of Earth's ocean, atmosphere, and biosphere in the past, and contain important information about tectonic events as recorded in depositional basins. An essential part of this task is determining reliable absolute ages for boundaries so that this framework can be used for understanding events that occur at and between those boundaries.

U–Pb geochronology has played a large role in calibrating the absolute geologic timescale. The majority of stage boundaries are now defined by U–Pb dates and, prior to the Cretaceous, boundary ages are almost exclusively defined by U–Pb dates [*Gradstein et al.*, 2012]. Given that the stage and higher order boundaries are defined by events recorded in the stratigraphic record, such as biologic extinctions and/or proliferations, and that the discussion above about U–Pb geochronology focuses entirely on dating metamorphic and igneous minerals, a reasonable question is how U–Pb geochronology is applied to the stratigraphic record and what are the assumptions and uncertainties that underpin these methods?

The answer is that many volcanic ash beds, whose deposition represent a moment in geologic time, contain high-U minerals, such as zircon, sourced from the magmatic system that produced the ash. Dating these high-U minerals thereby gives an estimate of the timing of ash-bed deposition, provided that the zircons crystallized immediately prior to eruption [*Tucker et al.*, 1990; *Tucker*, 1992 *Mundil et al.*, 1996 *Bowring et al.*, 1998]. As discussed in section 8.6.4, however, zircon can crystallize and retain age information at high temperatures within magmatic systems for thousands to hundreds of thousands, or even millions of years prior to eruption. The precision of ID-TIMS geochronology has increased over the past decades (Fig. 8.11), such that it has become increasingly clear that volcanic ash beds often contain a heterogeneous population of zircons, including many which crystallized before the eruption of interest [*Schoene et al.*, 2010a, 2015; *Sageman et al.*, 2014; *Deering et al.*, 2016]. U-series geochronology of zircons from very young eruptions (<100 ka) more often than not show that ash beds contain zircons with growth ages spanning over tens of thousands to hundreds of thousands of years [*Wilson and Charlier*, 2009; *Charlier and Wilson*, 2010; *Schmitt et al.*, 2010b, 2011]. Whether or not these zircons crystallized within a magma chamber prior to eruption or were entrained within the eruptive column is important for understanding plutonic and volcanic processes, but is somewhat irrelevant for the purposes of dating ash-bed deposition: the best U–Pb estimate of deposition of a primary, nonreworked ash bed is that of the youngest closed-system zircons (or other high-U minerals, though zircon is by far most utilized [*Schoene et al.*, 2010a]).

Verifying that the youngest U–Pb date from an ash bed actually overlaps within uncertainty of the eruption date is difficult given the propensity of zircon to retain magmatic dates. Dating eruptions by the $^{40}Ar/^{39}Ar$ technique provides an alternative means of dating ash beds, which has the advantage that Ar diffuses out of magmatic minerals at high-T and therefore only becomes a closed system at the time of eruption [*Renne*

et al., 1995; *Smith et al.*, 2010; *Sageman et al.*, 2014; *Sprain et al.*, 2014]. The downside of $^{40}Ar/^{39}Ar$, similar to U–Pb, is that crystals inherited immediately before eruption or during eruption may contain excess (not totally outgassed) Ar. Dating of low-K minerals that will produce large age uncertainties may mask this problem [*Renne*, 1995; *Kelley*, 2002; *Bachmann et al.*, 2010; *Hora et al.*, 2010; *Rivera et al.*, 2014; *Jicha et al.*, 2016]. By analogy, dating U–Pb thermochronometers such as apatite or titanite (see section 8.6.7) provides an alternative means to zircon of dating eruption because they should record the time at which they cooled below magmatic temperatures [*Schmitz and Bowring*, 2001]. However, generally these minerals are lower in U and higher in initial Pb than zircon and thus the resulting age precision and accuracy suffer [*Schoene and Bowring*, 2006]. Similarly, applying *in situ* dating techniques gives both the ability to analyze a large number of zircons in search of the youngest and also for targeting young rims in grains, but these methods are often not precise enough to solve relevant geologic problems without using large-*n* weighted means, which can mask subtle inheritance or Pb loss [*Ireland and Williams*, 2003; *Gehrels et al.*, 2008; *Horstwood*, 2008; *Košler et al.*, 2013].

Thus, U–Pb dates defining stratigraphic intervals through ash-bed geochronology rely on ID-TIMS U–Pb dates, and require that one or more zircons accurately record the eruption of that ash bed [*Bowring and Schmitz*, 2003]. The accuracy of U–Pb or U-series zircon dates in young ash beds can be evaluated indirectly by comparing them to other dating techniques with similar precision in that age range [*Bachmann et al.*, 2007; *Schmitt et al.*, 2010b; *Rivera et al.*, 2014; *Sageman et al.*, 2014]. Doing so reveals that nearly all cases where the eruption age is determined independently, the youngest zircon ages from a complex population agree with the independent estimates to within thousands of years. Because independent estimates of eruption age are usually not available or too imprecise to be helpful in older samples (and hence the reason to use high-precision U–Pb geochronology), these observations imply that one should focus on the youngest population of zircons or youngest single zircon for a best estimate of the eruption age of some given ash bed. This approach dominates age interpretations of zircon data from ash beds in recent studies, but still suffers from somewhat arbitrary inclusion and exclusion of analyses. Recent attempts to enhance the ability to interpret zircon U–Pb dates from ash beds include using zircon geochemistry to guide selection of dates from which weighted means can be calculated, by inferring that cogenetic zircons should have both the same age and geochemistry [*Rivera et al.*, 2014; *Schoene et al.*, 2015]. Whatever approach is used, increasing the precision of age interpretations without sacrificing accuracy is an ongoing challenge [*Bowring et al.*, 2006; *Simon et al.*, 2008; *Schoene et al.*, 2013].

No events in Earth history have garnered as much attention as mass extinctions. The causes and consequences of these events continue to puzzle geoscientists, and understanding these events relies in part on a robust timeline for both time-sensitive processes such as carbon cycling and also to pinpoint the driving mechanisms leading to environmental change and extinction. A recent example is the application of U–Pb ID-TIMS geochronology towards understanding the largest mass extinction event in Earth history at the Permian–Triassic boundary [*Knoll et al.*, 2007]. The global stratotype section and point, in Meishan, China, records the marine expression of the mass extinction event and also contains zircon-bearing ash beds [*Shen et al.*, 2011]. This section has been the subject of numerous biostratigraphic, isotopic, and geochemical studies and thus paints the most detailed picture of the extinction event and the hosting environment and climate. Most workers agree that the eruption of the Siberian Traps large igneous province was a probable cause of extinction through environmental disturbance related to volatile emissions (e.g., CO_2, SO_2, Cl, F) during eruption of basaltic lavas and intrusion of gabbroic sills that totaled $> 3 \times 10^6 \, km^3$ of magma [*Self et al.*, 2014]. Recent application of ID-TIMS U–Pb geochronology to ashbeds in Meishan and basalts and gabbros in Siberia have shown that more than two-thirds of the erupted lava was emitted in the ~300 ka leading up to the extinction interval, which itself occurred in less than tens of thousands of years [*Burgess et al.*, 2014; *Burgess and Bowring*, 2015] (Fig. 8.23). Continued igneous activity for ~600 ka following the mass extinction can now be correlated with early Triassic carbon cycling and biotic recovery. Importantly, these studies were carried out using U–Pb ID-TIMS geochronology both in the same laboratory and with the same tracer solution for isotope dilution, thereby minimizing sources of systematic uncertainties in the data sets, which had hampered earlier correlations [*Kamo et al.*, 2003; *Renne et al.*, 1998; *Shen et al.*, 2011]. Furthermore, a challenge that faces workers attempting to date mafic magmatism by U–Pb geochronology lies in the rarity of zircon or other high-U phases in basalts. In the studies highlighted here, abundant zircon was recovered from more slowly cooled gabbroic sills, and perovskite was obtained from the lava flows themselves. While perovskite can be high-U, it also contains abundant common Pb, thereby requiring that care is taken to constrain the isotopic composition of the initial Pb. Other recent approaches to dating basalts by U–Pb geochronology include finding and dating baddeleyite (see section 8.6.9), and also targeting high-SiO_2 ashbeds deposited between basalt flows, much like dating other sedimentary successions [*Schoene et al.*, 2015].

8.6.6 Detrital zircon geochronology

LA-ICPMS provides a fast and affordable way to generate a huge amount of U–Pb isotopic data (Fig. 8.24), which is ideal for characterizing complex detrital zircon populations [*Fedo et al.*, 2003; *Gehrels*, 2014]. *Gehrels* [2011] outlines three main motivations for detrital zircon studies: (i) to characterize the provenance of sediment compared to known sources; (ii) to correlate sedimentary units, assuming identical provenance; and (iii) to quantify the maximum depositional age of strata in the absence of datable volcanic material.

Provenance studies are used for paleogeographic reconstructions, identifying drainage-pattern switches in the past, constraining uplift histories, and recognizing pulses of magmatism [e.g. *Rainbird et al.*, 1992; *Bruguier et al.*, 1997; *Ireland et al.*, 1998; *Stewart et al.*, 2001; *DeGraaff-Surples et al.*, 2002; *Dickinson and Gehrels*, 2003; *LaMaskin*, 2012]. One recent example of this type of study comes from the East African Rift where detrital zircon geochronology has been used, in part to show that the eastern and western rift segments initiated simultaneously, as opposed to the consensus paradigm that the east predated the west by 15 Ma [*Roberts et al.*, 2012].

Detrital zircons have also been used to correlate sedimentary strata [*Murphy et al.*, 2004]. By assuming that a single layer should contain the same detrital zircon age distribution (commonly called age spectra), one can correlate units thousands of kilometers away by quantitatively matching spectra. This approach has led to improved tectonic models for orogenic belts such as the Himalaya, where the difficulty of correlating sedimentary sequences along strike has hindered an understanding of pre-collision basin geometries [*DiPietro and Isachsen*, 2001; *Gehrels et al.*, 2003; *Myrow et al.*, 2009; *Long et al.*, 2011].

A related approach is to use detrital zircon dates to provide maximum ages for deposition of sedimentary strata [*Robb et al.*, 1990]. In this case, accuracy requires analyzing a large number of zircons to obtain a representative population (e.g., > 100 [*Hervé et al.*, 2003; *Dickinson and Gehrels*, 2009]), from which the youngest grains can be used as a maximum age. Estimates of maximum depositional ages can then be used to correlate strata and estimate depositional rates for stratigraphic successions. This technique is not adequate for precise determinations of depositional ages, but in the absence of primary volcanic airfall, they can sometimes provide the only means of estimating depositional ages.

Applications of detrital zircon geochronology often require quantification of age spectra, for which robust statistical approaches are still being developed. There is currently no consensus on the best way to interpret detrital zircon age spectra in terms of the significance of peak heights (when plotted on probability density function diagrams), differences in the relative abundances of peaks between samples in stratigraphic succession, or what statistics can be applied to spectra [*Gehrels*, 2011; *Pullen et al.*, 2014]. Discussions of different approaches and mathematical developments are ongoing [*Vermeesch*, 2012, 2013; *Saylor and Sundell*, 2016]. Furthermore, several recent studies on modern sediments highlight the impact that biased provenance sampling and grain-size sorting during sediment transport can have on depositional age interpretations [*Moecher and Samson*, 2006; *Hietpas et al.*, 2011; *Malusà et al.*, 2013]. As workers continue to decide how to interpret detrital zircon spectra quantitatively and apply statistical models to these data, equal effort could be applied to understanding these sources of "geologic" bias. The addition of other detrital minerals such as monazite and apatite will also likely play a role in deciphering ages and provenance of sedimentary units [*White et al.*, 2001; *Suzuki and Adachi*, 1994; *Hietpas et al.*, 2010; *Chew and Donelick*, 2012].

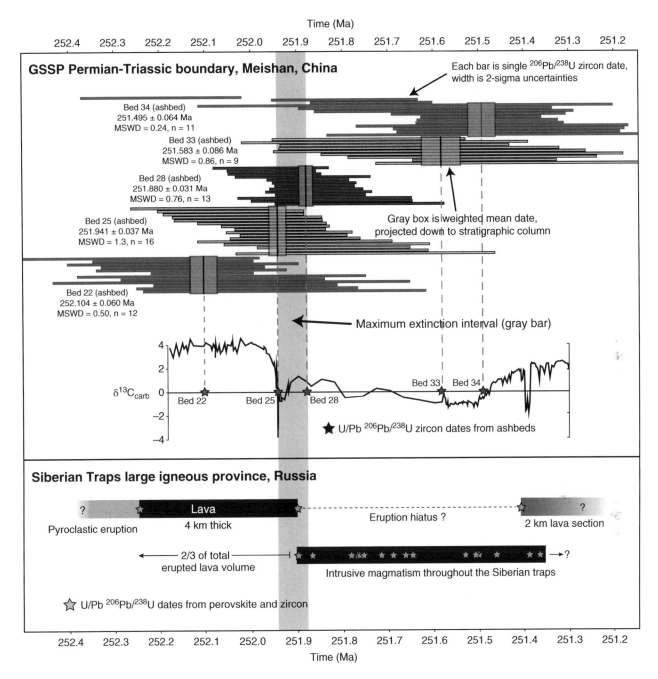

Fig. 8.23. U–Pb ID-TIMS geochronology linking the end-Permian mass extinction to the Siberian Traps large igneous province. (Source: Adapted from *Burgess et al.* [2014] and *Burgess and Bowring* [2015].)

A fourth application involves using detrital zircon records, coupled with Hf and O isotopic measurements from the same zircon grains, to estimate rates of continental growth over Earth history [*Kemp and Hawkesworth*, 2014; *Roberts and Spencer*, 2015]. Compilations of tens of thousands of zircon U–Pb and Hf analyses show peaks in zircon dates that roughly correspond to time periods of known or inferred supercontinent assembly, which are known in the case of Mesoproterozoic and younger supercontinents, but these reconstructions become more

difficult for older time periods (Fig. 8.24). These peaks in zircon dates can be interpreted as pulses in the production or growth of continental crust or as preservation bias [*Hawkesworth and Kemp*, 2006; *Condie et al.*, 2009, 2011; *Belousova et al.*, 2010; *Lancaster et al.*, 2011; *Voice et al.*, 2011; *Dhuime et al.*, 2012]. The combination of zircon dates with Hf and O isotopic compositions of the same grains, permitted through age and isotopic analysis *in situ*, has been used as a means of identifying "juvenile" zircons—i.e., those that have resulted from the extraction of basalt from the mantle and its subsequent fractionation rather than melting of older continental crust [*Kemp et al.*, 2007]. The argument is that if the initial Hf isotopic composition of a zircon matches that of estimates for the mantle at the same time, then this mass of zircon more likely records production of some new volume of continental crust. Similarly, the oxygen isotopic composition of melt having originated in the mantle (and

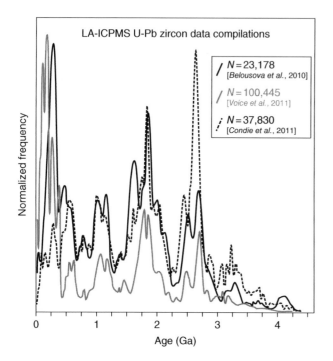

Fig. 8.24. Summary of recent zircon LA-ICPMS U–Pb date compilations illustrating the massive influx of such data. *N* is the number of analyses included in each compilation. Most of the data are for detrital zircon from young sediments, though several thousand from *Condie et al.* [2011] are orogenic granitoids of various ages. Data such as these are used to debate periodic continental growth versus preservation bias or crustal volume as a function of time, in addition to more focused studies on individual orogenic belts and sedimentary rocks.

subsequently differentiated to saturate zircon) will retain a mantle-like oxygen isotopic composition whereas melt generated within, or contaminated by, the continental crust will noticeably change the isotopic composition of zircon [*Hawkesworth et al.*, 2010].

These two methods for screening zircon dates used independently or in conjunction are useful because recycling or remelting of older continental crust does not result in net crustal growth. Numerous curves for crustal growth over the past 4 Ga have been published using detrital zircon records (Fig. 8.25), and all argue that crustal growth rates have decreased through time—in other words that the Archean eon saw rapid crustal growth and that crustal reworking (e.g., remelting of older crust) became more and more important through the Proterozoic and Phanerozoic eons [*Belousova et al.*, 2010; *Hawkesworth et al.*, 2010; *Dhuime et al.*, 2012]. Translating zircon abundance to crustal abundance, however, still requires assuming the mass of zircon available in the geologic record today tells us something about the mass of crust present at a given time in the past. Processes that recycle continental crust back into the mantle such as sediment subduction, crustal delamination, and subduction erosion are difficult to account for. Given our oldest Earth materials are detrital zircons found in younger sedimentary rocks, crustal recycling is certainly an important process that biases these crustal growth records, but we do not know to what degree.

8.6.7 U–Pb thermochronology

Volume diffusion of Pb through mineral lattices is generally sluggish compared to other thermochronometric techniques outlined in this book, such as those based on the diffusion of He and Ar. However, empirical, experimental, and theoretical studies show that Pb diffusion in titanite, apatite, rutile, and several other minerals is rapid enough that the U–Pb system can be a robust thermochronometer around 400–700 °C. This temperature range is useful for understanding the thermal evolution of the middle to lower crust for both active orogenic belts on

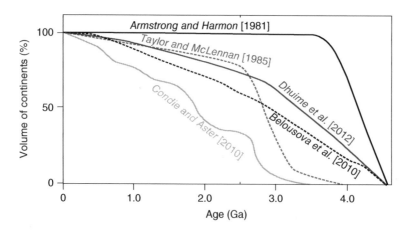

Fig. 8.25. Estimates of volume of continental crust over Earth history. *Armstrong and Harmon* [1981] and *Taylor and McLennan* [1985] rely on various petrologic and geophysical arguments, whereas the other three are based mostly on detrital zircon U–Pb data.

timescales of tens of millions of years and investigating billion-year thermal histories of cratonic lower crust. U–Pb thermochronology has been used to investigate the cooling histories of plutons in the middle and upper crust and also the exhumation timescales of high-grade metamorphic rocks.

The theory behind U–Pb thermochronology is identical to that underpinning other thermochronological techniques, and is outlined in Chapter 5. Experimental data for Pb diffusion is less abundant than for techniques based on gaseous species, however, due to the difficulty in conducting those measurements and the observation that U–Pb thermochronologists cannot typically do the diffusion experiments given existing equipment in their laboratories [*Cherniak et al.*, 1991; *Cherniak*, 1993; *Cherniak and Watson*, 2000, 2001] (Table 8.2). However, there is broad

Table 8.2 Experimentally determined diffusion kinetics for Pb in several high-U minerals used for U–Pb geochronology and thermochronology[a]

Mineral	E_a [kJ/mol]	D_0 [m²/s]	T_c [°C] (dT/dt = 1 °C/Ma; a = 10–1000 μm)
Zircon	544	7.8×10^{-3}	903–1127
Monazite	592	9.4×10^{-1}	907–1111
Titanite	331	1.1×10^{-4}	507–670
Rutile	220	2.1×10^{-11}	465–700
Apatite	230	1.3×10^{-8}	374–553

a is the effective diffusion dimension.
[a] Closure temperatures are calculated according to *Dodson* [1973], and spherical geometry is assumed. See text for data sources.

agreement between experiment-derived diffusion kinetics, ionic porosity theory, and empirical studies that compare relative dates of minerals with variable closure temperatures. These observations support nominal closure temperatures for titanite, apatite, and rutile from 400 to 700 °C and suggest that experimental kinetic data can be used quantitatively for modeling thermal histories of rocks (Fig. 8.26). Importantly, these same minerals can be involved in numerous metamorphic and/or hydrothermal reactions, so interpreting U–Pb dates as cooling dates must be done on a sample-by-sample basis. U–Pb thermochronometers are generally low U/Pb$_c$ compared to zircon and monazite, and therefore common-Pb corrections can hinder the precision obtained on single-grain analyses due to the uncertainty in that correction. Unfortunately, isochron techniques are only applicable to quickly cooled systems where all grains are closed to parent and daughter gain or loss simultaneously. Below, the existing empirical and experimental data are reviewed for the most often exploited U–Pb thermochronometers, titanite, apatite, and rutile, and then some applications of these systems are highlighted.

8.6.7.1 Titanite—CaTiSiO₅

Titanite was recognized decades ago as a candidate for U–Pb geochronology because it can have moderately high U/Pb$_c$ ratios and is common in silicic magmatic and mafic metamorphic rocks [*Tilton and Grunenfelder*, 1968; *Ishizaka and Yamaguchi*, 1969; *Hanson et al.*, 1971; *Corfu*, 1980, 1988; *Schärer*, 1980;

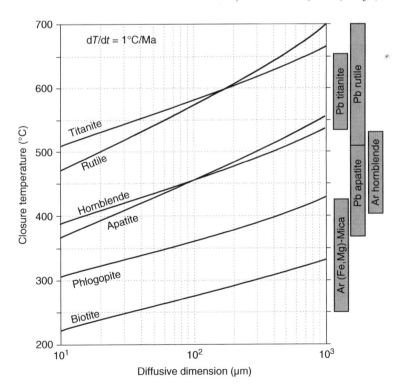

Fig. 8.26. Closure temperatures of U–Pb thermochronometers relative to some K–Ar based thermochronometers (Chapter 9). Kinetic data are located in Table 8.2, and closure temperatures are calculated using the *Dodson* [1973] formulation, assuming spherical geometry. (Source: *Schmitz and Bowring* [2003]. Reproduced with permission of Springer.)

Corfu et al., 1985]. These early studies commonly observed titanite dates that were younger than zircon from the same magmatic rocks, which *Mattinson* [1978] first hypothesized was due to diffusion of Pb at high temperatures, estimating a nominal closure temperature of 500 °C through comparison with K–Ar dates in hornblende. Subsequent multichronometer studies on slowly cooled metamorphic terranes are also consistent with nominal closure temperature similar to that of K–Ar in hornblende and slightly higher than U–Pb in apatite and rutile [*Mezger et al.*, 1991; *Cosca et al.*, 1995; *Hawkins and Bowring*, 1999; *Nemchin and Pidgeon*, 1999; *Schoene and Bowring*, 2007 *Blackburn et al.*, 2012b]. A single experimental study on diffusion of Pb in titanite [*Cherniak*, 1993] also suggests nominal closure temperatures in the range of 500–700 °C for spherical grain sizes between 0.01 and 1 mm at cooling rates of < 10 °C/Ma.

Several observations complicate the interpretation of U–Pb dates of titanite. The first is that, despite relatively good agreement between many empirical studies and experimental data for diffusion of Pb in titanite, there are several studies in which titanite records less diffusion of radiogenic Pb at high temperatures than would be predicted using the kinetics of *Cherniak* [1993]. These examples include partial to negligible resetting of titanite during ultrahigh-pressure (UHP) metamorphism at temperatures > 700–750 °C for millions of years in the western gneiss region, Norway [*Tucker et al.*, 1986; *Spencer et al.*, 2013]; nearly negligible diffusive reequilibration of Pb in titanite from partially melted metapelites in the Himalaya that reached temperatures of 700–775 °C [*Kohn and Corrie*, 2011]; titanite from the Dabie orogeny that retain preorogenic ages in magmatic cores despite having resided at > 800 °C for millions of years [*Gao et al.*, 2012]; and several examples of xenocrystic titanite into syenitic magmas that retain dates far predating the timing of magmatism [*Pidgeon et al.*, 1996; *Zhang and Schärer*, 1996]. Given these examples, further work is needed to better understand the controls on Pb diffusion in titanite and what controls whether a titanite date records cooling through the partial retention zone indicated by diffusion kinetics.

Also important for geochronology is that growth of titanite can occur near or well below its nominal closure temperature. Titanite is a common metamorphic phase in amphibolite-grade mafic and calc-silicate rocks [*Frost et al.*, 2000; *Spear*, 1993] and as a subsolidus alteration or metamorphic phase in granitoids [*Corfu*, 1988; *Corfu et al.*, 1994; *Verts et al.*, 1996; *Corfu and Stone*, 1998; *Storey et al.*, 2007; *Schoene et al.*, 2012). As such, growth of titanite at mid-temperature metamorphic conditions can be dated using U–Pb geochronology. In such cases, titanite would not be recording cooling from high temperatures, but if combined with petrographic and/or geochemical data, can constrain the prograde or retrograde *P–T–t* paths of rocks [*Corfu*, 1988; *Corfu and Stone*, 1998; *Aleinikoff et al.*, 2002; *Brewer et al.*, 2003; *Kylander-Clark et al.*, 2008; *Tanner and Evans*, 2003]. In many cases, these titanite contain complex growth zoning that is best identified by imaging or geochemical analyses, at which point *in situ* dating techniques may be preferable to TIMS

analysis due to the better spatial resolution, provided age differences are resolvable with these techniques.

8.6.7.2 Apatite—Ca(PO$_4$)$_3$(OH,F,Cl)

Use of apatite as a geochronometer initially focused on fission-track analysis (Chapter 10), presumably because of relatively low U/Pb$_c$ in apatite relative to some other common accessory minerals. However, an early attempt at dating apatite by the U–Pb method in radiogenic Archean grains was highly successful, yielding more concordant results than zircons from the same rocks [*Oosthuyzen and Burger*, 1973]. *Mattinson* [1978] suggested its use as a thermochronometer with a closure temperature similar to that of titanite, though this estimate has been modified based on empirical comparison with U–Pb dates from titanite and rutile as well as ^{40}Ar/^{39}Ar dates from various phases [*Frost et al.*, 2000; *Krogstad and Walker*, 1994; *Schoene and Bowring*, 2007; *von Blanckenburg*, 1992; *Willigers et al.*, 2001]. One experimental study [*Cherniak et al.*, 1991] indicates a closure temperature of ~450–550 °C for grains < 500 μm and cooling rates of 1–2 °C/Ma, which is in good agreement with these empirical calibrations.

Although multiple, temporally distinct, generations of apatite have been identified in single samples [*Schoene and Bowring*, 2007], there is good evidence that apatite dates often record cooling through the partial retention zone of Pb [*Chew and Spikings*, 2015]. Examples include the reproducible sequence of closure dates relative to other thermochronometers cited above, but also data sets that show clear correlations between date and grain size for single, whole apatite grains dated by ID-TIMS [*Schoene and Bowring*, 2007; *Cochrane et al.*, 2014] (Fig. 8.27). Age-gradients within single apatites have been measured by LA-ICPMS in slowly cooled rocks from the Andes [*Cochrane et al.*, 2014], and these profiles can be inverted for unique thermal histories recording mid-crustal exhumation.

Despite the successes and great potential for future application of U–Pb thermochronology of apatite, it is not uncommon for apatite dates to be reversely discordant (i.e., due to Pb gain or U loss), which is less common in other high-U accessory minerals [*Schoene et al.*, 2008; *Schoene and Bowring*, 2010]. The physical mechanisms of reverse discordance in apatite are not clear, but may reflect late-stage fluid leaching, dissolution/reprecipitation of apatite, or other mechanisms of fractionating U from Pb*. In some cases, spurious discordance could be related to the large correction for Pb$_c$, requiring either isochron techniques or leaching studies of cogenetic U/Pb phases [*Mattinson*, 1978; *Schoene and Bowring*, 2006; *Chew et al.*, 2014]. Using isochrons in slowly cooled rocks, however, violates the assumption of isochroneity unless identical grain sizes are used; estimation of Pb$_c$ isotopic composition by measuring coexisting low-U phases such as feldspar [*Frost et al.*, 2000] is time consuming. Because diffusion kinetics of Pb in feldspar are similar to those of apatite [*Cherniak*, 1995], the latter approach could be a reasonable approximation of the reservoir of Pb$_c$ most appropriate for apatite in some cases (cf. [*Schoene and Bowring*, 2006]). Whether or not the Pb$_c$ is

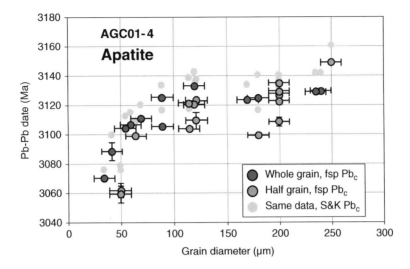

Fig. 8.27. Grain size versus age diagram for apatite from the eastern Kaapvaal Craton, Swaziland. ID-TIMS U–Pb dates for whole or half grains of apatite plotted against their diameter. Strong correlation between date and grain size is interpreted to reflect thermally activated Pb diffusion with the grain diameter as the effective diffusion dimension. Data are shown with multiple estimates for Pb_c isotopic composition. "fsp Pb_c" is the isotopic composition derived from leaching K-feldspar from the same rock and using the lowest-U aliquots to estimate apatite Pb_c. "S&K" refers to the Pb_c approximated by the *Stacey and Kramers* [1975] two-stage Pb evolution model (Fig. 8.13), evaluated at 3.2 Ga.

detrimental to the accuracy of a given data set depends in part on the question being asked (Fig. 8.27).

8.6.7.3   Rutile—TiO$_2$

Early U–Pb analysis on rutile focused on low-temperature growth in hydrothermal and ore deposits [*Corfu and Muir*, 1989; *de Ronde et al.*, 1991; *Richards et al.*, 1988; *Schandl et al.*, 1990; *Wong et al.*, 1991], though *Schärer et al.* [1986] speculated that differences between postmetamorphic rutile and titanite dates from the Grenville Province may have been the result of protracted cooling and differing closure temperatures. *Mezger et al.* [1989] documented the first detailed U–Pb study of rutile from a thermochronological point of view, in which they measured correlations between age and grain size that were used to calculate Paleoproterozoic cooling rates in the Pikwitonei granulite terrane. That study suggested that volume diffusion was an important control on the high-temperature Pb distribution in rutile and that grain size was a reasonable estimate of the effective diffusion domain. Subsequent work on rutile supports these conclusions [*Blackburn et al.*, 2011, 2012a, b; *Kooijman et al.*, 2010; *Schmitz and Bowring*, 2003; *Smye and Stockli*, 2014], though there is disagreement between empirical and experimental estimates of the diffusion kinetics of rutile. *Cherniak and Watson* [2000] report a set of experimentally determined kinetic data whose mean values correspond to a closure temperature of > 500 °C for a 100 μm grain cooling at 1 °C/Ma (Fig. 8.26). This is similar to that estimated for titanite (see above), though empirical estimates suggest rutile closure temperatures closer to that of apatite and well below experimental data for titanite [*Mezger et al.*, 1991; *Blackburn et al.*, 2012b; *Schmitz and Bowring*, 2003]. This apparent discrepancy may be

resolved using values at the upper end of the range of experimentally determined diffusion kinetics, which yields self-consistent estimates between Pb and Zr diffusion in rutile [*Blackburn et al.*, 2012a].

Regardless of the absolute values of diffusion kinetics for Pb in rutile, both grain-size–date correlations measured by ID-TIMS [*Blackburn et al.*, 2012b; *Mezger et al.*, 1991; *Schmitz and Bowring*, 2003] and age gradients consistent with diffusion profiles in single grains measured by depth profiling by LA-ICPMS [*Kooijman et al.*, 2010; *Smye and Stockli*, 2014; *Vry and Baker*, 2006] provide powerful tools for determining temperature–time paths of rocks exhumed from the middle crust (Fig. 8.28). These two different approaches provide complementary information, in that ID-TIMS provides higher precision dates whereas depth profiling accesses a greater range of apparent closure temperatures than whole grains alone.

8.6.7.4   Applications of U–Pb thermochronology

As mentioned above, U–Pb thermochronology provides access to higher temperature thermal histories than other methods such as K–Ar, fission track, or U–Th/He. Two examples where U–Pb thermochronology has yielded interesting tectonic implications include, first, in middle to lower crustal terranes that are now exposed at the surface, and second, rocks from similar depths that are still at depth and exhumed as xenoliths in volcanic rocks. Note that although the examples below focus on U–Pb thermochronology, these techniques are often used in conjunction with and/or interpreted within the context of other mid-range thermochronometers such as the K–Ar system.

U–Pb thermochronology has shown that in many cases, exhumation of granulite terranes from lower to middle crustal depths

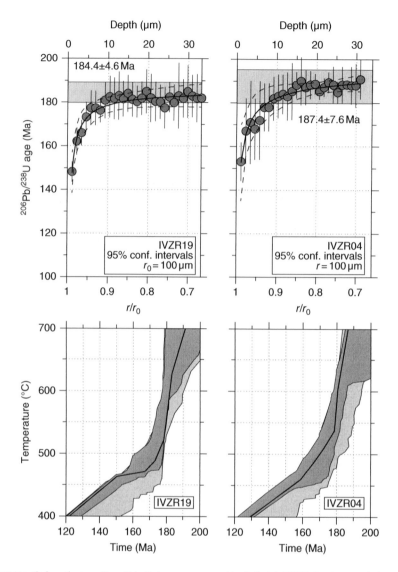

Fig. 8.28. U–Pb isotopic gradients in rutile from the Ivrea Zone, Italy. Dates were measured *in situ* by LA-ICPMS. Top two panels show data from two different grains; gray bars represent weighted average dates, while black curves are power-law fits to the date profiles, with 95% confidence intervals outlined by dashed lines. Lower two panels are temperature–time histories calculated through inversion of diffusion profiles in grains shown above (see Chapter 5 for inversion methods). (Source: Smye and Stockli [2014]. Reproduced with permission of Elsevier.)

occurred over tens to hundreds of millions of years at rates of < 2 °C/Ma. In the Adirondack mountains cooling rates following the Grenvillian Orogeny recorded by U–Pb systematics of titanite, garnet and rutile were < 2 °C/Ma and inferred to reflect post-orogenic cooling and isostatically compensated erosion [*Mezger et al.*, 1991]. A similar story emerges from rutile thermochronology of Paleoproterozoic Pikwitonei granulite terrane in Manitoba [*Mezger et al.*, 1989]. In contrast, *Mezger et al.* [1993] measured differences in cooling histories across major shear zones in the southeast Grenville orogeny using U–Pb thermochronometry, which they attribute to post-tectonic reactivation along those shear zones. In the southwest United States, cooling rates following the Mesoproterozoic Yavapai-Mazatzal orogeny have been inferred from apatite U–Pb thermochronology [*Chamberlain and Bowring*, 2001], in conjunction with

^{40}Ar/^{39}Ar thermochronology [*Hodges and Bowring*, 1995; *Heizler et al.*, 1997], to show slow postorogenic cooling at rates of < 1 °C/Ma. One explanation for differential cooling rates across blocks of crust in the southwest United States is due to gradients in radiogenic heat production [*Flowers et al.*, 2004]; another interpretation involves partial resetting of apatite and other moderate temperature thermochronometers due to crustal heating and metamorphism ca. 1.4 Ga [*Shaw et al.*, 2004].

In order to address the ambiguity of slow cooling versus reheating of thermochronometers, *Schoene and Bowring* [2007] employed U–Pb thermochronology of apatite and titanite and exploited grain-size versus age relationships in conjunction with forward thermal modeling to show that unique thermal histories can be obtained by using multiple thermochronometers. Using this method in conjunction with zircon

U–Pb dating of shear-zone movement, *Schoene and Bowring* [2007] documented differential unroofing of Archean terranes ca. 3.2–3.1 Ga in the eastern Kaapvaal craton and were able to "see through" the potential effects of thermal resetting by ca. 3.1 Ga magmatism. The potential for similar approaches using depth profiling *in situ* by LA-ICPMS [*Cochrane et al.*, 2014; *Smye and Stockli*, 2014] or SIMS [*Grove and Harrison*, 1999], but over the larger temperature range accessible via *in situ* analysis, provides a complementary way to constrain thermal histories of exhumed middle to lower crustal terranes.

In contrast to slowly cooled granulite terranes, U–Pb thermochronology has also been applied to more rapidly exhumed metamorphic terranes. In one example from the western gneiss terranes in Norway, U–Pb data from titanite and rutile, combined with thermobarometry and $^{40}Ar/^{39}Ar$ thermochronlogy, have been used to indicate very rapid exhumation (e.g., 10–30 mm/year) following burial of eclogite-grade metamorphic rocks to depths of 200 km [*Kylander-Clark et al.*, 2009]. These thermochronological data are thus crucial in building tectonic models for the generation and exhumation of UHP terranes during orogenesis and thus for how the continental crust is built and preserved.

U–Pb thermochronology has also been used to investigate the thermal history of cratonic lower crust, for example by dating rutile from lower crustal xenoliths in the Kaapvaal craton [*Schmitz and Bowring*, 2003] (Fig. 8.29). In that study rutile U–Pb data were used to calibrate the relaxation of cratonic geotherms following mid-Proterozoic thermal perturbation and subsequent Mesozoic lithospheric heating coincident with kimberlite eruption. *Blackburn et al.* [2011] conducted a similar study on kimberlite-borne lower crustal xenoliths from the Rocky Mountain region, USA. Using rutile U–Pb dates from three xenoliths, each representing different crustal depths, they employed a finite difference diffusion model to show that systematic discordance spanning > 1 Ga is inconsistent with Pb loss from reheating events. Instead, they fit *T–t* paths to the rutile data to illustrate that ≤ 0.1 °C/Ma cooling in the lower crust is required and that the results fit the analytical solution for diffusive Pb loss derived by *Tilton* [1960]. A subsequent contribution coupled these and other U–Pb thermochronometric data to model extremely long-term cratonic exhumation rates of < 2 m/Ma [*Blackburn et al.*, 2012b].

8.6.8 Carbonate geochronology by the U–Pb method

The ubiquity of carbonate rocks in the geologic record, and their importance in recording ocean–atmosphere processes through geologic time, make it a desirable target for geochronology. Although U-series geochronology of carbonates is a widely used tool for dating a variety of processes in the ocean–atmosphere–climate system, this technique is limited to materials < 800 ka. The modest U contents in carbonates that make them amenable to U-series geochronology also opens the door for directly dating carbonates by U–Pb geochronology [*Rasbury and Cole*, 2009].

Calcium carbonate can incorporate ≤ 10 ppm U into its crystal structure upon crystallization [*Richards and Dorale*, 2003; *Woodhead et al.*, 2012], due to the solubility of the uranyl ion (UO_2^{2+}) in oxidizing hydrous fluids (such as water at Earth's surface) and in turn the compatibility of the uranyl ion into the carbonate site of calcium carbonate [*Kelly et al.*, 2003, 2006]. The presence of carbonate in a wide variety of geologic environments, from soils to speleothems and of course the ocean, therefore presents an opportunity to apply the U–Pb method to calibrating the timescales of often difficult to date geologic processes. Below, applications to marine carbonates and speleothems are highlighted, but U–Th–Pb geochronology has been applied to a wide range of carbonates [*Brasier*, 2011], in particular, pedogenic carbonate often found at depositional hiatuses in sedimentary successions [*Hoff et al.*, 1995; *Winter and Johnson*, 1995; *Rasbury et al.*, 1998, 2000; *Wang et al.*, 1998], lacustrine carbonates [*Cole et al.*, 2005], and hydrothermal carbonates that can be associated with ore deposits [*Brannon et al.*, 1996; *Coveney et al.*, 2000; *Grandia et al.*, 2000].

The most voluminous carbonate deposits by far in the geologic record are in marine limestones and dolostones derived from biotic and abiotic carbon fixation. These rocks are unique recorders of the ocean–atmosphere–biosphere system through Earth history. Therefore, understanding the rates of chemical and biologic change in these systems, combined with the task of calibrating the geologic timescale, are ongoing goals that require robust time constraints from carbonate sequences. The most popular methods of providing such data are through U–Pb and Ar–Ar geochronology of interspersed volcanic ash beds (section 8.6.5), but this approach is inherently limited in that many stratigraphic sequences do not contain volcanic material. Direct dating of marine carbonates is therefore a tantalizing prospect. Attempts at this approach have found that uranium contents in marine limstones and dolostones are often < < 1 ppm and similar to that of common Pb, thus requiring U–Pb or Pb–Pb isochron techniques with a small range in U/Pb [*Jahn and Cuvellier*, 1994; *Rasbury and Cole*, 2009]. The resulting uncertainties are generally large (millions to tens of millions of years in Phanerozoic carbonates [*Smith et al.*, 1994]), and given the maturity of Phanerozoic biostratigraphic age control, such dates will usually not improve knowledge of the age of the rocks.

Furthermore, the ubiquity of diagenetic alteration in many if not most marine carbonates, which can occur immediately or tens of millions of years after deposition, presents an accuracy problem if interpreted in terms of depositional ages [*Smith et al.*, 1991]. If diagenesis can be demonstrated to be negligible, such as in one example from a Miocene aragonitic coral [*Denniston et al.*, 2008], both precise (± < 150 ka) and accurate ages for coral formation can be obtained. A combination of microanalytical techniques geared at evaluating diagenetic alteration in corals may provide a means of extending this technique further back into the Cenozoic [*Gothmann et al.*, 2015]. A recent attempt using LA-ICPMS U–Pb dating of high-U carbonate cements within Jurassic ammonites [*Li et al.*, 2014] highlights the

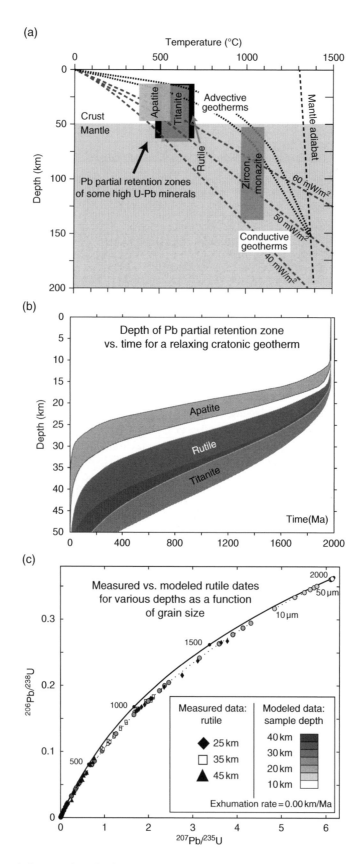

Fig. 8.29. U–Pb thermochronology applied to craton thermal evolution. (a) Temperature versus depth through a generic cratonic lithosphere illustrating advective "hot" geotherms compared to steady-state conductive "cold" cratonic geotherms for different surface-heat-flow values (mW/m²). Intersection of geotherms with mantle adiabat defines the base of the thermal lithosphere. Nominal closure temperatures for U–Pb thermochronometers (Table 8.8.2, Fig. 8.26) shown as shaded boxes with upper and lower bounds defined by the intersection with the geotherms. (b) An example showing how the ²⁰⁶Pb/²³⁸U partial retention zones (PRZ) of each thermochronometer move downward through the crust as a function of time during relaxation from a hot geotherm to a cold geotherm over 2 Ga. Limits of shaded envelopes encompass the PRZ for 10–50 μm grains using diffusion kinetics cited in the text. (c) Actual rutile U–Pb data from middle to lower crustal xenoliths from *Blackburn et al.* [2011] compared to the results of a numerical diffusion model. Black and open symbols are measured data from samples originating from various crustal depths, and shaded circles show the range of closure times for 10–50 μm grains from variable crustal depth for a conductively relaxing geotherm given no surface erosion [*Blackburn et al.*, 2011, 2012]. Note the good agreement between measured rutile dates and modeled dates, which preclude significant reheating as a source of discordance of the real data. (Source: *Schoene* [2014]. Reproduced with permission of Elsevier.)

possibility of dating diagenesis of carefully characterized carbonate fossils into the Mesozoic.

An increased interest in establishing climate proxy records beyond the upper limit of U-series geochronology (~800 ka at present) has driven some recent interest U–Pb geochronology to speleothem records [*Richards et al.*, 1998; *Walker et al.*, 2006; *Woodhead and Pickering*, 2012]. Uranium concentration in speleothems is highly variable (but often <1 ppm), so prescreening of samples for U content is useful to enable targeting of high-U zones for geochronology [*Pickering et al.*, 2010]. The application of LA-ICPMS U–Pb geochronology to speleothems, given the rapidity of throughput, facilitates generation of a large amount of data in different zones of the speleothem, rapidly identifying zones of high U/Pb$_c$ while performing isotopic measurements [*Woodhead and Pickering*, 2012]. Due to variable U/Pb$_c$, data are usually visualized on a Tera–Wasserburg concordia diagram and Pb$_c$-corrected using an isochron intercept. Multiple isochron ages can then be used to develop an age model for a speleothem (Fig. 8.30). An additional challenge for U–Pb geochronology of Quaternary speleothems lies in the correction for isotopic disequilibrium of intermediate daughter isotopes, resulting from both fractionation during calcite precipitation as well as substantial ^{234}U/^{238}U disequilibrium in meteoric water products [*Walker et al.*, 2006; *Meyer et al.*, 2009; *Pickering et al.*, 2010]. A rich literature from the U-series community has grown around methods of addressing this problem [*Edwards et al.*, 2003; *Dorale et al.*, 2004; *Yuan et al.*, 2004] (see Chapter 12). Given the interest in obtaining young time series for climatic and archeological records, the application of U–Pb geochronology to cave materials older than 500 ka is sure to increase.

8.6.9 U–Pb geochronology of baddeleyite and paleogeographic reconstructions

Understanding plate tectonics and the accurate reconstruction of continental movements into the Proterozoic and Archean eons requires a combination of robust paleomagnetic poles and reliable geochronology [*Ernst et al.*, 2008; *Evans and Pisarevsky*, 2008; *Li et al.*, 2008; *Evans*, 2013]. Shallowly emplaced mafic dikes and lava flows are ideal recorders of Earth's magnetic field but are notoriously difficult to date with high precision by any geochronologic method. Furthermore, obtaining rocks back in time with both undisturbed paleomagnetic signatures and closed-system geochronometers is difficult [*de Kock et al.*, 2009; *Evans*, 2013]. U–Pb geochronology on zircon is an ideal way to assess open-system behavior in Archean rocks using the dual U decay scheme as visualized on concordia diagrams, but zircon is rare to absent in quickly cooled mafic rocks.

Alternatively, the mineral baddeleyite (ZrO$_2$) has high U/Pb$_c$ and is not uncommon in Si-undersaturated mafic rocks [*Heaman and LeCheminant*, 1993]. Following abundant studies of baddeleyite geochronology beginning a few decades ago [*Andersen*

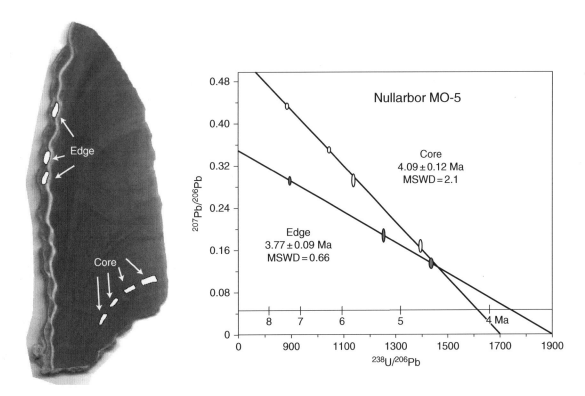

Fig. 8.30. U–Pb isochron diagram for carbonate speleothem. Core and rim analyses form two separate arrays whose lower intercept with the Tera–Wasserburg concordia diagram retains anticipated relative age progression, despite having different Pb$_c$ compositions defined by the y-intercept. Note that data are corrected for initial ^{234}U/^{238}U disequilibrium. (Source: *Woodhead et al.* [2006]. Reproduced with permission of Elsevier.)

and Hinthorne, 1972; *Krogh et al.*, 1987; *Dunning and Hodych*, 1990; *Heaman and Tarney*, 1989; *Heaman et al.*, 1992; *Heaman*, 1997; *Harlan et al.*, 2003], the accessibility of this technique has increased due to new methods for separating baddeleyite efficiently using standard mineral separation tools [*Söderlund and Johansson*, 2002; *Olsson et al.*, 2010], as well as the wider application of techniques for dating *in situ* small baddeleyite that may be lost during standard mineral separation [*Andersen and Hinthorne*, 1972; *Wingate and Compston*, 2000; *Chamberlain et al.*, 2010; *Schmitt et al.*, 2010a; *Souders and Sylvester*, 2012]. Advances in ID-TIMS and SIMS dating of high Pb_c phases, such as perovskite, in mafic rocks [*Kamo et al.*, 1996; *Kinny et al.*, 1997; *Burgess and Bowring*, 2015], and recognition of other micron-scale U-bearing phases such as zirconolite have amplified the use of U–Pb geochronology in mafic rocks [*Rasmussen and Fletcher*, 2004].

In addition to combining more robust geochronology with paleomagnetic results, an alternative use of dike orientation and precise geochronology called "barcoding" has been used to reconstruct relative continent positions through piecing together radiating dike swarms like puzzle pieces [*Bleeker and Ernst*, 2006; *Söderlund et al.*, 2010; *Ernst et al.*, 2013]. This method assumes a radial geometry for dike swarms associated with large igneous provinces, commonly also associated with continental breakup, and permits one to line up continents that were next to each other at the time indicated by the dates of the dike swarms. Because large igneous province eruptions are generally short (< 2 Ma) relative to the speed of plate tectonics (~1–5 cm/year), having U–Pb dates with uncertainties of < 2 Ma is sufficient to correlate dike swarms that barcode cratonic juxtaposition into the Archean without the inherent uncertainties in paleomagnetic data. Taken together, a growing database of dike-swarm correlations, paleomagnetic data, and geochronlogic data is playing an important role in piecing together past plate motions [*O'Neill et al.*, 2007] and supercontinent reconstructions [*Evans*, 2013] that are critical to understanding the relationship between continents, the mantle, oceans and atmosphere and the biosphere through Earth history.

8.7 CONCLUDING REMARKS

Despite being the oldest radioisotopic dating method, U–Th–Pb geochronology remains on the cutting edge and has proven to be one of the most widely used and versatile dating techniques. As analytical approaches improve and diversify, application of U–Th–Pb geochronology has spread to new minerals, smaller submineral domains, and all with increasing precision and accuracy.

What does the future hold for U–Th–Pb geochronology and thermochronology? High-spatial-resolution techniques such as SIMS and LA-ICPMS continue to strive for higher precision dates through understanding and minimizing elemental and isotopic fractionation during analysis. Geochronologists using these techniques also continue to seek more homogeneous and better-characterized mineral standards, and strive to mediate matrix effects to increase age accuracy. Even higher spatial resolution will continue to make accessible smaller domains of single crystals as well as microcrystals of high-U minerals that permit dating of a wider range of lithologies and geologic processes. Increasing accessibility to data through opening of new laboratories and decreasing analysis time in LA-ICPMS has changed the geochronologic landscape and is helping to make geochronologic data as common as geochemical and geologic data.

High-precision ID-TIMS U–Th–Pb geochronology continues to probe finer and finer timescales of geologic processes. While it was once sufficient to to correlate events such as mass extinctions and their causes to million-year precision, geologists, stratigraphers, and paleontologists are now asking questions that require ten-thousand-year precision or better. Obtaining dates that are precise and accurate to that level requires interpreting dates in terms of mineral formation processes and connecting those processes to the geologic event one is actually interested in dating. Thus, accurate dates are dependent on interpretation of those dates.

To achieve the goal of higher accuracy and precision, some of the most exciting research in U–Th–Pb geochronology is that capitalizing on the benefits of all analytical techniques as well as complementary geochemical, petrographic, and geologic information. Given the ubiquity of high-U and Th minerals in many geologic environments, we can look forward to the development of creative new approaches of applying U–Th–Pb geochronology with an ever finer lense to calibrate Earth history and the rates of geologic and solar system processes.

8.8 REFERENCES

Abe, M., Suzuki, T., Fujii, Y., Hada, M., and Hirao, K. (2008) An *ab initio* molecular orbital study of the nuclear volume effects in uranium isotope fractionations. *Journal of Chemistry and Physics* **129**(16), 164309.

Ahrens, L. (1955a) The convergent lead ages of the oldest monazites and uraninites (Rhodesia, Manitoba, Madagascar, and Transvaal). *Geochimica et Cosmochimica Acta* **7**(5), 294–300.

Ahrens, L. H. (1955b) Implications of the Rhodesia age pattern. *Geochimica et Cosmochimica Acta* **8**(1), 1–15.

Albarède, F. (2003) *Geochemistry: an Introduction*. Cambridge University Press.

Aleinikoff, J. N., Wintsch, R. P., Fanning, C. M., and Dorais, M. J. (2002) U–Pb geochronology of zircon and polygenetic titanite from the Glastonbury Complex, Connecticut, USA: an integrated SEM, EMPA, TIMS, and SHRIMP study. *Chemical Geology* **188**, 125–147.

Allègre, C. J., Albarède, F., Grünenfelder, M., and Köppel, V. (1974) $^{238}U/^{206}Pb$–$^{235}U/^{207}Pb$–$^{232}Th/^{208}Pb$ zircon geochronology in alpine and non-alpine environment. *Contributions to Mineralogy and Petrology* **43**(3), 163–194.

Allen, C. M. and Campbell, I. H. (2012) Identification and elimination of a matrix-induced systematic error in LA–ICP–MS $^{206}Pb/^{238}U$ dating of zircon. *Chemical Geology* **332**, 157–165.

Amelin, Y. and Zaitsev, A. N. (2002) Precise geochronology of phoscorites and carbonatites: The critical role of U-series disequilibrium in age interpretations. *Geochimica et Cosmochimica Acta* **66**(13), 2399–2419.

Amelin, Y., Lee, D.-C., Halliday, A. N., and Pidgeon, R. T. (1999) Nature of the Earth's earliest crust from hafnium isotopes in single detrital zircons. *Nature* **399**(6733), 252–255.

Amelin, Y., Lee, D. C., and Halliday, A. N. (2000) Early–middle Archaean crustal evolution deduced from Lu–Hf and U–Pb isotopic studies of single zircon grains. *Geochimica et Cosmochimica Acta.* **64**(24), 4205–4225.

Amelin, Y., Krot, A. N., Hutcheon, I. D., and Ulyanov, A. A. (2002) Lead isotopic ages of chondrules and calcium–aluminum-rich inclusions. *Science* **297**(5587), 1678–1683.

Amelin, Y., Ghosh, A., and Rotenberg, E. (2005) Unraveling the evolution of chrondrite parent asteroids by precise U–Pb dating and thermal modeling. *Geochimica et Cosmochimica Acta* **69**(2), 505–518.

Amelin, Y., Connelly, J., Zartman, R. E., Chen, J. H., Gopel, C., and Neymark, L. A. (2009) Modern U–Pb chronometry of meteorites: advancing to higher time resolution reveals new problems. *Geochimica et Cosmochimica Acta* **73**(17), 5212–5223.

Amelin, Y., Kaltenbach, A., Iizuka, T., *et al.* (2010) U–Pb chronology of the solar system's oldest solids with variable $^{238}U/^{235}U$. *Earth and Planetary Science Letters* **300**(3–4), 343–350.

Anczkiewicz, R., Oberli, F., Burg, J. P., Villa, I. M., Gunther, D., and Meier, M. (2001) Timing of normal faulting along the Indus Suture in Pakistan Himalaya and a case of major $^{231}Pa/^{235}U$ initial disequilibrium in zircon. *Earth and Planetary Science Letters* **191**, 101–114.

Andersen, C. A. and Hinthorne, J. R. (1972) U, Th, Pb and REE abundances and $^{207}Pb/^{206}Pb$ ages of individual minerals in returned lunar material by ion microprobe mass analysis. *Earth and Planetary Science Letters* **14**(2), 195–200.

Andersen, M., Romaniello, S., Vance, D., Little, S., Herdman, R., and Lyons, T. (2014) A modern framework for the interpretation of $^{238}U/^{235}U$ in studies of ancient ocean redox. *Earth and Planetary Science Letters* **400**, 184–194.

Andersen, T. (2002) Correction of common lead in U–Pb analyses that do not report ^{204}Pb. *Chemical Geology* **192**(1–2), 59–79.

Annen, C. (2011) Implications of incremental emplacement of magma bodies for magma differentiation, thermal aureole dimensions and plutonism–volcanism relationships. *Tectonophysics* **500**, 3–10. DOI: 10.106/j.tecto.2009.2004.2010

Armstrong, R. and Harmon, R. (1981) Radiogenic isotopes: the case for crustal recycling on a near-steady-state no-continental-growth Earth. *Philosophical Transactions of the Royal Society of London. Series A, Mathematical and Physical Sciences* **301** (1461), 443–472.

Asael, D., Tissot, F. L., Reinhard, C. T., *et al.* (2013) Coupled molybdenum, iron and uranium stable isotopes as oceanic paleoredox proxies during the Paleoproterozoic Shunga Event. *Chemical Geology* **362**, 193–210.

Bachmann, O. and Bergantz, G. W. (2003) Rejuvenation of the Fish Canyon magma body: A window into the evolution of large-volume silicic magma systems. *Geology* **31**(9), 789–792.

Bachmann, O., Dungan, M. A., and Lipman, P. W. (2002) The Fish Canyon Magma Body, San Juan Volcanic Field, Colorado: Rejuvenation and Eruption of an Upper-Crustal Batholith. *Journal of Petrology* **43**(8), 1469–1503.

Bachmann, O., Charlier, B. L. A., and Lowenstern, J. B. (2007) Zircon crystallization and recycling in the magma chamber of the rhyolitic Kos Plateau Tuff (Aegean arc). *Geology* **35**(1), 73–76. DOI: 10.1130/G23151A.23151.

Bachmann, O., Schoene, B., Schnyder, C., and Spikings, R. (2010) The 40Ar/39Ar and U/Pb dating of young rhyolites in the Kos-Nisyros volcanic complex, Eastern Aegean Arc, Greece: age discordance due to excess 40Ar in biotite. *Geochemistry, Geophysics, Geosystems* **11**, Q0AA08. DOI: 10.1029/2010gc003073.

Baker, J., Peate, D., Waight, T., and Meyzen, C. (2004) Pb isotopic analysis of standards and samples using a $^{207}Pb-^{204}Pb$ double spike and thallium to correct for mass bias with a double-focusing MC-ICP-MS. *Chemical Geology* **211**(3), 275–303.

Baldwin, J. A., Bowring, S. A., Williams, M. L., and Mahan, K. H. (2006) Geochronological constraints on the evolution of high-pressure felsic granulites from an integrated electron microprobe and ID-TIMS geochemical study. *Lithos* **88**(1–4), 173–200.

Baldwin, J., Brown, M., and Schmitz, M. (2007) First application of titanium-in-zircon thermometry to ultrahigh-temperature metamorphism. *Geology* **35**(4), 295–298.

Barboni, M., and Schoene, B. (2014) Short eruption window revealed by absolute crystal growth rates in a granitic magma. *Nature Geoscience* **7**, 524–528.

Barboni, M., Annen, C., and Schoene, B. (2015) Evaluating the construction and evolution of upper crustal magma reservoirs with coupled U/Pb zircon geochronology and thermal modeling: a case study from the Mt. Capanne pluton (Elba, Italy). *Earth and Planetary Science Letters* **432**, 436–448.

Bartley, J. M., Coleman, D. S., and Glazner, A. E. (2008) Incremental pluton emplacement by magmatic crack-seal. *Transactions of the Royal Society of Edinburgh: Earth Sciences* **97**, 383–396. DOI:310.1017/S0263593300001528

Bateman, H. (1910) Solution of a system of differential equations occurring in the theory of radioactive transformations. *Proceedings of the Cambridge Philosophical Society* **15**, 423–427.

Becquerel, A. (1896a) Émission de radiations nouvelles par l'uranium métallique. *Comptes Rendus de l' Academie des Sciences de Paris* **122**, 1086.

Becquerel, A. H. (1896b) Sur les radiations invisibles émises par les corps phosphorescents. *Comptes Rendus de l' Academie des Sciences de Paris* **122**, 501.

Begemann, F., Ludwig, K. R., Lugmair, G. W., *et al.* (2001) Call for an improved set of decay constants for geochronological use. *Geochimica et Cosmochimica Acta* **65**(1), 111–121.

Bell, E. A., Harrison, T. M., McCulloch, M. T., and Young, E. D. (2011) Early Archean crustal evolution of the Jack Hills Zircon source terrane inferred from Lu–Hf, $^{207}Pb/^{206}Pb$, and δ 18 O systematics of Jack Hills zircons. *Geochimica et Cosmochimica Acta* **75**(17), 4816–4829.

Bell, E. A., Harrison, T. M., Kohl, I. E., and Young, E. D. (2014) Eoarchean crustal evolution of the Jack Hills zircon source and loss of Hadean crust. *Geochimica et Cosmochimica Acta* **146**, 27–42.

Bell, E. A., Boehnke, P., Harrison, T. M., and Mao, W. L. (2015) Potentially biogenic carbon preserved in a 4.1 billion-year-old zircon. *Proceedings of the National Academy of Sciences* **112** (47), 14518–14521.

Belousova, E. A., Griffin, W. L., O'Reilly, S. Y., and Fisher, N. I. (2002) Igneous zircon: trace element composition as an indicator of source rock type. *Contributions to Mineralogy and Petrology* **143**, 602–622. DOI: 610.1007/s00410-00002-00364-00417.

Belousova, E. A., Griffin, W. L., and O'Reilly, S. Y. (2006) Zircon crystal morphology, trace element signatures and Hf isotope composition as a tool for petrogenetic modelling: examples from eastern Australian granitoids. *Jour. Pet.* **47**, 329–353. DOI:310.1093/petrology/egi1077.

Belousova, E. A., Kostitsyn, Y. A., Griffin, W. L., Begg, G. C., O'Reilly, S. Y., and Pearson, N. J. (2010) The growth of the continental crust: Constraints from zircon Hf-isotope data. *Lithos* **119**(3), 457–466.

Belyatsky, B., Rodionov, N., Antonov, A., and Sergeev, S. (2011) The 3.98–3.6 Ga zircons as indicators of major processes operating in the ancient continental crust of the east Antarctic shield (Enderby Land). *Proceedings Doklady Earth Sciences* **438**, 770–774.

Bickford, M. E., Chase, R. B., Nelson, B. K., Shuster, R. D., and Arruda, E. C. (1981) U–Pb studies of zircon cores and overgrowths, and monazite: implications for age and petrogenesis of the northeastern Idaho Batholith. *The Journal of Geology* **89** (4), 433–457.

Black, L., Williams, I., and Compston, W. (1986) Four zircon ages from one rock: the history of a 3930 Ma-old granulite from Mount Sones, Enderby Land, Antarctica. *Contributions to Mineralogy and Petrology* **94**(4), 427–437.

Black, L. P., Kamo, S., Allen, C. M., *et al.* (2004) Improved ^{206}Pb/^{238}U microprobe geochronology by the monitoring of a trace-element-related matrix effect; SHRIMP, ID-TIMS, ELA-ICP-MS and oxygen isotope documentation for a series of zircon standards. *Chemical Geology* **205**, 115–140.

Blackburn, T., Bowring, S., Schoene, B., Mahan, K., and Dudas, F. (2011) U–Pb thermochronology: creating a temporal record of lithosphere thermal evolution. *Contributions to Mineralogy and Petrology* **162**(3), 479–500.

Blackburn, T., Shimizu, N., Bowring, S. A., Schoene, B., and Mahan, K. H. (2012a) Zirconium in rutile speedometry: New constraints on lower crustal cooling rates and residence temperatures. *Earth and Planetary Science Letters* **317–318**, 231–240.

Blackburn, T. J., Bowring, S. A., Perron, J. T., Mahan, K. H., Dudas, F. O., and Barnhart, K. R. (2012b) An exhumation history of continents over billion-year time scales. *Science* **335**(6064), 73–76.

Bleeker, W. and Ernst, R. (2006) Short-lived mantle generated magmatic events and their dyke swarms: the key unlocking Earth's paleogeographic record back to 2.6 Ga. In *Dyke Swarms—Time Markers of Crustal Evolution*, Hanski, E., Mertanen, S., Rämö, T., and Vuollo, J. (eds), 3–26. Taylor & Francis, London.

Blichert-Toft, J. and Albarède, F. (2008) Hafnium isotopes in Jack Hills zircons and the formation of the Hadean crust. *Earth and Planetary Science Letters* **265**(3), 686–702.

Boehnke, P., and Harrison, T. M. (2014) A meta-analysis of geochronologically relevant half-lives: what's the best decay constant? *International Geology Review* **56**(7), 905–914.

Boehnke, P., Watson, E. B., Trail, D., Harrison, T. M., and Schmitt, A. K. (2013) Zircon saturation re-revisited. *Chemical Geology* **351**, 324–334.

Boehnke, P., Barboni, M., and Bell, E. (2016) Zircon U/Th model ages in the presence of melt heterogeneity. *Quaternary Geochronology* **34**, 69–74.

Boltwood, B. B. (1907) On the ultimate disintegration products of the radioactive elements. *Part II. The disintegration products of uranium. American Journal of Science* **23**, 77–88.

Bopp, C. J., Lundstrom, C. C., Johnson, T. M., and Glessner, J. J. (2009) Variations in ^{238}U/^{235}U in uranium ore deposits: Isotopic signatures of the U reduction process? *Geology* **37**(7), 611–614.

Bourdon, B., Turner, S., Henderson, G. M., and Lundstrom, C. C. (2003) Introduction to U-series geochemistry. *Reviews in Mineralogy and Geochemistry* **52**(1), 1–21.

Bouvier, A. and Wadhwa, M. (2010) The age of the solar system redefined by the oldest Pb–Pb age of a meteoritic inclusion. *Nature Geoscience* **3**(9), 637–641.

Bowring, S. A. and Housh, T. (1995) The Earth's early evolution. *Science* **269**, 1535–1540.

Bowring, S. A., and Schmitz, M. D. (2003) High-precision U–Pb zircon geochronology and the stratigraphic record. In Zircon, *Vol. 53*, Hanchar, J. M. and Hoskin, P. W. O. (eds), 305–326. Mineralogical Society of America, Washington, DC.

Bowring, S. A. and Williams, I. S. (1999a) Priscoan (4.00–4.03 Ga) orthogneisses from northwestern Canada. *Contributions to Mineralogy and Petrology* **134**(1), 3–16.

Bowring, S. A. and Williams, I. S. (1999b) Priscoan (4.00–4.03 Ga) orthogneisses from northwestern Canada. *Contributions to Mineralogy and Petrology* **134**(1), 3–16.

Bowring, S. A., King, J. E., Housh, T. B., Isachsen, C. E., and Podosek, F. A. (1989a) Neodymium and lead isotope evidence for enriched early Archaean crust in North America. *Nature* **340**, 222–225

Bowring, S. A., Williams, I. S., and Compston, W. (1989b) 3.96 Ga gneisses from the Slave province, Northwest Territories, Canada. *Geology* **17**(11), 971–975.

Bowring, S., Housh, T., and Isachsen, C. (1990) The Acasta gneisses: remnant of Earth's early crust. *Origin of the Earth* **1**, 319–343.

Bowring, S. A., Erwin, D. H., Jin, Y. G., Martin, M. W., Davidek, K., and Wang, W. (1998) U/Pb zircon geochronology and tempo of the end-Permian mass extinction. *Science* **280**, 1039–1045.

Bowring, S. A., Schoene, B., Crowley, J. L., Ramezani, J., and Condon, D. C. (2006) High-precision U–Pb zircon geochronology and the stratigraphic record: progress and promise. In *Geochronology: Emerging Opportunities, Paleontological Society Short Course*, Vol. **12**, Olszewski, T. (ed.), 25–45. The Paleontological Society, Philidelphia, PA.

Brannon, J. C., Cole, S. C., Podosek, F. A., *et al.* (1996) Th–Pb and U–Pb dating of ore-stage calcite and Paleozoic fluid flow. *Science* **271**(5248), 491–493.

Brasier, A. T. (2011) Searching for travertines, calcretes and speleothems in deep time: processes, appearances, predictions and the impact of plants. *Earth-Science Reviews* **104**(4), 213–239.

Brennecka, G. A., Weyer, S., Wadhwa, M., Janney, P. E., Zipfel, J., and Anbar, A. D. (2010) ^{238}U/^{235}U Variations in Meteorites: Extant 247Cm and Implications for Pb–Pb Dating. *Science* **327**(5964), 449–451.

Brennecka, G. A., Wasylenki, L. E., Bargar, J. R., Weyer, S., and Anbar, A. D. (2011) Uranium Isotope Fractionation during Adsorption to Mn-Oxyhydroxides. *Environmental Science and Technology* **45**(4), 1370–1375.

Brewer, T. S., Storey, C. D., Parrish, R. R., Temperley, S., and Windley, B. F. (2003) Grenvillian age decompression of eclogites in the Glenelg–Attadale Inlier, NW Scotland. *Journal of the Geological Society* **160**(4), 565–574.

Bruguier, O., Lancelot, J. R., and Malavieille, J. (1997) U–Pb dating on single detrital zircon grains from the Triassic Songpan-Ganze flysch (Central China): provenance and tectonic correlations. *Earth and Planetary Science Letters* **152**(1), 217–231.

Budzyń, B., Jastrzębski, M., Kozub-Budzyń, G. A., and Konečný, P. (2015) Monazite Th–U–total Pb geochronology and PT thermodynamic modelling in a revision of the HP-HT metamorphic record in granulites from Stary Gierałtów (NE Orlica-Śnieżnik Dome, SW Poland). *Geological Quarterly* **59**(4). DOI: 10.7306/gq. 1232.

Burgess, S. D. and Bowring, S. A. (2015) High-precision geochronology confirms voluminous magmatism before, during, and after Earth's most severe extinction. *Science Advances* **1**(7), e1500470.

Burgess, S. D. and Miller, J. S. (2008) Construction, solidification and internal differentiation of a large felsic arc pluton: Cathedral Peak granodiorite, Sierra Nevada Batholith. *Geological Society, London, Special Publications* **304**(1), 203–233.

Burgess, S. D., Bowring, S., and Shen, S.-Z. (2014) High-precision timeline for Earth's most severe extinction. *Proceedings of the National Academy of Sciences* **111**(9), 3316–3321.

Burnham, A. D. and Berry, A. J. (2012) An experimental study of trace element partitioning between zircon and melt as a function of oxygen fugacity. *Geochimica et Cosmochimica Acta* **95**, 196–212.

Carley, T. L., Miller, C. F., Wooden, J. L., *et al.* (2014) Iceland is not a magmatic analog for the Hadean: evidence from the zircon record. *Earth and Planetary Science Letters* **405**, 85–97.

Catanzaro, E. (1963) Zircon ages in southwestern *Minnesota: Journal of Geophysical Research* **68**(7), 2045–2048.

Catanzaro, E. J. and Hanson, G. N. (1971) U–Pb ages for sphene from Early Precambrian igneous rocks in northeastern Minnesota, northwestern Ontario. *Canadian Journal of Earth Sciences* **8**(10), 1319–1324.

Cavosie, A. J., Wilde, S. A., Liu, D., Weiblen, P. W., and Valley, J. W. (2004) Internal zoning and U–Th–Pb chemistry of Jack Hills detrital zircons: a mineral record of early Archean to Mesoproterozoic (4348–1576 Ma) magmatism. *Precambrian Research* **135**, 251–279.

Chamberlain, K. R. and Bowring, S. A. (2001) Apatite-feldspar U–Pb thermochronometer: a reliable mid-range (∼450 °C), diffusion controlled system. *Chemical Geology* **172**(1–2), 173–200.

Chamberlain, K. R., Schmitt, A. K., Swapp, S. M., *et al.* (2010) *In situ* U–Pb SIMS (IN-SIMS) micro-baddeleyite dating of mafic rocks: method with examples. *Precambrian Research* **183**(3), 379–387.

Chamberlain, K. J., Morgan, D. J., and Wilson, C. J. (2014) Timescales of mixing and mobilisation in the Bishop Tuff magma body: perspectives from diffusion chronometry. *Contributions to Mineralogy and Petrology* **168**(1), 1–24.

Charlier, B. L. A. and Wilson, C. J. N. (2010) Chronology and Evolution of Caldera-forming and Post-caldera Magma Systems at Okataina Volcano, *New Zealand from Zircon U–Th Model-age Spectra: Journal of Petrology* **51**(5), 1121–1141.

Charlier, B. L. A., Wilson, C. J. N., Lowenstern, J. B., Blake, S., Van Calstren, P. W., and Davidson, J. P. (2005) Magma generation at a large hyperactive silicic volcano (Taupo, New Zealand) revealed by U–Th and U–Pb systematics in zircons. *Journal of Petrology* **46**, 3–32. DOI:10.1093/petrology/egh1060.

Chen, J. H. and Wasserburg, G. J. (1981) Isotopic determination of uranium in picomole and subpicomol quantities. Anal. Chem. 53), 2060–2067.

Cheng, H., Lawrence Edwards, R., Shen, C.-C., *et al.* (2013) Improvements in ^{230}Th dating, ^{230}Th and ^{234}U half-life values, and U–Th isotopic measurements by multi-collector inductively coupled plasma mass spectrometry. *Earth and Planetary Science Letters* **371–372**, 82–91.

Cherniak, D. J. (1993) Lead diffusion in titanite and preliminary results on the effects of radiation damage on Pb transport. *Chemical Geology* **110**, 177–194.

Cherniak, D. J. (1995) diffusion of Pb in plagioclase and K-feldspar investigated using Rutherford backscatter and resonant nuclear reaction analysis. *Contributions to Mineralogy and Petrology* **120**, 358–371.

Cherniak, D. J., and Watson, E. B. (2000) Pb diffusion in rutile. *Contributions to Mineralogy and Petrology* **139**, 198–207.

Cherniak, D. J., and Watson, E. B. (2001) Pb diffusion in zircon: *Chemical Geology* **172**, 5–24.

Cherniak, D. J., Lanford, W. A., and Ryerson, F. J. (1991) Lead diffusion in apatite and zircon using ion implantation and Rutherford Backscattering techniques. *Geochimica et Cosmochimica Acta* **55**, 1663–1673.

Chew, D. M. and Donelick, R. A. (2012) Combined apatite fission track and U–Pb dating by LA-ICP-MS and its application in apatite provenance analysis. In *Quantitative Mineralogy and Microanalysis of Sediments and Sedimentary Rocks. Mineralogical Association of Canada, Short Course* **42**, 219–247.

Chew, D. M. and Spikings, R. A. (2015) Geochronology and thermochronology using apatite: time and temperature, lower crust to surface. *Elements* **11**(3), 189–194.

Chew, D., Petrus, J., and Kamber, B. (2014) U–Pb LA–ICPMS dating using accessory mineral standards with variable common Pb. *Chemical Geology* **363**, 185–199.

Christensen, J. N. and DePaolo, D. J. (1993) Time scales of large volume silicic magma systems: Sr isotopic systematics of phenocrysts and glass from the Bishop Tuff, Long Valley, California. *Contributions to Mineralogy and Petrology* **113**(1), 100–114.

Claiborne, L. L., Miller, C. F., Flanagan, D. M., Clynne, M. A., and Wooden, J. L. (2010) Zircon reveals protracted magma storage and recycling beneath Mount St. *Helens. Geology* **38**(11), 1011–1014.

Cocherie, A., Legendre, O., Peucat, J. J., and Kouamelan, A. N. (1998) Geochronology of polygenetic monazites constrained by *in situ* electron microprobe Th–U–total lead determination: implications for lead behaviour in monazite. *Geochimica et Cosmochimica Acta* **62**(14), 2475–2497.

Cochrane, R., Spikings, R. A., Chew, D., *et al.* (2014) High temperature (> 350 °C) thermochronology and mechanisms of Pb loss in apatite. *Geochimica et Cosmochimica Acta* **127**, 39–56.

Cole, J. M., Rasbury, E. T., Hanson, G. N., Montañez, I. P., and Pedone, V. A. (2005) Using U–Pb ages of Miocene tufa for correlation in a terrestrial succession, Barstow Formation, California. *Geological Society of America Bulletin* **117**(3–4), 276–287.

Coleman, D. S. and Glazner, A. F. (1997) The Sierra Crest magmatic event: rapid formation of juvenile crust during the Late Cretaceous in California. *International Geology Review* **39**(9), 768–787.

Coleman, D. S., Gray, W., and Glazner, A. F. (2004) Rethinking the emplacement and evolution of zoned plutons: geochronologic evidence for incremental assembly of the Tuolumne Intrusive Suite, California. *Geology* **32**(5), 433–436.

Coleman, D. S., Mills, R. D., and Zimmerer, M. J. (2016) The pace of plutonism. *Elements* **12**(2), 97–102.

Compston, W. and Pidgeon, R. T. (1986) Jack Hills, evidence of more very old detrital zircons in Western Australia. *Nature* **321**(6072), 766–769.

Compston, W., Williams, I. S., and Meyer, C. (1984) U–Pb Geochronology of Zircons From Lunar Breccia 73217 Using a Sensitive High Mass-Resolution Ion Microprobe. *Journal of Geophysics Research* **89**(S2), B525–B534.

Condie, K. C. and Aster, R. C. (2010) Episodic zircon age spectra of orogenic granitoids: the supercontinent connection and continental growth. *Precambrian Research* **180**(3), 227–236.

Condie, K. C., Belousova, E., Griffin, W. L., and Sircombe, K. N. (2009) Granitoid events in space and time: Constraints from igneous and detrital zircon age spectra. *Gondwana Research* **15** (3–4), 228–242.

Condie, K. C., Bickford, M. E., Aster, R. C., Belousova, E., and Scholl, D. W. (2011) Episodic zircon ages, Hf isotopic composition, and the preservation rate of continental crust. *Geological Society of America Bulletin* **123**(5–6), 951–957.

Condon, D., Schoene, B., McLean, N., Bowring, S., and Parrish, R. (2015) Metrology and traceability of U–Pb isotope dilution geochronology (EARTHTIME Tracer Calibration Part I). *Geochimica et Cosmochimica Acta* **164**, 464–480.

Connelly, J. N. and Bizzarro, M. (2009) Pb–Pb dating of chondrules from CV chondrites by progressive dissolution. *Chemical Geology* **259**(3–4), 143–151.

Connelly, J. N., Amelin, Y., Krot, A. N., and Bizzarro, M. (2008) Chronology of the solar system's oldest solids. *The Astrophysical Journal Letters* **675**(2), L121.

Connelly, J. N., Bizzarro, M., Krot, A. N., Nordlund, Å., Wielandt, D., and Ivanova, M. A. (2012) The absolute chronology and thermal processing of solids in the solar protoplanetary disk. *Science* **338**(6107), 651–655.

Coogan, L. A. and Hinton, R. W. (2006) Do the trace element compositions of detrital zircons require Hadean continental crust? *Geology* **34**(8), 633–636.

Coombs, M. L. and Vazquez, J. A. (2014) Cogenetic late Pleistocene rhyolite and cumulate diorites from Augustine Volcano revealed by SIMS ^{238}U–^{230}Th dating of zircon, and implications for silicic magma generation by extraction from mush. *Geochemistry, Geophysics, Geosystems* **15**(12), 4846–4865.

Cooper, K. M. and Kent, A. J. (2014) Rapid remobilization of magmatic crystals kept in cold storage. *Nature* **506**(7489), 480–483.

Corfu, F. (1980) U–Pb and Rb–Sr systematics in a polyorogenic segment of the Precambrian shield, central southern Norway. *Lithos* **13**(4), 305–323.

Corfu, F. (1988) Differential response of U–Pb systems in coexisting accessory minerals, Winnepeg River Subprovince, Canadian Shield: implications for Archean crustal growth and stabilization. *Contributions to Mineralogy and Petrology* **98**, 312–325.

Corfu, F. (2013) A century of U–Pb geochronology: the long quest towards concordance. *Geological Society of America Bulletin* **125** (1–2), 33–47.

Corfu, F. and Ayres, L. D. (1984) U–Pb age and genetic significance of heterogeneous zircon populations in rocks from the favorable lake area, nortwestern Ontario. *Contributions to Mineralogy and Petrology* **88**(1–2), 86–101. DOI: 110.1007/BF00371414.

Corrie, S. L. and Kohn, M. J. (2007) Resolving the timing of orogenesis in the Western Blue Ridge, southern Appalachians, via *in situ* ID-TIMS monazite geochronology. *Geology* **35**(7), 627–630. DOI: 610.1130/G23601A.23601.

Corfu, F. and Muir, T. (1989) The Hemlo–Heron Bay greenstone belt and Hemlo A–Mo deposit, Superior Province, Ontario, Canada 2. Timing of metamorphism, alteration and Au mineralization from titanite, rutile, and monazite U–Pb geochronology. *Chemical Geology: Isotope Geoscience Section* **79**(3), 201–223.

Corfu, F. and Stone, D. (1998) The significance of titanite and apatite U–Pb ages: Constraints for the post-magmatic thermal-hydrothermal evolution of a batholithic complex, Berens River area, northwestern Superior Province, Canada. *Geochimica et Cosmochimica Acta* **62**(17), 2979–2995.

Corfu, F., Krogh, T. E., and Ayres, L. D. (1985) U–Pb zircon and sphene geochronology of a composite Archean granitoid batholith, Favourable Lake area, nortwestern Ontario. *Canadian Journal of Earth Science* **22**, 1436–1451.

Corfu, F., Heaman, L. M., and Rogers, G. (1994) Polymetamorphic evolution of the Lewisian complex, NW Scotland, as recorded by U–Pb isotopic compositions of zircon, titanite and rutile. *Contributions to Mineralogy and Petrology* **117**(3), 215–228.

Corfu, F., Hanchar, J. M., Hoskin, P. W. O., and Kinny, P. (2003) Atlas of zircon textures. In Zircon, *Vol. 53*, Hanchar, J. M. and Hoskin, P. W. O. (eds), 468–500. Mineralogical Society of America, Washington, DC.

Cosca, M. A., Essene, E. J., Mezger, K., and van der Pluijm, B. A. (1995) Constraints on the duration of tectonic processes:

protracted extension and deep-crustal rotation in the Grenville orogen. *Geology* **23**(4), 361–364.

Costa, F., Dohmen, R., and Chakraborty, S. (2008) Time Scales of Magmatic Processes from Modeling the Zoning Patterns of Crystals. *Reviews in Mineralogy and Geochemistry* **69**(1), 545–594.

Cottle, J. M., Horstwood, M. S. A., and Parrish, R. R. (2009a) A new approach to single shot laser ablation analysis and its application to *in situ* Pb/U geochronology. *Journal of Analytical Atomic Spectrometry* **24**(10).

Cottle, J. M., Searle, M. P., Horstwood, M. S. A., and Waters, D. J. (2009b) Timing of midcrustal metamorphism, melting, and deformation in the Mount Everest region of southern Tibet revealed by U–Th–Pb geochronology. *The Journal of Geology* **117**(6), 643–664.

Coveney, R. M., Ragan, V. M., and Brannon, J. C. (2000) Temporal benchmarks for modeling Phanerozoic flow of basinal brines and hydrocarbons in the southern Midcontinent based on radiometrically dated calcite. *Geology* **28**(9), 795–798.

Crowley, J. L., Myers, J. S., Sylvester, P. J., and Cox, R. A. (2005) Detrital zircon from the Jack Hills and Mount Narryer, Western Australia: evidence for diverse > 4.0 Ga source rocks. *The Journal of Geology* **113**(3), 239–263.

Crowley, J. L., Schoene, B., and Bowring, S. A. (2007) U–Pb dating of zircon in the Bishop Tuff at the millennial scale. *Geology* **35** (12), 1123–1126. DOI: 1110.1130/G24017A.

Crowley, J. L., Waters, D., Searle, M., and Bowring, S. (2009) Pleistocene melting and rapid exhumation of the Nanga Parbat massif, Pakistan: age and P–T conditions of accessory mineral growth in migmatite and leucogranite. *Earth and Planetary Science Letters* **288**(3), 408–420.

Crowley, Q. G., Heron, K., Riggs, N., *et al.* (2014) Chemical abrasion applied to LA-ICP-MS U–Pb zircon geochronology. *Minerals* **4**(2), 503–518.

Cumming, G. L. and Richards, J. R. (1975) Ore lead isotope ratios in a continuously changing earth. *Earth and Planetary Science Letters* **28**(2), 155–171.

Curie, P. and Sklodowska-Curie, M. (1898) Sur une substance nouvelle radio-active, contenue dans la pechblende. *Comptes Rendus de l' Academie des Sciences de Paris* **127**, 175–178.

Dahl, T. W., Boyle, R. A., Canfield, D. E., *et al.* (2014) Uranium isotopes distinguish two geochemically distinct stages during the later Cambrian SPICE event. *Earth and Planetary Science Letters* **401**, 313–326.

Dalrymple, G. B. (1994) *The Age of the Earth*. Stanford University Press.

Davis, D. W. and Krogh, T. E. (2000) Preferential dissolution of ^{234}U and radiogenic Pb from alpha-recoil-damaged lattive sites in zircon: implications for thermal histories and Pb isotopic fractionation in the near surface environment. *Chemical Geology* **172**, 41–58.

Davis, D. W., Williams, I. S., and Krogh, T. E. (2003) Historical development of zircon geochronology. In Zircon, *Vol. 53*, Hanchar, J. M. and Hoskin, P. W. O. (eds), 145–181. Mineralogical Society of America, Washington, DC.

Davydov, V. I., Crowley, J. L., Schmitz, M. D., and Poletaev, V. I. (2010) High-precision U–Pb zircon age calibration of the global Carboniferous time scale and Milankovitch band cyclicity in the Donets Basin, eastern Ukraine. *Geochemistry, Geophysics, Geosystems* **11**(2), Q0AA04.

De Kock, M. O., Evans, D. A. D., and Beukes, N. J. (2009) Validating the existence of Vaalbara in the Neoarchean. *Precambrian Research* **174**(1–2), 145–154.

De Ronde, C. E. J., Kamo, S., Davis, D. W., de Wit, M. J., and Spooner, E. T. C. (1991) Field, geochemical and U–Pb isotopic constraints from hypabyssal felsic intrusions within the Barberton greenstone belt, South Africa: implications for tectonics and the timing of gold mineralization. *Precambrian Research* **49**, 261–280.

Deering, C. D., Keller, B., Schoene, B., Bachmann, O., Beane, R., and Ovtcharova, M. (2016) Zircon record of the plutonic-volcanic connection and protracted rhyolite melt evolution. *Geology* **44**(4), 267–270.

DeGraaff-Surpless, K., Graham, S. A., Wooden, J. L., and McWilliams, M. O. (2002) Detrital zircon provenance analysis of the Great Valley Group, California: Evolution of an arc-forearc system. *Geological Society of America Bulletin* **114**(12), 1564–1580.

Deliens, M., Delhal, J. and Tarte, P. (1977) Metamictization and U–Pb systematics – a study by infrared absorption spectrometry of Precambrian zircons. *Earth and Planetary Science Letters* **33** (3), 331–344.

Denniston, R. F., Asmerom, Y., Polyak, V. Y., *et al.* (2008) Caribbean chronostratigraphy refined with U–Pb dating of a Miocene coral. *Geology* **36**(2), 151–154.

DesOrmeau, J. W., Gordon, S. M., Kylander-Clark, A. R. C., *et al.* (2015) Insights into (U)HP metamorphism of the Western Gneiss Region, Norway: a high-spatial resolution and high-precision zircon study. *Chemical Geology* **414**, 138–155.

Dhuime, B., Hawkesworth, C. J., Cawood, P. A., and Storey, C. D. (2012) A change in the geodynamics of continental growth 3 billion years ago. *Science* **335**(6074), 1334–1336.

Dickinson, W. R. and Gehrels, G. E. (2003) U–Pb ages of detrital zircons from Permian and Jurassic eolian sandstones of the Colorado Plateau, USA: paleogeographic implications. *Sedimentary Geology* **163**(1–2), 29–66.

Dickinson, W. R. and Gehrels, G. E. (2009) Use of U–Pb ages of detrital zircons to infer maximum depositional ages of strata: A test against a Colorado Plateau Mesozoic database. *Earth and Planetary Science Letters* **288**(1–2), 115–125.

DiPietro, J. A. and Isachsen, C. E. (2001) U–Pb zircon ages from the Indian plate in northwest Pakistan and their significance to Himalayan and pre-Himalayan geologic history. *Tectonics* **20**(4), 510–525.

Diwu, C., Sun, Y., Wilde, S. A., *et al.* (2013) New evidence for ~4.45 Ga terrestrial crust from zircon xenocrysts in Ordovician ignimbrite in the North Qinling Orogenic Belt, China. *Gondwana Research* **23**(4), 1484–1490.

Dodson, M. H. (1973) Closure temperature in cooling geochronological and petrological systems. *Contributions to Mineralogy and Petrology* **40**, 259–274.

Dorale, J. A., Edwards, R. L., Alexander Jr, E. C., Shen, C.-C., Richards, D. A., and Cheng, H. (2004) Uranium-series dating of speleothems: current techniques, limits, and applications. In

Studies of Cave Sediments, Sasowsky, I. D. and Mylroie, J. (eds), 177–197. Springer,

Druitt, T. H., Costa, F., Deloule, E., Dungan, M., and Scaillet, B. (2012) Decadal to monthly timescales of magma transfer and reservoir growth at a caldera volcano. *Nature* **482**(7383), 77–80.

Dumond, G., McLean, N., Williams, M. L., Jercinovic, M. J., and Bowring, S. A. (2008) High-resolution dating of granite petrogenesis and deformation in a lower crustal shear zone: Athabasca granulite terrane, western Canadian Shield. *Chemical Geology* **254**(3–4), 175–196.

Dumond, G., Goncalves, P., Williams, M., and Jercinovic, M. (2015) Monazite as a monitor of melting, garnet growth and feldspar recrystallization in continental lower crust. *Journal of Metamorphic Geology* **33**(7), 735–762.

Dunning, G. R. and Hodych, J. P. (1990) U/Pb zircon and baddeleyite ages for the Palisades and Gettysburg sills of the northeastern United States: implications for the age of the Triassic/Jurassic boundary. *Geology* **18**, 795–798.

Eddy, M. P., Bowring, S. A., Miller, R. B., and Tepper, J. H. (2016) Rapid assembly and crystallization of a fossil large-volume silicic magma chamber. *Geology* **44**(4), 331–334.

Edwards, R. L., Gallup, C. D., and Cheng, H. (2003) Uranium-series dating of marine and lacustrine carbonates. *Reviews in Mineralogy and Geochemistry* **52**(1), 363–405.

Ernst, R., Wingate, M., Buchan, K., and Li, Z.-X. (2008) Global record of 1600–700 Ma large igneous provinces (LIPs): implications for the reconstruction of the proposed Nuna (Columbia) and Rodinia supercontinents. *Precambrian Research* **160**(1), 159–178.

Ernst, R. E., Bleeker, W., Söderlund, U., and Kerr, A. C. (2013) Large Igneous Provinces and supercontinents: Toward completing the plate tectonic revolution. *Lithos* **174**, 1–14.

Evans, D. A. D. (2013) Reconstructing pre-Pangean supercontinents. *Geological Society of America Bulletin* **125**(11–12), 1735–1751.

Evans, D. A. D., and Pisarevsky, S. A. (2008) Plate tectonics on early Earth? Weighing the paleomagnetic evidence. *Geological Society of America Special Papers* **440**, 249–263.

Fedo, C. M., Sircombe, K. N., and Rainbird, R. H. (2003) Detrital Zircon Analysis of the Sedimentary Record. *Reviews in Mineralogy and Geochemistry* **53**(1), 277–303.

Feng, R., Machado, N., and Ludden, J. (1993) Lead geochronology of zircon by laser probe-inductively coupled plasma mass spectrometry (LP-ICPMS). *Geochimica et Cosmochimica Acta* **57**(14), 3479–3486.

Ferry, J. and Watson, E. (2007) New thermodynamic models and revised calibrations for the Ti-in-zircon and Zr-in-rutile thermometers. *Contributions to Mineralogy and Petrology* **154**(4), 429–437.

Fletcher, I. R., McNaughton, N. J., Davis, W. J., and Rasmussen, B. (2010) Matrix effects and calibration limitations in ion probe U–Pb and Th–Pb dating of monazite. *Chemical Geology* **270**(1–4), 31–44.

Flowers, R. M., Royden, L. H., and Bowring, S. A. (2004) Isostatic constraints on the assembly, stabilization, and preservation of cratonic lithosphere. *Geology* **32**(4), 321–324.

Foster, G., Parrish, R. R., Horstwood, M. S., Chenery, S., Pyle, J., and Gibson, H. (2004) The generation of prograde P–T–t points and paths; a textural, compositional, and chronological study of metamorphic monazite. *Earth and Planetary Science Letters* **228**(1), 125–142.

Fraser, G., Ellis, D., and Eggins, S. (1997) Zirconium abundance in granulite-facies minerals, with implications for zircon geochronology in high-grade rocks. *Geology* **25**(7), 607–610.

Friend, C. R. L. and Kinney, P. D. (1995) New evidence for protolith ages of Lewisian granulites, northwest Scotland. *Geology* **23**(11), 1027–1030.

Frost, B. R., Chamberlain, K. R., and Schumacher, J. C. (2000) Sphene (titanite): phase relations and role as a geochronometer. *Chemical Geology* **172**, 131–148.

Froude, D. O., Ireland, T. R., Kinny, P. D., *et al.* (1983) ion microprobe identification of 4,100–4,200 Myr-old terrestrial zircons. *Nature* **304**(5927), 616–618.

Fryer, B. J., Jackson, S. E., and Longerich, H. P. (1993) The application of laser ablation microprobe-inductively coupled plasma-mass spectrometry (LAM-ICP-MS) to *in situ* (U)–Pb geochronology. *Chemical Geology* **109**(1–4), 1–8.

Fu, B., Page, F., Cavosie, A., *et al.* (2008) Ti-in-zircon thermometry: applications and limitations. *Contributions to Mineralogy and Petrology* **156**(2), 197–215.

Galer, S. and Abouchami, W. (1998) Practical application of lead triple spiking for correction of instrumental mass discrimination. *Mineralogy Magazine Series A* **62**, 491–492.

Gao, X. Y., Zheng, Y. F., Chen, Y. X., and Guo, J. (2012) Geochemical and U–Pb age constraints on the occurrence of polygenetic titanites in UHP metagranite in the Dabie orogen. *Lithos* **136–139**, 93–108.

Gehrels, G. (2011) *Detrital Zircon U–Pb Geochronology: Current Methods and New Opportunities, Tectonics of Sedimentary Basins.* John Wiley & Sons, 45–62.

Gehrels, G. (2014) Detrital zircon U–Pb geochronology applied to tectonics. *Annual Review of Earth and Planetary Sciences* **42**, 127–149.

Gehrels, G., DeCelles, P. G., Martin, A., Ojha, T. P., and Pinhassi, G. (2003) Initiation of the Himalayan orogen as an early Paleozoic thin-skinned thrust belt. *GSA Today* **13**(9), 4–9.

Gehrels, G. E., Valencia, V. A., and Ruiz, J. (2008) Enhanced precision, accuracy, efficiency, and spatial resolution of U–Pb ages by laser ablation-multicollector-inductively coupled plasma-mass spectrometry. *Geochemistry, Geophysics, Geosystems* **9**(3), Q03017.

Geisler, T., Pidgeon, R. T., van Bronswijk, W., and Kurtz, R. (2002) Transport of uranium, thorium, and lead in metamict zircon under low-temperature hydrothermal conditions. *Chemical Geology* **191**, 141–154.

Geisler, T., Pidgeon, R. T., Kurtz, R., van Bronswijk, W., and Schleicher, H. (2003) Experimental hydrothermal alteration of partially metamict zircon. *American Mineralogist* **88**(10), 1496–1513.

Gelman, S. E., Gutiérrez, F. J., and Bachmann, O. (2013) On the longevity of large upper crustal silicic magma reservoirs. *Geology* **41**(7), 759–762.

Gerling, E. (1942) Age of the Earth according to radioactivity data. *Proceedings Doklady (Proceedings of the Russian Academy of Science)* **34**, 259–261.

Gibson, H. D., Carr, S. D., Brown, R. L., and Hamilton, M. A. (2004) Correlations between chemical and age domains in monazite, and metamorphic reactions involving major pelitic phases: an integration of ID-TIMS and SHRIMP geochronology with Y–Th–U X-ray mapping. *Chemical Geology* **211**(3–4), 237–260.

Glazner, A. F., Bartley, J. M., Coleman, D. S., Gray, W., and Taylor, R. Z. (2004) Are plutons assembled over millions of year by amalgamation from small magma chambers?. *GSA Today* **14**(4), 4–11.

Goldich, S. and Mudrey, M. (1972) Dilatancy model for discordant U–Pb zircon ages. In *Contributions to Recent Geochemistry and Analytical Chemistry*, Tugarinov, A. I. (ed.), 415–418. Nauka.

Gordon, S. M., Bowring, S. A., Whitney, D. L., Miller, R. B., and McLean, N. (2010) Time scales of metamorphism, deformation, and crustal melting in a continental arc, North Cascades, USA. *Geological Society of America Bulletin*. DOI: 10.1130/B30060.1.

Gothmann, A. M., Stolarski, J., Adkins, J. F., *et al.* (2015) Fossil corals as an archive of secular variations in seawater chemistry since the Mesozoic. *Geochimica et Cosmochimica Acta* **160**, 188–208.

Gradstein, F. M., Ogg, J. G., and Smith, A. G. (2004) *A Geologic Time Scale 2004*. Cambridge University Press.

Gradstein, F. M., Ogg, J. G., Schmitz, M. D., and Ogg, G. M. (eds). (2012) *The Geologic Time Scale 2012*, Vols I and II. Elsevier.

Grandia, F., Asmerom, Y., Getty, S., Cardellach, E., and Canals, A. (2000) U–Pb dating of MVT ore-stage calcite: implications for fluid flow in a Mesozoic extensional basin from Iberian Peninsula: Journal of Geochemical Exploration 69), 377–380.

Grange, M. L., Nemchin, A. A., Timms, N., Pidgeon, R. T., and Meyer, C. (2011) Complex magmatic and impact history prior to 4.1 Ga recorded in zircon from Apollo 17 South Massif aphanitic breccia 73235. *Geochimica et Cosmochimica Acta* **75**, 2213–2232.

Grimes, C. B., John, B. E., Kelemen, P. B., *et al.* (2007) Trace element chemistry of zircons from oceanic crust: A method for distinguishing detrital zircon provenance. *Geology* **35**, 643–646. DOI: 610.1130/G23603A.23601.

Grove, M. and Harrison, T. M. (1999) Monazite Th–Pb age depth profiling. *Geology* **27**(6), 487–490.

Gualda, G. A., Pamukcu, A. S., Ghiorso, M. S., Anderson Jr, A. T., Sutton, S. R., and Rivers, M. L. (2012) Timescales of quartz crystallization and the longevity of the Bishop giant magma body. *PLoS One* **7**(5), e37492.

Guitreau, M., Blichert-Toft, J., Mojzsis, S. J., *et al.* (2014) Lu–Hf isotope systematics of the Hadean–Eoarchean Acasta Gneiss Complex (Northwest Territories, Canada). *Geochimica et Cosmochimica Acta* **135**, 251–269.

Hanchar, J. M. and Miller, C. F. (1993) Zircon zonation patterns as revealed by cathodoluminescence and backscattered electron images: implications for interpretation of complex crustal histories. *Chemical Geology* **110**, 1–13.

Hanchar, J. M. and Rudnick, R. (1995) Revealing hidden structures: the application of cathodoluninescence and back-scattered electron imaging to dating zircons from lower crustal xenoliths. *Lithos* **36**, 289–303.

Hanchar, J. M. and van Westrenen, W. (2007) Rare earth element behavior in zircon-melt systems. *Elements* **3**(1), 37–42.

Hanson, G. N., Catanzaro, E. J., and D.H., A. (1971) U–Pb for sphene in a contact metamorphic zone. *Earth and Planetary Science Letters* **12**, 231–237.

Harlan, S. S., Heaman, L., LeCheminant, A. N., and Premo, W. R. (2003) Gunbarrel mafic magmatic event: a key 780 Ma time marker for Rodinia plate reconstructions. *Geology* **31**(12), 1053–1056.

Harland, W. B. (1990) *A geologic time scale 1989*, Cambridge University Press.

Harley, S. L. and Kelly, N. M. (2007) The impact of zircon-garnet REE distribution data on the interpretation of zircon U–Pb ages in complex high-grade terrains: an example from the Rauer Islands, East Antarctica. *Chemical Geology* **241**(1–2), 62–87.

Harley, S. L., Kelly, N. M., and Möller, A. (2007) Zircon behaviour and the thermal history of mountain belts. *Elements* **3**(1), 25–30. DOI: 10.2113/gselements.2113.2111.2125.

Harrison, T. M. (2009) The Hadean crust: evidence from > 4 Ga zircons. *Annual Review of Earth and Planetary Sciences* **37**, 479–505.

Harrison, T., Blichert-Toft, J., Müller, W., Albarede, F., Holden, P., and Mojzsis, S. (2005) Heterogeneous Hadean hafnium: evidence of continental crust at 4.4 to 4.5 Ga. *Science* **310**(5756), 1947–1950.

Harrison, T. M., Schmitt, A. K., McCulloch, M. T., and Lovera, O. M. (2008) Early (≥ 4.5 Ga) formation of terrestrial crust: Lu–Hf, $\delta^{18}O$, and Ti thermometry results for Hadean zircons. *Earth and Planetary Science Letters* **268**(3–4), 476–486.

Harrison, T.M., Bell, E.A., and Boehnke, P. (2017) Hadean zircon petrochronology. *Reviews in Mineralogy and Geochemistry* **83**(1), 329–363.

Hawkesworth, C. J. and Kemp, A. I. S. (2006) Evolution of the continental crust. *Nature* **443**(19), 811–817.

Hawkins, D. P. and Bowring, S. A. (1999) U–Pb monzaite, xenotime and titanite geochronological constraints on the prograde to post-peak metamorphic thermal history of Paleoproterozoic migmatites from the Grand Canyon, Arizona. *Contributions to Mineralogy and Petrology* **134**, 150–169.

Hawkesworth, C. J., Dhuime, B., Pietranik, A. B., Cawood, P. A., Kemp, A. I. S., and Storey, C. D. (2010) The generation and evolution of the continental crust. *Journal of the Geological Society* **167**(2), 229–248.

He, S., Li, R., Wang, C., *et al.* (2011) Discovery of ∼ 4.0 Ga detrital zircons in the Changdu Block, North Qiangtang, Tibetan Plateau. *Chinese Science Bulletin* **56**(7), 647–658.

Heaman, L. M. (1997) Global mafic magmatism at 2.45 Ga: remnants of an ancient large igneous province? *Geology* **25**(4), 299–302.

Heaman, L. M. and LeCheminant, A. N. (1993) Geochemistry of Accessory Minerals Paragenesis and U–Pb systematics of baddeleyite (ZrO_2). *Chemical Geology* **110**(1), 95–126.

Heaman, L. and Tarney, J. (1989) U–Pb baddeleyite ages for the Scourie dyke swarm, Scotland: evidence for two distinct intrusion events. *Nature* **340**, 705–708

Heaman, L. M., LeCheminant, A. N., and Rainbird, R. H. (1992) Nature and timing of Franklin igneous events, Canada: implications for a Late Proterozoic mantle plume and the break-up of Laurentia. *Earth and Planetary Science Letters* **109**(1), 117–131.

Heizler, M. T., Ralser, S., and Karlstrom, K. E. (1997) Late Proterozoic (Grenville?) deformation in central New Mexico determined from single-crystal muscovite $^{40}Ar/^{39}Ar$ age spectra. *Precambrian Research* **84**(1–2), 1–15.

Hermann, J., Rubatto, D., Korsakov, A., and Shatsky, V. S. (2001) Multiple zircon growth during fast exhumation of diamondiferous, deeply subducted continental crust (Kokchetav Massif, Kazakhstan). *Contributions to Mineralogy and Petrology* **141**(1), 66–82.

Hervé, F., Fanning, C. M., and Pankhurst, R. J. (2003) Detrital zircon age patterns and provenance of the metamorphic complexes of southern Chile. *Journal of South American Earth Sciences* **16**(1), 107–123.

Hiess, J., Condon, D. J., McLean, N., and Noble, S. R. (2012) $^{238}U/^{235}U$ systematics in terrestrial uranium-bearing minerals. *Science* **335**(6076), 1610–1614.

Hietpas, J., Samson, S., Moecher, D., and Schmitt, A. K. (2010) Recovering tectonic events from the sedimentary record: Detrital monazite plays in high fidelity. *Geology* **38**(2), 167–170.

Hietpas, J., Samson, S., Moecher, D., and Chakraborty, S. (2011) Enhancing tectonic and provenance information from detrital zircon studies: assessing terrane-scale sampling and grain-scale characterization. *Journal of the Geological Society* **168**(2), 309–318.

Hinthorne, J. R., Andersen, C. A., Conrad, R. L., and Lovering, J. F. (1979) Single-grain $^{207}Pb^{206}Pb$ and U/Pb age determinations with a 10-μm spatial resolution using the ion microprobe mass analyzer (IMMA). *Chemical Geology* **25**(4), 271–303.

Hirata, T. and Nesbitt, R. W. (1995) U–Pb isotope geochronology of zircon: evaluation of the laser probe-inductively coupled plasma mass spectrometry technique. *Geochimica et Cosmochimica Acta* **59**(12), 2491–2500.

Hodges, K. V. and Bowring, S. A. (1995) 40Ar/39Ar thermochronology of isotopically zoned micas: Insights from the southwestern USA Proterozoic orogen. *Geochimica et Cosmochimica Acta* **59**(15), 3205–3220.

Hoff, J. A., Jameson, J., and Hanson, G. N. (1995) Application of Pb isotopes to the absolute timing of regional exposure events in carbonate rocks: An example from U–rich dolostones from the Wahoo Formation (Pennsylvanian), Prudhoe Bay, Alaska. *Journal of Sedimentary Research* **65**(1).

Hofmann, A. W. (2007) Sampling mantle heterogeneity through oceanic basalts: isotopes and trace elements. In Treatise on Geochemistry, *Vol. 2*, Carlson, R. W. (ed.), 1–44. Oxford, Pergamon.

Holden, P., Lanc, P., Ireland, T. R., Harrison, T. M., Foster, J. J., and Bruce, Z. (2009) Mass-spectrometric mining of Hadean zircons by automated SHRIMP multi-collector and single-collector U/Pb zircon age dating: the first 100,000 grains. *International Journal of Mass Spectrometry* **286**(2), 53–63.

Holder, R. M., Hacker, B. R., Kylander-Clark, A. R., and Cottle, J. M. (2015) Monazite trace-element and isotopic signatures of (ultra) high-pressure metamorphism: examples from the Western Gneiss Region, Norway. *Chemical Geology* **409**, 99–111.

Holmes, A. (1911) The association of lead with uranium in rock-minerals and its application to the measurement of geological time. *Proceedings of the Royal Society of London* **85**, 248–256. DOI: 210.1098/rspa.1911.0036.

Holmes, A. (1946) An estimate of the age of the earth. *Nature* **157**, 680–684.

Holmes, A. (1947) VII—the construction of a geological time-scale. *Transactions of the Geological Society of Glasgow* **21**(1), 117–152.

Holmes, A. (1959) A revised geological time-scale. *Transactions of the Edinburgh Geological Society* **17**(3), 183–216.

Hopkins, M., Harrison, T. M., and Manning, C. E. (2008) Low heat flow inferred from > 4 Gyr zircons suggests Hadean plate boundary interactions. *Nature* **456**(7221), 493–496.

Hopkins, M. D., Harrison, T. M., and Manning, C. E. (2010) Constraints on Hadean geodynamics from mineral inclusions in > 4 Ga zircons. *Earth and Planetary Science Letters* **298**(3–4), 367–376.

Hora, J. M., Singer, B. S., Jicha, B. R., *et al.* (2010) Volcanic biotite-sanidine $^{40}Ar/^{39}Ar$ age discordances reflect Ar partitioning and pre-eruption closure in biotite. *Geology* **38**(10), 923–926.

Horn, I. and von Blanckenburg, F. (2007) Investigation on elemental and isotopic fractionation during 196 nm femtosecond laser ablation multiple collector inductively coupled plasma mass spectrometry. *Spectrochimica Acta Part B: Atomic Spectroscopy* **62**(4), 410–422.

Horstwood, M. S. A. (2008) Data reduction strategies, uncertainty assessment and resolution of LA-(MC-)ICP-MS isotope data. In *Laser Ablation ICP-MS in the Earth Sciences: Current Practives and Outstanding Issues*, Vol.40, Sylvester, P. J. (ed.), 283–303. Short Course Series, Mineralogical Association of Canada, Quebec.

Horstwood, M. S. A., Foster, G. L., Parrish, R. R., Noble, S. R., and Nowell, G. M. (2003) Common-Pb corrected *in situ* U–Pb accessory mineral geochronology by LA-MC-ICP-MS. *Journal of Analytical Atomic Spectrometry* **18**(8).

Horstwood, M. S. A., Košler, J., Gehrels, G., *et al.* (2016) Community-derived standards for LA-ICP-MS U–Th–Pb geochronology—uncertainty propagation, age interpretation and data reporting. *Geostandards and Geoanalytical Research* **40**(3), 311–332.

Hoskin, P. W. O. and Ireland, T. R. (2000) Rare earth element chemistry of zircon and its use as a provenance indicator. *Geology* **28**, 627–630. DOI: 610.1130/0091-7613(2000)1128 < 1627: REECOZ > 1132.1130.CO;1132.

Housh, T., and Bowring, S. A. (1991) Lead isotopic heterogeneities within alkali feldspars: implications for the determination of initial lead isotopic compositions. *Geochimica et Cosmochimica Acta* **55**, 2309–2316.

Houtermans, F. G. (1946) Die Isotopenhäufigkeiten im natürlichen Blei und das Alter des Urans. *Naturwissenschaften* **33**(6), 185–186.

Huber, C., Bachmann, O., and Dufek, J. (2012) Crystal-poor versus crystal-rich ignimbrites: a competition between stirring and reactivation. *Geology* **40**(2), 115–118.

Ickert, R. B., Mundil, R., Magee, C. W., and Mulcahy, S. R. (2015) The U–Th–Pb systematics of zircon from the Bishop Tuff: a case study in challenges to high-precision Pb/U geochronology at the millennial scale. *Geochimica et Cosmochimica Acta* **168**, 88–110.

Iizuka, T., Horie, K., Komiya, T., *et al.* (2006) 4.2 Ga zircon xenocryst in an Acasta gneiss from northwestern Canada: evidence for early continental crust. *Geology* **34**(4), 245–248.

Iizuka, T., Komiya, T., Johnson, S. P., Kon, Y., Maruyama, S., and Hirata, T. (2009) Reworking of Hadean crust in the Acasta gneisses, northwestern Canada: evidence from *in-situ* Lu–Hf isotope analysis of zircon. *Chemical Geology* **259**(3–4), 230–239.

Ireland, T. R., and Williams, I. S. (2003) Considerations in zircon geochronology by SIMS. In Zircon, *Vol. 53*, Hanchar, J. M. and Hoskin, P. W. O. (eds), 215–241. Mineralogical Society of America, Washington, DC.

Ireland, T. R., Flöttmann, T., Fanning, C. M., Gibson, G. M., and Preiss, W. V. (1998) Development of the early Paleozoic Pacific margin of Gondwana from detrital-zircon ages across the Delamerian orogen. *Geology* **26**(3), 243–246.

Ishizaka, K. and Yamaguchi, M. (1969) U–Th–Pb ages of sphene and zircon from the Hida metamorphic terrain, Japan. *Earth and Planetary Science Letters* **6**(3), 179–185.

Jackson, M., Hart, S., Konter, J., Kurz, M., Blusztajn, J., and Farley, K. (2014) Helium and lead isotopes reveal the geochemical geometry of the Samoan plume. *Nature* **514**, 355–358.

Jaffey, A. H., Flynn, K. F., Glendenin, L. E., Bentley, W. C., and Essling, A. M. (1971) Precision measurement of half-lives and specific activities of ^{235}U and ^{238}U. *Physics Reviews* **C4**, 1889–1906.

Jahn, B.-M. and Cuvellier, H. (1994) Pb–Pb and U–Pb geochronology of carbonate rocks: an assessment. *Chemical Geology* **115**(1), 125–151.

Jeon, H. and Whitehouse, M. J. (2015) A Critical Evaluation of U–Pb Calibration Schemes Used in SIMS Zircon Geochronology. *Geostandards and Geoanalytical Research* **39**(4), 443–452.

Jicha, B. R., Singer, B. S., and Sobol, P. (2016) Re-evaluation of the ages of ^{40}Ar/^{39}Ar sanidine standards and supereruptions in the western US using a Noblesse multi-collector mass spectrometer. *Chemical Geology* **431**, 54–66.

Johnstone, J. H. L. and Boltwood, B. B. (1920) The relative activity of radium and the uranium with which it is in radioactive equilibrium: *American Journal of Science* **295**, 1–19.

Kamo, S. L., Czamanske, G. K., and Krogh, T. E. (1996) A minimum U–Pb age for Siberian flood-basalt volcanism. *Geochimica et Cosmochimica Acta* **60**(18), 3505–3511.

Kamo, S., Czamanske, G. K., Amelin, Y., Fedorenko, V. A., Davis, D. W., and Trofimov, V. R. (2003) Rapid eruption of Siberian flood-volcanic rocks and evidence for coincidence with the Permian–Triassic boundary and mass extinction at 251 Ma. *Earth and Planetary Science Letters* **214**, 75–91.

Katayama, I., Maruyama, S., Parkinson, C. D., Terada, K., and Sano, Y. (2001) Ion micro-probe U–Pb zircon geochronology of peak and retrograde stages of ultrahigh-pressure metamorphic rocks from the Kokchetav massif, northern Kazakhstan. *Earth and Planetary Science Letters* **188**(1–2), 185–198.

Kelley, S. (2002) Excess argon in K–Ar and Ar–Ar geochronology. *Chemical Geology* **188**(1), 1–22.

Kelly, N. M. and Harley, S. L. (2005) An integrated microtextural and chemical approach to zircon geochronology: refining the Archaean history of the Napier Complex, east Antarctica.

Contributions to Mineralogy and Petrology **149**, 57–84. DOI: 10.1007/s00410-00004-00635-00416.

Kelly, S. D., Newville, M. G., Cheng, L., *et al.* (2003) Uranyl incorporation in natural calcite. *Environmental Science and Technology* **37**(7), 1284–1287.

Kelly, S. D., Rasbury, E. T., Chattopadhyay, S., Kropf, A. J., and Kemner, K. M. (2006) Evidence of a stable uranyl site in ancient organic-rich calcite. *Environmental Science and Technology* **40**(7), 2262–2268.

Kemp, A. I. S. and Hawkesworth, C. J. (2014) Growth and differentiation of the continental crust from isotope studies of accessory minerals A2. In *Treatise on Geochemistry*, 2nd edn, Turekian, K. K. (ed.), 379–421. Elsevier, Oxford.

Kemp, A., Hawkesworth, C., Foster, G., *et al.* (2007) Magmatic and crustal differentiation history of granitic rocks from Hf–O isotopes in zircon. *Science* **315**(5814), 980–983.

Kemp, A. I. S., Wilde, S. A., Hawkesworth, C. J., *et al.* (2010) Hadean crustal evolution revisited: new constraints from Pb–Hf isotope systematics of the Jack Hills zircons. *Earth and Planetary Science Letters* **296**(1–2), 45–56.

Ketcham, R. A., Donelick, R. A., and Carlson, W. D. (1999) Variability of apatite fission-track annealing kinetics; III, extrapolation to geological time scales. *American Mineralogist* **84**(9), 1235–1255.

Kimura, J.-I., Chang, Q., and Tani, K. (2011) Optimization of ablation protocol for 200 nm UV femtosecond laser in precise U–Pb age dating coupled to multi-collector ICP mass spectrometry. *Geochemical Journal* **45**(4), 283–296.

Kinny, P., Griffin, B., Heaman, L., and Spetsius, Z. (1997) SHRIMP U–Pb ages of perovskite from Yakutian kimberlites. *Russian Geology and Geophysics C/C of Geologiia I Geofizika* **38**, 97–105.

Knoll, A. H., Bambach, R. K., Payne, J. L., Pruss, S., and Fischer, W. W. (2007) Paleophysiology and end-Permian mass extinction. *Earth and Planetary Science Letters* **256**(3), 295–313.

Kohn, M. J. and Corrie, S. L. (2011) Preserved Zr-temperatures and U–Pb ages in high-grade metamorphic titanite: evidence for a static hot channel in the Himalayan orogen. *Earth and Planetary Science Letters* **311**(1–2), 136–143.

Kohn, M. J., and Malloy, M. A. (2004) Formation of monazite via prograde metamorphic reactions among common silicates: implications for age determinations. *Geochimica et Cosmochimica Acta* **68**(1), 101–113.

Kooijman, E., Mezger, K., and Berndt, J. (2010) Constraints on the U–Pb systematics of metamorphic rutile from in situ LA-ICP-MS analysis. *Earth and Planetary Science Letters* **293**(3–4), 321–330.

Košler, J. and Sylvester, P. J. (2003) Present trends and the future of zircon in geochronology: laser ablation ICPMS. In Zircon, *Vol. 53*, Hanchar, J. M. and Hoskin, P. W. O. (eds), 243–275. Mineralogical Society of America, Washington, DC.

Košler, J., Sláma, J., Belousova, E., *et al.* (2013) U-Pb detrital zircon analysis–results of an inter-laboratory comparison. *Geostandards and Geoanalytical Research* **37**(3), 243–259.

Kramers, J. D. and Tolstikhin, I. N. (1997) Two terrestrial lead isotope paradoxes, forward transport modelling, core formation and the history of the continental crust. *Chemical Geology* **139**(1), 75–110.

Krogh, T. E. (1982) Improved accuracy of U–Pb zircon ages by the creation of more concordant systems using an air abrasion technique. *Geochimica et Cosmochimica Acta* **46**, 637–649.

Krogh, T. and Davis, G. (1974) Alteration in zircons with discordant U–Pb ages. *Carnegie Institution of Washington Yearbook* **73**, 560–567.

Krogh, T. E. and Davis, G. L. (1975) Alteration in zircons and differential dissolution of altered and metamict zircon. *Carnegie Institution of Washington Yearbook* **74**, 619–623.

Krogh, T., Corfu, F., Davis, D., *et al.* (1987) Precise U–Pb isotopic ages of diabase dykes and mafic to ultramafic rocks using trace amounts of baddeleyite and zircon. In *Mafic Dyke Swarms*, Halls, H. C. and Fahrig, W. F. (eds), 147–152. Special Paper 34, Geological Association of Canada

Krogstad, E. J. and Walker, R. J. (1994) High closure temperatures of the U–Pb system in large apatites from the Tin Mountain pegmatite, Black Hills, South Dakota, USA. *Geochimica et Cosmochimica Acta* **58**(18), 3845–3853.

Kryza, R., Crowley, Q. G., Larionov, A., Pin, C., Oberc-Dziedzic, T., and Mochnacka, K. (2012) Chemical abrasion applied to SHRIMP zircon geochronology: an example from the Variscan Karkonosze Granite (Sudetes, SW Poland). *Gondwana Research* **21**(4), 757–767.

Kylander-Clark, A. R. C., Hacker, B. R., and Mattinson, J. M. (2008) Slow exhumation of UHP terranes: titanite and rutile ages of the Western Gneiss Region, Norway. *Earth and Planetary Science Letters* **272**(3–4), 531–540.

Kylander-Clark, A. R., Hacker, B. R., Johnson, C. M., Beard, B. L., and Mahlen, N. J. (2009) Slow subduction of a thick ultrahigh-pressure terrane. *Tectonics* **28**(2). DOI:10.1029/2007TC002251.

Kylander-Clark, A. R. C., Hacker, B. R., and Cottle, J. M. (2013) Laser-ablation split-stream ICP petrochronology. *Chemical Geology* **345**, 99–112.

LaMaskin, T. A. (2012) Detrital zircon facies of Cordilleran terranes in western North America. *GSA Today* **22**(3), 4–11. DOI: 10.1130/GSATG1142A.1131.

Lancaster, P. J., Storey, C. D., Hawkesworth, C. J., and Dhuime, B. (2011) Understanding the roles of crustal growth and preservation in the detrital zircon record. *Earth and Planetary Science Letters* **305**(3–4), 405–412.

Langmuir, D. (1978) Uranium solution-mineral equilibria at low temperatures with applications to sedimentary ore deposits. *Geochimica et Cosmochimica Acta* **42**(6), 547–569.

Le Roux, L. J. and Glendenin, L. E. (1963) Half-life of [232]Th. *Proceedings of the National Meeting on Nuclear Energy*, Pretoria, South Africa; 83–94.

Lee, J. K. W. (1995) Multipath diffusion in geochronology. *Contributions to Mineralogy and Petrology* **120**, 60–82.

Lee, J. K. W. (1997) Pb, U, and Th diffusion in natural zircon. *Nature* **390**, 159–162.

Lewis, C. (2002) *The Dating Game: One Man's Search for the Age of the Earth*. Cambridge University Press.

Li, Q., Parrish, R. R., Horstwood, M. S. A., and McArthur, J. M. (2014) U–Pb dating of cements in Mesozoic ammonites. *Chemical Geology* **376**, 76–83.

Li, Z.-X., Bogdanova, S., Collins, A., *et al.* (2008) Assembly, configuration, and break-up history of Rodinia: a synthesis. *Precambrian Research* **160**(1), 179–210.

Liati, A. and Gebauer, D. (1999) Constraining the prograde and retrograde *P–T–t* path of Eocene HP rocks by SHRIMP dating of different zircon domains: inferred rates of heating, burial, cooling and exhumation for central Rhodope, northern Greece. *Contributions to Mineralogy and Petrology* **135**(4), 340–354.

Lissenberg, C. J., Rioux, M., Shimizu, N., Bowring, S. A., and Mével, C. (2009) Zircon dating of oceanic crustal accretion. *Science* **323**(5917), 1048–1050. DOI: 1010.1126/science.1167330.

Lo Pò, D., Braga, R., Massonne, H. J., Molli, G., Montanini, A., and Theye, T. (2016) Fluid-induced breakdown of monazite in medium-grade metasedimentary rocks of the Pontremoli basement (Northern Apennines, Italy). *Journal of Metamorphic Geology* **34**(1), 63–84.

Long, S., McQuarrie, N., Tobgay, T., Rose, C., Gehrels, G., and Grujic, D. (2011) Tectonostratigraphy of the Lesser Himalaya of Bhutan: Implications for the along-strike stratigraphic continuity of the northern Indian margin. *Geological Society of America Bulletin* **123**(7–8), 1406–1426.

Ludwig, K. R. (1980) Calculation of uncertainties of U–Pb isotope data. *Earth and Planetary Science Letters* **46**, 212–220.

Ludwig, K. R. (1998) On the treatment of concordant uranium-lead ages. *Geochimica et Cosmochimica Acta* **62**(4), 665–676.

Lundstrom, C. C. and Glazner, A. F. (eds). (2016) Enigmatic relationship between silicic volcanic and plutonic rocks: silicic magmatism and the volcanic–plutonic connection. *Elements*, **12**(2).

Maas, R., Kinny, P. D., Williams, I. S., Froude, D. O., and Compston, W. (1992) The Earth's oldest known crust: a geochronological and geochemical study of 3900–4200 Ma old detrital zircons from Mt. Narryer and Jack Hills, Western Australia. *Geochimica et Cosmochimica Acta* **56**(3), 1281–1300.

Mahan, K. H., Bartley, J. M., Coleman, D. S., Glazner, A. F., and Carl, B. S. (2003) Sheeted intrusion of the synkinematic McDoogle pluton, Sierra Nevada, California. *Geological Society of America Bulletin* **115**(12), 1570–1582.

Mahan, K. H., Goncalves, P., Williams, M. L., and Jercinovic, M. J. (2006) Dating metamorphic reactions and fluid flow: application to exhumation of high-P granulites in a crustAl–scale shear zone, western Canadian Shield. *Journal of Metamorphic Geology* **24**(3), 193–217.

Maier, A. C., Cates, N. L., Trail, D., and Mojzsis, S. J. (2012) Geology, age and field relations of Hadean zircon-bearing supracrustal rocks from Quad Creek, eastern Beartooth Mountains (Montana and Wyoming, USA). *Chemical Geology* **312**, 47–57.

Malusà, M. G., Carter, A., Limoncelli, M., Villa, I. M., and Garzanti, E. (2013) Bias in detrital zircon geochronology and thermochronometry. *Chemical Geology* **359**, 90–107.

Marsh, B. D. (1982) On the mechanics of igneous diapirism, stoping, and zone melting. *American Journal of Science* **282**(6), 808–855.

Mattinson, J. M. (1973) Anomalous isotopic composition of lead in young zircons: *Carnegie Institute of Washington Yearbook* **72**, 613–616.

Mattinson, J. M. (1978) Age, origin, and thermal histories of some plutonic rocks from the Salinian block of California. *Contributions to Mineralogy and Petrology* 67(3), 233–245.

Mattinson, J. M. (1987) U–Pb ages of zircons: a basic examination of error propagation. *Chemical Geology* 66, 151–162.

Mattinson, J. M. (2000) Revising the "gold standard"—the uranium decay constants of Jaffey *et al.* (1971). *Eos, Transactions of the American Geophysical Union, Spring Meeting Supplement,* Abstract V61A-02.

Mattinson, J. M. (2005) Zircon U–Pb chemicAl–abrasion ("CA-TIMS") method: combined annealing and multi-step dissolution analysis for improved precision and accuracy of zircon ages. *Chemical Geology* 220(1–2), 47–56.

Mattinson, J. M. (2010) Analysis of the relative decay constants of ^{235}U and ^{238}U by multi-step CA-TIMS measurements of closed-system natural zircon samples. *Chemical Geology* 275 (3–4), 186–198.

Mattinson, J. M. (2013) Revolution and evolution: 100 years of U–Pb geochronology. *Elements* 9(1), 53–57.

Matzel, J. P., Bowring, S. A., and Miller, R. B. (2006) Timescales of pluton construction at differing crustal levels: examples from the Mount Stuart and Tenpeak intrusions, North Cascades, WA. *Geological Society of America Bulletin* 118(11), 1412–1430. DOI: 1410.1130/B25923.25921.

McLean, N. M., Bowring, J. F., and Bowring, S. A. (2011) An algorithm for U–Pb isotope dilution data reduction and uncertainty propagation. *Geochemistry, Geophysics, Geosystems* 12, Q0AA18.

Meldrum, A., Boatner, L. A., Weber, W. J., and Ewing, R. C. (1998) Radiation damage in zircon and monazite. *Geochimica et Cosmochimica Acta* 62(14), 2509–2520.

Memeti, V., Paterson, S., Matzel, J., Mundil, R., and Okaya, D. (2010) Magmatic lobes as "snapshots" of magma chamber growth and evolution in large, composite batholiths: an example from the Tuolumne intrusion, Sierra Nevada, California. *Geological Society of America Bulletin* 122, 1912–1931. DOI: 1910.1130/B30004.

Meyer, M., Cliff, R., Spötl, C., Knipping, M., and Mangini, A. (2009) Speleothems from the earliest Quaternary: snapshots of paleoclimate and landscape evolution at the northern rim of the Alps. *Quaternary Science Reviews* 28(15), 1374–1391.

Mezger, K. and Krogstad, E. J. (1997) Interpretation of discordant U–Pb zircon ages. *An evaluation. Journal of Metamorphic Geology* 15, 127–140.

Mezger, K., Hanson, G. N., and Bohlen, S. R. (1989) High-precision U–Pb ages of metamorphic rutile: application to the cooling history of high-grade terranes. *Earth and Planetary Science Letters* 96, 106–118.

Mezger, K., Rawnsley, C. M., Bohlen, S. R., and Hanson, G. N. (1991) U–Pb garnet, sphene, monazite, and rutile ages: implications for the duration of high-grade metamorphism and cooling histories, Adirondack, Mts., New York. *Journal of Geology* 99, 415–428.

Mezger, K., Essene, E., Van der Pluijm, B., and Halliday, A. (1993) U–Pb geochronology of the Grenville Orogen of Ontario and New York: constraints on ancient crustal tectonics. *Contributions to Mineralogy and Petrology* 114(1), 13–26.

Miller, C. F. and Wark, D. A. (eds) (2008) Supervolcanoes. *Elements,* **4**.

Miller, J. S., Matzel, J. P., Miller, C. F., Burgess, S. D., and Miller, R. B. (2007) Zircon growth and recycling during the assembly of large, composite arc plutons. *Journal of Volcanology and Geothermal Research* 167, 282–299.

Miller, R. B. and Paterson, S. R. (1999) In defense of magmatic diapirs. *Journal of Structural Geology* 21, 1161–1173.

Mills, R. D. and Coleman, D. S. (2013) Temporal and chemical connections between plutons and ignimbrites from the Mount Princeton magmatic center. *Contributions to Mineralogy and Petrology* 165(5), 961–980.

Min, K., Mundil, R., Renne, P. R., and Ludwig, K. R. (2000) A test for systematic errors in 40Ar/39Ar geochronology through comparison with U–Pb analysis of a 1.1 Ga rhyolite. *Geochimica et Cosmochimica Acta* 64, 73–98.

Moecher, D. P. and Samson, S. D. (2006) Differential zircon fertility of source terranes and natural bias in the detrital zircon record: Implications for sedimentary provenance analysis. *Earth and Planetary Science Letters* 247(3–4), 252–266.

Mojzsis, S. J. and Harrison, T. M. (2002) Establishment of a 3.83-Ga magmatic age for the Akilia tonalite (southern West Greenland). *Earth and Planetary Science Letters* 202(3–4), 563–576.

Mojzsis, S. J., Harrison, T. M., and Pidgeon, R. T. (2001) Oxygen-isotope evidence from ancient zircons for liquid water at the Earth's surface 4,300 Myr ago. *Nature* 409(6817), 178–181.

Mojzsis, S. J., Cates, N. L., Caro, G., *et al.* (2014) Component geochronology in the polyphase ca. 3920 Ma Acasta Gneiss. *Geochimica et Cosmochimica Acta* 133, 68–96.

Montel, J.-M., Foret, S., Veschambre, M. l., Nicollet, C., and Provost, A. (1996) Electron microprobe dating of monazite. *Chemical Geology* 131(1–4), 37–53.

Moorbath, S., Whitehouse, M. J., and Kamber, B. S. (1997) Extreme Nd-isotope heterogeneity in the early Archaean—fact or fiction? Case histories from northern Canada and West Greenland. *Chemical Geology* 135(3–4), 213–231.

Mueller, P., Wooden, J., and Nutman, A. (1992) 3.96 Ga zircons from an Archean quartzite, Beartooth Mountains, Montana. *Geology* 20(4), 327–330.

Mundil, R., Brack, P., Meier, M., Rieber, H., and Oberli, F. (1996) High resolution U–Pb dating of Middle Triassic volcaniclastics: time-scale calibration and verification of tuning parameters for carbonate sedimentation. *Earth and Planetary Science Letters* 141(1), 137–151.

Mundil, R., Ludwig, K. R., Metcalfe, I., and Renne, P. R. (2004) Age and timing of the Permian mass extinctions: U/Pb dating of closed-system zircons. *Science* 305(5691), 1760–1763.

Murphy, J. B., Fernandez-Suarez, J., Keppie, J. D., and Jeffries, T. E. (2004) Contiguous rather than discrete Paleozoic histories for the Avalon and Meguma terranes based on detrital zircon data. *Geology* 32(7), 585–588.

Myrow, P. M., Hughes, N. C., Searle, M. P., Fanning, C. M., Peng, S. C., and Parcha, S. K. (2009) Stratigraphic correlation of Cambrian-Ordovician deposits along the Himalaya: Implications for the age and nature of rocks in the Mount Everest region. *Geological Society of America Bulletin* 121(3–4), 323–332.

Nadeau, S., Chen, W., Reece, J., *et al.* (2013) Guyana: the Lost Hadean crust of South America? *Brazilian Journal of Geology* **43**(4), 601–606.

Nasdala, L., Pidgeon, R. T., and Wolf, D. (1996) Heterogeneous metamictization of zircon on a microscale. *Geochimica et Cosmochimica Acta* **60**(6), 1091–1097.

Nasdala, L., Pidgeon, R. T., Wolf, D., and Irmer, G. (1998) Metamictization and U–PB isotopic discordance in single zircons: a combined Raman microprobe and SHRIMP ion probe study. *Mineralogy and Petrology* **62**(1), 1–27.

Nebel, O., Scherer, E. E., and Mezger, K. (2011) Evaluation of the ^{87}Rb decay constant by age comparison against the U–Pb system. *Earth and Planetary Science Letters* **301**(1–2), 1–8.

Nemchin, A. A. and Pidgeon, R. T. (1999) U–Pb ages on titanite and apatite from the Darling Range granite: Implications for Late Archaean history of the southwestern Yilgarn Craton. *Precambrian Research* **96**(1–2), 125–139.

Nemchin, A., Pidgeon, R., and Whitehouse, M. (2006) Re-evaluation of the origin and evolution of > 4.2 Ga zircons from the Jack Hills metasedimentary rocks. *Earth and Planetary Science Letters* **244**(1), 218–233.

Nier, A. O., Thompson, R. W., and Murphey, B. F. (1941) The isotopic constitution of lead and the measurement of geological time. III. *Physical Review* **60**(2), 112.

O'Neill, C., Lenardic, A., Moresi, L., Torsvik, T. H., and Lee, C. T. A. (2007) Episodic Precambrian subduction. *Earth and Planetary Science Letters* **262**(3–4), 552–562.

Oberli, F., Meier, M., Berger, A., Rosenberg, C. L., and Giere, R. (2004) U–Th–Pb and ^{230}Th/^{238}U disequilibrium isotope systematics: Precise accessory mineral chronology and melt evolution tracing in the Alpine Bergell intrusion. *Geochimica et Cosmochimica Acta* **68**(11), 2543–2560.

Olsson, J. R., Söderlund, U., Klausen, M. B., and Ernst, R. E. (2010) U–Pb baddeleyite ages linking major Archean dyke swarms to volcanic-rift forming events in the Kaapvaal craton (South Africa), and a precise age for the Bushveld Complex. *Precambrian Research* **183**(3), 490–500.

Oosthuyzen, E. J., and Burger, A. J. (1973) The suitability of apatite as an age indicator by the uranium-lead isotope method. *Earth and Planetary Science Letters* **18**, 29–36.

Paquette, J. L., Barbosa, J. S. F., Rohais, S., *et al.* (2015) The geological roots of South America: 4.1 Ga and 3.7 Ga zircon crystals discovered in N.E. Brazil and N.W. Argentina. *Precambrian Research* **271**, 49–55.

Parrish, R. R. (1990) U–Pb dating of monazite and its application to geological problems. *Canadian Journal of Earth Science* **27**, 1431–1450.

Parrish, R. R. and Noble, S. R. (2003) Zircon U–Th–Pb geochronology by isotope dilution – thermal ionization mass spectrometry (ID-TIMS). In Zircon, *Vol. 53*, Hanchar, J. M. and Hoskin, P. W. O. (eds), 183–213. Mineralogical Society of America, Washington, DC.

Paterson, S. R. and Ducea, M. N.(eds). (2015) Arc magmatic tempos: gathering the evidence. *Elements* **11**.

Patterson, C. (1956) Age of meteorites and the earth. *Geochimica et Cosmochimica Acta* **10**(4), 230–237.

Patterson, C., Tilton, G., and Inghram, M. (1955) Age of the earth. *Science* **121**(3134), 69–75.

Peck, W. H., Valley, J. W., Wilde, S. A., and Graham, C. M. (2001) Oxygen isotope ratios and rare earth elements in 3.3 to 4.4 Ga zircons: ion microprobe evidence for high δ^{18}O continental crust and oceans in the Early Archean. *Geochimica et Cosmochimica Acta* **65**(22), 4215–4229.

Peterman, E. M., Mattinson, J. M., and Hacker, B. R. (2012) Multi-step TIMS and CA-TIMS monazite U–Pb geochronology. *Chemical Geology* **312–313**, 58–73.

Petford, N., Kerr, R. C., and Lister, J. R. (1993) Dike transport of granitoid magmas. *Geology* **21**(9), 845–848.

Petford, N., Cruden, A. R., McCaffrey, K. J. W., and Vigneresse, J.-L. (2000) Granite magma formation, transport and emplacement in the Earth's crust. *Nature* **408**, 669–673.

Pidgeon, R. T., O'Neil, J. R., and Silver, L. T. (1966) Uranium and lead isotopic stability in a metamict zircon under experimental hydrothermal conditions. *Science* **154**, 1538–1540.

Pidgeon, R. T., Bosch, D., and Bruguier, O. (1996) Inherited zircon and titanite U–Pb systems in an Archaean syenite from southwestern Australia: Implications for U–Pb stability of titanite. *Earth and Planetary Science Letters* **141**(1–4), 187–198.

Pickering, R., Kramers, J. D., Partridge, T., Kodolanyi, J., and Pettke, T. (2010) U–Pb dating of calcite–aragonite layers in speleothems from hominin sites in *South Africa by MC-ICP-MS: Quaternary Geochronology* **5**(5), 544–558.

Prowatke, S. and Klemme, S. (2005) Effect of melt composition on the partitioning of trace elements between titanite and silicate melt. *Geochimica et Cosmochimica Acta* **69**(3), 695–709.

Prowatke, S. and Klemme, S. (2006a) Rare earth element partitioning between titanite and silicate melts: Henry's law revisited. *Geochimica et Cosmochimica Acta* **70**(19), 4997–5012.

Prowatke, S. and Klemme, S. (2006b) Trace element partitioning between apatite and silicate melts. *Geochimica et Cosmochimica Acta* **70**(17), 4513–4527.

Pullen, A., Ibáñez-Mejía, M., Gehrels, G. E., Ibáñez-Mejía, J. C., and Pecha, M. (2014) What happens when $n = 1000$? Creating large-n geochronological datasets with LA-ICP-MS for geologic investigations. *Journal of Analytical Atomic Spectrometry* **29**(6), 971–980.

Pupin, J. (1980) Zircon and granite petrology. *Contributions to Mineralogy and Petrology* **73**(3), 207–220.

Rainbird, R. H., Heaman, L. M., and Young, G. (1992) Sampling Laurentia: detrital zircon geochronology offers evidence for an extensive Neoproterozoic river system originating from the Grenville orogen. *Geology* **20**(4), 351–354.

Rasbury, E. T. and Cole, J. M. (2009) Directly dating geologic events: U–Pb dating of carbonates. *Review of Geophysics* **47**(3). DOI: 10.1029/2007RG000246.

Rasbury, E. T., Hanson, G. N., Meyers, W. J., Holt, W. E., Goldstein, R. H., and Saller, A. H. (1998) U–Pb dates of paleosols: constraints on late Paleozoic cycle durations and boundary ages. *Geology* **26**(5), 403–406.

Rasbury, E. T., Meyers, W. J., Hanson, G. N., Goldstein, R. H., and Saller, A. H. (2000) Relationship of uranium to petrography of caliche paleosols with application to precisely dating the time of sedimentation. *Journal of Sedimentary Research* **70**(3), 604–618.

Rasmussen, B., and Fletcher, I. R. (2004) Zirconolite: A new U–Pb chronometer for mafic igneous rocks. *Geology* **32**(9), 785–788.

Reddy, S. M., Timms, N. E., Trimby, P., Kinny, P. D., Buchan, C., and Blake, K. (2006) Crystal-plastic deformation of zircon: A defect in the assumption of chemical robustness. *Geology* **34**(4), 257–260.

Reid, M. R. (2008) How long does it take to supersize an eruption? *Elements* **4**(1), 23–28.

Reid, M. R. and Coath, C. D. (2000) In situ U–Pb ages of zircons from the Bishop Tuff: no evidence for long crystal residence times. *Geology* **28**, 443–446.

Renne, P. R. (1995) Excess ^{40}Ar in biotite and hornblende from the Noril'sk 1 intrusion, Siberia: implications for the age of the Siberian Traps. *Earth and Planetary Science Letters* **131**(3), 165–176.

Renne, P. R. (2000) 40Ar/39Ar age of plagioclase from Acapulco meteorite and the problem of systematic errors in cosmochronology. *Earth and Planetary Science Letters* **175**, 13–26.

Renne, P. R., Zichao, Z., Richards, M. A., Black, M. T., and Basu, A. R. (1995) Synchrony and causal relations between Permian-Triassic boundary crises and Siberian flood volcanism. *Science* **269**, 1413–1416.

Renne, P. R., Karner, D. B., and Ludwig, K. R. (1998) Absolute ages aren't exactly. *Science* **282**, 1840–1841.

Renne, P. R., Mundil, R., Balco, G., Min, K., and Ludwig, K. R. (2010) Joint determination of ^{40}K decay constants and ^{40}Ar/^{40}K for the Fish Canyon sanidine standard, and improved accuracy for ^{40}Ar/^{39}Ar geochronology. *Geochimica et Cosmochimica Acta* **74**(18), 5349–5367.

Richards, D. A., Bottrell, S. H., Cliff, R. A., Ströhle, K., and Rowe, P. J. (1998) U–Pb dating of a speleothem of Quaternary age. *Geochimica et Cosmochimica Acta* **62**(23), 3683–3688.

Richards, D. A. and Dorale, J. A. (2003) Uranium-series chronology and environmental applications of speleothems. *Reviews in Mineralogy and Geochemistry* **52**(1), 407–460.

Richards, J., Krogh, T., and Spooner, E. (1988) Fluid inclusion characteristics and U–Pb rutile age of late hydrothermal alteration and veining at the Musoshi stratiform copper deposit, Central African copper belt, Zaire. *Economic Geology* **83**(1), 118–139.

Richter, S., Alonso-Munoz, A., Eykens, R.,*et al.* (2008) The isotopic composition of natural uranium samples – measurements using the new $n(^{233}$U)$/n(^{236}$U) double spike IRMM-3636. *International Journal of Mass Spectrometry* **269**(1–2), 145–148.

Rioux, M., Bowring, S., Dudás, F., and Hanson, R. (2010) Characterizing the U–Pb systematics of baddeleyite through chemical abrasion: application of multi-step digestion methods to baddeleyite geochronology. *Contributions to Mineralogy and Petrology* **160**(5), 777–801.

Rioux, M., Bowring, S., Cheadle, M., and John, B. (2015) Evidence for initial excess ^{231}Pa in mid-ocean ridge zircons. *Chemical Geology* **397**, 143–156.

Rivera, T. A., Storey, M., Schmitz, M. D., and Crowley, J. L. (2013) Age intercalibration of ^{40}Ar/^{39}Ar sanidine and chemically distinct U/Pb zircon populations from the Alder Creek Rhyolite Quaternary *Geochronology* standard. *Chemical Geology* **345**, 87–98.

Rivera, T. A., Schmitz, M. D., Crowley, J. L., and Storey, M. (2014) Rapid magma evolution constrained by zircon petrochronology and ^{40}Ar/^{39}Ar sanidine ages for the Huckleberry Ridge Tuff, Yellowstone, USA. *Geology* **42**(8), 643–646.

Robb, L. J., Davis, D. W., and Kamo, S. L. (1990) U–Pb Ages on single detrital zircon grains from the Witwatersrand Basin, South Africa: constraints on the age of sedimentation and on the evolution of granites adjacent to the basin. *The Journal of Geology* **98**(3), 311–328.

Roberts, E. M., Stevens, N., O'connor, P., *et al.* (2012) Initiation of the western branch of the East African Rift coeval with the eastern branch. *Nature Geoscience* **5**(4), 289–294.

Roberts, M. P. and Finger, F. (1997) Do U–Pb zircon ages from granulites reflect peak metamorphic conditions?. *Geology* **25**(4), 319–322.

Roberts, N. M. and Spencer, C. J. (2015) The zircon archive of continent formation through time. *Geological Society, London, Special Publications* **389**(1), 197–225.

Roddick, J. C. (1987) Generalized numerical error analysis with applications to geochronology and thermodynamics. *Geochimica et Cosmochimica Acta* **51**(8), 2129–2135.

Roddick, J. C., Loveridge, W. D., and Parrish, R. R. (1987) Precise U–Pb dating of zircon at the sub-nanogram Pb level. *Chemical Geology: Isotope Geoscience Section* **66**(1–2), 111–121.

Rollinson, H. R. (2014) *Using Geochemical Data: Evaluation, Presentation*, Interpretation. Routledge.

Roth, A. S., Bourdon, B., Mojzsis, S. J., Rudge, J. F., Guitreau, M., and Blichert-Toft, J. (2014) Combined ^{147,146}Sm-^{143,142}Nd constraints on the longevity and residence time of early terrestrial crust. *Geochemistry, Geophysics, Geosystems* **15**(6), 2329–2345.

Rubatto, D. (2002) Zircon trace element geochemistry: partitioning with garnet and the link between U–Pb ages and metamorphism. *Chemical Geology* **184**, 123–138.

Rubatto, D. and Hermann, J. (2007a) Experimental zircon/melt and zircon/garnet trace element partitioning and implications for the geochronology of crustal rocks. *Chemical Geology* **241**(1–2), 38–61.

Rubatto, D. and Hermann, J. (2007b) Zircon behaviour in deeply subducted rocks. *Elements* **3**(1), 31–35.

Russell, A. (1923) LXXII. Radio-active disintegration series and the relation of actinium to uranium. *The London, Edinburgh, and Dublin Philosophical Magazine and Journal of Science* **46**(274), 642–656.

Rutherford, E. and Soddy, F. (1902a, LXXXIV—the radioactivity of thorium compounds. II. The cause and nature of radioactivity. *Journal of the Chemical Society, Transactions* **81**, 837–860.

Rutherford, E. and Soddy, F. (1902b) XLI. The cause and nature of radioactivity—part I. *The London, Edinburgh, and Dublin Philosophical Magazine and Journal of Science* **4**(21), 370–396.

Rutherford, E. and Soddy, F. (1902c) XXXIII—the radioactivity of thorium compounds. I. An investigation of the radioactive emanation. *Journal of the Chemical Society, Transactions* **81**, 321–350.

Sageman, B. B., Singer, B. S., Meyers, S. R., *et al.* (2014) Integrating 40Ar/39Ar, U–Pb, and astronomical clocks in the Cretaceous Niobrara Formation, Western Interior Basin, USA. *Geological Society of America Bulletin.*

Samperton, K. M., Schoene, B., Cottle, J. M., Brenhin Keller, C., Crowley, J. L., and Schmitz, M. D. (2015) Magma emplacement, differentiation and cooling in the middle crust: Integrated zircon geochronological–geochemical constraints from the Bergell Intrusion, Central Alps. *Chemical Geology* **417**, 322–340.

Saylor, J. E. and Sundell, K. E. (2016) Quantifying comparison of large detrital geochronology data sets. *Geosphere* **12**(1), 203–220.

Schaltegger, U., Fanning, C. M., Günther, D., Maurin, J. C., Schulmann, K., and Gebauer, D. (1999) Growth, annealing and recrystallization of zircon and preservation of monazite in high-grade metamorphism: conventional and in-situ U–Pb isotope, cathodoluminescence and microchemical evidence. *Contributions to Mineralogy and Petrology* **134**(2), 186–201.

Schaltegger, U., Brack, P., Ovtcharova, M., *et al.* (2009) Zircon and titanite recording 1.5 million years of magma accretion, crystallization and initial cooling in a composite pluton (southern Adamello batholith, northern Italy). *Earth and Planetary Science Letters* **286**(1–2), 208–218.

Schaltegger, U., Schmitt, A., and Horstwood, M. (2015) U–Th–Pb zircon geochronology by ID-TIMS, SIMS, and laser ablation ICP-MS: recipes, interpretations, and opportunities. *Chemical Geology* **402**, 89–110.

Schandl, E., Davis, D., and Krogh, T. (1990) Are the alteration halos of massive sulfide deposits syngenetic? Evidence from U–Pb dating of hydrothermal rutile at the Kidd volcanic center, Abitibi subprovince, Canada. *Geology* **18**(6), 505–508.

Schärer, U. (1980) U–Pb and Rb–Sr dating of a polymetamorphic nappe terrain: the Caledonian Jotun Nappe, southern Norway. *Earth and Planetary Science Letters* **49**(2), 205–218.

Schärer, U. (1984) The effect of initial 230Th disequilibrium on young U–Pb ages: the Makalu case, Himalaya. *Earth and Planetary Science Letters* **67**, 191–204.

Schärer, U., Krogh, T. E., and Gower, C. R. (1986) Age and evolution of the Grenville Province in eastern Labrador from U–Pb systematics in accessory minerals. *Contributions to Mineralogy and Petrology* **94**, 438–451.

Schärer, U., Tapponnier, P., Lacassin, R., Leloup, P. H., Zhong, D., and Ji, S. (1990) Intraplate tectonics in Asia: a precise age for large-scale Miocene movement along the Ailao Shan-Red River shear zone, China. *Earth and Planetary Science Letters* **97** (1–2), 65–77.

Schauble, E. A. (2007) Role of nuclear volume in driving equilibrium stable isotope fractionation of mercury, thallium, and other very heavy elements. *Geochimica et Cosmochimica Acta* **71**(9), 2170–2189.

Schenk, V. (1980) U–Pb and Rb–Sr radiometric dates and their correlation with metamorphic events in the granulite-facies basement of the serre, Southern Calabria (Italy). *Contributions to Mineralogy and Petrology* **73**(1), 23–38.

Scherer, E., Münker, C., and Mezger, K. (2001) Calibration of the Lutetium-Hafnium clock. *Science* **293**, 683–687.

Schmitt, A., Danišík, M., Evans, N., *et al.* (2011) Acigöl rhyolite field, Central Anatolia (part 1): high-resolution dating of eruption episodes and zircon growth rates. *Contributions to Mineralogy and Petrology* **162**(6), 1215–1231.

Schmitt, A. K. (2007) Ion microprobe analysis of (231Pa)/(235U) and an appraisal of protactinium partitioning in igneous zircon. *American Mineralogist* **92**(4), 691–694.

Schmitt, A. K. and Zack, T. (2012) High-sensitivity U–Pb rutile dating by secondary ion mass spectrometry (SIMS) with an O2+ primary beam. *Chemical Geology* **332**, 65–73.

Schmitt, A. K., Chamberlain, K. R., Swapp, S. M., and Harrison, T. M. (2010a) *In situ* U–Pb dating of micro-baddeleyite by secondary ion mass spectrometry. *Chemical Geology* **269**(3), 386–395.

Schmitt, A. K., Stockli, D. F., Lindsay, J. M., Robertson, R., Lovera, O. M., and Kislitsyn, R. (2010b) Episodic growth and homogenization of plutonic roots in arc volcanoes from combined U–Th and (U–Th)/He zircon dating. *Earth and Planetary Science Letters* **295**(1–2), 91–103.

Schmitz, M. D. and Bowring, S. A. (2001) U–Pb zircon and titanite sytematics of the Fish Canyon Tuff: an assessment of high-precision U–Pb geochronology and its application to young volcanic rocks. *Geochimica et Cosmochimica Acta* **65**(15), 2571–2587.

Schmitz, M. D. and Bowring, S. A. (2003) Constraints on the thermal evolution of continental lithosphere from U–Pb accessory mineral thermochronometry of lower crustal xenoliths, southern Africa. *Contributions to Mineralogy and Petrology* **144**, 592–618.

Schmitz, M. D. and Bowring, S. A. (2004) Lower crustal granulite formation during Mesoproterozoic Namaqua–Natal collisional orogenesis, southern Africa. *South African Journal of Geology* **107**(1–2), 261–284.

Schmitz, M. D., and Schoene, B. (2007) Derivation of isotope ratios, errors, and error correlations for U–Pb geochronology using 205Pb-235U-(233U)-spiked isotope dilution thermal ionization mass spectrometric data. *Geochemistry, Geophysics, Geosystems* **8** (8), Q08006.

Schoene, B. (2007) Determining accurate temperature-time paths in U–Pb thermochronology: an example from the SE Kaapvaal craton, southern Africa. *Geochimica et Cosmochimica Acta* **71**), 165–185.

Schoene, B. (2010) Rates and mechanisms of Mesoarchean magmatic arc construction, eastern Kaapvaal craton, Swaziland. *Geological Society of America Bulletin* **122**), 408–429. DOI: 410.1130/B26501.26501.

Schoene, B. (2014) U–Th–Pb geochronology, in Rudnick, R., ed., Treatise on Geochemistry, Volume 4.10: Oxford, U.K., Elsevier), 341–378.

Schoene, B., and Bowring, S. A. (2006) U–Pb systematics of the McClure Mountain syenite: thermochronological constraints on the age of the 40Ar/39Ar standard MMhb. *Contributions to Mineralogy and Petrology* **151**(5), 615–630.

Schoene, B. and Bowring, S. A. (2007) Determining accurate temperature–time paths in UPb thermochronology: an example from the SE Kaapvaal craton, southern Africa. *Geochimica et Cosmochimica Acta* **71**, 165–185.

Schoene, B. and Bowring, S. A. (2010) Rates and mechanisms of Mesoarchean magmatic arc construction, eastern Kaapvaal craton, Swaziland. *Geological Society of America Bulltin* **122**, 408–429. DOI: 410.1130/B26501.26501.

Schoene, B., Crowley, J. L., Condon, D. C., Schmitz, M. D., and Bowring, S. A. (2006) Reassessing the uranium decay constants for geochronology using ID-TIMS U–Pb data. *Geochimica et Cosmochimica Acta* **70**), 426–445.

Schoene, B., de Wit, M. J., and Bowring, S. A. (2008) Mesoarochean assembly and stabilization of the eastern Kaapvaal craton: A structural-thermochronological perspective. Tectonics 27), TC5010, DOI:5010.1029/2008TC002267.

Schoene, B., Guex, J., Bartolini, A., Schaltegger, U., and Blackburn, T. J. (2010a) Correlating the end-Triassic mass extinction and flood basalt volcanism at the 100,000-year level. *Geology* 38), 387–390. DOI: 310.1130/G30683.30681.

Schoene, B., Latkoczy, C., Schaltegger, U., and Gunther, D. (2010b) A new method integrating high-precision U–Pb geochronology with zircon trace element analysis (U–Pb TIMS-TEA). *Geochimica et Cosmochimica Acta* **74**(24), 7144–7159.

Schoene, B., Schaltegger, U., Brack, P., Latkoczy, C., Stracke, A., and Günther, D. (2012) Rates of magma differentiation and emplacement in a ballooning pluton recorded by U–Pb TIMS-TEA, Adamello batholith, Italy. *Earth and Planetary Science Letters* **355–356**, 162–173.

Schoene, B., Condon, D. J., Morgan, L., and McLean, N. (2013) Precision and accuracy in geochronology. *Elements* **9** (1), 19–24.

Schoene, B., Samperton, K. M., Eddy, M. P., *et al.* (2015) U–Pb geochronology of the Deccan Traps and relation to the end-Cretaceous mass extinction. *Science* **347**(6218), 182–184.

Schön, R., Winkler, G., and Kutschera, W. (2004) A critical review of experimental data for the half-lives of the uranium isotopes ^{238}U and ^{235}U. *Applied Radiation and Isotopes* **60**, 263–273.

Selby, D., Creaser, R. A., Stein, H. J., Markey, R. J., and Hannah, J. L. (2007) Assessment of the ^{187}Re decay constant by cross calibration of Re–Os molybdenite and U–Pb zircon chronometers in magmatic ore systems. *Geochimica et Cosmochimica Acta* **71**(8), 1999–2013.

Self, S., Schmidt, A., and Mather, T. A. (2014) Emplacement characteristics, time scales, and volcanic gas release rates of continental flood basalt eruptions on Earth. *Geological Society of America Special Papers* **505**.

Shaw, C. A., Heizler, M., and Karlstrom, K. E. (2004) ^{40}Ar/^{39}Ar thermochronologic record of 1.45–1.35 Ga intracontinental tectonism in the southern Rocky Mountains: interplay of conductive and advective heating with intracontinental deformation. In *The Rocky Mountain Region: an Evolving Lithosphere*, Karlstrom, K. E. and Keller, G. R. (eds), 163–184. Monograph 154, American Geophysical Union, Washington, DC.

Shen, S.-Z., Crowley, J. L., Wang, Y., *et al.* (2011) Calibrating the end-Permian mass extinction. *Science* **334**(6061), 1367–1372.

Shirey, S. B., Kamber, B. S., Whitehouse, M. J., Mueller, P. A., and Basu, A. R. (2008) A review of the isotopic and trace element evidence for mantle and crustal processes in the Hadean and Archean: implications for the onset of plate tectonic subduction. *Geological Society of America Special Papers* **440**, 1–29.

Silver, L. T. and Deutsch, S. (1963) Uranium–lead isotopic variations in zircons: a case study. *The Journal of Geology* **71**(6), 721–758.

Simon, J. I. and Reid, M. R. (2005) The pace of rhyolite differentiation and storage in an 'archetypical' silicic magma system, Long Valley, California. *Earth and Planetary Science Letters* **235**(1–2), 123–140.

Simon, J. I., Renne, P. R., and Mundil, R. (2008) Implications of pre-eruptive magmatic histories of zircons for U–Pb geochronology of silicic extrusions. *Earth and Planetary Science Letters* **266** (1–2), 182–194.

Simonetti, A., Heaman, L. M., Hartlaub, R. P., Creaser, R. A., MacHattie, T. G., and Bohm, C. (2005) U–Pb zircon dating by laser ablation-MC-ICP-MS using a new multiple ion counting Faraday collector array. *Journal of Analytical Atomic Spectrometry* **20**(8).

Slama, J., Košler, J., Condon, D. J., *et al.* (2008) Plesovice zircon—a new natural reference material for U–Pb and Hf isotopic microanalysis. *Chemical Geology* **249**(1–2), 1–35. DOI:10.1016/j. chemgeo.2007.1011.1005.

Smith, M. E., Chamberlain, K. R., Singer, B. S., and Carroll, A. R. (2010) Eocene clocks agree. Coeval ^{40}Ar/^{39}Ar, U–Pb, and astronomical ages from the Green River Formation. *Geology* **38**(6), 527–530.

Smith, P. E., Farquhar, R. M., and Hancock, R. G. (1991) Direct radiometric age determination of carbonate diagenesis using U–Pb in secondary calcite. *Earth and Planetary Science Letters* **105**(4), 474–491.

Smith, P. E., Brand, U., and Farquhar, R. M. (1994) U–Pb systematics and alteration trends of Pennsylvanian-aged aragonite and calcite. *Geochimica et Cosmochimica Acta* **58**(1), 313–322.

Smye, A. J. and Stockli, D. F. (2014) Rutile U–Pb age depth profiling: a continuous record of lithospheric thermal evolution. *Earth and Planetary Science Letters* **408**, 171–182.

Soddy, F. (1913a) Intra-atomic charge. *Nature* **92**, 399–400.

Soddy, F. (1913b) Radioactivity. *Annual Reports on the Progress of Chemistry* **10**, 262–288.

Söderlund, U. and Johansson, L. (2002) A simple way to extract baddeleyite (ZrO$_2$). *Geochemistry, Geophysics, Geosystems* **3**(2).

Söderlund, U., Hofmann, A., Klausen, M. B., Olsson, J. R., Ernst, R. E., and Persson, P.-O. (2010) Towards a complete magmatic barcode for the Zimbabwe craton: Baddeleyite U–Pb dating of regional dolerite dyke swarms and sill complexes. *Precambrian Research* **183**(3), 388–398.

Souders, A. and Sylvester, P. (2012) Use of MC-ICPMS for laser ablation U/Pb geochronology of baddeleyite. *Mineralogical Magazine* **76**, 2397.

Spear, F. S. (1993) *Metamorphic Phase Equilibria and Pressure–Temperature–Time Paths*. Monograph, Mineralogical Society of America, Washington, DC, 799 pp.

Spencer, K., Hacker, B., Kylander-Clark, A., *et al.* (2013) Campaign-style titanite U–Pb dating by laser-ablation ICP: implications for crustal flow, phase transformations and titanite closure. *Chemical Geology* **341**, 84–101.

Sprain, C. J., Renne, P. R., Wilson, G. P., and Clemens, W. A. (2014) High-resolution chronostratigraphy of the terrestrial Cretaceous-Paleogene transition and recovery interval in the Hell Creek region, Montana. *Geological Society of America Bulletin* **127**(3-4), 393–409.

Stacey, J. C. and Kramers, J. D. (1975) Approximation of terrestrial lead isotope evolution by a two-stage model. *Earth and Planetary Science Letters* **26**, 207–221.

Steiger, R. H. and Jäger, E. (1977) Subcommission on Geochronology: convention on the use of decay constants in geo- and cosmochronology. *Earth and Planetary Science Letters* **36**, 359–362.

Stern, R. A. and Amelin, Y. (2003) Assessment of errors in SIMS zircon U–Pb gechronology using a natural zircon standard and NIST SRM 610 glass. *Chemical Geology* **197**, 111–142.

Stern, R. A., Bodorkos, S., Kamo, S. L., Hickman, A. H., and Corfu, F. (2009) Measurement of SIMS instrumental mass fractionation of Pb isotopes during zircon dating. *Geostandards and Geoanalytical Research* **33**(2), 145–168.

Stewart, J. H., Gehrels, G. E., Barth, A. P., Link, P. K., Christie-Blick, N., and Wrucke, C. T. (2001) Detrital zircon provenance of Mesoproterozoic to Cambrian arenites in the western United States and northwestern Mexico. *Geological Society of America Bulletin* **113**(10), 1343–1356.

Stirling, C. H., Andersen, M. B., Potter, E.-K., and Halliday, A. N. (2007) Low-temperature isotopic fractionation of uranium. *Earth and Planetary Science Letters* **264**(1–2), 208–225.

Storey, C. D., Smith, M. P., and Jeffries, T. E. (2007) *In situ* LA-ICP-MS U–Pb dating of metavolcanics of Norrbotten, Sweden: records of extended geological histories in complex titanite grains. *Chemical Geology* **240**(1–2), 163–181.

Stracke, A., Scherer, E., and Reynolds, B. (2014) Application of isotope dilution in geochemistry. In *Treatise on Geochemistry* (2nd edn), 71–86. Elsevier, Oxford,

Suzuki, K. and Adachi, M. (1994) Middle Precambrian detrital monazite and zircon from the hida gneiss on Oki-Dogo Island, Japan: their origin and implications for the correlation of basement gneiss of Southwest Japan and Korea. *Tectonophysics* **235**(3), 277–292.

Tanner, P. W. G. and Evans, J. A. (2003) Late Precambrian U–Pb titanite age for peak regional metamorphism and deformation (Knoydartian orogeny) in the western Moine, Scotland. *Journal of the Geological Society* **160**(4), 555–564.

Tapster, S., Condon, D., Naden, J., *et al.* (2016) Rapid thermal rejuvenation of high-crystallinity magma linked to porphyry copper deposit formation; evidence from the Koloula Porphyry Prospect, Solomon Islands. *Earth and Planetary Science Letters* **442**, 206–217.

Taylor, S. R. and McLennan, S. M. (1985) *The Continental Crust: Its Composition and Evolution*. Blackwell, Oxford.

Tera, F. and Wasserburg, G. J. (1972a) U–Th–Pb systematics in lunar highland samples from the Luna 20 and Apollo 16 missions. *Earth and Planetary Science Letters* **17**(1), 36–51.

Tera, F. and Wasserburg, G. J. (1972b) U–Th–Pb systematics in three Apollo 14 basalts and the problem of initial Pb in lunar rocks. *Earth and Planetary Science Letters* **14**(3), 281–304.

Thirlwall, M. (2000) Inter-laboratory and other errors in Pb isotope analyses investigated using a ^{207}Pb–^{204}Pb double spike. *Chemical Geology* **163**(1), 299–322.

Till, C. B., Vazquez, J. A., and Boyce, J. W. (2015) Months between rejuvenation and volcanic eruption at Yellowstone caldera, Wyoming. *Geology* **43**(8), 695–698.

Tilton, G. R. (1960) Volume diffusion as a mechanism for discordant lead ages. *Journal of Geophysics Research* **65**(9), 2933–2945.

Tilton, G. R. and Grunenfelder, M. H. (1968) Sphene: uranium–lead ages. *Science* **159**(3822), 1458–1461.

Tissot, F. L. and Dauphas, N. (2015) Uranium isotopic compositions of the crust and ocean: age corrections, U budget and global extent of modern anoxia. *Geochimica et Cosmochimica Acta* **167**, 113–143.

Tissot, F. L. H., Dauphas, N., and Grossman, L. (2016) Origin of uranium isotope variations in early solar nebula condensates. *Science Advances* **2**(3).

Todt, W., Cliff, R. A., Hanser, A., and Hofmann, A. (1996) Evaluation of a ^{202}Pb–^{205}Pb double spike for high-precision lead isotope analysis. In *Earth Processes: Reading the Isotopic Code*, 429–437. Monograph, American Geophysical Union, Washington, DC

Tomkins, H., Powell, R., and Ellis, D. (2007) The pressure dependence of the zirconium-in-rutile thermometer. *Journal of Metamorphic Geology* **25**(6), 703–713.

Tornos, F. and Chiaradia, M. (2004) Plumbotectonic evolution of the Ossa Morena zone, Iberian Peninsula: tracing the influence of mantle-crust interaction in ore-forming processes. *Economic Geology* **99**(5), 965–985.

Trail, D., Mojzsis, S. J., Harrison, T. M., Schmitt, A. K., Watson, E. B., and Young, E. D. (2007) Constraints on Hadean zircon protoliths from oxygen isotopes, Ti-thermometry, and rare earth elements. *Geochemistry, Geophysics, Geosystems* **8**(6).

Trail, D., Watson, E. B., and Tailby, N. D. (2011) The oxidation state of Hadean magmas and implications for early Earth/'s atmosphere. *Nature* **480**(7375), 79–82.

Tucker, R. D. (1992) U–Pb dating of plinian-eruption ashfalls by the isotopic dilution method: a reliable and precise tool for time-scale calibration and biostratigraphic correlation. *Geological Society of America Abstracts with Programs* **24**, A192.

Tucker, R. D., Raheim, A., Krogh, T. E., and Corfu, F. (1986) Uranium–lead zircon and titanite ages from the northern portion of the Western Gneiss Region, south-central Norway. *Earth and Planetary Science Letters* **81**, 203–211.

Tucker, R. D., Krogh, T. E., Ross Jr, R. J., and Williams, S. H. (1990) Time-scale calibration by high-precision U–Pb zircon dating of interstratified volcanic ashes in the Ordovician and Lower Silurian stratotypes of Britain. *Earth and Planetary Science Letters* **100**(1–3), 51–58.

Vavra, G., Gebauer, D., Schmid, R., and Compston, W. (1996) Multiple zircon growth and recrystallization during polyphase Late Carboniferous to Triassic metamorphism in granulites of the Ivrea Zone (Southern Alps): and ion microprobe (SHRIMP) study. *Contributions to Mineralogy and Petrology* **122**, 337–358.

Vazquez, J. A. and Reid, M. R. (2004) Probing the accumulation history of the voluminous Toba magma. *Science* **305**(5686), 991–994.

Vermeesch, P. (2012) On the visualisation of detrital age distributions. *Chemical Geology* **312**, 190–194.

Vermeesch, P. (2013) Multi-sample comparison of detrital age distributions. *Chemical Geology* **341**, 140–146.

Verts, L. A., Chamberlain, K. R., and Frost, C. D. (1996) U–Pb sphene dating of metamorphism: the importance of sphene growth in the contact aureole of the Red Mountain pluton, Laramie Mountains, Wyoming. *Contributions to Mineralogy and Petrology* **125**, 186–199.

Vervoort, J. D. and Kemp, A. I. S. (2016) Clarifying the zircon Hf isotope record of crust–mantle evolution. *Chemical Geology* **425**, 65–75.

Villeneuve, M., Sandeman, H. A., and Davis, W. J. (2000) A method for intercalibration of U–Th–Pb and ^{40}Ar–^{39}Ar ages in the Phanerozoic. *Geochimica et Cosmochimica Acta* **64**(23), 4017–4030.

Voice, P. J., Kowalewski, M., and Eriksson, K. A. (2011) Quantifying the timing and rate of crustal evolution: global compilation of radiometrically dated detrital zircon grains. *The Journal of Geology* **119**(2), 109–126.

Von Blanckenburg, F. (1992) Combined high-precision chronometry and geochemical tracing using accessory minerals: applied to the CentrAl–Alpine Bergell intrusion (central Europe). *Chemical Geology* **100**, 19–40.

Von Quadt, A., Gallhofer, D., Guillong, M., Peytcheva, I., Waelle, M., and Sakata, S. (2014) U–Pb dating of CA/non-CA treated zircons obtained by LA-ICP-MS and CA-TIMS techniques: impact for their geological interpretation. *Journal of Analytical Atomic Spectrometry* **29**(9), 1618–1629.

Vry, J. K., and Baker, J. A. (2006) LA-MC-ICPMS Pb–Pb dating of rutile from slowly cooled granulites: confirmation of the high closure temperature for Pb diffusion in rutile. *Geochimica et Cosmochimica Acta* **70**(7), 1807–1820.

Walker, J., Cliff, R. A., and Latham, A. G. (2006) U–Pb isotopic age of the StW 573 hominid from Sterkfontein, South Africa. *Science* **314**(5805), 1592–1594.

Wang, H., Chen, L., Sun, Y., *et al.* (2007) ~ 4.1 Ga xenocrystal zircon from Ordovician volcanic rocks in western part of North Qinling Orogenic Belt. *Chinese Science Bulletin* **52**(21), 3002–3010.

Wang, Z., Rasbury, E., Hanson, G., and Meyers, W. (1998) Using the U–Pb system of calcretes to date the time of sedimentation of clastic sedimentary rocks. *Geochimica et Cosmochimica Acta* **62**(16), 2823–2835.

Wasserburg, G. J. (1963) Diffusion Processes in Lead-Uranium Systems. *Journal of Geophysics Research.* **68**(16), 4823–4846.

Watson, E. B. and Harrison, T. M. (1983) Zircon saturation revisited: temperature and composition effects in a variety of crustal magma types. *Earth and Planetary Science Letters* **64**(2), 295–304.

Watson, E. and Harrison, T. (2005) Zircon thermometer reveals minimum melting conditions on earliest Earth. *Science* **308**(5723), 841–844.

Watson, E., Wark, D., and Thomas, J. (2006) Crystallization thermometers for zircon and rutile. *Contributions to Mineralogy and Petrology* **151**(4), 413–433.

Wayne, D., Sinha, A., and Hewitt, D. (1992) Differential response of zircon U – Pb isotopic systematics to metamorphism across a lithologic boundary: an example from the Hope Valley Shear Zone, southeastern Massachusetts, USA. *Contributions to Mineralogy and Petrology* **109**(3), 408–420.

Wayne, D. M. and Sinha, A. K. (1988) Physical and chemical response of zircons to deformation. *Contributions to Mineralogy and Petrology* **98**(1), 109–121.

Wendt, I. (1984) A three-dimensional U/Pb discordia plane to evaluate samples with common lead of unknown isotopic composition. *Chemical Geology* **46**(1), 1–12.

Wetherill, G. W. (1956a) Discordant uranium–lead ages. *Transactions of the American Geophysical Union* **37**, 320–326.

Wetherill, G. W. (1956b) An interpretation of the Rhodesia and Witwatersrand age patterns. *Geochimica et Cosmochimica Acta* **9**(5), 290–292.

Weyer, S., Anbar, A. D., Gerdes, A., Gordon, G. W., Algeo, T. J., and Boyle, E. A. (2008) Natural fractionation of ^{238}U/^{235}U. *Geochimica et Cosmochimica Acta* **72**(2), 345–359.

White, L. T. and Ireland, T. R. (2012) High-uranium matrix effect in zircon and its implications for SHRIMP U–Pb age determinations. *Chemical Geology* **306–307**, 78–91.

White, N. M., Parrish, R. R., Bickle, M. J., Najman, Y. M. R., Burbank, D., and Maithani, A. (2001) Metamorphism and exhumation of the NW Himalaya constrained by U–Th–Pb analyses of detrital monazite grains from early foreland basin sediments. *Journal of the Geological Society* **158**(4), 625–635.

White, W. M. (2013) *Geochemistry*. John Wiley & Sons.

Whitehouse, M. J., Nägler, T. F., Moorbath, S., Kramers, J. D., Kamber, B. S., and Frei, R. (2001) Priscoan (4.00–4.03 Ga) orthogneisses from northwestern Canada by Samuel A. Bowring and Ian S. Williams: discussion. *Contributions to Mineralogy and Petrology* **141**(2), 248–250.

Wilde, S. A., Valley, J. W., Peck, W. H., and Graham, C. M. (2001) Evidence from detrital zircons for the existence of continental crust and oceans on the Earth 4.4 Gyr ago. *Nature* **409**, 175–178.

Williams, I. S. (1998) U–Th–Pb geochronology by ion microprobe. In *Applications of Microanalytical Techniques to Understanding Mineralizing Processes*, Vol. 7, McKibben, M. A., Shanks III, W. C., and Ridley, W. I. (eds), 1–35. Reviews in Economic Geology.

Williams, M. L. and Jercinovic, M. J. (2002) Microprobe monazite geochronology: putting absolute time into microstructural analysis. *Journal of Structural Geology* **24**(6–7), 1013–1028.

Williams, M. and Jercinovic, M. (2012) Tectonic interpretation of metamorphic tectonites: integrating compositional mapping, microstructural analysis and in situ monazite dating. *Journal of Metamorphic Geology* **30**(7), 739–752.

Williams, M., Jercinovic, M., Goncalves, P., and Mahan, K. (2006) Format and philosophy for collecting, compiling, and reporting microprobe monazite ages. *Chemical Geology* **225**(1), 1–15.

Williams, M. L., Jercinovic, M. J., and Hetherington, C. J. (2007) Microprobe monazite geochronology: understanding geologic processes by integrating composition and chronology. *Annual Review of Earth and Planetary Sciences* **35**(1), 137–175.

Willigers, B. J. A., Krogstad, E. J., and Wijbrans, J. R. (2001) Comparison of thermochronometers in a slowly cooled granulite terrain: Nagssugtoqidian Orogen, *West Greenland: Journal of Petrology* **42**(9), 1729–1749.

Wilson, C. J. N. and Charlier, B. L. A. (2009) Rapid rates of magma generation at contemporaneous magma systems, Taupo Volcano, New Zealand: insights from U–Th model-age spectra in zircons. *Journal of Petrology* **50**(5), 875–907.

Wingate, M. T. D. and Compston, W. (2000) Crystal orientation effects during ion microprobe U–Pb analysis of baddeleyite. *Chemical Geology* **168**(1–2), 75–97.

Winter, B. L. and Johnson, C. M. (1995) U–Pb dating of a carbonate subaerial exposure event. *Earth and Planetary Science Letters* **131**(3), 177–187.

Wong, L., Davis, D. W., Krogh, T. E., and Robert, F. (1991) U–Pb zircon and rutile chronology of Archean greenstone formation and gold mineralization in the Val d'Or region, Quebec. *Earth and Planetary Science Letters* **104**(2), 325–336.

Wood, B. J. and Blundy, J. D. (2003) Trace element partitioning under crustal and uppermost mantle conditions: the influences of ionic radius, cation charge, pressure, and temperature. In *Treatise on Geochemistry*, 395–424. Pergamon, Oxford.

Woodhead, J. and Pickering, R. (2012) Beyond 500 ka: progress and prospects in the UPb chronology of speleothems, and their application to studies in palaeoclimate, human evolution, biodiversity and tectonics. *Chemical Geology* **322–323**, 290–299.

Woodhead, J., Hellstrom, J., Pickering, R., Drysdale, R., Paul, B., and Bajo, P. (2012) U and Pb variability in older speleothems and strategies for their chronology: *Quaternary Geochronology* **14**, 105–113.

Wotzlaw, J.-F., Schaltegger, U., Frick, D. A., Dungan, M. A., Gerdes, A., and Günther, D. (2013) Tracking the evolution of large-volume silicic magma reservoirs from assembly to supereruption. *Geology* **41**(8), 867–870.

Wotzlaw, J.-F., Bindeman, I. N., Watts, K. E., Schmitt, A. K., Caricchi, L., and Schaltegger, U. (2014) Linking rapid magma reservoir assembly and eruption trigger mechanisms at evolved Yellowstone-type supervolcanoes. *Geology* **42**(9), 807–810.

Wotzlaw, J.-F., Bindeman, I. N., Stern, R. A., D'Abzac, F.-X., and Schaltegger, U. (2015) Rapid heterogeneous assembly of multiple magma reservoirs prior to Yellowstone supereruptions. *Scientific Reports* **5**.

Xing, G.-F., Wang, X.-L., Wan, Y., *et al.* (2014) Diversity in early crustal evolution: 4100 [emsp14] Ma zircons in the Cathaysia Block of southern China. *Scientific Reports* **4**.

Yuan, D., Cheng, H., Edwards, R. L., *et al.* (2004) Timing, duration, and transitions of the last interglacial Asian monsoon. *Science* **304**(5670), 575–578.

Yuan, H.-L., Gao, S., Dai, M.-N., *et al.* (2008) Simultaneous determinations of U–Pb age, Hf isotopes and trace element compositions of zircon by excimer laser-ablation quadrupole and multiple-collector ICP-MS. *Chemical Geology* **247**(1–2), 100–118.

Zartman, R. E. and Doe, B. R. (1981) Evolution of the upper mantle plumbotectonics—the model. *Tectonophysics* **75**(1), 135–162.

Zartman, R. E. and Haines, S. M. (1988) The plumbotectonic model for Pb isotopic systematics among major terrestrial reservoirs—a case for bi-directional transport. *Geochimica et Cosmochimica Acta* **52**(6), 1327–1339.

Zhang, L. S. and Schärer, U. (1996) Inherited Pb components in magmatic titanite and their consequence for the interpretation of U–Pb ages. *Earth and Planetary Science Letters* **138**(1–4), 57–65.

Zheng, Y.-F. (1992) The three-dimensional U/Pb method: Generalized models and implications for U–Pb two-stage systematics. *Chemical Geology* **100**, 3–18.

Fig. 1.3. The 1903 Nobel Prize winners Marie and Pierre Curie as depicted on the French 500-Franc note.

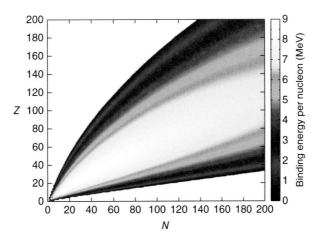

Fig. 1.4. The 1908 Nobel Prize winner Ernest Rutherford as depicted on the New Zealand 100-dollar note. Source: https://commons.wikimedia.org/wiki/File:100Neuseeland-Dollar_vorderseite_21585256953_02d6c65788_o.jpg. Used under CC BY SA 3.0.)

Fig. 2.3. Nuclear binding energies per nucleon as a function of the number of neutrons (N) versus protons (Z) in the nucleus. The binding energies are calculated using the Bethe–Weizsacker equation described above. (Source: Courtesy of Larry Nittler.)

Geochronology and Thermochronology, First Edition. Peter W. Reiners, Richard W. Carlson, Paul R. Renne, Kari M. Cooper, Darryl E. Granger, Noah M. McLean, and Blair Schoene.
© 2018 John Wiley & Sons Ltd. Published 2018 by John Wiley & Sons Ltd.

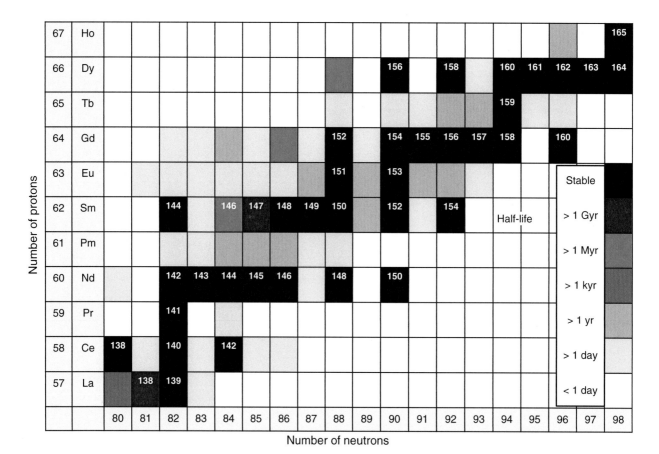

Fig. 2.7. Chart of the nuclides in the mass region of the rare earth elements. Black blocks denote isotopes with half-lives greater than 10^{12} years, dark gray are isotopes with half-lives between 10^6 and 10^{12} years, light gray are isotopes with half-lives of 1 to 10^6 years, and white blocks are isotopes with half-lives of less than a year. Numbers in the blocks give the atomic mass of the isotope.

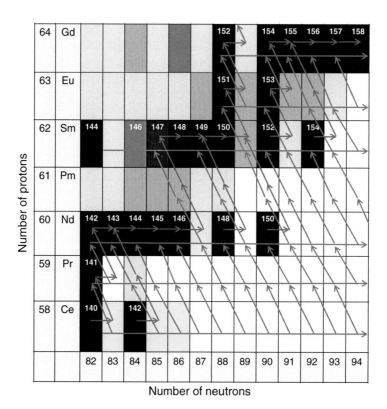

Fig. 2.15. Portion of the chart of the nuclides in the light rare earth element mass range showing the path of *s*- and *r*-process nucleosynthesis. The gray arrows show sequential neutron additions via the *s*-process followed by the β-decay that decreases N, and increases Z, by 1. The black arrows show *r*-process nucleosynthetic paths that rapidly increase N creating a number of very neutron-rich isotopes that then β-decay back to stable isotopes.

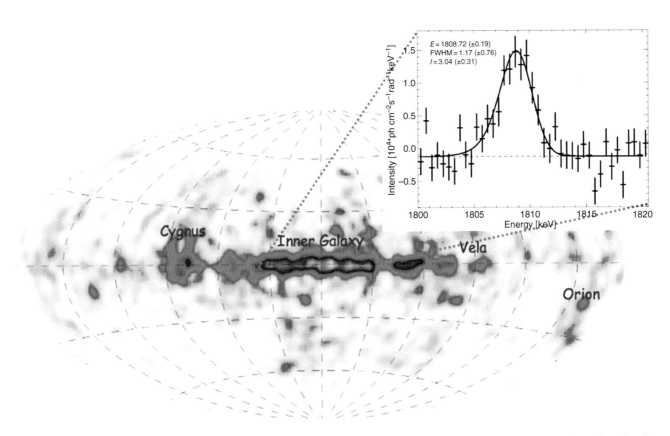

Fig. 2.17. COMPTEL map of ^{26}Al abundance and the INTEGRAL/SPI ^{26}Al gamma-ray spectrum from the inner Milky Way galaxy. Yellow and red colors reflect the highest abundances of ^{26}Al near the galactic center and other areas of active star formation, with gray and white reflecting lower abundances at the outskirts of the galaxy. The graph in the upper left shows the distribution of gamma ray intensity versus energy. *E* shows the peak energy, FWHM is the "full-width, half-maximum" of the distribution of energy, and *I* is the standard deviation of the data. (Source: Courtesy of Roland Diehl.)

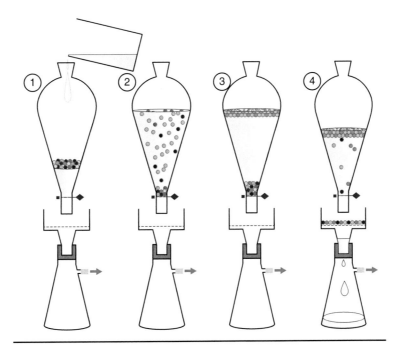

Fig. 3.1. Schematic of heavy liquid separation. Dense liquid is added to a separatory funnel (1) to which crushed and sieved minerals are added (2). The dense grains (black) sink in the liquid while the less dense grains (gray) float (3). The lower layer of the liquid is then drained through a filter (4) that captures the dense grains. The filter can then be replaced allowing the remaining low density grains to be removed from the funnel and captured by the filter paper. (Source: image from http://www.bgs.ac.uk/scienceFacilities/laboratories/mpb/media.html, reproduced by permission of the British Geological Survey.)

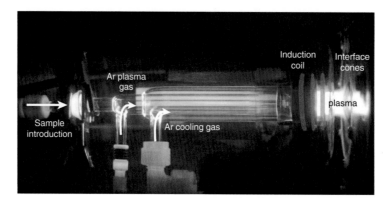

Fig. 3.6. Photograph of the ICP torch. In this image, argon flows through the three concentric quartz tubes from left to right. The small central tube carries the sample (introduced as aerosol at left) mixed with a small Ar gas flow. The Ar is turned into a plasma by a strong radiofrequency field inside the copper helical coil at the end of the torch. The sample flows through the plasma and is evaporated and ionized before it impacts the entrance cones to the mass spectrometer on the right-hand side of the photograph. (Source: Courtesy of Brad Hacker.)

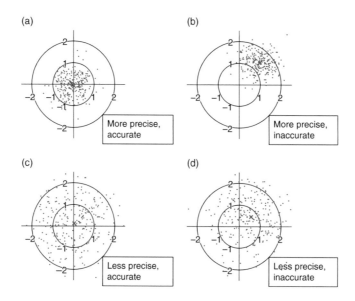

Fig. 4.1. The conventional understanding of accuracy and precision. Measured data, plotted as 250 points, have a true mean at the origin of each plot, and the mean of the measured data is plotted with an "x". Circles centered on the true value are drawn for reference. If the data sets are unbiased as in (a) and (c), the mean of a data set with lower precision is likely to be farther from the true value than the mean of a more precise data set. Because the accuracy depends on the difference between the measured and true values, in the language of the VIM this means a less precise data set is likely to be less accurate.

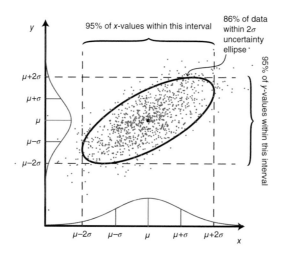

Fig. 4.4. Anatomy of an uncertainty ellipse. One thousand bivariate, normally distributed random numbers are plotted, along with the 2σ ellipse that illustrates the bivariate analog of their mean and standard deviation. On the x-axis and y-axis, the probability distributions for the x and y variables are plotted. They are univariate normal distributions, and known as marginal distributions. Note that the horizontal and vertical tangents of the ellipse correspond to the 2σ uncertainty envelopes for the x and y variables. The ellipse encloses a smaller number of points than the 95% confidence intervals, including only 86% of the simulated data. Therefore, a 2σ uncertainty ellipse is not a 95% uncertainty envelope.

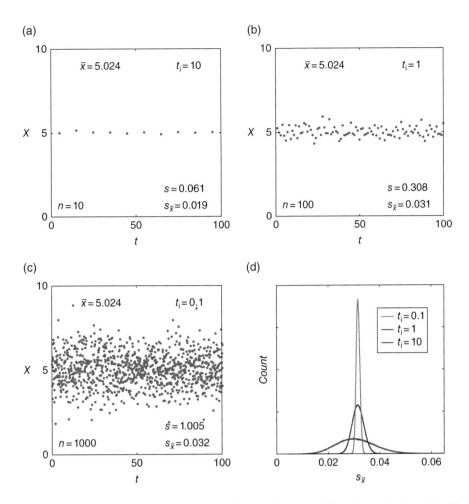

Fig. 4.6. Simulated mass spectrometer measurements of an isotope ratio, x, using the same data set and different integration times t_i. (a–c) Data generated with a true mean of $\bar{x} = 5$ and a true standard error of $s_{\bar{x}} = 0.032$. The estimated mean is exactly the same for each scenario because the underlying data are the same. Decreasing the integration time increases the standard deviation, but the standard error is approximately the same for each scenario. (d) Histograms showing variability of the calculated standard error for each integration time over 10^6 random simulations. Note that higher values of n yield better uncertainty estimates.

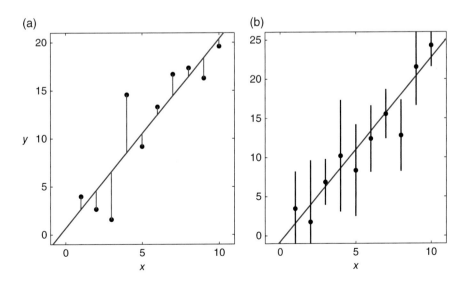

Fig. 4.9. Ordinary least squares and weighted least squares fits to $n = 10$ random, linearly related data points. Following convention, the independent variable is plotted on the horizontal (x-)axis and the dependent variable on the vertical (y-)axis. The best fit line is plotted for each. The dashed vertical lines in (a) are the normally distributed *residuals*, or differences between the measurements and the best fit model. The black vertical bars in (b) represent 2σ uncertainties in the measurements of y.

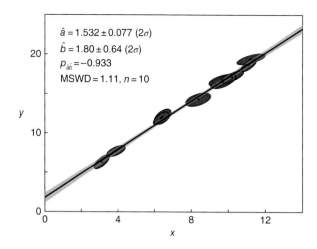

Fig. 4.10. Weighted least-squares linear regression (a "York fit") with correlated uncertainties in the x-axis and y-axis variables. The best fit line is shown in black and its 2σ uncertainty envelope is shaded. The best fit slope $\hat{a}$ and y-intercept $\hat{b}$ have uncertainties, and these uncertainties are correlated, with correlation coefficient $\rho_{\hat{a}\hat{b}}$. The correlation coefficient is needed to propagate uncertainties in any function of $\hat{a}$ and $\hat{b}$, such as the uncertainty envelope for the best fit line.

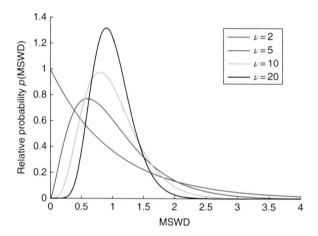

Fig. 4.12. Probability distribution functions for the reduced chi-squared distribution, illustrated for several degrees of freedom, ν. Each distribution is asymmetric, with a mode < 1 and mean of 1. The greater the degrees of freedom, the more the distribution resembles a normal distribution, with mean 1 and standard deviation $\sqrt{2/\nu}$.

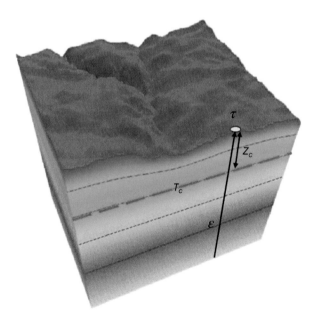

Fig. 5.28. Schematic landscape and underlying crust depicting a sample at surface today with age τ, which passed through its closure temperature T_c at its closure depth Z_c. The rate of erosion is simply the closure depth divided by the age, but closure depth depends on several parameters including erosion rate, and its determination may involve varying degrees of sophistication and precision depending on the application. (Source: *Braun et al.* [2006]. Reproduced with permission of Cambridge University Press.)

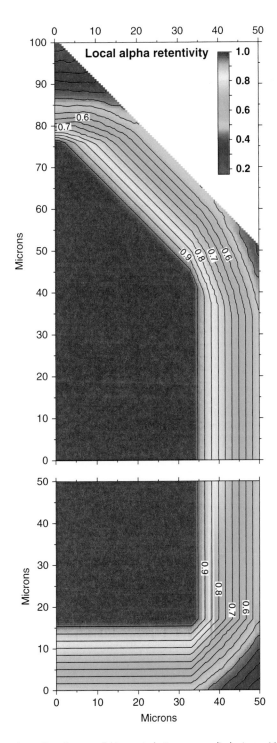

Fig. 11.4. Two-dimensional corner and basal quarter sections (top, parallel to c-axis; bottom, perpendicular to c-axis) of a model zircon crystal, showing contours of modeled fraction of ^{4}He retained within the crystal. (Source: *Hourigan et al.* [2005]. Reproduced with permission of Elsevier.)

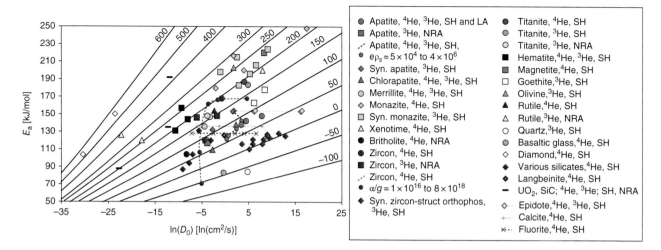

Fig. 11.38. Activation energy (E_a) as a function of natural logarithm of frequency factor [$\ln(D_0)$] for ^{4}He and ^{3}He diffusion in a variety of phases, from Table 11.2. Red and blue dashed lines connect zircon and apatite experiments with varying levels of radiation damage. Horizontal dashed lines for calcite, fluorite, and epidote represent uncertainties in diffusion domain size that allow for a wide range of frequency factors if diffusion domain size is significantly smaller than physical grain size.

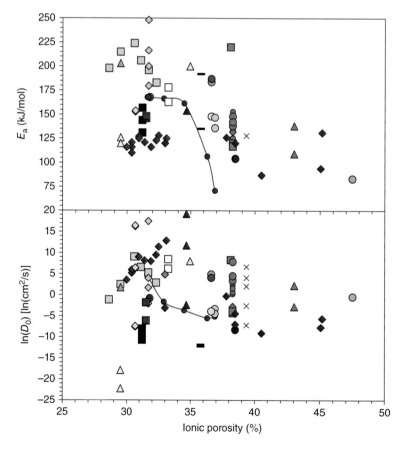

Fig. 11.39. Activation energies and frequency factors for He diffusion for minerals in Table 11.2, as a function of ionic porosity. Symbols as in Fig. 11.38.

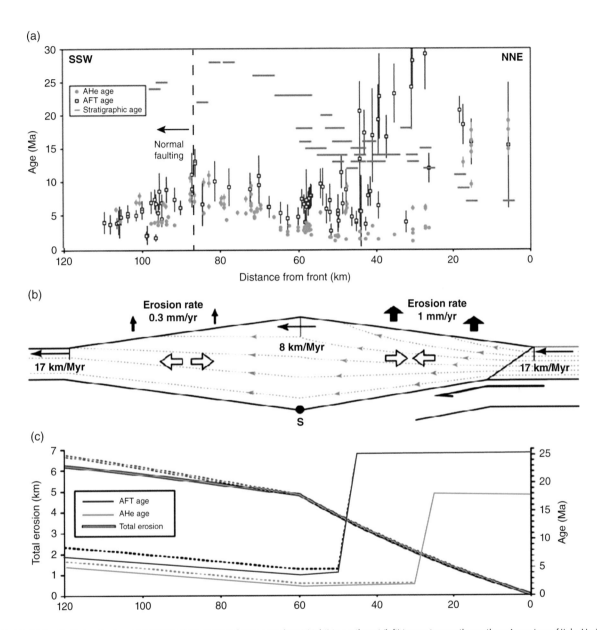

Fig. 11.48. (a) Apatite fission-track and (U–Th)/He dates projected onto a northeast (right) to southwest (left) transect across the northern Apennines of Italy. Horizontal dashes show stratigraphic ages of the greywackes and other low-grade metasedimentary host rocks. Note (i) systematic trends from old "unreset" dates on far "pro-wedge" flank of range, (ii) horizontally offset transitions to reset and younger dates farther west, (iii) increasing dates west of ~60 km distance from the front, and (iv) westward-decreasing dates in the region of active normal faulting. (b) Generalized orogen-scale kinematic model for the range, distances scaled to those in (a), showing horizontal and vertical velocity components across the range, and S-point of diverging velocities representing simplified downward velocity of subducting lithosphere converging from the east side. (c) Predicted cumulative erosion and thermochronologic dates of resulting from model in (b). (Source: *Thomson et al.* [2010]. Reproduced with permission of Geological Society of America.)

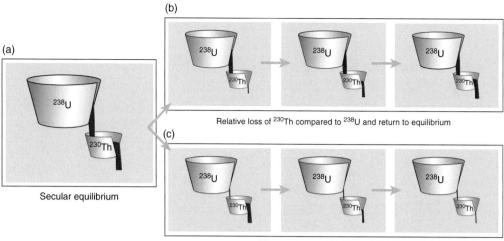

Fig. 12.2. Cartoon illustrating the return of the ^{238}U–^{230}Th parent–daughter pair to secular equilibrium after a disturbance (intermediate daughters are omitted in this diagram). Note that in these diagrams, the opening in each bucket is V-shaped in order to schematically represent the dependence of activity on the number of atoms present, and the different widths of the openings for ^{238}U and ^{230}Th represent the different half-lives. (a) The system at secular equilibrium, where the rate of decay of ^{230}Th matches that of ^{238}U. (b) Evolution of a system in which ^{230}Th has preferentially been lost (or excluded) relative to ^{238}U. (c) Evolution of a system in which ^{238}U has been preferentially lost (or excluded) relative to ^{230}Th. (Source: *Reid* [2008]. Reproduced with permission of Mineralogical Society of America.)

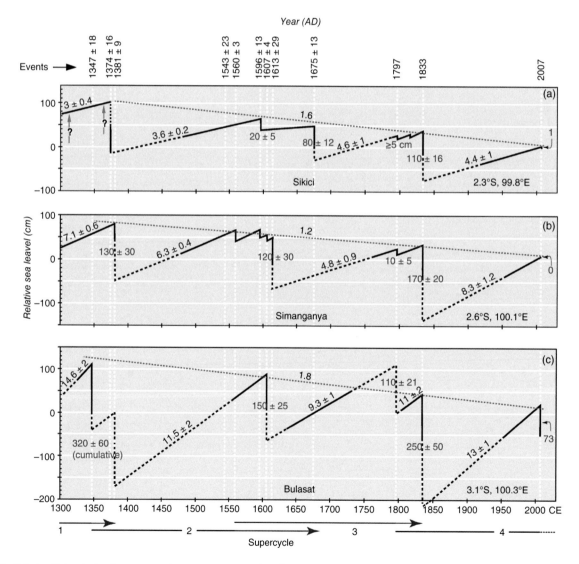

Fig. 12.8. Histories of interseismic submergence and coseismic emergence through seven centuries at three sites along the Sunda megathrust. Data constrain solid parts of the curves well; dotted portions are inferred. Emergence values (in centimeters ±2σ) are red. Interseismic submergence rates (in millimeters per year, ±2σ) are blue. Millennial emergence rates are black. Vertical dashed white lines mark dates of emergences. Red arrows at bottom highlight the timing of the failure sequence for each supercycle. (Source: *Sieh et al.* [2008]. Reproduced with permission of The American Association for the Advancement of Science.)

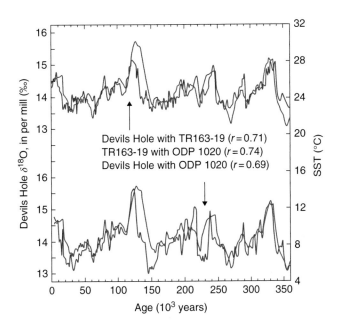

Fig. 12.9. Devils Hole δ¹⁸O (in black) and sea surface temperature (SST; in gray) time series, for two different SST records: marine core TR 163-19 and ODP 1020. Correlation coefficients (*r*) calculated for the period 360,000 to 4500 years ago. (Source: *Winograd et al.* [2006]. Reproduced with permission of Elsevier.)

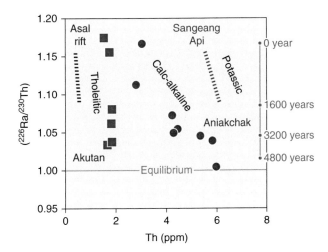

Fig. 12.16. Diagram showing (^{226}Ra)/(^{230}Th) as a function of the differentiation of the magma. Here, concentration of thorium, an incompatible element, is shown as an index of differentiation, with higher concentrations of Th indicating more differentiated magma. This pattern of decreasing disequilbria with increasing differentiation can be interpreted as a timescale of differentiation (axis on right side of diagram), but only in the case of closed-system evolution. (Source: *Turner and Costa* [2007]. Reproduced with permission of Mineralogical Society of America.)

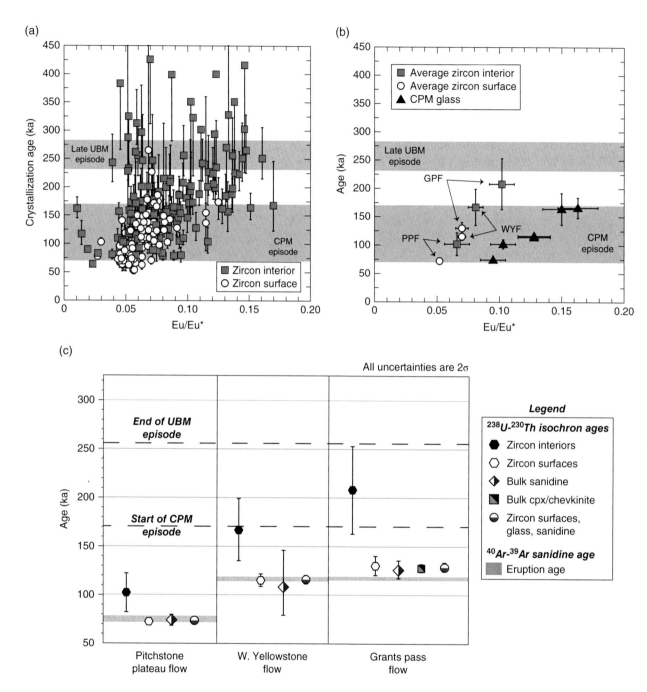

Fig. 12.19. Summary of U-series age data for Yellowstone case study; diagrams modified from *Stelten et al.* [2015]. (a) $^{238}U–^{230}Th$ zircon spot age versus Eu/Eu* for the same spots. Eu/Eu* is the europium anomaly (the deviation of Eu measured from that expected based on neighboring rare earth element concentrations), which is related to feldspar fractionation and magma differentiation. Spots located on zircon surfaces are shown as white circles, those located on polished interiors are shown as orange squares. (b) Average ages and Eu/Eu* of all zircon interior (orange squares) and surface (white circles) spot ages for individual eruptions studied, compared to glass data (black triangles) for the same eruptions. (c) Average zircon spot ages compared to sanidine ages and eruption ages (grey bars) for individual samples. (Source: *Stelten et al.* [2015]. Reproduced with permission of Oxford University Press.)

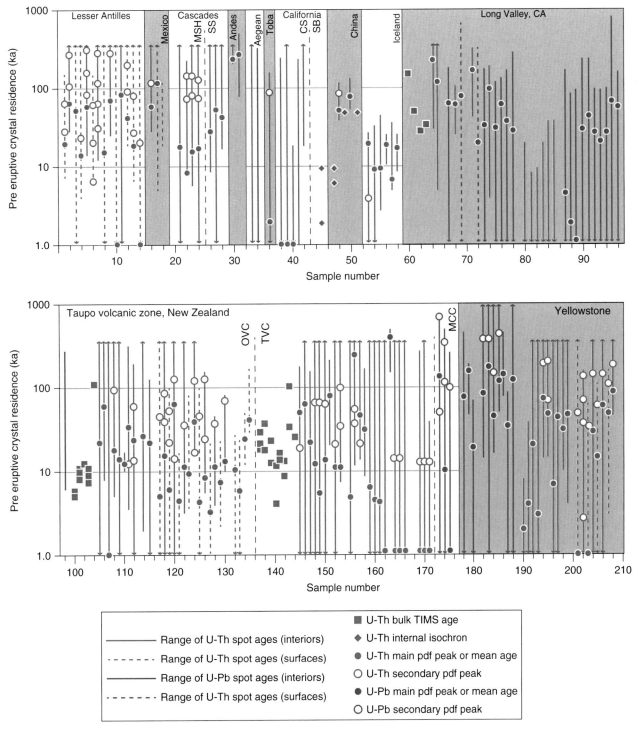

Fig. 12.20. Compilation of ^{238}U–^{230}Th and U–Pb ages of accessory minerals in volcanic rocks from *Cooper* [2015], expressed as preeruptive crystal residence ages (crystal age minus eruption age). Most ages shown are model ages based on ^{238}U–^{230}Th analyses *in situ* compared to host glasses (orange circles), with some U–Pb analyses *in situ* (green circles). Also shown are some internal isochron ages based on multiple spot analyses (orange diamonds), and a few bulk analyses of zircon by TIMS (orange squares). Each vertical array shows data for a single sample (or a group of closely related samples). Mean age or the dominant peak on probability density functions (PDF) is shown by solid circles; secondary PDF peaks are indicated by open circles. No circles are shown if no mean age or PDF is calculated in the original reference. Lines indicate range of ages for individual spot analyses; solid for interior analyses and dashed for unpolished surface analyses. Upward-pointing arrowheads on lines indicate analyses within error of secular equilibrium (or off-scale for U–Pb ages); downward-pointing arrowheads represent analyses within error of eruption age. Sample numbers are arbitrary. Geographic regions are separated by shaded bars, individual eruptive centers within these areas are separated by dashed black lines and labeled as follows: MSH, Mount St. Helens; SS, South Sister; CS, Coso; SB, Salton Buttes; OVC, Okataina Volcanic Center; TVC, Taupo Volcanic Center; MCC, Mangakino Caldera Center. (Source: *Cooper* [2015]. Reproduced with permission of The Geological Society.)

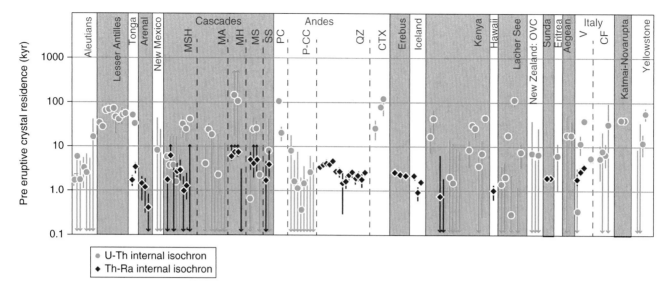

Fig. 12.21. Compilation of ^{238}U–^{230}Th and ^{230}Th–^{226}Ra ages of bulk mineral separates of major phases, expressed as the pre-eruptive residence age (bulk mineral age minus the eruption age). Error bars for U-series crystal residence ages indicate maximum and minimum residence ages calculated by combining the 2σ uncertainty on the crystal age with the 2σ uncertainty on the eruption. Arrows on error bars indicate samples with ages within error of eruption age (downward-pointing arrows) or indeterminate age maxima (upward-pointing arrows). Mean or best-estimate ages are shown by light blue filled circles (^{238}U–^{230}Th ages) or dark blue filled diamonds (^{230}Th–^{226}Ra ages); lines without symbols indicate samples for which only a maximum and minimum age estimate were available. Regions are labeled, and individual volcanic centers within each region are abbreviated as follows: MSH, Mount St. Helens; MA, Mound Adams; MH, Mount Hood; MS, Mount Shasta; SS, South Sister; PC, Parinacota; P-CC, Puyehue-Cordon Caulle; QZ, Quizapu; CTX, Cotopaxi; V, Vesuvius; CF, Campei Flegrei. (Source: *Cooper* [2015]. Reproduced with permission of The Geological Society.)

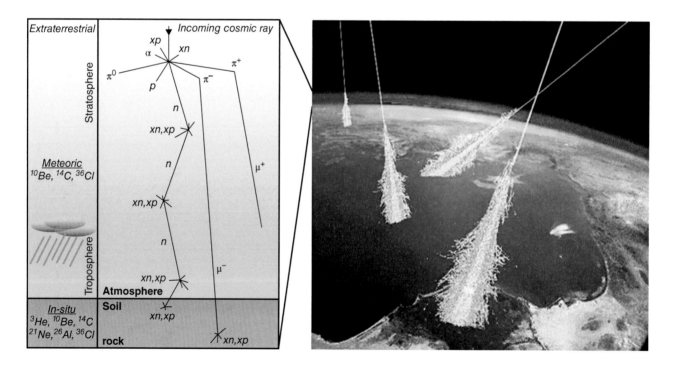

Fig. 13.1. A schematic cosmic-ray cascade generating cosmogenic nuclides in the atmosphere, soil, and rock. The incoming primary cosmic ray strikes an atmospheric nucleus and breaks it into smaller fragments. The high-energy fragments then travel through the atmosphere and generate further nuclear reactions, causing a cascade. Short-lived pions in the upper atmosphere decay into muons that travel long distances through the atmosphere and into rock. The products of these nuclear reactions are called cosmogenic nuclides. Those in the atmosphere are referred to as meteoric, those in rock soil are referred to as *in situ*-produced, and those produced in meteorites outside Earth's atmosphere are referred to as extraterrestrial.

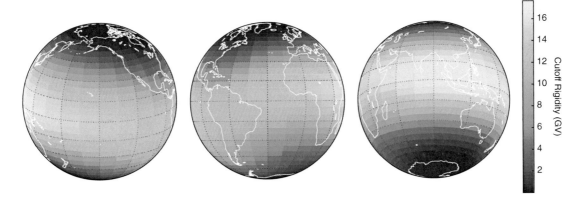

Fig. 13.2. Effective vertical cutoff rigidity (R_C). Lower energy cosmic rays are funneled to the poles, while only higher energy cosmic rays are able to penetrate the geomagnetic field to enter the atmosphere at lower latitudes. Cosmogenic nuclide-production rates are therefore partly a function of latitude. Variations in cutoff rigidity account for about a factor of two in cosmogenic nuclide production rates at the ground surface from high to low latitudes. (Source: *Sato et al.* [2008].)

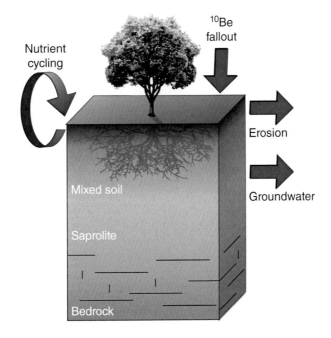

Fig. 13.10. A simplified $^{10}Be_{met}$ budget in a soil column. Meteoric ^{10}Be accumulates from wet and dry deposition on the surface. It quickly adsorbs to soil particles and is removed over time by particulate erosion. Depending on local rainfall and soil pH, a fraction is also carried away in groundwater. Beryllium can also be passively incorporated into the nutrient cycle, although this remains a little studied aspect of cosmogenic nuclides.

Fig. 13.11. Sampling a bouldery glacial moraine. It is a good idea to sample multiple boulders that protrude from the moraine surface to ensure that individual samples do not suffer from erosion, prior inheritance, or snow cover. For this boulder the researchers are recording the topographic shielding of low-angle secondary cosmic rays by measuring the angle to the horizon. Such topographic shielding must be applied as a correction factor to the cosmogenic nuclide-production rate. (Source: Courtesy of Marc Caffee.)

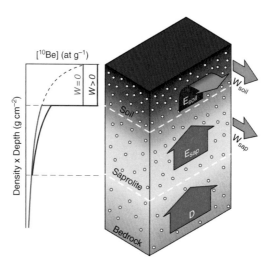

Fig. 13.12. Erosion and weathering convert bedrock to saprolite and soil. The total denudation rate (D) is equal to the sum of erosion (E) and chemical weathering (W). Weathering of minerals during the conversion of bedrock to soil leaves behind resistant minerals such as quartz, illustrated by the white circles. Additional weathering within the soil increases the quartz concentration. This results in a higher cosmogenic nuclide concentration in the quartz than would otherwise be present without chemical weathering. The curve at left shows the expected ^{10}Be concentration as a function of depth within the column, expressed in units of g/cm^2. The dashed line shows the ^{10}Be production rate. Vertical mixing within the soil homogenizes the upper part of the profile, leading to the gray curve for the case where weathering is negligible (equation (13.17)). Chemical weathering leads to higher concentrations in the profile, illustrated as the black curve (equation (13.18)).

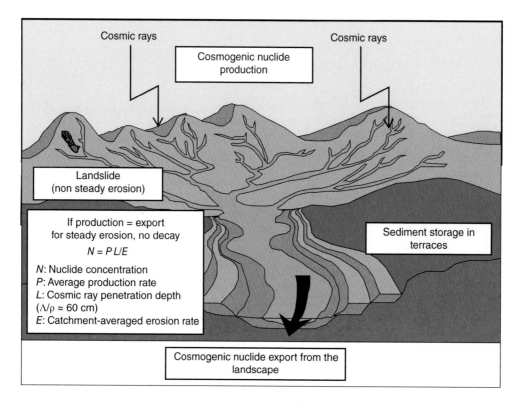

Fig. 13.13. Streams mix sediment that is eroded from throughout the landscape. The average ^{10}Be concentration in stream sediment can be used to estimate the spatially averaged erosion rate in the watershed, assuming steady state as in equation (13.21). Complications can arise due to nonsteady erosion processes such as landslides, or due to sediment storage and remobilization in terraces. Adapted from. (Source: *Granger and Schaller* [2014]. Reproduced with permission of Mineralogical Society of America.)

CHAPTER 9

The K–Ar and ^{40}Ar/^{39}Ar systems

9.1 INTRODUCTION AND FUNDAMENTALS

The K–Ar isotopic system was one of the first to be used for geochronology and it has been applied to diverse problems for more than 60 years. K–Ar dating provided the initial chronology of geomagnetic polarity reversals, which proved key to early development of the theory of plate tectonics. It was the first method to be applied systematically to the calibration of biochronological timescales [*Evernden et al.*, 1964] and to establish the antiquity of human evolution [*Leakey et al.*, 1961]. One of the early recognized limitations of the K–Ar geochronometer, the tendency of the radiogenic daughter ^{40}Ar* to "leak" from minerals at elevated temperatures, has subsequently been turned to advantage in thermochronology, an application that blossomed with the development of the derivative ^{40}Ar/^{39}Ar method (see Chapter 5). The K–Ar method has been increasingly supplanted by the more analytically complex but far more powerful ^{40}Ar/^{39}Ar method, which is based on the same radioactive decay scheme. We nonetheless discuss the K–Ar method in some detail, partly for historical reasons but more importantly because it forms the basis of the ^{40}Ar/^{39}Ar method and thus provides important groundwork for understanding the latter.

Potassium has a solar abundance of 4 ppm and is a strongly lithophilic and relatively volatile element with large ionic radius and 1^+ charge. It is commonly a major element in terrestrial igneous rocks, in which the oxide K_2O normally composes 0.1–10 wt%. Many minerals contain greater than 1 wt% K_2O, and in most planetary environments this endowment of parent element is the greatest of all radioisotopic systems. The naturally radioactive isotope ^{40}K has low abundance (0.0117%) compared with the other two natural isotopes of potassium ^{39}K (93.258%) and ^{41}K (6.730%), but the large abundance of potassium overall translates to unusually high concentrations of parent nuclide, approximately 10–1500 ppm in typical terrestrial igneous minerals. Accordingly, the enrichment in radiogenic daughter (^{40}Ar*) per unit time tends to be high, allowing relatively young events to be dated, although this is partially limited by the abundance of nonradiogenic ^{40}Ar in the atmosphere, which comprises 0.93% by dry volume.

The K–Ar system is distinctive compared with most other radioisotopic geochronometers in that the parent isotope (^{40}K) undergoes branched decay, to both ^{40}Ca (by β^- emission) and ^{40}Ar (primarily by electron capture, with a negligible contribution from β^+ decay) (Fig. 9.1). Thus the parent/daughter ratio (^{40}Ar*/^{40}K) changes over time at a faster rate than if all ^{40}K decayed to ^{40}Ar, and this must be accounted for. Use of the β^- decay to ^{40}Ca for geochronometry will be discussed later; here we focus on the ^{40}Ar-producing branch.

$$\frac{d(^{40}K)}{dt} = -\left[\frac{d(^{40}Ca)}{dt} + \frac{d(^{40}Ar)}{dt}\right] = -\lambda(^{40}K) \quad (9.1)$$

A second important feature of the K–Ar system is that the daughter isotope is a noble gas. This fact means that the daughter cannot be chemically bonded into mineral lattices after its formation by spontaneous decay, and is prone to migration by diffusion. Argon diffusion is thermally activated, and its rate increases with increasing temperature. The loss of ^{40}Ar* from K-bearing minerals by diffusion is both a liability and a benefit, much as in the case of ^{4}He* in U–Th-bearing minerals. The main liability associated with diffusion is that materials can lose ^{40}Ar* if they are subjected to moderate temperatures, and the K–Ar age will underestimate the age of formation. On the other hand, one of the benefits of diffusion is that minerals crystallizing from magma (generally at temperatures > 600 °C) tend to lose ^{40}Ar* rapidly, and do not begin to accumulate ^{40}Ar* quantitatively until cooling to much lower temperatures, i.e., upon eruption. This property makes the K–Ar method exceptionally useful for dating volcanic materials. A second important benefit of diffusion is that if the rate parameters for diffusion are known or can be determined experimentally, the propensity of ^{40}Ar* to diffuse from minerals forms the basis of Ar thermochronology (Chapter 5).

The K–Ar and ^{40}Ar/^{39}Ar dating methods are applicable to a uniquely large variety of materials and geologic environments

Geochronology and Thermochronology, First Edition. Peter W. Reiners, Richard W. Carlson, Paul R. Renne, Kari M. Cooper, Darryl E. Granger, Noah M. McLean, and Blair Schoene.
© 2018 John Wiley & Sons Ltd. Published 2018 by John Wiley & Sons Ltd.

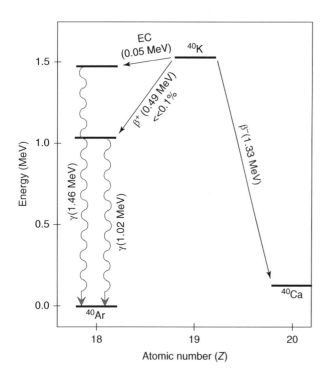

Fig. 9.1. Decay modes and energy levels for the radioactive isotope ^{40}K. The electron capture (EC) mode to an excited state of ^{40}Ar reduces the energy by 0.05 MeV, followed by emission of a 1.46 MeV gamma ray to the ground state of ^{40}Ar.

due to the ubiquity of K. For volcanic materials and some authigenic minerals, these methods may date the time of eruption or formation, respectively. For metamorphic and plutonic materials, these methods are invaluable tools for quantitifying thermal histories. The age range over which the K–Ar and $^{40}Ar/^{39}Ar$

methods are useful is unsurpassed by any radioisotopic method, ranging from the early solar system to less than 1000 years ago.

9.2 HISTORICAL PERSPECTIVE

The natural radioactivity of potassium was first demonstrated by *Campbell and Wood* [1907], before the concept of isotopes was established. The existence of potassium's two most abundant isotopes, ^{39}K and ^{41}K, was demonstrated by *Aston* [1921]. The existence of the rare radioactive isotope ^{40}K, predicted to decay by β− emission [*Klemperer*, 1935; *Newman and Walke*, 1935] was confirmed by *Nier* [1935], who determined a natural $^{39}K/^{40}K$ ratio of 8600 ± 10%. *Newman and Walke* [1935] noted the possibility of a dual decay of ^{40}K to ^{40}Ar and ^{40}Ca, and the former was supported by *von Weisacker* [1937] based on the observation of excess of ^{40}Ar in Earth's atmosphere relative to expectations about nuclear synthesis. *Bramley* [1937] provided further support for an electron-capture decay mode in addition to **β−** emission.

Von Weisacker [1937] made the prescient suggestion that his inference of an electron capture decay mode of ^{40}K to ^{40}Ar could be tested by analyzing old, K-rich rocks. A decade passed before this was accomplished by *Aldrich and Nier* [1948]. In their landmark study, Aldrich and Nier (Fig. 9.2) reported $^{40}Ar/^{36}Ar$ ratios for four K-rich minerals, and calculated K–Ar ages for them based on assumed stoichiometry of K and what was then known of the decay constants. This study launched the K–Ar dating method, and further applications quickly followed. By 1960, several laboratories devoted to K–Ar dating existed.

With increasing attention devoted to the topic, many fundamental refinements to the K–Ar method occurred in the 1950s and 1960s. Measurements of the β− and γ activity of K led to

Fig. 9.2. Lyman Thomas Aldrich (left) and Alfred Otto Carl Nier (right, showing off his homemade mass spectrometer) determined the first K–Ar ages in 1948. (Source: Aldrich: https://carnegiescience.edu/news/geophysicist-lyman-thomas-aldrich-dies; Nier: https://en.wikipedia.org/wiki/Alfred_O._C._Nier)

improved estimates of the decay constants for the β^- and electron-capture decay modes (respectively), as summarized by *Beckinsale and Gale* [1969] and revisited more objectively by *Min et al.* [2000]. The isotopic composition of argon in the terrestrial atmosphere, used to correct for ubiquitous nonradiogenic ^{40}Ar in samples, was determined by [*Nier*, 1950], whose results stood unsurpassed for a remarkable 56 years.

As of this writing, the K–Ar and ^{40}Ar/^{39}Ar dating methods have been in use for 65 and 49 years, respectively. K–Ar dating in many ways revolutionized paleontology; it was the first method to document the antiquity of human evolution [e.g., *Leakey et al.* 1961] and shortly thereafter it was the first to be applied systematically to the calibration of biochronological timescales [*Evernden et al.*, 1964]. K–Ar dating provided the first chronology of geomagnetic polarity reversals [*Cox et al.*, 1963; *McDougal and Tarling*, 1963], which proved key to the nascent theory of plate tectonics. Over the past 20 years, the venerable K–Ar method has given way to the more powerful ^{40}Ar/^{39}Ar technique. The ^{40}Ar/^{39}Ar technique is really just a modification of the K–Ar method, and although much more complex and time-consuming, it provides a plethora of tools to surmount the limitations of K–Ar dating. Unsurprisingly, the number of laboratories practicing K–Ar has dwindled dramatically over the past two decades, while those practicing ^{40}Ar/^{39}Ar has grown commensurately, but there are several advantages (discussed below) to the K–Ar method that ensure a viable role in the geochronology arsenal. There are approximately 50 ^{40}Ar/^{39}Ar laboratories in the world today, about 20 of which are in the United States.

9.3   K–AR DATING

The K–Ar age (t_u) of a sample is given by a slight modification of the standard age equation which accounts for the fact that the parent isotope ^{40}K decays to ^{40}Ca as well as ^{40}Ar, as shown by equation (9.1).

$$t_u = \frac{1}{\lambda}\ln\left[\left(\frac{\lambda}{\lambda_e}\right)\left(\frac{^{40}\text{Ar}^*}{^{40}\text{K}}\right) + 1\right] \tag{9.2}$$

in which the decay constants λ and λ_e are known, and the ratio of ^{40}Ar* to ^{40}K must be measured. This is accomplished by determining the concentrations of ^{40}Ar* to ^{40}K in the sample separately.

9.3.1   Determining ^{40}Ar*

The total abundance of ^{40}Ar is generally determined by isotope dilution (Chapter 3) using a ^{38}Ar tracer. There is always some nonradiogenic ^{40}Ar present in samples, and there is no a priori way to distinguish radiogenic from nonradiogenic ^{40}Ar. In terrestrial materials, atmospheric ^{40}Ar is always present as some combination of Ar that is initially trapped or is introduced during sample preparations. If all nonradiogenic Ar in a sample has

the isotopic composition of the modern atmosphere, a simple air correction can be made such that

$$^{40}\text{Ar}^* = {}^{40}\text{Ar}_t - {}^{36}\text{Ar}_t\left(\frac{^{40}\text{Ar}}{^{36}\text{Ar}}\right)_A \tag{9.3}$$

where ^{40}Ar$_t$ and ^{36}Ar$_t$ are the total measured quantities and $(^{40}\text{Ar}/^{36}\text{Ar})_A$ is the modern atmospheric ratio. This so-called "air correction" is ubiquitously applied in K–Ar dating.

The value of $(^{40}\text{Ar}/^{36}\text{Ar})_A$ used by convention since 1977 [*Steiger and Jager*, 1977] has been 295.5 based on the pioneering work of *Nier* [1950], but is now increasingly accepted to be equal to 298.56 $\pm$ 0.17 [*Lee et al.*, 2006]. The value of $(^{40}\text{Ar}/^{36}\text{Ar})_A$ is commonly used not only for the air correction but also to determine the measurement bias (mass discrimination) of the mass spectrometer, as discussed below. *Renne et al.* [2009a] showed that different values of $(^{40}\text{Ar}/^{36}\text{Ar})_A$ yield indistinguishable ages as long as the same value is used consistently for all computations. Nonetheless, the value of *Lee et al.* [2006] is recommended.

9.3.1.1   Initial ^{40}Ar

A serious limitation of the K–Ar method is that there are no intrinsic features of the data which allow testing of the assumption that all nonradiogenic Ar in a sample has the isotope composition of modern atmosphere. A majority of historic (hence, lacking detectable ^{40}Ar*) lava flows seem to support this assumption for volcanic rocks [*Dalrymple*, 1969]. Nonetheless, numerous cases are documented wherein samples have nonradiogenic Ar with $(^{40}\text{Ar}/^{36}\text{Ar})$ greater than atmospheric [e.g., *Kelley*, 2002]; this is especially common in deep-seated metamorphic or plutonic rocks subjected to high Ar pressures in the presence of old, K-rich crust. Such occurrences are commonly referred to as "excess Ar". *Dalrymple and Lanphere* [1969] present a more detailed nomenclature for various sources of initial ^{40}Ar. Less commonly, nonradiogenic Ar with $(^{40}\text{Ar}/^{36}\text{Ar})$ less than atmospheric are known in certain volcanic rocks, particularly obsidians, where they may arise from kinetic fractionation of atmospheric Ar [*Renne et al.*, 2009a].

A further complication is that there is no reason to believe that atmospheric Ar has been constant over geologic time. If rocks or minerals trap ambient atmosphere when they form, age calculations based on an air correction assuming modern atmospheric isotope composition will be biased. Several studies have supported the reasonable assumption that the ^{40}Ar/^{36}Ar of terrestrial atmospheric Ar has increased over time due to outgassing of radiogenic ^{40}Ar from the solid Earth [*Cadogan*, 1977; *Hanes et al.*, 1985; *Bender et al.*, 2008; *Pujol et al.*, 2013]. The exact form of the evolution of atmospheric ^{40}Ar/^{36}Ar over time is elusive because it depends on the still largely unknown history of degassing of the Earth. Constraining this history, especially whether it is episodic or continuous, remains a significant frontier in Earth science [*Parman*, 2007]. In sum, although there is evidence that the air correction as commonly applied is flawed, there is no clear alternative. The use of isochrons—especially practical in ^{40}Ar/^{39}Ar dating, as discussed below—offers the best solution to this conundrum.

9.3.2 Determining ^{40}K

^{40}K is determined using various methods. Most commonly, the total K concentration is measured using flame photometry, but other methods (X-ray fluorescence, atomic absorption) have been used. The ^{40}K concentration is then determined assuming the proportion ^{40}K/K of natural potassium today. Since 1977, this value has been widely accepted [*Steiger and Jager*, 1977] to be 0.01167% based on the isotope dilution measurements of *Garner et al.* [1975]. Isotope dilution is sometimes used to directly measure ^{40}K concentrations, but this is uncommon due to the more laborious nature of the technique.

The validity of assuming a constant value of ^{40}K/K for terrestrial materials was supported by *Humayun and Clayton* [1995] who showed that no variation in ^{41}K/^{39}K could be detected at the ~0.5% level. However, recent work with more sensitive techniques has discovered variations up to 0.13% in terrestrial ^{41}K/^{39}K [*Li et al.*, 2016; *Wang and Jacobsen*, 2016]. It is easy to show that variations at this level are insignificant in K–Ar dating, where accuracy better than 0.5% of the age is normally not expected. However, we will explore this further below in the context of ^{40}Ar/^{39}Ar dating.

9.4 ^{40}AR/^{39}AR DATING

In the simplest terms, the ^{40}Ar/^{39}Ar method is merely a specialized variant of the K–Ar method in which the ^{40}K concentration is measured by a form of neutron activation. Irradiating a sample with fast neutrons induces a neutron capture reaction whereby a neutron is absorbed by ^{39}K, and a proton is released to form ^{39}Ar. A convenient shorthand version of this reaction is ^{39}K(n,p)^{39}Ar. As ^{39}K/^{40}K is essentially constant at any time since nucleosynthesis, ^{39}K is a reasonable proxy for ^{40}K and thus ^{39}Ar is related straightforwardly to the parent isotope ^{40}K in the K–Ar system. The validity of this fundamental assumption is discussed later in this section.

A precursor of the ^{40}Ar/^{39}Ar method using counting techniques was proposed by *Wanke and Konig* [1959], whose approach required knowledge of the neutron energy distribution in the reactor, and allowed no provision for corrections for non-radiogenic ^{40}Ar. These limitations were surmounted by the seminal work of *Merrihue and Turner* [1966], who measured Ar isotopes with a mass spectrometer and introduced the concepts of (i) using a standard of known age to effectively determine K concentration, (ii) stepwise heating to reveal open system behavior and non-radiogenic ^{40}Ar, and (iii) measuring all relevant data from a single sample rather than using separate aliquots to determine ^{40}K and ^{40}Ar concentrations. Thus the important foundations of the ^{40}Ar/^{39}Ar method, and their advantages relative to conventional K–Ar dating, were established by *Merrihue*

Fig. 9.3. Grenville Turner made many fundamental contributions to development of the ^{40}Ar/^{39}Ar dating method and applied these to deducing thermal histories of meteorites and lunar rocks. Turner's account of this work in its broader context is chronicled in an oral history [*Sears*, 2012]. (Source: https://royalsociety.org/people/grenville-turner-12442/)

and Turner [1966]. Craig Merrihue died tragically in a mountain climbing accident before the publication of this seminal paper; he was honored posthumously by having the mineral Merrihueite named after him [*Dodd et al.*, 1965]. Grenville Turner (Fig. 9.3) went on to make many fundamental contributions to ^{40}Ar/^{39}Ar geochronology and cosmochronology, with emphasis on the thermal histories of meteorites and lunar rocks.

Unbeknownst to Merrihue and Turner, an unpublished manuscript had been written in Icelandic by *Sigurgeirsson* [1962], who outlined the theoretical basis for ^{40}Ar/^{39}Ar dating and its advantages for dating of young basalts. Sigurgeirsson's approach built upon the ideas of *Wanke and Konig* [1959] but encompassed many of the improvements adopted by *Merrihue and Turner* [1966], including the applicability of stepwise degassing by incremental heating. Unfortunately, Sigurgeirsson was unable to implement his proposal experimentally, and this important conceptual breakthrough was delayed for several years.

9.4.1 Neutron activation

The probability of the ^{39}K(n,p)^{39}Ar neutron capture reaction is measured by the capture cross-section σ, which is a function of neutron energy ε. In testimony to the humor of nuclear physicists, the units of σ are called barns. Neutron irradiation for ^{40}Ar/^{39}Ar dating is generally conducted in the core of a ^{235}U fission reactor, which produces a characteristic spectrum $\phi(\varepsilon)$ of neutron energies, where $\phi(\varepsilon)$ is the flux of neutrons of energy ε. The typical fission spectrum of energies is modified by materials in the reactor core, hence the neutron energy spectrum varies between different reactors and may even vary between irradiations in the same reactor depending on the placement of samples.

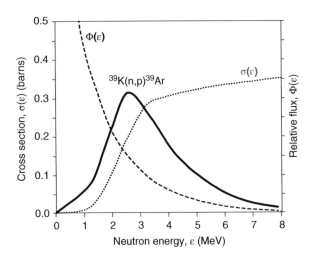

Fig. 9.4. Cross-section $\sigma(\varepsilon)$ for the reaction ^{39}K(n,p)^{39}Ar as a function of neutron energy ε. Neutron flux $\phi(\varepsilon)$ for a ^{235}U fission spectrum is shown in arbitrary units as this depends on the operating power of the reactor. The solid curve is the product $\sigma(\varepsilon)\phi(\varepsilon)$ and shows the relative proportion of ^{39}Ar produced as a function of neutron energy. The total amount of ^{39}Ar produced is equal to the area under the solid curve.

As shown in Fig. 9.4 the amount of ^{39}Ar produced from ^{39}K by the neutron-capture process in an irradiation of duration Δt is given by

$$^{39}\text{Ar}_\text{K} = {}^{39}\text{K} \cdot \Delta t \cdot \int_0^\infty \sigma(\varepsilon)\phi(\varepsilon) \tag{9.4}$$

The subscript K specifies the portion of ^{39}Ar produced from K, as distinct from other possible sources. Its usage is not consistent in the literature, but is often implied in the quantity ^{40}Ar*/^{39}Ar. The amount of radiogenic ^{40}Ar produced in a sample of age t is given by solving equation (9.2)

$$^{40}\text{Ar}^* = {}^{40}\text{K}\left(\frac{\lambda_\varepsilon}{\lambda}\right)\left(e^{\lambda t} - 1\right) \tag{9.5}$$

where $\lambda = \lambda_\text{b} + \lambda_\text{e}$ is the total decay constant of ^{40}K. Combining the above

$$\frac{^{40}\text{Ar}^*}{^{39}\text{Ar}_\text{K}} = \left(\frac{\lambda_\varepsilon}{\lambda}\right)\left(\frac{^{40}\text{K}\left(e^{\lambda t}-1\right)}{^{39}\text{K} \Delta t \int_0^\infty \sigma(\varepsilon)\phi(\varepsilon)}\right) \tag{9.6}$$

This expression could be solved for t in principle, but evaluating the neutron flux integral directly is prohibitively difficult because neither $\sigma(\varepsilon)$ nor $\phi(\varepsilon)$ are known quantitatively over the relevant portion of the neutron energy spectrum. A clever alternative was devised by *Mitchell* [1968], who defined a quantity J such that

$$J \equiv \left(\frac{^{39}\text{K}}{^{40}\text{K}}\right)\left(\frac{\lambda}{\lambda_\varepsilon}\right)\Delta t \int_0^\infty \sigma(\varepsilon)\phi(\varepsilon) \tag{9.7}$$

Strictly speaking, J has the units of ^{39}K atoms/^{40}K atoms, but since ^{39}K/^{40}K is normally expressed on an atomic basis it is generally considered to be dimensionless. The value of J is then determined by analyzing material of known age (t_S) referred to as a standard or neutron fluence monitor, which had the same irradiation history as the sample. Conceptually, J is a relative measure of the extent of conversion of ^{39}K to ^{39}Ar$_\text{K}$, and its value increases with irradiation duration. For a given reactor operating at a given power, J is a linear function of the time duration of the irradiation. Combining equations (9.6) and (9.7) yields

$$J = \left(\frac{^{39}\text{Ar}_\text{K}}{^{40}\text{Ar}^*}\right)\left(e^{\lambda t_\text{S}} - 1\right) \tag{9.8}$$

where t_S is the age of the standard. Rearranging this equation and substituting the age and isotope composition of the unknown for that of the standard yields the classic age equation:

$$t = \frac{1}{\lambda}\ln\left[1 + J\frac{^{40}\text{Ar}^*}{^{39}\text{Ar}_\text{K}}\right] \tag{9.9}$$

For many purposes it is convenient to express the definition of J explicitly to yield

$$t = \frac{1}{\lambda}\ln\left[1 + \left(\frac{^{39}\text{Ar}_\text{K}}{^{40}\text{Ar}^*}\right)_\text{s}\left(e^{\lambda \tau} - 1\right)\left(\frac{^{40}\text{Ar}^*}{^{39}\text{Ar}_\text{K}}\right)_\text{u}\right] \tag{9.10}$$

where U denotes the unknown. Defining an intercalibration factor [*Renne et al.*, 1998] as

$$R_\text{s}^\text{u} \equiv \left(\frac{^{40}\text{Ar}^*}{^{39}\text{Ar}_\text{K}}\right)_\text{u} \bigg/ \left(\frac{^{40}\text{Ar}^*}{^{39}\text{Ar}_\text{K}}\right)_\text{s} \tag{9.11}$$

allows this expanded age equation to be written

$$t = \frac{1}{\lambda}\ln\left[1 + \left(e^{\lambda \tau} - 1\right)R_\text{s}^\text{u}\right] \tag{9.12}$$

Several issues follow from the forgoing. Among these is the role of standards and how well their ages are known. This a rich topic and detailed discussion is reserved for the section on Accuracy. However, it is appropriate to comment here on the popular characterization of the ^{40}Ar/^{39}Ar technique as a relative dating method, due to the use of standards. Considering that standards are used only to determine the parent isotope concentration, it should be clear that the ^{40}Ar/^{39}Ar method is no more a relative dating technique than other methods that require use of a tracer to determine parent and daughter concentrations by isotope dilution.

9.4.1.1 Neutron fluence gradients

Various schemes are used for irradiating samples and standards. Because the neutron flux in reactors is not uniform, the value of J assigned to a given sample may be derived from analysis of standards that bracket the samples spatially. In such cases average or interpolated values of J may be used. For studies requiring utmost age precision, the number of standards analyzed may approach the number of unknowns. Fig. 9.5 shows two commonly used irradiation schemes. Scheme A, permitting irradiation of large (tens of mg) samples, is declining in use as it permits only characterization of vertical neutron fluence gradients. Scheme B allows much more precise control of neutron fluence gradients and the smaller amount of samples (and standards) irradiated is enabled by the increasingly high sensitivity of mass spectrometers and cleanliness of extraction systems.

A typical variation in neutron fluence for vertically arrayed samples is shown in Fig. 9.6a. To reduce neutron fluence gradients in the plane of disks such as those shown in Fig. 9.6b, the entire package may be rotated about the cylindrical axis in some

reactors set up to do so. While rotation does homogenize neutron fluence gradients to some extent, radial gradients of up to 2.2%/cm have been reported for samples rotated during irradiation [*Rutte et al.*, 2015].

9.4.1.2 Effects of variable ⁴⁰K/K

Application of equation (9.9) or (9.12) implicitly assumes that the $^{40}K/^{39}K$ of the sample is the same as that of the unknown. How much bias in age would be introduced if this assumption is invalid due to natural fractionation processes? To address this question we start with equation (9.12) to calculate the apparent age t_U' of an unknown whose $^{40}K/^{39}K$ may be different from that of the standard:

$$t_u' = \frac{1}{\lambda}\ln\left[1 + \left(e^{\lambda t} - 1\right)R_s^u\right] \tag{9.13}$$

Substitute in the definition of R_s^u:

$$t_u' = \frac{1}{\lambda}\ln\left[\left(\frac{\lambda}{\lambda_\varepsilon}\right)\left(\frac{^{40}Ar^*}{^{40}K}\right)_s\left(\frac{^{40}Ar^*}{^{39}Ar_K}\right)_u\left(\frac{^{39}Ar_K}{^{40}Ar^*}\right)_s + 1\right] \tag{9.14}$$

and the equations for production of $^{40}Ar^*$ and $^{39}Ar_K$:

$$^{40}Ar_u{}^* = {}^{40}K_u\left(\frac{\lambda_\varepsilon}{\lambda}\right)\left(e^{\lambda t_u} - 1\right) \tag{9.15}$$

$$^{39}Ar_{K,u} = {}^{39}K_u\Delta t\int_0^\infty \sigma(\varepsilon)\phi(\varepsilon) \tag{9.16}$$

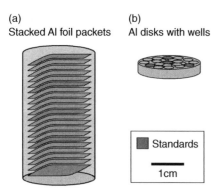

(a) Stacked Al foil packets

(b) Al disks with wells

Standards

1cm

Fig. 9.5. Two schemes for irradiating samples and standards. Scheme A allows detection only of vertical neutron fluence gradients and its use is declining as demands for precise knowledge of J-values increases. Scheme B allows tight control of horizontal neutron fluence by close bracketing or interpolation of J-values.

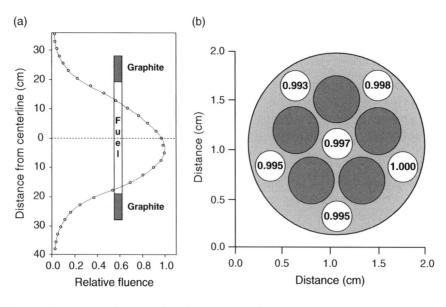

Fig. 9.6. Typical neutron fluence gradients corresponding to sample irradiation geometries shown in Fig. 9.5. (a) Vertical fluence gradient for an irradiation in the US Geological Survey TRIGA reactor. (Source: adapted from *Dalrymple et al.* [1981].) (b) Lateral fluence gradient for an irradiation in the Oregon State University TRIGA reactor. (Source: *Renne et al.* [2015]. Reproduced with permission of The American Association for the Advancement of Science.) Both panels show the parameter J normalized to its maximum value.

$$^{39}Ar_{s*} = {}^{40}K_s\left(\frac{\lambda_\varepsilon}{\lambda}\right)\left(e^{\lambda t_s} - 1\right) \tag{9.17}$$

$$^{39}Ar_{K,s} = {}^{39}K_s\Delta t\int_0^\infty \sigma(\varepsilon)\phi(\varepsilon) \tag{9.18}$$

to obtain

$$t'_u = \frac{1}{\lambda}\ln\left[\left(\frac{\lambda}{\lambda_\varepsilon}\right)\left(\frac{{}^{39}K_s\left(\frac{\lambda}{\lambda_\varepsilon}\right)\left(e^{\lambda t_s}-1\right)}{{}^{40}K_s}\right)\right.$$
$$\left.\left(\frac{{}^{40}K_u\left(\frac{\lambda_\varepsilon}{\lambda}\right)\left(e^{\lambda t_u}-1\right)}{{}^{39}K_u\Delta t\int_0^\infty \sigma(\varepsilon)\phi(\varepsilon)}\right)\left(\frac{{}^{39}K_s\Delta t\int_0^\infty \sigma(\varepsilon)\phi(\varepsilon)}{{}^{40}K_s\left(\frac{\lambda_\varepsilon}{\lambda}\right)\left(e^{\lambda t_s}-1\right)}\right)+1\right] \tag{9.19}$$

which reduces to

$$t'_u = \frac{1}{\lambda}\ln\left[\left(\frac{{}^{40}K_u}{{}^{39}K_u}\right)\left(\frac{{}^{39}K_s}{{}^{40}K_s}\right)\left(e^{\lambda t_s}-1\right)+1\right] \tag{9.20}$$

From this relationship it is easily shown that a maximum age bias of about 0.02% would arise from the 0.04% variation in ^{41}K/^{39}K observed by [*Li et al.*, 2016; *Wang and Jacobsen*, 2016], assuming that the variation in ^{40}K/^{39}K is half that of ^{41}K/^{39}K.

9.4.2   Collateral effects of neutron irradiation

In addition to the motivating production of ^{39}Ar from ^{39}K via the ^{39}K(n,p)^{39}Ar neutron activation process, there are additional consequences of neutron irradiation that are important for ^{40}Ar/^{39}Ar dating.

9.4.2.1   Interfering reactions
Neutrons interacting with elements having proton numbers near that of Ar (p = 18) may produce isotopes of Ar that must be accounted for. For example, a ^{40}K(n,p)^{40}Ar reaction produces excess ^{40}Ar beyond radiogenic and atmospheric contributions, and will cause spuriously old ages unless accounted for. On the other hand, the ^{42}Ca(n,a)^{39}Ar and ^{43}Ca(n,na)^{39}Ar reactions produce excess ^{39}Ar beyond that produced from ^{39}K, and will cause spuriously young ages unless accounted for. The ^{40}Ca(n,na)^{36}Ar and ^{35}Cl(n,g)^{36}Cl → ^{36}Ar produce excess ^{36}Ar, which leads to overcorrection for atmospheric ^{40}Ar, producing spuriously young ages, unless corrected for.

The extent to which these interfering reactions may bias ages depends on the chemical composition of a sample, the energy distribution of the irradiating neutrons, and the duration of the irradiation. The duration of irradiation is the most straightforward parameter to control, but most be considered in light of the desirability of producing an ^{40}Ar/^{39}Ar close enough to unity that this measurement is not compromised by the spectrometer's mass resolution. In general, a balance must be struck between minimizing irradiation duration (to minimize the extent of the ^{40}K

(n,p)^{40}Ar and ^{40}Ca(n,na)^{36}Ar reactions), and maintaining a ^{40}Ar/^{39}Ar ratio suitably close to unity. The important topic of optimizing irradiation duration in this context is discussed at length by *Turner* [1971] and *Dalrymple et al.* [1981].

An important contribution to minimizing the ^{40}K(n,p)^{40}Ar reaction was made by *Tetley et al.* [1980], who used cadmium shielding to absorb the low-energy neutrons for which the reaction has the highest cross-section. Use of Cd shielding is now commonplace, and some reactors have built in special Cd-lined vessels for this purpose. Cd shielding has the additional benefit of reducing the total radioactivity of samples, thus reducing the radiologic hazards associated with ^{40}Ar/^{39}Ar dating. The only downside to this practice aside from the generation of mixed (toxic and radioactive) waste is that filtering out low-energy neutrons also reduces the ^{37}Cl(n,g)^{38}Cl → ^{38}Ar production, which may be useful for geochemical inference about Ar sources.

Fortunately, correcting for the so-called interfering nuclear reactions is possible if we know the production rate from a given element of an interfering isotope relative to one that does not appear explicitly or implicitly in the age equation. For example, if we know the rate at which the ^{40}Ca(n,na)^{36}Ar reaction proceeds relative to the ^{40}Ca(n,a)^{37}Ar reaction, we can define the production ratio of ^{36}Ar/^{37}Ar from Ca, denoted by $(^{36}Ar/^{37}Ar)_{Ca}$, which can be used to correct for Ca-derived ^{36}Ar before making the ^{36}Ar-based atmospheric correction on ^{40}Ar. If there are no other sources of ^{36}Ar, this simplifies to:

$$^{36}Ar_A = {}^{36}Ar_T - {}^{36}Ar_{Ca} = {}^{37}Ar_{Ca}\left(\frac{{}^{36}Ar}{{}^{37}Ar}\right)_{Ca} \tag{9.21}$$

where $^{36}Ar_A$ is atmospheric ^{36}Ar and $^{36}Ar_T$ is the total measured ^{36}Ar. Eight production ratios may be important in ^{40}Ar/^{39}Ar geochronology, depending on chemical composition of samples and standards, duration of irradiation, and delay between irradiation and analysis. These production ratios are:

$$\left(\frac{{}^{40}Ar}{{}^{39}Ar}\right)_K;\left(\frac{{}^{38}Ar}{{}^{39}Ar}\right)_K;\left(\frac{{}^{37}Ar}{{}^{39}Ar}\right)_K;\left(\frac{{}^{36}Ar}{{}^{39}Ar}\right)_K;\left(\frac{{}^{39}Ar}{{}^{37}Ar}\right)_{Ca};$$
$$\left(\frac{{}^{38}Ar}{{}^{37}Ar}\right)_{Ca};\left(\frac{{}^{36}Ar}{{}^{37}Ar}\right)_{Ca};\left(\frac{{}^{36}Ar}{{}^{38}Ar}\right)_{Cl}$$

These production ratios are normally measured on simple compounds containing only one of the target elements Ca, K, and Cl, and irradiated under similar conditions as the samples. In some cases, such as the $(^{40}Ar/^{39}Ar)_K$, the two Ar isotopes are produced from different isotopes of K and therefore the ^{40}K/^{39}K of the material used to measure this production ratio is assumed to be the same as in samples and unknowns. In most cases, this assumption is valid within the limits of significance. An exception is the $(^{39}Ar/^{37}Ar)_{Ca}$ production ratio, which involves mainly the ^{42}Ca(n,a)^{39}Ar reaction in the numerator and the ^{40}Ca (n, a)^{37}Ar reaction in the denominator, hence is a function of ^{42}Ca/^{40}Ca of the material. Insofar as this ratio varies with age and K content (due to the β$^-$ decay of ^{40}K to ^{40}Ca) the ^{42}Ca/^{40}Ca may vary between samples, standards, and the compound used to measure the $(^{39}\text{Ar}/^{37}\text{Ar})_{\text{Ca}}$ production ratio. A similar caveat applies to the $(^{38}\text{Ar}/^{37}\text{Ar})_{\text{Ca}}$ production ratio, although this is less important due to the small cross-section of the $^{42}\text{Ca}(n,a)^{39}\text{Ar}$ reaction. Fortunately, variations in Ca isotopic composition are largest in old, K-rich materials, and these are exactly the materials that are least sensitive to Ca-based interference corrections. Also fortunate is the fact that the Ca-based production ratio that generally has the largest effect on age calculations, $(^{36}\text{Ar}/^{37}\text{Ar})_{\text{Ca}}$, involves both Ar isotopes derived from ^{40}Ca, hence any variations in Ca-isotope composition are irrelevant.

Two of the interfering isotope production pathways result from radioactive decay of unstable neutron activation products. $^{36}\text{Ar}_{\text{Cl}}$ is produced indirectly by the $^{35}\text{Cl}(n,g)^{36}\text{Cl}$ reaction, whereby the ^{36}Cl subsequently decays by β^- emission to ^{36}Ar. This process can be written in shorthand as $^{35}\text{Cl}(n,g)^{36}\text{Cl} \to {}^{36}\text{Ar}$. Analogously, $^{36}\text{Ar}_{\text{K}}$ is produced by the $^{39}\text{K}(n,a)^{36}\text{Cl} \to {}^{36}\text{Ar}$ reaction. Fortunately, the long half-life (3×10^8 years) of ^{36}Cl means that ingrowth of $^{36}\text{Ar}_{\text{Cl}}$ and $^{36}\text{Ar}_{\text{K}}$ are negligible unless samples are analyzed very long (i.e., years) after irradiation [*Renne et al.,* 2008], and their effects are generally ignored although, as seen below, the corrections are straightforward. Many samples of interest are effectively Cl-free, further minimizing $^{36}\text{Ar}_{\text{Cl}}$. However, because K is present by design, often at concentrations of order 10 atom%, the $^{39}\text{K}(n,a)^{36}\text{Cl} \to {}^{36}\text{Ar}$ reaction is more likely to be significant on normal laboratory timescales. The ratio of ^{235}U fission spectrum average cross-sections for the $^{39}\text{K}(n,a)^{36}\text{Cl}$ to the $^{39}\text{K}(n,p)^{39}\text{Ar}$ reactions is 41.1mb/113.3mb = 0.36, i.e. the production ratio $(^{36}\text{Cl}_{\text{K}}/^{39}\text{Ar}_{\text{K}}) = 0.36$. It is easy to show that the ratio $(^{36}\text{Ar}/^{39}\text{Ar})_{\text{K}}$ grows to $\sim 10^{-6}$ in 1 year, a value that may be $\sim 10\%$ of the total measured $^{36}\text{Ar}/^{39}\text{Ar}$ in many samples. The $(^{36}\text{Cl}_{\text{K}}/^{39}\text{Ar}_{\text{K}})$ production ratio for the irradiations experienced by $^{40}\text{Ar}/^{39}\text{Ar}$ dating samples, which are generally modified relative to the ^{235}U fission spectrum, has not been thoroughly investigated.

The time-dependent production ratios are given by:

$$\left(\frac{^{36}\text{Ar}}{^{38}\text{Ar}}\right)_{\text{Cl}} = P\left(\frac{^{36}\text{Cl}}{^{38}\text{Cl}}\right)_{\text{Cl}} \left(1 - e^{\lambda_{36} \, _{Cl}\Delta t}\right) \quad (9.22\text{a})$$

$$\left(\frac{^{36}\text{Ar}}{^{39}\text{Ar}}\right)_{\text{K}} = P\left(\frac{^{36}\text{Cl}}{^{38}\text{Ar}}\right)_{\text{K}} \left(1 - e^{\lambda_{36} \, _{Cl}\Delta t}\right) \quad (9.22\text{b})$$

where P refers to the production ratio in the reactor, λ is the decay constant for the β^- decay of ^{36}Cl, and Δt is the time interval between irradiation and analysis. For piecewise irradiations a slightly more complex decay correction should be used [*Wijbrans and McDougall,* 1987], as adapted for $^{36}\text{Ar}_{\text{Cl}}$ by *Renne et al.* [2008].

The number of interfering nuclear reactions is a consequence of the fact that the energy spectrum of neutrons produced by the ^{235}U fission process is continuous and broad. The use of moderators such as Cd to reduce the low-energy neutron flux, hence

minimizing $(^{40}\text{Ar}/^{39}\text{Ar})_{\text{K}}$ [*Tetley et al.,* 1980], has been successful in reducing undesirable reactions, but it would be desirable to also reduce the flux of higher energy (i.e., > 5 MeV) neutrons. In principle, neutrons produced by a deuterium fusion process, with 2.45 MeV energy, would greatly reduce undesirable nuclear reactions on K, Ca and Cl, and would have the radiological and environmental benefits of limited reactions on other targets [*Renne et al.,* 2005].

9.4.2.2 Apportioning Ar isotopes between sources

The contributions of various sources to the Ar isotope budget of an irradiated sample can be summarized by the following mass balance statements:

$$^{40}\text{Ar}_{\text{T}} = {}^{40}\text{Ar}^* + {}^{36}\text{Ar}_{\text{A}} + {}^{36}\text{Ar}_{\text{K}} \quad (9.23\text{a})$$

$$^{39}\text{Ar}_{\text{T}} = {}^{39}\text{Ar}_{\text{K}} + {}^{39}\text{Ar}_{\text{Ca}} \quad (9.23\text{b})$$

$$^{38}\text{Ar}_{\text{T}} = {}^{38}\text{Ar}_{\text{A}} + {}^{38}\text{Ar}_{\text{K}} + {}^{38}\text{Ar}_{\text{Ca}} + {}^{38}\text{Ar}_{\text{Cl}} \quad (9.23\text{c})$$

$$^{37}\text{Ar}_{\text{T}} = {}^{37}\text{Ar}_{\text{K}} + {}^{40}\text{Ar}_{\text{Ca}} \quad (9.23\text{d})$$

$$^{36}\text{Ar}_{\text{T}} = {}^{36}\text{Ar}_{\text{A}} + {}^{36}\text{Ar}_{\text{K}} + {}^{36}\text{Ar}_{\text{Ca}} + {}^{36}\text{Ar}_{\text{Cl}} \quad (9.23\text{e})$$

where $^x\text{Ar}_{\text{T}}$ refers to the total measured relative abundance of Ar isotope of mass x, corrected for mass discrimination, background, and decay where relevant; $^x\text{Ar}_{\text{A}}$ refers to the contribution from atmosphere; $^x\text{Ar}_{\text{K}}$, $^x\text{Ar}_{\text{Ca}}$, and $^x\text{Ar}_{\text{Cl}}$ refer to contributions from neutron reactions on K, Ca, and Cl respectively.

From this information, we need to infer $^{40}\text{Ar}^*$ and $^{39}\text{Ar}_{\text{K}}$ in order to solve the age equation. It is also often desirable to know $^{37}\text{Ar}_{\text{Ca}}$ and $^{38}\text{Ar}_{\text{Cl}}$, in order to infer the Ca:K:Cl composition of the source of $^{40}\text{Ar}^*$. However, equations (9.23a)–(9.23e) appear to pose a severely underconstrained system of five equations in 14 unknowns whose unique solution for the variables of interest is therefore impossible. Fortunately, we can make use of several relationships between the variables based on the neutron production and atmospheric abundance ratios, which can be measured.

$$^{40}\text{Ar}_{\text{A}} = {}^{36}\text{Ar}_{\text{A}} \left(\frac{^{40}\text{Ar}}{^{36}\text{Ar}}\right)_{\text{A}}; \, ^{40}\text{Ar}_{\text{K}} = {}^{39}\text{Ar}_{\text{K}} \left(\frac{^{40}\text{Ar}}{^{36}\text{Ar}}\right)_{\text{K}} \quad (9.24\text{a})$$

$$^{39}\text{Ar}_{\text{Ca}} = {}^{37}\text{Ar}_{\text{Ca}} \left(\frac{^{36}\text{Ar}}{^{37}\text{Ar}}\right)_{\text{Ca}} \quad (9.24\text{b})$$

$$^{38}\text{Ar}_{\text{A}} = {}^{36}\text{Ar}_{\text{A}} \left(\frac{^{38}\text{Ar}}{^{36}\text{Ar}}\right)_{\text{A}}; \, ^{38}\text{Ar}_{\text{K}} = {}^{39}\text{Ar}_{\text{K}} \left(\frac{^{38}\text{Ar}}{^{39}\text{Ar}}\right)_{\text{K}};$$

$$^{38}\text{Ar}_{\text{Ca}} = {}^{37}\text{Ar}_{\text{Ca}} \left(\frac{^{38}\text{Ar}}{^{39}\text{Ar}}\right)_{\text{Ca}} \quad (9.24\text{c})$$

$$^{37}\text{Ar}_{\text{K}} = {}^{39}\text{Ar}_{\text{K}} \left(\frac{^{37}\text{Ar}}{^{39}\text{Ar}}\right)_{\text{Ks}} \quad (9.24\text{d})$$

$$^{36}\text{Ar}_{\text{Ca}} = {}^{37}\text{Ar}_{\text{Ca}}\left(\frac{^{36}\text{Ar}}{^{37}\text{Ar}}\right)_{\text{Ca}}; \quad {}^{36}\text{Ar}_{\text{Cl}} = {}^{38}\text{Ar}_{\text{Cl}}\left(\frac{^{36}\text{Ar}}{^{38}\text{Ar}}\right)_{\text{Cl}};$$

$$^{36}\text{Ar}_{\text{K}} = {}^{39}\text{Ar}_{\text{K}}\left(\frac{^{36}\text{Ar}}{^{39}\text{Ar}}\right)_{\text{K}} \tag{9.24e}$$

Substituting these identities into the Ar source budget equations [equations (9.23a)–(9.23e)] yields a system of five equations in five unknowns. As structured here, the five unknowns are $^{40}\text{Ar}^*$, $^{39}\text{Ar}_{\text{K}}$, $^{38}\text{Ar}_{\text{Cl}}$, $^{37}\text{Ar}_{\text{Ca}}$, and $^{36}\text{Ar}_{\text{A}}$.

$$^{40}\text{Ar}_{\text{T}} = {}^{40}\text{Ar} + {}^{36}\text{Ar}_{\text{A}}\left(\frac{^{40}\text{Ar}}{^{36}\text{Ar}}\right)_{\text{A}} + {}^{39}\text{Ar}_{\text{K}}\left(\frac{^{40}\text{Ar}}{^{36}\text{Ar}}\right)_{\text{K}} \tag{9.25a}$$

$$^{39}\text{Ar}_{\text{T}} = {}^{39}\text{Ar}_{\text{K}} + {}^{37}\text{Ar}_{\text{Ca}}\left(\frac{^{39}\text{Ar}}{^{37}\text{Ar}}\right)_{\text{Ca}} \tag{9.25b}$$

$$^{38}\text{Ar}_{\text{T}} = {}^{36}\text{Ar}_{\text{A}}\left(\frac{^{38}\text{Ar}}{^{36}\text{Ar}}\right)_{\text{A}} + {}^{39}\text{Ar}_{\text{K}}\left(\frac{^{38}\text{Ar}}{^{39}\text{Ar}}\right)_{\text{K}}$$
$$+ {}^{37}\text{Ar}_{\text{Ca}}\left(\frac{^{38}\text{Ar}}{^{39}\text{Ar}}\right)_{\text{Ca}} + {}^{38}\text{Ar}_{\text{Cl}} \tag{9.25c}$$

$$^{37}\text{Ar}_{\text{T}} = {}^{37}\text{Ar}_{\text{Ca}} + {}^{39}\text{Ar}_{\text{K}}\left(\frac{^{36}\text{Ar}}{^{37}\text{Ar}}\right)_{\text{K}} \tag{9.25d}$$

$$^{36}\text{Ar}_{\text{T}} = {}^{36}\text{Ar}_{\text{A}} + {}^{37}\text{Ar}_{\text{Ca}}\left(\frac{^{36}\text{Ar}}{^{37}\text{Ar}}\right)_{\text{Ca}} + {}^{38}\text{Ar}_{\text{Cl}}\left(\frac{^{36}\text{Ar}}{^{37}\text{Ar}}\right)_{\text{Cl}}$$
$$+ {}^{39}\text{Ar}_{\text{K}}\left(\frac{^{36}\text{Ar}}{^{39}\text{Ar}}\right)_{\text{K}} \tag{9.25e}$$

This system of equations can be written in matrix form as

$$\begin{pmatrix} ^{40}\text{Ar}_{\text{T}} \\ ^{39}\text{Ar}_{\text{T}} \\ ^{38}\text{Ar}_{\text{T}} \\ ^{37}\text{Ar}_{\text{T}} \\ ^{36}\text{Ar}_{\text{T}} \end{pmatrix} = \begin{pmatrix} 1 & \left(\frac{^{40}\text{Ar}}{^{36}\text{Ar}}\right)_{\text{A}} & \left(\frac{^{40}\text{Ar}}{^{39}\text{Ar}}\right)_{\text{K}} & 0 & 0 \\ 0 & 0 & 1 & \left(\frac{^{39}\text{Ar}}{^{37}\text{Ar}}\right)_{\text{Ca}} & 0 \\ 0 & \left(\frac{^{38}\text{Ar}}{^{36}\text{Ar}}\right)_{\text{A}} & \left(\frac{^{38}\text{Ar}}{^{39}\text{Ar}}\right)_{\text{K}} & \left(\frac{^{38}\text{Ar}}{^{37}\text{Ar}}\right)_{\text{Ca}} & 1 \\ 0 & 0 & \left(\frac{^{37}\text{Ar}}{^{39}\text{Ar}}\right)_{\text{K}} & 1 & 0 \\ 0 & 1 & \left(\frac{^{36}\text{Ar}}{^{39}\text{Ar}}\right)_{\text{K}} & \left(\frac{^{36}\text{Ar}}{^{37}\text{Ar}}\right)_{\text{Ca}} & \left(\frac{^{36}\text{Ar}}{^{38}\text{Ar}}\right)_{\text{Cl}} \end{pmatrix} \times \begin{pmatrix} ^{40}\text{Ar}^* \\ ^{36}\text{Ar}_{\text{A}} \\ ^{39}\text{Ar}_{\text{K}} \\ ^{37}\text{Ar}_{\text{Ca}} \\ ^{38}\text{Ar}_{\text{Cl}} \end{pmatrix} \tag{9.26}$$

The 5×5 matrix of coefficients (nuclear production and atmospheric ratios) is invertible and can be solved to yield unique values of $^{40}\text{Ar}^*$, $^{39}\text{Ar}_{\text{K}}$, $^{38}\text{Ar}_{\text{Cl}}$, $^{37}\text{Ar}_{\text{Ca}}$, and $^{36}\text{Ar}_{\text{A}}$. Uncertainties in these five key parameters can then be estimated using Monte Carlo simulation.

It should be noted that the above approach for attributing components of the Ar isotopes fails to account for cosmogenic ^{38}Ar and ^{36}Ar [*Niedermann et al.*, 2007; *Renne et al.*, 2001a], whose common presence in extraterrestrial materials would invalidate the application of equation (9.26). It should also be noted that some of the production ratios, e.g., $(^{36}\text{Ar}/^{39}\text{Ar})_{\text{K}}$, used in equation (9.26) are not well known for all reactors. Finally, it is worth emphasizing that equation (9.26) attributes all supra-atmospheric ^{40}Ar, i.e. anything with $^{40}\text{Ar}/^{36}\text{Ar}$ greater than atmospheric, to $^{40}\text{Ar}^*$ and thus fails to correct for excess ^{40}Ar (see below).

A simplified solution of equations (9.25a)–(9.25e), which assumes that no ^{37}Ar is produced from K (i.e., $(^{37}\text{Ar}/^{39}\text{Ar})_{\text{K}} = 0$) and neglecting Cl-produced Ar, gives a reasonable approximation in most cases to the fundamental parameter that is fed into the age equation:

$$\frac{^{40}\text{Ar}^*}{^{39}\text{Ar}_{\text{K}}} = \frac{^{40}\text{Ar}_{\text{T}} - 298.56\left[^{36}\text{Ar}_{\text{T}} - {}^{37}\text{Ar}_{\text{T}}\left(\frac{^{36}\text{Ar}}{^{37}\text{Ar}}\right)_{\text{Ca}}\right] - \left(\frac{^{40}\text{Ar}}{^{39}\text{Ar}}\right)_{\text{K}}\left[^{39}\text{Ar}_{\text{T}} - {}^{37}\text{Ar}\left(\frac{^{39}\text{Ar}}{^{37}\text{Ar}}\right)_{\text{Ca}}\right]}{^{39}\text{Ar}_{\text{T}} - {}^{37}\text{Ar}\left(\frac{^{39}\text{Ar}}{^{37}\text{Ar}}\right)_{\text{Ca}}} \tag{9.27}$$

Once the relative abundances of the three neutron-produced isotopes $^{39}Ar_K$, $^{38}Ar_{Cl}$, and $^{37}Ar_{Ca}$ have been determined, the relative abundances of their parent elements (K, Cl, and Ca) may be determined by analysis of a sample whose composition is known independently. For example, *Renne et al.* [2010] derived the following relationships from electron microprobe data reported by *Jourdan and Renne* [2007] for the hornblende standard Hb3gr irradiated in the Oregon State University TRIGA reactor:

$$\frac{Ca}{K} = (1.92 \pm 0.05)\left(\frac{^{37}Ar_{Ca}}{^{39}Ar_K}\right) \tag{9.28a}$$

$$\frac{Ca}{Cl} = (0.122 \pm 0.007)\left(\frac{^{37}Ar_{Ca}}{^{38}Ar_{Cl}}\right) \tag{9.28b}$$

The values of the conversion factors depend on the specific neutron-energy spectrum during the irradiation, and this may differ between reactors or even between different irradiations in the same reactor. This is especially true for Ca/Cl or K/Cl due to the large difference in $\sigma(\varepsilon)$ for the relevant neutron capture reactions.

9.4.2.3 Recoil effects

The $^{39}K(n,p)^{39}Ar$ neutron-capture reaction imparts *recoil energy* to the nascent ^{39}Ar atom, which causes it to be displaced from the original site of the parent ^{39}K atom. As a result, ^{39}Ar produced near the surfaces of mineral grains may be expelled from the grains, inducing a spuriously elevated $^{40}Ar^*/^{39}Ar_K$ (hence, apparent age) in the near-surface portions of the grains. The distribution of recoil energies, hence of displacements, is a function of how anisotropic the irradiating neutron flux is. Subject to some simplifying assumptions, *Turner and Cadogan* [1974] determined a mean depth of the depletion layer to be 0.08 mm. Thus the magnitude of ^{39}Ar recoil artifacts is expected to be a strongly dependent on grain size, as has been verified by numerous experiments [*Huneke and Villa*, 1981; *Foland et al.*, 1984,1992, 1993; *Hess and Lippolt*, 1986; *Villa*, 1997; *Paine et al.*, 2006; *Jourdan et al.*, 2007; *Hall*, 2014]. In many cases the inferred ^{39}Ar recoil loss is larger than predicted by *Turner and Cadogan* [1974], leading *Hall* [2014] to conclude that mineral defects or other microstructures may enhance the recoil distance. K–Ar dating is immune to recoil artifacts and is thus better suited to dating fine-grained materials, one of the major advantages of this method compared with $^{40}Ar/^{39}Ar$.

^{37}Ar produced by the $^{40}Ca(n,a)^{37}Ar$ reaction also undergoes recoil displacement, with about three times the energy of ^{39}Ar recoil [*Onstott et al.*, 1995]. *Jourdan et al.* [2006] and *Jourdan and Renne* [2014] found ^{37}Ar losses from plagioclase and hornblende approximately an order of magnitude larger than expected, with mean partial depletion depths of 3–4 mm. The greater energy of ^{37}Ar recoil than ^{39}Ar recoil implies that inferred Ca/K of the outermost several millimeters are biased towards lower values. A further complication is that the $^{40}Ca(n,na)^{36}Ar$ reaction has negligible recoil energy compared with the $^{40}Ca(n,a)^{37}Ar$ reaction, which means that recoil fractionates the $(^{36}Ar/^{37}Ar)_{Ca}$ production ratio. Thus in regions affected by recoil loss, excess ^{36}Ar would be perceived and a standard air correction would overcorrect the ^{40}Ar, resulting in a spuriously young age.

The biasing effect of ^{39}Ar recoil loss was exemplified in studies of calcium–aluminum-rich inclusions in the Allende carbonaceous chondrite, some of which yielded apparent ages > 5 Ga [*Jessberger et al.*, 1980] interpreted as possible evidence of presolar material or isotopically anomalous $^{39}K/^{40}K$. This was disputed by *Villa et al.* [1983], who showed that the results were consistent with ^{39}Ar recoil-loss artifacts. ^{39}Ar recoil redistribution and/or loss also complicates the inference of discrete intracrystalline domains at very small (e.g., < 10 μm) spatial scales, a point that has been raised [*Onstott et al.*, 1995; *Villa*, 1997] against the ability to infer thermochronometric data from comparably small diffusion domains via incremental heating, as is required in the multidomain diffusion theory [*Lovera et al.*, 1989]. Further implications of ^{39}Ar recoil are discussed below in the context of incremental heating and age spectra. Meanwhile, we note that recoil phenomena, particularly of ^{37}Ar, remain incompletely understood and further research would be fruitful.

9.4.3 Appropriate materials

An unusually large variety of materials can be dated by K–Ar and $^{40}Ar/^{39}Ar$ methods due to the relatively large terrestrial abundance of K compared with other radionuclides. The main considerations determining utility depend on the type of information being sought, but generally they are K concentration, Ca/K ratio, tendency to include initial (trapped) Ar, Ar retentivity, and grain size. Several materials dominate the menu, as summarized below. Approximate Dodsonian closure temperatures (see Chapter 5) are given for selected minerals; these vary with diffusion length scale (which is equal to or less than grain dimensions), cooling rate, and specific composition in the case of some solid solutions, and the values given do not necessarily encompass the full possible range of these variables.

9.4.3.1 Alkali feldspars

The volcanic alkali feldspars sanidine and anorthoclase are the mainstay of high-precision $^{40}Ar/^{39}Ar$ geochronology due to their high K concentrations (generally > 5 wt% K_2O), low Ca/K, generally low initial Ar concentrations, and moderate Ar retentivity. These minerals are common as phenocrysts in siliceous magmas, and are especially useful in rhyolite tuffs interbedded in sedimentary sequences for chronostratigraphy. Alkali feldspars in plutonic and metamorphic rocks are commonly used in thermochronometry due to their low to moderate Ar retentivity (closure temperatures 200–400 °C), but the common presence of complex microstructures and alteration may inhibit their reliability for this purpose.

9.4.3.2 Micas

Trioctahedral micas loosely termed biotites that are mainly solid solutions of annite and phlogopite, but ubiquitously contain other components, have been a very important mineral for both K–Ar and $^{40}Ar/^{39}Ar$ geochronology. The advantages are high

K concentrations (generally > 7 wt% K_2O) and low Ca/K, as well as ease of fusion and complete extraction of Ar, as is necessary for K–Ar. A disadvantage is the relatively high concentration of initial Ar. Moderate Ar retentivity (closure temperatures 250–400 °C) means that biotites from plutonic and metamorphic rocks generally record a cooling age rather than a crystallization age. The high surface/volume ratio of biotites' tabular habit makes them particularly prone to ^{39}Ar recoil artifacts.

Dioctahedral micas, mainly comprising muscovite and phengite, share most properties with biotites except that that they tend to have slightly higher Ar retention properties, with closure temperatures often assumed to be ~350 °C, though some data indicate values > 400 °C [*Harrison et al.*, 2009]. These minerals are almost exclusively found in plutonic and metamorphic rocks, in which they tend to record older ages than biotites due to their greater Ar retention.

9.4.3.3 Amphiboles

Amphiboles whose most common igneous varieties are loosely termed hornblende are commonly dated by K–Ar and ^{40}Ar/^{39}Ar methods. The main advantage of hornblendes relative to micas or feldspars is their high Ar retentivity (closure temperatures 450–600 °C), meaning that they record higher temperatures (older ages) than micas or feldspars in plutonic and metamorphic rocks. Disadvantages include their relatively low K concentration (generally < 2 wt% K_2O), high Ca/K, and moderate tendency to incorporate initial Ar. These latter factors weigh against hornblende as sources of high-precision ages as is often desired in chronostratigraphic studies, but the high Ar retentivity makes it desirable for thermochronometry.

9.4.3.4 Plagioclase feldspars

Plagioclases share many of the properties of alkali feldspars, except that they (by definition) have lower K concentrations (generally < 1 wt% K_2O) and higher Ca/K. For high-precision work, they are subordinate to alkali feldspars in desirability for these reasons. In low-K volcanic rocks such as basalts, they are often the only K-bearing mineral (i.e., as phenocrysts) available for extraction in practical grain size, hence the only game in town.

9.4.3.5 Whole rock

Occasionally samples are dated which are not pure mineral separates. This is most common in lavas that either are aphyric or lack K-rich phenocrysts. Often such samples are processed to remove K-poor phenocrysts such as olivine and pyroxenes, which dilute K concentrations at best, and may host excess ^{40}Ar at worst, in which case the samples are sometimes termed "groundmass" samples. In any case, only very fresh material, i.e., lacking alteration of interstitial glass, is appropriate for K–Ar or ^{40}Ar/^{39}Ar dating, and grain size limits the applicability of ^{40}Ar/^{39}Ar methods due to ^{39}Ar recoil. Whole-rock analysis of metamorphic, plutonic, or pyroclastic rocks is not recommended.

9.4.3.6 Glass

Magma quenched to glass, such as obsidian, has been successfully dated by the K–Ar and ^{40}Ar/^{39}Ar methods. Compositions vary widely, hence generalizations on this basis are difficult. The main limitations are that glasses are subject to alteration in surficial environments, and may contain substantial amounts of initial Ar. They are also prone to hydration, which may mobilize K. There is some evidence for kinetic isotope fractionation of atmospheric Ar dissolved in some glasses, which tends to enrich the ^{36}Ar/^{40}Ar of the initial trapped Ar component and thereby bias an air correction. Very fine grain sizes, such as are common in glass shards of vitroclastic tuffs, are prone to ^{39}Ar recoil and thus are generally unsuitable for ^{40}Ar/^{39}Ar dating. Tektites, glasses quenched from impact melts, often yield highly reproducible data.

9.4.3.7 Other

Various other minerals have been used for K–Ar and ^{40}Ar/^{39}Ar dating, including various authigenic oxides and sulfates, evaporates, and clays. Certain K-bearing Mn oxides have proven useful in dating weathering events [*Vasconcelos et al.*, 1994b]. The K-bearing phosphates alunite and jarosite have been used extensively to date weathering events and supergene ore deposits [*Alpers and Brimhall*, 1988; *Vasconcelos et al.*, 1994a]. Many studies have been conducted on various K-bearing evaporate minerals, including the pioneering study of *Aldrich and Nier* [1948], but most are prone to open-system behavior and yield ages that are significantly younger than deposition. An exception is the sulfate langbeinite, which in some cases has been shown to yield plausible deposition ages [*Leost et al.*, 2001; *Renne et al.*, 2001b]. The mineraloid glaucony, which forms authigenically in some marine environments, has been used extensively to date deposition and/or diagenesis [*Fiet et al.*, 2006; *Odin*, 1986]; its typically small grain size renders it subject to extensive recoil effects [*Foland et al.*, 1984], hence K–Ar is preferred to ^{40}Ar/^{39}Ar dating although some successful applications of the latter have been reported [*Smith et al.*, 1998]. Similarly, clay minerals such as illite have been used to date both sedimentary and tectonic (i.e., faulting) paragenesis, in some cases from the same sample using mineralogical data to deconvolve the different generations [*van der Pluijm et al.*, 2001]. As with glaucony, the diminutive size of clay minerals renders them highly susceptible to ^{39}Ar recoil loss, hence K–Ar is the preferred analytical approach although vacuum-encapsulation methods [*Dong et al.*, 1995] have been developed to mitigate ^{39}Ar recoil loss and essentially reduce the ^{40}Ar/^{39}Ar method to K–Ar.

These and other unspecified, and doubtless many yet untried, materials are useful for specific applications. Their utility depends on the criteria discussed above, in addition to whether or not their paragenesis can be straightforwardly linked to the event whose age is sought. Creativity and need will undoubtedly spawn the application of K–Ar and ^{40}Ar/^{39}Ar to additional materials in the future.

9.5 EXPERIMENTAL APPROACHES AND GEOCHRONOLOGIC APPLICATIONS

One of the profound advantages of the ^{40}Ar/^{39}Ar technique compared with K–Ar is that all of the relevant information is derived from a single sample using a single analytical method (mass spectrometry) rather than making separate measurements of ^{40}K and ^{40}Ar on separate subsamples. This allows analysis of arbitrarily small amounts of material, limited only by the instrumental sensitivity and background levels of Ar isotopes. Coupled with low-background methods for degassing samples, such as laser heating, the ability to date individual crystals has revolutionized the ability to date pyroclastic deposits accurately. Such deposits frequently contain xenocrystic contaminants, which may bias apparent ages when multigrained aliquots are analyzed.

9.5.1 Single crystal fusion

This mode of analysis was enhanced by the advent of process automation [*Layer et al.*, 1987]. With these capabilities, it is not uncommon to analyze tens or even hundreds of individual crystals from tephras when high precision is desired and the presence of xenocrysts is possible, as in many chronostratigraphic studies.

The importance of single crystal analysis was showcased in the landmark study of *Lobello et al.* [1987], who showed that Mesozoic apparent ages determined for multigrained sanidine samples from a Quaternary pumice flow were artifacts due to xenocrysts. When single crystals were analyzed a plausible age of 580 ± 20 ka was determined. Myriad subsequent studies have been similarly successful in recognizing and obviating biases due to xenocrysts. Data from this mode of analysis are commonly displayed as age–probability diagrams [*Deino and Potts*, 1990], in which individual ages are assumed to have Gaussian distributions, and the probabilities are summed over an appropriate age range using a suitable partition.

Extracting a single age from a number of analyses is generally carried out using a weighted mean calculation, where the weighting factor of each analysis is the inverse variance. This approach accounts for variable precision of the individual analyses such that imprecise single-crystal ages affect the weighted mean age less than more precise ones. A fundamental consideration in applying a weighted mean calculation is that the individual ages can be said to represent a homogeneous population such that their scatter is consistent with their analytical uncertainties as represented by parameters such as the mean square weighted deviation (MSWD; see Chapter 4). Excess scatter, i.e. MSWD > 1, can arise from various issues including xenocrysts, variable alteration, or underestimation of uncertainties. For the purposes of testing whether a set of analyses is homogeneous, and thus that their weighted mean is meaningful, systematic uncertainties (discussed further below) must be ignored.

A typical age–probability diagram is depicted in Fig. 9.7, which displays the results of single-crystal analyses from a tuff in the Shungura Formation of northern Kenya [*McDougall et al.*, 2012].

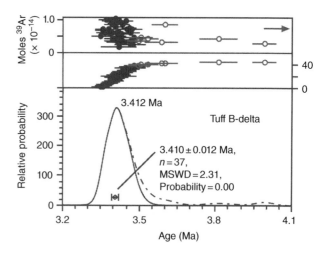

Fig. 9.7. Age–probability diagram [*Deino and Potts*, 1990] from *McDougall et al.* [2012] displaying the results of 45 laser fusion analyses of single anorthoclase crystals from a tuff in the Shungura Formation of northern Kenya. Shaded symbols show analyses rejected from a weighted mean age calculation. (Source: *McDougall et al.* [2012]. Reproduced with permission of the Geological Society, London.)

Elsewhere, this tuff is associated with fossils of *Australopithecus afarensis*. Three obvious xenocrysts > 3.8 Ma can be straightforwardly rejected; the remaining five deletions rely on more subjective criteria. The stratigraphic section contains numerous hominid fossils, and the ages of interbedded tuffs is the principal means of determining the ages of the hominids. In general, the ^{40}Ar/^{39}Ar method is the most important tool for establishing the timeline of human evolution beyond the range of ^{14}C, i.e., ~50 ka.

^{40}Ar/^{39}Ar ages from tuffs or lavas interbedded in stratigraphic sequences have proven invaluable to constraining features such as sediment accumulation rates (which may have paleoclimate or tectonic implications), and rates of biotic processes such as extinctions and radiations. ^{40}Ar/^{39}Ar dating in the stratigraphic context, along with U/Pb dating of zircons (see Chapter 8) provides the majority of radioisotopic age constraints on calibration of the geologic timescale. A good example of this application is the study of *Sprain et al.* [2015], who dated single crystals of sanidine from nine tuffs interbedded with terrestrial sediments spanning the Cretaceous–Paleogene boundary (KPB) in eastern Montana (Fig. 9.8). This study dated the KPB at 66.043 ± 0.010 Ma (analytical precision) or ± 0.043 Ma (full uncertainty including systematic sources), and constrained the post-KPB disaster fauna to a duration of 70 ka, with full faunal recovery within ~900 ka.

Another important application of grain-specific ^{40}Ar/^{39}Ar dating enabled by laser fusion is to dating detrital minerals in clastic sedimentary rocks. The age distributions of various minerals can provide valuable information about sediment provenance [*Renne et al.*, 1990]. Combined with knowledge about Ar closure temperatures and depositional ages, powerful constraints can be placed on tectonics of sediment source regions [*Renne et al.*,

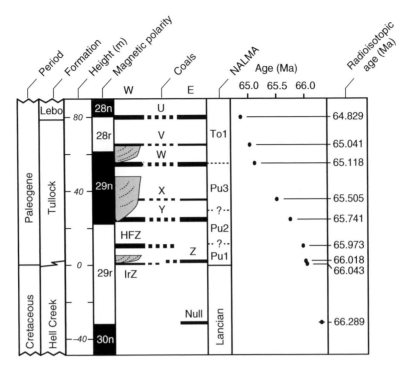

Fig. 9.8. Single crystal ^{40}Ar/^{39}Ar dates of sanidine from silicic tuffs interbedded with terrestrial sedimentary rocks spanning the Cretaceous–Paleogene boundary (KPB) in eastern Montana [*Sprain et al.*, 2015]. The dates define a smoothly varying sediment accumulation rate and constrain the timing and pace of recovery from the KPB mass extinction. NALMA is the North American Land Mammal Age. Nomenclature of the coal beds is described by *Sprain et al.* [2015].

1990; *Carrapa et al.*, 2003; *Vermeesch*, 2015]. This latter topic is thoroughly reviewed by *Hodges et al.* [2005].

9.5.2 Intragrain age gradients

Lasers enable fusing small portions of single crystals, extending the ^{40}Ar/^{39}Ar technique to the intracrystalline scale. The first application to ^{40}Ar/^{39}Ar dating was by *Megrue* [1973], who used a pulsed laser to fuse 75–150 mm diameter spots in a polymict lunar breccia to determine distinct age components of 3.7 Ga and 2.9 Ga amongst the clasts. Subsequently, many studies have been conducted on individual crystals to characterize internal Ar isotopic gradients. Such studies have been used to test Ar diffusion models [*Phillips and Onstott*, 1988], to evaluate diffusion geometry [*Kelley and Turner*, 1991; *Hames and Bowring*, 1994], and to infer thermal histories [*Dejong et al.*, 1992].

9.5.3 Incremental heating

A further advantage of the ^{40}Ar/^{39}Ar method compared with K–Ar is the ability to release argon incrementally from a sample by heating it in steps and analyzing the gas released in each heating step. Heating of samples may be achieved by furnaces based on electrical resistance or radiofrequency induction, or by continuous wave laser.

9.5.3.1 Age spectrum

The sequence of model ages calculated from incremental heating steps as a function of the fractional degassing of the sample is termed an age spectrum or apparent age spectrum, the latter term reflecting the fact that individual step ages are generally model ages based on an air correction or other assumptions about the isotopic composition of nonradiogenic ^{40}Ar in the sample. The age spectrum was first employed by *Turner* [1970] and has become the standard means to depict incremental-heating or step-heating data. A flat age spectrum, i.e. the model ages are statistically indistinguishable throughout an incremental heating experiment, is strong evidence that (i) the sample has not suffered partial loss of ^{40}Ar* and (ii) the assumed isotopic composition of nonradiogenic Ar is correct. Such age spectra (e.g., Fig. 9.9) are considered to be *concordant*.

An age spectrum with a significant number of contiguous steps having mutually indistinguishable ages is said to contain an (apparent) age plateau. A concordant age spectrum yields a plateau age by definition. Age spectra that are not concordant are common, and these are termed *discordant*. Discordant age spectra commonly yield plateau ages, although the definition of a plateau is somewhat arbitrary and none of those in common use are grounded in statistical theory. One the earliest and still widely used definitions of an age plateau is due to *Fleck et al.* [1977], who stipulated that a plateau should comprise at least 50% of the ^{39}Ar released in contiguous steps, with none of the plateau steps being distinguishable at the 95% confidence level. Countless variations of this definition have been used, often seemingly to accommodate special cases.

Several generic patterns of discordance occur frequently enough to be discussed categorically. One of the most common

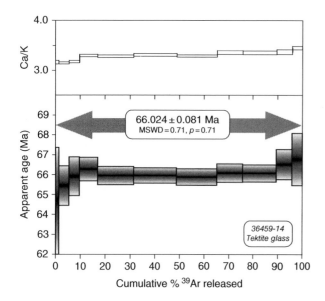

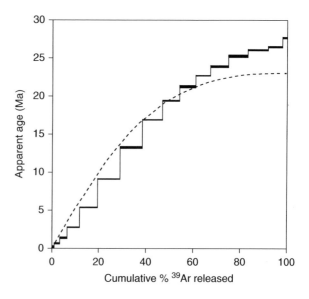

Fig. 9.9. A concordant age spectrum (lower panel) derived by incremental heating of a tektite from the Cretaceous–Paleogene boundary [*Renne et al.*, 2013]. Uncertainties (2σ) in step model ages are shown as the vertical heights of boxes. The individual step model ages are mutually indistinguishable at the 95% confidence level. The upper panel shows Ca/K atomic ratios for each step, derived from $^{37}Ar_{Ca}/^{39}Ar_{K}$ as explained in the text. (Source: *Renne et al.* [2013]. Reproduced with permission of the American Association for the Advancement of Science.)

Fig. 9.10. Discordant apparent age spectrum produced by laboratory partial degassing of sanidine from the Fish Canyon tuff [*Zeitler*, 1987]. Dotted curve shows a theoretical age spectrum for a population of uniform spheres with homogeneous initial $^{40}Ar^*$ and K concentrations, degassed by 40% at zero age.

cases is due to partial loss of $^{40}Ar^*$, resulting in anomalously low step ages in the initial increments of ^{39}Ar released followed by increasing step ages that may or may not eventually yield a plateau (Fig. 9.10). Some of the earliest applications of $^{40}Ar/^{39}Ar$ were focused on the significance of such spectra derived from meteorites, which were interpreted to reflect loss of $^{40}Ar^*$ by diffusion due to an episodic heating event [e.g., Turner, 1970]. The basic idea is that diffusive loss of $^{40}Ar^*$ from minerals produces age gradients such that $^{40}Ar^*$ is preferentially lost from grain surfaces, and progressively more retained in grain interiors. If incremental heating progressively taps Ar from surficial regions to interiors of grains with increasing proportions of ^{39}Ar released, then the age spectrum can be interpreted as a rendition of a diffusion profile imposed on the initial distribution of ^{40}Ar and K. Further discussion of such spectra and their implications for thermochonology is given in Chapter 5.

Interpretation of discordant age spectra of the form illustrated in Fig. 9.10 as due to either episodic ^{40}Ar loss or slow cooling through the partial retention zone are subject to the requirement that the discordance is not due to the degassing of younger minerals in the sample. For example, intergrowths of phyllosilicates in hornblende produced discordance reminiscent of loss profiles, but simply represent two-phase mixtures in which the two phases have distinct Ar retention ages and laboratory degassing kinetics [*Onstott and Peacock*, 1987; *Ross and Sharp*, 1988]. Many discordant hornblende age spectra interpreted as loss profiles may in fact be due to this phenomenon. The younger

contaminating phase is often much more K-rich than hornblende, hence has strong leverage on the age spectrum, but this allows use of the $^{37}Ar_{Ca}/^{39}Ar_{K}$ spectrum to diagnose the cause of discordance [*Wang and Jacobsen*, 2016].

Another common type of discordant age spectrum reflects ^{39}Ar recoil artifacts, often manifest in age spectra for fine-grained lavas, clays or clay-like minerals, or other samples whose grain-size approaches ^{39}Ar recoil distances. In such cases step ages generally rise to unrealistically old values within the first 10–30% of ^{39}Ar released (i.e., steps A and B in Fig. 9.11), then progressively decrease through the remainder of the spectrum or eventually stabilize to a plateau-like sequence of steps. The normal interpretation of such spectra is that ^{39}Ar is preferentially depleted from grain surfaces by recoil, such that early heating steps have spuriously high $^{40}Ar^*/^{39}Ar_{K}$ and therefore old apparent ages. Then as recoil depleted near-surface regions of grains are progressively degassed, apparent ages decrease and may approach or yield realistic values.

A commonly encountered problem with such spectra is that spurious age plateaus can result in which successive steps reveal continuously decreasing ages. If analytical uncertainties are large enough that the step ages are mutually indistinguishable, the formal definition of a plateau may be achieved despite an obvious decreasing trend which negates the interpretation that a homogeneous reservoir is being degassed.

Another common type of age spectrum is the so-called saddle (Fig. 9.12), in which anomalously old apparent ages occur in the earliest and latest fractions of ^{39}Ar released, and an intermediate (saddle) portion of the spectrum yields lower apparent ages. Ages defining the saddle are generally older than, but often approach, the true Ar retention age of the sample. This pattern is common

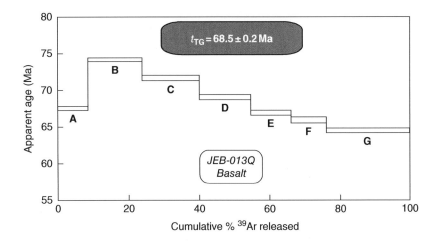

Fig. 9.11. A discordant age spectrum for a basalt from the Thakurvadi Formation of the Deccan Traps [*Baksi*, 1994] is typical of whole-rock samples with fine-grained, altered groundmass. This behavior is commonly a result of ^{39}Ar recoil redistribution and possible net loss from the sample. Although no plateau is evident, the total gas age (t_{TG}) may be geologically meaningful if there is no net loss of ^{39}Ar from the sample. In this case, the total gas age is older than expected as compared with plateau ages of 66.20 ± 0.13 and 66.24 ± 0.08 Ma obtained from plagioclase separates from two lavas in the Thakurvadi Formation [*Renne et al.*, 2015], hence net loss of ^{39}Ar is implied.

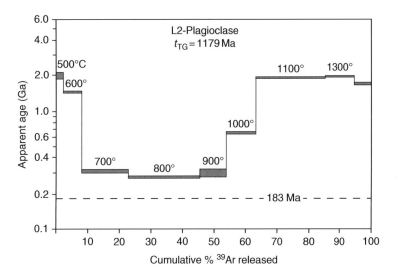

Fig. 9.12. A typical saddle-shaped apparent age spectrum for plagioclase from a dolerite dike intruded into Proterozoic crystalline rocks [*Lanphere and Dalrymple*, 1976]. The youngest apparent ages approach the expected age (183 Ma, dashed line). Note logarithmic scale on vertical (age) axis. Extraction temperatures (°C) are shown for each step.

in cases where there is initial nonradiogenic ^{40}Ar with ^{40}Ar/^{36}Ar greater than atmospheric, and is generally interpreted as an indication of the presence of excess Ar [e.g., *Lanphere and Dalrymple*, 1976]. Saddle-shaped age spectra have been reported from volcanic sanidines lacking any indication of excess ^{40}Ar [*Phillips and Matchan*, 2013]; no definitive explanation for this behavior is known.

Saddle-shaped age spectra are common in metamorphic and plutonic rocks that likely crystallized under high partial pressures of Ar. A noteworthy category of this is in basaltic bodies intruded into much older, K-rich crystalline rocks, in which cases the heat of the basaltic magma outgases ^{40}Ar* from the basement and may be incorporated into the basalt. An early

example of this phenomenon was shown by *Lanphere and Dalrymple* [1976] in both plagioclase and pyroxene in a dolerite dike intruded into Proterozoic basement gneiss in Liberia (Fig. 9.12). In this case, other dikes from the same swarm that intrude Paleozoic sedimentary rocks yielded whole-rock K–Ar ages of 173–192 Ma [*Lanphere and Dalrymple*, 1976], consistent with their origin as part of the ∼201 Ma Central Atlantic Magmatic Province [*Marzoli et al.*, 1999], with some later ^{40}Ar* loss as is common in whole-rock K–Ar dates from such rocks. As these latter dikes appear to lack excess ^{40}Ar despite having ascended through crystalline basement underlying the Paleozoic sediments, it is implied that uptake of excess ^{40}Ar by the dike via degassing of the country rocks occurred dominantly *in situ*, after the dike magma ceased to flow. An analogous situation has been reported from the Ponta Grossa dike swarm of Brazil [*Renne et al.*, 1996].

Various permutations of the saddle-shaped spectrum occur; some with pronounced asymmetry grading into cases with anomalously old ages only at high or low fractional ^{39}Ar release; some with multiple age minima. In all such cases, a discordant age spectrum is a priori evidence that a simple air correction is inadequate, and one or more components of excess ^{40}Ar are present. When these are manifest in different portions of the age spectrum, generally derived at different extraction temperatures, it is implied that the excess ^{40}Ar is stored in distinct crystal-chemical environments with distinct thermal release thresholds. The nature of these energetically distinct environments is not generally clear, but may include distinct phases due to exsolution, and/or some excess ^{40}Ar being sited in anion vacancies with distinct activation energy for diffusion [*Harrison and McDougall*, 1981]. Alternatively, excess ^{40}Ar in the phase of interest is frequently associated with inclusions, either of different minerals, fluids, or of quenched melt in the case of volcanic rocks [*Kelley*, 2002]. An excellent example of excess ^{40}Ar hosted by melt inclusions was documented by *Esser et al.* [1997] in anorthoclase phenocrysts from a phonolite lava erupted in 1984 from Erebus volcano in Antarctica, which yield apparent K–Ar and ^{40}Ar/^{39}Ar ages as old as 700 ka. In this case, the excess ^{40}Ar was shown to correlate with elevated ^{38}Ar$_{Cl}$/^{39}Ar$_K$ due to the high concentration of Cl in the melt inclusions relative to the anorthoclase (Fig. 9.13). Correlation between excess ^{40}Ar and ^{38}Ar$_{Cl}$ has also been noted in fluid inclusions [*Harrison et al.*, 1994].

9.5.3.2 Isochrons

Some insight into the various causes of discordant age spectra, especially those arising from excess ^{40}Ar, can be gained from casting step-heating data on an isotope correlation diagram, commonly plotting ^{40}Ar/^{39}Ar$_K$ versus ^{39}Ar$_K$/^{36}Ar or ^{36}Ar/^{40}Ar versus ^{39}Ar$_K$/^{40}Ar, also termed isochron and inverse isochron diagrams, respectively. The former (Fig. 9.14a) is directly analogous to isochrons used in other systems (e.g., Rb–Sr), as discussed in Chapter 4. For a collinear array of data representing binary mixtures of radiogenic and nonradiogenic Ar, each of which have a single value, an isochron is defined whose slope is proportional to an age and whose y-axis intercept gives the initial isotopic composition of nonradiogenic Ar. In the so-called inverse isochron diagram (Fig. 9.14b), the x-axis intercept gives the reciprocal of ^{40}Ar*/^{39}Ar$_K$, hence is a function of age, and the y-axis intercept gives the isotopic composition of nonradiogenic Ar, which may contain excess ^{40}Ar, atmospheric Ar, and/or some other components. Note that the presence of any diffusion gradients, such as may arise from slow cooling or episodic partial degassing, will produce nonlinear arrays in isochron diagrams.

The computation of isochron slopes and intercepts accounting for error correlations between the variables [*York*, 1969] is discussed in Chapter 4. It has been shown [*Dalrymple et al.*, 1988] that both types of isochron yield identical information if they are correctly calculated, and the choice between the two is essentially esthetic. Inverse isochrons have steadily gained popularity, probably because they tend to spread the data more uniformly for Phanerozoic samples, which dominate the sphere of ^{40}Ar/^{39}Ar geochronology, but they are not intrinsically superior in statistical terms.

The profound advantage of isochrons is that no a priori assumptions are required about the isotopic composition of nonradiogenic argon. Recall that an age spectrum is just a sequence of model ages derived from incremental heating in which the nonradiogenic Ar is generally assumed to have modern atmospheric composition. A meaningful isochron can be determined from data whose age spectrum is significantly discordant, as shown in Figure 9.15.

Isotope correlation diagrams are very useful in qualitatively evaluating discordant age spectra, especially those due to excess ^{40}Ar contamination. Unfortunately it is relatively rare that statistically valid isochrons, i.e. with MSWD ~1.0, are obtained from such data, presumably because excess ^{40}Ar has variable isotopic composition, or initial trapped (excess) ^{40}Ar is mixed with modern atmospheric ^{40}Ar, or some combination of these. In rare cases, multiple isochrons have been determined from a single step-heating experiment [*Heizler and Harrison*, 1988]. In one such case (Fig. 9.16) the data define two isochrons with a common age but distinct trapped ^{40}Ar/^{36}Ar, suggesting the presence of two components of excess ^{40}Ar with distinct thermal retentivities that were incorporated at the same time. In another example [*Renne*, 1995], two components of trapped excess ^{40}Ar with

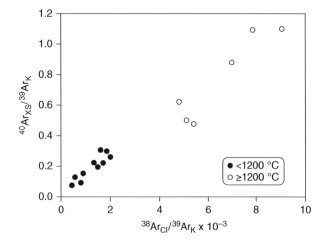

Fig. 9.13. Analysis of chlorine-free anorthoclase phenocrysts with up to 30% chlorine-rich melt inclusions shows a correlation of excess ^{40}Ar (^{40}Ar$_{xs}$) with chlorine-derived ^{38}Ar (^{38}Ar$_{Cl}$) [*Esser et al.*, 1997]. Chlorine concentrations were not measured independently, but are known to be high in alkalic melts and negligible in feldspars. The amount of excess ^{40}Ar was calculated using the method of *Harrison et al.* [1993], who used the technique to detect and correct for excess ^{40}Ar in fluid inclusions.

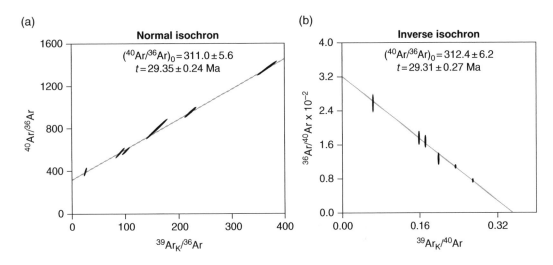

Fig. 9.14. (a) "Normal" and (B) "inverse" isochrons determined from the same Ar isotope data derived from step-heating of a hawaiite lava [*Dalrymple et al.*, 1988]. Both approaches in general yield indistinguishable values for trapped $^{40}Ar/^{36}Ar$ and $^{40}Ar^*/^{39}Ar_K$ (age) if computed correctly.

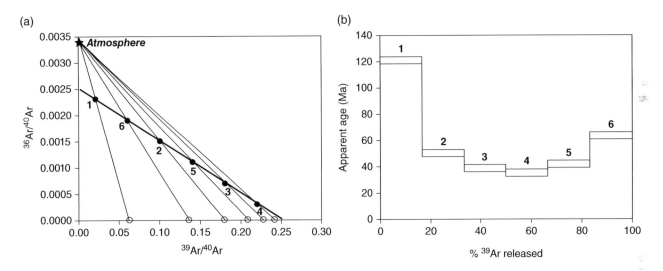

Fig. 9.15. Relationship between (a) an inverse isochron and (b) the corresponding age spectrum. Individual steps are plotted as solid circles along the isochron in (a). The ages for steps 1–6 in the age spectrum (b) are model ages based on an air correction, graphically corresponding to projections from atmosphere through each data point to the open circles on the horizontal ($^{39}Ar/^{40}Ar$) axis in (a).

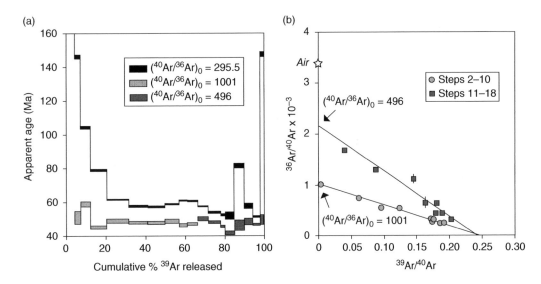

Fig. 9.16. (a) Step heating of a multigrained K-feldspar sample [*Heizler and Harrison*, 1988] yields a discordant apparent age spectrum based on a standard air correction. (b) The data cast on an isotope correlation diagram reveal two components of trapped (excess) ^{40}Ar as shown by distinct inverse isochrons. Both isochrons have a common $^{39}Ar/^{40}Ar$ intercept, interpreted to reflect the Ar retention age of the K-feldspar.

indistinguishable ages were inferred for biotite, but high MSWD values of 8.3 and 2.9 for the low-temperature and high-temperature components, respectively, weaken the validity of the isochrons and it is possible that recoil artifacts contribute to the scatter of data if not to the very appearance of two distinct components of excess ^{40}Ar.

Finally, we note that isotope correlation diagrams are particularly useful in interpreting data from extraterrestrial samples, which a priori cannot be assumed to have a nonradiogenic component of Ar dominated by modern terrestrial atmosphere. Additionally, meteorites and lunar samples commonly have significant concentrations of ^{38}Ar and ^{36}Ar produced by cosmic radiation and solar wind implantation [*Turner et al.*, 1971], and to further complicate matters the general antiquity of such materials necessitates relatively long irradiations, which may also produce significant $^{38}Ar_{Cl}$ and $^{36}Ar_{Cl}$. A method for deconvolving the various Ar isotopes from these sources under ideal circumstances has been developed [*Cassata et al.*, 2010] and the reader is directed to this paper for details.

9.6 CALIBRATION AND ACCURACY

The accuracy of K–Ar and $^{40}Ar/^{39}Ar$ ages can be viewed as having three distinct components, each with distinct limitations. In the following, we use the term "uncertainty" as a quantification of doubt about the accuracy of measurements or results. We reserve the term "error" to designate the difference between a measured value or result and the true value of the quantity in question.

(1) First, the accuracy of Ar isotope data is governed by measurement precision and corrections for background concentrations and mass biases. Mass biases arise from both the ionization source and the detector(s), whose efficiencies are generally mass-dependent. This is especially true for electron multiplier detectors, and less so for Faraday-type detectors. In the case of multicollector mass spectrometers, some uncertainty is associated with intercalibration of the detectors. Mass bias (usually termed mass discrimination) is generally determined by normalizing measured relative abundances to a reference gas, most commonly purified air Ar whose isotopic composition is well-known [*Lee et al.*, 2006]. Unfortunately only two isotopes (^{40}Ar and ^{36}Ar) in air are generally measurable with suitable precision and accuracy, and the form of the mass bias function for intermediate masses (^{39}Ar, ^{38}Ar, and ^{37}Ar) is not generally well-determined as a consequence [*Renne et al.*, 2009a]. The various corrections required to produce accurate relative abundances of Ar isotopes are nonsystematic in nature, and in modern mass spectrometers typically introduce uncertainties in the most important ratios ($^{40}Ar/^{39}Ar$ and $^{36}Ar/^{39}Ar$) that are better than 0.1 and 0.5%, respectively.

(2) The second category affecting accuracy concerns the operations that transform corrected Ar isotope data into the parameters needed to solve the age equation, namely $^{40}Ar^*$, $^{36}Ar_A$, $^{39}Ar_K$, $^{37}Ar_{Ca}$, and $^{38}Ar_{Cl}$, as given, for example, by equation (9.26). The uncertainties arising here are introduced by those in the nucleogenic production ratios (also known as nuclear interference corrections) such as $(^{40}Ar/^{39}Ar)_K$, $(^{36}Ar/^{37}Ar)_{Ca}$, and $(^{39}Ar/^{37}Ar)_{Ca}$, as shown in equations (9.24a)–(9.24e) and those in the isotopic composition of air argon $(^{40}Ar/^{36}Ar)_A$ and $(^{38}Ar/^{36}Ar)_A$. The uncertainties introduced by these parameters may be regarded as pseudosystematic insofar as errors in their values have variable effects on the age equation, depending on sample composition. This poses complications for comparing results from individual analyses (e.g., steps in incremental heating experiments) neglecting systematic errors as appropriate, because compositional heterogeneity due to mineral zoning, for example, may exist.

(3) The third and most important category governing ultimate accuracy of $^{40}Ar/^{39}Ar$ ages arises from the variables associated with the ages of standards and the ^{40}K decay constants. For standards whose age is determined by K–Ar dating, age of the standard is not an independent variable and the relevant variables are the $^{40}Ar^*/^{40}K$ of the standard and the ^{40}K decay constants.

Uncertainties arising from the first and second categories above are complex due to the large number and extent of error correlations. For example, both blank and discrimination corrections commonly involve the application of results from multiple measurements to each unknown, which means that the corrected relative abundances are not mutually independent. For a comprehensive treatment of uncertainties and their propagation in $^{40}Ar/^{39}Ar$ dating, the reader is referred to *Scaillet* [2000] and *Vermeesch* [2015], the latter being especially thorough concerning the importance of "error" (uncertainty) correlations or covariances between the many variables.

Due to increasing awareness of the importance of the accuracy-limiting nature of systematic uncertainties (i.e., category 3 above), it is becoming common practice to report both the uncertainties arising only from categories 1 and 2 (commonly referred to as analytical or intralaboratory uncertainties) and those arising from all sources (categories 1–3; commonly called systematic or interlaboratory uncertainties). A common format for this is to report an age as $X \pm Y/Z$, where X is the age, and Y and Z are the analytical analytical (intralaboratory) and systematic (interlaboratory) uncertainties, respectively.

9.6.1 ^{40}K decay constants

From 1977 until 2000, the ^{40}K decay constants in nearly universal use were those recommended by *Steiger and Jager* [1977], which were based on a compilation of ^{40}K activity data [*Beckinsale and Gale*, 1969] and a determination of the isotopic abundance of ^{40}K (i.e., ^{40}K/K) [*Garner et al.*, 1975]. Unfortunately, *Steiger and Jäger* [1977] did not recommend values for the uncertainties associated with their constants, consequently the propagation of appropriate uncertainties into the age equation was extremely rare until analytical precision in Ar isotope measurements improved to the point that decay constant uncertainties became the conspicuously dominant limitation to accuracy.

A critical reevaluation of the same ^{40}K activity data used by *Beckinsale and Gale* [1969] led to revised ^{40}K decay constants [*Min et al.*, 2000], which are in widespread usage today. A third set of ^{40}K decay constants results from a statistical optimization approach [*Renne et al.*, 2011] that is based on (i) the λ_e and λ_b values reported by *Min et al.* [2000] combined with (ii) ^{40}Ar*/^{40}K of a widely used standard, sanidine from the Fish Canyon Tuff (FCs), and (iii) carefully selected pairs of ^{238}U/^{206}Pb–^{40}Ar/^{39}Ar dates for 16 volcanic rocks whose closure age is expected to be identical for both systems. Through this latter component, the optimization method implicitly includes constraints from the relatively well-determined ^{238}U decay constant. *Renne* [2014] provides further discussion of the appropriate use of the optimization approach. A summary of the ^{40}K decay constants most widely used today is given in Table 9.1.

9.6.2   Standards

Standards are necessary in ^{40}Ar/^{39}Ar dating to calibrate the amount of ^{40}K that can be related to a given amount of ^{40}Ar in an analysis. In this sense, ^{40}Ar/^{39}Ar standards are analogous to tracers used for isotope dilution in thermal ionization mass spectrometers (TIMS) analysis.

As of today, the ages of standards and their uncertainties are widely viewed as a bellwether of the accuracy of the ^{40}Ar/^{39}Ar dating method. This view is somewhat naïve insofar as it fails to acknowledge the more fundamental importance of uncertainties in the ^{40}K decay constants.

9.6.2.1   Calibration by K–Ar
Historically, the ages of ^{40}Ar/^{39}Ar standards were determined by K–Ar analysis of so-called *first-principles standards*, which tended to be micas and amphiboles because these minerals could be quantitatively degassed of Ar—a fundamental requirement for K–Ar analysis. The age of an unknown sample is then given by

$$t_u = \left(\frac{1}{\lambda_\varepsilon + \lambda_\beta}\right) \ln\left[\left(\frac{\lambda_\varepsilon + \lambda_\beta}{\lambda_\varepsilon}\right)\left(^{40}\text{Ar}^{*40}\text{K}\right)R_s^u + 1\right] \quad (9.29)$$

where R is as defined in equation (9.11). The ages of samples may be referred to *secondary standards*, which are calibrated relative to first-principles standards, using the following identity:

$$R_{s0}^u \equiv R_{s1}^u R_{s0}^{s1} \quad (9.30)$$

Equation (9.30) may be extended to an arbitrary number N of stepwise intercalibrations between different standards such that the generalized form of equation (9.29) becomes:

$$t_u = \left(\frac{1}{\lambda_\varepsilon + \lambda_\beta}\right) \ln\left[\left(\frac{\lambda_\varepsilon + \lambda_\beta}{\lambda_\varepsilon}\right)\left(\frac{^{40}\text{Ar}^*}{^{40}\text{K}}\right)R_{i+1}^u \prod_{i=0}^{N} R_i^{i+1} + 1\right] \quad (9.31)$$

Equation (9.31) enables recalculating an age based on any given standard to an age based on another standard, or to a different age for the same standard) using another identity:

$$R_j^i = \frac{\left(e^{\lambda t_i} - 1\right)}{\left(e^{\lambda t_j} - 1\right)} \quad (9.32a)$$

and the corollary

$$R_j^i \equiv \frac{1}{R_j^i} \quad (9.32b)$$

Box 9.1   Example 1

Suppose we want to recalculate the age of the Permo-Triassic boundary (PTB) as dated [*Renne et al.*, 1995] at 249.91 Ma using the secondary standard FCs at 28.03 Ma, to an age based on the first principles standard GA-1550 biotite, which has been intercalibrated with FCs [*Renne et al.*, 1998], and the decay constants of [*Min et al.*, 2000]. Application of equation (9.32b) yields:

$$R_{GA-1550}^{FCs} = \frac{1}{R_{FCs}^{GA-1550}} = \frac{1}{3.5957} = 0.2781$$

Using the ^{40}K decay constants of *Steiger and Jäger* [1977] from Table 9.1, as these were used to obtain the original ages, we calculate the R-value between the PTB and FCs using equation (9.32a):

$$R_{FCs}^{PTB} = \frac{\left(e^{(5.543E-04/Ma)(249.91Ma)} - 1\right)}{\left(e^{(5.543E-04/Ma)(28.03Ma)} - 1\right)} = 9.4888$$

Combining equations (9.29), (9.32), and (9.33)

$$R_{GA-1550}^{PTB} = R_{FCs}^{PTB} R_{GA-1550}^{FCs} = 9.4888 \cdot 1.2781 = 26388$$

Applying these results to equation (9.29) or equivalently, equation (9.31), now using the decay constants of *Min et al.* [2000], and the value of ^{40}Ar*/^{40}K = 0.00590 for GA-1550 biotite [*Renne et al.*, 1998] we obtain:

$$t_{PTB} = \left(\frac{1}{5.463 \times 10^{-10}}\right)$$
$$\ln\left[\left(\frac{5.463 \times 10^{-10}}{0.580 \times 10^{-10}}\right)(0.00590)(2.688) + 1\right] = 250.48\,Ma$$

9.6.2.2   Calibration by independent means
An increasingly popular means of determining the ages of ^{40}Ar/^{39}Ar standards is to use dating methods that are completely independent of the K–Ar system. Most common among these methods is to use astronomically tuned ages for tephra that are interbedded with sediments that record climate cycles which

Table 9.1 Summary of modern ^{40}K decay constants

Parameter	Steiger and Jäger [1977]	Min et al. [2000]	Renne et al. [2011]
λ_{EC}	0.581×10^{-10}/year	$(0.580 \pm 0.007) \times 10^{-10}$/year	$(5.757 \pm 0.016) \times 10^{-11}$/year
λ_b	4.962×10^{-10}/year	$(4.884 \pm 0.049) \times 10^{-10}$/year	$(4.9548 \pm 0.0134) \times 10^{-10}$/year
λ_{Total}	5.543×10^{-10}/year	$(5.463 \pm 0.054) \times 10^{-10}$/year	$(5.531 \pm 0.0135) \times 10^{-10}$/year

can be correlated with periodic fluctuations in Earth's orbital parameters of eccentricity, precession, and obliquity, and their ages are determined by reference to models of solar system dynamics [*Hinnov and Hilgen*, 2012]. The astronomical age (t_{ast}) is then attributed to a dated tephra, and the age (t_s) of a co-irradiated standard is determined by forcing the unknown's age to equal the astronomical age [*Renne et al.*, 1994; *Hilgen et al.*, 1997; *Kuiper et al.*, 2008; *Rivera et al.*, 2011].

By whatever independent means are used to determine the age (t_s) of a ^{40}Ar/^{39}Ar standard, the age (t_u) of an unknown relative to that standard is given by:

$$t_u = \frac{1}{\lambda} \ln\left[\left(e^{\lambda t_s} - 1\right) R_s^u + 1\right] \tag{9.33}$$

Note that this equation requires only the use of the total ^{40}K decay constant. For ages near the age of the standard, this age equation is relatively insensitive to uncertainties in the age of the standard [*Kuiper et al.*, 2008].

Box 9.2 Example 2

Let us recalculate the age of the Permo-Triassic boundary (see Example 1 above) using the astronomically tuned age of 28.201 Ma for the FCs standard, which also uses (by definition) the same ^{40}K decay constant as before. Substituting appropriate values into equation (9.33) and propagating uncertainties, we obtain:

$$t_{PTB} = \frac{1}{5.563 \times 10^{-4}/\text{Ma}} \ln\left[\left(e^{\left(5.463 \times 10^{-4}/\text{Ma}\right)(28.201\,\text{Ma})} - 1\right)(9.4888) + 1\right]$$
$$= 251.56 \pm 0.03/0.36\,\text{Ma}$$

which is more than a million years older than calculated in Example 1.

Box 9.3 Example 3

Now let us calculate the age of the Permo-Triassic boundary using the optimization calibration [*Renne et al.*, 2011], which returns both intralaboratory and interlaboratory uncertainties:

$$t_{PTB} = \frac{1}{5.531 \times 10^{-4}/\text{Ma}} \ln\left[\left(e^{\left(5.531 \times 10^{-4}/\text{Ma}\right)(28.294\,\text{Ma})} - 1\right)(9.4888) + 1\right]$$
$$= 252.20 \pm 0.03/0.15\,\text{Ma}$$

The results of Examples 2 and 3 differ by 0.64 Ma, which is insignificant at 95% confidence considering systematic uncertainties, and both are indistinguishable from ^{238}U/^{206}Pb zircon ages of 251.94 ± 0.14 Ma [*Burgess et al.*, 2014] or 252.40 ± 0.17 Ma [*Mundil et al.*, 2004] for the same tuff bed at the PTB.

9.6.3 So which is the best calibration?

This question is the subject of ongoing investigation. The examples above illustrate that different calibrations can yield significantly different results. Most workers who are familiar with the details of the topic agree that the values of λ_e and λ_b reported by *Min et al.* [2000] are more accurate than those of *Steiger and Jäger* [1977], particularly when an estimate of the full uncertainty is desired, as in cases where ^{40}Ar/^{39}Ar ages will be compared with ages derived from other methods. This leaves us realistically with a choice between the optimization calibration [*Renne et al.*, 2011], which incorporates diverse physical measurements with readily quantifiable uncertainties, and an astronomical calibration such as that of *Kuiper et al.* [2008].

From the standpoint of propagated uncertainty, the optimization approach yields much lower uncertainties as shown in Fig. 9.17, and would appear on this basis to be the more accurate calibration. On the other hand, consider the ages indicated for the FCs standard itself by the two approaches (Table 9.2) compared with various ^{238}U/^{206}Pb zircon ages for the Fish Canyon tuff, from which the FCs is derived. The ^{40}Ar retention age of sanidine is expected to be younger than the Pb retention age of the zircon, hence the ^{40}Ar/^{39}Ar age of FCs should be younger than the youngest credible ^{238}U/^{206}Pb zircon age. By invoking credibility, we dismiss the results of *Lanphere and Baadsgaard* [2001] on the grounds discussed by *Schmitz et al.* [2003]. The results of *Schmitz and Bowring* [2001] are now clearly understood to be biased by pre-eruptive zircon crystallization (see Chapter 8) based on single zircon age distributions [*Bachmann et al.*, 2007; *Wotzlaw et al.*, 2013]; values for the youngest zircon measured in the latter two studies are those shown in Table 9.2.

A critical evaluation of the results shown in Table 9.2 clearly indicates that the youngest zircon ^{238}U/^{206}Pb ages are around 28.2 Ma, and the orbital (astronomical) tuning ages are slightly younger, with some overlap. Thus it is tempting to infer that the optimization-based ^{40}Ar/^{39}Ar sanidine age of 28.294 ± 0.036 Ma is slightly too old and therefore that this calibration, although yielding the smallest propagated uncertainties, is inaccurate. On the other hand, good agreement between this calibration of the ^{40}Ar/^{39}Ar system and the ^{238}U/^{206}Pb system is evident in older samples, including the one illustrated in Example 3 above. It is possible that misfit between the optimization-calibrated ^{40}Ar/^{39}Ar system and the ^{238}U/^{206}Pb system at younger ages is due to unmitigated residence time in some of the ^{238}U/^{206}Pb zircon ages used in the former. However, it is also possible that ^{238}U/^{206}Pb ages for younger zircons, which

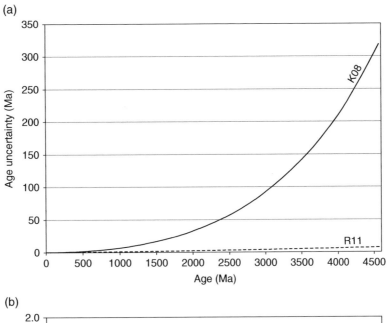

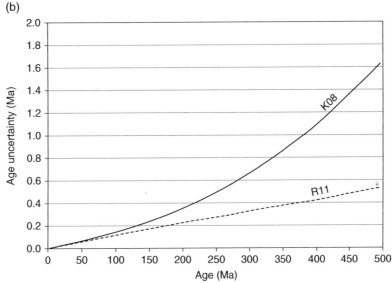

Fig. 9.17. (a) Comparison of propagated uncertainty in ages calculated by the optimization (R11) [*Renne et al.*, 2011a] and astronomical (K08) calibrations of *Kuiper et al.* [2008]. (b) Inset showing detail for the past 500 Ma. Values shown are for an *R*-value of an unknown relative to the FCs standard, as defined by equation (9.11), assuming a 0.1% relative precision for this value.

Table 9.2 Selected inferred eruption ages for the Fish Canyon tuff

Method	Age (Ma)	±σ (Ma)	Source
^{238}U/^{206}Pb zircon	28.498	0.018	*Schmitz and Bowring* [2001]
^{238}U/^{206}Pb zircon	28.196	0.019	*Wotzlaw et al.* [2013]
^{238}U/^{206}Pb zircon	28.18	0.20	*Bachmann et al.* [2007]
^{238}U/^{206}Pb zircon	27.52	0.09	*Lanphere and Baadsgaard* [2001]
Orbital tuning	28.201	0.023	*Kuiper et al.* [2008]
Orbital tuning	28.172	0.014	*Rivera et al.* [2011]
Orbital tuning	28.03	0.18	*Renne et al.* [1994]
Orbital tuning	28.15	0.19	*Hilgen et al.* [1997]
^{40}Ar/^{39}Ar sanidine[a]	28.294	0.036	*Renne et al.* [2011]
^{40}Ar/^{39}Ar sanidine[b]	28.02	0.28	*Renne et al.* [1998]
^{40}Ar/^{39}Ar sanidine[b]	27.62	0.13	*Lanphere and Baadsgaard* [2001]

[a] Based on the optimization calibration.
[b] Based on constants of *Steiger and Jager* [1977] and intercalibration with first-principles K–Ar standards.

are especially sensitive to corrections for blank and/or common Pb and initial Th/U isotopic disequilibrium, tend to be biased. Whatever the cause, the matter remains unresolved at present.

The existence and usage of several absolute calibrations of $^{40}Ar/^{39}Ar$ dates is somewhat confusing to nonspecialists, and even to some specialists. Fortunately, it is straightforward to recalculate ages from one calibration to another using methods outlined in this section, provided that the relevant metadata are provided [*Renne et al.*, 2009b]. Ideally, there will eventually be convergence such that a single calibration will be adopted by all laboratories. A particularly promising approach to improving $^{40}Ar^*$ concentration measurements for standards, i.e., towards improved first-principles K–Ar calibrations, is outlined by *Morgan et al.* [2011].

9.6.4 Interlaboratory issues

As noted above, inconsistent use of absolute calibrations is confusing but at least we have straightforward computational tools to enable conversion of ages *ex post facto*, to convert ages from one calibration to another and compare them on a common basis. Perhaps a greater problem is posed by interlaboratory variations in measured Ar isotope compositions. For example, Table 9.3 shows results of different published values of the intercalibration factor or *R*-value, as defined by equation (9.11) between the Alder Creek sanidine (ACs) and Fish Canyon sanidine (FCs) standards. The ACs was proposed as an $^{40}Ar/^{39}Ar$ standard by *Turrin et al.* [1994], partly because it is tied to the Cobb Mountain geomagnetic polarity event.

These *R*-values are independent of calibrations and are presumed to represent measurements of a single value, which is the age of eruption of the Alder Creek Rhyolite relative to that of the Fish Canyon Tuff. Thus the *R*-values in Table 9.3 are expected to be indistinguishable within the stated uncertainties. However, the three values published prior to 2011 (all measured in a single laboratory) are distinctly larger than those published subsequently, and the latter data yield consistent values. A subset of these was combined by *Niespolo et al.* [2017] to yield a weighted mean of 0.041707 ± 0.000011, which was recommended as an interlaboratory value.

Table 9.3 Intercalibration factors between Fish Canyon and Alder Creek sanidine standards

Source	R^{ACs}_{FCs}	$\pm\sigma_R$
Turrin et al. [1994]	0.04229	0.00021
Renne et al. [1998]	0.04229	0.00006
Nomade et al. [2005]	0.04230	0.00006
Coble et al. [2011]	0.04182	0.00021
McDougall et al. [2012]	0.04189	0.00060
Rivera et al. [2013]	0.04175	0.00003
Phillips and Matchan [2013]	0.04169	0.00002
Singer [2014]	0.04185	0.00011
Jicha et al. [2016]	0.04176	0.00004
Niespolo et al. [2016]	0.04170	0.00001

The manifest convergence of the more recently published values presumably reflects community-wide adoption of best practices in irradiation strategies and mass spectrometry; interlaboratory intercalibration of additional standards at similar levels of precision are expected to follow in the near future. Ideally, a suite of standards should be developed whose ages are related by relatively small multiples in order to minimize the difference in Ar isotope ratios to be measured between standards and unknowns. For example, a suite of eight standards with ages of 0.33, 1, 3, 9, 27, 81, 240, and 722 Ma, i.e., related by a factor of three, would tightly bracket any Phanerozoic unknown sample in terms of $^{40}Ar^*/^{39}Ar_K$ to be measured. Arguably, the 1 Ma and 27 Ma place in the hypothetical progression above are already established by the ACs and FCs standards. The relationships between the various standards are logically summarized in matrix form, as shown by *Niespolo et al.* [2017]. Aside from relative ages of standards, other important criteria in their selection for interlaboratory use are those of reproducibility and availability.

9.7 CONCLUDING REMARKS

The classical K–Ar technique maintains some relevance for two main reasons: (i) for dating fine-grained materials subject to ^{39}Ar recoil artifacts during neutron irradiation; and (ii) as a means of providing first-principles calibration of $^{40}Ar^*$ concentrations in standards. In nearly all other respects, the $^{40}Ar/^{39}Ar$ variant of this method offers powerful advantages that can be summarized as follows:

- The ability to conduct stepwise heating experiments, which permits age spectrum and isochron analyses that frequently allow recognition of open-system behavior, excess $^{40}Ar^*$, and/or sample homogeneity.
- The ability to analyze arbitrarily small samples, at the scale of single crystals or portions thereof, limited only by procedural backgrounds and/or mass spectrometer sensitivity, allows recognition of xenocrysts and age gradients due to open-system behavior.
- The ability to automate the entire analysis process allows rapid data acquisition and standardization of analytical conditions.

9.7.1 Remaining challenges

The $^{40}Ar/^{39}Ar$ method currently has the broadest scope of any radioisotopic geochronometer in terms of its age range of applicability and diversity of materials that can be usefully dated, and it is capable of unsurpassed precision. Nonetheless, there are several areas in which the method can be improved.

The issues of ultimate calibration arising from decay constants and standards, as discussed in the previous section, are under ongoing investigation and will continue to improve accuracy of the $^{40}Ar/^{39}Ar$ method. Similarly, the issue of interlaboratory reproducibility as manifest in Table 9.3 and related discussion will be improved with ongoing efforts of the $^{40}Ar/^{39}Ar$ community. This effort will focus on technical developments and analytical protocols such as the treatment of background measurements, intercalibration of detectors for multicollector mass spectrometers, evaluating the linearity of detector sensitivity and mass discrimination, and improving the signal to noise ratio of low-level ion beam measurement.

Further benefits to accuracy, precision, and scope of the method will be realized by other developments, including the following:

- Improved knowledge of nuclear reactions and recoil phenomena, including factors governing self-shielding and detailed knowledge of cross-sections of interfering reactions as a function of neutron energy. Development of optimized neutron sources will have both scientific and radiological benefits.
- Improved knowledge of the evolution of ^{40}Ar/^{36}Ar in paleoatmosphere over time. This is important in its own right as a constraint on outgassing history of the solid Earth, but also to better inform the application of an air correction to determine relative concentrations of ^{40}Ar*.
- Improved understanding of the mechanism(s) for incorporation and uptake of nonradiogenic Ar (e.g., excess ^{40}Ar) by various materials, including relevant Ar solubilities and diffusion kinetics. Included here are the possible effects of isotopic fractionation of Ar during uptake of nonradiogenic components such as atmospheric Ar.

9.8 REFERENCES

Aldrich, L. T. and Nier, A. O. (1948) Argon-40 in potassium minerals. *Physical Review* **74**, 876–877.

Alpers, C. N. and Brimhall, G. H. (1988) Middle Miocene climatic-change in the Atacama desert, northern Chile—evidence from supergene mineralization at La-Escondida. *Geological Society of America Bulletin* **100**, 1640–1656.

Aston, F. W. (1921) Mass-spectra and atomic weights—A lecture delivered before the chemical society on April 7th 1921. *Journal of the Chemical Society* **119**, 677–687.

Bachmann, O., Oberli, F., Dungan, M. A., Meier, M., Mundil, R., and Fischer, H. (2007) ^{40}Ar/^{39}Ar and U-Pb dating of the Fish Canyon magmatic system, San Juan Volcanic field, Colorado: Evidence for an extended crystallization history. *Chemical Geology* **236**, 134–166.

Baksi, A. K. (1994) Geochronological studies on whole-rock basalts, Deccan Traps, India—evaluation of the timing of volcanism relative to the K–T boundary. *Earth and Planetary Science Letters* **121**, 43–56.

Beckinsale, R. D. and Gale, N. H. (1969) A reappraisal of decay constants and branching ratio of 40K. *Earth and Planetary Science Letters* **6**, 289–294.

Bender, M. L., Barnett, B., Dreyfus, G., Jouzel, J., and Porcelli, D. (2008) The contemporary degassing rate of ^{40}Ar from the solid Earth. *Proceedings of the National Academy of Sciences of the United States of America* **105**, 8232–8237.

Bramley, A. (1937) The potassium-argon transformation. *Science* **86**, 424–425.

Burgess, S. D., Bowring, S., and Shen, S.-Z. (2014) High-precision timeline for Earth's most severe extinction. *Proceedings of the National Academy of Sciences of the United States of America* **111**, 3316–3321.

Cadogan, P. H. (1977) Paleoatmospheric argon in Rhynie Chert. *Nature* **268**, 38–41.

Campbell, N. R. and Wood, A. (1907) Radio-activity of the alkali metals. *Proceedings of the Cambridge Philosophical Society* **14**, 15–21.

Carrapa, B., Wijbrans, J., and Bertotti, G. (2003) Episodic exhumation in the Western Alps. *Geology* **31**, 601–604.

Cassata, W. S., Shuster, D. L., Renne, P. R., and Weiss, B. P. (2010) Evidence for shock heating and constraints on Martian surface temperatures revealed by ^{40}Ar/^{39}Ar thermochronometry of Martian meteorites. *Geochimica et Cosmochimica Acta* **74**, 6900–6920.

Coble, M. A., Grove, M., and Calvert, A. T. (2011) Calibration of Nu-Instruments Noblesse multicollector mass spectrometers for argon isotopic measurements using a newly developed reference gas. *Chemical Geology* **290**, 75–87.

Cox, A., Dalrymple, G. B., and Doell, R. R. (1963) Geomagnetic polarity epochs and Pleistocene geochronometry. *Nature* **198**, 1049–1051.

Dalrymple, G. B. (1969) ^{40}Ar/^{36}Ar analyses of historic lava flows. *Earth and Planetary Science Letters* **6**, 47–54.

Dalrymple, G. B. and Lanphere, M. A. (1969) *Potassium-argon Dating*. W. H. Freeman and Co., San Francisco, 258 pp.

Dalrymple, G. B., Alexander, E. C. Jr., Lanphere, M. A., and Kraker, G. P. (1981) Irradiation of samples for ^{40}Ar/^{39}Ar dating using the Geological Survey TRIGA reactor. *U.S. Geological Survey Professional Paper* **1176**

Dalrymple, G. B., Lanphere, M. A., and Pringle, M. S. (1988) Correlation diagrams in ^{40}Ar/^{39}Ar dating—is there a correct choice. *Geophysical Research Letters* **15**, 589–591.

Deino, A. and Potts, R. (1990) Single-crystal ^{40}Ar/^{39}Ar dating of the Olorgesailie Formation, southern Kenya rift. *Journal of Geophysical Research—Solid Earth and Planets* **95**, 8453–8470.

Dejong, K., Wijbrans, J. R., and Feraud, G. (1992) Repeated thermal resetting of phengites in the Mulhacen Complex (Betic zone, southeastern Spain) shown by ^{40}Ar/^{39}Ar step heating and single grain laser probe dating. *Earth and Planetary Science Letters* **110**, 173–191.

De Laeter, J. and Kurz, M. D. (2006) Alfred Nier and the sector field mass spectrometer. *Journal Mass Spectrometry* **41**(7), 847–854.

Dodd, R. T., van Schmu, W. R., and Marvin, U. B. (1965) Merrihueite a new alkali-ferromagnesian silicate from Mezo-Madaras chondrite. *Science* **149**, 972–974.

Dong, H. L., Hall, C. M., Peacor, D. R., and Halliday, A. N. (1995) Mechanisms of argon retention in clays revealed by laser ^{40}Ar–^{39}Ar dating. *Science* **267**, 355–359.

Esser, R. P., McIntosh, W. C., Heizler, M. T., and Kyle, P. R. (1997) Excess argon in melt inclusions in zero-age anorthoclase feldspar from Mt. Erebus, Antarctica, as revealed by the ^{40}Ar/^{39}Ar method. *Geochimica et Cosmochimica Acta* **61**, 3789–3801.

Evernden, J. F., Savage, D. E., Curtis, G. H., and James, G. T. (1964) Potassium-argon dates and the Cenozoic mammalian chronology of north america. *American Journal of Science* **262**, 145–198.

Fiet, N., Quidelleur, X., Parize, O., Bulot, L. G., and Gillot, P. Y. (2006) Lower Cretaceous stage durations combining radiometric data and orbital chronology: Towards a more stable relative time scale? *Earth and Planetary Science Letters* **246**, 407–417.

Fleck, R. J., Sutter, J. F., and Elliot, D. H. (1977) Interpretation of discordant ^{40}Ar–^{39}Ar age-spectra of Mesozoic tholeiites from Antarctica. *Geochimica et Cosmochimica Acta* **41**, 15–32.

Foland, K. A., Linder, J. S., Laskowski, T. E., and Grant, N. K. (1984) ^{40}Ar/^{39}Ar dating of glauconites—measured ^{39}Ar recoil loss from well-crystallized specimens. *Isotope Geoscience* **2**, 241–264.

Foland, K. A., Hubacher, F. A., and Arehart, G. B. (1992) ^{40}Ar/^{39}Ar dating of very fine-grained samples—an encapsulated-vial procedure to overcome the problem of ^{39}Ar recoil loss. *Chemical Geology* **102**, 269–276.

Foland, K. A., Fleming, T. H., Heimann, A., and Elliot, D. H. (1993) Potassium argon dating of fine-grained basalts with massive Ar loss—application of the ^{40}Ar/^{39}Ar technique to plagioclase and glass from the Kirkpatrick basalt, Antarctica. *Chemical Geology* **107**, 173–190.

Garner, E. L., Murphy, T. J., Gramlich, J. W., Paulsen, P. J., and Barnes, I. L. (1975) Absolute isotopic abundance ratios and atomic weight of a reference sample of potassium. *Journal of Research of the National Bureau of Standards Section A—Physics and Chemistry* **79**, 713–725.

Hall, C. M. (2014) Direct measurement of recoil effects on ^{40}Ar/^{39}Ar standards. In *Advances in ^{40}Ar/^{39}Ar Dating: From Archaeology to Planetary Sciences, Jourdan, F., Mark, D. M., and Verati, C. (eds)*, 53–62. Special Publication 378, Geological Society, London.

Hames, W. E. and Bowring, S. A. (1994) An empirical-evaluation of the argon diffusion geometry in muscovite. *Earth and Planetary Science Letters* **124**, 161–167.

Hanes, J. A., York, D., and Hall, C. M. (1985) An ^{40}Ar/^{39}Ar geochronological and electron-microprobe investigation of an Archean pyroxenite and its bearing on ancient atmospheric compositions. *Canadian Journal of Earth Sciences* **22**, 947–958.

Harrison, T. M. and McDougall, I. (1981) Excess ^{40}Ar in metamorphic rocks from Broken Hill, New-South-Wales—implications for ^{40}Ar–^{39}Ar age spectra and the thermal history of the region. *Earth and Planetary Science Letters* **55**, 123–149.

Harrison, T. M., Heizler, M. T., and Lovera, O. M. (1993) Invacuo crushing experiments and K-feldspar thermochronometry. *Earth and Planetary Science Letters* **117**, 169–180.

Harrison, T. M., Heizler, M. T., Lovera, O. M., Chen, W. J., and Grove, M. (1994) A chlorine disinfectant for excess argon released from K-feldspar during step heating. *Earth and Planetary Science Letters* **123**, 95–104.

Harrison, T. M., Celerier, J., Aikman, A.B., Hermann, J., and Heizler, M. T. (2009) Diffusion of ^{40}Ar in muscovite. *Geochimica et Cosmochimica Acta* **73**, 1039–1051.

Heizler, M. T. and Harrison, T. M. (1988) Multiple trapped argon isotope components revealed by ^{40}Ar/^{39}Ar isochron analysis. *Geochimica et Cosmochimica Acta* **52**, 1295–1303.

Hess, J. C. and Lippolt, H. J. (1986) Kinetics of Ar isotopes during neutron-irradiation—Ar-39 loss from minerals as a source of error in ^{40}Ar/^{39}Ar dating. *Chemical Geology* **59**, 223–236.

Hilgen, F. J., Krijgsman, W., and Wijbrans, J. R. (1997) Direct comparison of astronomical and ^{40}Ar/^{39}Ar ages of ash beds: Potential implications for the age of mineral dating standards. *Geophysical Research Letters* **24**, 2043–2046.

Hinnov, L. A. and Hilgen, F. J. (2012) Cyclostratigraphy and astrochronology. In *Geologic Time Scale 2012*, Vols **1 and 2**, Gradstein, F. M., Ogg, J. G., Schmitz, M. D., and Ogg, G. M. (eds), 63–83. Elsevier.

Hodges, K. V., Ruhl, K. W., Wobus, C. W., and Pringle, M. S. (2005) ^{40}Ar/^{39}Ar thermochronology of detrital minerals. In *Low-Temperature Thermochronology: Techniques, Interpretations, and Applications*, Vol. **58**, Reiners, P. W. and Ehlers, T. A. (eds), 239–257. Mineralogical Society of America.

Humayun, M. and Clayton, R. N. (1995) Precise determination of the isotopic composition of potassium—application to terrestrial rocks and lunar soils. *Geochimica et Cosmochimica Acta* **59**, 2115–2130.

Huneke, J. C. and Villa, I. M. (1981) Ar-39 loss during neutron-irradiation and the aging of Allende inclusions. *Meteoritics* **16**, 329–330.

Jessberger, E. K., Dominik, B., Staudacher, T., and Herzog, G. F. (1980) ^{40}Ar–^{39}Ar ages of Allende. *Icarus* **42**, 380–405.

Jicha, B. R., Singer, B. S., and Sobol, P. (2016) Re-evaluation of the ages of ^{40}Ar/^{39}Ar sanidine standards and supereruptions in the western US using a Noblesse multi-collector mass spectrometer. *Chemical Geology* **431**, 54–66.

Jourdan, F. and Renne, P. R. (2007) Age calibration of the Fish Canyon sanidine ^{40}Ar/^{39}Ar dating standard using primary K–Ar standards. *Geochimica et Cosmochimica Acta* **71**, 387–402.

Jourdan, F. and Renne, P. R. (2014) Neutron-induced Ar-37 recoil ejection in Ca-rich minerals and implications for ^{40}Ar/^{39}Ar dating. In *Advances in ^{40}Ar/^{39}Ar Dating: From Archaeology to Planetary Sciences, Jourdan, F., Mark, D. M., and Verati, C. (eds)*, 33–52. Special Publication 378, Geological Society, London.

Jourdan, F., Matzel, J. P., and Renne, P. R. (2006) Ar-39 and Ar-37 recoil ejection during and plagioclase crystals. *Geochimica et Cosmochimica Acta* **70**, A299–A299.

Jourdan, F., Matzel, J. P., and Renne, P. R. (2007) ^{39}Ar and ^{37}Ar recoil loss during neutron irradiation of sanidine and plagioclase. *Geochimica et Cosmochimica Acta* **71**, 2791–2808.

Kelley, S. (2002) Excess argon in K–Ar and Ar/Ar geochronology. *Chemical Geology* **188**, 1–22.

Kelley, S. P. and Turner, G. (1991) Laser probe ^{40}Ar–^{39}Ar measurements of loss profiles within individual hornblende grains from the Giants Range Granite, northern Minnesota, USA. *Earth and Planetary Science Letters* **107**, 634–648.

Klemperer, O. (1935) Radioactivity of potassium and rubidium. *Proceedings of the Royal Society of London, Series A (Mathematical and Physical Sciences)* **148**, 638–648.

Kuiper, K. F., Deino, A., Hilgen, F. J., Krijgsman, W., Renne, P. R., and Wijbrans, J. R. (2008) Synchronizing rock clocks of Earth history. *Science* **320**, 500–504.

Lanphere, M. A. and Baadsgaard, H. (2001) Precise K–Ar, ^{40}Ar/^{39}Ar, Rb–Sr and U/Pb mineral ages from the 27.5 Ma Fish Canyon Tuff reference standard. *Chemical Geology* **175**, 653–671.

Lanphere, M. A. and Dalrymple, G. B. (1976) Identification of excess ^{40}Ar by ^{40}Ar–^{39}Ar age spectrum technique. *Earth and Planetary Science Letters* **32**, 141–148.

Layer, P. W., Hall, C. M., and York, D. (1987) The derivation of ^{40}Ar/^{39}Ar age spectra of single grains of hornblende and biotite by laser step-heating. *Geophysical Research Letters* **14**, 757–760.

Leakey, L. S., Curtis, G. H., and Evernden, J. F. (1961) Age of bed 1, Olduvai Gorge, Tanganyika. *Nature* **191**, 478–479.

Lee, J. Y., Marti, K., Severinghaus, J. P., *et al.* (2006) A redetermination of the isotopic abundances of atmospheric Ar. *Geochimica et Cosmochimica Acta* **70**, 4507–4512.

Leost, I., Feraud, G., Blanc-Valleron, M. M., and Rouchy, J. M. (2001) First absolute dating of Miocene langbeinite evaporites by ^{40}Ar/^{39}Ar laser step-heating: $K_2Mg_2(SO_4)_3$ Stebnyk mine (Carpathian Foredeep Basin). *Geophysical Research Letters* **28**, 4347–4350.

Li, W., Beard, B. L., and Li, S. (2016) Precise measurement of stable potassium isotope ratios using a single focusing collision cell multi-collector ICP-MS. *Journal of Analytical Atomic Spectrometry* **31**, 1023–1029.

Lobello, P., Feraud, G., Hall, C. M., York, D., Lavina, P., and Bernat, M. (1987) ^{40}Ar–^{39}Ar step-heating and laser fusion dating of a quaternary pumice from Neschers, Massif-Central, France—the defeat of xenocrystic contamination. *Chemical Geology* **66**, 61–71.

Lovera, O. M., Richter, F. M., and Harrison, T. M. (1989) The ^{40}Ar–^{39}Ar thermochronometry for slowly cooled samples having a distribution of diffusion domain sizes. *Journal of Geophysical Research—Solid Earth and Planets* **94**, 17917–17935.

Marzoli, A., Renne, P. R., Piccirillo, E. M., Ernesto, M., Bellieni, G., and De Min, A. (1999) Extensive 200-million-year-old continental flood basalts of the Central Atlantic Magmatic Province. *Science* **284**, 616–618.

McDougal, I. and Tarling, D. H. (1963) Dating of polarity zones in Hawaiian islands. *Nature* **200**, 54–56.

McDougall, I., Brown, F. H., Vasconcelos, P. M., Cohen, B. E., Thiede, D. S., and Buchanan, M. J. (2012) New single crystal ^{40}Ar/^{39}Ar ages improve time scale for deposition of the Omo Group, Omo-Turkana Basin, East Africa. *Journal of the Geological Society* **169**, 213–226.

Megrue, G. H. (1973) Spatial-distribution of ^{40}Ar/^{39}Ar ages in lunar breccia 14301. *Journal of Geophysical Research* **78**, 3216–3221.

Merrihue, C. and Turner, G. (1966) Potassium-argon dating by activation with fast neutrons. *Journal of Geophysical Research* **71**, 2852–2857.

Min, K. W., Mundil, R., Renne, P. R., and Ludwig, K. R. (2000) A test for systematic errors in ^{40}Ar/^{39}Ar geochronology through comparison with U/Pb analysis of a 1.1-Ga rhyolite. *Geochimica et Cosmochimica Acta* **64**, 73–98.

Mitchell, J. G. (1968) ^{40}Ar/^{39}Ar method for potassium-argon age determination. *Geochimica et Cosmochimica Acta* **32**, 781–790.

Morgan, L. E., Postma, O., Kuiper, K. F., *et al.* (2011) A metrological approach to measuring Ar-40* concentrations in K–Ar and ^{40}Ar/^{39}Ar mineral standards. *Geochemistry, Geophysics, Geosystems* **12**.

Mundil, R., Ludwig, K. R., Metcalfe, I., and Renne, P. R. (2004) Age and timing of the Permian mass extinctions: U/Pb dating of closed-system zircons. *Science* **305**, 1760–1763.

Newman, F. H. and Walke, H. J. (1935) Radioactivity of potassium. *Nature* **135**, 98–98.

Niedermann, S., Schaefer, J. M., Wieler, R., and Naumann, R. (2007) The production rate of cosmogenic ^{38}Ar from calcium in terrestrial pyroxene. *Earth and Planetary Science Letters* **257**, 596–608.

Nier, A. O. (1935) Evidence for the existence of an isotope of potassium of mass 40. *Physical Review* **48**, 283–284.

Nier, A. O. (1950) A redetermination of the relative abundances of the isotopes of carbon, nitrogen, oxygen, argon, and potassium. *Physical Review* **77**, 789–793.

Niespolo, E. M., Rutte, D., Deino, A. L., and Renne, P. R. (2017) Intercalibration and age of the Alder Creek sanidine ^{40}Ar/^{39}Ar standard. *Quaternary Geochronology* **39**, 205–213.

Nomade, S., Renne, P. R., Vogel, N., *et al.* (2005) Alder Creek sanidine (ACs-2): a Quaternary ^{40}Ar/^{39}Ar dating standard tied to the Cobb Mountain geomagnetic event. *Chemical Geology* **218**, 315–338.

Odin, G. S. (1986) Recent advances in Phanerozoic time-scale calibration. *Chemical Geology* **59**, 103–110.

Onstott, T. C. and Peacock, M. W. (1987) Argon retentivity of hornblendes—a field experiment in a slowly cooled metamorphic terrane. *Geochimica et Cosmochimica Acta* **51**, 2891–2903.

Onstott, T. C., Miller, M. L., Ewing, R. C., Arnold, G. W., and Walsh, D. S. (1995) Recoil refinements—implications for the ^{40}Ar/^{39}Ar dating technique. *Geochimica et Cosmochimica Acta* **59**, 1821–1834.

Paine, J. H., Nomade, S., and Renne, P. R. (2006) Quantification of ^{39}Ar recoil ejection from GA1550 biotite during neutron irradiation as a function of grain dimensions. *Geochimica et Cosmochimica Acta* **70**, 1507–1517.

Parman, S. W. (2007) Helium isotopic evidence for episodic mantle melting and crustal growth. *Nature* **446**, 900–903.

Phillips, D. and Matchan, E. L. (2013) Ultra-high precision ^{40}Ar/^{39}Ar ages for Fish Canyon Tuff and Alder Creek Rhyolite sanidine: New dating standards required? *Geochimica et Cosmochimica Acta* **121**, 229–239.

Phillips, D. and Onstott, T. C. (1988) Argon isotopic zoning in mantle phlogopite. *Geology* **16**, 542–546.

Pujol, M., Marty, B., Burgess, R., Turner, G., and Philippot, P. (2013) Argon isotopic composition of Archaean atmosphere probes early Earth geodynamics. *Nature* **498**, 87.

Renne, P. R. (1995) Excess ^{40}Ar in biotite and hornblende from the Norilsk-1 intrusion, Siberia—implications for the age of the Siberian traps. *Earth and Planetary Science Letters* **131**, 165–176.

Renne, P. R. (2014) Some footnotes to the optimization-based calibration of the ^{40}Ar/^{39}Ar system. In *Advances in ^{40}Ar/^{39}Ar Dating: From Archaeology to Planetary Sciences, Jourdan, F.,*

Mark, D. M., and Verati, C. (eds), 21–31. Special Publication 378, Geological Society, London.

Renne, P. R., Becker, T. A., and Swapp, S. M. (1990) $^{40}Ar/^{39}Ar$ laser-probe dating of detrital micas from the montgomery creek formation, northern california—clues to provenance, tectonics, and weathering processes. *Geology* **18**, 563–566.

Renne, P. R., Deino, A. L., Walter, R. C., et al. (1994) Intercalibration of astronomical and radioisotopic time. *Geology* **22**, 783–786.

Renne, P. R., Zhang, Z.C., Richards, M. A., Black, M. T., and Basu, A.R. (1995) Synchrony and causal relations between Permian-Triassic boundary crises and Siberian flood volcanism. *Science* **269**, 1413–1416.

Renne, P. R., Deckart, K., Ernesto, M., Feraud, G., and Piccirillo, E. M. (1996) Age of the Ponta Grossa dike swarm (Brazil), and implications to Parana flood volcanism. *Earth and Planetary Science Letters* **144** 199–211.

Renne, P. R., Swisher, C. C., Deino, A. L., Karner, D. B., Owens, T. L., and DePaolo, D. J. (1998) Intercalibration of standards, absolute ages and uncertainties in $^{40}Ar/^{39}Ar$ dating. *Chemical Geology* **145**, 117–152.

Renne, P. R., Farley, K. A., Becker, T. A., and Sharp, W. D. (2001a) Terrestrial cosmogenic argon. *Earth and Planetary Science Letters* **188**, 435–440.

Renne, P. R., Sharp, W. D., Montanez, I. P., Becker, T. A., and Zierenberg, R. A. (2001b) $^{40}Ar/^{39}Ar$ dating of Late Permian evaporites, southeastern New Mexico, USA. *Earth and Planetary Science Letters* **193**, 539–547.

Renne, P. R., Knight, K. B., Nomade, S., Leung, K. N., and Lou, T. P. (2005) Application of deuteron–deuteron (D–D) fusion neutrons to $^{40}Ar/^{39}Ar$ geochronology. *Applied Radiation and Isotopes* **62**, 25–32.

Renne, P. R., Sharp, Z. D., and Heizler, M. T. (2008) Cl-derived argon isotope production in the CLICIT facility of OSTR reactor and the effects of the Cl-correction in $^{40}Ar/^{39}Ar$ geochronology. *Chemical Geology* **255**, 463–466.

Renne, P. R., Cassata, W. S., and Morgan, L. E. (2009a) The isotopic composition of atmospheric argon and $^{40}Ar/^{39}Ar$ geochronology: Time for a change? *Quaternary Geochronology* **4**, 288–298.

Renne, P. R., Deino, A. L., Hames, W. E., et al. (2009b) Data reporting norms for $^{40}Ar/^{39}Ar$ geochronology. *Quaternary Geochronology* **4**, 346–352.

Renne, P. R., Schwarcz, H. P., Kleindienst, M. R., Osinski, G. R., and Donovan, J. J. (2010) Age of the Dakhleh impact event and implications for Middle Stone Age archeology in the Western Desert of Egypt. *Earth and Planetary Science Letters* **291** 201–206.

Renne, P. R., Balco, G., Ludwig, K. R., Mundil, R., and Min, K. (2011) Response to the comment by W. H. Schwarz et al. on "Joint determination of ^{40}K decay constants and $^{40}Ar^*/^{40}K$ for the Fish Canyon sanidine standard, and improved accuracy for $^{40}Ar/^{39}Ar$ geochronology" by P. R. Renne et al. (2010). *Geochimica et Cosmochimica Acta* **75**, 5097–5100.

Renne, P. R., Sprain, C. J., Richards, M. A., Self, S., Vanderkluysen, L., and Pande, K. (2015) State shift in Deccan volcanism at the Cretaceous–Paleogene boundary, possibly induced by impact. *Science* **350**, 76–78.

Rivera, T. A., Storey, M., Zeeden, C., Hilgen, F. J., and Kuiper, K. (2011) A refined astronomically calibrated $^{40}Ar/^{39}Ar$ age for Fish Canyon sanidine. *Earth and Planetary Science Letters* **311**, 420–426.

Rivera, T. A., Storey, M., Schmitz, M. D., and Crowley, J.L. (2013) Age intercalibration of $^{40}Ar/^{39}Ar$ sanidine and chemically distinct U/Pb zircon populations from the Alder Creek Rhyolite Quaternary geochronology standard. *Chemical Geology* **345**, 87–98.

Ross, J. A. and Sharp, W. D. (1988) The effects of sub-blocking temperature metamorphism on the K/Ar systematics of hornblendes—$^{40}Ar/^{39}Ar$ dating of polymetamorphic garnet amphibolite from the Franciscan Complex, California. *Contributions to Mineralogy and Petrology* **100**, 213–221.

Rutte, D., Pfander, J. A., Koleska, M., Jonckheere, R., and Unterricker, S. (2015) Radial fast-neutron fluence gradients during rotating $^{40}Ar/^{39}Ar$ sample irradiation recorded with metallic fluence monitors and geological age standards. *Geochemistry, Geophysics, Geosystems* **16**, 336–345.

Scaillet, S. (2000) Numerical error analysis in $^{40}Ar/^{39}Ar$ dating. *Chemical Geology* **162**, 269–298.

Schmitz, M. D. and Bowring, S. A. (2001) U–Pb zircon and titanite systematics of the Fish Canyon Tuff: an assessment of high-precision U–Pb geochronology and its application to young volcanic rocks. *Geochimica et Cosmochimica Acta* **65**, 2571–2587.

Schmitz, M. D., Bowring, S. A., Ludwig, K. R., and Renne, P. R. (2003) Comment on "Precise K–Ar, $^{40}Ar–^{39}Ar$, Rb–Sr and U–Pb mineral ages from the 27.5 Ma Fish Canyon Tuff reference standard" by M. A. Lanphere and H. Baadsgaard. *Chemical Geology* **199**, 277–280.

Sears, D. W. G (2012) Oral histories in meteoritics and planetary science—XVI: Grenville Turner. *Meteoritics and Planetary Science* **47**, 434–448.

Sigurgeirsson, T. (1962) *Age Dating of Young Basalts with the Potassium–Argon Method* (in Icelandic). Unpublished report, Physics Laboratory, University of Iceland. (English translation by L. Kristjansson, University of Iceland (1973.)

Singer, B. S. (2014) A Quaternary geomagnetic instability time scale. *Quaternary Geochronology* **21**, 29–52.

Smith, P. E., Evensen, N. M., York, D., and Odin, G. S. (1998) Single-grain $^{40}Ar–^{39}Ar$ ages of glauconites: implications for the geologic time scale and global sea level variations. *Science* **279**, 1517–1519.

Sprain, C. J., Renne, P. R., Wilson, G. P., and Clemens, W.A. (2015) High-resolution chronostratigraphy of the terrestrial Cretaceous-Paleogene transition and recovery interval in the Hell Creek region, Montana. *Geological Society of America Bulletin* **127**, 393–409.

Steiger, R. H. and Jager, E. (1977) Subcommission on geochronology—convention on use of decay constants in geochronology and cosmochronology. *Earth and Planetary Science Letters* **36**, 359–362.

Tetley, N., McDougall, I., and Heydegger, H. R. (1980) Thermal-neutron interferences in the $^{40}Ar–^{39}Ar$ dating technique. *Journal of Geophysical Research* **85**, 7201–7205.

Turner, G. (1970) ^{40}Ar/^{39}Ar dating of lunar rock samples. *Science* **167**, 466–468.

Turner, G. (1971) ^{40}Ar–^{39}Ar dating—optimization of irradiation parameters. *Earth and Planetary Science Letters* **10**, 227–234.

Turner, G. and Cadogan, P. (1974) Possible effects of ^{39}Ar recoil in ^{40}Ar–^{39}Ar dating. *Proceedings of the 5th Lunar and Planetary Science Conference*; 1601–1615.

Turner, G., Huneke, J. C., Podosek, F. A., and Wasserburg, G. J. (1971) ^{40}Ar–^{39}Ar ages and cosmic-ray exposure ages of Apollo-14 samples. *Earth and Planetary Science Letters* **12** 19–35.

Turrin, B. D., Donnelly-Nolan, J. M., and Hearn, B. C. (1994) ^{40}Ar–^{39}Ar ages from the rhyolite of Alder Creek, California—age of the Cobb-Mountain normal-polarity subchron revisited. *Geology* **22**, 251–254.

Van der Pluijm, B. A., Hall, C. M., Vrolijk, P. J., Pevear, D. R., and Covey, M. C. (2001) The dating of shallow faults in the Earth's crust. *Nature* **412**, 172–175.

Vasconcelos, P. M., Brimhall, G. H., Becker, T. A., and Renne, P. R. (1994a) ^{40}Ar/^{39}Ar analysis of supergene jarosite and alunite—implications to the paleoweathering history of the western USA and west-Africa. *Geochimica et Cosmochimica Acta* **58**, 401–420.

Vasconcelos, P. M., Renne, P. R., Brimhall, G. H., and Becker, T. A. (1994b) Direct dating of weathering phenomena by ^{40}Ar/^{39}Ar and K–Ar analysis of supergene K–Mn oxides. *Geochimica et Cosmochimica Acta* **58**, 1635–1665.

Vermeesch, P. (2015) Revised error propagation of ^{40}Ar/^{39}Ar data, including covariances. *Geochimica et Cosmochimica Acta* **171**, 325–337.

Villa, I. M. (1997) Direct determination of ^{39}Ar recoil distance. *Geochimica et Cosmochimica Acta* **61**, 689–691.

Villa, I. M., Huneke, J. C., and Wasserburg, G. J. (1983) ^{39}Ar recoil losses and presolar ages in Allende inclusions. *Earth and Planetary Science Letters* **63**, 1–12.

Von Weisacker, C. F. (1937) The possibility of a dual ss-disintegration of potassium. *Physikalische Zeitschrift* **38**, 623–624.

Wang, K. and Jacobsen, S.B. (2016) An estimate of the Bulk Silicate Earth potassium isotopic composition based on MC-ICPMS measurements of basalts. *Geochimica et Cosmochimica Acta* **178**, 223–232.

Wanke, H. and Konig, H. (1959) Eine neue methode zur kalium-argon-altersbestimmung und ihre anwendung auf steinmeteorite. *Zeitschrift Für Naturforschung Part A-Astrophysik Physik Und Physikalische Chemie* **14**, 860–866.

Wijbrans, J. R. and McDougall, I. (1987) On the metamorphic history of an Archean granitoid greenstone terrane, east Pilbara, Western Australia, using the ^{40}Ar–^{39}Ar age spectrum technique. *Earth and Planetary Science Letters* **84**, 226–242.

Wotzlaw, J.-F., Schaltegger, U., Frick, D. A., Dungan, M. A., Gerdes, A., and Günther, D. (2013) Tracking the evolution of large-volume silicic magma reservoirs from assembly to supereruption. *Geology* **41**, 867–870.

York, D. (1969) Least squares fitting of a straight line with correlated errors. *Earth and Planetary Science Letters* **5**, 320–324.

Zeitler, P. K. (1987) Argon diffusion in partially outgassed alkali feldspars—insights from ^{40}Ar/^{39}Ar analysis. *Chemical Geology* **65**, 167–181.

CHAPTER 10

Radiation-damage methods of geochronology and thermochronology

10.1 INTRODUCTION

Radiation-damage-based methods of geochronology rely on the same principles as other radioisotopic techniques except that daughter products are not nuclides but physical or electronic damage (stored energy) within a crystal. Radioactive decay of parent nuclides releases energy imparted to nuclei, electrons, and other types of ionizing radiation that propagate through crystal lattices, causing a range of damage types including atomic vacancies and interstitial impurities, amorphous zones, lattice strain, local charge imbalance, and electrons trapped at high-energy levels. Observable manifestations of this damage includes pleochroic haloes around high U–Th inclusions, tracks created by alpha recoil, tracks created by moving fission fragments, and luminescence and electron spin signatures created by trapping of electrons in metastable high-energy defect sites.

Analogous to diffusive loss of daughter products in other systems, these types of radiation damage are subject to annealing as a result of energy input. This energy may be imparted through exposure to solar radiation, as in the case of optically stimulated luminescence. More typically it is thermal, so as with other thermochronometers the effects of annealing are a function of time and temperature. Of the various damage dating approaches, the fission-track technique stands out in the extensive knowledge of annealing kinetics and wide thermochronometric applications, but other approaches have also been used in analogous ways. It should be mentioned that quantification of the extent of cumulative crystallographic radiation damage by macroscopic or spectroscopic properties [e.g., *Hurley and Fairburn*, 1953; *Holland*, 1954; *Hurley et al.*, 1956; *Holland and Gottfried*, 1955; *Nasdala et al.*, 2004; *Pidgeon*, 2014] also pose potential for geochronologic use.

10.2 THERMAL AND OPTICALLY STIMULATED LUMINESCENCE

10.2.1 Theory, fundamentals, and systematics

At temperatures above absolute zero all materials emanate radiation over a broad range of wavelengths, with the spectrum of wavelengths described by Planck's Law (and the peak of this distribution by Wien's Law). A very different type of radiation, characterized by emission at discrete wavelengths, is also emitted by nonconducting crystalline solids. This radiation corresponds to the transition of electrons from metastable high energy states to lower energy states. Electrons within crystals attain these high energy states (e.g., by movement from the valence to conduction bands) through interaction with ionizing radiation (Fig. 10.1). In most cases this radiation comes from alpha, beta, and gamma decay of the principal radioactive elements U, Th, K, and Rb, in and around geologic samples, though cosmic rays also contribute. Most electrons affected by this interaction immediately decay back to their original, stable energy level, but some are displaced and migrate through the lattice to crystal defects associated with locally high concentrations of positive charge, where they become trapped. The electronic vacancy, or "hole," left by the electron may also migrate, in this case to defects associated with locally high negative charge (Fig. 10.1). Defects may be vacancies, as depicted in Fig. 10.1, or ionic substitutions that lead to local charge imbalances. The metastable excited state of electrons occupying defects is relieved when electrons recombine with holes and return to their original energy level. This requires energy input to overcome the activation energy of the recombination process, and the magnitude of the activation energy primarily depends on the type of defect the hole occupies. This energy may be provided by either heat, in which case the resulting radiation is called thermoluminescence (TL),

Geochronology and Thermochronology, First Edition. Peter W. Reiners, Richard W. Carlson, Paul R. Renne, Kari M. Cooper, Darryl E. Granger, Noah M. McLean, and Blair Schoene.
© 2018 John Wiley & Sons Ltd. Published 2018 by John Wiley & Sons Ltd.

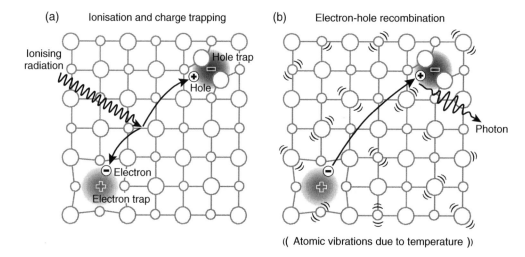

(a) Ionisation and charge trapping

Ionising radiation

Hole trap

Hole

Electron

Electron trap

(b) Electron-hole recombination

Photon

((Atomic vibrations due to temperature))

Fig. 10.1. Principles of charge trapping and release due to ionizing radiation. Left: schematic of electron excitation by ionizing radiation, and the subsequent mobilization of both electron and the valence "hole" left behind, to lower energy positions in the crystal lattice, which are defects associated with locally high positive or negative charge, respectively. Right: thermal (vibrational motion) or optical stimulation of the crystal lattice may overcome the activation energy required to "detrap" an electron and combine it with a trapped hole, releasing radiation of a characteristic energy, a luminescent photon. (Source: *Guralnik et al.* [2015]. Reproduced with permission of Elsevier.)

or by light (optically stimulated luminescence, OSL) or infrared radiation (IRSL).

Trapped-charge dating methods are often explained with the aid of a battery analogy (Duller, 2008), in which charges in a battery build up over time (as a result of energy from radioactive decay), and are discharged by energy input from exposure to light or heat. The return of electrons to lower energy levels releases radiation of a characteristic wavelength that corresponds to both the activation energy required to free it, and the type of trap it was held in. Release of electrons from distinct types of traps with distinct energy can be seen in a spectrum of luminescence intensity as a function of energy input (sample temperature). Comparison of a natural "glow curve" to one for a sample to which additional radiation was added, provides a measure of the fraction of the total luminescence energy that can be stored in a crystal. Figure 10.2 shows an example of this for thermoluminescence, comparing luminescence for a natural sample (curve N) and an analogous one with a known, artificially generated dose (curve $N + ß$).

Different minerals have different charge-accumulation responses to radiation dose, and different kinetics of luminescence annealing. When subject to variable exposure to energy input in discrete wavelengths in the visual or infrared parts of the energy spectrum, quartz releases UV luminescence through traps with a relatively simple distribution of kinetic controls that additively behave as a simple exponential function, whereas feldspar has a more complex continuum of traps types (Fig. 10.3).

The age equation for TL and OSL (here IRSL will be subsumed into OSL) can be expressed as the ratio of the accumulated dose [or total dose, paleodose (the true accumulated dose in nature), equivalent dose (the laboratory estimation of paleodose from a single type of radiation), or accumulated energy per unit mass since a hypothetical "resetting" event], and the dose rate, as in

$$t(\text{date}) = \frac{\text{accumulated dose (Gy)}}{\text{doserate}\left(\frac{\text{Gy}}{a}\right)}$$

where a is years. Dose can be described by the Kinetic Energy Released per unit MAss (KERMA) by ionizing radiation. Paleodosimetry methods generally use the SI version of gray (Gy), or one joule of ionizing radiation per one kilogram of matter. The principal challenge of luminescence dating (and other paleodosimetry methods) is thus to characterize:

(1) the luminescence present in a sample;
(2) the relationship between dose and luminescence in the sample (e.g., curve $N + ß$ in Fig. 10.2, or the black and gray curves in Fig. 10.3b,c);
(3) the dose rate in the sample over time.

Luminescence dating has been used for many decades, particularly in archeological and geomorphological applications [*Huntley et al.*, 1985; *Aitken*, 1998; *Rhodes*, 2011], but has also been used as a thermochronometer for a variety of planetary and tectonic applications (e.g., *Guralnik et al.* [2015] and references therein). State of the art approaches include higher precision and accuracy techniques, such as single aliquot regenerative (SAR) dose protocol [*Murray and Wintle*, 2000, 2003], and calibration and detailed understanding of kinetic behaviors of discrete types of luminescence in particular minerals [*Herman et al.*, 2010; *Guralnik et al.*, 2013, 2015].

10.2.2 Analysis

Thermal and optically stimulated luminescence dating employ a wide range of analytical methods, but the essential elements involve:

(1) measurement of the total luminescence contained in a sample upon heating (TL) or exposure to intense light (OSL) or infrared radiation (IRSL);

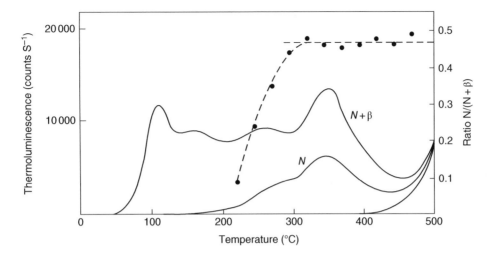

Fig. 10.2. Representative thermoluminescence glow curves. Emitted thermoluminescence is measured as a function of temperature in a natural sample (N) as well as an aliquot of the same sample that was also subject to a known natural dose of radiation (N + ß). The N + ß aliquot shows several distinct but overlapping peaks of luminescence that correspond to multiple types of traps. The lower temperature peaks are not observed in the natural glow curve (N) because electrons were liberated from the corresponding traps over the time and temperature history of the sample prior to sampling and analysis. The dashed line shows the ratio of luminescence values between the two samples, and its plateau can be used to estimate the paleodose. (Source: *Aitken* [1985]. Reproduced with permission of Elsevier.)

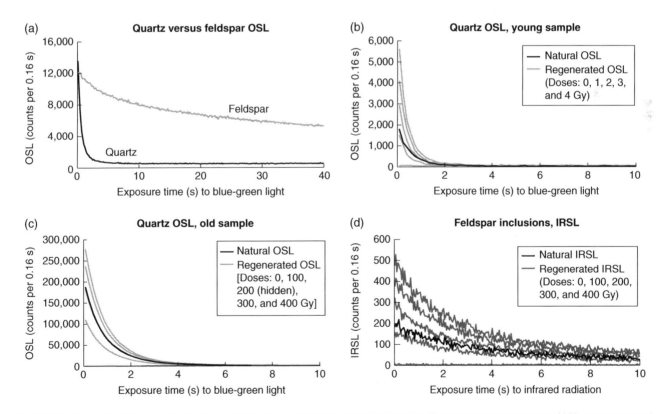

Fig. 10.3. (a–c) Examples of the decay of optically stimulated luminescence (OSL) and (d) infrared stimulated luminescence (IRSL) in quartz and feldspar. OSL is emitted upon exposure to intense light, typically in the blue-green spectrum (~470 ± 20 nm), and IRSL by exposure to radiation at 880 ± 80 nm. (a) Distinct decay curves of OSL in quartz and feldspar, pointing to a wider range of kinetics for charge release in feldspar. (b) Natural OSL in a young quartz grain (black) compared with OSL generated in similar aliquots that were completely annealed and subsequently irradiated (regenerative-dose curves), to determine dose-OSL response curves (gray). (c) Similar curves for a quartz grain that was isolated in a sand dune for roughly 200 ka. (d) IRSL luminescence from feldspar inclusions in quartz grains. (Source: *Rhodes* [2011]. Reproduced with permission of Annual Review of 39, Volume © by Annual Reviews, http://www.annualreviews.org.)

(2) the relationship between radiation dose and accumulation of luminescence (the dose-response curve);

(3) the paleodose rate.

The first two are typically measured on samples of a uniform grain sizes of a single mineral and repeated on multiple aliquots. The commonly used SAR protocol is typically used for quartz and feldspar but can also be adapted to mixed mineral aliquots in some cases. The dose rate (or paleodose rate) is measured on the sample's surroundings as well as the sample itself (microdosimetry).

10.2.2.1 Equivalent dose measurements (total luminescence and dose response)

Two general procedures are used to measure the total luminescence and the dose-response curve. In the first, the sample is heated or stimulated by light to release all its luminescence, which is then integrated as a glow curve (Fig. 10.4a). Similarly annealed aliquots of the sample are then subject to irradiation to generate artificial glow curves that establish the dose–response curve (Fig. 10.4b). An equivalent dose D_e can be measured in this method by comparing the dose required to obtain the equivalent natural luminescence observed in the samples.

An alternative procedure, the addition method, subjects the untreated sample to irradiation, and then the resulting luminescence is measured to produce a correlation that is extrapolated to zero luminescence (Fig. 10.4). This extrapolation thus provides a measure of the equivalent dose, D_e, contained in the sample, assuming the same dose–response curve at low doses and over the natural accumulation of dose.

As implied schematically in Fig. 10.4a, the luminescence-radiation relationship is, especially over larger ranges, typically convex upward, reflecting an approach to saturation of the traps with electrons. Saturation also limits the dateable age range

for luminescence (as well as ESR) techniques. Curved dose–response curves arising from saturation are typically fit using the relationship

$$I(D) = I_0 \left(1 - e^{-D/D_0} \right) \tag{10.1}$$

where $I(D)$ is the luminescence as a function of dose D, I_0 is the luminescence at saturation, and D_0 describes the shape of the curve.

Aliquots from a given sample do not always show a single reproducible equivalent dose or even a narrow range of doses. This may arise from incomplete resetting of the luminescence and multiple types of charge centers among the grains, inhomogeneities in paleodose rate, or in the case of sedimentary grains, mixing of grains with distinct exposure and/or resetting histories. In some cases distinct peaks of equivalent dose, corresponding to mixed populations, may be detected, as shown in the radial plot [*Galbraith and Green*, 1990] examples of Fig. 10.5.

10.2.2.2 Dose-rate measurements/estimates

The dose rate affecting a sample in nature (sometimes called the paleodose or natural dose rate) must be determined for the region around a sample where it resided during the time being dated. This can be done either by directly measuring radioactivity, or by measuring the concentrations of radioactive elements and converting these to activities. In both cases measurements should be made in the sample-specific region. In practice, multiple methods are often combined to characterize relative contributions of proximal alpha, beta, gamma, and cosmic radiation. The effective sphere of influence and therefore length scale over which radiation should be measured depends on the radiation type. Gamma radiation has a sphere of influence of approximately

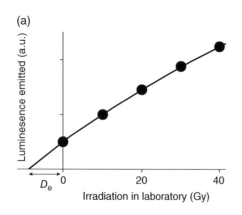
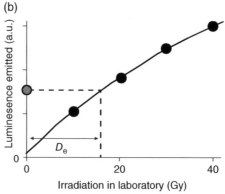

Fig. 10.4. The (a) addition and (b) regenerative methods for determining paleodose of a sample. (a) Aliquots of a sample with initial luminescence (point at zero irradiation), are subject to varying radiation dose and the corresponding luminescence measurements at each dose lead to a relationship between the two whose absolute value of the x-intercept corresponds to the paleodose, or equivalent dose D_e. (b) The luminescence-dose relationship established on aliquots that have been completely annealed or zeroed of luminescence, and the luminescence observed in another aliquot (gray symbol), or a single aliquot in the case of a single-aliquot regenerative dose protocol, is fit to the relationship to determine the equivalent dose. (Source: Adapted from *Wintle* [2008]. Reproduced with permission of John Wiley & Sons.)

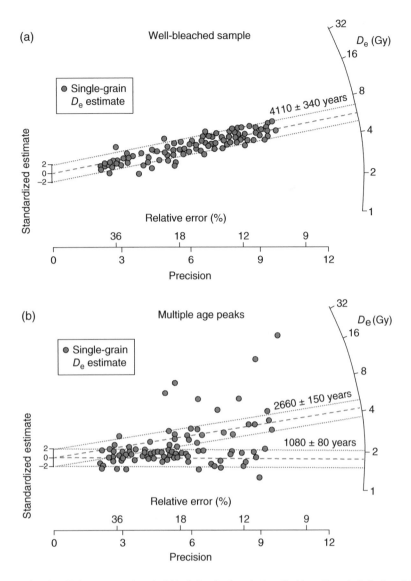

Fig. 10.5. Radial plots of OSL equivalent dose (D_e) measurements on individual aliquots of quartz, from fluvial sand from Australia. In radial plots such as these, which are often used in fission-track dating as well, the horizontal axis corresponds to inverse uncertainty on the measurement (points to the right are more precise, and thus have greater leverage on the estimated value of a variable that is plotted radially, increasing counterclockwise). In (a) example, nearly all aliquots fall within two-sigma of a trend pointing at an equivalent dose of a little more than 6 Gy. Together with an estimate of paleodose rate (see text), this yields the apparent age of all the aliquots. In (b), significant scatter in the D_e measurements precludes identification of a single D_e peak, but peak deconvolution protocols identified two statistically robust populations with distinct apparent ages. (Source: *Rhodes* [2011]. Reproduced with permission of Annual Review of 39, Volume © by Annual Reviews, http://www.annualreviews.org.)

30 cm, so this is often measured in the field (rather than collecting a sample with a 30-cm radius). Approximate spheres of influence for beta and alpha radiation are much shorter, at ∼3 mm and ∼30 μm, respectively, so measurements can be made on aliquots in the laboratory. Of course, cosmic-ray contributions have a much larger range. They vary relatively predictably as a function of latitude (above about 60° their influence increases strongly) and altitude (increasing with elevation). For mid- or low latitude sites near sea level they contribute on order 0.01–0.1 Gy/ka, which is typically about 5–10% of the paleodose. If the depth of burial of a sample has varied significantly in the past, however,

this can strongly affect the cosmogenic contribution, because of the approximately meter-scale attenuation depth of cosmic rays.

In some cases radioactivity measurements can be made *in situ* by burying an artificial phosphor at a site and collecting radiation over a period of months to years, or by using a portable gamma ray spectrometer; these approaches have the theoretical advantage of integrating a radiation dose over a longer length scale and, at least in principle, accounting for compositional heterogeneity in the medium that is more difficult to sample directly. Alternatively samples may be collected and radioactivity counted by spectroscopic methods such as thick-source alpha spectroscopy.

One of the most common modern methods employed is chemical analysis of the naturally radioactive elements in the vicinity of the sample: U, Th, K, and Rb. Knowledge of the energies associated with decay of each of these nuclides (and in the case of U and Th, the intermediate daughters in their decay series) can be converted to doses using data such as those provided by *Nambi and Aitken* [1986] and *Guérin et al.* [2011, 2012] for U, Th, Rb, K, and intermediate daughters in the case of U and Th.

For U and Th, assumptions of secular equilibrium or more detailed measurements of specific disequilibrium (and assumptions of constancy in time) must be made [e.g., *Olley et al.*, 1999]. The potential for loss of radon gas is one of the particular concerns in the U and Th series. Many environments are also prone to changing amounts of U and Th and the intermediate daughters, particularly those with high organic contents. Water content of the material surrounding the sample is an additional concern, because of its ability to absorb differently from the rock or soil matrix surrounding it. While variable, a general rule of thumb is that a 1% increase in water content will affect ages by about 1%.

Once dose rates are converted from measurements of radionuclide concentrations, these are combined with the sensitivities of luminescence to different types of radiation, though typically only alpha is distinguished from the other types. For example, equation (10.1) is typically modified to:

$$t = \text{natural luminescence}/\left[X_a D_a + X_b\left(D_b + D_g + D_c\right)\right] \quad (10.2)$$

where X denotes luminescence sensitivities for each radiation type and D denotes dose rates for each type. Alternatively, if the effective dose D_e can be constrained as described above, this can be combined as a ratio with the dose rate to determine the age.

10.2.3 Fundamental assumptions and considerations for interpretations

The robustness of OSL and TL dating results depends on many sample-specific and site-specific factors but some of the most general considerations include the following:

- *Spatial and temporal uniformity of the natural dose rate.* Natural radioactive element concentrations may be heterogeneously distributed in the vicinity of a sample, and this heterogeneity may be expressed over a variety of length scales that impact the different sources of radioactivity differently. Concentrations of radioactive sources may also vary with time, for example due to groundwater–rock reactions. Of particular concern in at least some circumstances is the potential for secular disequilibrium among the U-decay and Th-decay series. Movement of Rn is often considered as a possible violation of this assumption.

- *Moisture content.* As mentioned above, interstitial water may strongly affect the effective length scales over which ionizing radiation affects neighboring grains. Water content may be measured in representative samples, but the question of whether this represents the time-averaged conditions over the appropriate duration of time may be difficult to address.

- *The dose–luminescence relationship.* The relationship between luminescence and radiation dose is sample specific and requires a functional characterization over an appropriate range of variation. In addition, it must be assumed that the this laboratory-established relationship, measured over very short timescales and high dose rates, can be extrapolated to natural conditions of much longer timescales and lower dose rates. This rate-extrapolation assumption is in fact characteristic of any thermochronologic technique that constrains the kinetics of annealing, diffusion, or signal fading over laboratory timescales.

- *Kinetics and variation in "zeroing."* Many TL or OSL studies involve relatively simple interpretations that assume an initial state of complete lack of luminescence, caused by complete annealing (thermal) or bleaching (sunlight). However, the kinetics of this "blanking" or "zeroing" in nature are not always clear, and grains may be reset to varying degrees by the same heating or exposure event for reasons that are not understood.

- *Saturation.* As suggested in the curvature of the luminescence–dose relationship in Fig. 10.4, the capacity for minerals to trap electrons and therefore produce luminescence is finite (and this finite limit is itself assumed to be unchanging with time, which may not be true under some conditions). Therefore TL and OSL are subject to saturation, whereby a sample cannot accumulate luminescence beyond that corresponding to a certain age. The saturation limits are sample and method specific but in general ages beyond about 1 Ma are not achievable. However, new approaches and understanding of saturation are pushing this limit [e.g., *Rhodes*, 2011].

- *"Open systems."* In many cases it is advantageous to consider luminescence dates not as on-off switches in which a grain's signal is either fully reset by heating or exposure, or is "charging up" at a maximum rate while completely protected from heat or light. This situation is analogous to the simultaneous growth and loss of daughter products with a partial retention zone for a fission-track-based or noble-gas-based thermochronometer.

As an example of useful interpretation of open-system behavior, *Herman et al.* [2010] adapted the quartz OSL method as a thermochronometer by modeling the combined effects of progressive trapping and detrapping with a finite activation energy, with consideration of a saturation age. Assuming a single type of trap for electrons, the change in the number of trapped electrons with time follows a first-order Arrhenius law as described by *Aitken et al.* [1978]

$$-\frac{dn}{dt} = ns \exp\left[-\frac{E_a}{kT}\right] \quad (10.3)$$

where dn/dt is the change in number of filled traps, n is the number of occupied traps, s is a frequency factor related to escape-attempt frequency, E_a is the activation energy for detrapping, k is the Boltzmann constant, and T is temperature. For reference, the Boltzmann constant is typically expressed as 8.62×10^{-5} eV/K, and converting a kinetic expression of this type to one involving the universal gas constant R can be done by using 1 eV = $1.60217646 \times 10^{-19}$ J and accounting for the conversion from a single electron or atom to moles using Avogadro's number. Although equation (10.3) describes the kinetics of electron release from a single type of trap with particular energy, real luminescence behavior shows a broad distribution of intensities as a function of input energy (e.g., temperature), consistent with electron transitions from a range of types of traps.

Herman et al. [2010] and *King et al.* [2016] extended this expression to combine the rates of trapping and detrapping, as well as saturation, into a single expression,

$$\frac{dn}{dt} = \frac{\dot{D}}{D_0}(N-n) - ns \exp\left[-\frac{E_a}{kT}\right] \qquad (10.4)$$

where $\dot{D}$ is the dose rate, and D_0 is a characteristic or saturation dose.

Using these relationships *Herman et al.* interpreted OSL dating results from rapidly exhuming bedrock in the Southern Alps of New Zealand (Fig. 10.6) in terms of cooling ages, using a modified version of Dodson's (1973) closure temperature equations [*Herman et al.*, 2010; *Guralnik et al.*, 2013]. They found that erosion rates over 10^5-year timescales were 8–10 mm/year, but even though these are among the highest on Earth, and glacial activity likely modified the landscape significantly, overall there has been little change in topographic relief over the past few hundred thousand years.

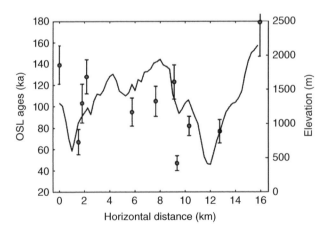

Fig. 10.6. OSL ages (filled circles) of quartz in bedrock, from an orogen-parallel transect along the Southern Alps of New Zealand with mean topography represented by the solid line. The overall correlation between mean elevation and age, combined with a more detailed spectral analysis of the age–elevation relationship, led *Herman et al.* [2010] to infer erosion rates of 8–10 mm/year but little or no topographic change over 10^5-year timescales. (Source: *Herman et al.* [2010]. Reproduced with permission of Elsevier.)

10.2.4 Applications

Until recently, most luminescence dating has focused on applications involving relatively straightforward interpretations of time elapsed since "zeroing" of the luminescence signal. Rhodes [2011] provides a good overview of applications, as well as principles and procedures, applied to sediment burial. In many cases, uncertainties on the order of 5–10% (1σ) are achievable. Notable examples include the use of luminescence to eolian transport processes, ages of marine and lacustrine shoreline materials, alluvial and fluvial processes, and rates of soil creep. Luminescence dating also holds considerable potential for thermochronometric applications and is the focus of much ongoing work.

Guralnik et al. [2015] provide an example of the use of thermochronologic application of luminescence dating. They addressed the low-temperature thermal history of the upper part of the crystalline rocks in the KTB borehole in Germany. After finding that more standard quartz OSL dating yielded complex results, in part compromised by strong signals from small grains of interstitial and included feldspar, they turned to IRSL of feldspar, performing luminescence and dose experiments at 50 °C in order to avoid unstable luminescence signals at lower temperatures. Luminescence and dose–response curves were established for 12 oligoclase feldspar samples in the upper 2.5 km of the borehole (Fig. 10.7).

These samples are clearly not ones for which luminescence accumulation at a fixed rate followed complete "zeroing." Instead they are open systems. That is their luminescence is characterized by a combination of growth and decay, where the growth rate depends on the paleodose rate and decay rate depends on the temperature of the sample, or more specifically, its thermal history. Thus in addition to the paleodose rate, the luminescence decay rate must be determined. For these samples the paleodose rate was determined from U, Th, and K in whole-rock aliquots surrounding each sample, as well as the K content of the feldspar itself. The kinetics of luminescence decay were characterized by a six-parameter model [similar to that described in equations (10.3) and (10.4)] fit to observations of isothermal decay as a function of time at different temperatures (Fig. 10.8).

Guralnik et al. [2015] converted the observed luminescence, dose–response curves, paleodose rates, and kinetics of signal decay, to apparent ages and "storage temperatures," for samples up to about 2.4 km depth (Fig. 10.8). Samples in the uppermost kilometer displayed saturated luminescence, limiting results to minimum apparent ages and maximum apparent storage temperatures. But deeper samples show a familiar trend of decreasing apparent ages with depth, with ages about two orders of magnitude younger than those from the apatite fission-track and (U–Th)/He systems on the same samples. The apparent age trend also displays an apparent partial retention zone corresponding to the limits of 10–90% retention of the age, analogous to those of other thermochronometers. The apparent storage temperatures also show a consistent increase with depth implying a paleogeothermal gradient of about 29 °C/km that is indistinguishable

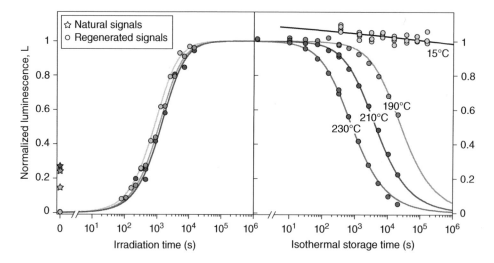

Fig. 10.7. IRSL for feldspar from 2.1 km depth in the KTB borehole. Left: observed luminescence (stars) and luminescence dose–response curves (circles) for a feldspar aliquot (compare this panel to (b) in Fig. 10.4). Three sets of filled circles and lines in response curves correspond to different dose rates ranging from 0.15 to 0.25 Gy/s. Right: decay of luminescence signal in the same feldspar sample after isothermal holding at four different temperatures for varying durations. The observed signal decay (circles) can be fitted by a kinetic model similar to that in equations (10.3) and (10.4), but involving six empirical parameters including an activation energy and frequency factor. (Source: *Guralnik et al.* [2015]. Reproduced with permission of Elsevier.)

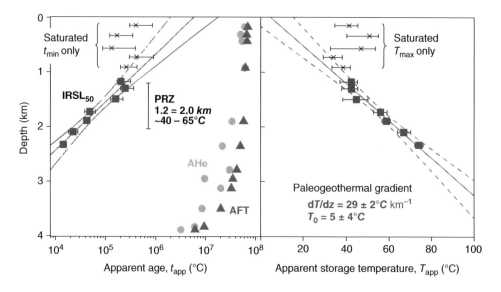

Fig. 10.8. Left: apparent ages of feldspar from the KTB borehole using the IRSL technique, compared with other thermochronometric techniques (AHe, apatite (U–Th)/He and AFT, apatite fission track). IRSL results in the shallowest 1 km of the core (crosses and large error bars) indicate saturation of the luminescence signal, providing only minimum apparent ages. Right: apparent storage temperatures of the same feldspar samples as a function of depth. The saturated samples provide only a maximum temperature estimate. The apparent paleogeothermal gradient of 29 °C/km is indistinguishable from the modern one, implying little or no change in the thermal field in the uppermost 2.5 km of the crust in this region over the past ~300 ka. (Source: *Guralnik et al.* [2015]. Reproduced with permission of Elsevier.)

from the modern one. These results do not imply a significantly different thermal field or exhumation rate from the modern ones over the past ~300 ka, but they do suggest potential for application of these approaches to other settings where temperature changes in the range of 40–65 °C might be expected on 10^4–10^5 year timescales.

10.3 ELECTRON SPIN RESONANCE

10.3.1 Theory, fundamentals, and systematics

The basis of ESR dating is very similar to that of luminescence dating in that it relies on measurement of a proxy for the abundance of electrons occupying defect traps, and this abundance is proportional to the integrated radiation dose of the crystal. Thus the age is determined from the ratio of an estimate of the total radiation dose to the dose rate. One advantage of this method, relative to luminescence methods, is that measurement does not liberate electrons from the traps or destroy the traps, so multiple measurements can be made on the same sample. ESR as a geochronologic tool was first proposed in the late 1960s by Zeller and coworkers, and explored by Ikeya in the late 1970s, who obtained the first ESR date, on a speleothem, in 1975. Besides important applications in understanding paramagnetic defects in crystals and in dosimetry, ESR has been used in a wide range of dating applications, primarily in Quaternary studies and archeology (e.g., see review by *Rink* [1997]). The best reviews of the principles and procedures of the method are those of *Grün* (1989) and *Rink* [1997].

The basis of ESR relies on the quantum-mechanical principles describing the spin of electrons. Paired electrons occupying the same orbital have opposite spins, canceling their magnetic moments and making their host material diamagnetic. With time, however, as described in the luminescence dating section, radiation excites electrons from the valence to conduction band, and some of these are trapped by crystal defects, where they become unpaired, generating a paramagnetic signal.

Defects or centers typically responsible for this type of trapping include local charge imbalances created by ionic substitutions such as CO_3^{2-} in apatite, and Ge, Al, and Ti in quartz that create local regions of high positive charge. The paramagnetic characteristics of such defects or centers have a distinctive signature specific to the center that contains the electron occupying it, which depends on the ratio of the magnetic moment and angular momentum of the electron. This is the Lande factor, or g-value.

Measuring ESR signals requires applying a combination of a magnetic field and microwave radiation. When a magnetic field is applied to the crystal, unpaired electrons acquire a precession (like a wobbling top) of a certain frequency and which is aligned with the magnetic field. When subjected to an electromagnetic radiation with the same frequency as the precession (in the microwave band), this causes the spin to resonate, inducing a reversal. Resonance occurs at the condition where

$$h\nu = g\mu_B H \qquad (10.5)$$

where h is Planck's constant divided by 2π, ν is the frequency, g is the Lande factor or g-value, μ_B is the physical constant Bohr magneton, and H is the magnetic field.

Resonance also causes absorption of the microwave energy, permitting detection of the phenomenon. Thus the abundance of unpaired electrons occupying traps with a particular paramagnetic signature can be measured by monitoring the absorption of microwave energy across a spectrum of magnetic field intensity. At the field value corresponding to resonance and absorption, a large change in microwave intensity corresponds to a particular g-value for the trapped electrons (Fig. 10.9). Assuming no other effects such as resonance of organic radicals in this part of the spectrum, the magnitude of this intensity change is then

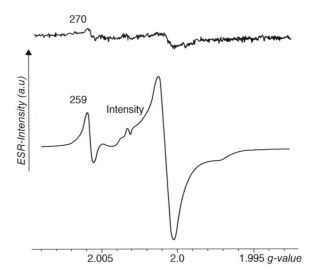

Fig. 10.9. Representative ESR spectra for two speleothem samples with strongly contrasting radiation doses caused by different ages (young on top; old on bottom). The y-axis is equivalent to the derivative of the microwave radiation intensity with respect to magnetic field; the x-axis is the magnetic field, represented as equivalent g-value following equation (10.5). Each of these spectra show three clearly observable resonance conditions with distinct g-values, with their relative amplitudes potentially proportional to the abundance of electrons trapped in each. The amplitude of the top spectrum is magnified ten times for clarity. (Source: Adapted from *Grün* [1989]. Reproduced with permission of Elsevier.)

proportional to the abundance of trapped electrons and therefore integrated radiation dose.

10.3.2 Analysis

As in luminescence dating, an accumulated or equivalent dose of a sample must be determined in order to calculate an age, in this case by the addition method, for example by irradiation with ^{60}Co or another source. As shown in Fig. 10.10, in some cases changes to the ESR spectra upon irradiation can be complex. Some of this is due to the presence of overlapping peaks of multiple origin, some of which respond to irradiation differently. For example, some peaks may represent organic radicals in the crystal; these generally do not respond to increasing dose, though they may interfere with detection of desired ESR signals.

Deriving the accumulated dose and the dose–response curve for ESR may be complicated by saturation, as in the luminescence case. ESR intensity as a function of dose is often fitted with a logarithmic function related to the signal intensity at saturation (Fig. 10.11b, c). An additional effect is supralinearity of the dose–response curve at low dose, which leads to an underestimation of accumulated dose (Fig. 10.11a).

As with luminescence methods, it should be remembered that accumulated dose and dose–response-curve measurements involve dose rates that are roughly nine orders of magnitude higher than the natural dose rates.

ESR dose rates (the denominator in the basic age equation) are calculated in the same way as for luminescence methods, from measured concentrations of natural radionuclides (with consideration of potential U-series and Th-series disequilibria), gamma radiation measurements over the same ~30 cm sphere of influence, consideration of integrated cosmic-ray doses and water contents of the material, and conversions to dose rates [e.g., *Guérin et al.*, 2011, 2012].

10.3.3 Fundamental assumptions and considerations for interpretations

Although ESR is less sensitive to problems associated with sample preparation than luminescence methods, some ESR signals in some minerals are affected by sample grinding. Procedures exist for minimizing these effects [*Grün*, 1989]. However, undesirable ESR signals in geological materials can also arise from other effects, including the presence of unpaired electrons associated with organic radicals. In some cases these may swamp observation of desired ESR signals.

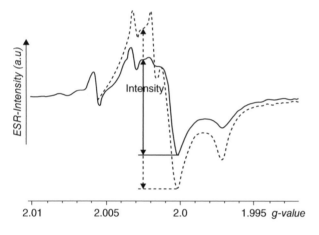

Fig. 10.10. ESR spectra of an untreated mollusc (solid line) and the same sample irradiated for addition-method determination of accumulated dose. Five distinct signals are observed in the region of interest, which respond differently to irradiation. Vertical arrows denote the magnitude of the pre-irradiation and postirradiation signals, whose difference is used to calibrate the dose–response relationship. (Source: Adapted from *Grün* [1989]. Reproduced with permission of Elsevier.)

Spatial heterogeneities in radionuclides on the scale of the sample as well as in the larger surroundings may contribute to large variations and local saturation in some samples. In addition, as with luminescence, variations in dose rate with time may arise from a variety of factors affecting local radionuclide concentrations, including U uptake [e.g., *Grün*, 1989], mobilization of Ra and Rn, and other temporal variations in disequilibria.

ESR dates are usually interpreted in the context of representing a time elapsed since a "zeroing" event. This is analogous to exposure to sunlight and/or heat for sufficient durations to erase OSL or TL signals. In ESR, a zeroing event may be mineral formation, heating, and cataclastic deformation or recrystallization [e.g., *Fukuchi et al.*, 2007; *Lee and Schwarcz*, 1994].

Aside from saturation, which limits the maximum age accessible to both ESR and luminescence dating techniques, the other primary limitation is natural loss of signal. The extent to which electron detrapping and loss of luminescence or ESR signal is due to thermal effects is generally not well known. OSL is thought to be affected by both athermal and thermal loss, and distinguishing between these phenomena and calibrating their kinetics is still in its infancy [e.g., *Guralnik et al.*, 2015; *King et al.*, 2016].

In ESR, fading is typically described by a mean life of the ESR signal, which is temperature dependent. As is essentially the same for OSL and TL, assuming first-order kinetics of electron detrapping, a mean life τ can be written as

$$\frac{1}{\tau} = v_0 e^{-E/kT} \tag{10.6}$$

where v_0 and E are the frequency factor and activation energy of detrapping, k is the Boltzmann constant, and T is temperature. The mean life at a particular temperature can be measured by the decrease in ESR signal as a function of time (Fig. 10.12). Doing this at multiple temperatures leads to a pseudo-Arrhenius trend, as is used in calibration of fission-track annealing, from which activation energy and frequency factor can be derived (Fig. 10.12b).

The mean-life τ of an ESR signal may be sufficiently short that, combined with a given dose D, it leads to a steady-stage ESR

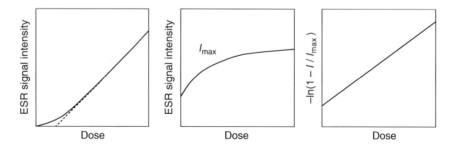

Fig. 10.11. Complexities in ESR dose–response curves. (a) Supralinearity of the dose–response curve, which is observed at low dose (solid line). This leads to an underestimate of accumulated dose from a linear extrapolation. (b) Samples with higher natural accumulated dose often show a convex-upward dose–response curve, reflecting saturation of the ESR signal by accumulation of a finite number of electrons in a given center. (c) This can be accommodated by a logarithmic fit to the curve; sometimes called a supersaturating exponential fit. (Source: Adapted from *Grün* [1989]. Reproduced with permission of Elsevier.)

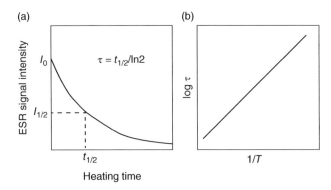

Fig. 10.12. Principle of mean life measurement for ESR signal and determination of fading kinetics. (a) ESR signal as a function of time at a given temperature. (b) Pseudo-Arrhenius trend relating mean life and inverse temperature, from which activation energy and frequency factor could be extracted. (Source: Adapted from *Grün* [1989]. Reproduced with permission of Elsevier.)

signal that is unrelated to saturation. For this condition, which is reached in about 5τ, the accumulated (or equivalent) dose D_e of a sample [*Grün*, 1989] is

$$D_e = D\tau\left(1 - e^{-t/\tau}\right) \tag{10.7}$$

More recent attention to ESR as a potential thermochronometer has shown that ESR signals of distinct defects or centers in quartz do indeed respond to temperature in systematic ways that may form the basis of a thermochronometric system. Toyoda and Ikeya (1991) calibrated the time-temperature characteristics of "annealing" of ESR signals associated with Al and Ti substitution in quartz, as well as oxygen vacancies (E' centers) (Fig. 10.13).

Toyoda and Ikeya showed that the behavior exhibited by these centers required a second-order kinetic model, consistent with the centers combining with one another rather than diffusing or otherwise annealing away independently. They showed that the logarithm of the fundamental parameter of this kinetic model, the decay factor λ, showed systematic Arrhenius trends with distinct activation energies and preexponential factors (Fig. 10.13). The preexponential factors and activation energies determined in this study imply Dodsonian [*Dodson*, 1973] closure temperatures for cooling rates of 10 K/Ma ranging from 31 to 91 °C.

The preliminary kinetics calibrations discussed above hold considerable promise for thermochronometric applications, but thus far nearly all ESR applications have focused on interpretations involving clear "zeroing" of ages [*Grün*, 1989; *Rink*, 1997].

In general, ages of specimens up to around 1–2 Ma can be achieved in ESR dating, with uncertainties similar to those of luminescence (~5–15%, 1σ). In some cases ages as young as a few thousand years may be measured. In some circumstances, including for environments with low dose rates so saturation is not achieved, much older dates potentially could be measured.

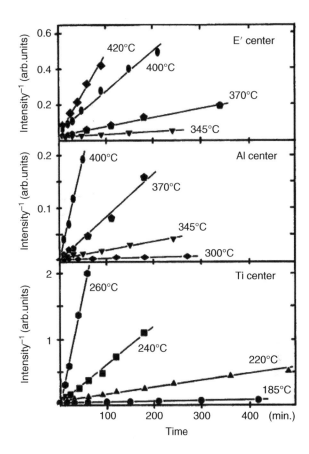

Fig. 10.13. Reciprocal intensity of ESR signals for three different centers in quartz with distinct *g*-values. (Source: *Toyoda and Ikeya* [1991]. Reproduced with permission of Geochemical Society of Japan.)

10.3.4 Applications

Three of the most common minerals routinely dated by ESR include apatite, calcite/aragonite, and quartz, though some work has been done on other phases including gypsum and bone. Most apatite studies focus on tooth enamel (e.g., *Ikeya* [1982]; references in *Rink* [1997]), where it is the dominant phase, and ages of samples of interest are typically less than a few million years, precluding saturation limitations. The primary source of uncertainty in these applications is the history of diagenetic U uptake (and in some cases loss) in fossil tooth enamel [*Grün and Invernati*, 1985], and most studies rely on assumptions for the U-uptake history.

Many ESR studies focus on both biogenic and abiogenic calcite and aragonite, for example in speleothems, travertine, corals, foraminifera, molluscs, and caliche. The first ESR date, by *Ikeya* [1975] was on a speleothem. As in other geochronologic and geochemical studies of carbonates, one of the principal concerns in many ESR applications is recrystallization, especially of aragonite to calcite. Other concerns include U-uptake histories, and the particular sensitivity of carbonate ESR signals to alpha radiation, as opposed to the beta or gamma typically used to measure equivalent dose and dose–response curves.

ESR study of quartz has found widespread use in studies of volcanic rocks, rocks heated by fire, and fault zones. *Ikeya et al.* [1982] were the first to date volcanic rocks. Many applications of ESR on either phenocrystic or xenocrystic quartz yield relatively problematic results for reasons that are not well understood, but generally seem to work better for samples younger than about 60 ka and which have not spent considerable time very close to the surface where diurnal heating may be important [*Rink*, 1997].

Ikeya et al. [1982] and *Miki and Ikeya* [1982] were the first to explore the use of ESR in dating fault activity. Resetting of quartz ESR signals in such environments is often considered to result from deformation, particularly crushing and comminution, mirroring the effects seen on ESR signals from some sample-preparation approaches. Thermal effects are also important [e.g., *Fukuchi*, 2012]. Other minerals including clay and gypsum precipitated in fault zones have also been the focus of attention [*Ikeda and Ikeya*, 1992; *Fukuchi*, 1996].

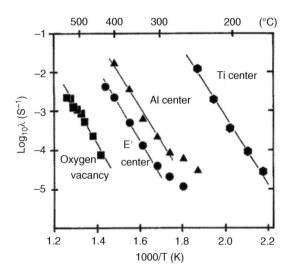

Fig. 10.14. Arrhenius trends for decay of three distinct ESR signals in quartz, along with that for oxygen vacancies. Lambda is the decay factor in a second-order model for signal decay as a function of time and temperature. (Source: *Toyoda and Ikeya* [1991]. Reproduced with permission of Geochemical Society of Japan.)

10.4 ALPHA DECAY, ALPHA-PARTICLE HALOES, AND ALPHA-RECOIL TRACKS

10.4.1 Theory, fundamentals, and systematics

10.4.1.1 Alpha decay

Although for ESR and luminescence methods alpha decay produces a disproportionately small amount of "equivalent daughter product," it is by far the most frequent and abundant source of radiation and damage in most minerals. Alpha decay describes the spontaneous emission of a ^{4}He nucleus from a radioactive parent. Several heavy isotopes including ^{147}Sm and ^{190}Pt decay in this way, but the most frequent are from ^{238}U, ^{235}U, ^{232}Th, and their intermediate daughter products in their decay chains (see Figs 8.1 and 12.1). Alpha decays in these chains have energies ranging from about 4–9 MeV (from a minimum of 4.270 MeV for ^{238}U to a maximum of 8.995 MeV for ^{212}Po). In most minerals these energies correspond to stopping distances of ~10–35 μm for the alpha particle itself, which carries the bulk of the energy, although decay of exceptionally high-energy parents like ^{212}Po can have larger ranges.

10.4.1.2 Alpha-particle damage

One source of radiation damage resulting from alpha decay is the damage created by the alpha particle itself. Using compiled data and the resources of the SRIM (stopping and range of ions in matter) project and web site (http://www.srim.org [*Ziegler et al.*, 2008]), *Ketcham et al.* [2011] derived empirical polynomial relationships between particle energies and ranges in a variety of minerals (Fig. 10.15; also see Table 11.1).

Although they travel relatively long distances, alpha particles do relatively little damage along their path. The bulk of their damage is thought to be at the end points of their paths, where they produce on order of tens to hundreds of Frenkel defects through collision cascade and ionization of the crystal lattice.

This damage can be seen in Raman band broadening and shift in zircon [*Nasdala et al.*, 2005], and for many years the effects of alpha particle stopping has been recognized as producing gradients in optical and cathodoluminescence in regions of heterogeneous U–Th concentrations in a variety of minerals [*Joly*, 1907; *Joly and Rutherford*, 1913; *Henderson*, 1934, 1939; *Henderson and Bateson*, 1934; *Henderson and Turnbull*, 1934; *Henderson and Sparks*, 1939; *Voznyak et al.*, 1996; *Nasdala et al.*, 2001, 2002, 2006]. These features, called alpha-particle haloes, radiohaloes, or pleochroic haloes if visible in optical microscopy (Figs 10.16 and 10.17), have been used to estimate ages of the rocks containing them [*Henderson*, 1934].

The major limitations in this method are development of a robust proxy for the degree of damage in the halo region (e.g., a reliable spectroscopic index; e.g., *Moazed et al.* [1975]), a reliable proxy for the relationship between damage as a function of dose (probably also accounting for different initial spectroscopic conditions), a measurement technique appropriate for the relevant parent nuclides in the source region, and a kinetic model for the annealing of the haloes as a function of time and temperature [e.g., *Laney and Laughlin*, 1981] that would be required for thermochronologic interpretations. These limitations are conceivably determinable, which might someday lead to a useful thermochronometer or geochronometer.

10.4.1.3 Alpha-recoil tracks

The heavy nuclide complement to the ^{4}He nucleus in alpha decay carries only a small amount of the total energy of the decay, traveling roughly 20–40 nm in the opposite direction as the alpha particle. However, the damage from the massive particle is relatively large, creating thousands of Frenkel defects and an

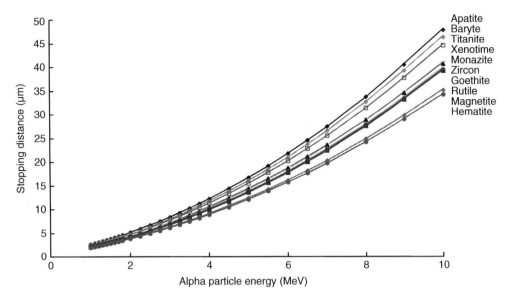

Fig. 10.15. Stopping distance, or range, of alpha particles as a function of energy in several minerals with appreciable concentrations of the parent nuclides that produce most alpha particles, U and Th. Energies are shown for various parents in the ^{238}U, ^{235}U, ^{232}Th decay chains as well as for ^{147}Sm. (Source: *Ketcham et al.* [2011]. Reproduced with permission of Elsevier.)

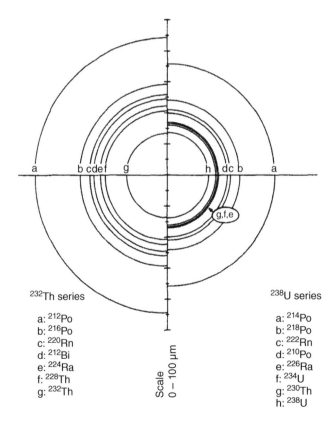

Fig. 10.16. Schematic radiation-damage haloes resulting from ionization produced by each of the subparent nuclei in the ^{232}Th and ^{238}U series. Although hints of discrete halos of distinct radii are seen in some samples, in reality broad and diffuse halos are generally seen. (Source: *Owen* [1988]. Reproduced with permission of Geological Society of America.)

essentially amorphous region in the lattice tens of nanometers across. Some molecular dynamics simulations and transmission electron microscopy (TEM) images of alpha-recoil damage zones suggest roughly spherical or ovoid damage zones [*Gögen and Wagner*, 2000; *Trachenko et al.*, 2002, 2003], but consideration of particle energies and stopping ranges suggests that alpha-recoil tracks should be more elongated, with lengths of ~25 nm and cross-sectional widths about ten times lower [*Ketcham et al.*, 2013]. Many studies have offered direct observations of alpha-recoil tracks (ARTs) on chemically etched surfaces [*Huang and Walker*, 1967; *Katcoff*, 1969; *Hashimoto et al.*, 1980; *Hashemi-Nezhad and Durrani*, 1981, 1983; *Gögen and Wagner*, 2000; *Stubner et al.*, 2005].

Three-dimensional imaging of features that may be fossil alpha-recoil tracks have been observed by *Valley et al.* [2014], using atom-probe tomography of zircon to image preferential partitioning of incompatible trace elements Y and Pb in ovoid clusters with dimensions and spacing consistent with the sizes of alpha-recoil tracks (Fig. 10.18). These clusters, which were presumably filled by diffusing trace elements that became trapped in the defect damage zones, are slightly smaller than might be predicted from particle energies and stopping distances, and do not show the expected clustering of linked decay chains. But the damage zones seen by *Valley et al.* [2014] may have been annealed over time [*Yuan et al.*, 2009], leaving the question of the dimensions of the original ARTs unclear.

Monte Carlo models of ART accumulation suggest that they should form a bimodal distribution of sizes, corresponding to lone tracks and their clusters (Fig. 10.19). A method that could ratio the abundances of alpha-recoil tracks to U and Th would

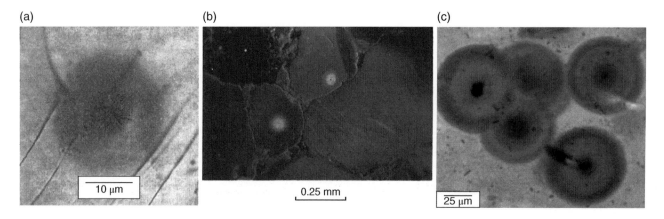

Fig. 10.17. Alpha-particle haloes, also known as radiohaloes (or -halos). (a) Halo in biotite, viewed in visible spectrum. Fission tracks also emanate from the center, which must host a source rich in ^{238}U. (Source: *Price and Walker* [1963a]. Reproduced with permission of John Wiley & Sons.) (b) Cathodoluminescence contrasts showing radiohalos in quartz caused by U–Th inclusions therein. (Source: *Owen* [1988]. Reproduced with permission of Geological Society of America.) (c) Visible-light haloes in chamosite originating from uraninite inclusions. (Source: *Nasdala et al.* [2006]. Reproduced with permission of Springer.)

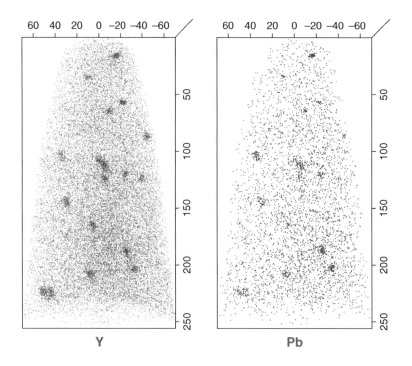

Fig. 10.18. Clusters of high Y and Pb, interpreted as alpha subparent recoil-damage-zone traps in Archean zircon. (Source: *Valley et al.* [2015]. Reproduced with permission of Mineralogical Society of America.)

provide radiation damage ages in similar way as fission-track dating. Although many studies have outlined the potential for this approach, few have demonstrated its use. One approach is to quantify accumulated alpha damage by a bulk spectroscopic measurement like changes in Raman bands in zircon [e.g., *Pidgeon*, 2014]. Like fission tracks, individual unetched ARTs can also be seen by TEM microscopy, but observing them at a scale that also allows estimation of their overall abundance (i.e., areal density) in a crystal requires etching. Numerous studies have

demonstrated the observability of etched ARTs in various minerals, particularly mica, where low U concentrations (typically ppb) lead to a manageably low track density when etched with HF. These studies also showed a linear dependence between density of visible tracks and etch time. *Gögen and Wagner* [2000] showed that etch rates are anisotropic, and the density–duration relationship of etching actually flattens out at high densities due to overlapping tracks. They also posited that individual ARTs from all of the intermediate daughters in the U and Th decay

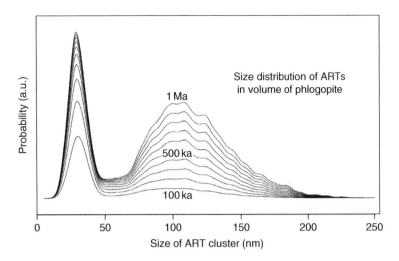

Fig. 10.19. Expected size distributions of alpha recoil tracks in phlogopite as observed volumetrically. (Source: *Jonckheere and Gögen* [2001]. Reproduced with permission of Elsevier.)

chains will not register as individually recognizable tracks, due to proximity of formation and overlap created during etching, resulting in composite ARTs more than 100 nm in size (Fig. 10.19). They also contend that only considering the long-lived nuclei ^{238}U, ^{234}U, ^{235}U, ^{230}Th, and ^{232}Th in transforming ART densities to age results simplifies the problem significantly with less than about 2% inaccuracy. Their age equation, assuming secular equilibrium among the intermediate daughters, and relating the total areal density of ARTs ρ_v to time t is:

$$\rho_v = {}^{238}U_a e^{\lambda_8 t} \cdot \left[\left(1 - e^{-\lambda_8 t}\right) + \frac{\lambda_8}{\lambda_4}\left(1 - e^{-\lambda_4 t}\right) + D\frac{\lambda_8}{\lambda_0}\left(1 - e^{-\lambda_0 t}\right) \right.$$
$$\left. + I\left(1 - e^{-\lambda_5 t}\right) + D\left(\frac{Th}{U}\right)_m \left(1 - e^{-\lambda_2 t}\right) \right] \cdot \eta_{tot} \qquad (10.8)$$

where $^{238}U_a$ is the areal abundance of ^{238}U, λ_8, λ_4, λ_0, λ_5, and λ_2 are the decay constants for the nuclides with the respective numerals in the last digits of their atomic weights, $(Th/U)_m$ is the Th/U of the magma (or fluid) in equilibrium with the mineral being dated, I is the ratio of ^{235}U/^{238}U abundances and decay constants, η_{tot} is the total efficiency coefficient for ART revelation ($\sim$1), and D is the partition coefficient for Th/U between the mineral and the magma/fluid from which it precipitated. All these parameters are known or measurable except D, which can be estimated for the system of interest. *Gögen and Wagner* [2000] used these relationships to determine ART ages on phlogopite samples with ages ranging from $\sim$10–400 ka, with precision of about 20%.

The kinetics of annealing of ARTs are poorly known. *Gögen and Wagner* [2000] suggest an apparent closure temperature of about 50 °C for phlogopite, and *Yuan et al.* [2009] suggest even lower closure temperatures of only 26–37 °C for cooling rates of 10–100 K/Ma. As with fission-tracks, these annealing kinetics refer to the retention of etchable and observable ARTs only, not necessarily the retention of alpha-recoil damage defects themselves.

10.5 FISSION TRACKS

10.5.1 History

Fission-track analysis, including dating and track-length measurements, provides one of the most powerful and versatile geochronologic, and especially thermochronologic, methods available. *Price and Walker* [1962a,b,c, 1963a,b] first recognized the potential for and pioneered the measurement of radiation damage tracks in solids. Since then, numerous studies have established robust understanding and calibrations of the method through development of analytical methods for counting and measuring track lengths [*Fleischer and Price*, 1964; *Naeser*, 1967; *Gleadow et al.*, 1986, 2009; *Dumitru*, 1993], statistical approaches for representing and understanding the unique constraints fission-track analysis presents [*Galbraith*, 2005], experiments and models of the kinetics of annealing and therefore thermal sensitivity of fission tracks [*Green et al.*, 1986; 1989; *Laslett et al.*, 1987; *Green*, 1988; *Duddy*, 1988; *Carlson et al.*, 1999; *Ketcham*, 2005; *Tagami*, 2005), and numerous applications to diverse geologic problems. Several publications also address the subject of fission-track analysis and modeling, particularly in apatite, in detail [e.g., *Gallagher et al.*, 1998; *Ketcham*, 2005; *Ketcham et al.*, 2007; *Donelick et al.*, 2005; *Wagner and van den Haute*, 2012], and this section only scratches the surface of this large and impactful subdiscipline of geochronology.

10.5.2 Theory, fundamentals, and systematics

Uranium-238 decays not only by alpha emission with λ_a of 1.55125×10^{-10} year^{-1}, but also by spontaneous fission, with a much more infrequent λ_{sf} of approximately 8.46×10^{-17} year^{-1}. Although much rarer than alpha decay, spontaneous fission of ^{238}U releases about 50–100 times more energy than the 4–8 MeV released by U-series and Th-series alpha decays. This energy is roughly split between two nuclei (Fig. 10.20) that recoil from

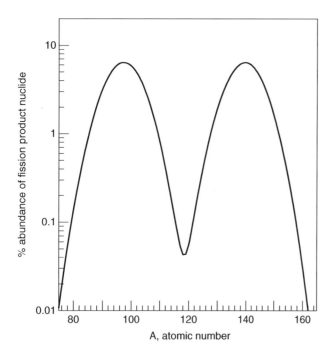

Fig. 10.20. Approximate probability distribution of nuclei yields from spontaneous fission of ^{238}U. Most decay produces two nuclei with ~3/2 mass ratios.

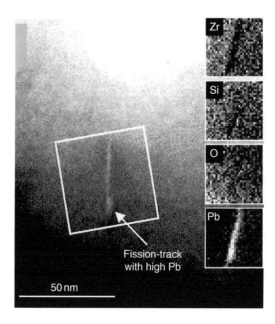

Fig. 10.21. HAADF-STEM (high-angle annular dark-field scanning transmission electron microscope) image of region surrounding a spontaneous fission-track in Archean Jack Hills zircon, showing Pb partitioning into the track. (Source: *Utsunomiya et al.* [2004]. Reproduced with permission of Elsevier.)

one another with stopping distances of about 8–12 µm per particle. In an insulating solid, these nuclei create a large linear damage zone about twice this length and only a few nanometers wide called a fission track. A fission track is characterized by a high degree of crystal lattice disorder, forming an amorphous, and in some minerals, a possibly partly porous channel [e.g., *Li et al.*, 2011]. However, the mechanisms by which the track is registered in the crystal are debated. The most widely accepted model, the "ion explosion spike" mechanism, calls on ionization of atoms adjacent to the track caused by the movement of the highly charged particles [*Fleischer et al.* 1965, 1975]. The fact that poorly insulating solids do not appear to retain or at least register fission tracks has been used to support the ion spike model, and the maintenance of tracks as local regions of intense charge gradients [*Fleischer et al.*, 1975]. In contrast, the "thermal spike" model [*Seitz*, 1949; *Bonfiglioli et al.*, 1961; *Chadderton and Montagu-Pollock*, 1963] proposes that damage to the lattice is primarily due to extreme heating caused by rapid motion of the particle through the region and the intense energy release. Both mechanisms may contribute to the formation and persistence of the fission-track [*Chadderton*, 2003].

Fission tracks that have not been enlarged by chemical etching for easier observation and counting are called latent fission tracks, and have been observed by TEM, atomic-force microscopy (AFM), and scanning-electron microsopy (SEM). As with ARTs [*Valley et al.*, 2014], in at least some cases they have been associated with locally high concentrations of incompatible trace elements in minerals (Fig. 10.21), suggesting that they may act as traps for nonstoichiometric ions.

Fission tracks are a daughter product of the decay of parent ^{238}U by spontaneous fission. Thus, the measured ratio of the two, as in most radioisotopic systems, can be used to date geologic materials. This means track abundances must be measured in a way that is directly related to the U concentration in the same region of a mineral. This relies on the assumption that fission tracks are not removed following their formation. However, fission tracks are metastable in insulating solids, meaning they fade, anneal, shorten, or otherwise disappear over time. In practice understanding the kinetics of fission-track annealing with respect to time, temperature, and composition has shown fission-track dating to be a very useful thermochronometer in a wide range of applications. This has resulted in a great deal of work being applied to experimental development of the technique and to quantifying fission-track annealing.

Price and Walker [1962a,b,c, 1963a,b] pioneered the recognition of the potential of radiation damage tracks in minerals to yield geochronologically useful information. Some of the first geochronologic ages using the technique were presented in 1964, dating tektites and related materials from Libya, and numerous studies followed in the late 1960s and through the 1970s that greatly expanded the technique. Studies of fission-track annealing proliferated in the middle to late 1970s and 1980s, as thermochronologic applications became more widespread.

10.5.3 Analyses

In order for volume or areal densities of fission tracks (Fig. 10.22) to be useful for geochronologic purposes, tracks must be observable in abundances that permit counting and statistical

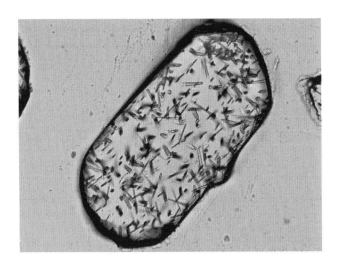

Fig. 10.22. Etched spontaneous tracks in apatite. (Source: courtesy of Stuart Nigel Thomson.)

interpretations robust enough to produce ages sufficiently precise to be geologically useful. A large number of minerals and phases contain sufficient U for fission-track dating in various applications, including glasses, epidote, and titanite, but by far the largest share of analytical and application attention has focused on the minerals apatite and zircon. These typically contain tens to thousands of ppm U, generating track densities in grains of typical dimensions to yield relatively precise ages, especially when multiple grains from the same sample are dated. The higher U concentrations in zircon are actually an impediment to dating many older samples, as track densities can be too high to allow counting, and the accumulated radiation damage may not allow etching to reveal the tracks.

The fundamental ingredients of the fission-track dating procedure include measurement of natural, or spontaneous, track densities and the concentration of ^{238}U in the dated grain. Many studies also complement these data with the frequency distribution of the lengths of tracks meeting certain criteria, which, as will be discussed, can provide powerful constraints on a sample's thermal history aside from the date itself.

Numerous permutations of fission-track dating procedures exist, but most fall into two categories: grain population methods, where spontaneous tracks and uranium concentrations are measured on separate grain aliquots; and grain-by-grain methods where spontaneous tracks and uranium concentrations are measured on the same grain. In the latter approach, two common methods are used to quantify uranium concentrations: the external detector method, and a newer approach involving laser-ablation inductively coupled plasma mass spectrometry (LA-ICP-MS). All methods require increasing the size and contrast of latent fission-tracks with the surrounding crystal. This is done by chemical etching with acids or bases, which preferentially dissolve the track and region around it, enlarging it to the point where it is easily visible using optical microscopy. Some newer techniques have used SEM [e.g., *Gombosi et al.*, 2014], enabling counting at much higher track densities.

The population method, which has largely been supplanted by the external detector method described below, is nonetheless historically important and provides a basis for understanding newer approaches. It involves measuring spontaneous track densities in one aliquot by revealing interior surfaces of the crystal through polishing and etching the tracks on the surface, and counting the areal abundance of tracks. The U measurements are then made in a separate aliquot with assumed identical U concentration, but which has first been subject to high-temperature heating for sufficiently long durations to anneal all spontaneous fission tracks. The aliquot is then irradiated with thermal neutrons in a nuclear reactor, which induces fission of only ^{235}U. These induced fission tracks can then be exposed, etched, and counted just as the spontaneous ones were, and the U concentration can be calculated from the natural ^{238}U/^{235}U and knowledge of the efficiency of induced fission of ^{235}U during irradiation. In practice, and analogous to procedures used in ^{40}Ar/^{39}Ar dating with fast neutrons, calculating this efficiency from the flux of neutrons with various energies, the ^{235}U capture cross-section, and duration of irradiation is much harder and less precise than simply measuring it based on a standard glass (or other phase) of known U concentration subject to the same irradiation.

The external detector method (Fig. 10.23) has the advantage of measuring track densities and U concentrations in regions that are much more closely related to one another, and over much smaller length scales (i.e., in virtually the same aliquot, rather than separate aliquots). The first step (Fig. 10.23) is polishing and etching of spontaneous tracks in interior crystal surfaces, as in the population method. Typically before the tracks are counted, however, the "external detector" is attached to the polished surface of the grain. In practice this is performed on many grains exposed in a teflon or epoxy mount. The package of grains and the external detector affixed to the surfaces of all the grains is then irradiated in a nuclear reactor as in the population method. Induced fission of ^{235}U in the dated grains is then registered in the external detector material by fission particles emanating from the grain and recoiling into the detector. Because the detector only receives fission fragments from within half a track distance of the polished grain surface, and only from one side, this geometry of implantation must be taken into account when calculating the U concentration of the grain; this is referred to as the 2π geometry factor. The most common material used for the external detector is muscovite mica, because of its very low U concentration (typically 1 ppb) and perfect cleavage allowing extensive flat planes to be easily affixed to polished grain mounts. Following irradiation, the grain and the external detector are separated and mounted "face up" as mirror images of one another, with fiduciary marks used to colocate regions within grains even over submicron scales for precise correlation of spontaneous and induced track densities. Spontaneous track densities are then counted in the grains and induced track densities are counted in the mica, and the ratio between the two is used to calculate the age.

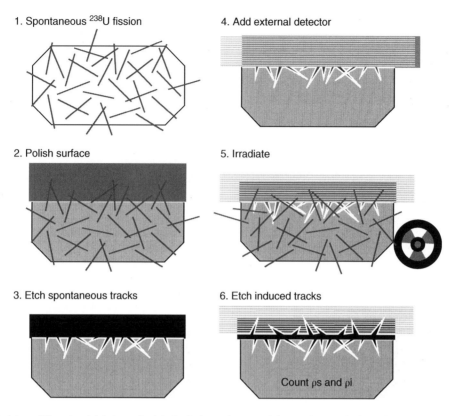

1. Spontaneous ^{238}U fission

2. Polish surface

3. Etch spontaneous tracks

4. Add external detector

5. Irradiate

6. Etch induced tracks

Count ρs and ρi

Fig. 10.23. The analytical steps of the external detector method. In the final step, the external detector (horizontal rule) and grain (below the detector) in panel 6 are removed from one another, laid facing up, and tracks in corresponding regions of each are counted. (Source: Courtesy of Andrew Gleadow.)

More recently, many fission-track dating studies have measured U concentrations in dated grains by LA-ICP-MS, after counting spontaneous track densities [*Hasebe et al.*, 2004]. This avoids irradiation, whose main disadvantages are the time required to wait for the decrease in induced radioactivity upon return from the reactor, time involved in counting of the induced tracks, and the inconveniences of handling and shipping radioactive materials. Uranium measurement by LA-ICP-MS also has the advantage of allowing simultaneous measurement of other useful compositional and isotopic information in dated grains such as U–Pb ages and trace element compositions that may be correlated with provenance or fission-track annealing characteristics [e.g., *Chew and Donelick*, 2012; *Hasebe et al.*, 2013].

Chemical etching of fission-tracks is required regardless of the parent measurement technique or whether tracks are counted by optical microscopy or SEM. Tagami and O'Sullivan [2005] provide a thorough review of track-etching procedures and considerations. Various etchants ranging from acids to bases are typically used, and the temperature and duration of exposure is well controlled to ensure systematic behavior. For example, apatite is typically etched with 5.5 N HNO₃ for 20 s at ~21 °C, zircon with a molten eutectic mixture of KOH and NaOH at 220 °C, with etch time dependent on accumulated radiation damage of the zircon varying from 1 h (high damage) to over 50 h (zero damage), and the muscovite detector with 48% HF for 20 min at 20 °C. Standards of known age are also etched in the same experimental batch, with the idea that they have the same etching behavior as the unknowns. An additional concern is the effect of crystallographically anisotropic etching, the magnitude of which also varies with radiation damage. In general, the objective of etching is to reveal etched density and track lengths that are as similar as possible to those of the latent tracks. Etching behavior depends on both the rate of track etching and the somewhat slower bulk-etching rate of the undamaged polished mineral surface. These combine in such a way to yield a relationship between etched track length and the duration of etchant exposure characterized by a rapid rise and then slow increase in etched track length. Ideal etching therefore involves targeting the rollover in this track length curve.

10.5.4 Fission-track age equations

As with any radioisotopic age measurement, the fundamental requirements for an age are the abundance of the daughter product, in this case fission tracks F_s, the abundance of the parent nuclide, ^{238}U, and the rate constants governing the daughter's production λ_{sf} and parent's decay $\lambda_\alpha \approx \lambda_\alpha + \lambda_{sf}$:

$$F_S = \frac{\lambda_{sf}}{\lambda_\alpha} {}^{238}U \left(e^{\lambda_\alpha t} - 1 \right) \tag{10.9}$$

To account for incomplete revelation of tracks by etching, an etching efficiency factor q relates the actual spontaneous track density ρ_s to the abundance counted in analysis F_s

$$F_s = \frac{\rho_s}{q} \tag{10.10}$$

In reality q is often paired with an additional factor accounting for track-length differences among samples η, because both etching efficiency and track length affect the measured track density. Here we simplify the derivation with a single variable q. Combining equations (10.9) and (10.10) yields

$$\rho_S = q \frac{\lambda_{sf}}{\lambda_\alpha} {}^{238}U \left(e^{\lambda_\alpha t} - 1 \right) \tag{10.11}$$

For the external detector method, the density of induced tracks ρ_i is

$$\rho_i = q {}^{235}U \phi \sigma \tag{10.12}$$

where ϕ is the thermal neutron flux per unit volume of sample and σ is the neutron capture cross-section of ${}^{235}U$ for induced fission.

Assuming that q is the same in dated crystal and the external mica detector, dividing the spontaneous track equation by the induced track equation yields

$$\frac{\rho_S}{\rho_i} = \frac{\lambda_{sf}}{\lambda_\alpha} {}^{238}U {}^{235}U \frac{1}{\phi\sigma} \left(e^{\lambda_\alpha t} - 1 \right) \tag{10.13}$$

which can then be solved for t,

$$t = \frac{1}{\lambda_\alpha} \ln \left(1 + \frac{\rho_S}{\rho_i} \frac{\lambda_\alpha}{\lambda_{sf}} \frac{\phi\sigma}{137.88} \right) \tag{10.14}$$

assuming a given ${}^{238}U / {}^{235}U$.

This equation suffers from impracticality because λ_{sf}, ϕ, and σ are not well known or hard to predict for a given irradiation. Similar to the approach used for ${}^{40}Ar/{}^{39}Ar$ dating, these variables can be eliminated by dating a standard of known age as well as measuring the induced track density ρ_d in a glass dosimeter with known U concentration.

The neutron flux is replaced by

$$\phi = B\rho_d \tag{10.15}$$

where B is a constant, leading to

$$\frac{\rho_S}{\rho_i} = \frac{\lambda_{sf}}{\lambda_\alpha} \frac{137.88}{B\rho_d\sigma} \left(e^{\lambda_\alpha t} - 1 \right) \tag{10.16}$$

Finally, a new constant ζ, for each irradiation batch is assigned to be

$$\zeta = \frac{B\sigma}{\lambda_{sf} \cdot 137.88} \tag{10.17}$$

so that

$$\zeta = \frac{B\sigma}{\lambda_{sf} \cdot 137.88} = \frac{\rho_i}{\rho_s} \cdot \frac{\left(e^{\lambda_\alpha t} - 1 \right)}{\lambda_\alpha \rho_d} \tag{10.18}$$

The right-hand side of equation (10.18) can be measured in a standard of known age and for a glass dosimeter irradiated in the same way as the standard and unknowns. This then represents ζ, so that the final age equation for an unknown sample is

$$t = \frac{1}{\lambda_\alpha} \ln \left(1 + \frac{\rho_S}{\rho_i} \lambda_\alpha \rho_d \zeta \right) \tag{10.19}$$

Other variations on this equation for LA-ICP-MS-based analysis of ${}^{238}U$ are provided in a number of recent publications, including *Hasebe et al.* [2004, 2013] and *Soares et al.* [2014].

Like any technique, the robustness of fission-track dating results depends on the analytical precision. But unlike many techniques that measure extremely large amounts of parent and daughter nuclides in a given aliquot, the essential practice of literally counting numbers of tracks introduces distinctive data interpretation challenges for the technique [*Galbraith*, 2005], particularly in cases of low U concentrations and/or young cooling ages. Because the uncertainty on the date determined a single grain may be relatively large, most studies combine observations from many grains from a single sample. The observations are often combined in three different ways to determine an age and uncertainty.

(1) *Pooled age*: the ratio of spontaneous-track to induced-track densities in equation (10.19) is based on the sum of each from all grains in a sample;

(2) *Arithmetic mean age*: the average ratio of spontaneous tracks to induced tracks among all grains is used in equation (10.19);

(3) *Central age*: all single-grain ages are combined and the central age is the weighted mean of their log-normal distribution [*Galbraith and Laslett*, 1993]. The likelihood that all grains compose a single population, based on their observed dispersion and estimated analytical precision, is often quantified with the chi-square statistic, similar to the MSWD.

In the case that central, pooled, or mean ages actually comprise multiple populations, the variation in single-grain ages may contain valuable information in itself. In some samples it may reflect variable annealing kinetics of component populations that may be used to constrain thermal histories. In detrital samples that have not been post-depositionally heated sufficiently to reset source cooling ages, this variation may reflect important provenance information (also see Chapter 5 on interpreting thermochronologic results), particularly if component peaks can be deconvoluted from probability distributions of single-grain ages [*Brandon*, 1992; *Brandon and Vance*, 1992; *Garver et al.*, 1999].

Another method for distinguishing component populations within samples is the radial plot, an example of which was also shown above in the context of luminescence dating. The radial, or Galbraith, plot [*Galbraith*, 1988, 1990] provides a graphical way to interpret the significance of analyses with different uncertainties. In a fission-track radial plot, each single-grain age is assigned an error bar of the same size, but the distance of the point away from the origin on the horizontal axis is inversely

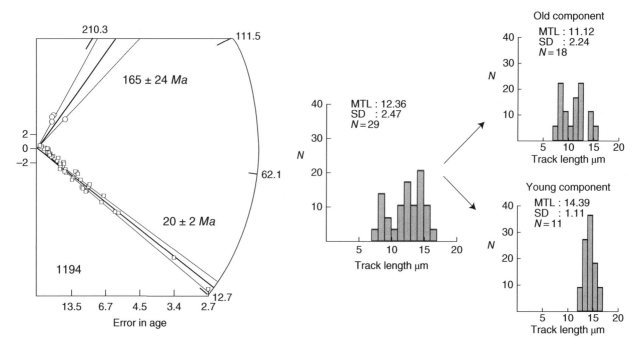

Fig. 10.24. Left: example radial plot of single-grain apatite fission-track analyses. Two distinct populations are resolved: an older one with relatively poor precision, and a younger one with higher precision, including a few grains with extremely low SE. Right: histograms of track-length distributions, showing that when the two populations are separated, the younger one has a more uniform distribution of longer track lengths, consistent with rapid cooling through the AFT closure temperature at the age indicated in the radial plot. (Source: *Gallagher et al.* [1998]. Reproduced with permission of Annual Review of 26, Volume © by Annual Reviews, http://www.annualreviews.org.)

proportional to the standard error (SE) of the analysis. The radial axis is then the age. One way to think about this is that lines drawn from the origin through the maximum and minimum of the error bars will intersect or "cast a shadow" along a large arc of the age axis (large age range) if the point is near the origin (high SE) but a small arc (small age range) if the point is far from the origin (small SE). An example is shown in Fig. 10.24, in which two distinct populations with distinct ages also have track-length distributions reflecting different thermal histories.

10.5.5 Fission-track annealing

As with other radiation-damage techniques, fission tracks are not permanently preserved in crystals but fade, shorten, segment, and disappear in a process called annealing. The full range of conditions or phenomena that may affect this annealing is not well known, but the most important effect in most circumstances is caused by heating. Thermal annealing of damage tracks has been recognized since almost the inception of the fission-track technique [*Fleischer et al.*, 1975], and incorporation of the kinetics of this annealing into interpretation of measured ages played an important role in defining modern thermochronology.

Observations of annealing of latent tracks during heating provide information about the physical processes involved. TEM observations of tracks in apatite [e.g., *Paul and Fitzgerald*, 1992; *Paul*, 1993; *Li et al.*, 2011, 2012] show that tracks segment

into regions with shorter aspect ratios, pinching off into discontinuous damage zones that eventually disappear with further heating. However, most understanding of annealing properties is actually based on changes visible in etched tracks using standard techniques, not in latent tracks. Thus even in "fully annealed" tracks, as determined from their lack of revelation by etching, radiation damage of some type may be preserved. Indeed, recent work by *Li et al.* [2011] shows that, in apatite, remnant discontinuous latent fission tracks are clearly visible by TEM even after heating to much higher temperatures for longer durations than required for the effective complete disappearance of etchable tracks. In contrast, zircon fission tracks appear not to segment during annealing, such that latent and etchable tracks anneal with roughly similar kinetics [*Li et al.* 2011]. This may suggest that tracks anneal in these two minerals by very different mechanisms.

Because of the practical value of fission-track thermochronology for understanding time–temperature histories of rocks, a great deal of experimental work has been carried out to characterize the annealing kinetics of etchable fission tracks, particularly in apatite, and to a lesser degree in zircon and other phases. Although some of the models arising from these experiments contain at least elements of mechanistic physical descriptions of rate laws associated with diffusion, release of strain energy, and topotactic crystallization [e.g., *Carlson*, 1990], most fission-track models are largely empirical. Normalized track length, l/l_0, rather than density, is the primary observation in experiments

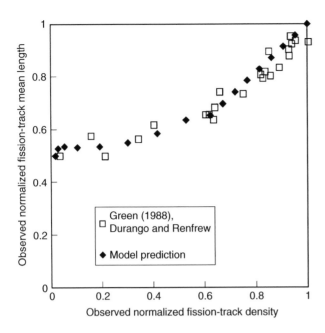

Fig. 10.25. Relationship between normalized fission-track mean length (normalized to starting length) and normalized fission-track density. (Source: *Donelick et al.* [1999]. Reproduced with permission of Mineralogical Society of America.)

varying isothermal holding at different temperatures for varying durations. Length can then be transformed to track density and thus age, as in Fig. 10.24. *Ketcham* [2005] provides examples of these relationships for apatite, in which, owing to segmentation of annealed tracks, normalized length reductions approaching ~0.55 correspond to essentially zero normalized density and complete resetting of the apatite fission-track system (Fig. 10.25). Because of anisotropic track annealing, observations of length reduction also need to be crystallographically oriented and either mean or c-axis projected lengths consistently used for modeling.

Once a suite of experiments has provided quantitative relationships between temperature, time, and fission-track length, contours of constant length reduction are related to time and temperature of annealing through the "pseudo-Arrhenius" plot (Fig. 10.26). Several models have been proposed to relate normalized track-length reduction ($r = l/l_0$) to time and temperature through empirical relationships involving three or more fitted constants. Here we review three basic forms proposed for apatite. *Green et al.* [1985] found that the following relationship provided a good fit:

$$\ln(1-r) = c_0 + c_1 \ln(t) + \frac{c_2}{T} \tag{10.20}$$

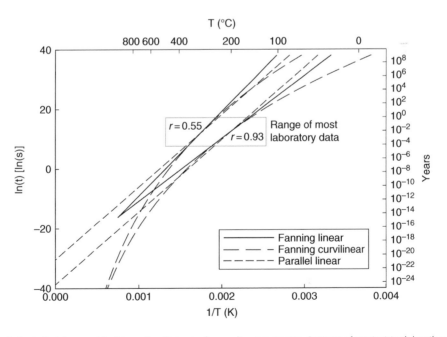

Fig. 10.26. The "pseudo-Arrhenius" plot as used for interpreting fission-track annealing experiments. Contours of constant track-length reduction are plotted as a function of (log) duration (*y*-axis) and (inverse) temperature (*x*-axis) of isothermal holding producing each length reduction. In this example, each of three possible model forms is characterized by two contours of track-length reduction, one for 0.93 (the estimated reduction that occurs for holding at Earth surface temperatures over geologic timescales) and 0.55 (the approximate reduction at which track density approaches zero; i.e., full resetting). Fanning, parallel, linear, and curvilinear denote model forms that have been proposed to fit experimental annealing data. In the temperature and time range of most experimental data, differences between these models are relatively small, but when extrapolated to geologic timescales they can be very significant. (Source: *Ketcham* [2005]. Reproduced with permission of Mineralogical Society of America.)

where c_i are fitted parameters. This relationship results in the "parallel linear" model shown in Fig. 10.26. Using annealing data on Durango apatite, *Laslett et al.* [1987] proposed a slightly more complicated model involving additional fitted parameters α, β, and c_3, the latter causing the contours of constant annealing to "fan."

$$\frac{\left[(1-r^\beta/\beta)^\alpha\right]-1}{\alpha} = c_0 + c_1\left[\frac{\ln(t)-c_2}{(1/T)-c_3}\right] \quad (10.21)$$

This "fanning linear" model is shown graphically in Fig. 10.26. Motivated in part by observations from geologic samples and extrapolating to timescales complementing conventional experimental data, *Ketcham et al.* [1999] proposed a variation to *Laslett et al.*'s model

$$\frac{\left[(1-r^\beta/\beta)^\alpha\right]-1}{\alpha} = c_0 + c_1\left[\frac{\ln(t)-c_2}{\ln(1/T)-c_3}\right] \quad (10.22)$$

This results in the "fanning curvilinear" model in Fig. 10.26. *Ketcham* [2005] also discussed the potential relationships between these empirical models and the mechanistic aspects of track annealing. Although the constants in equation (10.22) vary for different types of apatite, one set that is appropriate for c-axis projected tracks in a fairly representative specimen, Renfrew apatite, are $\alpha = -0.35878$, $\beta = -2.9633$, $c_0 = -61.311$, $c_1 = 1.292$, $c_2 = -100.53$, and $c_3 = -8.7225$. To then relate the normalized track-length reduction to fractional track density, *Ketcham* [2005] used two different equations for length reductions above and below 0.624 (later changed to 0.612): if.

$$\rho_s \geq 0.612, \ r = (\rho_s + 0.6)/1.6 \quad (10.23)$$

if

$$\rho_s < 0.612, \ r = 0.4974 + \frac{\sqrt{0.30642 + 36.82\rho_s}}{18.41} \quad (10.24)$$

Apatite is also known to show annealing kinetic variations that depend on a variety of characteristics, only one of which may be chemistry. These characteristics can be constrained by the dimensions of etch pits created during etching of fission tracks (D_{per} and D_{par}), and the inferred kinetic variations can be accounted for in models of track annealing through comparing fitted parameters with those of more or less annealing-resistant apatite specimens [*Ketcham et al.*, 1999].

Compared with apatite, fewer experimental constraints exist on the annealing kinetics of fission tracks in zircon. A complicating factor in zircon is the wide range of radiation damage present in grains from different samples and settings [*Garver et al.*, 2005], caused by both the wider range of effective uranium (eU) concentrations as well as the fact alpha damage in zircon does not anneal at low temperature the way it does in apatite. *Yamada et al.* [1995] and *Tagami et al.* [1998] proposed fanning linear and parallel models. Tagami's fanning linear model has the form:

$$r = l/l_0 = c_0 + c_1\left[\frac{\ln(t)+c_2}{(10^3/T)-c_3}\right] \quad (10.25)$$

Ketcham [in preparation; also described in *Guenthner et al.*, 2013] also proposed a fanning curvilinear form:

$$r = l/l_0 = \left[\left(c_0 + c_1\frac{\ln(t)+c_2}{\ln(\frac{1}{T})-c_3}\right)^{\frac{1}{\beta}} + 1\right]^{-1} \quad (10.26)$$

where the constants are: $\beta = -0.05721$, $c_0 = 6.24534$, $c_1 = -0.11977$, $c_2 = -314.937$, and $c_3 = -14.2868$. This model uses the relationship between reduced track length r and reduced track density r/r_0 from *Tagami et al.* [1990], which applies down to r/r_0 of 0.36:

$$\rho_r = \frac{\rho}{\rho_0} = 1.25(r-0.2) \quad (10.27)$$

Although the model described in equations (10.26) and (10.27) yield an effective closure temperature around 275 °C (for a cooling rate of 10 °C/Ma), field-based, empirical constraints on zircon fission-track closure temperatures predict significantly lower closure temperatures (~200–240 °C) than most laboratory based methods (~320–340 °C) (Fig. 10.27; *Zaun and Wagner* [1985]; *Hurford* [1986]; *Brandon et al.*, [1998]; *Bernet* [2009]). This is likely due to effects of natural radiation damage on annealing properties. *Rahm et al.* [2004] incorporated constraints from a variety of experiments and damage extents to develop a baseline model for zircon fission-track annealing kinetics in the case of zero-damage.

10.5.6 Track-length analysis

Although tracks shorten when their host grains are subjected to elevated temperatures, new unannealed tracks are forming continuously, meaning that not only the track density but also their length distribution can provide valuable thermal history information. Track-length analysis [*Gleadow et al.*, 1986] is usually performed on apatite, because the kinetics of length reduction are best understood in this mineral, providing most robust time–temperature constraints.

Tracks that intersect the polished surface of a crystal at any angle other than perfectly parallel to the polished surface are obviously shorter than they were prior to polishing. So track lengths reflecting the full length as they existed in the crystal prior to polishing can be measured only on confined tracks—those still embedded within a crystal. Confined tracks can be etched and therefore made visible only if they intersect another track or fracture that in turn intersects the surface and acts as a pathway for etchant to reach it. These kinds of tracks are sometimes called TINT (track in track) or TINCle (track in cleavage).

Track-length distributions vary from largely unimodal, with a mean length equivalent to that of fresh, unannealed tracks, to broadly distributed between this value and very short tracks, to bimodal. A sample that experienced rapid cooling to low temperatures and subsequently remained at low temperatures for its entire history will have exclusively long tracks (near the maximum of ~14.5 mm characteristic of fission tracks in apatite).

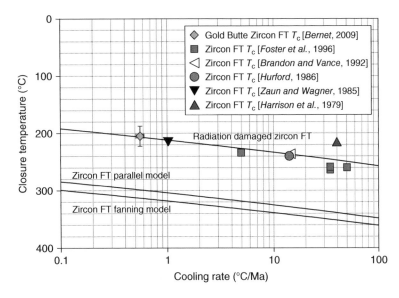

Fig. 10.27. Estimates of the zircon fission-track closure temperature (as a function of cooling rate) for the experimentally determined parallel and fanning models of *Yamada et al.* [1995] and *Tagami et al.* [1998], compared with empirical closure temperature estimates from geologic studies. (Source: *Bernet* [2009]. Reproduced with permission of Elsevier.)

In contrast, a sample recovered from a location where the temperature is near that of the closure temperature for the apatite fission-track system ($\sim$90–115 °C), such as in a deep borehole, will have tracks that are mostly short, because they anneal immediately after formation (at least over geologic time periods in excess of ca. 10^6 years). In a sample undergoing slow cooling, tracks that have formed most recently will be longer because they form at lower temperatures and have had less time to anneal (Fig. 10.28). Samples with more complex histories involving reheating may have mixed, skewed, or bimodal track-length populations, with the form of this population dependent on the thermal history. Thus, track-length distributions provide potentially detailed, though nonunique, constraints on the time–temperature path at temperatures and over durations where fission tracks anneal (e.g., in the partial annealing zone), while track densities and ages scale and position this path in time (Fig. 10.28).

10.5.7 Applications

10.5.7.1 The Otway Basin, Australia

One of the most historically important applications of apatite fission-track thermochronology is understanding the burial and exhumation history of sedimentary basins, and in some cases, the thermal histories of source terranes that supplied the detrital grains. Work in the Otway Basin of southeastern Australia is an excellent example of this type of study and demonstrates the utility of the fission-track dating and track-length analysis methods and their integration with paleotemperature estimates from vitrinite reflection methods [*Gleadow and Duddy*, 1981; *Green et al.*, 1989a,b; *Brown et al.*, 1994; *Dumitru*, 2000]. Most of the apatite grains in Otway basin sediments were derived from volcanic rocks

with an age of about 120 Ma, buried fairly quickly in the basin during rifting. Following recovery from elevated geothermal gradients after rifting, they experienced a fairly stable geologic and thermal history through the Cenozoic, with a modern geothermal gradient of about 35 $\pm$ 2 °C. Apatite grains in these samples have track densities yielding ages ranging from the detrital age of 120 Ma at shallow depths to close to zero at $\sim$3-km depths where temperatures are close to the approximate nominal closure temperature of the apatite fission-track system. Track-length distributions become broader and mean track length decreases with depth, reflecting partial annealing of older tracks at higher temperatures (Fig. 10.29). Radial plots also show expected shifts with depth from relatively high-precision and unimodal age near the surface, to increasing age dispersion at increasing depths, but a relatively uniform and near-zero age at the deepest levels.

10.5.7.2 The Wasatch Front, Utah

As also shown in Chapter 11, low-temperature thermochronologic methods, including fission-track analysis, have proven to be extremely useful for understanding the timing, rates, and depths of tectonic exhumation by normal faulting [e.g., *Stockli*, 2005]. One illustrative example from a region still posing seismic hazard to its inhabitants, as tectonic exhumation continues, is the Wasatch front. *Armstrong et al.* [2003] and *Ehlers et al.* [2003] presented numerous apatite, and some zircon, fission-track data, along with complementary apatite (U–Th)/He dates on the same samples, from a transect perpendicular to the strike of the normal fault scarp forming the west flank of the Wasatch Mountains east of Salt Lake City. Both zircon and apatite fission-track ages show fairly systematic increases away from the fault trace, with a change in slope in a region where several ancillary

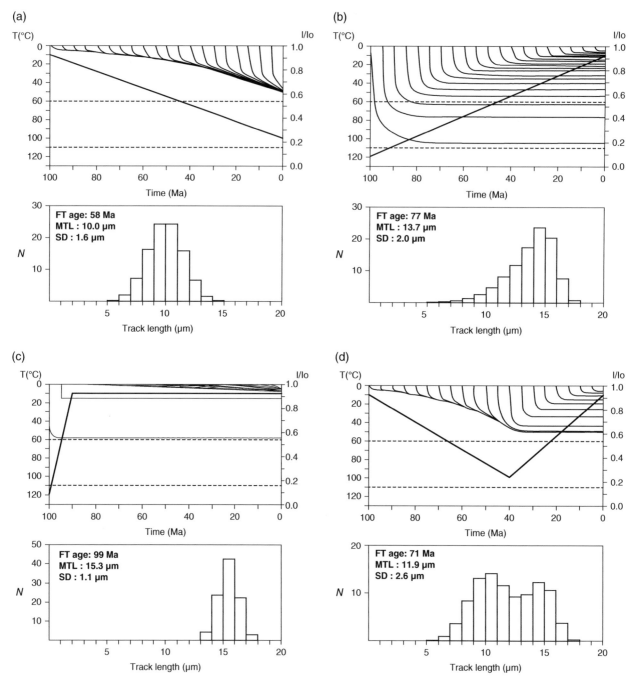

Fig. 10.28. (a–d) Schematic examples of the evolution of track-length distributions and corresponding fission-track ages, for the apatite fission-track system, for four different thermal histories. In each case, the upper panel shows the thermal history (bold line and left-side y-axis) as well as the normalized length (l/l_0) evolution of each of 20 distinct representative tracks formed at different times in the grain history (fine lines and right-side y-axis). The lower panel in each example shows the histogram of track lengths, as well as the apparent fission-track age, mean track length (MTL), and standard deviation (SD) of track lengths. (a) Slow heating, resulting in a unimodal but fairly broad distribution of lengths with a mean length significantly lower than that of a freshly formed track in apatite. In this case, the age does not correspond to any particular event. (b) Slow cooling, resulting in a skewed distribution with mostly long tracks but also containing a long tail of shortened tracks formed while the grain was still at high temperature. In this case, the age corresponds approximately to the time when the sample was at the closure temperature of the apatite fission-track system for this particular cooling rate. (c) Rapid cooling followed by isothermal holding at low temperature, resulting in a unimodal and narrow distribution of relatively long track lengths and an age corresponding to the timing of rapid cooling. (d) Heating followed by cooling, producing a bimodal distribution of track lengths and age that does not correspond to a particular event. (Source: *Gallagher et al.* [1998]. Reproduced with permission of Annual Review of 26, Volume © by Annual Reviews, http://www.annualreviews.org.)

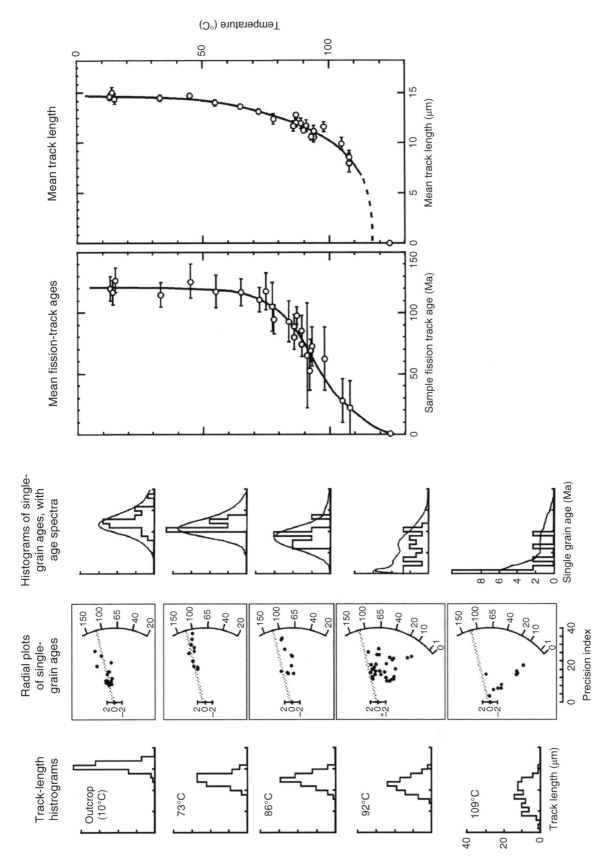

Fig. 10.29. Age, track-length, and other information from samples taken from the well-studied Otway Basin boreholes off southern Australia. Otway basin sediments were derived from volcanic rocks with a unimodal age of about 120 Ma, and buried fairly quickly in the basin where they resided at temperatures approximately as shown on right-hand side of figure until they were recently recovered by drilling. Apatite grains in these samples have track densities yielding ages ranging from the detrital age of 120 Ma at shallow depths to close to zero at depths where temperatures are greater than the approximate closure temperature of the apatite fission-track system. As shown at right, mean track length distributions systematically increase with age, and as shown at far left, the track length distribution becomes increasingly broad and skewed to short lengths with depth in the borehole as well. Also shown are radial plots and histograms of age distributions, showing the increasing dispersion of ages with depth, except for the deepest sample, for which most grains point to an extremely young age. (Source: *Gallagher et al.* [1998]. Reproduced with permission of Annual Reviews, http://www.annualreviews.org. Data from [*Gleadow and Duddy* 1981; *Green et al.*, 1989a,b; *Brown et al.*, 1994, *Dumitru*, 2000]).

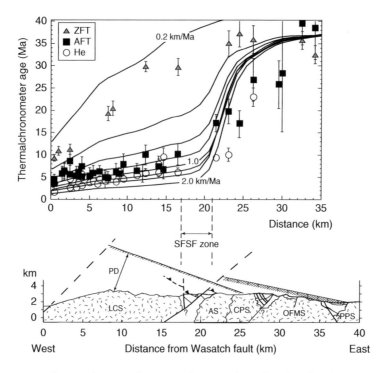

Fig. 10.30. Zircon fission-track (ZFT), apatite fission-track (AFT), and apatite (U–Th)/He (He) dates of samples collected in a transect perpendicular to the trace of the Wasatch normal fault, on the west side of the Wasatch Mountain range. Some degree of structural complexity (from the Silver Fork and Superior Faults [SFSF] zone) complicates what is otherwise a very systematic trend of thermochronologic ages that decrease with proximity to the fault and show system offsets consistent with the relative closure temperatures of each system. The model curves correspond to predictions from a coupled tectonic–thermal model for different rates of exhumation as shown, with initial ages of 35 Ma (from Eocene intrusions), a fault dip of 45°, and tilt hinge position 25 km from the fault trace. (Sources: *Ehlers et al.* [2003]. Reproduced with permission of John Wiley & Sons.; *Armstrong et al.* [2003]. Reproduced with permission of John Wiley & Sons.)

faults cut through the range. A similar trend, offset to younger ages and consistent with the lower closure temperature of the system, is seen for apatite (U–Th)/He ages. Using a two-dimensional thermokinematic model, *Ehlers et al.* [2003] used these relationships to solve for best-fit solutions for the rate of exhumation along the fault (0.8–1.2 km/Ma), the geometry of tilting in the footwall block (hinge minimum of 20–25 km from fault), the fault dip angle (45° or 60°), and age of onset of faulting (12 ± 2 Ma). The trends of the thermochronologic data are also most consistent with a >30% decrease in faulting rate at about 5 Ma (Fig. 10.30).

10.6 CONCLUSIONS

Radiation damage methods of dating are similar to other systems except that the daughter product is a physical or electronic manifestation of radiation damage instead of a nuclide. In most cases these manifestations anneal or otherwise disappear as a function of time and temperature with kinetics that make them useful for either low-temperature thermochronology or dating very near-surface processes.

Optically stimulated luminescence (OSL) and thermoluminescence (TL) rely on excitation and crystal defect trapping of

electrons of by ionizing radiation from radioactive decay originating in or near a sample. Over time, trapped charge accumulates in a naturally irradiated crystal, whereas exposure to heat or light releases or discharges it. Discharging crystals emit a characteristic wavelength of radiation that can be detected as TL or OSL. Different minerals have different charge-accumulation responses to radiation dose and kinetics and different kinetics of discharge in response to time and temperature. The principal challenge of luminescence (and other paleodosimetry methods) dating is thus to characterize (i) the luminescence present in a sample, (ii) the relationship between dose and luminescence, and (iii) the dose rate in the sample in its natural environment. Luminescence dating is often used for applications assuming an initial condition of complete resetting (zeroing, or bleaching), followed by simple accumulation with no annealing/discharge. However, with experimental calibration of the kinetics of luminescence signal decay, it can be used as a low-temperature thermochronometer.

Electron spin resonance (ESR) relies on paramagnetic signals caused by unpaired electrons excited by radiation and trapped in crystal defects. Defects typically responsible for this type of trapping include local charge imbalances created by ionic substitutions. Each defect has a distinctive signature, characterized by a *g*-value, related to the magnetic moment and angular momentum of electrons that become trapped in it, and the abundance of trapped electrons can be associated with the intensity of a change in microwave absorption across a spectrum of magnetic field intensity. As with luminescence, ESR dating involves estimating the ESR intensity at a particular *g*-value, the dose–response relationship, and the dose rate in the sample's natural environment. The kinetics of ESR signal decay have been determined for various types of centers and minerals for use in low-temperature thermochronometry.

Energetic alpha decay of U and Th causes several types of radiation damage, including that from stopping of alpha particles themselves, at distances $\sim$10–20 µm from their sources, as well as from recoil of the parent nuclide, which has a much lower stopping distance but creates many more atomic displacements in a small volume. Alpha particle damage is evident in alpha particle haloes (also known as radiohaloes) around small inclusions with high U–Th concentrations. These hold potential for geochronology but currently lack of robust quantification methods and uncertainty in annealing kinetics are limitations to use. Alpha recoil (of parent nuclides) has also shown promise using both microspectroscopic or track-counting measurements.

Fission-track dating is relatively well-developed technique based on spontaneous fission of ^{238}U. This type of decay is more energetic but much less frequent than alpha decay, and produces a damage track in insulating solids that is roughly 15–20 µm in length. The analytical procedure for most applications involves optically counting natural tracks in a thin layer of a polished exposed interior of a crystal, and relating this to the U concentration in the same region, either by laser ablation or inducing tracks from ^{235}U via neutron irradiation. Fission-track dating is most commonly applied to apatite and zircon, but other minerals and materials have also been dated using the technique. In apatite, track-length distributions are also commonly measured. Several different ways of calculating fission track dates from samples exist, and variation among ages of single crystals may provide additional provenance or thermal-history information. This is particularly true for detrital samples, but even samples from crystalline or reset sedimentary rocks can show multiple date populations that may relate to variation in fission-track annealing kinetics. These may be revealed in radial plots. Annealing of fission tracks is crystallographically anisotropic, and the relationship of length reduction to measured track density is not linear. Much experimental work has been done to calibrate the annealing kinetics of fission tracks with respect to temperature and time, particularly in apatite and zircon. The form of the annealing equations is empirical, with six or more fit parameters, but can be related to mechanistic principals. Annealing kinetics are typically represented as contours of constant length reduction in log-time versus inverse-temperature space, and both fanning linear and curvilinear models, both with varying activation energy, have been proposed. Effective closure temperatures for the apatite and zircon systems (for cooling rates of 10 K/Ma) are approximately 90–115 °C and 250–350 °C, respectively. At least for zircon, annealing kinetics also depend on the accumulated radiation damage. Track-length frequency distributions in apatite provide a powerful constraint on thermal histories near the temperature range of the partial annealing zone. Broadly speaking, track-length distributions that are unimodal and long typically indicate rapid cooling through partial annealing zone temperatures, whereas bimodal or widely varying length distributions may indicate reheating to, or slow cooling through, partial annealing-zone temperatures.

10.7 REFERENCES

Aitken, M. J. (1985) *Thermoluminescence Dating*. Academic Press.

Aitken, M. J. (1998) *An Introduction to Optical Dating: the Dating of Quaternary Sediments by the Use of Photon-stimulated Luminescence*. Oxford University Press.

Aitken, J. M., Young, D. R., and Pan. K. (1978) Electron trapping in electron-beam irradiated SiO$_2$. *Journal of Applied Physics* **49**(6), 3386–3391.

Armstrong, P. A., Ehlers, T. A., Chapman, D. S., Farley, K. A., and Kamp, P. J. (2003) Exhumation of the central Wasatch Mountains, Utah: 1. Patterns and timing of exhumation deduced from low-temperature thermochronology data. *Journal of Geophysical Research: Solid Earth* **108**(B3).

Bernet, M. (2009) A field-based estimate of the zircon fission-track closure temperature. *Chemical Geology* **259**(3), 181–189.

Bonfiglioli, G., Ferro, A., and Mojoni, A. (1961) Electron microscope investigation on the nature of tracks of fission products in mica. *Journal of Applied Physics* **32**(12), 2499–2503.

Brandon, M. T. (1992) Decomposition of fission-track grain-age distributions. *American Journal of Science* **292**(8), 535–564.

Brandon, M. T. and Vance, J. A. (1992) Tectonic evolution of the Cenozoic Olympic subduction complex, Washington State, as deduced from fission track ages for detrital zircons. *American Journal of Science* **292**, 565–565.

Brandon, M. T., Roden-Tice, M. K., and Garver, J. I. (1998) Late Cenozoic exhumation of the Cascadia accretionary wedge in the Olympic Mountains, northwest Washington State. *Geological Society of America Bulletin* **110**(8), 985–1009.

Brown, R.W., Summerfield, M.A., and Gleadow, A.J.W. (1994) Apatite fission track analysis: its potential for the estimation of denudation rates and implications of long-term landscape development. In *Process Models and Theoretical Geomorphology*, Kirkby, M. J. (ed.), 23–53. J. Wiley & Sons, New York.

Carlson, W. D. (1990) Mechanisms and kinetics of apatite fission-track annealing. *American Mineralogist* **75**(9–10).

Carlson, W. D., Donelick, R. A., and Ketcham, R. A. (1999) Variability of apatite fission-track annealing kinetics: I. Experimental results. *American Mineralogist* **84**(9), 1213–1223.

Chadderton, L. T. (2003) Nuclear tracks in solids: registration physics and the compound spike. *Radiation Measurements* **36**(1), 13–34.

Chadderton, L. T. and Montagu-Pollock. H. M. (1963) Fission fragment damage to crystal lattices: heat-sensitive crystals. *Proceedings of the Royal Society of London A: Mathematical, Physical and Engineering Sciences* **274**(1357), 239–252.

Chew, D. M. and Donelick, R. A. (2012) Combined apatite fission-track and U–Pb dating by LA-ICP-MS and its application in apatite provenance analysis. *Quantitative Mineralogy and Microanalysis of Sediments and Sedimentary Rocks: Mineralogical Association of Canada, Short Course* **42**, 219–247.

Dodson, M. H. (1973) Closure temperature in cooling geochronological and petrological systems. *Contributions to Mineralogy and Petrology* **40**(3), 259–274.

Donelick, R., Ketcham, R., and Carlson, W. (1999) Variability of apatite fission-track annealing kinetics: II. Crystallographic orientation effects. *American Mineralogist* **84**(9), 1224–1234.

Donelick, R. A., O'Sullivan, P. B., and Ketcham, R. A. (2005) Apatite fission-track analysis. *Reviews in Mineralogy and Geochemistry* **58**(1), 49–94.

Duller, G. A. T. (2008) *Luminescence Dating: Guidelines in Using Luminescence Dating in Archaeology.* English Heritage, Swindon.

Duddy, I. R., Green, P. F., and Laslett, G. M. (1988) Thermal annealing of fission tracks in apatite 3. Variable temperature behaviour. *Chemical Geology: Isotope Geoscience Section* **73**(1), 25–38.

Dumitru, T. A. (1993) A new computer-automated microscope stage system for fission-track analysis. *Nuclear Tracks and Radiation Measurements* **21**(4), 575–580.

Dumitru, T. A. (2000) Fission-Track Geochronology. Quaternary geochronology: Methods and applications, 131–155.

Ehlers, T. A., Willett, S. D., Armstrong, P. A., and Chapman, D. S. (2003) Exhumation of the central Wasatch Mountains, Utah: 2. Thermokinematic model of exhumation, erosion, and thermochronometer interpretation. *Journal of Geophysical Research: Solid Earth* **108**(B3).

Fleischer, R. L. and Price, P. B. (1964) Techniques for geological dating of minerals by chemical etching of fission fragment tracks. *Geochimica et Cosmochimica Acta* **28**(10), 1705–1714.

Fleischer, R. L., Price, P. B., and Walker, R. M. (1965) Solid-state track detectors: applications to nuclear science and geophysics. *Annual Review of Nuclear Science* **15**(1), 1–28.

Fleischer, R. L., Price, P. B., and Walker, R. M. (1975) *Nuclear Tracks in Solids: Principles And Applications.* University of California Press.

Foster, D. A., Kohn, B. P., & Gleadow, A. J. W. (1996) Sphene and zircon fission track closure temperatures revisited; empirical calibrations from $^{40}Ar/^{39}Ar$ diffusion studies of K-feldspar and biotite. *In International Workshop on Fission Track Dating, Abstracts, Gent,* **37**. August.

Fukuchi, T. (1996) Quartet ESR signals detected from natural clay minerals and their applicability to radiation dosimetry and dating. *Japanese Journal of Applied Physics* **35**(Part 1: 3).

Fukuchi, T. (2012) *ESR Techniques for the Detection of Seismic Frictional Heat.* INTECH Open Access Publisher.

Fukuchi, T., Yurugi, J.-I., and Imai, N. (2007) ESR detection of seismic frictional heating events in the Nojima fault drill core samples, Japan. *Tectonophysics* **443**(3), 127–138.

Galbraith, R. F. (1988) Graphical display of estimates having differing standard errors. *Technometrics* **30**(3).

Galbraith, R. F. (1990) The radial plot: graphical assessment of spread in ages. *Nuclear Tracks* **17**, 207–214.

Galbraith, R. F. (2005) *Statistics for Fission Track Analysis.* CRC Press.

Galbraith, R. F. and Green, P. F. (1990) Estimating the component ages in a finite mixture. *Nuclear Tracks and Radiation Measurements* **17**, 196–206.

Galbraith, R. F. and Laslett., G. M. (1993) Statistical models for mixed fission track ages. *Nuclear Tracks and Radiation Measurements* **21**(4), 459–470.

Gallagher, K., Brown, R., and Johnson, C. (1998) Fission track analysis and its applications to geological problems. *Annual Reviews in Earth and Planetary Science* **26**, 519–72.

Garver, J. I., Brandon, M. T., Roden-Tice, M., and Kamp, P. J. (1999) Exhumation history of orogenic highlands determined by detrital fission-track thermochronology. *Geological Society, London, Special Publications* **154**(1), 283–304.

Garver, J. I., Reiners, P. W., Walker, L. J., Ramage, J. M., and Perry, S. E. (2005) Implications for timing of Andean uplift from thermal resetting of radiation-damaged zircon in the Cordillera Huayhuash, Northern Peru. *The Journal of Geology* **113**(2), 117–138.

Gleadow, A. J. W., Duddy, I. R., Green, P. F., and Lovering, J. F. (1986) Confined fission track lengths in apatite: a diagnostic tool for thermal history analysis. *Contributions to Mineralogy and Petrology* **94**(4), 405–415.

Gleadow, A. J., Gleadow, S. J., Belton, D. X., Kohn, B. P., Krochmal, M. S., and Brown, R. W. (2009) Coincidence mapping-a key strategy for the automatic counting of fission tracks in natural minerals. *Geological Society, London, Special Publications* **324** (1), 25–36.

Gleadow, A. J. W. and Duddy. I. R. (1981) A natural long-term track annealing experiment for apatite. *Nuclear Tracks* **5**(1–2), 169–174.

Gögen, K. and Wagner. G. A. (2000) Alpha-recoil track dating of Quaternary volcanics. *Chemical Geology* **166**(1), 127–137.

Gombosi, D. J., Garver, J. I., and Baldwin, S. L. (2014) On the development of electron microprobe zircon fission-track geochronology. *Chemical Geology* **363**, 312–321.

Green, P. F. (1988) The relationship between track shortening and fission track age reduction in apatite: combined influences of inherent instability, annealing anisotropy, length bias and system calibration. *Earth and Planetary Science Letters* **89**(3–4), 335–352.

Green, P. F., Duddy, I. R., Gleadow, A. J. W., Tingate, P. R., and Laslett, G. M. (1985) Fission-track annealing in apatite: track length measurements and the form of the Arrhenius plot. *Nuclear Tracks and Radiation Measurements (1982)* **10**(3), 323–328.

Green, P. F., Duddy, I. R., Gleadow, A. J. W., Tingate, P. R., and Laslett, G. M. (1986) Thermal annealing of fission tracks in apatite: 1. *A qualitative description. Chemical Geology: Isotope Geoscience Section* **59**, 237–253.

Green, P. F., Duddy, I. R., Laslett, G. M., Hegarty, K. A., Gleadow, A. W., and Lovering, J. F. (1989a) Thermal annealing of fission tracks in apatite 4. Quantitative modelling techniques and extension to geological timescales. *Chemical Geology: Isotope Geoscience Section* **79**(2), 155–182.

Green, P. F., Duddy, I.R., Gleadow, A. J. W., and Lovering, J. F. (1989b) Apatite fission track analysis as palaeotemperature indicator for hydrocarbon exploration. In *Thermal History of Sedimentary Basins*, Naeser, N. D. and McCulloh, T. H. (eds), 181–95. Springer-Verlag. New York.

Grün, R. (1989) Electron spin resonance (ESR) dating. *Quaternary International* **1**, 65–109.

Grün, R. and Invernati, C. (1985) Uranium accumulation in teeth and its effect on ESR dating—a detailed study of a mammoth tooth. *Nuclear Tracks and Radiation Measurements (1982)* **10** (4), 869–877.

Guenthner, W. R., Reiners, P. W., Ketcham, R. A., Nasdala, L., and Giester, G. (2013) Helium diffusion in natural zircon: radiation damage, anisotropy, and the interpretation of zircon (U–Th)/He thermochronology. *American Journal of Science* **313**(3), 145–198.

Guérin, G., Mercier, N., and Adamiec, G. (2011) Dose-rate conversion factors: update. *Ancient TL* **29**(1), 5–8.

Guérin, G., Mercier, N., Nathan, R., Adamiec, G., and Lefrais, Y. (2012) On the use of the infinite matrix assumption and associated concepts: a critical review. *Radiation Measurements* **47**(9), 778–785.

Guralnik, B., Jain, M., Herman, F., *et al.* (2013) Effective closure temperature in leaky and/or saturating thermochronometers. *Earth and Planetary Science Letters* **384**, 209–218.

Guralnik, B., Jain, M., Herman, F., *et al.* (2015) OSL-thermochronometry of feldspar from the KTB borehole, Germany. *Earth and Planetary Science Letters* **423**, 232–243.

Harrison, T. M., Armstrong, R. L., Naeser, C. W., and Harakal, J. E. (1979) Geochronology and thermal history of the Coast Plutonic Complex, near Prince Rupert, British Columbia. *Canadian Journal of Earth Sciences* **16**, 400–410

Hasebe, N., Barbarand, J., Jarvis, K., Carter, A., and Hurford, A. J. (2004) Apatite fission-track chronometry using laser ablation ICP-MS. *Chemical Geology* **207**(3), 135–145.

Hasebe, N., Tamura, A., and Arai, S. (2013) Zeta equivalent fission-track dating using LA-ICP-MS and examples with simultaneous U–Pb dating. *Island Arc* **22**(3), 280–291.

Hashimoto, T., Hirokazu Sugiyama, H., and Sotobayashi, T. (1980) Alpha-recoil track formation on muscovite and measurement of recoil-range using ^{252}Cf-sources. *Nuclear Tracks* **4**(4), 263–269.

Hashemi-Nezhad, S. R. and Durrani. S. A. (1981) Registration of alpha-recoil tracks in mica: the prospects for alpha-recoil dating method. *Nuclear Tracks* **5**(1), 189–205.

Hashemi-Nezhad, S. R. and Durrani. S. A. (1983) Annealing behaviour of alpha-recoil tracks in biotite mica: implications for alpha-recoil dating method. *Nuclear Tracks and Radiation Measurements (1982)* **7**(3), 141–146.

Henderson, W. J. (1934) The upper limits of the continuous β-ray spectra of thorium C and C″. *Proceedings of the Royal Society of London. Series A, Mathematical and Physical Sciences* **147** (862), 572–582.

Henderson, G. H. (1939) A quantitative study of pleochroic haloes. V. The genesis of haloes. *Proceedings of the Royal Society of London. Series A, Mathematical and Physical Sciences* **173**(953), 250–264.

Henderson, G. H. and Bateson. S. (1934) A quantitative study of pleochroic haloes. I. *Proceedings of the Royal Society of London. Series A, Containing Papers of a Mathematical and Physical Character* **145**(855), 563–581.

Henderson, G. H. and Sparks, F. W. (1939) A quantitative study of pleochroic haloes. IV. New types of haloes. *Proceedings of the Royal Society of London. Series A, Mathematical and Physical Sciences* **173**(953), 238–249.

Henderson, G. H. and Turnbull, L. G. (1934) A quantitative study of pleochroic haloes. II. *Proceedings of the Royal Society of London. Series A, Containing Papers of a Mathematical and Physical Character* **145**(855), 582–591.

Herman, F., Rhodes, E. J., Jean Braun, J., and Heiniger, L. (2010) Uniform erosion rates and relief amplitude during glacial cycles in the Southern Alps of New Zealand, as revealed from OSL-thermochronology. *Earth and Planetary Science Letters* **297**(1), 183–189.

Holland, H. D. (1954) Radiation damage and its use in age determination. In *Nuclear Geology*, Faul, H. (ed.), 175–179. Wiley, New York.

Holland, H. D. and Gottfried, D. (1955) The effect of nuclear radiation on the structure of zircon. *Acta Crystallographica* **8**(6), 291–300.

Huang, W. H. and Walker, R. M. (1967) Fossil alpha-particle recoil tracks: a new method of age determination. *Science* **155**(3766), 1103–1106.

Huntley, D. J., Godfrey-Smith, D. I., and Thewalt, M. L. W. (1985) Optical dating of sediments. *Nature* **313**, 105–107.

Hurford, A. J. (1986) Cooling and uplift patterns in the Lepontine Alps South Central Switzerland and an age of vertical movement on the Insubric fault line. *Contributions to Mineralogy and Petrology* **92**(4), 413–427.

Hurley, P. M. and Fairbairn, H. W. (1953) Radiation damage in zircon: a possible age method. *Geological Society of America Bulletin* **64**(6), 659–673.

Hurley, P. M., Larsen, E. S., and Gottfried, D. (1956) Comparison of radiogenic helium and lead in zircon. *Geochimica et Cosmochimica Acta* **9**(1–2), 98–102.

Ikeda, S. and Ikeya, M. (1992) Electron spin resonance (ESR) signals in natural and synthetic gypsum: an application of ESR to the age estimation of gypsum precipitates from the San Andreas Fault. *Japanese Journal of Applied Physics* **31**(2A), L136.

Ikeya, M. (1975) Dating a stalactite by electron paramagnetic resonance. *Nature* **255**, 48–50.

Ikeya, M. (1982) A model of linear uranium accumulation for ESR age of Heidelberg (Mauer) and Tautavel bones. *Japanese Journal of Applied Physics* **21**(11A), L690.

Ikeya, M., Miki, T., and Tanaka, K. (1982) Dating of a fault by electron spin resonance on intrafault materials. *Science* **215**(4538), 1392–1393.

Joly, J. (1907) XXIX. Pleochroic halos. *The London, Edinburgh, and Dublin Philosophical Magazine and Journal of Science* **13**(75), 381–383.

Joly, J. and Rutherford, E. (1913) LXIII. The age of pleochroic haloes. *The London, Edinburgh, and Dublin Philosophical Magazine and Journal of Science* **25**(148), 644–657.

Jonckheere, R. and Gögen, K. (2001) A Monte-Carlo calculation of the size distribution of latent alpha-recoil tracks. *Nuclear Instruments and Methods in Physics Research Section B: Beam Interactions with Materials and Atoms* **183**(3), 347–357.

Katcoff, S. (1969) Alpha-recoil tracks in mica: registration efficiency. *Science* **166**(3903), 382–384.

Ketcham, R. A. (2005) Forward and inverse modeling of low-temperature thermochronometry data. *Reviews in Mineralogy and Geochemistry* **58**(1), 275–314.

Ketcham, R. A., Donelick, R. A., and Carlson, W. D. (1999) Variability of apatite fission-track annealing kinetics: III. Extrapolation to geological time scales. *American Mineralogist* **84**, 1235–1255.

Ketcham, R. A., Carter, A., Donelick, R. A., Barbarand, J., and Hurford, A. J. (2007) Improved modeling of fission-track annealing in apatite. *American Mineralogist* **92**(5–6), 799–810.

Ketcham, R. A., Gautheron, C., and Tassan-Got, L. (2011) Accounting for long alpha-particle stopping distances in (U–Th–Sm)/He geochronology: refinement of the baseline case. *Geochimica et Cosmochimica acta* **75**(24), 7779–7791.

Ketcham, R. A., Guenthner, W. R., and Reiners, P. W. (2013) Geometric analysis of radiation damage connectivity in zircon, and its implications for helium diffusion. *American Mineralogist* **98**(2–3), 350–360.

King, G. E., Guralnik, B., Valla, P. G., and Herman, F. (2016) Trapped-charge thermochronometry and thermometry: A status review. *Chemical Geology* **446**, 3–17.

Laney, R. and Laughlin, A. W. (1981) Natural annealing of pleochroic haloes in biotite samples from deep drill holes, Fenton Hill, New Mexico. *Geophysical Research Letters* **8**(5), 501–504.

Laslett, G. M., Green, P. F., Duddy, I. R., and Gleadow, A. J. W. (1987) Thermal annealing of fission tracks in apatite 2. A quantitative analysis. *Chemical Geology: Isotope Geoscience Section* **65**(1), 1–13.

Lee, H. K. and Schwarcz, H. P. (1994) ESR plateau dating of fault gouge. *Quaternary Science Reviews* **13**(5), 629–634.

Li, W., Wang, L., Lang, M., Trautmann, C., and Rodney C. Ewing, R. C. (2011) Thermal annealing mechanisms of latent fission tracks: Apatite vs. zircon. *Earth and Planetary Science Letters* **302**(1), 227–235.

Li, W., Lang, M., Gleadow, A. J. W., Zdorovets, M. V., and Ewing, R. C. (2012) Thermal annealing of unetched fission tracks in apatite. *Earth and Planetary Science Letters* **321**, 121–127.

Miki, T. and Ikeya, M. (1982) Physical basis of fault dating with ESR. *Naturwissenschaften* **69**(8), 390–391.

Moazed, C., Overbey, R., and Spector, R. M. (1975) Precise determination of critical features in radiohalo-type coloration of biotite. *Nature* **258**, 315–317.

Murray, A. S. and Wintle, A. G. (2000) Luminescence dating of quartz using an improved single-aliquot regenerative-dose protocol. *Radiation measurements* **32**(1), 57–73.

Murray, A. S. and Wintle, A. G. (2003) The single aliquot regenerative dose protocol: potential for improvements in reliability. *Radiation Measurements* **37**(4), 377–381.

Naeser, C. W. (1967) The use of apatite and sphene for fission track age determinations. *Geological Society of America Bulletin* **78**(12), 1523–1526.

Nambi, K. S. V. and Aitken, M. J. (1986) Annual dose conversion factors for TL and ESR dating. *Archaeometry* **28**(2), 202–205.

Nasdala, L., Wenzel, M., Vavra, G., Irmer, G., Wenzel, T., and Kober, B. (2001) Metamictisation of natural zircon: accumulation versus thermal annealing of radioactivity-induced damage. *Contributions to Mineralogy and Petrology* **141**(2), 125–144.

Nasdala, L., Lengauer, C. L., Hanchar, J. M., *et al.* (2002) Annealing radiation damage and the recovery of cathodoluminescence. *Chemical Geology* **191**(1), 121–140.

Nasdala, L., Götze, J. E. N. S., Hanchar, J. M., Gaft, M., and Krbetschek. M. R. (2004) Luminescence techniques in earth sciences. *Spectroscopic methods in mineralogy* **6**, 43–91.

Nasdala, L., Wildner, M., Wirth, R., Groschopf, N., Pal, D.C., and Moller, A. (2006) Alpha particle haloes in chlorite and cordierite. *Mineralogy and Petrology* **86**, 1–27.

Nasdala, L., Hanchar, J. M., Kronz, A., and Whitehouse, M. J. (2005) Long-term stability of alpha particle damage in natural zircon. *Chemical Geology* **220**(1), 83–103.

Olley, J. M., Caitcheon, G. G., and Roberts, R. G. (1999) The origin of dose distributions in fluvial sediments, and the prospect of dating single grains from fluvial deposits using optically stimulated luminescence. *Radiation Measurements* **30**(2), 207–217.

Owen, M. R. (1988) Radiation-damage halos in quartz. *Geology* **16**(6), 529–532

Paul, T. A. (1993) Transmission electron microscopy investigation of unetched fission tracks in fluorapatite—physical process of annealing. *Nuclear Tracks and Radiation Measurements* **21**(4), 507–511.

Paul, T. A. and Fitzgerald, P. G. (1992) Transmission electron microscopic investigation of fission tracks in fluorapatite. *American Mineralogist* **77**(3-4), 336–344.

Pidgeon, R. T. (2014) Zircon radiation damage ages. *Chemical Geology* **367**, 13–22.

Price, P. B. and Walker, R. M. (1962a) Electron microscope observation of etched tracks from spallation recoils in mica. *Physical Review Letters* **8**(5), 217.

Price, P. B. and R. M. Walker. (1962b) Observation of fossil particle tracks in natural micas. *Nature* **196**, 732–734.

Price, P. B. and Walker. R. M. (1962c) Chemical etching of charged-particle tracks in solids. *Journal of Applied Physics* **33**(12), 3407–3412.

Price, P. B. and Walker, R. M. (1963a) Fossil tracks of charged particles in mica and the age of minerals. *Journal of Geophysical Research* **68**(16), 4847–4862.

Price, P. B. and Walker, R. M. (1963b) A simple method of measuring low uranium concentrations in natural crystals. *Applied Physics Letters* **2**, 23–25.

Rahn, M. K., Brandon, M. T. G., Batt, E., and Garver, J. I. (2004) A zero-damage model for fission-track annealing in zircon. *American Mineralogist* **89**(4), 473–484.

Rhodes, E. J. (2011) Optically stimulated luminescence dating of sediments over the past 200,000 years. *Annual Review of Earth and Planetary Sciences* **39**, 461–488.

Rink, W. J. (1997) Electron spin resonance (ESR) dating and ESR applications in Quaternary science and archaeometry. *Radiation Measurements* **27**(5), 975–1025.

Seitz, F. (1949) On the disordering of solids by action of fast massive particles. *Discussions of the Faraday Society* **5**, 271–282.

Soares, C. J., Guedes, S., Hadler, J. C., Mertz-Kraus, R., Zack, T., and Iunes, P. J. (2014) Novel calibration for LA-ICP-MS-based fission-track thermochronology. *Physics and Chemistry of Minerals* **41**(1), 65–73.

Stockli, D. F. (2005) Application of low-temperature thermochronometry to extensional tectonic settings. *Reviews in Mineralogy and Geochemistry* **58**(1), 411–448.

Stübner, K., Jonckheere, R. C., and Ratschbacher, L. (2005) Alpha-recoil track densities in mica and radiometric age determination. *Radiation Measurements* **40**(2), 503–508.

Tagami, T. (2005) Zircon fission-track thermochronology and applications to fault studies. *Reviews in Mineralogy and Geochemistry* **58**(1), 95–122.

Tagami, T. and O'Sullivan, P. B. (2005) Fundamentals of fission-track thermochronology. *Reviews in Mineralogy and Geochemistry* **58**(1), 19–47.

Tagami, T., Galbraith, R. F., Yamada, R., and Laslett, G. M. (1998) Revised annealing kinetics of fission tracks in zircon and geological implications. In *Advances in Fission-track Geochronology*, 99–112. Springer.

Tagami, T., Ito, H., and Nishimura, S. (1990) Thermal annealing characteristics of spontaneous fission tracks in zircon: *Chemical Geology: Isotope Geoscience Section*, **80**, 159–169.

Toyoda, S. and Ikeya, M. (1991) Thermal stabilities of paramagnetic defect and impurity centers in quartz: basis for ESR dating of thermal history. *Geochemical Journal* **25**(6), 437–445.

Trachenko, K., Dove, M. T., and Salje, E. K.H. (2002) Structural changes in zircon under α-decay irradiation. *Physical Review B* **65**(18), 180102.

Trachenko, K., Dove, M. T., and Salje, E. K.H. (2003) Large swelling and percolation in irradiated zircon. *Journal of Physics: Condensed Matter* **15**(2), L1.

Utsunomiya, S., Palenik, C. S., Valley, J. W., Cavosie, A. J., Simon A. Wilde, S. A., and Ewing, R. C. (2004) Nanoscale occurrence of Pb in an Archean zircon. *Geochimica et Cosmochimica Acta* **68**(22), 4679–4686.

Valley, J. W., Cavosie, A. J., Ushikubo, T., *et al.* (2014) Hadean age for a post-magma-ocean zircon confirmed by atom-probe tomography. *Nature Geoscience* **7**(3), 219–223.

Valley, J. W., Reinhard, D. A., Cavosie, A. J., *et al.* (2015) Presidential address. Nano- and micro-geochronology in Hadean and Archean zircons by atom-probe tomography and SIMS: New tools for old minerals. *American Mineralogist* **100**(7), 1355–1377.

Voznyak, D. K., Pavlishin, V. I., Bugaenko, V. N., and Galaburda, Y. A. (1996) Nature, origin and geochronology of radiogenic haloes in minerals of the Polokhivske deposit (Ukrainian Shield). *Mineralogicheskii Zhurnal* **18**(5), 3–17.

Wagner, G. and Van den Haute, P. (2012) Fission-track *Dating*, Vol. **6**. Springer Science and Business Media.

Wintle, A. G. (2008) Luminescence dating: where it has been and where it is going. *Boreas* **37**(4), 471–482.

Yamada, R., Tagami, T., Nishimura, S., and Ito, H. (1995) Annealing kinetics of fission tracks in zircon: an experimental study. *Chemical Geology* **122**(1), 249–258.

Yuan, W., Ketcham, R. A., Gao, S., Dong, J., Bao, Z., and Deng, J. (2009). Annealing behavior of alpha recoil tracks in phlogopite. *Chemical Geology* **266**(3), 343–349.

Zaun, P. E. and Wagner, G. A. (1985) Fission-track stability in zircons under geological conditions. *Nuclear Tracks and Radiation Measurements (1982)* **10**(3), 303–307.

Ziegler, J. F., Biersack J. P., and Ziegler, M. D. (2008) *SRIM The Stopping and Range of Ions in Matter*. SRIM Co., Chester, MD.

CHAPTER 11

The (U–Th)/He system

11.1 INTRODUCTION

Among radioisotopic systems in which daughter products are nuclides (as opposed to radiation damage features) no other system is characterized by parent and daughter elements with more different chemical behaviors than (U–Th)/He. This provides unique opportunities for geochronologic understanding of processes not accessible by other techniques, but also poses analytical and interpretational challenges. (U–Th)/He dating (often called He dating) saw a resurgence of development and use in recent years, but this is partly due to the fact that it was all but abandoned for several decades after being used as the first radioisotopic system to derive a geochronologic date [*Rutherford*, 1905, 1906]. This is testament to some of its idiosyncrasies, but also to the fact that appreciation for the kinds of understanding it can provide in Earth and planetary science was delayed until the basics of thermochronology, rather than geochronology, had been established for other systems. Modern (U–Th)/He dating is most often used for low-temperature thermochronology in ways similar to fission-track dating, typically for understanding thermal histories associated with burial and exhumation from depths in the uppermost few kilometers of the Earth's crust. But the system is also applied to a wide variety of minerals for a wide range of purposes, and although it is rarely a high-precision technique, its versatility and distinctive characteristics provide unique opportunities for innovative applications and complementary geologic constraints.

11.2 HISTORY

The first U/He (the contribution of Th was not yet recognized) date was also the first radioisotopic date of any sort. Although it was one of his relatively minor contributions, in 1905 Ernest Rutherford kicked off the entire field of quantitative geochronology by publishing an apparent date of a sample of fergusonite (a Nb–Ta oxide). This date was actually based on the relative concentrations of Ra and He, as measured by the contemporary noble gas giants William Ramsay and Morris Travers. Rutherford himself had only recently discovered that He was a decay product of Ra, and that Ra itself was an intermediate daughter of U. Assuming secular equilibrium between the Ra and its parental U, and a rate of He production (which he assumed to be constant even though he was well aware of the exponential nature of radioactive growth and decay), he calculated a date of 40 Ma for the fergusonite. One year later Rutherford revised this estimate to 500 Ma, and added another date calculation for uraninite from Connecticut. This sample had been analyzed by William Hillebrand and found to contain 72% U and about 2.4% of an inert gas Hillebrand interpreted to be N_2. Assuming that this N_2 was actually He, Rutherford calculated a similar age of 500 Ma for this sample as well. Although it would take geologists decades to lose their initial distrust of these new geochronologic methods, these first U–He dates nonetheless provided them with something of a vindication because they generally felt that intervals at least this long were required for development of geologic features on Earth.

One of the most prolific early practitioners of He dating was R. J. Strutt, whose many age determinations reflected the then primary objective of establishing reproducible ages for stratigraphic horizons by any means possible. *Strutt* [1908a,b,c, 1910a,b] analyzed a huge number and variety of materials, including phosphatic nodules, fossil bones and teeth, zircon, iron-ores, halides, titanites, fluorite, beryl, and others. Unfortunately, more often than not the efforts produced inconsistent results when samples from the same stratigraphic horizon but different locations were dated. Given the relative chemical behaviors of the parent and daughter elements involved in the system, and the fact that ages often appeared to be "too-young," the obvious explanation was He loss after formation of the specimen. As Strutt [1910b, p. 388] noted, "[these dates] are minimum values, because helium leaks out from the mineral, to what extent it is impossible to say." The same dour assessments were expressed by Patrick Hurley [1954, p. 302] in the middle of the 20th century, who denounced He dating's geochronological

Geochronology and Thermochronology, First Edition. Peter W. Reiners, Richard W. Carlson, Paul R. Renne, Kari M. Cooper, Darryl E. Granger, Noah M. McLean, and Blair Schoene.
© 2018 John Wiley & Sons Ltd. Published 2018 by John Wiley & Sons Ltd.

potential with "…at best, only minimum results could be obtained." The upshot was that because of the presumably unreliable and flighty nature of the daughter product, He dating was deemed unreliable for most of the twentieth century, a sentiment persisting in textbooks through the mid-1990s. In the early to middle part of the century most geochronologic attention turned towards U/Pb and other methods, which seemed to provide more consistent stratigraphic results, and (U–Th)/He dating received little attention until decades thereafter. This is not to say that radiogenic He was not used for geochronologic purposes during this time. It continued to be explored as a dating technique for zircon [*Damon*, 1957], and to find important applications in the study of groundwater flow, and more specialized and experimental studies of a wide range of materials, including magnetite [*Fanale and Kulp*, 1962], corals [*Bender*, 1973] fossils [*Fanale and Schaeffer*, 1965; *Turekian*, 1970], volcanic rocks [*Leventhal*, 1975; *Ferreira et al.*, 1975], and hematite [e.g., *Lippolt and Weigel*, 1988; *Bähr et al.*, 1993; *Lippolt et al.*, 1993, 1994; *Wernicke and Lippolt*, 1994a]. Applications outside groundwater studies met with varying degrees of success, however, until the late 20th century, when the potential of the method as a thermochronometer was realized.

A breakthrough in He dating came with the realization that, contrary to Stutt's assertion, it is possible, at least in some cases, to estimate the extent of He loss, or more precisely, to derive geologically useful information from He contents that are lower than would be expected in the case of full retention since formation. Specifically, in cases where the system cannot be used as a geochronometer, it may provide valuable thermochronometric information. For the case of (U–Th)/He dating this insight came with *Zeitler et al.*'s [1987] paper showing that He ages of apatite from a well-known near-surface magmatic rock from Durango, Mexico, yielded the known formation age of about 31 Ma. More importantly, they also showed that He loss from the specimen basically followed the principles of thermally activated volume diffusion, opening up the possibility of using the system as a low-temperature thermochronometer. The principles of radioisotopic thermochronology (see Chapter 5) had already been worked out in the preceding several decades [e.g., *Nicolayson*, 1957], largely by the fission-track and K/Ar and Ar/Ar dating communities, so it was simply a matter of transferring the approaches to the new system. In detail other considerations, including the long alpha stopping distances, radiation damage, and other aspects of He dating, complicate this simple technological transfer. But one of the primary hurdles in interpreting (U–Th)/He dates has always been development of a quantitative understanding of the kinetics of He mobility within and loss from minerals.

11.3 THEORY, FUNDAMENTALS, AND SYSTEMATICS

Uranium and thorium are highly refractory lithophilic actinides that are among the heaviest naturally occurring nuclides in the cosmos. Their solar abundances are only about 8 and 29 ppb

for U and Th respectively, but their concentrations in continental crust are about 1–9 ppm. U and Th typically form ions with high charge (usually 4^+ or 6^+) and are relatively immobile in most common rock-forming minerals. In high temperature systems pertinent to magmatic and metamorphic rock formation, U and Th are usually incompatible with respect to the volumetrically dominant minerals and are therefore enriched in silicate liquids produced from differentiation or small degrees of melting. Their concentrations in granitoids, for example, are roughly 2–5 ppm for U and ~15–20 ppm for Th, whereas in mid-ocean ridge basalts they are usually about one to two orders of magnitude less abundant. Uranium and Th are compatible in several geochemically important accessory minerals, including zircon, titanite, apatite, monazite, and others. These are often targeted for (U–Th)/He dating applications.

In near-surface systems at lower temperature and in aqueous solutions the behavior of U and Th is more complex and depends on complexation with other species. In a nutshell, U^{6+} can be highly soluble, whereas U^{4+} and Th are insoluble, but such straightforward simplifications are complicated by solution chemistry, adsorption/desorption, and complexation phenomena.

The only other significant source of radiogenic ^{4}He besides U and Th is ^{147}Sm. As a refractory lithophile lanthanide, Sm has a chemical behavior similar to that of Th, despite the fact that it forms Sm^{3+} in contrast to Th^{4+}. The chemical similarity of Sm and Th means that natural samples with Sm concentrations much higher than that of Th (and U) are rare. Combined with the relatively low alpha productivity of Sm, this explains why the (U–Th)/He system often neglects mention of Sm in the name, despite the fact that Sm is often measured and included in date calculations.

Helium occupies a unique niche in the elemental inventory of the universe as one of three primordial elements to condense into nuclear material within minutes of the Big Bang. It composes nearly a quarter of baryonic matter in the universe (together with H it composes 99%), and it is at least two orders of magnitude more abundant than the next most abundant elements (C and O). Helium's inert nature is the result of its stable electron configuration conferring it with a one full orbital (the simplest noble gas), but its abundance is due to the relatively high nuclear binding energy of its most abundant isotope ^{4}He. The other stable isotope of He is ^{3}He, which has a much lower abundance in most materials, including Earth's atmosphere where the ^{4}He/^{3}He ratio is 1.34×10^6. Helium also has several radioactive isotopes with atomic mass numbers five through eight, and half-lives ranging from 0.8 to 10^{-22} s.

Helium is endowed with almost opposite chemical behaviors of its refractory, highly charged, and often immobile parents. Although the controls on its diffusivity in minerals are complex in detail, to first order it is highly mobile in most natural materials at temperatures far below those affecting the parent isotope contents. Partly because of this mobility, but also because of variable production rates in Earth reservoirs, He is highly variable in concentration among reservoirs. Compared to its solar concentration of about 25%, it is highly depleted in the atmosphere at 5.24 ppm,

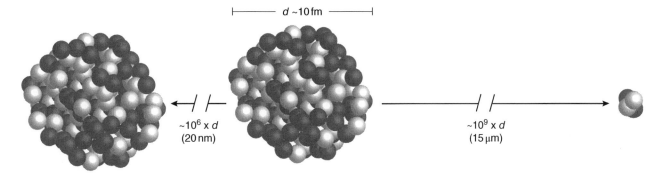

Fig. 11.1. Schematic depiction of the relative length scales of recoil involved in alpha decay. The nucleus of the parent nuclide is approximately 10 fm in diameter. The heavy daughter nucleus recoils about one million times this distance, while the alpha particle (^{4}He nucleus) recoils about a billion times this distance in the opposite direction.

due to gravitational escape. Its abundance in the Earth varies from ~0.1 ppb in the depleted mantle to as much as 8% in natural gas reservoirs.

Although most He in the universe is primordial, it is also produced by stellar fusion in stars, cosmogenic and nucleogenic reactions, and by alpha decay. Several important reactions produce ^{3}He alone, including beta decay of ^{3}H and spallation on a wide variety of targets, but ^{4}He is by far the dominant product of nuclear reactions that produce He. The most productive of all these is alpha decay, or the spontaneous emission of a ^{4}He nucleus from a radioactive parent. Several heavy isotopes including ^{147}Sm and ^{190}Pt decay in this way, but most radiogenic ^{4}He is produced by ^{238}U, ^{235}U, ^{232}Th, and their intermediate daughter products in the decay chain leading to stable ^{206}Pb, ^{207}Pb, and ^{208}Pb, respectively (see Chapter 12). Roughly 99% of terrestrial He comes from these reactions. Neglecting all the intermediate daughter products, they can be written as:

$$^{238}\text{U} \rightarrow {}^{206}\text{Pb} + 8 \cdot {}^4\text{He} + 6 \cdot \beta^- + Q \quad \lambda = 1.55125 \times 10^{-10} \text{year}^{-1}$$
(11.1)

$$^{235}\text{U} \rightarrow {}^{207}\text{Pb} + 7 \cdot {}^4\text{He} + 4 \cdot \beta^- + Q \quad \lambda = 9.84850 \times 10^{-10} \text{year}^{-1}$$
(11.2)

$$^{232}\text{Th} \rightarrow {}^{206}\text{Pb} + 6 \cdot {}^4\text{He} + 4 \cdot \beta^- + Q \quad \lambda = 4.9475 \times 10^{-11} \text{year}^{-1}$$
(11.3)

The Q-values (total energy released) for each decay chain can be estimated by the mass difference of the daughter and parent nuclides and its energy equivalent. For example, the 0.0555 u (9.22×10^{-29} kg) mass difference between ^{238}U and its ultimate daughters ^{206}Pb and eight ^{4}He nuclides leads to a total Q-value of 51.7 MeV. Each of the eight alpha decays have energies ranging from 4 to 8 MeV (from a minimum of 4.270 MeV for ^{238}U to a maximum of 8.995 MeV for ^{212}Po), adding up to a total of 43.7 MeV. The difference between this and the total Q-value is carried by the six beta decays in the series, with energies ranging from 0.3 to 3.3 MeV.

The kinetic energy in each alpha decay (Fig. 11.1) is apportioned between the alpha particle and the heavy daughter nuclide, and conservation of momentum dictates that they move apart from one another with kinetic energies proportional to the ratio of their masses. For example, the kinetic energy of the alpha particle from decay of ^{238}U is about $234/(4 + 234)$ of the total, and the energy imparted to the recoiling heavy daughter is only a few percent of this. This means that the alpha particle travels a relatively long distance, typically tens of micrometers, whereas the heavy daughter only recoils a few tens of nanometers. Despite the smaller recoil distance of the larger nuclide, its large size and highly ionized nature causes much more lattice damage than the alpha particle, displacing approximately 3000 atoms in a billiard-like cascade, compared with only 100 or so at the end of the alpha-particle track.

Another source of ^{4}He is the ^{147}Sm to ^{143}Nd system, which is an important chronometer in its own right, even if its contribution to ^{4}He inventories of minerals is typically minor.

$$^{147}\text{Sm} \rightarrow {}^{143}\text{Nd} + {}^4\text{He} + Q \quad \lambda = 6.54 \times 10^{-12} \text{year}^{-1}$$
(11.4)

The Q-value for this decay is only 2.3 MeV, compared with the 4–7 MeV for the U-series and Th-series isotopes, and the recoil distances of the alpha particle and ^{143}Nd nuclides are correspondingly shorter.

The equation for ingrowth of radiogenic ^{4}He, including ^{147}Sm, is written

$$^4\text{He} = 8 \cdot {}^{238}\text{U}\left(e^{\lambda 238 t} - 1\right) + 7 \cdot {}^{235}\text{U}\left(e^{\lambda 235 t} - 1\right) + 6 \cdot {}^{232}\text{Th}\left(e^{\lambda 232 t} - 1\right) + {}^{147}\text{Sm}\left(e^{\lambda 147 t} - 1\right)$$
(11.5)

where nuclides are expressed in moles or atoms, the λs are the decay constants of the subscripted nuclides, and t is time.

More often than not, the central challenges and rewards for (U–Th)/He dating are that t in this equation may represent something other than a simple formation age. In most cases the accumulation and loss of He, and therefore the apparent t, are influenced by other factors, for example temperature, or more precisely thermal history, although for at least some minerals radiation damage also plays a significant role. The significance of the date can be interpreted only within this context, and it is this context that provides the power of the technique.

11.4 ANALYSIS

Because of the large chemical behavior differences between parents and daughter products in the (U–Th)/He system, most analytical methods for "conventional" (U–Th)/He dating employ a two-stage procedure involving initial He measurement, followed by U–Th (and Sm) measurement. In most cases analyses are made on the same aliquot so as to avoid complications arising from non-uniform parent nuclide concentrations in the sample. At present, *in situ* methods [*Boyce et al.*, 2006; *Vermeesch et al.*, 2012; *Tripathy-Lang et al.*, 2013] require separate parent and daughter aliquots.

11.4.1 "Conventional" analyses

(U–Th)/He dating of one sort or another has been around since the beginning of the 20th century, so it has been applied to a wide range of materials using a wide range of analytical approaches. Helium has been extracted by fusion, resistance furnace and laser heating, and measured by quadrupole and sector gas-source mass spectrometry, not to mention the pre-mass-spectrometric techniques like optical-emission spectrometry [e.g., *Strutt*, 1908a]. Uranium and Th have been measured using intermediate daughter isotopes as proxies, alpha counting, radiography, thermal ionization mass spectrometry (TIMS), and inductively coupled plasma mass spectrometry (ICP-MS).

A diversity of analytical approaches for this technique exists today, and new ones are currently in development that have the potential to increase sample throughput, spatial resolution, and potentially improve analytical and date uncertainty. For commonly dated minerals such as apatite and zircon, however, at least since the mid-1990s, most analyses have followed fairly standardized approaches described in *Wolf et al.* [1996, 1997] and *House et al.* [2000]. The following descriptions focus largely on standard methods used for typical apatite and zircon analyses. Other approaches are adapted for different mineral types. Routine approaches for apatite and zircon can be divided roughly into four parts: (i) grain selection and characterization, (ii) ^{4}He extraction, purification, and measurement, (iii) U–Th (and sometimes Sm) measurement, and (iv) correction for alpha ejection, if necessary.

11.4.1.1 Grain selection

Although some applications call for aliquots comprising multiple crystals, especially those for very young or low-U and –Th materials, most procedures for dating apatite and zircon use single-crystal aliquots. Grains are typically hand-picked with fine-tipped tweezers from mineral concentrates, often in liquids to reduce refractive index contrast and allow better internal inspection, and selected on the basis of morphology, size, optical clarity, and, ideally, compositional uniformity. Often euhedral crystals or grains with morphologies that can be geometrically approximated as one of several typical morphologies, such as hexagonal or orthorhombic prisms with pryamidal terminations, are favored, because these simplify alpha-ejection corrections. Highly irregular or rounded grains can also be dated if such corrections are not needed or can be applied to prolate spheroids.

Especially in the case of apatite, larger grains are generally favored over smaller ones because this provides higher analyte contents that often, but not always, lead to better analytical precision and minimize consequences of inaccurate alpha-ejection corrections. Selection of grains based on these criteria is usually accomplished with a high-powered (~120–160×) stereo-zoom microscope and/or scanning-electron microsope (SEM), often with the aid of digital photographic documentation and measurement (e.g., Fig. 11.2). Finally, grains may be selected on the basis of compositional uniformity, particularly with respect to U and Th, to minimize complications related to alpha-ejection corrections, perturbations to He diffusion profiles, and potential intragrain He diffusivity variations arising from radiation damage [*Farley et al.*, 2011]. Grain selection on this basis may be accomplished using electron microbeam or cathodoluminescence (CL) imaging, or laser ablation (LA) ICP-MS, including on grain interiors [e.g., *Hourigan et al.*, 2005].

11.4.1.2 He measurement

In most cases conventional (U–Th)/He dating does not require measurements of concentrations of parent or daughter nuclides. Instead, ages are generally calculated by molar ratios of total nuclides in an aliquot. This obviates the need for weighing very small samples such as single crystal aliquots. Helium is typically extracted from aliquots using furnace or laser heating. Laser heating generally has advantages of lower He blanks and higher sample throughput due to shorter temperature ramp times. This approach typically uses relatively long-wavelength lasers such as those with diode (~800–900 nm), fundamental Nd:YAG (1064 nm), or CO_2 (10.6 μm) sources, that efficiently couple with, but do not photodissociate or ablate, the metal foils used to contain the minerals. Refractory and malleable metals such as Pt or Nb are typically used as "microfurnaces" to contain

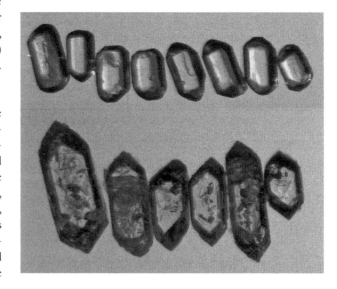

Fig. 11.2. Apatite (top) and zircon (bottom) crystals with sizes and morphologies suitable for conventional (U–Th)/He dating procedures. These crystals have c-axis-perpendicular widths of 75–125 μm.

analyzed minerals, because direct laser heating of some minerals causes parent nuclide volatilization.

Crystal-bearing foil tubes (and also empty ones to check ^{4}He blanks) are individually heated under a laser beam to about 900–1300 °C for several minutes in an ultrahigh vacuum chamber typically pumped to < 10^{-8} Torr using turbomolecular and/or ion pumps. The UHV line itself is typically made of polished stainless steel rather than glass to minimize He transfer through the walls, and to minimize internal surface area. Likewise, valves to control gas flow through the line typically have Cu or other metal seats, to prevent cross-seat He transfer. After release from the sample, the gas can either be spiked with ^{3}He for measurement by isotope dilution, or it can proceed to the purification step directly if it is to be peak-height or ion-counting quantified using an external ^{4}He standard.

In most laboratories, the released gas (plus or minus the ^{3}He spike) is purified to some extent through the use of heated alloy-metal getters, or a temperature-cycled cryogenic cold-head in the UHV line [*Lott and Jenkins*, 1984]. The main purpose of this step is to remove some fraction of the hydrogen and other reactive species, to improve ionization and resolution of the ^{3}He, and in some cases ^{4}He, peaks. Once purified, the gas is released into the ionizing source of a gas-source mass spectrometer. Both sector and quadrupole spectrometers are routinely used for this step. Sector instruments generally have the advantages of resolving ^{3}He from HD and H_3^+ (if isotope dilution by ^{3}He is involved), as well as higher ^{4}He sensitivity (counts per second or volts per mole) than quadrupoles, but in most cases the lower limit to ^{4}He sample size is not sensitivity but ^{4}He blank, which may be as low as 0.01–0.1 fmol. In quadrupole systems using ^{3}He isotope dilution, the intensity of the mass 3 peak is corrected for isobaric interferences from HD and H_3^+ based on correlations between mass 1 (inferred to be H^+) and mass 3 intensities when no He is present in the system. Because of the large mass fractionation characteristic of gas-source spectrometers, measured ^{4}He/^{3}He ratios cannot generally be directly converted to ^{4}He contents simply using an estimate of the spike's ^{3}He content and ^{4}He/^{3}He ratio. A common way to quantify ^{4}He, therefore, is to first measure ^{4}He/^{3}He on a ^{3}He-spiked reference sample of ^{4}He representing a known amount ^{4}He. Assuming a linear response between He moles and ratios across some range of confidence, the molar ^{4}He content of an unknown can then be calculated as the ratio of the ^{4}He/^{3}He measurements of the sample and spiked ^{4}He reference standard, multiplied by the moles of ^{4}He delivered in the reference standard. Typically this reference standard is analyzed, using the same procedures as unknown samples, several times during an analytical session, to account for changes in isotopic fractionation or sensitivity bias with a variety of causes.

As in many noble gas applications that do not rely solely on isotopic ratios, quantification of molar quantities of He requires calibration of a molar ^{4}He reference standard. This in turn requires establishing a system for delivering reproducible and predictable amounts of ^{4}He. In most laboratories, ^{4}He is delivered through a small (0.1–1.0 cm^3) pipette attached to a larger volume tank of ^{4}He. With knowledge of the volumes of the standard tank and pipette, as well as the initial molar ^{4}He content inside the standard volume, the ^{4}He delivery of the reference standard and its depletion as a function of number of shots taken can be calculated. Measurement of these reference volumes is performed by gas expansion and absolute pressure measurement using capacitance manometry, and correcting for thermal transpiration effects [*Setina*, 1999].

11.4.1.3 U–Th–Sm measurement

There are several ways to quantify the molar content of parent nuclides in apatite and zircon following ^{4}He extraction and measurement. The most common, however, is via isotope-dilution using ICP-MS. As outlined in *House et al.* [2000], post-heating, crystal-bearing metal tubes are removed from the UHV line, placed in plastic containers, and spiked with nitric acid solutions of U and Th (and sometimes Sm) isotopes enriched in ^{233}U or ^{235}U and ^{229}Th or ^{230}Th (and sometimes ^{147}Sm or ^{149}Sm). Acid solutions are then added and heat is applied to dissolve the apatite or zircon contents of the tubes and mix the aliquot with the isotopically enriched spike. Apatite can be dissolved by relatively dilute HNO$_3$ at temperatures less than ~100 °C for an hour or so. Zircon and some other minerals require more aggressive techniques, such as serial HF-HNO$_3$ and HCl solutions in high-pressure dissolution vessels at temperatures above 200 °C. Another approach for dissolving zircon is melting with borate flux, followed by less onerous dissolution methods. Isotope ratio measurements made by ICP-MS can be converted to molar contents of U and Th (and Sm) in aliquots using standard isotope dilution equations (see Chapter 3). In some cases Ca and Zr (or other elements) may be quantified by isotope dilution or peak height standardization in order to stoichiometrically estimate the mass of crystal matrix (e.g., apatite or zircon) in each aliquot. This is not necessary for calculating dates, as this is done on a molar basis for parent and daughters. But knowledge of crystal matrix mass allows for calculation of parent and daughter concentrations in aliquots, which can provide complementary constraints bearing on provenance, petrogenesis, or radiation damage effects. Other trace elements are sometimes measured for similar purposes. If this is performed on the same aliquot from which He was extracted, these must be interpreted with the possibility for element volatilization during heating in mind.

Some minerals commonly analyzed for the (U–Th)/He technique, such as apatite, can have Sm concentrations much higher than those of U or Th. However, the much longer half-life and relatively low atomic abundance of ^{147}Sm, as well as the fact that it produces a single ^{4}He compared with the multiple decays in the actinide chains, relegate it to a minor contributor (typically less than 0.5% to a few percent in rare cases, of ^{4}He). Figure 11.3 shows the fractional ^{4}He contribution of ^{147}Sm compared with that of U and Th, with the latter quantified by the index equivalent U, or eU, which combines ^{238}U, ^{235}U, ^{232}Th, and ^{147}Sm in a weighted average in proportion to their alpha productivity. For example, an approximate eU concentration based on U, Th, and Sm elemental concentrations would be eU = U + 0.234 · Th + 0.00463 · Sm.

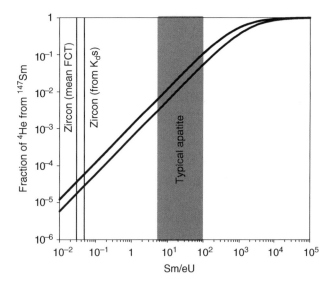

Fig. 11.3. The relative contribution of ^{147}Sm to the ^{4}He inventory, calculated for samples with dates of 0.1 and 4500 Ma (the latter has a lower contribution to total ^{4}He at a given Sm/eU due to higher relative production of U and Th in the past and is represented by the lower bold line), and the typical range of Sm/eU for apatite. Zircon values are for relative zircon/granite partition coefficients of Sm and U–Th, from *Nardi et al.* [2013], and as measured on a suite of Fish Canyon Tuff zircons. The apatite band is from about 4000 apatite analyses from typical granitoid rocks.

11.4.1.4 Dealing with long alpha-stopping distances

One of the most important differences between He dating and most other techniques is the relatively long distance that daughter nuclides travel away from their parents when they form. Because of their high energies and tiny size, alpha particles like those produced from U and Th have stopping distances on the order of millimeters in water and centimeters in air. Alphas produced from decay of U and Th parent isotopes and their intermediate daughters have mean stopping distances from about 5 to 22 µm in apatite and zircon (and most other minerals) [*Farley et al.*, 1996; *Ketcham et al.*, 2011]. The fact that most alpha particles end up on order tens of micrometers away from their birthplace presents several challenges (and perhaps not even a single advantage) for (U–Th)/He dating, as described here.

Alpha ejection corrections

For samples that are internal fragments of much larger domains with reasonably uniform parent nuclide concentrations (e.g., Durango apatite, though see *Boyce et al.* [2006] for documentation of internal parent nuclide heterogeneity), long alpha stopping distances may be ignored. This is because any ^{4}He produced within the aliquot but "ejected" out of it would presumably be balanced by "implantation" of ^{4}He from adjacent regions. However, typical apatite or zircon crystals have U and Th concentrations higher than adjacent minerals in their host rocks. They also have dimensions (~50–200 µm) that are not many times the length scale of alpha stopping distances. Therefore some significant fraction of alpha particles created within them will have been ejected from the crystal, and any age

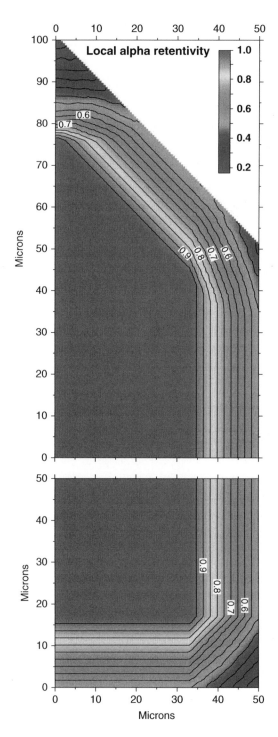

Fig. 11.4. Two-dimensional corner and basal quarter sections (top, parallel to c-axis; bottom, perpendicular to c-axis) of a model zircon crystal, showing contours of modeled fraction of ^{4}He retained within the crystal. (Source: *Hourigan et al.* [2005]. Reproduced with permission of Elsevier.) (*See insert for color representation of the figure.*)

calculated directly from the measured (U–Th)/He will be younger than the "true" formation or cooling age, regardless of whether ^{4}He was lost via diffusion (Fig. 11.4).

In most situations, the most straightforward way to deal with this is to "correct" the measured (U–Th)/He age in a way that effectively adds back in the ^{4}He that has been lost to extragranular ejection. In detail there are differing views on whether this is exactly appropriate in all cases, as discussed below. Because alpha stopping distances are known or can be calculated quite precisely for each nuclide and for each mineral [*Zeigler et al.*, 2008], the appropriate

correction for each nuclide can be calculated as a function of the size and morphology of the dated aliquot. For parent nuclides with multiple intermediate daughter products, a mean stopping distance can be assumed (*Ketcham et al.* [2011]; Table 11.1).

For an idealized spherical crystal, *Farley et al.* (1996) showed that the fraction of ^{4}He with stopping distance S that come to rest within a sphere of radius R (i.e., the fraction of the total produced, hence F_T) will be

$$F_T = 1 - \left(\frac{3}{4}\right)\left(\frac{S}{R}\right) + \left(\frac{1}{16}\right)\left(\frac{S^3}{R^3}\right) \tag{11.6}$$

for large spheres ($R >> S$) this simplifies to

$$F_T = 1 - \left(\frac{S}{4}\right)\beta \tag{11.7}$$

where β is the surface area to volume ratio of the sphere (because for a sphere $\beta = 3/R$). For more complex and realistic morphologies, Monte Carlo models can be used to fit polynomial expressions for F_T values for a given nuclide and a given grain morphology, as a function of β, as shown in Fig. 11.5 (see Box 11.1).

Table 11.1 Mean stopping distances of α particles produced from the U, Th, and Sm decay chains in several minerals used for (U–Th)/He dating

Mineral	Formula	r (g/cm^3)	Mean stopping distance (μm)			
			^{238}U	^{235}U	^{232}Th	^{147}Sm
Apatite	$Ca_5(PO_4)_3F$	3.20	18.81	21.80	22.25	5.93
Zircon	$ZrSiO_4$	4.65	15.55	18.05	18.43	4.76
Titanite	$CaTiSiO_5$	3.53	17.46	20.25	20.68	5.47
Monazite	$CePO_4$	5.26	16.18	18.74	19.11	4.98
Xenotime	YPO_4	4.75	15.20	17.63	17.99	4.68
Rutile	TiO_2	4.25	15.30	17.76	18.14	4.77
Magnetite	Fe_3O_4	5.18	13.97	16.16	16.49	4.51
Hematite	Fe_2O_3	5.26	13.59	15.72	16.04	4.39
Goethite	$FeO(OH)$	4.28	15.54	18.00	18.38	4.95
Barite	$BaSO_4$	4.5	18.14	21.05	21.50	5.54

Source: *Ketcham et al.* [2011]. Reproduced with permission of Elsevier.

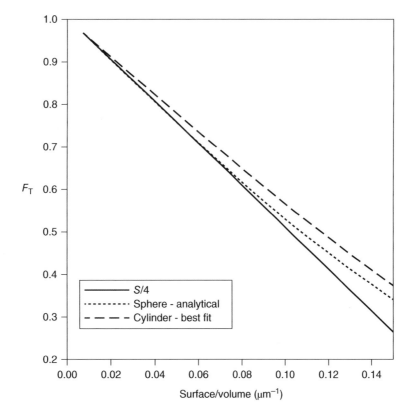

Fig. 11.5. F_T for ^{238}U (based on the average stopping distance of all decays in that chain) in apatite with spherical and cylindrical morphologies, as well as for the simpler sphere case that is a function only of surface-area to volume ratio and S/4 (equation 11.7). For all morphologies, F_T is strongly correlated with surface-area to volume ratio (β). Only for extremely small β do F_T values for different morphologies deviate significantly from one another and from predictions for a simple sphere. (Source: *Farley et al.* [1996]. Reproduced with permission of Elsevier.)

Box 11.1 Correcting for alpha-ejection effects in realistic crystals

Although calculating and correcting for the fraction of ^{4}He ejected from a sphere with a certain surface-area to volume ratio is simple, real crystals are not spheres. Morphologic differences for crystals of typical dimensions cause alpha ejection correction differences that are usually less than 5–10% different from the spherical approximation. These adjustments are not difficult to make and are routine in most (U–Th)/He dating procedures.

The simplest *Farley et al.*'s [1996] polynomial expression for F_T for a given nuclide i is

$$F_T^i = 1 + A_1\beta + A_2\beta^2 \tag{11.B1.1}$$

where A_1 and A_2 are coefficients derived from generalized fits to alpha ejection observed in Monte Carlo models for a grain of a given morphology. Coefficients for apatite [*Farley*, 2002] and zircon [*Hourigan et al.*, 2005] commonly used in this approach are shown in Table 11.B1.1, and surface-area and volume formulae for idealized morphologies commonly used for routine apatite and zircon crystals are given in Table 11.B1.2.

The *Farley* [2002] and *Hourigan et al.* [2005] approaches described in Table 11.B1.2 assume a single set of coefficients for ^{232}Th and ^{235}U, because the mean alpha stopping distances in their decay chains are the same to within 2% in both apatite and zircon. F_T for ^{147}Sm can be approximated by scaling the F_T values for ^{238}U and ^{147}Sm in the Ketcham model (see below) for typical apatite and zircon dimensions:

apatite:

$$F_T^{^{147}Sm} = -0.09158 \cdot \left(F_T^{^{238}U}\right)^2 + 0.46974 \cdot F_T^{^{238}U} + 0.61895 \tag{11.B1.2}$$

zircon:

$$F_T^{^{147}Sm} = -0.00779 \cdot \left(F_T^{^{238}U}\right)^2 + 0.45193 \cdot F_T^{^{238}U} + 0.62526 \tag{11.B1.3}$$

Natural apatite crystals are often cleavage-fractured perpendicular to the c-axis, which may result in loss of the original c-axis-perpendicular terminations. Assuming this fracturing happened after the thermal history recorded in the He content of the grain (e.g., during mineral separation), and that it removed more than one alpha-stopping distance, the surface-area to volume ratio of the grain should be modified to reflect the loss of some of the alpha-ejection-affected surface. *Farley* [2002] showed that the alpha-ejection-related consequences (but not the diffusion-profile consequences) of this can be conveniently determined by multiplying the measured length by 1.5 or 2, for one or two missing tips, respectively, in the equation for the surface-area to volume ratio,

$$\beta = \frac{8}{\sqrt{3}w} + \frac{2}{l} \tag{11.B1.4}$$

Analogously, zircon crystals approximating orthorhombic prisms that have lost their (usually pyramidal) terminations, have surface-area to volume ratios expressed by:

$$\beta_{1m} = \frac{1}{l} + \frac{2}{w_1} + \frac{2}{w_2} \tag{11.B1.5}$$

$$\beta_{2m} = \frac{2}{w_1} + \frac{2}{w_2} \tag{11.B1.6}$$

Where the subscripts 1 m and 2 m refer to the number of missing terminations, l is the c-axis-parallel length, and w_1 and w_2 are the c-axis-perpendicular widths of the prism.

A different but similar adjustment to the surface-area to volume ratio must be made to grains that have lost significant surface area along a face parallel to the c-axis. This may occur naturally or upon mineral separation, but more commonly it arises with grains plucked from grain mounts as might be made for fission-track dating or microanalytical techniques such as ion probe or LA-ICP-MS. For apatite, a modified surface-area to volume ratio can be estimated assuming that polishing was subparallel to the c-axis, and removed more than one alpha-stopping distance and less than half of the entire width of the crystal [*Reiners et al.* 2007]. Under these conditions, the modified surface-area to volume ratio of the polished grain (for an assumed prepolishing cylindrical geometry with pinacoidal terminations and not including the surface area of the polished face) is

$$\beta = \frac{2}{l} + \frac{2r\omega}{r^2\omega + (r-d)\sqrt{2rd - d^2}} \tag{11.B1.7}$$

where d is polishing depth, l is crystal length, r is c-axis-perpendicular half-width (radius), and

$$\omega = \pi - \cos^{-1}\left(1 - \frac{d}{r}\right) \tag{11.B1.8}$$

A similar and simpler approach is also possible for zircon, if the pyramidal terminations are ignored, so that

$$\beta = \frac{2}{l} + \frac{1}{r_1} + \frac{1}{2r_2 - d} \tag{11.B1.9}$$

where r_1 and r_2 are the c-axis-perpendicular half-widths of the crystal parallel and normal to the polishing directions, respectively.

Ketcham et al.'s [2011] more recent and slightly more generalized treatment of alpha-ejection correction casts the general equation for the numerical fits to specific morphologies somewhat differently than *Farley et al.* (1996):

$$F_T = 1 - \left(\frac{3}{4}\right)\left(\frac{R}{R_S}\right) + b_2\left(\frac{R}{R_S}\right)^2 + b_3\left(\frac{R}{R_S}\right)^3 \tag{11.B1.10}$$

where in this case R is the stopping distance and R_s is the radius of a sphere with the equivalent β of the grain of interest. As in the previous cases, the polynomial coefficients are determined from Monte Carlo models for specific morphologies. The *Ketcham et al.* approach has the advantage of casting the formulation as one in which the stopping distance R is explicit in the expression so that it may be refined or adapted to any nuclide, even if the expressions for the combined coefficients and R_s can become a little complicated. For example, for apatite with a generalizeable morphology as shown in Fig. 11.B1.1 (and pyramidal terminations, if present, constrained to be at 45° angles from the c-axis), the F_T equation is

$$F_T = 1 - \left(\frac{3}{4}\right)\left(\frac{R}{R_S}\right) + \left[(0.2093 - 0.465\,N_P)\left(W + \frac{L}{\sqrt{3}}\right) + \left(0.1062 + \frac{0.2234R}{R + 6(W\sqrt{3} - L)}\right)\left(H - N_P\frac{W\sqrt{3}/2 + L}{4}\right)\right]\frac{R^2}{V} \tag{11.B1.11}$$

where N_P is the number of pyramidal terminations, V is the volume of the crystal, and other variables are as shown in Fig. 11.B1.1. For zircon with morphology of an orthorhombic prism with or without pyramidal terminations (constrained to be at 45° angles from the c-axis), F_T can be adequately described as

$$F_T = 1 - \left(\frac{3}{4}\right)\left(\frac{R}{R_S}\right) + \left(0.2095(a + b + c) - \left(0.096 - 0.013\frac{a^2 + b^2}{c^2}\right)(a + b)N_P\right)\frac{R^2}{V} \tag{11.B1.12}$$

where the variables are as in the previous case and shown in Fig. 11.B1.1.

Figure 11.B1.2 shows F_T values for apatite and zircon for a range of crystal sizes, calculated using the *Farley* [2002], *Ketcham et al.* [2011], and *Hourigan et al.* [2005] methods.

A mean F_T for a specific aliquot with specific molar proportions of parent nuclides can be calculated from the weighted F_T values for each nuclide specific to the size and shape of the grain, based on the relative alpha productivities and the concentrations of each element,

$$F_T^m = \Sigma_k\, F_T^i M_k\, A_k^i\, \frac{\lambda_k^i}{\Sigma_k M_k\, A_k^i\, \lambda_k^i} \tag{11.B1.13}$$

where F_T^i, A_k^i, and λ_k^i are the mean F_T, atomic abundance, and decay constant of the nuclide of interest, respectively, and M_k is the number of moles or atoms of the element of the nuclide.

As an approximation, the alpha-ejection correction can be applied by simply dividing the raw (U–Th)/He date or ^{4}He content by F_T^m, but this is not accurate over long timescales, because of the exponential nature of daughter ingrowth. A more accurate approach is to consider each parent nuclide as capable of only producing the fraction of ^{4}He daughter that remains in the crystal, by multiplying each measured parent nuclide content by its nuclide-specific F_T, and recasting the age equation as

$$^4He = 8\cdot{}^{238}U\cdot F_T^{238}\left(e^{\lambda 8t} - 1\right) + 7\cdot{}^{235}U\cdot F_T^{235}\left(e^{\lambda 7t} - 1\right) + 6\cdot{}^{238}Th\cdot F_T^{232}\left(e^{\lambda 2t} - 1\right) + {}^{147}Sm\cdot F_T^{147}\left(e^{\lambda 147t} - 1\right). \tag{11.B1.14}$$

Table 11.B1.1 Factors A_1 and A_2 for calculating fraction of He retained in crystals from the ^{238}U and ^{232}Th decay series in zircon, for different assumed crystal geometries. In these formulations the ^{235}U and ^{232}Th stopping distances are considered similar enough to yield equal polynomial coefficients for F_T calculations

Parent nuclide	Tetragonal prism with pinacoid terminations [*Farley, 2002*]		Hexagonal prism with pinacoid terminations [*Farley* 2002]		Tetragonal prism[a] with pyramidal terminations [*Hourigan et al., 2005*]	
	A_1	A_2	A_1	A_2	A_1	A_2
^{238}U	−4.31	4.92	−5.13	6.78	−4.28	4.37
^{232}Th	−5.00	6.80	−5.90	8.99	−4.87	5.61

[a] Tetragonal prism morphology was used for Monte Carlo modeling but use of an orthorhombic prism with c-axis-perpendicular variation within the range of most zircons produced similar results.

Table 11.B1.2 Surface areas and volumes of idealized morphologies similar to those commonly encountered in routine apatite and zircon (U–Th)/He dating

Geometry	Volume	Surface area
Orthorhombic prism with pyramidal terminations	$V_z = 4r_1r_2\left[(l - h_1 - h_2) + \frac{1}{3}h_1h_2\right]$	$SA_z = 4(l - h_1 - h_2)(r_1 + r_2) + 2r_1a + 2r_2b$ $a = \sqrt{h_1^2 + r_2^2} + \sqrt{h_2^2 + r_2^2}$ $b = \sqrt{h_1^2 + r_1^2} + \sqrt{h_2^2 + r_1^2}$
Prolate Spheroid	$V_{ps} = \frac{2}{3}\pi r^2 l$	$SA_{ps} = 2\pi r^2 + \left[\frac{2\pi r(^1/_2)^2}{\sqrt{(^1/_2)^2 - r^2}}\right]\sin^{-1}\left[\frac{\sqrt{(^1/_2)^2 - r^2}}{(^1/_2)}\right]$
Hexagonal prism	$V_{hp} = \frac{3\sqrt{3}}{8}w^2 l$	$SA_{hp} = 3lw + \frac{3\sqrt{3}}{4}w^2$

l, c-axis-parallel length; h_1, h_2, pyramidal termination lengths; r_1, r_2, mutually perpendicular prism half-widths or average equatorial radius; w, distance between mutually opposed apices parallel to c-axis in hexagonal prism.

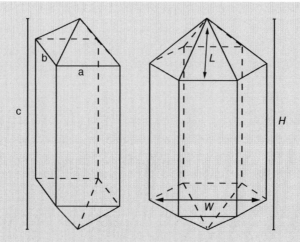

Fig. 11.B1.1. Schematic of idealized crystal morphologies for zircon and apatite, showing dimensional parameters used in example F_T equations (11.B1.11) and (11.B1.12). (Source: *Ketcham et al.* [2011]. Reproduced with permission of Elsevier.)

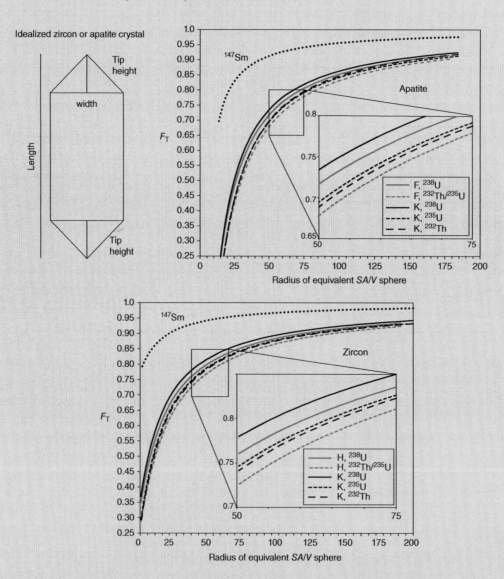

Fig. 11.B1.2. F_T values for apatite and zircon grains with dimensions on an idealized crystal morphology as shown on left side. Trend labels are shown in inset: F, *Farley* [2002]; H, *Hourigan et al.* [2005]; K, *Ketcham et al.* [2011]. These examples show results for grains with $L/W = 2.67$ (near the midpoint of the distribution shown in *Farley et al.* [1996]) for typical apatite grains. The x-axis is R_s, the radius of a sphere with an equivalent surface-area to volume ratio as the grain. In these examples, apatite c-axis-perpendicular width is $R_s/0.5786$; zircon c-axis-perpendicular width is $R_s/0.6319$. Apatite and zircon in the *Ketcham et al.* [2011] model require assumptions of pyramidal tips at 45° from prisms, but can accommodate grains with one or no pyramidal tips. *Farley et al.* [2002] and *Hourigan et al.* [2005] models do not explicitly calculate F_T values for ^{147}Sm, and assume the same F_T values for ^{235}U and ^{232}Th, based on the very similar average stopping distances of these two nuclides.

Complications and considerations related to alpha ejection correction

Several possible complications can arise in attempting to routinely treat the problem of long alpha-stopping distances in (U–Th)/He dating. Chief among these are the effects of: (i) interaction of the alpha ejection and diffusive concentration profiles in affecting diffusive loss; (ii) the presence of strong intragranular parent nuclide zonation; (iii) implantation of He into grains; and (iv) natural or artificial abrasion of grains that removes some of the alpha-ejection-affected profile.

Interaction of alpha ejection and diffusion   The presence of a ~10–20 μm zone of reduced ^{4}He concentration around grain rims means that natural grains have a lower ^{4}He concentration gradient near their margins than they would if the ^{4}He particles had zero stopping distances. This results in a decrease in the rate of diffusive loss, and an apparently higher retentivity and slightly higher closure temperature for the natural grain [*Meesters and Dunai*, 2002a, b]. This means that measured and alpha-ejection corrected He ages will always be older than the predicted age of a grain with a hypothetical zero alpha-stopping distance that experienced the same thermal history and had the same He diffusion kinetics. This discrepancy goes to zero for quickly cooled grains in which diffusion has not significantly modified the He concentration profile and inventory. But for thermal histories involving a relatively prolonged residence (relative to the bulk date of the grain) within a range of temperatures where He is partially retained (the partial retention zone), it can amount to several percent for apatite grains of typical size, with smaller grains generally being affected more strongly (Fig. 11.6) [*Gautheron et al.*, 2012].

This potential "overcorrection" effect has led some authors to suggest that raw, measured dates without alpha-ejection corrections should be the common currency of routine (U–Th)/He dating of apatite and zircon. In fact, predicted cooling age results of numerical models may be better compared with measured raw uncorrected dates for this reason, assuming model and measured dates are otherwise appropriately related by grain size, radiation damage, and other factors. However, when comparing measured (U–Th)/He dates among crystals or samples or regions, F_T-corrected dates should be used, because:

(1) raw uncorrected dates often vary widely due entirely to well-understood alpha-ejection processes;
(2) these differences are far greater than the relatively minor inaccuracies that may arise from overcorrection for samples with certain thermal histories;
(3) this facilitates comparisons with other lines of geologic evidence such as stratigraphic, magmatic, or climatic events.

Parent nuclide zonation   Intragranular parent nuclide zonation has at least three different effects on (U–Th)/He dates:

(1) perturbation of the He concentration profile and therefore the bulk retentivity;
(2) generation of spatially heterogeneous diffusivity properties through heterogeneous radiation damage;
(3) inaccuracy of the alpha-ejection correction [*Farley et al.*, 2011].

Except in cases where thermal histories involving long durations of partial retention or partial resetting, the last of these is probably the most significant.

Most routine alpha-ejection correction procedures assume a uniform distribution of parent nuclides within dated grains.

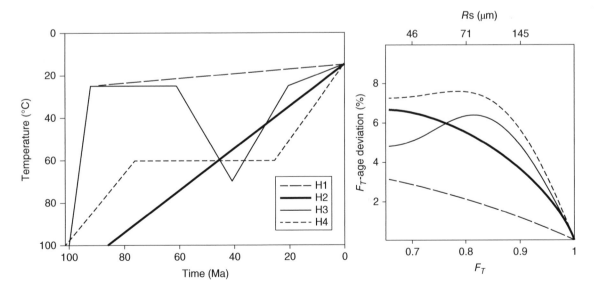

Fig. 11.6. Left: Thermal histories used for investigating the effect of alpha ejection on He diffusion. Right: Percent differences between model predicted age including the effects of alpha ejection relative to a theoretical one without alpha ejection (i.e., deviation % = (A–A$_{ef}$)/A$_{ef}$, where A is the date with alpha ejection and A$_{ef}$ is the theoretical age without it) for thermal histories shown at left. (Source: *Gautheron et al.* [2012]. Reproduced with permission of Elsevier.)

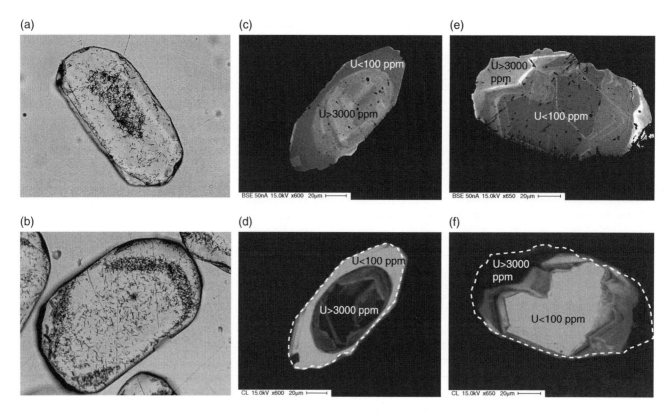

Fig. 11.7. Photomicrographs showing systematic core and rim enrichments in U of large magnitude that would cause significant He age inaccuracies through alpha-ejection correction assuming homogeneous U distribution. (a,b) U distributions, as indicated by fission-track densities, in apatite grains from Baffin Island. (Source: Courtesy of Stuart Nigel Thomson.) (c,d) Back-scattered electron and cathodoluminesence images of zircon grain from New Zealand, showing ~30-fold core-to-rim enrichment of U. (e,f) Similar concentration contrast as left, but with reverse sense of zonation. (Source: *Reiners et al.* [2004]. Reproduced with permission of Elsevier.)

However, natural grains will naturally have variable internal distributions of U, Th, and Sm (e.g., Fig. 11.7), which inevitably leads to some degree of inaccuracy in corrected (U–Th)/He dates [*Farley et al.*, 1996]. Even if it is extreme in magnitude, parent nuclide zonation does not always lead to significant age inaccuracies. Oscillatory or unsystematic patchy zonation, for example, may be negligible for the overall He retention properties.

What does produce significant effects are systematic core-to-rim contrasts in parent concentrations. As an extreme example, a grain with the entire inventory of parent nuclides sited more than ~20 μm corewards from the rim would result in an actual F_T of unity whereas a grain with all parents located precisely at the rim would result in an F_T of about 0.4–0.5, depending on morphology and size. In the absence of information about such zonation, application of a routine alpha ejection correction would result in ages that are too old and too young, respectively, with the magnitude depending on the size, morphology, and thermal history of the grain. This can be extended to the general conclusion that a naïve alpha-ejection correction (that made in the absence of knowledge or consideration of parent zonation) will result in "too-old" dates for grains with enriched cores and "too-young" dates for those with enriched rims. Also, without considering radiation-damage or thermal-history-specific effects, the magnitude of parent zonation on naïve alpha-

ejection-corrected He dates will be larger for smaller grains. Systematically enriched cores will also have the largest effect when they are 10–20 mm from grain boundaries, whereas strongly enriched rims will have more significant effects when they comprise the outermost few millimeters of the grain. Modeling of the type shown in Fig. 11.8 but even including zonation's effects on He diffusion and radiation damage, shows that for grains of typical dimensions, in most cases, even extreme core-to-rim zonation will not produce systematic (U–Th)/He date inaccuracies greater than approximately 20–35%.

Unfortunately, systematic core-to-rim parent zonation is not uncommon in natural apatite and zircon. This may be caused by multiple growth and redissolution events during crystal growth (and possibly aqueous fluid reactions). Several studies have shown that apatite commonly exhibits zonation of this type with magnitudes of differences about twofold to threefold [*Ault and Flowers*, 2011; *Farley et al.*, 2011]. Zircons, on the other hand, often display order-of-magnitude core–rim concentration contrasts (Fig. 11.7).

Other types of zonation, while less systematic and extreme, can also produce dispersion. Zircons from the well-known Fish Canyon Tuff, for example, exhibit ubiquitous heterogeneities that lead to a small predicted (~3–4%) inaccuracy and expected common dispersion of ~10% (Fig. 11.9). *Ault and Flowers* [2011]

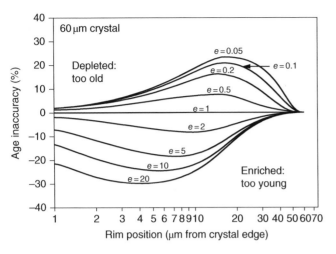
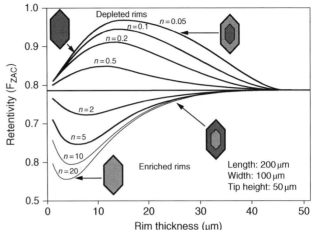

Fig. 11.8. Left: (U–Th)/He date inaccuracy predicted for an idealized 60-μm-radius spherical crystal with rims that are either depleted or enriched relative to the core of the grain, assuming a naïve (unzoned) F_T correction and no additional effects from concentration-profile or diffusivity variations. For typically sized grains, inaccurate F_T application to grains with ~20× core/rim concentration contrasts causes date inaccuracies up to about 30% "too-young" or ~20% "too-old." Right: analogous relationships for ideal zircon morphologies. (Source: *Hourigan et al.* [2005]. Reproduced with permission of Elsevier.)

showed that most apatite grains from granitoid rocks in a cratonic region displayed eU zonation that would predict age inaccuracies of < ~15%.

If estimates of the magnitude and style of zonation are available, as by LA-ICP-MS or fission-track mapping, then grain-specific alpha-ejection corrections (F_{ZAC}) may be numerically derived and applied. Ideally, zonation characterization is performed on the same grain that is dated, because zonation style and magnitude typically varies enormously from grain to grain even within the same igneous sample. Methods that characterize zonation while preserving a sufficient volume of a grain to later analyze can only constrain patterns on a two-dimensional plane (polished mount) or one-dimensional profile (depth profile) (Fig. 11.9). If individual grains cannot be directly characterized prior to dating, at least representative grains from a larger population can be, if the desired result is to explain observed date dispersion among a nondetrital population.

In detail, parent zonation affects observed (U–Th)/He dates not just through the alpha-ejection correction but also through modification of the He diffusion profile and through the influence of radiation damage in creating spatially heterogeneous diffusivity [*Farley et al.*, 2011]. The convolution of these effects, which sometimes are additive in the same sense and sometimes counteract one another, can produce complicated and in some cases counterintuitive results [*Farley et al.*, 2011; *Gautheron et al.*, 2012]. Although parent zonation is not typically characterized as part of routine conventional (U–Th)/He dating procedures, this discussion underscores the interpretational advantage of selecting grains with uniform compositions. Varying styles and magnitudes of parent zonation among grains in a single sample could have distinct sensitivities to certain thermal histories, raising the possibility of using this variation to benefit thermal history interpretations.

He implantation A fundamental assumption in whole-grain apatite or zircon (U–Th)/He dating is that no other source of ^{4}He resided within one alpha stopping distance of the crystal in its host rock or other media. To first order, the robustness of this assumption depends on the eU concentration of the grain compared with other minerals likely to be found near it. This means it is likely to be true for high-eU phases such as zircon, monazite, or titanite, and less likely to be true for low-eU phases like apatite or magnetite. The problem arises from a reverse alpha-ejection phenomenon: alpha implantation from eU-bearing material external to and not recovered with the dated crystal. The age inaccuracy arising from an attempt to date an apatite crystal that was adjacent to relatively high-eU phases depends on the thermal history of the rock, but can easily be several factors higher than the "true" age [*Gautheron et al.*, 2012].

Working with detrital apatite from deep-sea sediments, *Spiegel et al.* [2009] observed dates up to a factor of two older than geologically possible, and an inverse correlation between date and grain eU concentration (Fig. 11.10). This was attributed to implantation of He from the host-rock matrix and other minerals. Laboratory abrasion of the outermost ~20 μm of another suite of apatite crystals yielded more realistic dates, consistent with the implantation hypothesis.

In addition to primary phases and host matrix with relatively high eU, apatite may also be susceptible to implantation from secondary phases that often form in association with it. In some volcanic rocks, apatite crystals are coated over large fractions of their exterior margins in Fe-oxides and clay minerals. These secondary minerals can have eU concentrations as high and in some cases much higher than the apatite grains (Fig. 11.11).

The effects of He implantation from external phases adjacent to apatite depend on several factors: the size of the apatite grain, the thickness and eU concentration of the external phase relative

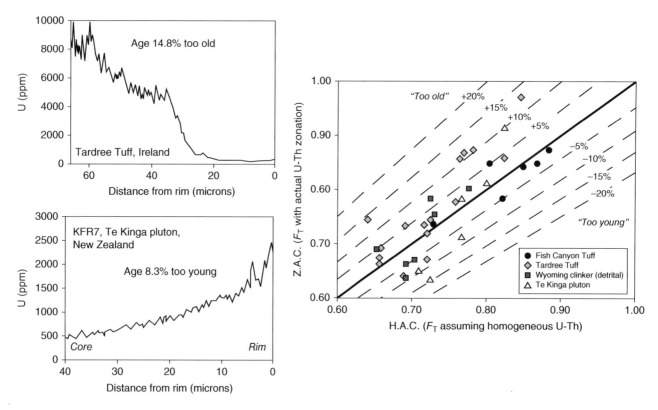

Fig. 11.9. Left: examples of extreme systematic core-to-rim zonation of contrasting types (Sources: adapted from *Hourigan et al.* [2005]; *Reiners et al.* [2004]. Reproduced with permission of Elsevier.) Right: F_{ZAC} versus F_T for zircons with U–Th measured by LA-ICP-MS and imposed concentrically throughout the crystal. Note that some samples are displaced preferentially one way or another relative to the 1:1 line, whereas others (like FCT) are more or less centered on it, due to variable zonation styles. Most of these grains in this suite of zircon grains examined by LA-ICP-MS exhibit zonation that generates less than ~10% inaccuracies.

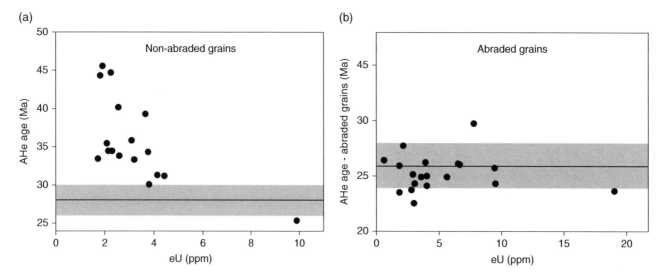

Fig. 11.10. (a) Inverse apatite (U–Th)/He date–eU correlations for single-grain analyses, caused by implantation of He from adjacent phases and rock matrix. (b) Dates obtained on a separate suite of aliquots that had had ~20 μm of the crystal rims removed by laboratory abrasion. (Source: *Spiegel et al.* [2009]. Reproduced with permission of Elsevier.)

to the apatite, the timing of formation of the external phase relative to the cooling age of the apatite, and whether the external phase is recovered and analyzed with the apatite. Various permutations of these factors lead to different types of correlations

between apparent age and eU concentration among single-grain apatite analyses (Fig. 11.12). In most cases, analytical procedures will favor selection of grains lacking significant external phases. If these phases formed earlier than, or near the same as the cooling

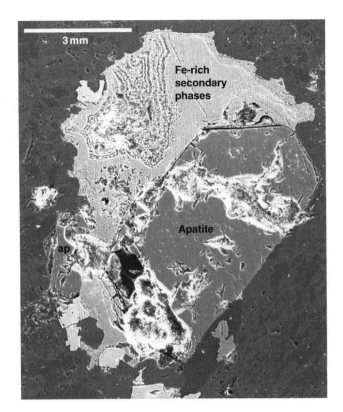

Fig. 11.11. Back-scattered electron image of apatite from the Henry Mountains (28 Ma trachyandesite) and surroundings. The eU concentration of the Fe-rich secondary phases adjacent to the apatite is approximately 80–90 ppm, whereas most regions of the apatite are about 25–50 ppm. (Source: *Murray et al.* [2014]. Reproduced with permission of Elsevier.)

age of the grain, then this produces an inverse age–eU correlation among single-grain aliquots (Fig. 11.12), with large-magnitude age inaccuracies for smaller and lower-eU grains.

As shown by *Spiegel et al.* [2009], one way to ameliorate and potentially avoid He implantation effects is to abrade away at least 20 μm of the grain rims prior to analyses. This can be done using an air abrasion apparatus similar to that designed by *Krogh* [1982] for zircons. This does not solve the problem of inward diffusion (or impeded rates of outward diffusion) of He that may come from implantation. Nor does it address the fact that internal portions of grains will always yield an older date than the bulk grain would, even in the absence of diffusion. Nonetheless, even for very slow cooling histories, the resulting age inaccuracies from these effects may be on the order of observed reproducibility (~8%), whereas inaccuracies arising from implantation may, under easily envisaged circumstances such as clustered or adjacent apatite (±zircon) crystals, be much larger [*Gautheron et al.*, 2012]. The success of this method requires that grains are large enough to yield sufficient material for analysis after abrasion. A grain with a relatively common equivalent-sphere radius of 40 μm, for example, will lose 90% of its volume for a 20-μm rim removal. Unless the remaining core is extremely high in eU and He, this is likely to have a significant negative impact on analytical precision.

Natural grain abrasion   While laboratory grain abrasion may, at least in theory, mitigate implantation problems, detrital grains may experience natural abrasion and rounding during transport, and this poses other complications for alpha ejection correction. Although it is often possible to tell if a given grain has experienced some abrasion, it is often not so clear if it has lost 20 μm or more from all surfaces, as would be required to remove the full alpha-ejection-affected rim. Furthermore, even if abrasion resulted in removal of the full ejection-affected rim, it is usually not possible to tell when this occurred relative to He ingrowth. If abrasion occurred prior to all He ingrowth, for example after deep reburial, then a full alpha-ejection correction should be applied that is appropriate to the observed, presumably rounded, morphology. If abrasion of the full ejection-affected rim occurred only recently, as might be the case for rounded grains emerging from rivers draining rapidly eroding crystalline basement, then no correction should be made. If abrasion occurred at a time after only some of the observed He had accumulated, as might occur for grains being a record of their source provenance, but which resided in sediment or sedimentary rocks that were never deeply buried, then an alpha-ejection correction should be made only for the duration that the grain resided in the sediment or sedimentary rock. Some types of detrital grains such as multicycle zircons may also experience multiple phases of ingrowth and abrasion, which further complicates matters.

For highly rounded grains recovered from modern sediment, so that abrasion may have occurred at any time up until yesterday, the most reasonable and straightforward solution is often to report the raw (U–Th)/He date, and an upper "error bar" that incorporates both analytical uncertainty and the maximum date for the grain if it were fully alpha-ejection corrected (i.e., if the abrasion actually occurred prior to all He ingrowth). The lower error bar would then be the analytical precision or some other measure of uncertainty on the raw date.

For highly rounded grains recovered from ancient sediment or sedimentary rock, but which retained all He from their previous history following deposition, an assumption that abrasion occurred just prior to deposition [*Rahl et al.*, 2003] leads to an alpha-ejection correction as

$$A_c = A_d(1 - F_T) + A_r \tag{11.8}$$

where A_c is the corrected date, A_d is the depositional age, and A_r is the measured (raw) date. Note that this correction does not account for the exponential growth of He through time, but is appropriate for dates less than a few hundred million years.

A similar approach, which does not explicitly assume a time of abrasion [*Thomson et al.*, 2013], is to let the maximum possible date be the fully alpha-ejection-corrected date plus analytical uncertainty, the initial minimum date be the raw date (minus uncertainty), and the "plot date" be a mean of the two, corrected for alpha ejection for the duration of residence in sediment or sedimentary rock at shallow depths. The total alpha-ejection correction (A_{tc}) then is

$$A_{tc} = A_{fc} - A_r \tag{11.9}$$

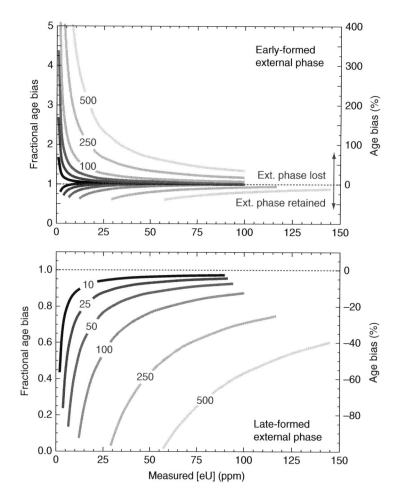

Fig. 11.12. Model apparent-age–eU correlations among apatite grains affected by implantation of He from adjacent phases. Upper panel: external phases formed before or approximately at the same time as the cooling age of the apatite. Lower panel: external phases formed much more recently than the apatite. Contours show eU concentrations (ppm) of the external phase. Fixed model parameters: equivalent spherical radius of apatite grain = 50 μm; thickness of external phase around apatite = 2 μm; eU concentrations of external phase are 10, 25, 50, 100, 250, and 500 ppm. In these examples, the eU concentration of the apatite grain varies from 1–100 ppm. (Source: *Murray et al.* [2014]. Reproduced with permission of Elsevier.)

where A_{fc} and A_r are the fully corrected date and the raw date, respectively. A_{tc} is then multiplied by A_{fr}, the ratio of the age of the sediment (A_s) to the cooling date (fully corrected), as

$$A_{fr} = A_s / A_{fc} \qquad (11.10)$$

which yields a new alpha-ejection-corrected minimum date (A_{min})

$$A_{min} = A_r + (A_{tc} A_{fr}) \qquad (11.11)$$

which replaces the raw date and is considered the minimum possible date or lower "error bar" extent.

11.4.1.5   Date calculation

Equations (11.B.14 and 11.5) are transcendental with respect to t so require either a numerical solution or some other analytical approximation. *Meesters and Dunai* [2005] suggested a method involving a weighted mean of (linearized) ^{4}He production rates

for a particular sample. Alternatively, iteration of a simple Newton–Raphson approach also provides a straightforward solution with minimal error, as:

$$t_{i+1} = t_i - \frac{\alpha(t) - {}^4\text{He}}{\alpha'(t)} \qquad (11.12)$$

In the first iteration t_i is an initial date guess, for example based on a linearized Taylor-Series approximation to equation (11.B14), $\alpha(t)$ is ^{4}He calculated by equation (11.12) using this date, and $\alpha'(t)$ is the derivative with respect to time. The number of iterations required for convergence to an accurate solution depends on the age. Five or ten iterations on a 1 Ga age yield accuracy to about one part in 10^3 or 10^8, respectively.

11.4.2   Other analytical approaches

Several studies have explored the use of laser ablation as a component of (U–Th)/He date measurements. In at least some

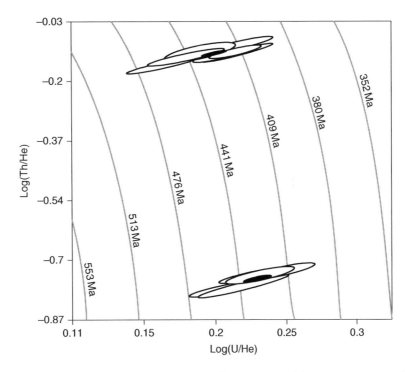

Fig. 11.13. (U–Th)/He dates of two Sri Lankan zircon specimens, shown as functions of log ratios U/He and Th/He. Contours represent loci of constant (U–Th)/He date. These analyses were performed by laser ablation *in situ* measurements and standardization to a third Sri Lankan zircon specimen with independently known U, Th, and He contents. (Source: *Vermeesch et al.* [2012]. Reproduced with permission of Elsevier.)

cases, this approach is capable of measuring spatially resolved dates in confined regions of larger crystals, obviating the need for alpha-ejection corrections (assuming compositional uniformity). It also may allow for the mapping of intragranular distribution of (U–Th)/He dates to provide valuable constraints on integrated thermal histories. Because this approach does not assume a correspondence of the molar parent and daughter contents in a single grain, it requires that parent and daughter contents are either converted to concentrations or are compared with standards of known concentration and/or age.

Using the concentration-measurement approach for He, *Boyce et al.* [2006, 2009] measured accurate and precise (U–Th)/He dates on monazites with U concentrations of 0.1–0.8% and Th concentrations of 4–8%. They extracted and measured He from pits about 25–30 μm in diameter and about 5–14 μm deep. Subsequent measurement of the ablation pit volume by scanning interferometry provided a conversion to He concentration. Uranium and Th concentrations were then measured in a separate region by electron microprobe. Assuming compositional uniformity at least over the scale of the distance between the He and U–Th measurement regions, these concentrations were then combined to calculate dates matching those generated by conventional methods for monazites with ages both as old as ~450 Ma and as young as 750 ka.

Using the standardization approach, *Vermeesch et al.* [2012] measured He and U–Th contents of (separate) laser-ablation pits in large internal fragments of homogeneous Sri Lankan zircon crystals. By normalizing these results to those on another sample with known concentrations, and well as normalizing U and Th LA-ICP-MS signals to the ^{29}Si peak, they converted their measured signal intensities to molar contents (assuming similar pit volumes for the standards), and derived (U–Th)/He dates that agree well with conventional analyses (Fig. 11.13). A similar approach was also taken by *Tripathy-Lang et al.* [2013] in dating detrital zircons.

11.4.3 Uncertainty and reproducibility in (U–Th)/He dating

The variety of analytical approaches used for (U–Th)/He dating in different laboratories make it difficult to generalize about the primary sources of analytical uncertainty inherent in routine (U–Th)/He dating. One important, though difficult to quantify, source of uncertainty comes from the procedure used for quantifying ^{4}He contents.

Many laboratories measure ^{4}He by either comparison to isotope dilution or peak height calibration to a ^{4}He reference aliquot. Either require a source of ^{4}He that delivers a known number of moles in each reference aliquot. This is typically a tank/pipette delivery system attached to the UHV analysis line whose volumes and initial ^{4}He pressure are calibrated manometrically by reference to a reference volume that is known precisely. *Min et al.* [2003] assumed a suite of uncertainties on parameters required for this calibration and calculated an estimated accuracy uncertainty of 0.53% on the ^{4}He pipette delivery system. In detail this probably varies, and may be reasonably estimated to be larger in many cases if other sources of uncertainty such as spatial and temporal temperature gradients, and thermal transpiration during manometry, are considered. Nonetheless, absolute accuracy on standard delivery is likely to be less than 1%.

Aside from the standard delivery calibration, analytical precision on ^{4}He contents of unknowns depends on the similarity of standard and unknown aliquots (e.g., differing total gas pressure and suppression of ionization by other matrix gases), appropriate corrections for isobars, drift in sensitivity and mass fractionation between reference and unknown aliquots (and reproducibility of reference aliquots), reproducibility or systematic behavior (e.g., for linear regression to time zero gas entry) of peak intensities during measurement, and appropriate correction for ^{4}He blanks. Propagations of uncertainties involving these and sometimes other factors typically leads to an estimated analytical precision for ^{4}He on representative apatite and zircon single-grain aliquots of approximately 1–3% (1σ). Due to blank corrections and signal instability for small samples this rises for smaller He contents; values of less than 1–2% being typical for He contents greater than about 10 fmol, and rising to ~4% for 1 fmol and ~10% for 0.1 fmol.

Sources of analytical uncertainty in the measurement of U, Th, and Sm also include the accuracy of the reference elemental concentration solutions used for either peak height comparisons or to calibrate the spike solutions for isotope dilution. Other sources are the reproducibility of measurements on spike isotopic composition and concentration standards, the value and reproducibility of blanks, matrix matching of standards and unknowns, and the uncertainty on measured isotope ratios or peak heights of unknown samples. Propagations of uncertainties involving these and sometimes other factors typically leads to an estimated analytical precision for U, Th, and Sm on representative apatite and zircon single grain aliquots of approximately 1–2% (1σ). Due to blank corrections and signal instability for small samples, this rises for smaller parent nuclide contents; values of less than 0.1–3% being typical for U, Th, or Sm contents from ~1–10 ng, and rising to ~1–2% for 20–50 pg and ~2–5% for 2–5 g.

Using estimated analytical uncertainties on parent and daughter nuclide measurements like those described above, formal analytical precision on raw dates from typical apatite and zircon single grains is usually less than about 2–4% (2σ). Although rapidly cooled grains with high eU sometimes yield observed reproducibility at this level, very few natural samples do. Typical samples from moderately to slowly cooled (~1–50 °C/Ma) crystalline bedrock, for example, yield multiple single-grain dates with reproducibilities in the range of about 6–10% (2σ). Some of the reason for this likely comes from uncertainties associated with the uniform eU distribution assumption inherent in the alpha-ejection correction. The ubiquitous parent zonation in at least some samples of Fish Canyon Tuff zircon alone should produce dispersion of about 10% [*Dobson et al.*, 2008] (Fig. 11.9). However, even samples that do not require alpha-ejection corrections, such as internal fragments of much larger crystals of

Durango apatite, typically yield dispersion of about 6% (2σ). At least some of this may be due to heterogeneous U and Th distributions in specimens of this type [*Boyce and Hodges*, 2005].

Aside from uncertainties associated with application of the alpha-ejection correction, date variation among single-crystal replicates of more typical apatite and zircon crystals arises from a range of factors and often produces dispersion that is much larger than analytical precision. Some of these, such as grain-size differences, radiation-damage-induced diffusivity variations, or the presence or absence of diffusion-profile-bearing grain tips, are potentially predictable and useful, as discussed elsewhere in this chapter. But in many cases, He date dispersion is larger and more difficult to understand than would be predicted by these factors alone. Likely nonanalytical sources for cryptic dispersion includes variable intragranular parent zonation, internal fractures that reduce diffusion domain size, He implantation, and possibly highly localized thermal disturbances such as wildfire [*Mitchell and Reiners*, 2003; *Fitzgerald et al.*, 2006; *Flowers and Kelly*, 2011]. The effects of these factors on measured He dates are, in some cases, likely to be emphasized in cases where grains have cooled slowly, resided in the partial retention zone for long periods, or experienced partial resetting events. In any case, estimating the uncertainty these factors introduce on an a priori, grain-by-grain basis is impractical and perhaps impossible in many cases.

Because of the difficulty in predicting uncertainty associated with these factors, some practitioners prefer to represent uncertainty on (U–Th)/He dates by a proxy of observed reproducibility on other samples considered "typical." Often Fish Canyon Tuff is used for zircon, which as discussed above has an observed reproducibility of about 10% (2σ), perhaps not coincidentally approximately the magnitude of dispersion predicted by intragranular zonation [*Dobson et al.*, 2008]. Durango is often used for apatite, despite the fact that it is far from typical of most apatite as a quickly cooled, minimally zoned, very high Th/U sample that is not subject to alpha-ejection correction or implantation.

11.4.3.1 A quick example

Given the range of sample types, thermal histories, petrographic context, and general objectives in (U–Th)/He thermochronology, there is unlikely to be much agreement about what constitutes a sample that is typical with respect to its date dispersion. Relatively sophisticated methods for both graphical presentation and statistical analysis of "overdispersed" (U–Th)/He data sets are available (e.g., Helioplot [*Vermeesch*, 2010]). The following simple example illustrates the type of date dispersion that is commonly observed in apatite (U–Th)/He dating and how it may relate to sources that arise from a variety of sources, some of which are useful and others of which are not understood.

One-hundred and ninety-six apatite He dates from a sample of granitoid from the northern Sierra Nevada (Fig. 11.14) vary from 53 to 96 Ma. As shown in Fig. 11.14 there is a first-order correlation between date and eU for these grains, and a variety of types of regressions show that more than half of the variation seen in the distribution can be explained by the eU concentration variation.

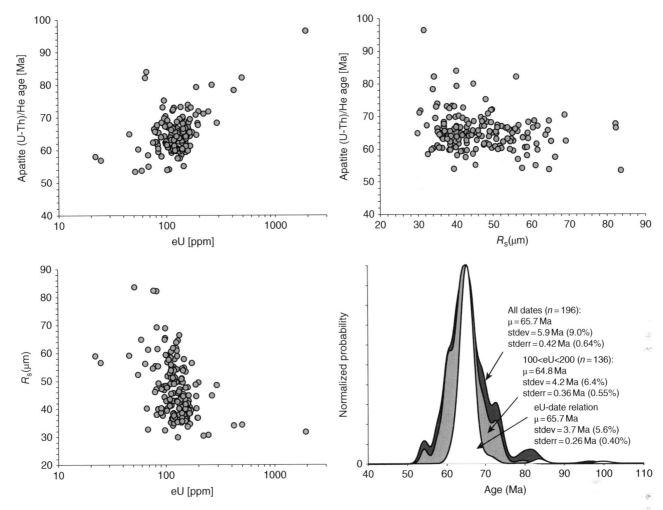

Fig. 11.14. Upper left: apatite date–eU relationship for 196 single-grain aliquots from a sample from the northern Sierra Nevada. Note the broad date–eU correlation, consistent with an effect of radiation-damage-induced diffusivity on dates (in apatite, increased damage leads to decreasing diffusivity). Upper right: date–grain-size covariation. These grains show a possible weak inverse (hyperbolic) correlation, which is the opposite of what would be expected for a simple grain-size (diffusion domain size) control on diffusivity. Lower left: the reason for the possible inverse grain-size–date correlation is likely the inverse correlation between grain-size and eU, a common feature in many igneous rocks. Lower right: probability density diagrams and statistical characterization of (i) the entire data set (dark gray), (ii) grains with 100 < eU < 200 (light gray), and (iii) dates predicted by a third-order polynomial fit to the distribution in the upper left. Comparison of these distributions indicates that other factors besides eU lead to an additional ~2–3% standard variation in this data set.

For example, a probability density distribution of dates predicted from a third-order polynomial fit to the date-eU correlation predicts a standard deviation of 5.6%, compared with the observed standard deviation of 9.0% for all grains and 6.4% for just grains with eU between 100 and 200 ppm.

This date–eU correlation is consistent with a significant influence on the dates by radiation-damage-induced He diffusivity variations, a phenomenon described later. This in turn requires a thermal history involving either slow cooling or long residence in the partial retention zone (or a shorter duration partial resetting event after development of the radiation-damage differences). As one example consistent with geologic constraints from this area, this date–eU correlation can be reproduced well by a steady cooling rate of ~1 °C/Ma from 100 Ma to the present. This tells us that more than half the observed date dispersion

is not meaningless scatter but a predictable and useful consequence of diffusivity variations among grains.

A secondary potential source of dispersion is variation in grain size. Varying grain size influences grain dates under the same circumstances the eU does (i.e., slow cooling or partial resetting), though generally to lesser extent. In this case, however, there is no noticeable correlation between grain size and date, except for possibly a broad hyperbolic inverse one; the opposite of what would be expected. This is likely due to the fact that grain size and eU concentration are inversely correlated in this sample. This is a common observation in many rocks and may be attributable to smaller crystals experiencing most of their growth later than larger ones, when incompatible element concentrations in the residual liquid are higher. In any case, for this sample the expected effect of grain size is outweighed by the anticorrelation

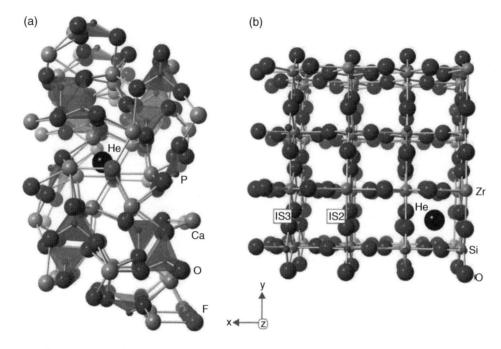

Fig. 11.15. Schematic view of atomic structures of (a) apatite and (b) zircon, and the interstitial gaps in which He may reside and migrate. (Source: *Bengtson et al.* [2012]. Reproduced with permission of Elsevier.)

of size with eU. The residuals of the fit to eU also show no correlation with grain size, suggesting that we are unable to evaluate the role that grain size may have on date dispersion in this sample.

Other likely sources of dispersion in this data set that cannot be easily evaluated are the presence or absence of broken grain tips, microfractures that reduce effective diffusion domain size, parent zonation, and the effects of implantation from adjacent He sources. Deducing the relative magnitudes of the contributions from these and other sources would require more work including careful grain morphology, fracturing, and internal zonation characterization, as well as examination of some representative population of the grains in their native petrographic habitat. Nonetheless we can speculate that these factors are responsible for the difference between the 9.0% standard deviation observed in the entire data set, and the 5.6% predicted for the regressed date–eU correlation.

11.5 HELIUM DIFFUSION

11.5.1 Introduction

One of the most distinctive and useful features of the (U–Th)/He system, as well as the feature that prevented its widespread use until the recent recognition of its utility as a thermochronometer, is the relatively high mobility of its daughter product. High He mobility can be thought of as arising from the fact that there are relatively low energy barriers to the movement of He atoms from one energy-minimum (presumably interstitial) site to another within minerals. In essence, He atoms are "small" compared with

the interatomic bond distances in minerals (Fig. 11.15), so their migration through the lattice does not require much energetic input. At the macroscopic scale, He mobility is thought to follow the principles of thermally activated volume diffusion of an interstitial impurity (see Chapter 5). There are complications to this, however, as described later.

Interpreting He diffusion in the context of Arrhenius behavior and deriving practically useful kinetic parameters requires estimating He diffusion rates over a range of temperatures. This can be done in several ways. One is computational, for example using density-functional theory (DFT) to estimate the energy barriers and the thermal energy needed to overcome them for migration of He along various crystallographic directions [*Reich et al.*, 2007; *Saadoune and de Leeuw*, 2009; *Saadoune et al.*, 2009; *Bengston et al.*, 2012]. These approaches have yielded insights such as the potential importance of crystallographically anisotropic diffusion. In practice, however, it is not clear how to scale theoretically derived diffusivities to those observed in real minerals, and the effects of realistic natural crystal defects, impurities, and radiation damage may have much larger effects than can be easily accounted for in such models. For example *Bengston et al.* [2012] predict higher He diffusion rates in zircon than in apatite, whereas empirical diffusion studies and relative (U–Th)/He dates from these minerals suggest the opposite. The reason for this discrepancy between predicted and observed behaviors is not known, but it may suggest the importance of crystal defects and radiation damage as controls on diffusion rates. This and other results from computational approaches provide insights into the nature of more complex and practically important behavior governing thermochronologic behavior of these systems.

Empirical methods for measuring He diffusivity (to be precise, measurements are made of mobility and/or loss, and this is interpreted as diffusivity) in minerals typically center on perturbation of concentration profiles or step-heating degassing experiments at different temperatures. Ideally these experiments would be carried out at temperatures and over timescales similar to those that are interpreted from (U–Th)/He dates (e.g., often millions of years and temperatures near T_c values). For most thermal histories of interest this is obviously impractical, so measurements are made at higher temperatures and over shorter timescales and extrapolated over many orders of magnitude to conditions of geological interest. The validity of using these laboratory-based calibrations to understand diffusion in nature is based on an assumption that the same mechanisms and kinetics apply over this extrapolated range.

The workhorse method for most noble gas diffusion studies, and for establishing the interpretational bases of (U–Th)/He thermochronology in general, has traditionally been in vacuo step-heating of specimens to sequentially release increasing fractions of a sample's inventory of either natural ^{4}He from radioactive decay, or artificially produced ^{3}He from proton bombardment. Readers seeking more background on the fundamentals of step heating experiments and Arrhenius trends are referred to Chapter 5. Some of the earliest attempts to do this were by *Gerling* [1939], who measured the "heat of diffusion" (i.e., activation energy) for He diffusion in several different minerals, including monazite and uraninite. Other methods for experimental measurement of He diffusion rates have recently come to the forefront though, including Rutherford backscattering spectroscopy (RBS) and nuclear reaction analysis (NRA), which provide complementary, and in some cases challenging, results to the more traditional approaches. All these empirical approaches provide useful comparisons to the computational ones described above.

Helium diffusion experiments that provide information useful for interpreting (U–Th)/He thermochronology have been published on upwards of 20 different minerals. Partly because of its utility in understanding thermal histories in the uppermost ~1–3 km of the Earth's crust, no mineral has received greater attention to its He diffusion kinetics than apatite. Zircon has also received considerable attention. Previous work and insights from these two minerals are reviewed in some detail here. Many of the issues raised and phenomena observed for these phases are pertinent to He diffusion in other minerals as well.

11.5.2 Apatite

Apatite occupies a special place in thermochronologists' hearts for several reasons. One is simply that it is one of the few U,Th-rich minerals that is abundant in many rock types at the Earth's surface. More important, however, is the fact that the kinetics of fission-track annealing and He diffusion in apatite are well suited for understanding thermal histories at temperatures less than about 120 °C, and less than ~80 °C, respectively. Temperatures this low are typical of the uppermost ~2–4 km of crust, so thermal histories in this range provide key understanding of the timing and rates of exhumation at shallow depths, which can be related to erosion, normal faulting, and other processes of interest to tectonic geomorphologists.

Intepreting apatite (U–Th)/He thermochronology requires an accurate knowledge of He diffusion kinetics and the factors that control them. Studies demonstrating consistency between apatite He dates and constraints from other thermochronometers and geologic constraints are reassuring, but confidently interpreting apatite He dates requires independent experiments that establish the fundamental kinetic controls on He diffusion. Most observational constraints on He diffusion kinetics in apatite come from step-heating experiments, but several other approaches are also used (see Box 11.2).

11.5.2.1 Step-heating experiments

As discussed in Chapter 5, thermally activated volume diffusion for simple systems characterized by a single diffusion domain, an initially uniform diffusant concentration, and Arrhenian kinetic behavior, should produce simple linear correlations between the logarithm of diffusivity and inverse temperature. For many thermochronometric systems it is conventional to cast diffusivity not just as D (which has units of length-squared per time), but as D/a^2, where a is the characteristic lengthscale of the domain through which the diffusant must travel before escaping from the mineral, or at least reaching a "fast pathway" in which escape is much more rapid. The reason for this is twofold. First, the fraction of daughter product retained or lost in any thermochronometer is a

---

Box 11.2   Other ways to measure He diffusion in apatite

Helium diffusion in apatite has also been measured by at least two other approaches aside from step heating. Using laser ablation, *van Soest et al.* [2011] directly measured ^{4}He concentration profiles over length scales of up to ~70 μm in large Durango apatite crystals subjected to laboratory heating of varying temperature and duration (Fig. 11.B2.1). The resulting Arrhenius trends are consistent with step-heating results from the same mineral, providing some confidence in both methods.

More direct observations of He concentration profiles in apatite have also been measured using ion implantation, followed by heating to disturb these initial concentration distributions, and then by ion beam microanalytical techniques to map the resulting diffusion profiles. *Ouchani et al.* [1998] used this approach with elastic recoil detection analysis and primary C^{4+} ions to measure He diffusivity in apatite, obtaining results that were broadly similar to the existing constraints at the time. A similar approach but using the ^{3}He(d,p)^{4}He reaction to map implanted ^{3}He (a procedure called nuclear-reaction analysis, NRA), *Cherniak et al.* [2009]

(Figs 11.B2.1 and 11.B2.2) found similar diffusion kinetics as *Ouchani et al.* [1998] for He in Durango apatite while also documenting modest anisotropy of diffusion. In detail, He diffusion kinetics determined from ion implantation techniques yield diffusivities significantly lower, and T_c approximately 10–20 °C higher, than those determined by both step-heating and direct laser-ablation analysis on the same types of samples. These discrepancies are not understood. *Shuster and Farley* [2009] observed relatively low diffusivity and high T_c (117 °C for zero-damage) for ^{3}He diffusion in a synthetic apatite crystal, which is also not understood.

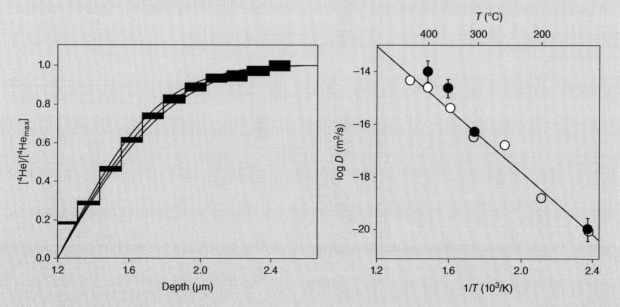

Fig. 11.B2.1. Left: depth profile of ^{4}He normalized concentration beneath surface of Durango apatite subjected to heating at 450 °C for 10 min, measured by laser ablation. (Source: *van Soest et al.* [2011]. Reproduced with permission of Elsevier.) Right: ^{3}He diffusion in apatite normal to c-axis (open symbols) and parallel to c-axis (filled symbols) as determined by NRA method. (Source: *Cherniak et al.* [2009]. Reproduced with permission of Elsevier.)

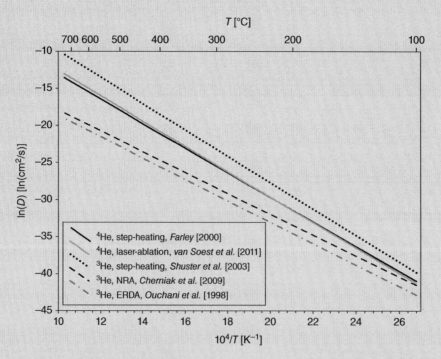

Fig. 11.B2.2. Helium diffusion in natural apatite specimens from radiogenic ^{4}He, proton-bombardment-induced ^{3}He, laser-ablation depth profiling, and ^{3}He implantation followed by heating, followed by NRA and ERDA. The latter methods yield consistently lower activation energies and lower D_0, but not lower closure temperatures, than the other methods.

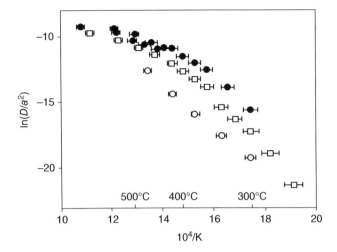

Fig. 11.16. *Zeitler et al.'s* [1987] Arrhenius trends for He diffusion in Durango apatite, based on step-heating experiments. Steps at higher temperatures exhibit increasing lower apparent slopes, inconsistent with strict Arrhenius behavior over the full range of temperatures examined. T_c values calculated from the low-temperature portions of these experiments average $105 \pm 30\,°C$. (Source: *Zeitler et al.* [1987]. Reproduced with permission of Elsevier.)

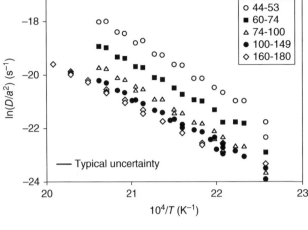

Fig. 11.17. *Farley's* [2000] Arrhenius trends for shards of Durango apatite of different size ranges. Note linearity of trends throughout the temperature range examined. Higher y-axis intercepts but similar slopes for each trend are consistent with the idea that the diffusion domain size (*a*) for apatite scales with physical grain size. Regression of these trends yields best-estimate kinetic parameters for natural ^{4}He in Durango apatite of $D_0 = 31.6^{+94}_{-24}$ cm^2/s and $E_a = 138 \pm 2$ kJ/mol. (Source: *Farley* [2000]. Reproduced with permission of John Wiley & Sons.)

function of D/a^2, not just D (i.e., a larger single-domain crystal will retain a larger fraction of its radiogenic daughter than a smaller one). Second, many experimental protocols to determine the Arrhenius parameters of thermally activated volume diffusion (activation energy E_a, and the preexponential term D_0/a^2) are not based on fractional loss measurements and therefore can only determine D/a^2.

The earliest studies of He diffusion in apatite are those of *Zeitler et al.* [1987], accompanying their groundbreaking paper outlining the bases for apatite (U–Th)/He thermochronology. *Lippolt et al.* [1994] also measured He diffusion in apatite grains, finding a wide range of kinetic parameters. The step-heating diffusion experiments of *Zeitler et al.* [1987] are shown in Fig. 11.16. They show a typical form of Arrhenius trends for He diffusion in apatite. Despite the complicating curvature in these initial results at high T, regressions through the He release data at lower temperature, which is arguably more appropriate for understanding He loss in conditions most relevant to most geologic applications, yielded kinetic parameters of about $\ln(D_0/a^2)$ of $16.4 \pm 2.8\,s^{-1}$ and activation energy (E_a) of 38.5 ± 8.1 kcal/mol, translating to an effective closure temperature of about $105 \pm 30\,°C$, slightly higher than, but not inconsistent with, later measurements. More recent analyses of He diffusion in apatite have characterized Arrhenius trends at lower temperatures (e.g., Fig. 11.17), where curvature in the step-heating trends is not present.

Detailed analysis of both ^{4}He and ^{3}He diffusion (the latter produced *in situ* in apatite grains by proton bombardment) from Durango apatite revealed several conclusions important both for apatite and other minerals. First, both ^{3}He and ^{4}He isotopes

yield concordant kinetic parameters. For an aliquot with restricted grain size (160–180 µm radius) of Durango apatite, *Shuster et al.* [2004] determined $\ln(D_0/a^2)$ of 16.0 ± 0.3 (ln [s^{-1}]) and E_a of 148 ± 1 kJ/mol for ^{3}He, and $\ln(D_0/a^2)$ of 15.8 ± 0.2 (ln[s^{-1}]) and E_a of 148 ± 1 kJ/mol for ^{4}He. Assuming a single domain size or 170 µm, the $\ln(D_0/a^2)$ terms reduce to D_0 of 2568 cm^2/s for ^{3}He and 2102 cm^2/s for ^{4}He. For a more typical (but still on the large side, compared with natural apatite grains) diffusion domain length scale of 75 µm, these yield T_c of ~67 and 69 °C, respectively, assuming a cooling rate of 10 °C/Ma. These are slightly higher than the calculated T_c from *Farley's* [2000] Durango data (Fig. 11.18) using a domain size of 75 µm (58 °C). Interestingly, identical diffusion kinetics for ^{3}He and ^{4}He is not entirely expected, if isotopes diffuse at rates following a simple inverse-square-of-mass-dependence law. The fact that these isotopes do seem to behave similarly means that either this law is not relevant to this type of diffusion, or possibly that the measured losses reflect the kinetics of some other process which is rate limiting for He diffusion.

Farley [2000] showed that He diffusivity in apatite is only slightly anisotropic with respect to the crystallographic axis, with c-axis parallel diffusion having a higher E_a by about 8 kJ/mol (~6%) and a higher D_0 by about a factor of 10–15. These relatively minor differences, compared with those associated with He diffusion in zircon, for example, are also evident in NRA experiments, as described later. The same study noted a systematic relationship between $\ln(D_0/a^2)$ and grain size of analyzed specimens (Fig. 11.18), consistent with the correspondence of diffusion domain size and grain size. This important result means that the effective dimension of the diffusion lengthscale a is the

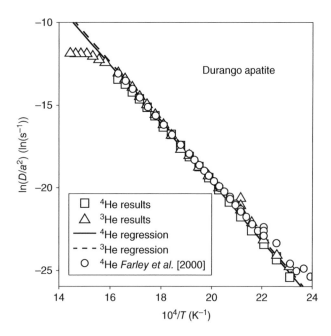

Fig. 11.18. Comparison of Arrhenius trends for ^{4}He and ^{3}He from Durango apatite [*Shuster et al.*, 2004]. Step-heating release patterns of ^{4}He and ^{3}He are identical in this sample and also match those of earlier experiments on nonproton-bombarded Durango apatite. Note the "rollover" to lower diffusivities seen at high temperatures, an effect that has been attributed to significant annealing of radiation damage within the apatite during the experiment.

grain itself, which, in some cases, has important implications for interpreting age results.

Helium diffusivity in apatite (and in several other minerals) is commonly observed to be higher, and initial apparent activation energies are lower, in the earliest initial stages of step-heating diffusion experiments. The origins of this behavior are not well understood, but may be related to intraaliquot or intragrain diffusion domain-size differences. However, the reappearance and/or persistence of this behavior for some step-heating schedules shows that a simple explanation involving step-wise exhaustion of progressively larger domains cannot explain it all.

Decreasing activation energy of He diffusion at high temperatures in apatite, first observed as the high-temperature rollover of *Zeitler et al.* [1987], actually occurs gradually at temperatures above about 265 °C, but the magnitude of the effect increases dramatically with temperatures above ~335 °C. Heating to such temperatures on laboratory timescales (e.g., minutes to hours) also produces an irreversible increase in diffusivity that persists at subsequent low-temperature steps. In detail, the magnitude of these changes in He diffusion kinetics appears to correlate with predictions of the extent of fission-track annealing in apatite, leading to the idea that radiation damage and perhaps other crystallographic defects in apatite may strongly influence diffusion. Following on earlier suggestions by *Trull et al.* [1991], *Farley* [2000] proposed that crystal defects may decrease bulk diffusivity

by effectively acting as energetically favorable traps for migrating He atoms. Using principles of equilibrium partitioning between these sites, he proposed that:

$$\frac{D}{a^2} = \frac{D_{\text{vol},0}}{a^2} \cdot \frac{e^{(-E_a/RT)}}{k_{t,0}e^{(-E_t/RT)} + 1} \tag{11.13}$$

where $D_{\text{vol},0}$ and E_a are the diffusivities at infinite temperature and activation energy for the defect-free apatite lattice, E_t is the energetic difference between trapped and free states of the He atom, and $k_{t,0}$ is a preexponential factor governing the Arrhenius relationship describing equilibrium partitioning of He between the lattice and the trap. Use of this model with $E_a = 25.4$ kcal/mol, $E_t = -7.5$ kcal/mol, $k_{t,0} = 2 \times 10^{-3}$, and $D_{\text{vol},0}/a^2 = 1500$ explains the position and curvature seen in the He-in-apatite Arrhenius trends, but only for prograde step-heating schedules. The fact that He diffusivity increases (and activation energy decreases) irreversibly after heating to high temperatures requires another explanation, as discussed in the radiation damage section below.

11.5.2.2 Radiation damage in apatite

In the mid-2000s, practitioners of apatite (U–Th)/He dating began recognizing cases in which model thermal histories derived from apatite fission-track (AFT) and (U–Th)/He data from the same rocks yielded inconsistent results [*Crowley et al.*, 2002; *Green and Duddy*, 2006; *Green et al.*, 2006; *Hansen and Reiners*, 2006]. Assuming accurate calibration of the annealing kinetics of fission tracks in apatite, these discrepancies could be explained if He retentivity were higher in crystals with higher eU and/or higher ^{4}He content. This speculation was confirmed when *Shuster et al.* [2006] demonstrated through numerous step-heating experiments that the diffusivity of both ^{3}He and ^{4}He were inversely correlated with He concentration (Fig. 11.19) apatite.

Hypothesizing that ^{4}He concentration was just a proxy for radiation damage, *Shuster et al.* [2006] expanded on the approach used by *Trull et al.* [1991] and *Farley* [2000], to develop the trapping model for He diffusion in apatite. The essence of the model is that He atoms migrating through a crystal lattice via diffusion will encounter localized zones, presumably radiation-damage-induced regions of high vacancies or disorder, which are energetically favorable for He residence. Migration of He out of these "traps" and back into the crystal lattice requires surmounting an additional activation energy E_t (Fig. 11.20).

Following the approach of *Farley* [2000], *Shuster et al.* [2006] expressed the trapping model for He diffusivity in apatite as

$$\frac{D(T, {}^4\text{He})}{a^2} = \frac{\dfrac{D_0}{a^2} \cdot e^{(-E_a/RT)}}{k_t^* + 1} = \frac{\dfrac{D_0}{a^2} \cdot e^{(-E_a/RT)}}{(k_0 \cdot \nu_{\text{rd}} \cdot e^{(E_t/RT)}) + 1} \tag{11.14}$$

Where the variables are the same as in equation (11.13), but here $k_{t,0}$ has been replaced by $k_0 \times \nu_{\text{rd}}$, where ν_{rd} is volume fraction of radiation-damage sites in a crystal, and k_0 is a constant that ends up being determined empirically. This expression also uses a positive sign for the energy of partitioning, E_t, describing the

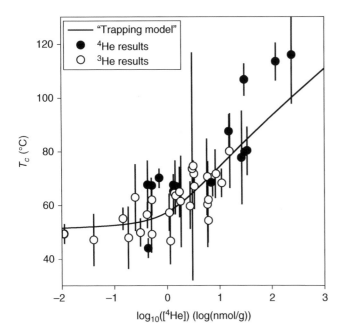

Fig. 11.19. Apatite (U–Th)/He T_c as a function of ^{4}He concentration in apatite, as measured by both ^{3}He and ^{4}He diffusion experiments. "Trapping model" is that of *Shuster et al.* [2006]. (Source: *Shuster et al.* [2006]. Reproduced with permission of Elsevier.)

energy required for movement of He from a trap to the crystal lattice.

Assuming that ^{4}He was a perfect proxy for radiation damage in a crystal, Shuster posited that

$$\nu_{rd} = \eta \left[^4\text{He}\right] \tag{11.15}$$

where η is a proportionality constant (with units g/nmol) relating ^{4}He concentration to the volume fraction of damage sites, so

$$\frac{D\left(T,{}^4\text{He}\right)}{a^2} = \frac{\dfrac{D_0}{a^2} \cdot e^{\left(-E_a/RT\right)}}{\left(k_0 \cdot \eta \cdot \left[^4\text{He}\right] \cdot e^{\left(-E_t/RT\right)}\right) + 1} \tag{11.16}$$

The constants were then grouped together to create a single fudge factor, ψ,

$$\psi = k_0 \eta \tag{11.17}$$

so that the model involved four free parameters: E_a, E_t, D_0 and ψ. Values of these were iteratively fit to the observed diffusion characteristics of the suite of samples shown in Fig. 11.19, and the variance on the observed E_a, D_0/a^2 of the suite was minimized.

The trapping model demonstrates the important effect of radiation damage (again assuming He concentration is a proxy for it) on the T_c and apparent apatite (U–Th)/He date. In practice, eU, not He content, is actually used as a proxy for relative radiation damage among grains that have experienced the same thermal history. For any given thermal history, apatite grains with higher eU are predicted to have higher T_c and older (U–Th)/He date (Fig. 11.21). From a broader perspective, this shows that the

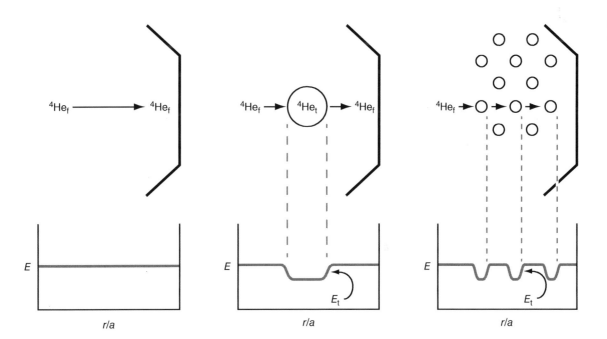

Fig. 11.20. Schematic model of the influence of radiation-damage traps on He diffusion. Left: free He atoms (He$_f$) migrating through a homogeneous, damage-free crystal with uniform activation energy. Middle: He atom may encounter and be partitioned into a relatively low-energy region, becoming trapped (He$_t$), and requiring an extra energetic bump to escape. Right: similar situation as middle, but with more trapping sites, demonstrating that at any time the He inventory of the grain may be partitioned into free and trapped sites. (Source: *Shuster et al.* [2006]. Reproduced with permission of Elsevier.)

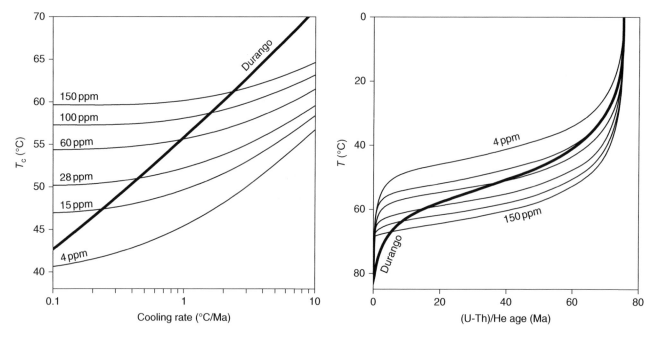

Fig. 11.21. Left: T_c (for a cooling rate of 10 °C/Ma) for Durango apatite [*Farley*, 2000] and grains with varying eU concentrations (ppm), and therefore ^{4}He contents, according to the trapping model of *Shuster et al.* [2006]. Right: apparent (U–Th)/He date of apatite with varying eU, held for 75 Ma at varying T, according to the trapping model. Note that grains with the identical thermal histories (e.g., isothermal holding at 50 °C for 75 Ma) but commonly observed eU variations (e.g., 4–150 ppm) may develop extremely different dates (e.g., 5–65 Ma). (Source: *Shuster et al.* [2006]. Reproduced with permission of Elsevier.)

kinetics of He loss from apatite evolve with the thermal history of the specimen in question, and for many combinations of thermal history and eU, are extremely different from those of Durango apatite.

The use of ^{4}He concentration as a proxy for the absolute amount of radiation damage is problematic in several ways. One is the likelihood that He may be lost and damage may be annealed over time; after all, (etchable) fission-tracks—one form of damage—is annealed with kinetics that are relatively well understood. Both of these are likely to depend on both time and temperature, but there is no reason to expect that they have the same kinetics. This led *Shuster and Farley* [2009] to compare the predictions of the trapping model with observations of the effect of direct radiation damage by neutron bombardment on He diffusivity. They showed that E_a, D_0, and T_c of apatite from samples with simple thermal histories changed with actual observed radiation dose (in kerma) in the same way as they would if ^{4}He accumulated proportionally as predicted with alpha-damage dose (Fig. 11.22).

Importantly, this same study also showed that not only does the accumulation of natural and artificial damage have the same effect on He diffusivity (and T_c), but annealing of damage by heating produced the opposite effect, essentially annealing away damage and reducing the T_c. This means that the He diffusion kinetics of apatite will evolve with time and temperature, up and down the trend in Fig. 11.22, as damage is annealed.

Once it was shown that radiation damage was the controlling parameter for He diffusivity, the question was how to quantify the extent of damage present in any given sample.

Shuster and Farley [2009] noted that among their experimental samples, fission-track lengths were well correlated with the (U–Th)/He T_c, suggesting that this parameter, which represents the combined effects of time and temperature associated with annealing, could be used to represent radiation damage.

The use of fission-track length as a proxy for radiation damage may be surprising given the fact that the vast majority of crystallographic defects in a crystal are likely to be caused by recoiling heavy nuclides following alpha decay. Alpha-recoil tracks, not fission-tracks, are generally thought to be responsible for changes in He diffusivity because they are volumetrically much more abundant than fission-tracks, and also because Durango apatite, which has an exceptionally low U/Th, displays the same relationship between diffusivity and He concentration as all other apatites, even though most of the He is produced from Th. The fact that Durango and other types of apatite change similarly with both increasing and decreasing damage extents is also consistent with the presumption that, at least to first order, fission-track annealing kinetics have similar kinetics to the damage that is responsible for He diffusivity changes.

Fission-track length is not a perfect proxy for use in a radiation-damage model, however. For one thing, only specimens with very simple thermal histories will have unimodal track lengths. More often they bear a distribution of lengths that reflects growth and annealing of tracks at various stages of a sample's thermal history. More importantly, natural fission tracks arise only from ^{238}U whereas alpha damage also comes from ^{235}U and ^{232}Th.

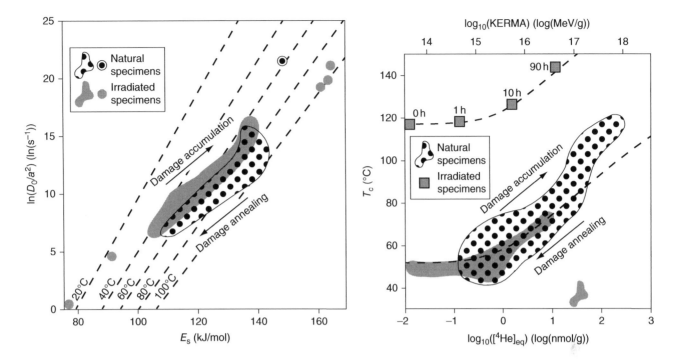

Fig. 11.22. Left: $\ln(D_0/a^2)$ and E_a of He diffusion in apatite, showing that natural specimens and specimens with varying amounts of artificially induced radiation damage follow similar trends of kinetic parameters and T_c (contours). Right: T_c of natural (unfilled symbols), and artificially irradiated samples, as a function of ^{4}He for natural samples and KERMA for neutron-irradiated samples (KERMA is the ionizing kinetic energy released per unit mass). Samples follow a similar trend suggesting that ^{4}He is indeed a proxy for radiation damage. Another suite of experiments showed that damage annealing through laboratory heating produced a reverse trend, suggesting that varying amounts of damage and annealing simply move individual samples up and down along the same trend. The gray circles in the upper left of the plot are samples of synthetic apatite irradiated to varying extents. Although they form a similar trend as the rest of the data, they plot at much higher T_c, and the zero-damage specimen has much higher retentivity than would be predicted based on the other samples. (Source: adapted from *Shuster and Farley* [2009].)

To serve as a damage proxy that is easy to quantify, has annealing properties that are well understood, and takes into account damage arising from both Th and ^{235}U, *Flowers et al.* [2009] recast the relationship between T_c and KERMA (KERMA is the ionizing kinetic energy released per unit mass by either artificial neutron irradiation or natural alpha-decay-induced irradiation) as one between T_c and effective fission-track density, $e\rho_s$ [tracks/cm^2], which is simply the measured fission-track density multiplied by the ratio of total alphas produced to those produced by only ^{238}U (Fig. 11.23).

In order to parameterize changes in He diffusion kinetics as a function of $e\rho_s$, *Flowers et al.* [2009] rewrote a version of the trapping equation (one last time, here) as

$$\frac{D(T,^4\text{He})}{a^2} = \frac{\dfrac{D_{0\text{L}}}{a^2}\cdot e^{(-E_\text{L}/RT)}}{\left(k_0\cdot\nu_\text{rd}\cdot e^{\left(E_\text{trap}/RT\right)}\right)+1} \tag{11.18}$$

and then instead of replacing the preexponential terms in the denominator of the right-hand side with a function of ^{4}He concentration ($\psi\,[^4\text{He}]$) as *Shuster et al.* [2006] did, they replaced them with a third-order polynomial function of $e\rho_s$

$$\psi = \left(\psi_\rho e\rho_s + \Omega_\rho e\rho_s^3\right) \tag{11.19}$$

The third-order polynomial in $e\rho_s$ was found to provide a much better fit to the data than any simpler functional form, even

though there is no particular mechanistic explanation for it. Indeed, a similar model was independently derived and calibrated by *Gautheron et al.* [2009]. The primary difference in the latter model is the linear dependence of the preexponential term in the denominator of equation (11.18), which they argued was more theoretically justified.

Flowers et al. [2009] then performed a fitting analysis to derive best-fit values of $D_{0\text{L}}$, E_L, E_trap, y_r and Ω_r, to explain the observed diffusion data at various levels of damage. Although the analysis produced several possible arrays of parameters that adequately described the data to one degree or another, their preferred parameter fit yielded: $\log(\Omega_\rho) = -22$, $\log(\psi_\rho) = -13$, E_trap (kJ/mol) $= 34$, $\ln(D_{0\text{L}}/a^2)$ ($\ln[\text{s}^{-1}]$) $= 9.733$, and E_L (kJ/mol) $= 122.3$.

Both the *Gautheron et al.* [2009] and *Flowers et al.* [2009] models account for annealing of radiation damage with time and temperature using the empirically based fission-track annealing model of *Ketcham et al.* [2007], which states that

$$r_{c,B2} = \left\{\left[C_0 + C_1\frac{\ln(t)-C_2}{\ln\left(^1/_T\right)-C_3}\right]^{1/_a}\right\}^{-1} \tag{11.20}$$

where C's and α are empirically derived parameters, and $r_{c,B2}$ is the reduced c-axis parallel mean fission-track length for a certain type of apatite (specimen B2; see Chapter 12). Because natural

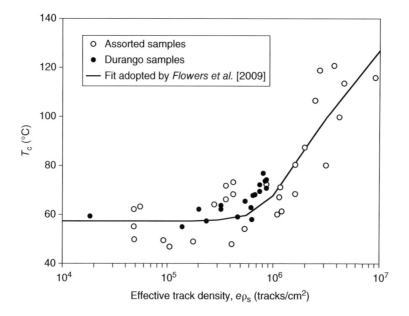

Fig. 11.23. T_c as a function of $e\rho_s$, a measure of extent of radiation damage, among apatite. (Source: adapted from *Flowers et al.* [2009].)

apatite specimens display a range of annealing kinetics related to composition and other poorly understood causes, the actual mean track length has to be adjusted using

$$r_{c,lr} = \left(\frac{r_{c,B2} - r_{mr0}}{1 - r_{mr0}}\right)^{\kappa} \tag{11.21}$$

where $r_{c,lr}$ is the reduced c-axis projected mean track length for an apatite of interest, and r_{mr0} and K are empirically fit parameters specific to the apatite of interest. In practice, these values are generally assigned by deriving K for an apatite of interest and setting $K \approx 1.04 - r_{mr0}$. For example, Durango and most fluorapatite specimens have r_{mr0} near 0.79–0.83.

Reduced track lengths are then converted to reduced track densities, ρ_r, using the empirical relationship of *Ketcham* [2005]

$$\rho_r = 1.600 r_c - 0.600, r_c \geq 0.765 \tag{11.22};$$

$$\rho_r = 9.250 r_c^2 - 9.157 r_c + 2.269, \ r_c < 0.765 \tag{11.23}$$

For modeling the evolution of track densities for arbitrary thermal histories, rather than a square heating pulse as equation (11.20) is explicitly written, a "reduced time" approach is taken to keep track of reduced track densities through time, which is necessary for predicting He diffusivity with time. A thermal history is discretized into segments and $r_{c,B2}$ is calculated for the first time step. For a second time step at a different temperature, the previous $r_{c,B2}$ value is used with equation (11.20) to calculate an equivalent time and the T of the second time step. Then this time, plus the time during the second time step at the same temperature, are summed, to derive the $r_{c,B2}$ after the second time step. This process is repeated for all subsequent time steps in the thermal history. Predicting a fission-track date at this step involves

summing all the reduced track densities from each step multiplied by the duration of each step (see *Ketcham* [2005], equation 14).

For predicting He diffusivity at each time step, however, the reduced areal track density used for fission-track dating from equation (11.20) must be converted to volumetric track density, and must also account for damage arising from ^{235}U and ^{232}Th. For a single time step of duration Δt then,

$$e\rho_s(\Delta t) = \frac{\lambda f}{\lambda_D} \rho_\nu(\Delta t, \text{U,Th}) \eta q L \rho_r(t, T) \tag{11.24}$$

where λ_f and λ_D are the spontaneous fission and total decay constants for ^{238}U, η and q are dimensionless constants related to fission-track etching efficiency whose product is estimated as ~0.91 [*Jonckheere and Van den Haute*, 2002], L is half the etchable length of a fission-track (~8.1 μm), and ρ_ν reflects the generation of new damage in each time step, as

$$\rho_\nu = \left[^{238}\text{U}\right]\left(e^{\lambda_{238} t_2} - e^{\lambda_{238} t_1}\right) + \frac{7}{8}\left[^{235}\text{U}\right]\left(e^{\lambda_{235} t_2} - e^{\lambda_{235} t_1}\right)$$
$$+ \frac{6}{8}\left[^{232}\text{Th}\right]\left(e^{\lambda_{232} t_2} - e^{\lambda_{232} t_1}\right) \tag{11.25}$$

Accounting for the $e\rho_s$ at any given time step then requires first calculating the reduced track densities using the equivalent time method, then calculating the $e\rho_s$ for the same step, and then summing for all previous time steps. For each time step then an $e\rho_s$ can be determined that governs the diffusivity of He as described in equation (11.18).

When combined with a model for He diffusivity in the grain at each time step, the end result of the radiation damage and accumulation model (RDAAM) is a prediction of the (U–Th)/He date as a function of the continuously evolving diffusivity

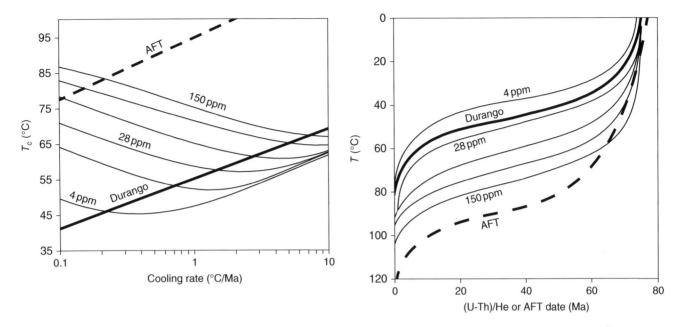

Fig. 11.24. Left: T_c (for canonical cooling rate of 10 °C/Ma) for Durango apatite [*Farley*, 2000] and grains with varying eU, and therefore varying amounts of radiation damage, according to the RDAAM of *Flowers et al.* [2009]. Right: apparent (U–Th)/He date of apatite with varying eU, held for 75 Ma at varying T, according to the RDAAM. Note that grains with identical thermal histories (e.g., isothermal holding at 50 °C for 75 Ma) but commonly observed eU variations (e.g., 4–150 ppm) may develop extremely different dates (e.g., 10 and 75 Ma). (Source: *Flowers et al.* [2009]. Reproduced with permission of Elsevier.)

resulting from a thermal history. As shown earlier, in practice, a commonly used proxy for relative amounts of radiation damage and thus diffusivity variations among grains that have experienced the same thermal history (e.g., in the same hand sample), is eU. Figure 11.24 shows the predicted T_c values and He dates for apatite grains with varying eU, analogous to the results shown for the simple trapping model of *Shuster et al.* [2006]. The RDAAM predicts a much larger range of T_c, and generally higher T_c as a function of cooling rate than the trapping model. It also predicts that very slowly cooled grains with relatively high eU will have higher T_c than that for the apatite fission-track system. This larger range is also reflected in the predicted He dates for apatite held isothermally at varying temperatures for 75 Ma (Fig. 11.24). This also shows that apatite (U–Th)/He dates can be significantly older than AFT dates in some cases.

The examples above show that apatite (U–Th)/He dates may be expected to be older than AFT dates from the same sample in cases of high eU grains and either slow monotonic cooling or prolonged isothermal residence at temperatures between about 30 and 60 °C (Fig. 11.25). Another scenario that can lead to strongly "inverted" apatite He and AFT dates are some kinds of nonmonotonic cooling paths. Reheating is a common and often necessary feature in many settings, including burial by sedimentary or volcanic rocks followed by erosional exhumation. In such settings RDAAM predicts that AFT dates will be younger than apatite (U–Th)/He dates for even moderate to low eU apatite grains, if burial and reheating did not result in full resetting of the apatite He system (i.e., if maximum burial temperatures

did not reach approximately 80–100 °C, depending on the eU of the grain; Fig. 11.25).

Although the effects of radiation damage complicate the interpretation of apatite (U–Th)/He thermochronology because of the coevolution of thermal history and diffusion kinetics, in cases where rocks (or localized regions that have experienced the same thermal history) contain apatite grains with a variety of eU concentrations, RDAAM actually provides a powerful tool for reconstructing past thermal histories. This is because grains with different eU contents experience different degrees of He loss for the same thermal history (i.e., they have different closure temperatures). By using eU as a proxy for relative amounts of radiation damage, observed date–eU correlations, if they exist, can be forward or inversely modeled for thermal histories that produce them. Figure 11.26 shows theoretical examples of correlations between (U–Th)/He age and eU concentration arising from different thermal histories, and a natural example is shown in section 11.7.

11.5.2.3 Crystal/grain size

As discussed in Chapter 5, for any arbitrary heat pulse or more complex thermal history, the fractional loss of daughter product from a diffusion domain will scale with the dimensionless parameter Dt/a^2 or its time-integrative equivalent, where a is a characteristic length scale of the diffusion domain. This means that larger diffusion domains will experience lower fractional losses than smaller domains and will therefore yield older dates. Similarly, for a constant temperature thermal history, a diffusion

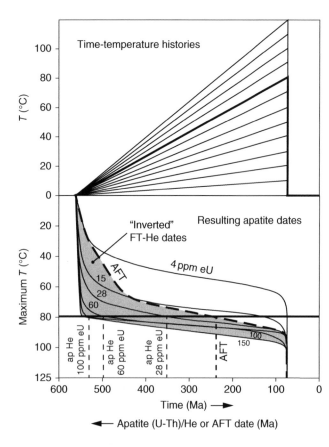

Fig. 11.25. Upper: model thermal histories simulating evolution expected for a typical basement-cored block buried by sediment and then rapidly exhumed. Each trend represents a history involving rapid cooling from high temperatures at 550 Ma, followed by gradual burial and heating to temperatures between 10 and 120 °C until 75 Ma, followed by rapid exhumation and cooling over about 1.7 Ma, followed by residence at the surface (assumed to be 0 °C). Lower: predicted apatite (U–Th)/He and AFT dates as a function of maximum temperature reached in thermal histories above. For all grains with eU > ~20 ppm, apatite He dates are older than AFT dates. Even grains with eU as low as 10–15 ppm are predicted to have "inverted" AFT-He dates if maximum temperatures did not exceed roughly 70 °C. Bold line in upper plot shows t–T history for maximum burial temperature of 80 °C. Horizontal bold line in lower plot shows intersections of AFT and apatite He dates for this t–T history. Vertical lines in lower plot show AFT date (243 Ma) predicted for this t–T history and apatite He dates for grains with 28, 60, and 100 ppm eU for the same t–T history (345, 491, and 523 Ma, respectively). It should be noted that although the AFT date for the $T_{max} = 80$ °C t–T history results in a partially reset AFT date (which would be evident in the track length distribution), even histories resulting in full resetting of the AFT dates (i.e., $T_{max} \sim 85$–100 °C) also result in AFT-apatite He date inversions if eU is greater than about 60 ppm.

domain of size a will approach an equilibrium date where diffusive loss and radiogenic production are balanced. *Wolf et al.* [1998] showed that this date t_{eq}, is approximately

$$t_{eq} \approx \frac{a^2}{D}\left(\frac{1}{15}\right) \qquad (11.26)$$

If natural grain sizes of apatite correspond to diffusion domain sizes, as suggested by step-heating experiments [*Farley*, 2000], then equation (11.25) shows that equilibrium dates will vary with

the square of their size. This also suggests that domains of different size will develop larger age differences for small D, or lower T. However, *Wolf et al.* [1998] also showed that e-folding time of the approach to t_{eq}, is approximately $1.5 \times t_{eq}$. The upshot of all this is that domains of different sizes will develop the largest age differences when they have resided at low temperatures for long periods of time, or, in the case of monotonic cooling histories, slow rates of cooling.

Because step-heating and laser-ablation diffusion experiments clearly indicate that in most cases the He diffusion domain for apatite scales with the crystal or grain size, the above analysis suggests that certain thermal histories should produce observable correlations between grain-size and He age. Although natural variations in crystal sizes and eU among apatites from typical hand-samples usually make the radiation damage effect more important for generating age variability, the grain-size effect can be important in cases where grains show little eU variation.

Figure 11.27 shows an example from the Bighorn Mountains, Wyoming, where multigrain aliquots of apatite crystals display clear age–size correlations with more than a factor-of-three age difference. As in the case of eU variations, each aliquot can be considered a thermochronometer with distinct bulk diffusivity (Dt/a^2), constraining the range of possible thermal histories more precisely than a single aliquot could.

When dealing with samples that have cooled slowly through, or resided for long durations (relative to their age), within their partial retention zone, intraaliquot date variations should be interpreted with both eU and grain-size effects in mind. Differences in the magnitude of variations of eU–date and grain-size–date relationships together may yield distinctive constraints, because the influence of eU coevolves with the thermal history and is affected by annealing, whereas the grain-size effect does not.

11.5.2.4 Intragrain diffusion profiles and crystal fragmentation

Another useful consequence of the grain-size to diffusion domain correspondence for apatite is that the intragranular distribution of He, if it can be determined, provides a much higher degree of constraint on a sample's thermal history than a bulk grain date. The general principles of this phenonomenon are described in detail in Chapter 5, and these form the basis for ^{4}He/^{3}He and some aspects of ^{40}Ar/^{39}Ar thermochronology. A less sophisticated, but practically useful way to recover information about the internal He concentration profile is to target analyses of broken grains that integrate the original He concentration profile over varying proportions of the original diffusion domain [*Beucher et al.*, 2013; *Brown et al.*, 2013]. He diffusion profiles will be expressed over a longer distance in directions parallel to the c-axis than they will in the c-axis radial direction. This means that an apatite grain that has lost one or both of its tips along c-axis perpendicular fractures (the natural preferred cleavage direction) will contain a higher He concentration, and therefore date, than the bulk grain would, assuming uniform intragranular and intergranular parent nuclide concentrations. By selectively analyzing a

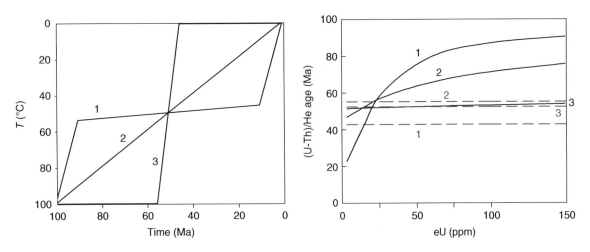

Fig. 11.26. The influence of thermal history on date–eU correlations, according to the RDAAM, for three thermal histories (left) whose date–eU correlations are shown at right. Bold numbers on right correspond to predicted ages from the RDDAM, whereas gray numbers (and dashed lines) are for the monokinetic model for Durango apatite. For the case of monotonic cooling paths such as these, cooling rate through a temperature range of about 40–80 °C is inversely proportion to the difference in age between high and low eU apatite grains. Note that model three predicts only a slight difference between ages across this eU range. (Source: *Flowers et al.* [2009]. Reproduced with permission of Elsevier.)

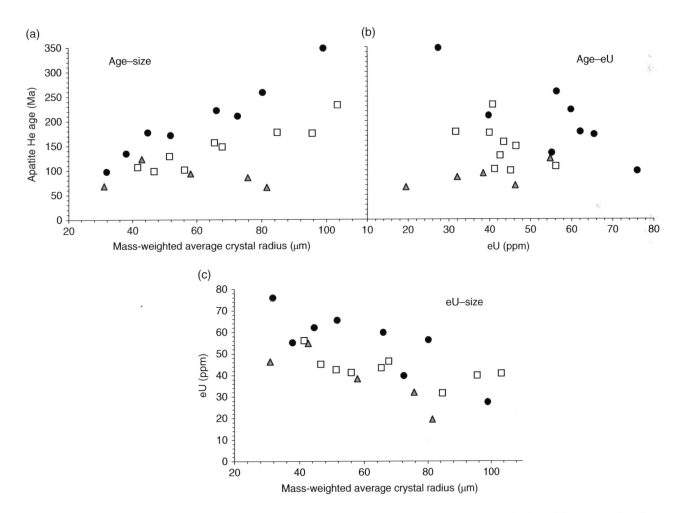

Fig. 11.27. Relationships between apatite He age, grain-size, and eU concentration in multigrain aliquots from three samples from Shell Canyon in the Bighorn Mountains of Wyoming. Filled circles and open squares were presented in *Reiners and Farley* [2001]. (a) Age–size correlations are seen for two samples but not a third (gray triangles). The age–grain-size trends are consistent with a thermal history involving rapid cooling at 600 Ma, followed by burial to depths to temperatures lower than about 90 °C, followed by exhumation to the surface between 120–55 Ma [*Reiners and Farley*, 2001]. (b) Age–eU correlation is seen for the one sample that does not show an age–size correlation. (c) These samples show a broad inverse correlation between eU and grain-size that is often seen among igneous apatite grains from a single hand-sample.

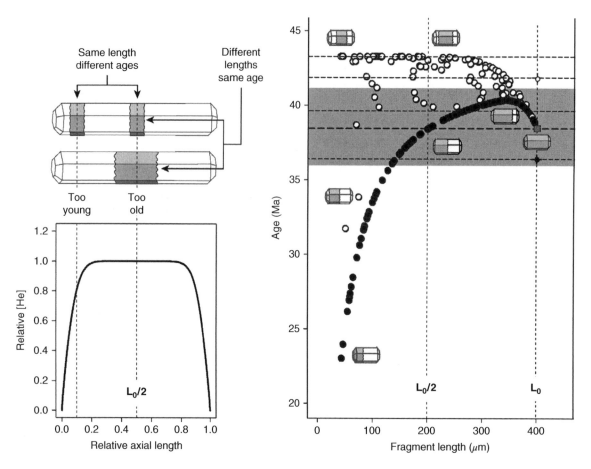

Fig. 11.28. The effect of intracrystalline He zonation on apatite (U–Th)/He ages measured on grains missing terminations due to breakage along planes perpendicular to the c-axis. The He concentration gradient and predicted ages as a function of fragment length are based on the "Wolf 5" thermal history of *Wolf et al.* (1998), which involves slow heating from 20–50 °C over 90 Ma, followed by cooling to 20 °C since 10 Ma. Left: normalized He concentration along a transect parallel to the c-axis, and schematic regions representing fragments that would yield bulk ages that are either "too-young" or "too-old," relative to the bulk age of a whole intact grain. Right: model results for He age as a function of fragment length; lower line delineates ages of grains with at least one original termination remaining. Upper line shows maximum ages of grains lacking original terminations. Square represents the age of the whole, intact crystal. All fragmented grains should occupy the shaded region, showing one possible origin of age dispersion in grains with thermal histories involving prolonged residence in the apatite He partial retention zone. (Source: *Brown et al.* [2013]. Reproduced with permission of Elsevier.)

range of apatite crystals with zero, one, or both c-axis perpendicular terminations removed, one can predict the distribution of this type of age dispersion as a function of grain morphology, for a given thermal history (Fig. 11.28). Grains from a sample having experienced slow cooling to surface temperatures recently will be characterized by large age differences between internal fragments and whole crystals, for example, whereas rapid cooling should result in little to no date variation among morphology types (beyond that predicted by the alpha-ejection phenomenon alone). This approach assumes that morphologic modifications to the crystals occurred only recently, exposing fresh interiors of He concentration profiles in fractured grains.

11.5.3 Zircon

After apatite, the mineral receiving the most attention in (U–Th)/He dating is zircon. The first published analyses are from *Strutt* [1910a, b], who dated a wide range of specimens and

reported dates as young as 100 ka from Vesuvian volcanic eruptions to as old as 565 Ma for zircon from Ontario. Strutt recognized the possibility, indeed likelihood in some cases, that these were not dates of formation but only minimum dates, due to post-formation He loss. Later analyses included He dates on xenocrystic zircons in kimberlites by *Holmes and Paneth* [1936], and zircons in granitoid rocks by *Larsen and Keevil* [1942] and *Keevil et al.* [1944]. All of these studies obtained dates thought to be much younger than the host rocks. In the 1950s, Hurley and coworkers measured a large number of zircon (U–Th)/He dates from specimens in many locations and tectonic settings [*Hurley*, 1952, 1954; *Hurley and Fairbarin*, 1953; *Hurley et al.*, 1956], including some of the first on an important Sri Lankan suite, on which they measured dates of 435–495 Ma for grains with low eU and 130 Ma for a grain with much higher eU. *Damon and Kulp* [1957] also measured (U–Th)/He dates on zircons from Ontario and Sri Lanka.

After several decades of zircon (U–Th)/He dates with mostly ambiguous significance, in the 1950s a good deal of attention turned to attempts to understand the origins of "too-young" dates. One of the potential culprits for He loss was (and still is) radiation damage. *Holland* [1954] noted that features of the zircon crystal structure and a variety of macroscopic properties appear to change relatively rapidly with increasing radiation dose, once dosages reach about 2×10^{18} α/g. Hurley and coworkers [*Hurley*, 1952, 1954; *Hurley and Fairbairn*, 1953; *Hurley et al.*, 1956] related these dosages and property changes to He dates, partly in an attempt to develop use of a radiation-damage proxy for age. This was not successful, but insights from Hurley's work foreshadowed modern (U–Th)/He thermochrology in several ways. For example, he showed that in some regions, a variety of minerals with a wide variety of radiation dosages appeared to record similar dates that were much younger than the known formation age, suggesting that they may "have accumulated He only since some period of metamorphism" [*Hurley*, 1954, p. 326].

More recent studies of zircon (U–Th)/He dating [*Reiners et al.*, 2002, 2004, 2005; *Tagami et al.*, 2003; *Wolfe and Stockli*, 2010] have generally followed approaches and perspectives used in the development of He dating for apatite [e.g., *Farley et al.*, 1996; *Wolf et al.*, 1996, 1998; *Farley*, 2000) and titanite [*Reiners and Farley*, 1999]. Recent work has focused on understanding the effects of radiation damage in zircon [*Guenthner et al.*, 2013], which is similar in some ways, but also more complex than in apatite [*Flowers et al.*, 2009; *Gautheron et al.*, 2009].

11.5.3.1 Step-heating experiments and other approaches

Figure 11.29 shows Arrhenius trends representing some of observed features of He step-heating diffusion experiments on

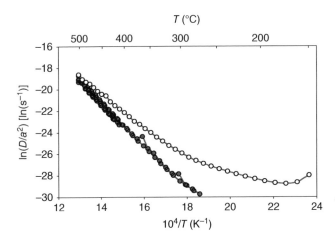

Fig. 11.29. Example of ⁴He-release step-heating results for zircon. These experiments start at low *T* (high 1/*T*) with an initial prograde cycle (open symbols), followed by several subsequent retrograde and prograde cycles (filled symbols). In most experiments, the first ~1% of gas release has higher diffusivity and lower apparent E_a than subsequent steps. Although the origin of this is not understood, most studies therefore only consider and regress results after the initial high-*T* heating step. (Source: *Guenthner et al.* [2013]. Reproduced with permission of American Journal of Science.)

zircon. In general, step-heating diffusion experiments on natural zircon are more complex than those for apatite in several ways. First, even when using slabs or fragments of compositionally homogeneous zircon with no preexisting He concentration profile, experiments typically exhibit progressive changes in the apparent activation energy and frequency factor of diffusion in the first ~1–2% of gas released (Fig. 11.29). It is not clear to what degree this feature is due to different step-heating schedules typically used for zircon experiments. In at least some cases this type of behavior can be removed by performing multiple isothermal steps to remove the first few percent of gas. This behavior has been attributed to several possible causes, including variable domain sizes due to microcracks and radiation damage effects, but the origin is not yet understood. It has been argued that this behavior is not important for bulk zircon (U–Th)/He dating because it is associated with only a small fraction of the gas released. Assuming that this is true and that the kinetic behavior relevant for interpreting lower temperature behavior on geologic timescales is represented by gas released after this first ~1%, most diffusion studies use only the heating steps following the initial high-temperature step of the prograde path of the experiment to derive kinetic parameters.

Other complexities in zircon He diffusion results include a return to slightly concave-up Arrhenius trends in some samples even after release of significant fractions of gas at high temperature, and also slight decreases in apparent D_0/a^2 in some experiments. None of these complexities are understood either, but one possible qualitative explanation is degassing, recharging, and exhaustion of domains with distinct diffusivity.

Ignoring these complexities, regressions of step-heating He diffusion data from natural zircons with intermediate extents of radiation damage (effective alpha dosages between $\sim 2 \times 10^{16}$ and 2×10^{18} α/g, which are approximately the doses for a typical zircon with 300 ppm U and 150 ppm Th with ages between ~20 Ma and 2.0 Ga) yield E_a values between ~160 and 170 kJ/mol and D_0 between ~0.1 and 1 cm²/s [*Reiners et al.*, 2004; *Guenthner et al.*, 2013]. The mean kinetic parameters E_a and D_0 for intermediate-damage zircons from these studies yield E_a of 168 ± 4.3 and D_0 of $0.172^{+0.56}/_{-0.13}$ (uncertainties as standard deviations of values from 11 different specimens). Using an equivalent sphere radius of 60 μm, and a cooling rate of 10° C/Ma, this corresponds to a T_c of 190 °C. This is near the upper end of the range and higher than the mean of T_c values determined by *Reiners et al.* [2004] (171–196 °C; 183 °C, respectively), likely because most specimens studied by *Guenthner et al.* [2013] had alpha dosage more than a factor of ten higher. Indeed, as discussed later, radiation damage plays an important role in He diffusion kinetics for zircon, although in a more complex way than in apatite.

In contrast to apatite, He diffusion in at least some varieties of zircon is significantly anisotropic. Computational models of ideal zircon crystals predict that different crystallographic orientations should have activation energies that differ by hundreds of kJ/mol [*Reich et al.*, 2007; *Bengston et al.*, 2012]. This was proposed as an explanation for the progressive change in step-heating behavior in the first few percent of gas released [*Reich et al.*, 2007]. However, anisotropy cannot explain the loss (or at least dramatic reduction) of this behavior at low temperature with progressive He extraction.

Both ion implantation/NRA and step-heating experiments indicate that anisotropy in zircon manifests itself not in E_a but in D_0 differences. This is not necessarily expected for a simple model of interstitial volume diffusion in which one would expect interionic lattice gaps in one direction to be manifest as higher activation energies. Using ion-implantation, heating, and NRA, *Cherniak et al.* [2009] measured diffusion in the c-axis parallel and perpendicular directions, finding identical E_a, but D_0 parallel to the c-axis two orders of magnitude higher than that in the c-axis perpendicular direction (Fig. 11.30). The diffusivities from this study match the E_a and range of D_0 measured in natural ^{4}He diffusion experiments, and neither show evidence that anisotropy is manifest as significantly different activation energies in natural zircon. Helium-3 diffusion experiments on synthetic orthophosphate crystals with zircon structure also indicate much faster diffusion in the c-axis-parallel direction.

Farley [2007] also found a strong cation compositional control on diffusivity, consistent with control by ionic porosity on bulk diffusivity and T_c. Another important result of the synthetic orthophosphate study was that apparent diffusivities for

phosphate isomorphs of zircon yielded much lower T_c for any composition than is observed in actual zircon. Assuming an approximate correspondence between diffusion domain length scale a and grain size, and assuming an equivalent spherical radius corresponding to that of a typical zircon (roughly 60 μm), the apparent T_c values for *Farley's* [2007] synthetic orthophosphates would range from –11 to 17 °C. Part of this T_c difference potentially can be explained by the lower ionic porosity and c-axis-parallel widths of actual zircon compared to its phosphate isomorphs. But extrapolating the results to values appropriate for zircon would still predict a T_c much lower than that observed in natural zircon. *Farley* [2007] speculated that this may be due to radiation damage that effectively "clogs up" c-axis-parallel channels, decreasing bulk diffusivity.

11.5.3.2 Anisotropic diffusion of He in zircon

Theoretical and experimental results suggest that He diffusion is strongly anisotropic with respect to crystallographic orientation in several minerals, including zircon, rutile, and monazite, although there is little consensus about the details of how this anisotropy is expressed in D_0/a^2 or E_a, and how important it is in natural specimens compared with other complications such as radiation damage and intragranular defects and structural domains. Some observations also suggest, and concord with simple crystallographic expectations, that Ar diffusion is strongly

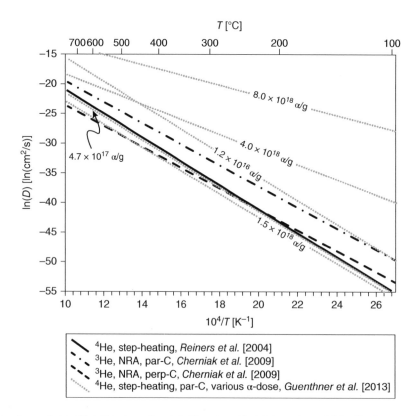

Fig. 11.30. Summary Arrhenius diagram for He diffusivity in zircon. Gray dotted lines show diffusivity in c-axis-parallel direction for compositionally uniform zircon grains with differing radiation doses as shown. Dashed and dash-dot lines show NRA experiments, and solid line shows trend corresponding to average kinetic parameters of several specimens with intermediate levels of radiation dosage.

anisotropic in micas, but as with He there is some question about the practical significance of this, as other evidence suggests an important role for structurally bound microdomains [*Harrison et al.*, 2009]. The largest departures from isotropic diffusion observed so far seem to be manifest between c-axis-parallel and c-axis-perpendicular directions. Helium in zircon, rutile, and orthophosphates appears to diffuse significantly faster in the c-axis parallel direction [e.g., *Farley*, 2007; *Cherniak et al.*, 2009; *Cherniak and Watson*, 2011], at least in theoretical predictions and in weakly radiation-damaged samples. In contrast, argon in micas appears to diffuse faster perpendicular to the c-axis, consistent with the greater lattice spacing or ionic porosity in this direction. Although the senses of anisotropy in these cases are antithetical, both cases motivate consideration of the diffusion domain as a cylinder. There are two main considerations for how anisotropy may be manifest in thermochronologic applications:

(1) the interpretation of experimental diffusion results, especially in the case of incremental step-heating to derive Arrhenius trends and kinetic parameters;

(2) the interpretation of the significance of observed dates themselves and their potential variation among specimens from the same sample.

One possible approach to dealing with anisotropy might be to ignore diffusion in the slow direction and assume all significant loss occurs either parallel or orthogonal to the long-axis direction. This is common practice for Ar diffusion in micas. Even though micas typically exhibit much larger c-axis-perpendicular than c-axis-parallel grain size dimensions, or perhaps more importantly the cleavage plane spacing in the same orientations, most Ar migration is thought to occur between and parallel to the crystallographic sheets. Thus the dimension of the diffusion domain is taken to be that in the c-axis-perpendicular direction, and its morphology is taken to be an infinite cylinder.

For He in zircon and rutile, theoretical computations suggest that anisotropy should be enormous [e.g., *Reich et al.*, 2007; *Bengston et al.*, 2012], which might warrant ignoring the c-axis perpendicular contribution and considering domains as infinite plane-sheets with dimensions equal to c-axis-parallel lengths. However, experiments on specimens of natural zircons suggest much less pronounced anisotropy, with c-axis-parallel and c-axis-perpendicular differences of about one order of magnitude and similar activation energies so that this difference does not change significantly as a function of temperature. Thus the effects of anisotropy in at least the case of zircon, and perhaps rutile as well, require more sophisticated treatments.

Watson et al. [2010] presented both analytical and numerical solutions for fractional losses and equivalent sphere diffusivities in circular axisymmetric cylinders where diffusion in the c-axis-parallel direction (D_{33}) is distinct from that in the c-axis-perpendicular direction (D_{11}). This morphology is obviously a simplification to the tetragonal and hexagonal cases of zircon, rutile, and orthophosphates, but the loss of accuracy is likely small. Using this formulation, and calling the c-axis-parallel

and c-axis-perpendicular directions z and r, respectively, the change in daughter concentration within the cylinder is

$$\frac{\partial C}{\partial t} = D_{11}\frac{1}{r}\frac{\partial}{\partial r}\left(r\frac{\partial C}{\partial r}\right) + D_{33}\frac{\partial^2 C}{\partial z^2} \tag{11.27}$$

Watson et al. [2010] also define dimensionless release coordinates $y_1(t)$ and $y_3(t)$ corresponding to the c-axis-perpendicular and c-axis-parallel directions, respectively, as

$$y_1(t) = \int_0^t \frac{D_{11}(t')}{a^2}dt \tag{11.28}$$

and

$$y_3(t) = \int_0^t \frac{D_{33}(t')}{h^2}dt \tag{11.29}$$

where a and h are the radius and height (length) of the cylinder respectively. Obviously $y_1(t)$ and $y_3(t)$ simplify to their respective Dt/a^2 values for square-pulse heating events. Assuming an initially uniform concentration of the diffusing species of C_0, after a heating event yielding $y_1(t)$ and $y_3(t)$, the concentration profile is

$$C(r,z,t) = C_0 R(r,t)Z(z,t) \tag{11.30}$$

where

$$R(r,t) = 2\sum_{m=1}^{\infty} \frac{J_0\left(\frac{\alpha_m r}{a}\right)}{\alpha_m J_1(\alpha_m)}\exp\left[-\alpha_m^2 y_1(t)\right] \tag{11.31}$$

and

$$Z(z,t) = \frac{4}{\pi}\sum_{n=0}^{\infty} \frac{(-1)^n}{(2n+1)}\cos[(2n+1)\pi z/h]\exp\left[-(2n+1)^2\pi^2 y_3(t)\right] \tag{11.32}$$

As in the solutions for the concentration profile in an isotropic cylinder earlier, a_m represents the mth root of the Bessel function of the first kind of order zero (J_0) or of the first order (J_1).

Watson et al. [2010] also derived expressions for fractional retention and loss of a diffusing species as a function of y_1 and y_3, the direction-specific measures of Dt/a^2. Fractional retention after events characterized by y_1 and y_3 will be a function of the retained fraction (i.e., not lost via one direction or the other) in the c-axis perpendicular and parallel directions, G_1 and G_3

$$G(y_1,y_3) = G_1(y_1)G_3(y_3) = \left[4\sum_{m=1}^{\infty}\frac{1}{\alpha_m^2}\exp\left[-\alpha_m^2 y_1(t)\right]\right]$$
$$\times \left[\frac{8}{\pi^2}\sum_{n=0}^{\infty}\frac{1}{(2n+1)^n}\exp\left[-(2n+1)^2\pi^2 y_3(t)\right]\right] \tag{11.33}$$

The fractional loss is then

$$F(y_1,y_3) = 1 - G(y_1,y_3) = 1 - G(y_1)G(y_3) = 1 - [1-F_1(y_1)][1-F_3(y_3)] \tag{11.34}$$

or

$$F(y_1,y_3) = F_1(y_1) + F_3(y_3) - F_1(y_1)F_3(y_3) \tag{11.35}$$

Equation (11.35) makes sense because small losses can be reasonably approximated by the sum of the losses from the sides (y_1 direction) and top/bottom (y_3) of the cylinder, but when these loss profiles begin to feel one anothers' effects at higher fractional losses, the cross-term is necessary.

As with fractional loss from isotropic domains, practically useful approximations to these fractional loss expressions can be derived that are suitable over certain ranges of loss. For the case of the diffusionally anisotropic cylinder, *Watson et al.* [2010] combined approximations to the infinite sheet and infinite cylinder to obtain

$$F(y_1, y_3) \cong \frac{4}{\sqrt{\pi}} \left(\sqrt{y_1} + \sqrt{y_3} \right) \tag{11.36}$$

for F_1 and F_3 less than about 0.1, and

$$F(y_1, y_3) \cong \frac{4}{\sqrt{\pi}} \left(\sqrt{y_1} + \sqrt{y_3} \right) - y_1 - \frac{16}{\pi} \sqrt{y_1 y_3} - \frac{y_1^{3/2}}{3\sqrt{\pi}} + 4 y_1 \sqrt{\frac{y_3}{\pi}} + \frac{4}{3\pi} y_1^{3/2} \sqrt{y_3} \tag{11.37}$$

for F_1 and F_3 up to about 0.5. Unfortunately simple approximations have not be found for larger fractional loss.

Watson et al. [2010] pointed out that the directional difference in fractional release, and thus the anisotropy of concentration profiles within a diffusionally anisotropic cylinder, reach a maximum at an intermediate value of $y_1 = y_3 = 0.12$. However, the actual fractional release and concentration profiles will depend on the actual D_{11} and D_{33} that compose y_1 and y_3. They also noted the similarity between the low-release fractional release equation (11.36) with that for a sphere

$$F_s(y_s) \cong 6 \sqrt{\frac{y_s}{\pi}} \tag{11.38}$$

where

$$y_s(t) = \int_0^t \frac{D_s(t')}{r^2} dt' \tag{11.39}$$

and D_s is the diffusivity in the sphere and r is its radius (i.e., the Dt/a^2 or Dt/r^2 for a square-pulse heating step). Equating the low-release fractional losses of the sphere and diffusionally anisotropic cylinder (equations (11.38) and (11.36)) yields

$$y_s \approx \frac{4}{9} \left(\sqrt{y_1} + \sqrt{y_3} \right)^2 \tag{11.40}$$

which shows that if $y_1 \gg y_3$ or $y_3 \gg y_1$, then the controlling parameter y (Dt/a^2, or Fo) for fractional loss is either y_1 or y_3, which, when multiplied by 4/9, can be used in the fractional loss equation for a sphere. In the case of a square-pulse heating step,

$$\frac{D_s}{r^2} = \frac{4}{9} \left(\frac{D_{11}}{a^2} + \frac{D_{33}}{b^2} + 2 \sqrt{\frac{D_{11} D_{33}}{a^2 b^2}} \right) \tag{11.41}$$

which relates the effective bulk diffusivity of the diffusionally anisotropic cylinder to that of an isotropic sphere with equivalent bulk effective diffusivity.

Several numerical methods for modeling diffusional anisotropy allow simulation of complex boundary and initial conditions, including the finite element CYLMOD of *Cherniak et al.* [2009] and *Watson et al.* [2010] for cylindrical morphology, the lattice-Boltzmann approach of *Huber et al.* [2011], and the Monte Carlo diffusion model of *Gautheron and Tassan-Got* [2010]. The latter models permit general solutions for the loss and ingrowth of daughter product in grains diffusional anisotropy combined with arbitrary shapes and diffusional anisotropy such as tetragonal prisms and prisms with pyramidal terminations. These models generally show that with the exception of minor differences arising from edge effects in complex grain shapes, the diffusional behavior of any morphology can be represented without large (~10%) inaccuracies by recasting a grain into some kind of equivalent sphere, for example by scaling diffusion rates in the different crystallographic directions according to the shape of the grain, to yield an effective $D(t)/a^2$, where a is the radius of the "active sphere."

Examples

Using the finite cylinder as an example provides several insights into the potential importance of diffusional anisotropy for the types of cases that might be encountered in real minerals. Consider a cylinder with radius a and height h, measured perpendicular and parallel to the c-axis, respectively, that exhibits c-axis-perpendicular diffusivity

$$D_{11} = D_{0_{11}} \exp[-E_{a_{11}}/RT] \tag{11.42}$$

and c-axis-parallel diffusivity

$$D_{33} = D_{0_{33}} \exp[-E_{a_{33}}/RT] \tag{11.43}$$

D_{11} may be greater than D_{33} or vice versa, depending on their relative D_0 or E_a values and, in the latter case, temperature. Here we assume for simplicity (and as we consider likely given that most isokinetic points for the same diffusant are at high T compared with temperatures of thermochronologic interest such as T_c) that the system with higher E_a will have lower D at $T < \sim 1000\,°C$.

Figure 11.31a shows the predicted Arrhenius trends as $\ln(D/\lambda^2)$, where $\lambda = a$ for D_{11}, h for D_{33}, and r for the active sphere as in equation (11.41), assuming a cylinder with $h = 200\,\mu m$ and $a = 50\,\mu m$, for a height to width aspect ratio $AR = 2$. Aspect ratios for typical apatite, zircon, rutile, and other tetragonal or hexagonal minerals are typically within 1.5–3.5. Micas often have much lower aspect ratios. In Fig. 11.31a–c, both directions of diffusion have the same E_a (150 kJ/mol), but $D_{0_{33}}$ is $10^4 \times D_{0_{11}}$, so D_{33} is ~9 ln units faster than D_{11} at any T. Because diffusional loss in this grain is dominated by migration in the D_{33} direction, the effective bulk diffusivity for the active equivalent sphere D_s/r^2 (equation 11.41]) is close to D_{33} although it is shifted slightly towards the slower D_{11} (Fig. 11.31a). T_c (for an assumed cooling rate of 10 °C/Ma) provides a useful index of the thermochronologic implications of this. For a grain with $h = 200\,\mu m$ and $a = 50\,\mu m$, the apparent T_c of the equivalent sphere is 155 °C,

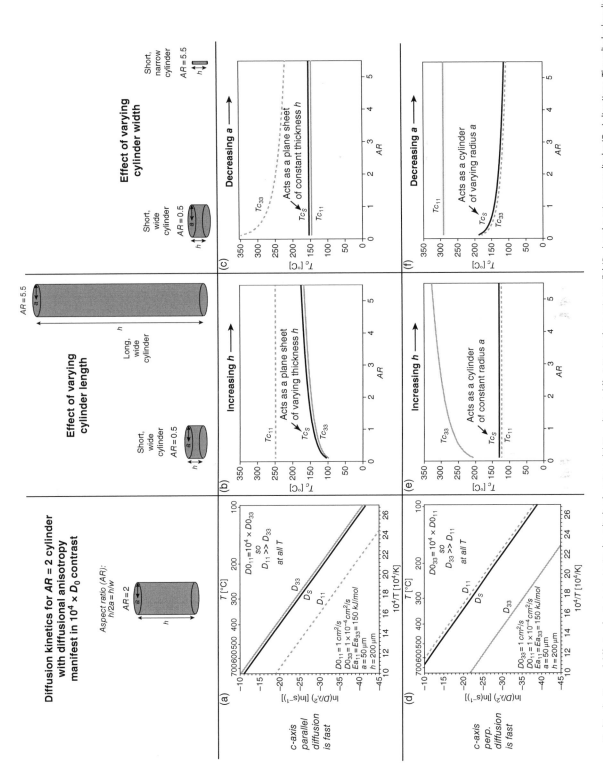

Fig. 11.31. Some practical implications of diffusional anisotropy in circular cylinder domains exhibiting distinct diffusivities in the c-axis-parallel (D_{33}) and c-axis-perpendicular (D_{11}) directions. The cylinder has dimension directions a and h as shown in the sketches at the top. Aspect ratio (AR) is the height to width ratio ($h/2a$), where λ is a for the radial (c-axis-perpendicular dimension), h for the c-axis-parallel direction, and r for the equivalent sphere radius (equation 11.41). In this example $D0_{11}$ is 10^4 lower than $D0_{33}$, so D_{33} is 10^4 times faster than D_{11}, so diffusive loss is nearly all out the ends of the prism rather than out the c-axis-parallel sides. This is reflected in the fact that D_s (the effective diffusivity of the equivalent sphere) is only slightly lower than D_{33}. (b,c) Apparent T_c of the equivalent sphere (T_c values) as a function of aspect ratio via changing h and a. The effective diffusivity of the grain changes only when h changes. Hypothetical T_c values of the D_{33} and D_{11} directions are also shown for reference. (d) Same as (a) except D_{11} and D_{33} are reversed (c-axis-perpendicular diffusion is faster). (e,f) T_c values and hypothetical T_c of the D_{11} and D_{33} directions corresponding to D.

close to the hypothetical T_c of the relatively fast D_{33}/h^2 dimension of 147 °C, and much less than the hypothetical T_c of the slow D_{11}/a^2 dimension. Varying the aspect ratio from 0.5 to 5.5 (by holding a constant at 50 μm and varying h from 50 to 550 μm) changes the length scale over which most of the diffusant is lost, resulting in a change in T_c of the equivalent sphere from 104 to 175 °C (Fig. 11.31b). In contrast, varying the aspect ratio by holding h constant at 200 μm and changing a from 200 to 18 μm does not change the length scale of fast diffusion, so the T_c of the equivalent sphere remains fixed at 155 °C (Fig. 11.31c). This type of behavior is qualitatively similar, although much more extreme, to that predicted for He diffusion in zircon, as discussed later. The predicted thermochronologic significance of this is that fractional losses, T_c values, and observed dates (for slowly cooled or partially reset samples) should vary only with c-axis-parallel length. Systems where c-axis parallel diffusivity is much faster than other directions could theoretically be treated as approximating infinite plane sheets with c-axis dimension h.

Reversing the relative diffusivities of the c-axis-parallel and c-axis-perpendicular directions (Fig. 11.31d,f) means that most diffusive loss occurs perpendicular to the c-axis, through the cylinder prism face. Changing aspect ratio by varying h (unless h reaches extremely small dimensions) therefore has no effect on the bulk diffusivity behavior of the system, and T_c for the equivalent sphere remains pinned at 130 °C regardless of the length of the cylinder (Fig. 11.31e). When aspect ratio is increased by decreasing cylinder radius, however, T_c of the equivalent sphere decreases as a becomes smaller (Fig. 11.31f).

As long as the isokinetic crossover for the Arrhenius trends of the c-axis-parallel and c-axis-perpendicular directions is well above the T_c for the equivalent sphere, differing diffusivities due to E_a differences has similar effects as due to D_0 values (Fig. 11.32). The case shown in Fig. 11.32a–c is qualitatively similar (but much exaggerated) to the behavior of He observed in some natural zircon crystals and synthetic isostructural analogues. If this applies to natural zircon, it implies that the c-axis-perpendicular width of grains is irrelevant to their thermochronologic properties, but the length of the c-axis-parallel dimension could exert a strong control on properties such as fractional loss and T_c. Figure 11.33 shows the predicted equivalent-sphere T_c for zircon using the anisotropically resolved NRA–^{3}He diffusion experiment results of *Cherniak et al.* [2009], for grains with a simplified circular cylinder morphology, cylindrical radius of 50 mm, and AR from 0.1–5.5 (dashed line). Over this range of prism lengths ($h = 10$–550 μm), the equivalent sphere T_c values vary from less than 120 °C to more than 170 °C. However, even if the dimensions of actual zircon grains correspond perfectly to diffusion-domain length scales and internal discontinuities do not reduce the effective diffusion-domain length scales, most natural zircons have a much narrower range of aspect ratios. In a database of several hundred example zircon grains from many different geologic settings, roughly two-thirds of the grains have AR of 2.2 ± 0.6 (one

standard deviation), corresponding to sphere-equivalent T_c values of 160–168 °C, and essentially all have AR between about 0.8 and 3.5 (T_c values of 148–171 °C [Fig. 11.33]); T_cs calculated using *Cherniak et al.'s* [2009] kinetics). In addition, radiation damage over commonly encountered ranges of dose ($\sim10^{16}$–10^{18} α/g) appear to have a much stronger control on He diffusivity in zircon than anisotropy (though as discussed below, it is possible that at least some of the effects of radiation damage on He diffusivity at low dose may be related to progressive destruction of anisotropy).

11.5.3.3 Radiation damage in zircon

As discussed earlier, the role of radiation damage on He diffusivity at relatively high doses in zircon was explored by several studies in the 1950s and by *Nasdala et al.* [2004]. These studies noted that He diffusivity appeared to increase dramatically, and He dates decrease dramatically, at dosages exceeding about 2×10^{18} α/g. *Guenthner et al.* [2013] expanded the investigation of radiation damage effects of He diffusion in zircon to both high and low extents of damage. They combined previously collected step-heating experiments with new ones on polished, crystallographically oriented slabs of homogeneous zircon with widely varying effective alpha doses. In this study effective alpha dose was calculated based on the eU concentration of the grain and an estimate of the time of cooling through temperatures of roughly 200–300 °C. This was undertaken in order to use effective alpha dose as a proxy for radiation damage, with the rationale that damage is annealed with kinetics similar to those of zircon fission-track annealing.

As in previous studies, *Guenthner et al.* [2013] found that zircon with effective doses higher than about 1–2×10^{18} α/g displayed dramatically lower retentivities. But at lower doses, they found a positive correlation between T_c and dose, similar to observations in apatite. Assuming effective alpha dose is indeed a proxy for radiation damage, radiation damage in zircon initially decreases diffusivity and increases T_c from about 125 °C at 1×10^{16} α/g, to ~190 °C at $\sim1.5 \times 10^{18}$ α/g. With higher dose, T_c then decreases rapidly, reaching apparent T_c values of $\sim$ –60 °C for fully amorphous zircon.

Guenthner et al. [2013] also found evidence for anisotropic He diffusion in at least one sample, again manifest as about a one-order-of-magnitude difference in D_0/a^2, with no difference in E_a. However, the magnitude of this effect on variation in T_c (~20 °C) is small compared with the effect of effective alpha dose (~65 °C below 1–2×10^{18} α/g, and much larger above this dose). Thus it is likely that zircon He date variations among single grains that have experienced the same thermal history (e.g., from the same crystalline rock) will be more strongly controlled by eU variations than by any index related to anisotropy, such as aspect ratio.

The initial increase in T_c with (inferred) radiation damage in zircon could, in principle, have the same origin as that ascribed to a similar behavior in apatite: trapping (favorable energetic partitioning) of ^{4}He into radiation damage regions, effectively

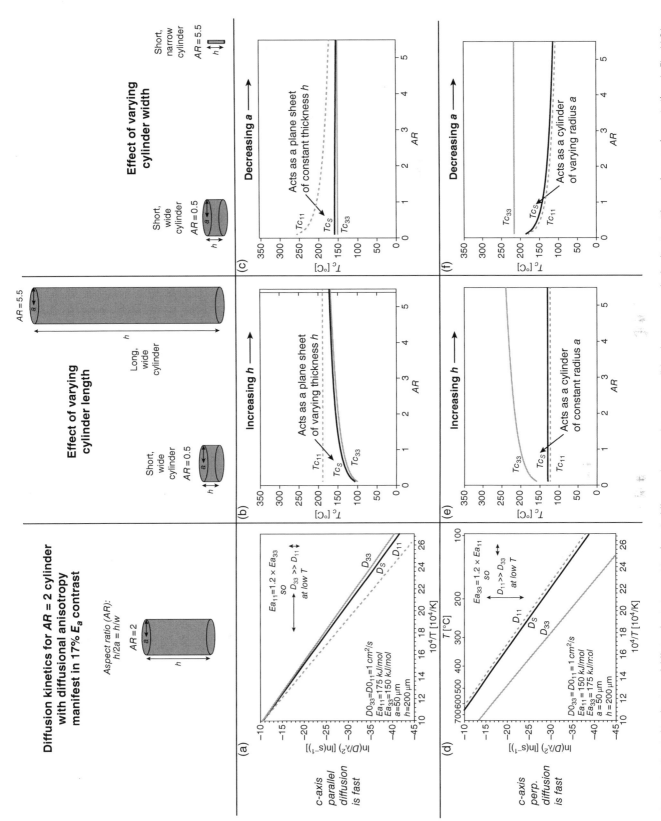

Fig. 11.32. Practical implications of diffusional anisotropy arising from E_a differences between the c-axis-parallel and c-axis-perpendicular directions. The panels are analogous to those in Fig. 11.31.

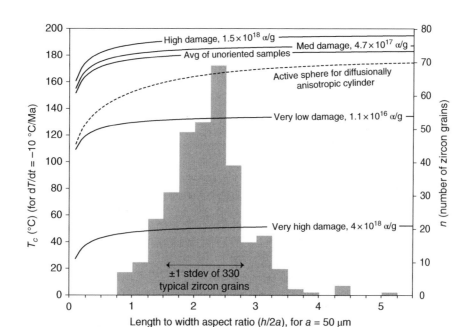

Fig. 11.33. Closure temperature of equivalent sphere zircon grains with simplified circular cylinder morphologies and aspect ratios 0.1–5.5 with $a = 50\ \mu m$. Dashed line: T_c of equivalent (active) sphere for diffusionally anisotropic zircon with orthogonal diffusivities as measured on ^{3}He by NRA [*Cherniak et al.*, 2009]. The increasing T_c is caused by increasing prism length in the fast-diffusing direction. Solid lines: T_c of equivalent sphere morphologies of zircon with diffusivities measured on zircon with varying radiation doses. Increasing radiation dose from 1.1×10^{16} α/g to 1.5×10^{18} α/g causes increasing T_c across the entire range of T_c for the active sphere for the diffusionally anisotropic case. Increasing dose from 1.5×10^{18} α/g then causes rapidly decreasing T_c, as shown in the 4×0^{18} α/g trend.

slowing the migration of diffusing ^{4}He. However, several observations, including (i) evidence for strong c-axis-parallel preferred He migration [*Farley*, 2007; *Cherniak et al.*, 2009; *Guenthner et al.*, 2013], (ii) the computational result that ideal zircon crystals should have lower T_c than apatite [*Bengston et al.*, 2012], and (iii) much lower T_c in zero-damage zircon-structure phosphate isomorphs [*Farley*, 2007], led *Guenthner et al.* to propose a different explanation. As suggested by previous studies, they reasoned that the initial decrease in diffusivity is due to progressive clogging of preferred c-axis-parallel channels by radiation damage, presumably alpha recoil tracks. The thinking behind this is that although alpha recoil tracks may have high-vacancy amorphous interiors, their "rinds" are expected to be relatively high-density defect regions, which could block access to large lattice channels by migrating ^{4}He (Fig. 11.34).

This increase in tortuosity can be described as

$$D_e = \frac{D_z}{\tau} \tag{11.44}$$

where D_e is the effective diffusivity in the bulk zircon grain, D_z is the diffusivity in the lattice of a minimally damaged endmember, and τ is a term reflecting tortuosity. Tortuosity could take a variety of forms, but *Guenthner et al.* [2013] chose to use an expression for the mean length a particle could travel in the lattice before encountering an alpha-recoil damage zone, l_{int}, as defined by *Ketcham et al.* [2013]. Hence

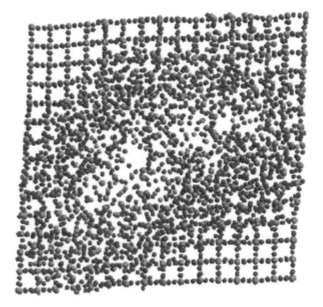

Fig. 11.34. Computed model image of an alpha-recoil track in zircon lattice resulting from recoil of a heavy parent nuclide that emitted an alpha particle. Note high-density rind and high-vacancy core of the track. View is down c-axis. (Source: *Trachenko et al.* [2002]. Reproduced with permission of American Physical Society.)

$$\tau = \left(\frac{l_{int0}}{l_{int}}\right)^2 \tag{11.45}$$

where l_{int0} is this distance in a minimally damaged zircon referenced to 1×10^{14} α/g. In *Guenthner et al.* [2013] l_{int0} = 45920 nm.

In detail, l_{int} values are computed by a geometrical model of alpha-recoil track growth and intersection, and depend on the surface-area-to-volume ratio of an alpha recoil track *SV*, and a term for the fraction of damage in a lattice, which is taken to be the amorphous fraction f_a, using the direct-impact model for zircon [*Gibbons*, 1972]. As determined by *Ketcham et al.* [2013]

$$l_{int} = \frac{4.2}{f_a SV} - 2.5 \tag{11.46}$$

where

$$f_a = 1 - \exp[-B_a\alpha] \tag{11.47}$$

and α is the effective alpha dose and B_a is the mass of amorphous material produced per alpha decay event (5.48×10^{-19} g/α-event).

The dramatic increase in diffusivity at doses above about $1–2 \times 10^{18}$ α/g, as well as the slowing of rate of increase of diffusivity approaching this value, are considered to result from progressively larger degrees of interconnection of high-vacancy cores of alpha recoil tracks. The diffusivity of He within the alpha recoil tracks themselves is assumed to be that of the completely amorphous specimen studied by *Guenthner et al.* [2013], the N17 glass and erstwhile zircon of *Nasdala et al.* [2004].

Combining the diffusivity of He in the remaining lattice with that in the amorphous alpha recoil tracks as a harmonic average,

and accounting for the changing effective length scales of both domains due to partitioning of the structure among them, yields

$$\frac{1}{D_c/a^2} = \frac{1}{\left(\frac{l_{int0}}{l_{int}}\right)^2} \times \frac{f_c'}{D_z/_{(a \cdot f_c')^2}} + \frac{f_{a'}}{D_{N17}/_{(a \cdot f_{a'})}} \tag{11.48}$$

where *a* is the equivalent sphere radius, and f_c' and f_a' are modified crystalline and amorphous fractions

$$f_a' = 1 - \exp[-B_a\alpha\Omega] \tag{11.49}$$

$$f_c' = 1 - f_a' \tag{11.50}$$

where Ω is an empirical parameter set to 3 in the model. The physical significance of Ω is not known but a value above unity could be thought of increasing the onset of interconnection of damage zones, possibly as a result of interaction between alpha recoil and fission tracks [*Ketcham et al.*, 2013].

Model predictions for changes in diffusivity and T_c with increasing effective alpha dose are shown in Fig. 11.35. As in the RDAAM for apatite, in order to put this model of the damage–diffusivity relationship to practical use, some knowledge, or at least assumption, about the kinetics of annealing of the radiation damage that affects He diffusivity is required. Unfortunately experiments demonstrating changes in He diffusivity resulting from controlled and reversible damage accumulation and annealing are not available for zircon, as they are for apatite [*Shuster and Farley*, 2009]. Empirical models for damage annealing have been developed for short durations and laboratory conditions, but at least those developed thus far are not appropriate for geologic timescales. Therefore the existing model for damage annealing in zircon assumes that pertinent damage anneals with the same kinetics as fission-tracks, analogous to the approach taken for apatite [*Guenthner et al.*, 2013].

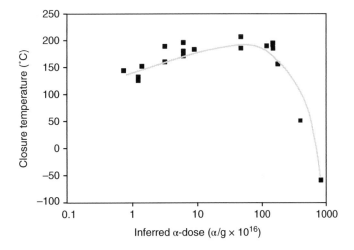

Fig. 11.35. Closure temperatures (for cooling rate of 10 °C/Ma) of zircon specimens (filled symbols) with varying alpha doses inferred from eU concentrations and cooling ages (either ZFT or (U–Th)/He). Also shown is the model prediction for zircon as a function of alpha dose predicted from the *Guenthner et al.* [2013] model. All closure temperatures are shown for *a* = 60 μm and d*T*/d*t* = 10 °C/Ma. Note gradual decrease in diffusivity (and increase in T_c) at low effective alpha dose, followed by rapid increase in diffusivity (and decrease in T_c) at effective alpha doses higher than about $1–2 \times 10^{18}$ α/g. (Source: *Guenthner et al.* [2013]. Reproduced with permission of American Journal of Science.)

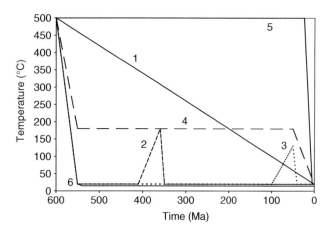
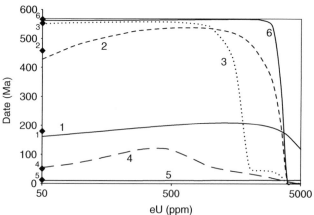

Fig. 11.36. Model predictions of date–eU correlations (right) for various thermal histories (left). Both positive and negative date–eU correlations are possible, depending on the timing of cooling events. Numbers next to y-axis represent dates predicted by a static kinetic model using kinetic parameters from *Reiners et al.* [2004]. (Source: *Guenthner et al.* [2013]. Reproduced with permission of American Journal of Science.)

Because the relationship between He diffusivity and effective alpha dose in zircon changes sign after a critical dosage, correlations between (U–Th)/He date and eU concentration (as proxy for relative radiation damage among grains with the same thermal history) can be complex, depending on the thermal history of a grain (Fig. 11.36).

11.5.4 Other minerals

A large number of minerals and bulk materials besides apatite and zircon have been the focus of attempted (U–Th)/He dating, with varying degrees of success. A smaller number, but at least a dozen or so, have also been the focus of He diffusion experiments, which is a necessary but not sufficient requirement for development of a practical thermochronometer. In principle (U–Th)/He dating might be expected to be more successful than it is generally found to be in experimental applications, given the expected low initial He concentrations and relatively high concentrations of U and Th (typically ppm) in many minerals. Potential complications likely arise from many sources, however, including U and Th mobility within and on grain-boundaries of crystals and complex and possibly evolving diffusion domain structure of minerals through time. The high mobility of He itself, which may be quantifiable and predictable under ideal conditions, may lead to complex loss and accumulation patterns in minerals with realistic defects, impurities, and damage. Nonetheless, numerous examples exist demonstrating promising results and insights, which we review briefly here.

11.5.4.1 Fe-oxides

Many early He dating studies [e.g., Strutt, 1910a; Hurley, 1954] analyzed Fe-oxides, with varying degrees of success. More recent studies have revived interest in Fe-oxide (U–Th)/He dating, due in part to their common occurrence, moderate to occasionally high eU contents, and interest in the diverse processes, including

crustal fluid flow and shallow weathering, leading to their formation. *Fanale and Kulp* [1962] and *Blackburn et al.* [2007] dated magnetite from hydrothermal and volcanic settings; the latter measured ^{4}He diffusion and proposed one of the highest E_a values found for He thus far (~218 kJ/mol) and an approximate T_c of ~250 °C, based on a relatively linear portion of an Arrhenius trend after ~16% of gas release. Several studies by Lippolt and coworkers [*Lippolt et al.*, 1993; *Bahr et al.*, 1994; *Wernicke and Lippolt*, 1994a,b] focused on hematite He dating, performing He diffusion experiments that yielded apparent closure temperatures of roughly 100–200 °C, and measuring (U–Th)/He dates on specular and botryoidal hematite from hydrothermal veins consistent with geologic constraints. Although the results of their He diffusion studies are not simple, linear regression of the steps below about 900 °C could be interpreted as indicating closure temperatures of about 100–200 °C. *Farley and Flowers* [2012] found evidence for multiple He diffusion domains in polycrystalline hematite (Fig. 11.37), which they proposed are caused by variations in crystal size over a factor of roughly 10^3–10^4. Perhaps not coincidentally this would be consistent with the size variation observed in the actual crystal sizes of the sample, leading to the tentative suggestion of grain size and diffusion domain size equivalency in hematite supported by other experiments that estimated D_0 [*Evenson et al.*, 2014]. Farley and Flowers interpreted the step-heating ^{4}He/^{3}He release of these samples, together with corresponding nucleogenic ^{21}Ne [(U–Th)/Ne] dates, in the context of a multidomain-diffusion model to reconstruct the cooling (and reheating) history of the sample.

Other Fe-oxide He dating and ^{4}He/^{3}He thermochronology studies have focused on pedogenic or lateritic minerals [*Pidgeon et al.*, 2004; *Shuster et al.*, 2005b, 2006, 2012; *Heim et al.*, 2006; *Vasconcelos et al.*, 2013]. Several of them have dated Cenozoic weathering and formation of residual Fe-oxides in Australia and Brazil, finding results consistent with geologic constraints, and in some cases allowing comparison between (U–Th)/He

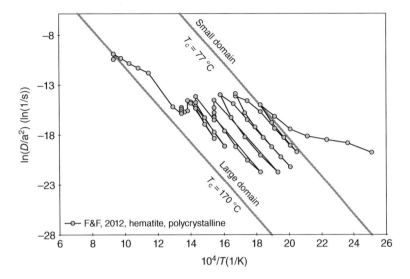

Fig. 11.37. Step-heating ^{3}He diffusion results from fine-grained polycrystalline hematite (gray circles; *Farley and Flowers* [2012]). The polycrystalline hematite-release pattern is interpreted to result from a multidomainal structure in which, assuming constant E_a and D_0, the sizes of the domains carrying > 95 % of the gas vary by a factor of about 10^3 and have T_c values ranging from about 60–170 °C. This size range is qualitatively consistent with observed crystal sizes in the sample, suggesting that crystal size scales with diffusion domain size in hematite.

and ^{40}Ar/^{39}Ar dates interpreted as formation ages, and cosmogenic ^{3}He dates interpreted as exposure ages.

11.5.4.2 Calcite

Several studies have attempted (U–Th)/He dating of calcite, aragonite, and dolomite (e.g., *Fanale and Kulp*, 1961; *Fanale and Schaeffer*, 1965; *Bender*, 1973; *Copeland et al.*, 2007; *Pik and Marty*, 2009; *Cros et al.*, 2014]. Many of these found inconsistent results, often attributed to loss as well as excess He, but open system U behavior was found in some cases, deduced from (^{234}U/^{238}U) excesses. Results from step-heating diffusion experiments on calcite have so far yielded a wide range of He diffusion kinetics (e.g., apparent E_a and D_0/a^2, with most samples between about 117 and 160 kJ/mol and 10^2 and 10^7 s^{-1}, respectively) and results suggesting subgrain domains [*Copeland et al.*, 2007]. NRA-based experiments [*Cherniak et al.*, 2015] on carbonates generally found low He retentivity (E_a of ~55–95 kJ/mol and $D_0 \sim 1e^{-4}$ cm^2/s). These results suggest that quite low fractional retention for He would occur in most carbonates for typical crystal sizes and near-surface conditions over geologic durations, consistent with previous results on ^{3}He from cosmogenic studies [*Amidon et al.*, 2015].

11.5.4.3 Titanite

Early studies using titanite (U–Th)/He dating found dates significantly less than presumed formation ages or known U/Pb dates [*Strutt*, 1910c; *Larsen and Keevil*, 1942; *Keevil et al.*, 1944]. Step-heating diffusion experiments [*Reiners and Farley*, 1999] showed that titanite has E_a of 187 ± 14 (2σ) kJ/mol and D_0 of ~60 cm^2/s, and a correspondence of grain size and diffusion domain size. For a cooling rate of 10 °C/Myr and grain sizes of several hundred microns (as typically observed in igneous

and metamorphic rocks), these kinetics yield apparent T_c values in the range of 191–218 °C. Other ^{4}He step-heating diffusion experiments [*Stockli and Farley*, 2004] from the KTB drillhole in Germany, and those using both ^{3}He and ^{4}He [*Shuster et al.*, 2004] yielded very similar kinetics and T_c values. ^{3}He diffusion measured by the NRA method [*Cherniak and Watson*, 2011] found no evidence of anisotropy, and as is the case with other minerals, lower apparent E_a of 148 ± 8 kJ/mol and D_0 of 0.0214 cm^2/s than step-heating diffusion experiments, yielding apparent T_c values about 20–30 °C lower for a given grain size.

With its ~190 °C T_c, titanite (U–Th)/He dating has been used to estimate the timing, rate, and depth of exhumation through depths of approximately 8–10 km. *Reiners et al.* [2000] compared titanite He dates with other thermochronometric systems to understand the depth of exhumation and structure of the Gold Butte block in the Basin and Range. *Spotila et al.* [2001] used contrasting titanite and apatite He dates to constrain the exhumation depths in an extremely rapidly uplifting block in the San Andreas fault zone, documenting an earlier "Laramide" phase of exhumation. *Pik et al.* [2003] used a similar approach to constrain incision depths in the Afar rift, and documented an earlier phase of Pan-African cooling recorded by the titanite He dates. *Stockli and Farley* [2004] documented a well-expressed titanite He partial retention zone in the KTB borehole in Germany. The depth–date relationships in this suite matched model predictions based on laboratory kinetics, and these authors also documented evidence for at least several kilometers of mid-Cretaceous exhumation.

11.5.4.4 Monazite

Boyce et al. [2005] performed step-heating ^{4}He diffusion experiments on monazite and interpreted E_a ranging from 180 to 248 kJ/mol and $\ln(D_0/a^2)$ from 2.2×10^3 to 5.3×10^{11} cm^2/s, for apparent T_c (for a cooling rate of 10 °C/Ma) of 206–286 °C. Boyce *et al.* speculated that the wide range of kinetics may be due to strong intragranular, and presumably intergranular, compositional variation, which could create complexities in release patterns but also manifest as real diffusivity variations if the large cationic variations in monazite affect He diffusivity. ^{3}He stepheating diffusion experiments in crystallographically oriented slabs of synthetic monazite specimens with end-member cation compositions ranging from La to Gd [*Farley*, 2007] showed no evidence for crystallographic anisotropy, and E_a and D_0 from ~183 to 224 kJ/mol and ~0.33 to 9.0×10^4 cm^2/s, respectively. Both E_a and D_0 were highest in the intermediate composition specimen (NdPO$_4$), and decreased towards both increasing and decreasing cationic atomic mass. Calculated T_c values, on the other hand, showed a systematic increase from La to Gd end members, following the trend of decreasing ionic porosity from 32 to 26%. When calculated using $a = 60$ μm, to represent an approximate size of typical "wild" monazite crystals, the end-member La and Gd specimens have T_c of about 184 and 263 °C, respectively.

As discussed above (see Chapter 3), *Boyce et al.* [2006, 2009] have measured apparently accurate and precise (U–Th)/He dates of monazites representing cooling ages of ~750 ka and 450 Ma.

11.5.4.5 Rutile

Step-heating ^{4}He diffusion experiments on rutile, some of which have complex non-Arrhenius behavior, yield E_a of 200–213 kJ/mol, and $\ln(D_0/a^2)$ of 6.5–7.1 s^{-1}, for a T_c of ~200–235 °C [*Stockli et al.*, 2007; *Wolfe*, 2009]. NRA-based experiments on rutile showed about a 100-fold anisotropy in ^{3}He diffusion, expressed in D_0 at 2.26×10^{-6} cm^2/s for the perpendicular to the c-axis direction, and 1.75×10^{-4} cm^2/s for the parallel to the c-axis direction. As with other minerals, the NRA results yield much lower E_a (similar in the c-axis-perpendicular and c-axis-parallel directions), at 126 ± 11 and 120 ± 7 kJ/mol, respectively, predicting T_c of 116 and 186 °C in the c-axis-parallel and c-axis-perpendicular directions, respectively.

11.5.4.6 Meteoritic phosphates

Early studies of He diffusion from stony meteorites (or silicate phases derived from them) demonstrated behavior consistent with thermally activated volume diffusion for both ^{3}He and ^{4}He [*Fechtig et al.*, 1963; *Huneke et al.*, 1969]. On the other hand, He loss from iron meteorites appears to be inconsistent with diffusion and instead may follow first-order loss or more complex laws describing defect-impurity interactions [*Levskiy and Aprub*, 1970; *Nyquist et al.*, 1972]. More recent work on ^{3}He diffusion from meteoritic phosphates showed diffusion kinetics of $E_a = 136 \pm 4.8$ kJ/mol and $\ln(D_0/a^2) = 5.83 \pm 0.66$ $\ln(s^{-1})$ for merrillite, and $E_a = 109 \pm 9.7$ kJ/mol and $\ln(D_0/a^2) = 8.15 \pm 1.93$ $\ln(s^{-1})$ for chlorapatite. Assuming grain dimensions characterize the diffusion domain size, and a grain size typical of

terrestrial apatite (60 μm), meteoritic merrillite and chlorapatite would have T_c (for dT/dt = 10 °C/Ma) of 126 °C and 38 °C, respectively.

(U–Th)/He dating has been applied to phosphates from a number of meteorites, in most cases yielding a wide range of dates that likely reflects sampling of fragments from original grain edges affected by diffusive loss and alpha ejection. An exception to this is dates of 3.2 and 2.3 Ma from merrillite and apatite, respectively, from Martian meteorite Los Angeles, which agree with cosmogenic exposure dates and presumably reflect resetting during the impact that launched this meteorite from Mars [*Min et al.*, 2004]. Even in cases where dates from all aliquots span a wide range, clusters of dates appear to provide significant and geologically reasonable constraints. For example, the population of oldest whitlockite and apatite dates from the Acapulco meteorite, are 4538 ± 32 Ma [*Min et al.*, 2003], and populations of the oldest merrillite and chlorapatite dates from the St. Severin LL6 chondrite are 4412 ± 75 Ma and 4152 ± 70 Ma, respectively (1σ uncertainties), displaying differences expected for these phases based on their apparent kinetics [*Min et al.*, 2013]. The Martian meteorite ALH84001 yielded a wide range of phosphate (U–Th)/He dates from 60 to 1800 Ma [*Min and Reiners*, 2007], which was interpreted to represent partial resetting of grains/domains with varying size by the impact event that launched the meteorite from Mars.

11.5.5 A compilation of He diffusion kinetics

Helium diffusion has been measured in a variety of other minerals not discussed above, mostly by step heating and/or NRA methods. In a few cases other techniques, such as elastic recoil detection analysis (ERDA) and laser-ablation depth profiling were used. Table 11.2 provides a compilation of kinetic data for ^{4}He and ^{3}He diffusion in representative specimens and experiments from a variety of minerals. Figures 11.35 and 11.36 show the relationships between apparent E_a, $\ln(D_0)$, T_c, and ionic porosity for He diffusion in most of these phases (a few are not shown in the figures because they plot far off scale relative to other phases).

Comparing He diffusion kinetics among different phases is most useful when the frequency factor or $\ln(D_0)$ can be directly compared, rather than $\ln(D_0/a^2)$, which convolves diffusion domain size effects. Several of the kinetics shown below derive from experiments observing apparent diffusion profiles in situ, such as those from NRA and laser-ablation profile methods. In these cases frequency factor can be estimated directly from the fit to the post-heating concentration profile. But most entries in Table 11.2 are from step-heating experiments, which require additional information on the diffusion domain size a, in order to derive frequency factor or $\ln(D_0)$ from the more direct observation of $\ln(D_0/a^2)$. In some cases, the equivalency, or at least proportionality, of experimental grain size and diffusion domain size has been demonstrated, as for apatite and titanite, so that dimensions of grains used in the experiments provide a simple way to

Table 11.2 He diffusion kinetics for various minerals

Mineral/isotope	Reference	Sample	Method	Comments	E_a [kJ/mol]	$\ln(D_0/a^2)$ [ln(s^{-1})]	a for D_0 (µm)	D_0 (cm^2/s)	a for T_c (µm)	T_c (°C)	T_c (°C) for a = 100 µm
Phosphates											
Apatite ^{4}He	Farley [2000]	Durango	Step heating	Radiogenic ^{4}He	138	na	na	3.16E+01	60	69	76
Apatite ^{3}He	Shuster et al. [2004]	Durango	Step heating	PB ^{3}He	148	16.00	170	2.57E+03	60	64	71
Apatite ^{4}He	Van Soest et al. [2011]	Durango	Loss profile	Laser-ablation	142	na	na	9.00E+01	60	72	79
Apatite ^{3}He	Ouchani et al. [1998]	Durango	ERDA	Implantation, heating, profile characterization	120	na	na	1.45E-02	60	79	88
Apatite ^{3}He	Cherniak et al. [2009]	Durango	NRA	Implantation, heating, profile characterization	117	na	na	2.10E-02	60	68	76
Apatite ^{4}He $e/_s \sim$ 5e4	Flowers et al. [2009]	DYJS5	Step heating	PB ^{3}He	122	10.50	60[a]	1.31E+00	60	50	57
Apatite ^{4}He $e/_s \sim$ 1.2e5	Flowers et al. [2009]	MCO1-11	Step heating	PB ^{3}He	125	10.90	60[a]	1.95E+00	60	57	64
Apatite ^{4}He $e/_s \sim$ 5.4e5	Flowers et al. [2009]	98MR-86	Step heating	PB ^{3}He	133	12.60	60[a]	1.07E+01	60	65	72
Apatite ^{4}He $e/_s \sim$ 1.6e6	Flowers et al. [2009]	MH96-17	Step heating	PB ^{3}He	142	13.70	60[a]	3.21E+01	60	80	87
Apatite ^{4}He $e/_s \sim$ 3.7e6	Flowers et al. [2009]	S089D49B	Step heating	Radiogenic ^{4}He	153	11.80	60[a]	4.80E+00	60	120	129
Apatite ^{3}He, synthetic fluorapatite	Shuster and Farley [2009]	"WSAa with no neutrons"	Step heating	PB ^{3}He	132	5.66	150	6.46E-02	60	100	109
Chlorapatite, ^{4}He	Min et al. [2013]	Chondritic meteorite	Step heating	Radiogenic ^{4}He	138	13.17	43	9.70E+00	60	78	85
Chlorapatite, ^{3}He	Min et al. [2013]	Chondritic meteorite	Step heating	PB ^{3}He	109	8.15	43	6.40E-02	60	38	45
Merrillite, ^{4}He	Min et al. [2013]	Chondritic meteorite	Step heating	Radiogenic ^{4}He	147	6.98	59	3.74E-02	60	146	156
Merrillite, ^{3}He	Min et al. [2013]	Chondritic meteorite	Step heating	PB ^{3}He	136	5.83	59	1.18E-02	60	126	136
Monazite ^{4}He ("grain 3")	Boyce et al. [2005]	Monazite 554	Step heating	Radiogenic ^{4}He	248	27.40	70	3.89E+07	60	203	211
Monazite ^{4}He ("grain 4")	Boyce et al. [2005]	Monazite 554	Step heating	Radiogenic ^{4}He	180	7.66	78	1.29E-01	60	224	235
Monazite ^{4}He ("grain 5")	Boyce et al. [2005]	Monazite 554	Step heating	Radiogenic ^{4}He	217	11.32	83	5.68E+00	60	277	289
Monazite ^{4}He	Farley and Stockli [2002][c]	Monazite 554	Step heating	Radiogenic ^{4}He	200	na	na	5.46E+01	60	214	224
Monazite ^{3}He, LaPO$_4$	Farley [2007]	Synthetic endmember	Step heating	PB ^{3}He	183	na	569	1.82E+01	60	183	193
	Farley [2007]		Step heating	PB ^{3}He	196	na	694	2.00E+02	60	193	202

(Continued)

Table 11.2 (Continued)

Mineral/isotope	Reference	Sample	Method	Comments	E_a [kJ/mol]	$\ln(D_0/a^2)$ [ln(s^{-1})]	a for D_0 (μm)	D_0 (cm^2/s)	a for T_c (μm)	T_c (°C)	T_c (°C) for a = 100 μm
Monazite ^{3}He, CePO$_4$	Farley [2007]	Synthetic endmember	Step heating	PB ^{3}He	206	na	850	7.35E+02	60	204	213
Monazite ^{3}He, PrPO$_4$	Farley [2007]	Synthetic endmember	Step heating	PB ^{3}He	224	na	850	8.96E+03	60	222	231
Monazite ^{3}He, NdPO$_4$	Farley [2007]	Synthetic endmember	Step heating	PB ^{3}He	215	na	850	1.22E+01	60	265	277
Monazite ^{3}He, SmPO$_4$	Farley [2007]	Synthetic endmember	Step heating	PB ^{3}He	198	na	850	3.33E-01	60	262	274
Monazite ^{3}He, GdPO$_4$	Farley [2007]	Synthetic endmember	Step heating	PB ^{3}He	200	na	na	2.98E+03	60	179	187
Xenotime ^{4}He	Farley and Stockli [2002][c]	QC-A (Miocene, Tibet)	Step heating	Radiogenic ^{4}He							
Britholite ^{3}He	Costantini et al. [2002]	Synthetic polycrystalline	NRA		104	na	na	2.60E-04	60	65	74
Zircon (and zircon structure orthophosphates)											
Zircon ^{4}He	Reiners et al. [2004]	Various natural zircon	Step heating	Radiogenic ^{4}He	168	na	na	4.60E-01[b]	60	180	190
Zircon ^{3}He, perpendicular to c-axis	Cherniak et al. [2009]	Mud tank and uknown sample	NRA	Implantation, heating, profile characterization	146	na	na	2.30E-03	60	172	183
Zircon ^{3}He, parallel to c-axis	Cherniak et al. [2009]	Mud tank and uknown sample	NRA	Implantation, heating, profile characterization	148	na	na	1.70E-01	60	136	145
Zircon ^{4}He, parallel to c-axis, 1.2e16 α/g	Guenthner et al. [2013]	Mud tank	Step heating	Radiogenic ^{4}He	168	15.30	50	1.11E+02[b]	60	132	140
Zircon ^{4}He, parallel to c-axis, 4.7e17 α/g	Guenthner et al. [2013]	RB140	Step heating	Radiogenic ^{4}He	166	10.83	20	2.01E-01[b]	60	184	194
Zircon ^{4}He, parallel to c-axis, 1.5e18 α/g	Guenthner et al. [2013]	M127	Step heating	Radiogenic ^{4}He	162	7.18	45	2.65E-02[b]	60	191	202
Zircon ^{4}He, parallel to c-axis, 4.0e18 α/g	Guenthner et al. [2013]	G3	Step heating	Radiogenic ^{4}He	107	6.06	34	4.19E-03[b]	60	50	58
Fully amorphous zircon ^{4}He, parallel to c-axis, 8.0e18 α/g	Guenthner et al. [2013]	N17	Step heating	Radiogenic ^{4}He	71	4.73	75	6.37E-03[b]	60	-58	-53
Zircon structure orthophosphate ^{3}He, TbPO$_4$	Farley [2007]	Synthetic endmember	Step heating	PB ^{3}He	125	na	na	4.00E+05	60	-12	-7
Zircon structure orthophosphate ^{3}He, DyPO$_4$	Farley [2007]	Synthetic endmember	Step heating	PB ^{3}He	128	na	na	9.48E+04	60	1	6

Zircon structure orthophosphate ^{3}He, YPO$_4$	Farley [2007]	Synthetic endmember	Step heating	PB ^{3}He	123	na	na	1.36E+04	60	0	5
Zircon structure orthophosphate ^{3}He, HoPO$_4$	Farley [2007]	Synthetic endmember	Step heating	PB ^{3}He	116	na	na	2.95E+03	60	-8	-3
Zircon structure orthophosphate ^{3}He, ErPO$_4$	Farley [2007]	Synthetic endmember	Step heating	PB ^{3}He	121	na	na	3.64E+03	60	2	7
Zircon structure orthophosphate ^{3}He, TmPO$_4$, pref. Parallel to c-axis	Farley [2007]	Synthetic endmember	Step heating	PB ^{3}He	127	na	na	8.27E+03	60	11	16
Zircon structure orthophosphate ^{3}He, TmPO$_4$	Farley [2007]	Synthetic endmember	Step heating	PB ^{3}He	125	na	na	8.10E+03	60	7	12
Zircon structure orthophosphate ^{3}He, YbPO$_4$, pref. Parallel to c-axis	Farley [2007]	Synthetic endmember	Step heating	PB ^{3}He	110	na	na	1.79E+02	60	-8	-3
Zircon structure orthophosphate ^{3}He, YbPO$_4$	Farley [2007]	Synthetic endmember	Step heating	PB ^{3}He	116	na	na	3.76E+02	60	2	8
Zircon structure orthophosphate ^{3}He, YbPO$_4$ pref. Perpendicular to c-axis	Farley [2007]	Synthetic endmember	Step heating	PB ^{3}He	121	na	na	3.88E+02	60	14	19
Zircon structure orthophosphate ^{3}He, LuPO$_4$	Farley [2007]	Synthetic endmember	Step heating	PB ^{3}He	116	na	na	3.55E+01	60	15	21
Titanite											
Titanite ^{4}He	Reiners and Farley [1999]		Step heating	Radiogenic ^{4}He	187	n/a	n/a	6.00E+01	250	209	191
Titanite ^{3}He	Shuster et al. [2004]		Step heating	PB ^{3}He	183.66	147.5	13.34	1.35E+02	250	193	176
Titanite ^{3}He	Cherniak and Watson [2011]		NRA	Implantation, heating; profile characterization	148	n/a	n/a	2.14E-02	250	185	165
Fe-oxides											
Hematite ^{4}He, radius = 200 μm	Lippolt et al. [1993]	Rimbach, vosges	Step heating	Radiogenic ^{4}He	144	na	200	3.34E-04	200	218	200
Hematite ^{4}He, radius = 60 μm	Lippolt et al. [1993]	Rimbach, vosges	Step heating	Radiogenic ^{4}He	131	na	60	2.31E-05	60	178	190
Hematite ^{3}He, polycrystalline	Farley and Flowers [2012]	Grand canyon	Step heating	PB ^{3}He	157	na	na	9.00E-05	1	154	259

(Continued)

Table 11.2 (Continued)

Mineral/isotope	Reference	Sample	Method	Comments	E_a [kJ/mol]	$\ln(D_0/a^2)$ [ln(s^{-1})]	a for D_0 (µm)	D_0 (cm^2/s)	a for T_c (µm)	T_c (°C)	T_c (°C) for a = 100 µm
Magnetite ^{4}He	Blackburn et al. [2007]	Bala kimberlite	Step heating	Radiogenic ^{4}He; first 16% of gas ignored; evidence for MD behavior	220	15.70	250	4.11E+03	250	247	229
Goethite ^{3}He, HRD, sample 111.4	Shuster et al. [2005b]	Bahia, Brazil	Step heating	PB ^{3}He	163	26.00	0.5	4.89E+02	0.5	51	117
Goethite ^{3}He, HRD, sample 114	Shuster et al. [2005b]	Bahia, Brazil	Step heating	PB ^{3}He	178	28.30	0.5	4.88E+03	0.5	68	134
Other minerals/phases											
Olivine ^{3}He	Shuster et al. [2004]	Guadalupe megacryst	Step heating	PB ^{3}He	154	3.00	690	9.56E-02	250	187	168
Olivine ^{4}He	Hart (1984)	Reunion dunite	Step heating	Magmatic/radiogenic ^{4}He	502.0920502	na	na	2.20E+08	250	697	671
Olivine ^{4}He	Trull and Kurz [1993]	Hualalai xenolith	Step heating	Magmatic/radiogenic ^{4}He	420	na	na	1.26E+05	250	648	619
Rutile ^{4}He	Wolfe et al. [2007]	KTB garnet amphibolite	Step heating	Radiogenic ^{4}He	203	na	na	5.79E+00	60	244	255
Rutile ^{3}He, parallel to a-axis	Cherniak and Watson [2011]	Synthetic crystal	NRA	Implantation, heating, profile characterization	126	na	na	2.26E-10	60	355	380
Rutile ^{3}He, parallel to c-axis	Cherniak and Watson [2011]	Synthetic crystal	NRA	Implantation, heating, profile characterization	120	na	na	1.75E-08	60	240	258
Quartz ^{3}He	Shuster and Farley [2005]	Minas Gerais quartz	Step heating	PB ^{3}He	84.5	11.10	430	1.22E+02	250	-55	-63
Basaltic glass ^{4}He	Kurz and Jenkins [1981]	Various MORB	Step heating	Magmatic/radiogenic ^{4}He	83.3	na	na	6.72E-01	250	-32	-42
Clinopyroxene ^{4}He	Trull and Kurz [1993]	Hualalai xenolith	Step heating		290	na	na	1.26E+02	250	456	431
Diamond ^{4}He	Zashu and Hiyagon [1995]	Ubangi carbonado	Step heating		103.5	na	na	6.31E-14	60	478	520
Diamond ^{3}He	Wiens et al. [1994]	Industrial specimens, loc unknown	Step heating		150	na	na	6.10E-11	60	507	539
Garnet ^{4}He	Dunai and Roselieb [1996]	Orissa, Py55Alm40Gr4Sp1	Sorption	Three T's examined	660	na	na	3.98E+15	250	735	713
Muscovite ^{4}He	Lippolt and Weigel [1988]	Bahrh., granite, Germany	Step heating		87	na	na	1.15E-04	250	41	25
Sanidine ^{4}He	Lippolt and Weigel [1988]	TSG, tuff, Switzerland	Step heating		94	na	na	4.57E-04	250	52	37
Hornblende ^{4}He	Lippolt and Weigel [1988]	LK-5, diorite, Austria	Step heating		104	na	na	9.55E-04	250	79	62

Mineral	Reference	Sample description	Method	Notes							
Hornblende 4He	Lippolt and Weigel [1988]	SAU B, eclogite, Germany	Step heating		120	na	na	1.23E-02	250	106	89
Augite 4He	Lippolt and Weigel [1988]	Fohren, tephrite, Germany	Step heating		120	na	na	4.57E-02	250	94	78
Nepheline 4He	Lippolt and Weigel [1988]	134, metasyenite, India	Step heating		131	na	na	3.63E-03	250	152	133
Langbeinite [(K₂Mg₂(SO₄)₃)] 4He	Lippolt and Weigel [1988]	PR3, salt rock, Germany	Step heating		126	na	na	7.59E-01	250	87	73
Uranium dioxide 3He	Garcia et al. [2012]	Synthetic polycrystalline UO₂	NRA	Implantation, heating, profile characterization	135	na	na	5.00E-06	250	254	226
Uranium dioxide 4He	Nakajima et al. [2011]	Synthetic UO₂	Sorption/infusion	Knudsen effusion MS	192	na	na	7.19E-06	250	459	421
Silicon carbide 4He	Pramono et al. [2004]	Synthetic sic	Step heating		88	na	na	1.38E-10	250	237	199
Tektite 3He	Reynolds [1960]	Kalgoorlie Australia, 100-15 mesh	Step heating	Three points; intercept estimated from figure in paper	26.3	na	na	1.50E-05	250	-168	-174
Epidote 4He (assume grain size = domain size)	Nicolescu and Reiners [2005][c]	Ocna de Fier	Step heating		153	20.06	1500[b]	1.16E+07[b]	1500[b]	63	34
Epidote 3He (assume grain size = domain size)	Nicolescu and Reiners [2005][c]	Ocna de Fier	Step heating	PB 3He	154	20.26	1500[b]	1.42E+07[b]	1500[b]	64	35
Epidote 4He (assume 1 μm = domain size)	Nicolescu and Reiners [2005][c]	Ocna de Fier	Step heating		153	20.06	1[b]	5.15E+00[b]	1[b]	63	128
Epidote 3He (assume 1 μm = domain size)	Nicolescu and Reiners [2005][c]	Ocna de Fier	Step heating	PB 3He	154	20.26	1[b]	6.29E+00[b]	1[b]	64	129
Epidote 4He (assume 10 μm = domain size)	Nicolescu and Reiners [2005][c]	Ocna de Fier	Step heating		153	20.06	10[b]	5.15E+02[b]	10[b]	63	93
Epidote 3He (assume 10 μm = domain size)	Nicolescu and Reiners [2005][c]	Ocna de Fier	Step heating	PB 3He	154	20.26	10[b]	6.29E+02[b]	10[b]	64	94
Fluorite 4He (assume grain size = domain size)	Evans et al. [2005]	Yucca Mountain ECRB25 + 30	Step heating	Assume grain size = domain size	128	11.28	275[b]	5.99E+01[b]	250[b]	60	47
Fluorite 4He (assume 1 μm = domain size)	Evans et al. [2005]	Yucca Mountain ECRB25 + 30	Step heating		128	11.28	1[b]	7.92E-04[b]	250[b]	159	139
Fluorite 4He (assume 10 μm = domain size)	Evans et al. [2005]	Yucca Mountain ECRB25 + 30	Step heating		128	11.28	10[b]	7.92E-02[b]	250[b]	112	96

(Continued)

Table 11.2 (Continued)

Mineral/isotope	Reference	Sample	Method	Comments	E_a [kJ/mol]	$\ln(D_0/a^2)$ [$\ln(s^{-1})$]	a for D_0 (µm)	D_0 (cm²/s)	a for T_c (µm)	T_c (°C)	T_c (°C) for a = 100 µm
Fluorite ^{4}He (assume 10 µm = domain size)											
Fluorite ^{4}He (assume 100 µm = domain size)	Evans et al. [2005]	Yucca Mountain ECRB25 + 30	Step heating		128	11.28	100[b]	7.92E+00[b]	250[b]	74	61
Fluorite ^{4}He (assume 1 µm = domain size)	Evans et al. [2005]	Yucca Mountain ECRB25 + 30	Step heating		128	11.28	1000[b]	7.92E+02[b]	250[b]	43	32
Calcite ^{4}He (assume 1 µm = domain size)	Copeland et al. [2007]	Various	Step heating	Evidence for MD behavior	136	12.52	1[b]	2.74E-03[b]	250[b]	172	152
Calcite ^{4}He (assume 10 µm = domain size)	Copeland et al. [2007]	Various	Step heating	Evidence for MD behavior	136	12.52	10[b]	2.74E-01[b]	250[b]	125	109
Calcite ^{4}He (assume 100 µm = domain size)	Copeland et al. [2007]	Various	Step heating	Evidence for MD behavior	136	12.52	100[b]	2.74E+01[b]	250[b]	86	73
Calcite ^{4}He (assume 1 µm = domain size)	Copeland et al. [2007]	Various	Step heating	Evidence for MD behavior	136	12.52	1000[b]	2.74E+03[b]	250[b]	55	44
		Various natural zircon	Step heating	Radiogenic ^{4}He	168	na	na	4.60E-01	60	180	190

na, not applicable; PB, proton bombardment, to create uniform intragranular ^{3}He distribution prior to step-heating.

[a] Other evidence [e.g., Farley, 2000] established the equivalence of experimental grain size and diffusion domain size in apatite, but grain sizes are not reported in these sources, so equivalent-sphere radius of typical igneous apatite assumed here for calculating D_0.

[b] Ambiguity about whether grain size in experiment is equivalent to diffusion domain size, but experimental grain size (half-width or radius of equivalent-sphere) is assumed here to be equivalent to diffusion domain size.

[c] Primary data not actually reported in source; values reported here were taken from abstracts or statements of kinetics reported in text of sources.

estimate a from $\ln(D_0/a^2)$. In other cases, experimental procedures were not sufficient, or were not sufficiently described, to confidently interpret diffusion domain size. In these cases values of $\ln(D_0)$ come from assuming a is a dimension approximating the experimental grain's half-width, or the radius of a sphere with equivalent surface-area to volume ratio of the experimental grain(s). In some cases, however, the effective diffusion domain size of a mineral is probably significantly less than the size of the experimental grain, as for calcite, fluorite, and possibly epidote. For these cases, the several possible values of $\ln(D_0)$ were estimated from a range of possible values for a, up to the size of the experimental grains.

Figure 11.38 shows activation energy and log frequency factor for most of the data shown in Table 11.2, along with contours of constant T_c assuming an arbitrary but uniform diffusion-domain length scale of 100 µm, spherical geometry, and a cooling rate of 10 °C/Ma. Most minerals have an apparent $\ln(D_0)$ between –10 and 10 ($\ln(\text{cm}^2/\text{s})$ [D_0 between 4.5×10^{-5} and $2.2 \times 10^4 \text{ cm}^2/\text{s}$, E_a between 100 and 220 kJ/mol, and T_c (for $a = 100$ µm)] between 0 and 250 °C. If minerals were randomly distributed across this full range of frequency factors and activation energies observed, the range of closure temperatures would be significantly higher than observed. The smaller range of T_c values is due to a weak correlation between E_a and $\ln(D_0)$. This is much stronger among specimens of similar species (e.g., among apatite, monazite, or zircon-structure orthophosphates), and also among phases of similar composition (e.g., silicates, phosphates). The largest outliers in this compilation are for He diffusion in diamond, UO_2, SiC, and hematite on the He-retentive side, and epidote, quartz, and basaltic glass on the low-retentivity side.

One observation from this compilation is the consistent offset between apparent diffusion kinetics from NRA and step-heating experiments. In all cases, NRA experiments show both lower E_a and $\ln(D_0)$ than step-heating experiments. These offsets are strongest for rutile, with E_a differences of about 75 kJ/mol and $\ln(D_0)$ differences of about 18 natural log units, but there are also strong contrasts in NRA versus step-heating results for titanite, apatite, and zircon. This cannot be due to anisotropic diffusion as He diffusion in distinct crystallographic orientations are consistently and significantly lower in NRA. Some of this discrepancy may be due to the compensation law expected for volume diffusion, which may or may not be an artifact of one or both of the methods. But because NRA results are based on interpretations of diffusion kinetics over 10–100 nm scales, whereas most other techniques interrogate diffusion over length scales about 1000 times larger, it is also possible that these discrepancies are due to differing kinetic mechanisms operating over these different length scales.

He diffusion kinetic parameters show only a weak correlation with ionic porosity (Fig. 11.39), although the correlation is somewhat stronger for E_a than for D_0. It is not clear whether this is due to analytical artifacts, interrogation of diffusion operating over very different length scales, or simply that other crystal chemical or structural aspects besides ionic porosity also strongly influence He migration kinetics. One possible explanation for the failure of ionic porosity to explain this variation is the role of variations in sizes of interatomic gaps in different crystallographic orientations. The molecular sieve model of *Cherniak and Watson* [2011] predicts that densities and sizes of minimum interionic "pores" in certain orientations play a strong role in crystallographic anisotropy of He diffusion, and similar differences that are perhaps only weakly related to overall three-dimensional porosity may also influence intermineral differences in He diffusion.

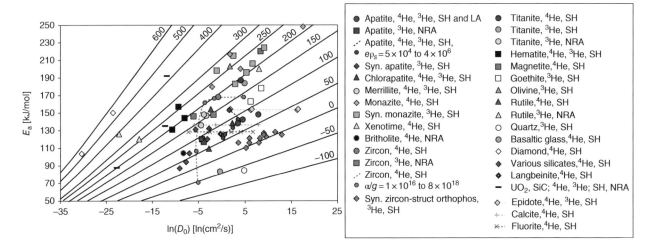

Fig. 11.38. Activation energy (E_a) as a function of natural logarithm of frequency factor [$\ln(D_0)$] for ^{4}He and ^{3}He diffusion in a variety of phases, from Table 11.2. Dashed lines connect zircon and apatite experiments with varying levels of radiation damage. Horizontal dashed lines for calcite, fluorite, and epidote represent uncertainties in diffusion domain size that allow for a wide range of frequency factors if diffusion domain size is significantly smaller than physical grain size. (*See insert for color representation of the figure.*)

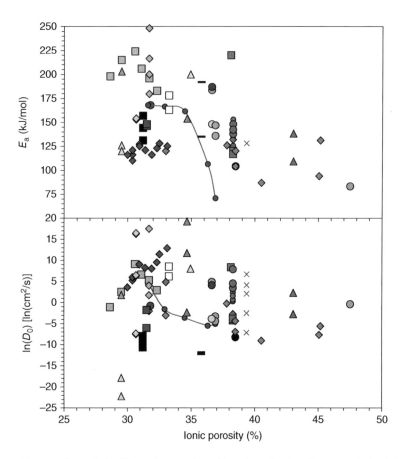

Fig. 11.39. Activation energies and frequency factors for He diffusion for minerals in Table 11.2, as a function of ionic porosity. Symbols as in Fig. 11.38. (*See insert for color representation of the figure.*)

11.6 ^{4}He/^{3}He THERMOCHRONOMETRY

Up this point, most of the theory and practice of (U–Th)/He chronometry described herein has focused on whole-grain contents of parent and daughter nuclides, which provide bulk grain dates. With certain assumptions or in concert with independent geologic constraints, a bulk-grain date may be associated with a T_c estimate to provide a single point in a monotonic cooling path. In some cases, multiple points on a cooling path or inferred thermal history may be established through dating multiple grains with varying grain size or radiation-damage-induced diffusivity, or, at least in principle, by dating multiple interior portions of a grain *in situ*.

A great deal more information about a mineral's thermal history exists in the intragrain distribution of He than can be recovered by bulk dating, or even spot analyses *in situ*. This approach has long been realized in ^{40}Ar/^{39}Ar chronometry [*Turner*, 1969; *Albarède*, 1978], made possible by the irradiation production of a uniform distribution of a second, nonradiogenic isotope, ^{39}Ar, within the grain. Comparison of the step-heating release patterns of both the radiogenic and irradiation-produced isotopes constrains the distribution of the radiogenic isotope and can be interpreted as an intragranular diffusion profile or as interdomain proportions. An additional benefit of this approach in ^{40}Ar/^{39}Ar dating is that the irradiation-induced isotope ^{39}Ar is derived from an isotope, ^{39}K, of the radioactive parent element, so that each step-heating release can be directly associated with a date.

The internal distribution of ^{4}He in minerals can be measured by an approach similar to, but not completely analogous to, the ^{40}Ar/^{39}Ar method. A uniform distribution of ^{3}He is produced with the grain, and the evolution of ^{4}He/^{3}He in sequential step-heating releases is converted to provide the initial intragranular ^{4}He distribution. This approach requires a mass spectrometer with sufficient mass resolution to resolve HD$^+$ and other isobaric interferences on small amounts of ^{3}He$^+$, as well as the ability to perform sufficiently accurate and precise heating steps on small samples. The two main benefits to this approach in (U–Th)/He dating are:

(1) the intragranular ^{4}He distribution can be interpreted as a continuous path in a time-temperature history (at least assuming monotonic cooling), rather than a single point as in the bulk-grain date;

(2) the grain-specific diffusion kinetics for He can be determined precisely because the initial distribution conforms precisely to the uniform distribution condition underlying step-heating theory [*Fechtig and Kalbitzer*, 1966].

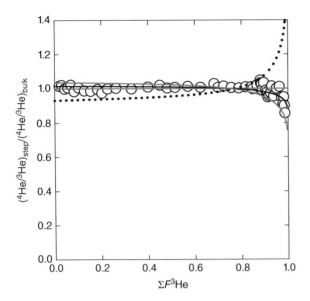

Fig. 11.40. Example presentation of results for ^{4}He and ^{3}He releases in a ^{4}He/^{3}He step-heating diffusion experiment on Durango apatite. Y-axis is R_{step}/R_{bulk}, or ^{4}He/^{3}He of each step normalized to that of the bulk aliquot, and x-axis is ΣF_{3He}, or the fractional release of ^{3}He in the corresponding step. Dotted curve shows predicted trend for $D_{4He}/D_{3He} = 0.868$ (inverse square of the masses of each isotope) assuming uniform ^{4}He distribution. Solid black curve shows best fit trend, with 95% confidence intervals shown as gray lines. Assuming a uniform ^{4}He distribution the solid black curve yields $D_{4He}/D_{3He} = 1.03$. This example suggests that ^{4}He and ^{3}He are released with approximately the same kinetic properties, and that the ^{4}He distribution within the crystal is spatially uniform. (Source: *Shuster and Farley* [2005]. Reproduced with permission of Elsevier.)

The latter provides especially important constraints for minerals with low ^{4}He contents due to high diffusivities at near-surface conditions.

Production of ^{3}He in minerals is performed by bombarding samples with a $\sim$150 MeV proton beam. At these energies, the cross sections for generation of spallogenic ^{3}He via X(p,x)^{3}He reactions are roughly constant across nuclides. Spallation is accomplished by reactions on many possible target nuclides that emit a ^{3}He nucleus upon deexcitation after receiving a proton and emitting another high-energy nucleon. Examples of reactions that may be important are ^{28}Si(p,p)^{28}Si* $\rightarrow$ ^{25}Mg + ^{3}He or ^{40}Ca(p,pn)^{39}Ca* $\rightarrow$ ^{36}Ar + ^{3}He. Integrated across the target nuclides, the proton bombardment process produces about 10^8–10^9 atoms of ^{3}He per milligram of sample. Similar to the phenomena of ^{4}He recoil during alpha decay, ^{3}He nuclei are scattered in random directions with stopping distances of $\sim$1–50 µm. Analyzed grains are surrounded by material that also produces ^{3}He, so homogeneous ^{3}He distributions are produced in samples as long as the length scales of the analyzed aliquots are small compared with sample/packaging unit subjected to irradiation. Proton bombardment also generates ^{4}He with a ^{4}He/^{3}He production ratio of about 10, but this is negligible compared to the radiogenic ^{4}He inventory in most samples of interest.

Experiments with proton bombardment [*Shuster et al.*, 2004; *Shuster and Farley*, 2005] have demonstrated, through comparison with conventional radiogenic ^{4}He experiments, that proton bombardment does not appear to affect He diffusion properties, at least among most minerals of interest that have been studied in detail (e.g., Fig. 11.B2.1); quartz is a possible exception [*Shuster and Farley*, 2005b]. The procedures also do not appear to cause significant heating, so natural He loss during bombardment is insignificant.

Another important observation from detailed step-heating diffusion experiments on both ^{3}He and ^{4}He is that diffusivity differences between the two isotopes appear negligible, as mentioned previously. This is not necessarily expected, however, even if He isotopes diffuse differentially at rates predicted by simple theory, the practical consequences for interpretations of ^{4}He diffusion profiles would likely be negligible. This is because the predicted departures in step-heating behavior appear only with isotopic distillation after large loss fractions, long after the critical information on the ^{4}He profile is extracted in the first few percent of loss [*Shuster and Farley*, 2005].

^{4}He/^{3}He diffusion experiments have been used for a number of purposes related to (U–Th)/He and in some cases cosmogenic ^{3}He dating purposes. Typically, diffusion experiment results are presented as shown in Fig. 11.37, with ^{4}He/^{3}He of each step as a ratio to the ^{4}He/^{3}He of the bulk aliquot (as determined at the end of the experiment) shown as a function of the total fraction of ^{3}He released from the aliquot. The critical information in this type of presentation is the R_{step}/R_{bulk} in the initial extractions. Assuming equivalent grain and diffusion-domain sizes, values less than unity indicate lower ^{4}He near grain margins than interior portions. After correction for the effects of alpha ejection, this can be inverted to provide a representative concentration profile across a grain, typically using the spherical approximation.

Figure 11.40 shows a ^{4}He/^{3}He ratio-evolution curve, in this case for an interior fragment of Durango apatite. The same data for which are also shown in Arrhenius trend in Fig. 11.B2.1, illustrating how these experiments produce both kinetic and concentration profile constraints. In this example there is no change in the observed ^{4}He/^{3}He through the experiment, as expected for a uniform distribution of ^{4}He across the grain and negligible diffusive fractionation of ^{4}He and ^{3}He. In detail, the data are consistent with a small diffusive fractionation, but with a 3% faster diffusion rate for ^{4}He than ^{3}He.

One important use of ^{4}He/^{3}He diffusion experiments is to provide an indication of the He retentivity of natural samples. If the thermal history of a particular sample is known precisely, and if ^{3}He is assumed to be produced and derived during step-heating from the same regions of the aliquot where natural radiogenic ^{4}He is expected to be (e.g., assuming uniform U–Th and He distributions), then comparison of ^{3}He and ^{4}He provides a fractional retention estimate for ^{4}He. For example, results from supergene goethite are consistent with loss of about 5–15% of radiogenic ^{4}He since formation, despite near-surface residence since that time [*Shuster et al.*, 2005]. *Shuster et al.* [2005b] and *Heim et al.* [2006] used deficit fractions calculated in this way to provisionally correct goethite (U–Th)/He dates from supergene weathering deposits.

The other primary utility of ^{4}He/^{3}He thermochronometry comes from inversion of ratio evolution diagrams (Fig. 11.40) into model profiles of ^{4}He concentration across dated grains. Together with the kinetic information from the same experiment (e.g., Fig. 11.B2.1), these can be used to model continuous time–temperature histories of samples [*Shuster and Farley*, 2004]. From a forward modeling perspective the principles of this method are fairly simple. The ^{4}He distribution in a grain reflects the time-integrated internal production from parent nuclides, minus loss from diffusion, and alpha ejection. Calculations are made for radial positions within a model spherical grain, using the equivalent radius sphere procedure. As shown by *Farley et al.* [2010] and other studies this spherical assumption introduces negligible inaccuracies compared to more complex grain morphologies such as cylinders, provided diffusion is isotropic.

For any arbitrary thermal history, the dimensionless parameter τ can be calculated as

$$\tau(T,t) = \int_0^t \frac{D(T,t')}{a^2} \cdot dt \tag{11.51}$$

as in Chapter 5. Following *Carslaw and Jaeger* [1959], assuming an initial concentration profile $C_0(r)$ and leaving out radiogenic ingrowth for the moment, the radial concentration profile $C(r)$ is:

$$C(r,\tau) = \frac{2}{ar} \sum_{k=1}^{\infty} e^{-k^2\pi^2\tau} \sin\left(\frac{k\pi r}{a}\right) \int_0^a r' C_0(r') \sin\left(\frac{k\pi r}{a}\right) dr' \tag{11.52}$$

The actual *t-T* history in equation (11.34) can be represented by a sequence of shorter duration step-wise portions of constant temperature,

$$\tau_i = \tau(T_i, t_i) \tag{11.53}$$

obviating the integral, and also allowing for incremental radiogenic ^{4}He ingrowth to occur between steps. Then the concentration profile after each step is

$$C_i(r) = C(r, \tau_i) \tag{11.54}$$

As an example, Fig. 11.41a shows three different thermal histories that, when combined with a He diffusion model for a 75 µm apatite with Durango kinetics, produce concentration profiles as shown in Fig. 11.41b. Each model grain in Fig. 11.45b would produce a bulk-grain (U–Th)/He date of 10 Ma, but only the internal distribution of ^{4}He within the grain can be used to infer the specific *t-T* histories depicted in Fig. 11.41a.

Figure 11.41c shows the ^{4}He/^{3}He ratio evolution diagram that would be measured on each of the grains shown in Figure 11.41b. The highly curved ^{4}He profile characteristic of steady-state diffusion production seen in the sample that

experienced isothermal holding at ~62 °C produces a ratio-evolution diagram characterized by the lowest ^{4}He/^{3}He in the early stages of the experiment. In contrast, the grain that cooled rapidly at 5 Ma shows relatively high ^{4}He/^{3}He in the earliest steps.

Essentially the same procedure used for generating predicted concentration profiles is used to predict ratio-evolution results like those in Fig. 11.40c [*Shuster and Farley*, 2004], except each step of the step-heating experiment is used as the input in equation (11.50), and the remaining amount of ^{3}He or ^{4}He in the grain after heating step *i* is calculated by integrating their concentrations over the spherical domain

$$N_i = \int_0^a 4\pi r^2 C_i(r) dr \tag{11.55}$$

so the fraction of ^{3}He or ^{4}He remaining in the grain at each step is

$$f_i = \frac{N_0 - N_i}{N_0} \tag{11.56}$$

where N_0 is the original amount before the experiment. For $\tau <$ 0.05, corresponding to fractional loss up to about 60%, fractional loss can be approximated as

$$f(\tau) = \int_0^1 K(x,\tau) b(x) dx + \varepsilon \tag{11.57}$$

where x is defined as the dimensionless parameter r/a (radial position over domain radius), $b(x) = xC(x)$, ε is analytical error, and

$$K(x,\tau) = 3x - 3\left(\text{erf}\frac{1+x}{2\sqrt{\tau}} - \text{erf}\frac{1-x}{2\sqrt{\tau}}\right) \tag{11.58}$$

As described by *Shuster and Farley* [2004], breaking up equation (11.40) into discrete steps with continuous τ then allows a user to associate $f(\tau_i)$ with a unique f_i for a given step in the experiment, and therefore reconstruct $C_0(x)$ from the experimental values of τ_i and f_i. This is undertaken using equation (11.56) to calculate the amount of ^{4}He and ^{3}He released from the sample during step heating, creating a synthetic ratio-evolution diagram. By changing the assumed starting ^{4}He concentration profile $C_0(x)$, the synthetic ratio-evolution diagram can be modified until it reproduces that observed to an acceptable degree, as a forward model matching procedure. Inverse methods of solving for $C_0(x)$ are also possible [*Shuster and Farley*, 2004; *Schildgen et al.*, 2010].

As mentioned previously, the ^{4}He/^{3}He method constrains the intragranular ^{4}He distribution (as well as the grain-specific He diffusion kinetics, assuming similar diffusivities of both isotopes), but in contrast to its analogue, the ^{40}Ar/^{39}Ar method, individual step-heating fractions or their inferred radial concentrations are not directly associated with a specific (U–Th)/He date. However, model (U–Th)/He dates can be estimated from individual steps. Figure 11.42 shows six model thermal histories that all result in the same bulk apatite He date of 5 Ma, and their corresponding model ratio-evolution diagrams. Analogous to intragranular ^{40}Ar profiles in single-domain minerals, the ^{4}He

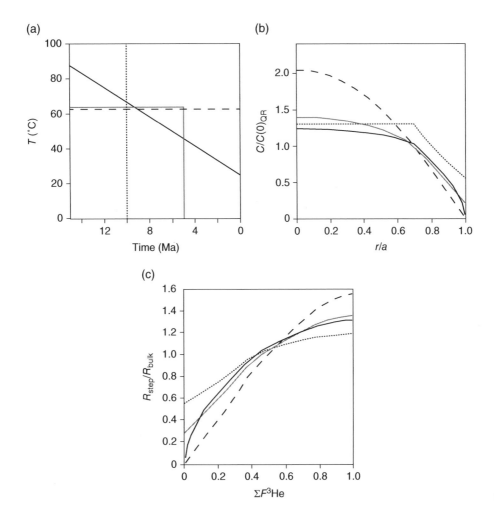

Fig. 11.41. (a) Three model thermal histories that all generate the same bulk-grain apatite (U–Th)/He date of 10 Ma. (b) Resulting internal radial distributions of ^{4}He concentrations (normalized to the concentration at the center of the grain with the same integrated concentration) in "equivalent sphere" grains from each of the histories in (a). "r/a" is radial distance normalized to radius of grain. All curves include the effect of alpha-ejection from the outermost $\sim$20 μm of the domain. (c) Ratio-evolution diagram for grains with corresponding profiles in (b). (Source: *Shuster and Farley* [2004]. Reproduced with permission of Elsevier.)

concentration at the edge of the grain, and inferred from the earliest steps of the ratio-evolution diagram, is proportional to the hypothetical date of a rapid cooling event that effectively ceased all He loss.

The edge model date T_e, can be estimated as a proportion of the alpha-ejection-corrected (U–Th)/He date, where the proportionality constant is the ratio of the observed $R_{initial}/R_{bulk}$ to the expected $R_{initial}/R_{bulk}$ in the case where the concentration profile only reflects alpha-ejection:

$$T_e = \frac{\left(\dfrac{R_{intial}}{R_{bulk}}\right)_{measured}}{\left(\dfrac{R_{intial}}{R_{bulk}}\right)_{\alpha-ref}} \cdot T_{\alpha-corr} \quad (11.59)$$

where $T_{\alpha-corr}$ is the alpha-ejection-corrected bulk grain He date, and

$$\left(\frac{R_{intial}}{R_{bulk}}\right)_{\alpha-ref} = \frac{F_{Tlocal}}{F_T}\sqrt{\frac{D^{4}{}^{He}}{D^{3}{}^{He}}} \quad (11.60)$$

where the ratio of the diffusivities allows for possible differences, assumed or determined, and the local fraction of alpha retention [*Farley et al.*, 1996; *Hourigan et al.*, 2005] is

$$F_T^{local}(r) = 1 - \frac{(R - S_m + r)(R + S_m - r)}{4rS_m} \quad (11.61)$$

where R is the grain radius, r is radial position, and S_m is the stopping distance of the mth parent nuclide. An effective F_T^{local} can be calculated by combining values for each parent nuclide in weighted proportions specific to the analyzed grain.

Similarly, other steps within the profile may also have date significance, and the observed ^{4}He/^{3}He for any step can be interpreted model date, at least relative to the bulk grain date. Thus a step date T_s may be calculated in a similar way by replacing $R_{initial}/R_{bulk}$ with $R_{initial}/R_{bulk}$, and comparing "measured" and "α-ref" values for the same step. These step dates may be combined to form a model date spectrum analogous to the date spectra commonly used in ^{40}Ar/^{39}Ar dating, but the important

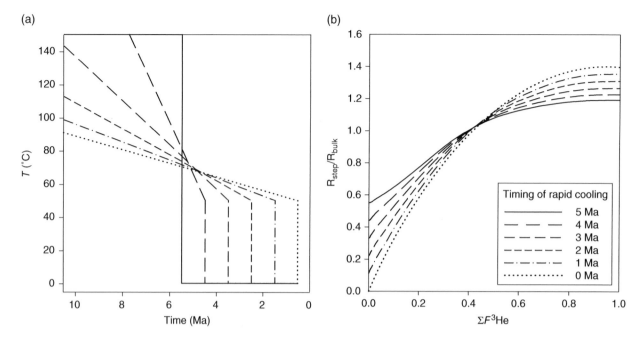

Fig. 11.42. (a) Six model thermal histories resulting in an identical bulk grain apatite He date of 5 Ma. (b) Ideal ^{4}He/^{3}He ratio-evolution diagrams that would be observed for grains from each history. Alpha-ejection effects have been included in these models. As suggested by the regular progression of increasing R_{step}/R_{bulk} in the earliest steps (lowest ΣF_{3He}), these differences are proportional to the time-differences in the rapid cooling events shown in (a), leading to concept of the edge date. (Source: *Shuster and Farley* [2005]. Reproduced with permission of Elsevier.)

assumption of uniformity of parent nuclides in the ^{4}He/^{3}He method must be recognized.

11.6.1 Method requirements and assumptions

Successful implementation of the ^{4}He/^{3}He method requires several specific conditions of sample characteristics and analytical methods are met. In some, but not all, cases problematic behavior may be evident in deviations of ratio-evolution trends into "forbidden zones," manifesting violations of underlying assumptions such as uniform intragranular eU concentrations.

As with bulk-grain (U–Th)/He dating, grains should be free of internal mineral or fluid inclusions, but this requirement is more stringent for interpretable ^{4}He/^{3}He ratio-evolution diagrams. Also, single grains must be analyzed for ratio-evolution diagrams, because grains of slightly varying characteristics such as He concentration, He diffusivities, or size may severely complicate the spectra [*Farley et al.*, 2010].

Currently, ^{4}He/^{3}He results are interpreted in the context of equivalent-sphere grains. For minerals like apatite that exhibit isotropic diffusion the deviation from expected behavior due to more realistic morphologies is negligible, but for diffusionally anisotropic minerals this is not true. Analyzed grains should have intact crystal surfaces and lack recent cracks or new surfaces, otherwise anomalously high ^{4}He release will result from fresh surfaces. The magnitude of this effect depends on thermal history and other factors, but *Farley et al.* [2010] showed that this is generally not a major source of error.

As with bulk-grain dating, intragranular parent zonation and implantation from external alpha sources is a major concern in ^{4}He/^{3}He thermochronometry, but it can be a significantly more serious problem in interpretations of ratio-evolution diagrams. As with bulk-grain dating, implantation is difficult to diagnose without independent observations on grains other than those analyzed. In the case of parent zonation, severely parent-enriched rims may signal their presence by generating step-heating spectra that fall well into the forbidden zone of ratio-evolution diagrams (Fig. 11.43), but enriched cores do not, and could easily be confused with realistic t-T histories requiring highly rounded profiles consistent with slow cooling (or partial resetting).

Work by *Farley et al.* [2010, 2011] has shown that intragranular zonation patterns can be documented by laser-ablation analysis, and the results can be used to predict or correct for zonation-induced modifications to ratio-evolution diagrams. As with bulk-grain dating, there are three interdependent effects of parent zonation that lead to deviations of expectations based on uniform eU distributions and can strongly affect ^{4}He/^{3}He interpretations: (i) the alpha-ejection profile; (ii) the ^{4}He diffusion profile; (iii) the equivalent radial profile of He diffusivity due to variable amounts of accumulated radiation damage [*Farley et al.*, 2011]. Figure 11.44 shows the observed ^{4}He/^{3}He ratio-evolution diagram, with ratios presented as normalized step dates, of an apatite grain from the western Grand Canyon, compared with expectations from a model presuming uniform eU and models of parent zonation actually measured in the same grain following

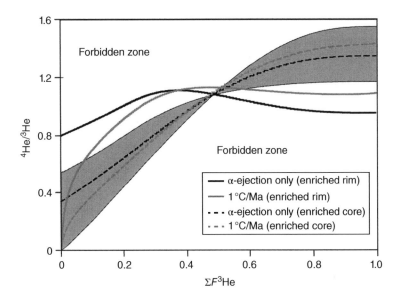

Fig. 11.43. Simulated ^{4}He/^{3}He spectra for apatite grains with eU concentration zonation (dashed and solid lines), compared with possible spectra for any possible internal distribution of ^{4}He resulting from any possible thermal history and a compositionally uniform grain (gray field). The model spherical apatite grains have a radius of 75 μm and a threefold contrast in eU concentration (either enriched core or rim) at a distance 60 μm from the core. Solid lines represent predicted ^{4}He/^{3}He spectra for the enriched rim grain; in this case both the rapid cooling (alpha-ejection only) and slow cooling (1 °C/Ma) produce ^{4}He/^{3}He spectra that fall in the forbidden zones, potentially allowing for identification of the zonation problem. However, core-enriched grains plot in acceptable regions regardless of the cooling rate, even though the actual ^{4}He/^{3}He spectra do not correspond to the actual *t–T* histories of the grains. (Source: adapted *Farley et al.* [2011].)

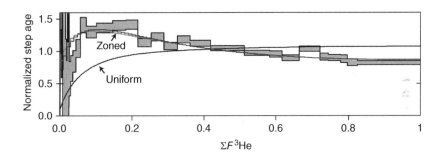

Fig. 11.44. ^{4}He/^{3}He ratio-evolution diagram using (normalized) step-date as in equations (11.59)–(11.61) for an apatite grain from the western Grand Canyon displaying core-to-rim eU enrichment. The observed ratio evolution (gray boxes) shows much higher ^{4}He/^{3}He in early release steps than expected for a uniform distribution of eU (solid line), but compares well with predictions based on the observed zonation, as measured by LA-ICP-MS following degassing. The two stepped-lines labeled "zoned" are a model of the predicted step-age change for this apatite using representative U and Th profiles with an approximately threefold concentration contrast from core to rim, combined with a thermal history independently determined for this region. (Source: *Farley et al.* [2011]. Reproduced with permission of Elsevier.)

step-heating. This exercise requires an assumption of a thermal-history, which comes from bulk-grain He dating of other grains from the same sample (and other independent constraints). This example shows that zonation will have large effects on ^{4}He/^{3}He step-heating data, but that with independent constraints on zonation patterns such as from laser-ablation data, and a model to account for the interacting effects, sensible thermal histories may still be constrained from even severely zoned samples.

Finally, as with any thermochronometric technique, it is worth bearing in mind that although thermal history interpretations are often restricted to paths involving cooling or possibly isothermal

histories, ^{4}He/^{3}He ratio-evolution data (as well as bulk dates and other thermochronologic data) may be equally well explained by nonmonotonic histories. Allowing candidate thermal histories involving reheating may lead to an ill-posed problem in interpreting thermochronologic data. Often, additional geologic constraints can be used to restrict the range of allowable possibilities. But constraints from ^{4}He/^{3}He ratio-evolution data, as well as other techniques that may permit finite *t-T* path estimations under the assumption of monotonic cooling, are subject to the same fundamental ambiguities surrounding potential reheating histories as bulk-grain dating.

11.7 APPLICATIONS AND CASE STUDIES

11.7.1 Tectonic exhumation of normal fault footwalls

As discussed earlier, some of the earliest attempts to apply the (U–Th)/He system focused on the general objective of establishing linchpins of absolute ages in the geologic timescale. As it gradually became evident that this was not an appropriate system to use for such purposes, the technique received little work until the 1990s when its thermochronologic utility began to become apparent. Many of the earliest applications in the renaissance of He dating focused on characterizing the timing, rate, and structural style of tectonic exhumation in normal faults in the Basin and Range of the western United States [*Wolfe et al.*, 1996; *Stockli et al.*, 2000]. Some of the principles in this approach are similar to those used in the "vertical transect" approach described in Chapter 5, except that normal-fault footwall blocks exhumed in extensional settings are usually tilted to high angles, so the paleovertical dimension is now expressed over a subhorizontal distance. As in vertical transect approaches, thermochronologic dates near what was formerly the top of the crustal section are usually oldest, and the date–paleodepth relationship shows a break in slope that marks the timing of onset of rapid exhumation of the block. At greater paleodepths the slope of the date–paleodepth relationship provides an estimate of the exhumation rate, which, with additional constraints, can be used to constrain the rate of slip on the fault that exhumed the block.

In some cases, more than one break in slope is evident, as in the example from the Wassuk Range, Nevada determined by *Stockli et al.* [2002] (Fig. 11.45). Here, an approximately 8.5-km-thick section of crust was exhumed by Basin and Range normal faulting and tilted about 60° to the west. Apatite He, apatite fission-track, and zircon He dates show similar patterns of decreasing date with increasing paleodepth, at least up to paleodepths of about 2.7, 3.6, and 6.5 km, respectively. These depths correspond to the predicted depths of the bases of the partial retention zones (PRZs) for each system, if the pre-exhumational geothermal gradient was 27 ± 5 °C/km. This "fossil PRZ" developed during a period of very slow (or zero) exhumation that likely began around 60–70 Ma, the end of rapid exhumation associated with Laramide orogeny in this region, as indicated by the oldest zircon He dates and converging dates for all three thermochronometers at the shallowest paleodepths. This interval of slow exhumation ended at about 15 Ma, the date corresponding to the "break in slope" beneath each fossil PRZ. Assuming the cooling represented by these identical 15 Ma dates for all three systems was caused entirely by tectonic exhumation (i.e., normal faulting), the ~2.5 km range of invariant dates requires an equivalent depth of exhumation over a short duration of time, likely less than 0.5–1.0 Ma. This is consistent with abundant geologic evidence from the region for widespread low-angle normal faulting throughout this part of the Basin and Range at this time. However, the trend to much younger apatite He and apatite FT dates in the deepest parts of this crustal block clearly show that exhumation slowed dramatically not long

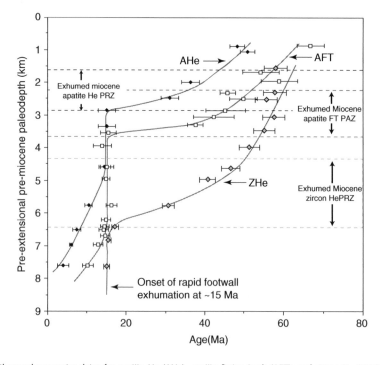

Fig. 11.45. Paleodepth versus thermochronometer dates for apatite He (AHe), apatite fission-track (AFT), and zircon He (ZHe) from the Wassuk Mountains, a tectonically exhumed normal-fault footwall in the Basin and Range Province of Nevada. All three systems preserve a "fossil PRZ" that developed from approximately 60 through 15 Ma in the shallower portions of the block. Approximately 2–3 km of rapid exhumation occurred at 15 Ma, as evidenced by the paleodepth-invariant dates at this time, but the shallower paleodepth-date slopes of the AHe and AFT systems at greater paleodepths require a subsequent slowdown in exhumation rates. (Source: *Stockli* [2005]. Reproduced with permission of Mineralogical Society of America.)

after it started. These parallel trends may indicate the development of another, post-15-Ma PRZ, or simply much slower exhumation after this time. Finally, although not explicitly represented in the dates from this particular block, the exhumation of rocks with these younger dates in the deep portions of this block require that another episode of rapid tectonic exhumation, later than the youngest exposed apatite He date of ~5 Ma, again exhumed the block. More direct evidence for this Pliocene event is found in the deepest portions of other exhumed normal fault footwalls in the region, where paleodepth-invariant apatite He dates of 3 Ma are seen at the structurally deepest levels of the White Mountains [*Stockli et al.*, 2003].

11.7.2 Paleotopography

11.7.2.1 Apatite He dating of Sierra Nevada paleotopography
The archetypal paleotopographic study using thermochronology is the pair of groundbreaking studies by *House et al.* [1998,

2001]. *House et al.* collected dozens of granitoid intrusive rocks in two parallel transects along strike of the Mesozoic Sierra Nevada batholith that crossed several major range-perpendicular canyons, including the San Joaquin and Kings Canyons, characterized by 1 km of relief over topographic wavelengths of about 50 km (Fig. 11.46). In order to compare apatite He dates of samples at the same elevation along strike of the orogen, *House et al.* removed the effects of small elevation differences by correcting their values to a common elevation using a vertical transect date–elevation relationship measured in another location (Fig. 11.46b). In the western transect, these dates show clear anticorrelations with the modern elevation, i.e., He dates are older in regions where the long-wavelength topography has low elevation. The simplest explanation for the phase and amplitude of these He date differences is that when these samples were erosionally exhumed through their closure depths at about 60–80 Ma, closure depths were much deeper for samples that are now near large valleys. Because isotherms at depth follow a

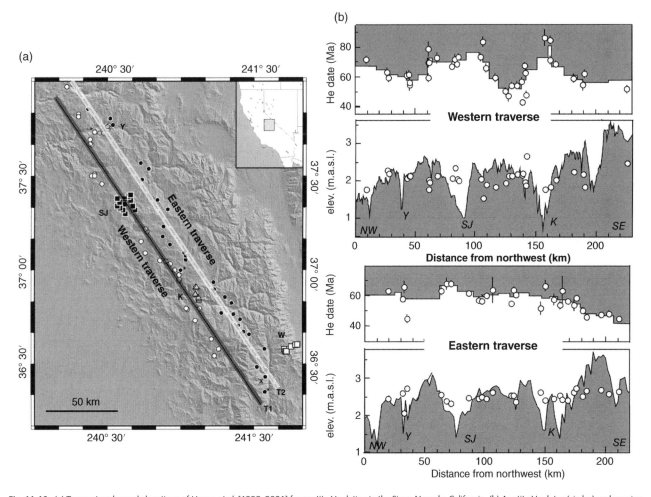

Fig. 11.46. (a) Transect and sample locations of *House et al.* [1998, 2001] for apatite He dating in the Sierra Nevada, California. (b) Apatite He dates (circles) and running means (joined-line segments under gray shading), and surface elevation with sample locations, for both transects. In the western transect, apatite He dates are anticorrelated with elevation in the vicinity of the largest canyons. *House et al.* [2001] used the He date differences, cooling/erosion rates at 60–80 Ma, and some assumptions about the shallow crustal thermal field to infer deep orogeny-perpendicular canyons in the vicinity of the western transect, but much lower relief only a few tens of kilometers to the east. (Source: *House et al.* [2001]. Reproduced with permission of American Journal of Science.)

muted form of the overlying topography, the most likely explanation for this is that 60–80 million years ago the valleys were located in the same positions they are now. Moreover, the magnitude of the date differences, combined with constraints on cooling and exhumation erosion rates in the late Cretaceous and early Paleogene, require that the Late Cretaceous–Early Paleogene valleys were significantly larger than they are today, with relief of 1.5 ± 0.5 km.

In contrast, the He dates in the eastern transect show no significant relationship with topography, suggesting that the canyon depths farther into the range were much less. This leads to a paleotopographic reconstruction something like the western edge of the north-central Andes, in which a high-elevation orogenic plateau is flanked by very deep canyons on the western edge. Other types of thermochronologic analysis of the Sierra have largely supported this interpretation [e.g., *Braun*, 2002].

11.7.2.2 ^{4}He/^{3}He thermochronometry of glacial incision

As shown by *Dodson* [1986], the concept of closure temperature has meaning not only for bulk grains, assuming grain and domain equivalency, but also for each radial position within a grain, so that the bulk grain T_c is actually the integrated closure profile across it. For an apatite grain of typical He diffusivity and size, T_c for its interior and exterior regions can range from as high as 80–90 °C to as low as 20–30 °C. By effectively characterizing (U–Th)/He dates across the entire radial domain of the grain, the ^{4}He/^{3}He method therefore provides a way to interpret a monotonic cooling history corresponding to crustal depths as shallow as 1 km or less. This provides a way to determine the timing, rate, and spatial patterns of topographic evolution of many landscapes with local relief on this scale, such as glacial landscapes.

The landscape of western British Columbia is characterized by numerous U-shaped valleys and fjords with vertical relief of 2–3 km over only slightly larger horizontal length scales. Just one example is the Kliniklini Valley, with a depth below sea level only about 20 km away from the highest point in British Columbia, Mount Waddington, at almost 4.5 km above sea level (Fig. 11.47). It is likely that glaciers were responsible for some of the incision that created this relief, but without robust temporal constraints it is impossible to tell if most of it was created by rivers long before the appearance of Plio-Pleistocene glaciers, or if most of the relief is much older and carved long ago by rivers. Similarly, if the relief is primarily due to glacial erosion, was it carved gradually over the past 2–3 Ma, or very quickly during the first appearance of alpine ice in the region?

^{4}He/^{3}He ratio-evolution plots for samples taken from Kliniklini Valley show that samples near the valley bottom have relatively high ^{4}He concentrations all the way across the grains (except for the alpha-ejection-affected grain rims), requiring rapid cooling (Fig. 11.47). In contrast, samples at high elevation appear to have strongly rounded ^{4}He concentration profiles, requiring slow cooling (assuming monotonic cooling). These data, in combination with bulk grain (U–Th)/He dates on the

same samples, showed that valley-bottom samples cooled rapidly from about 80 °C to less than 20 °C at 1.8 Ma, whereas ridge samples cooled much more slowly throughout the Late Miocene to present. Assuming a typical geothermal gradient of about 25 °C/km, these data require more than 2 km of incision within less than a few hundred thousand years at 1.8 Ma, highlighting the potential for rapid glacial modification of topography in regions with dynamic alpine glaciers.

Other thermochronologic studies of glacial landscapes have reached similar conclusions [e.g., *Ehlers et al.*, 2006]. Many of these studies reach a conclusion that the earliest phases of glacial erosion generate rapid and deep incision, even though the topographic lows may be occupied by ice long afterwards [e.g., *Shuster et al.*, 2011; *Valla et al.*, 2011; *Thomson et al.*, 2013].

11.7.3 Orogen-scale trends in thermochronologic dates

Several previous examples have highlighted the use of low-temperature thermochronometric applications using date variations across elevation, paleodepth, or multiple systems with different closure temperatures to infer exhumation rate changes with time. Variation in thermochronometric dates over large areas of the surface can also be used to infer the spatial patterns of exhumation. This is often used to infer spatial patterns of erosion [*Brandon et al.*, 1998; *Willett and Brandon*, 2013], which can in turn be interpreted as patterns in changing topography [e.g., *Ehlers et al.*, 2006; *Schildgen et al.*, 2010], or spatial patterns of rock uplift caused by long-term variations in isostatic rebound [*Braun and Robert*, 2005], crustal deformation [*Batt et al.*, 2001], or deeper-seated dynamic topography. Thermochronologic data may be especially useful for elucidating the large-scale kinematics of active orogens, where both horizontal and vertical components of advection and changing morphology of a mountain range may produce complex spatial patterns of dates on the surface [*Batt et al.*, 2001; *Batt and Brandon*, 2002].

Using a large data set of both apatite FT and apatite He dates from deformed metasedimentary rocks in the Northern Apennines of Italy, *Thomson et al.* [2010] observed systematic changes in cooling dates in a northeast–southwest transect across range and parallel to the convergence direction as indicated by GPS and kinematics of folds and thrust faults (Fig. 11.48). In the far northeast, cooling dates of apatites are older than the depositional ages of their host rocks. In the parlance of thermochronology of orogenic wedges this is called the "dead zone," because these rocks have never been deeply buried or exhumed, so dates are "unreset" and only bear information about the exhumation history of their distal sources, which in this case were in the Alps. Moving farther (west) into the range, however, first apatite He and then apatite FT dates become much younger, reaching minima of about 1–2 Ma and 3–5 Ma about 40 and 50 km from the range front. Farther west, dates of both systems gradually increase at the same rate until about 85 km from the range front. West of there, in a region characterized by widespread active normal faulting, dates decrease again.

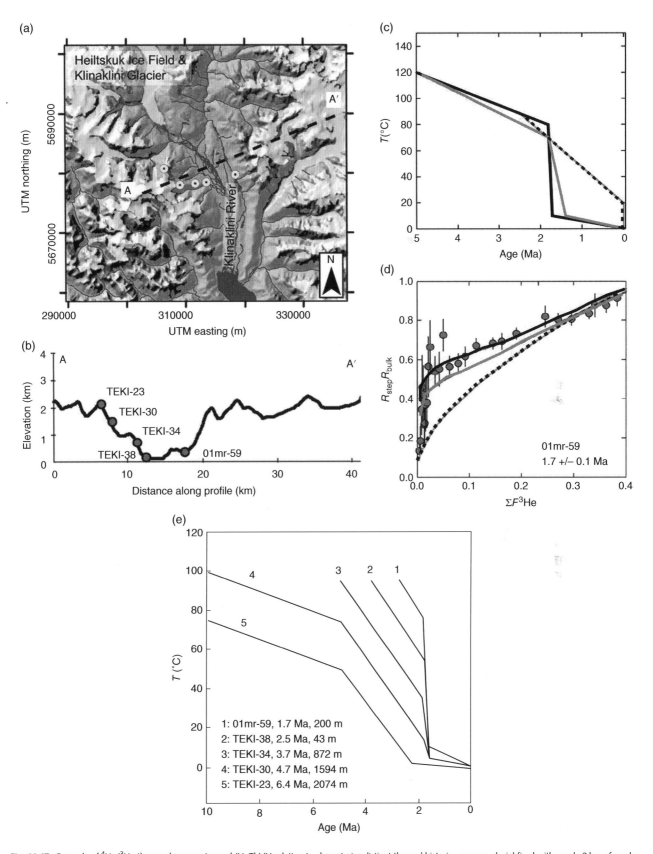

Fig. 11.47. Example of ^{4}He/^{3}He thermochronometry and (U–Th)/He dating to characterize distinct thermal histories across a glacial fjord with nearly 2 km of modern topographic relief. (a,b) Location map and topographic cross-section of Kliniklini valley, western British Columbia, and sample locations in *Shuster et al.* [2005]. (c,d) Three example *t–T* histories and their ratio-evolution predictions for ^{4}He/^{3}He step-heating measurements, and ratio-evolution diagram for sample 01mr-59, near the base of the valley. The results for this sample most closely match the predicted history involving very rapid cooling from ~80 °C at a time slightly older than the bulk date of the grain, 1.7 Ma. (e) Model thermal histories derived from ^{4}He/^{3}He results as in (d) for samples at different elevations across the Klinklini valley, showing progressive increase in amount of cooling experienced by samples at lower elevation. Together these data require rapid incision of the entire 2 km depth of the valley at 1.8 Ma. (Source: *Shuster et al.* [2005]. Reproduced with permission of The American Association for the Advancement of Science.)

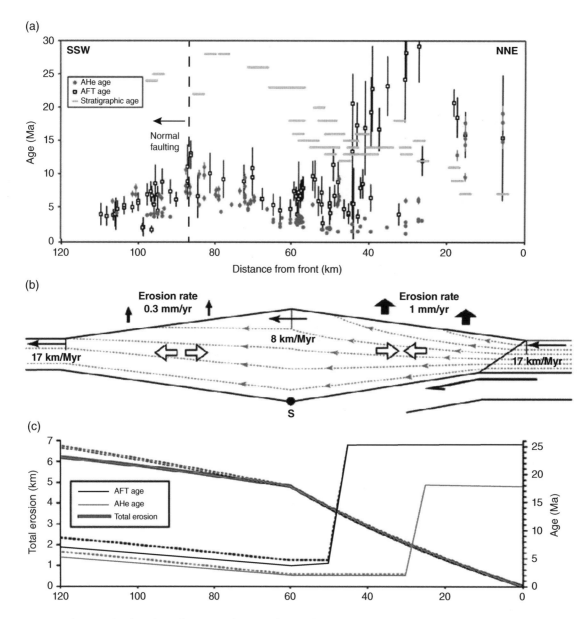

Fig. 11.48. (a) Apatite fission-track and (U–Th)/He dates projected onto a northeast (right) to southwest (left) transect across the northern Apennines of Italy. Horizontal dashes show stratigraphic ages of the greywackes and other low-grade metasedimentary host rocks. Note (i) systematic trends from old "unreset" dates on far "pro-wedge" flank of range, (ii) horizontally offset transitions to reset and younger dates farther west, (iii) increasing dates west of ~60 km distance from the front, and (iv) westward-decreasing dates in the region of active normal faulting. (b) Generalized orogen-scale kinematic model for the range, distances scaled to those in (a), showing horizontal and vertical velocity components across the range, and S-point of diverging velocities representing simplified downward velocity of subducting lithosphere converging from the east side. (c) Predicted cumulative erosion and thermochronologic dates of resulting from model in (b). (Source: *Thomson et al.* [2010]. Reproduced with permission of Geological Society of America.) (*See insert for color representation of the figure.*)

The young dates in the far western part of the range are due to tectonic exhumation along large normal faults (e.g., see earlier section), but the systematic date–distance trends east of there require an explanation involving both horizontal and vertical advection of rocks through the mountain range. *Thomson et al.* [2010] explain the first-order trends with a simple orogen-scale kinematic model involving equal accretion and excretion rates across a mountain range in flux steady-state but a faster erosion rate on the pro-wedge (east in this case) side. Rocks are accreted

to the wedge on the eastern thrust front and advected laterally through the wedge but only show "reset" dates younger than their host rock depositional ages once the cumulative erosional exhumation for any westward trajectory through the wedge exceeds the apatite He and FT closure depths. This occurs closer to the range from for the apatite He system because, for a given crustal thermal model, it has a shallower closure depth than the AFT system; the offset between the dead zone to reset dates for the two systems is proportional to the horizontal advection

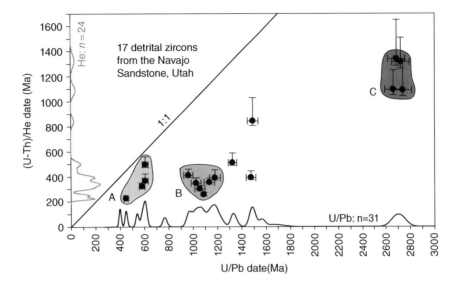

Fig. 11.49. Combined (U–Th)/He and U/Pb dates of 17 zircon grains from the Navajo Sandstone in southern Utah. Three dominant modes of combined crystallization (U/Pb) and cooling (He) dates were interpreted by *Rahl et al.* [2003] as requiring a sediment source for the Navajo Sandstone thousands of kilometers to the east, in the Appalachian/Caledonide chain. (Source: adapted from *Rahl et al.* [2003].)

velocity, here about 17 km/Ma. In turn, the reset dates of AFT and apatite He systems in the pro-wedge region just west of the dead zone are inversely proportional to the erosion rate in this region, here about 1 km/Ma. West of the topographic high of the range, dates increase with distance, approaching older dates corresponding to the slower erosion rate on the retrowedge side of the range, until extension results in sufficient tectonic exhumation to produce decreasing westward dates once again.

11.7.4 Detrital double-dating and sediment provenance

Thermochronologic dates of detrital minerals have long been used to elucidate sources and transport of sediment and orogenic histories of source terrains [e.g., *Garver et al.*, 1999; *Bernet and Spiegel*, 2004], and recent studies have advanced many aspects of the statistical and interpretational bases of detrital thermochronology [*Brandon and Vance*, 1992; *Stock and Montgomery*, 1996; *Vermeesch*, 2004; *Ruhl and Hodges*, 2005; *Amidon et al.*, 2005; *Rahl et al.*, 2007]. (U–Th)/He dating of both apatite and zircon have been used for detrital applications once restricted to fission-track or Ar/Ar methods, providing provenance constraints on both regional and local [*Stock and Montgomery*, 1996] scales, and constraining depositional ages and thermal histories of orogenic source terrains. Several studies have combined two or more thermochronometers and/or geochronometers on single detrital grains to deduce more precise provenance and thermal history constraints [*Campbell et al.*, 2005; *Reiners et al.*, 2005; *Carrapa et al.*, 2009; *Perry et al.*, 2009; *Saylor et al.*, 2012; *Thomson et al.*, 2013].

The first such "double-dating" application used U/Pb and (U–Th)/He dating of 17 grains of detrital zircon to trace the dominant sediment sources for the Navajo Sandstone in southern Utah. Although analysis of very few grains as in this study cannot

robustly rule out important contributions from other sources, the results of this study suggested that most zircon grains in this early Jurassic eolian sandstone were derived from three sources with distinct combinations of crystallization (U/Pb) dates and cooling (He) dates (Fig. 11.49). The most abundant of these has Grenville (~1.0–1.2 Ga) crystallization dates and "Appalachian" cooling dates of 200–400 Ma. The next most abundant populations have either Appalachian crystallization and cooling dates, or Archean crystallization dates combined with Grenvillian cooling dates. Together, these point to a source region in the Appalachian and/or Caledonide mountain ranges, thousands of kilometers east of the Navajo Sandstone and orthogonal to the proximal paleocurrent directions of the Navajo Sandstone. *Rahl et al.* [2003] interpreted these results to indicate westward continental-scale transport of Appalachian detritus over thousands of kilometers, and subsequent regional eolian transfer to the Navajo depocenter. Similar conclusions were derived from analysis of several thousand detrital zircon U/Pb dates from the same region [*Dickinson and Gehrels*, 2003].

11.7.5 Volcanic double-dating, precise eruption dates, and magmatic residence times

Obtaining precise dates on volcanic eruptions younger than about 1 Ma is often challenging for any geochronometer for several reasons, among them low daughter concentrations and inherited preeruptive daughter products. The dramatic increase in precision in U/Pb and ^{40}Ar/^{39}Ar in the past decades has also led to growing recognition of the abundance of inherited, preeruptive xenocrysts or antecrysts in many eruptions [*Charlier and Zellmer*, 2000]. Even eruptions unaffected by wallrock contamination of any kind may contain crystals formed over the lifetime of typical magma chambers: hundreds of thousands of years.

In theory, comparison of U/Pb and ^{40}Ar/^{39}Ar dates on many volcanic minerals should allow for the discrimination of preeruptive and eruptive ages, because the closure temperatures of the respective systems should bracket the preeruptive magma temperature. But in practice small but finite uncertainties on decay constants and branching ratios and differing standardization procedures and assumptions about decay systematics have made such comparisons difficult.

(U–Th)/He dating has been used in several cases to constrain ages of volcanic eruptions where other techniques are difficult to apply [e.g., *Farley et al.*, 2002; *Aciego et al.*, 2003, 2007; *Edgar et al.*, 2007; *Min et al.*, 2006; *Blondes et al.*, 2007; *Cooper et al.*, 2011], but one of the most useful approaches uses combined ^{230}Th/^{238}U disequilibria and (U–Th)/He methods on magmatic zircons. For eruptions on the order of 1 Ma or younger, either of these methods separately would yield unreliable eruption ages. The former would provide an estimate of the zircon's formation age, which may be significantly older than the eruption age, and the latter would be subject to significant U-series (and Th-series) disequilibria that would likely produce "too young" dates (because typical magmatic zircons have Th/U < 1). But

when combined, the extent of (^{230}Th/^{238}U) disequilibria in the zircon (together with the Th/U of the whole rock as a proxy for the initial magma composition, and other assumptions [*Farley et al.*, 2002]) can be used to correct the (U–Th)/He date, providing a more precise eruption age constraint, and the comparison between the ^{230}Th/^{238}U and He dates provides insights into magmatic residence time and assimilation of xenocrysts and antecrysts in the magmatic system [*Schmitt et al.*, 2006]. In some cases, erupted zircons appear to have formed in multiple plutonic episodes and can be interpreted with spectral approaches similar to those used in detrital studies of sediments [*Schmitt et al.*, 2010], in other cases, protracted durations of growth in single magma chambers have been interpreted from core to rim changes in Th/U ages, permitting estimates of magmatic residence times and crystal growth rates [*Schmitt et al.*, 2006, 2011]. The example shown in Figure 11.50 shows U-series disequilibria plots compared with (U–Th)/He dates for zircon grains from two eruptions in New Zealand previously interpreted to be older than ~60 ka. Zircon (U–Th)/He dates uncorrected for Th/U disequilibria range from 15 to 21 ka, but when corrected for disequilibrium using the whole-rock U/Th and

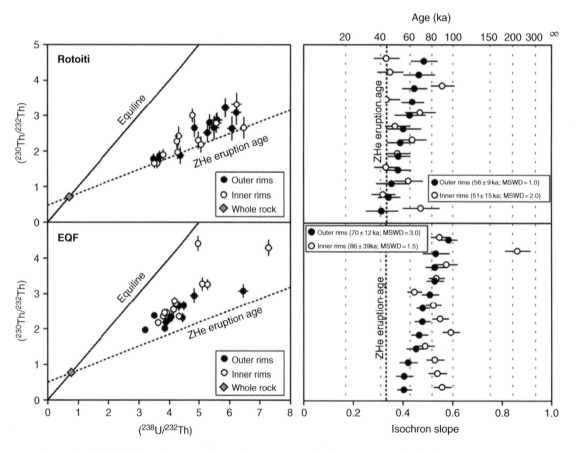

Fig. 11.50. Example of combined (^{230}Th/^{238}U) and (U–Th)/He dating of magmatic zircon, from the Rotoiti and Earthquake Flat (EQF) eruptions in northern New Zeland formerly interpreted as older than ~60 ka. Using the measured (^{230}Th/^{238}U) dates, which range from ~40–100 ka, and Th/U and (^{230}Th/^{232}Th) of the whole rock (inferred to be that of bulk magma), the measured (U–Th)/He dates were corrected by about 100% for the effects of (^{230}Th/^{238}U) disequilibrium. The resulting eruption dates from the corrected zircon He dates are then indistinguishable at 45.1 ka. (Source: *Danišík et al.* [2012]. Reproduced with permission of Elsevier.)

(^{230}Th/^{232}Th) and Th/U dates ranging from 41 to 96 ka, the disequilibrium-corrected zircon He dates yield precise constraints for the two eruptions of 45.1 ± 3.3 and 45.1 ± 2.9 ka, indistinguishable from high precision ^{14}C results from adjacent layers. Among other conclusions, this shows that two large (~100 km^3) eruptions can be produced from the same magmatic system within a short period of time (e.g., < 3000 years).

11.7.6 Radiation-damage-and-annealing model applied to apatite

As shown above, increasing radiation damage in apatite has the effect of decreasing He diffusivity and increasing closure temperature, at least up to radiation doses documented thus far in nature and in the lab. In general this means that the thermal sensitivity and closure temperature of the apatite He system continuously evolves with time, and that quantitative thermal history interpretations should use a kinetic model such as the RDAAM that explicitly accounts for this. It also follows that apatite grains with varying eU (i.e., a proxy for relative rate of radiation damage accumulation among related grains) from a small volume of rock (e.g., hand sample, outcrop, or larger-scale volume depending on the setting) that have experienced the same thermal history, will provide a range of dates corresponding to distinct T_c values. This can be used to constrain the thermal history more precisely than a single bulk He date and T_c. For example, grains with strongly contrasting eU but identical He dates require relatively rapid cooling, whereas correlations between He date and eU require slow cooling or an episode of partial resetting due to reheating.

Figure 11.51 shows an example of the RDAAM approach to constraining thermal histories [*Flowers*, 2009]. In this region of the Canadian shield, apatite grains from several rock samples from a single outcrop or nearby outcrops vary in He date by about 700 Ma, and this variation is well correlated with a range of eU concentrations over about 40 ppm. In this particular example, apatite fission-track dates and track-length models require the last cooling below the AFT T_c (~120 °C) in the Proterozoic, but geologic evidence suggests burial by Paleozoic sedimentary rocks up to several kilometers thick. Taken together, these constraints require two different episodes of heating to temperatures between ~60–80 °C, one in the Proterozoic and another in the early Paleozoic. Reconstructing nonmonotonic burial and exhumation histories using RDAAM in this way allows (U–Th)/He dating to resolve geologically complex (and presumably realistic) histories with important implications for a variety of purposes, not limited to cratonal uplift/subsidence and thermal maturation.

11.8 CONCLUSIONS

The (U–Th)/He system provided some of the first radioisotopic dates ever measured, and although the results confirmed that some rocks were at least hundreds of millions of years old, uncertainty surrounding the potential for post-crystallization loss of

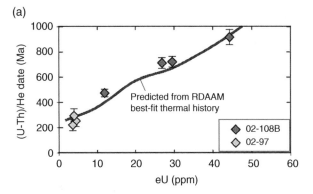

(a)

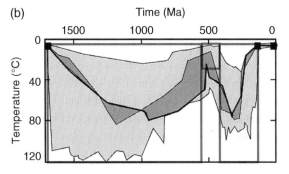

(b)

Fig. 11.51. (a) Correlation between single-grain apatite (U–Th)/He dates and bulk grain eU used, together with geologic constraints, to (b) deduce the thermal history for the Hearne region of the Canadian shield [*Flowers*, 2009]. In this case the thermal history requires two distinct periods of heating to temperatures between the nominal apatite He and apatite FT closure temperatures, though in detail, each apatite grain has an independent T_c that is a function of its eU concentration and which varies through time with accumulation (and potentially, though not in this example, annealing) of radiation damage. (Source: *Flowers et al.* [2009]. Reproduced with permission of Elsevier.)

^{4}He from samples prevented widespread use of the system until the later part of the 20th century, when it began to be used widely as a low-temperature thermochronometer. Most modern (U–Th)/He applications focus on apatite or zircon, but other minerals are often used as well, with iron and manganese oxides showing considerable potential. In zircon, most radiogenic ^{4}He is produced from ^{238}U, ^{235}U, and ^{232}Th, but ^{147}Sm may contribute a significant fraction of the daughter product in apatite (and other phosphates) in some cases.

Conventional analytical methods usually involve heating single crystals to release He, and measurement of that element using gas-source mass spectrometry. Parent nuclides are typically measured in the same aliquot using ICP-MS. (U–Th)/He analysis by other means, including laser ablation, have also been used and are a focus of ongoing development. The relatively high-energy alpha-decay events of the U and Th series that produce ^{4}He daughter nuclides lead to alpha-particle stopping distances that average ~15–20 μm in most minerals. The resulting physical separation of parent and daughter products thus requires considerations for accurate date interpretation. In the case of apatite and zircon crystals of typical size, a relatively large (~15–35%) correction is made to the measured parent/daughter ratio to account for ^{4}He ejected from the grain volume. Complications that may arise from long alpha-stopping distances include large and systematic intragranular parent nuclide heterogeneity (i.e., particular types of U–Th zonation), He implantation into analyzed aliquots from adjacent U–Th-bearing sources that are not accounted for, and, in detrital samples, grain abrasion during transport.

Helium migrates through and is lost from minerals primarily by thermally activated volume diffusion, so the bulk retentivity (and therefore closure temperature) of radiogenic He in an aliquot is a function of the diffusion domain size and He diffusion kinetics (e.g., activation energy and frequency factor). In the case of apatite, zircon, and several other minerals, the diffusion domain is the physical crystal or grain size, though fracturing, and radiation damage in the case of zircon, may modify this. This means that closure temperature, and therefore He age, depends in part on the grain/crystal size or size distribution in an aliquot, and relationships between grain/crystal size and He age among different aliquots from the same sample may be used to infer a sample's thermal history.

The kinetics of He diffusion within minerals are usually measured by step-heating release experiments on radiogenic ^{4}He or artificially introduced ^{3}He, but other methods including ERDA and NRA have also yielded important constraints over shorter length scales. Radiation damage has been shown to cause large changes in He mobility and retention in both apatite and zircon. In apatite, He diffusivity decreases (increasing effective closure temperature) with increasing accumulated radiation damage. This has been interpreted and quantitatively modeled as an increasing abundance of crystallographic defects that act as local energy minimum traps for diffusing He atoms. These radiation damage defect traps also anneal as a function of time and temperature. Thus the He retentivity and closure temperature in a given apatite crystal depend on both the abundance of parent nuclides that cause radiation damage, and the thermal history that affects both He loss and damage annealing. The bulk retentivity of He in zircon, which also has been shown to be highly crystallographically anisotropic, also increases (increasing closure temperature) with increasing radiation damage, but only up to a critical threshold of damage, after which He retentivity decreases. Because differences in eU concentrations among aliquots (typically single crystals) reflect differences in accumulated radiation damage, variations in He dates that are correlated with eU concentrations can be interpreted and quantitatively modeled to constrain thermal histories.

^{4}He/^{3}He thermochronology involves bombarding samples with high-energy protons to produce a homogeneous ^{3}He concentration in an aliquot, and then performing a step-heating degassing in which both the artificially produced ^{3}He and natural ^{4}He are released. This not only yields Arrhenius trends to estimate the kinetics of He diffusion in the sample, but the spectrum of ^{4}He/^{3}He evolution through the experiment can be interpreted as characterizing the concentration profile of ^{4}He within a single diffusion domain, or the relative ^{4}He concentrations among domains of differing retentivity.

A compilation of He diffusion kinetics from a variety of minerals shows significant differences in frequency factors and activation energies determined from step-heating and other types of experiments. He diffusion kinetics among minerals appear to show little correlation with ionic porosity.

Modern (U–Th)/He dating is a versatile approach with a wide range of applications but some of the most common in use today are:

(1) characterizing thermochronometric dates in samples collected across "vertical" or paleodepth transects to deduce erosional or tectonic exhumation histories;
(2) using spatial patterns of dates across landscapes to deduce paleotopographic evolution or large-scale erosion rate patterns related to rock uplift;
(3) combining He dating with other techniques (most commonly U/Pb) on single detrital grains to yield "double" or "multiple" dating constraints on provenance and/or depositional age;
(4) combining He dating with U-series dating on volcanic zircon to yield precise eruption ages;
(5) using He date-eU correlations to deduce thermal histories of rocks and regions.

11.9 REFERENCES

Aciego, S., Kennedy, B. M., DePaolo, D. J., Christensen, J. N., and Hutcheon, I. (2003) U–Th/He age of phenocrystic garnet from the 79 AD eruption of Mt. Vesuvius. *Earth and Planetary Science Letters* **216**(1), 209–219.

Aciego, S. M., DePaolo, D. J., Kennedy, B. M., Lamb, M. P., Sims, K. W., and Dietrich, W. E. (2007) Combining [^{3}He] cosmogenic dating with U–Th/He eruption ages using olivine in basalt. *Earth and Planetary Science Letters* **254**(3), 288–302.

Albarède, F. (1978) The recovery of spatial isotope distributions from stepwise degassing data. *Earth and Planetary Science Letters* **39**(3), 387–397.

Amidon, W. H., Burbank, D. W., and Gehrels, G. E. (2005) Construction of detrital mineral populations: insights from mixing of U–Pb zircon ages in Himalayan rivers. *Basin Research* **17**(4), 463–485.

Ault, A. K. and Flowers, R. M. (2011) Significance of U–Th zonation in cratonic apatites for interpretation of (U–Th)/He thermochronometry data: an example from the Slave craton. In *AGU Fall Meeting Abstracts*, Vol. **1**, 2549.

Bähr, R., Lippolt, H. J., and Wernicke, R. S. (1994) Temperature-induced ^{4}He degassing of specularite and botryoidal hematite: a ^{4}He retentivity study. *Journal of Geophysical Research: Solid Earth (1978–2012)* **99**(B9), 17695–17707.

Batt, G. E. and Brandon, M. T. (2002) Lateral thinking: 2-D interpretation of thermochronology in convergent orogenic settings. *Tectonophysics* **349**(1), 185–201.

Batt, G. E., Brandon, M. T., Farley, K. A., and Roden-Tice, M. (2001) Tectonic synthesis of the Olympic Mountains segment of the Cascadia wedge, using two-dimensional thermal and kinematic modeling of thermochronological ages. *Journal of Geophysical Research: Solid Earth (1978–2012)* **106**(B11), 26731–26746.

Bender, M. L. (1973) Helium–uranium dating of corals. *Geochimica et Cosmochimica Acta* **37**(5), 1229–1247.

Bengtson, A., Ewing, R. C., and Becker, U. (2012) He diffusion and closure temperatures in apatite and zircon: a density functional theory investigation. *Geochimica et Cosmochimica Acta* **86**, 228–238.

Bernet, M. and Spiegel, C. (2004) Introduction: detrital thermochronology. *Geological Society of America Special Papers* **378**, 1–6.

Beucher, R., Brown, R. W., Roper, S., Stuart, F., and Persano, C. (2013) Natural age dispersion arising from the analysis of broken crystals: Part II. Practical application to apatite (U–Th)/He thermochronometry. *Geochimica et Cosmochimica Acta* **120**, 395–416.

Blackburn, T. J., Stockli, D. F., and Walker, J. D. (2007) Magnetite (U–Th)/He dating and its application to the geochronology of intermediate to mafic volcanic rocks. *Earth and Planetary Science Letters* **259**(3), 360–371.

Blondes, M. S., Reiners, P. W., Edwards, B. R., and Biscontini, A. (2007) Dating young basalt eruptions by (U–Th)/He on xenolithic zircons. *Geology* **35**(1), 17–20.

Boyce, J. W. and Hodges, K. V. (2005) U and Th zoning in Cerro de Mercado (Durango, Mexico) fluorapatite: Insights regarding the impact of recoil redistribution of radiogenic ^{4}He on (U–Th)/He thermochronology. *Chemical Geology* **219**(1), 261–274.

Boyce, J. W., Hodges, K. V., Olszewski, W. J., and Jercinovic, M. J. (2005) He diffusion in monazite: implications for (U–Th)/He thermochronometry. *Geochemistry, Geophysics, Geosystems* **6**(12).

Boyce, J. W., Hodges, K. V., Olszewski, W. J., Jercinovic, M. J., Carpenter, B. D., and Reiners, P. W. (2006) Laser microprobe (U–Th)/He geochronology. *Geochimica et Cosmochimica Acta* **70**(12), 3031–3039.

Boyce, J. W., Hodges, K. V., King, D., *et al.* (2009) Improved confidence in (U–Th)/He thermochronology using the laser microprobe: an example from a Pleistocene leucogranite, Nanga Parbat, Pakistan. *Geochemistry, Geophysics, Geosystems* **10**(9).

Brandon, M. T. and Vance, J. A. (1992) Tectonic evolution of the Cenozoic Olympic subduction complex, Washington State, as deduced from fission track ages for detrital zircons. *American Journal of Science* **292**(8), 565–636.

Brandon, M. T., Roden-Tice, M. K., and Garver, J. I. (1998) Late Cenozoic exhumation of the Cascadia accretionary wedge in the Olympic Mountains, northwest Washington State. *Geological Society of America Bulletin* **110**(8), 985–1009.

Braun, J. (2002) Estimating exhumation rate and relief evolution by spectral analysis of age–elevation datasets. *Terra Nova* **14**(3), 210–214.

Braun, J., and Robert, X. [2005] Constraints on the rate of post-orogenic erosional decay from low-temperature thermochronological data: Application to the Dabie Shan, China. *Earth Surface Processes and Landforms* **30**(9), 1203–1225.

Brown, R. W., Beucher, R., Roper, S., Persano, C., Stuart, F., and Fitzgerald, P. (2013) Natural age dispersion arising from the analysis of broken crystals. Part I: Theoretical basis and implications for the apatite (U–Th)/He thermochronometer. *Geochimica et Cosmochimica Acta* **122**, 478–497.

Campbell, I. H., Reiners, P. W., Allen, C. M., Nicolescu, S., and Upadhyay, R. (2005) He–Pb double dating of detrital zircons from the Ganges and Indus Rivers: implication for quantifying sediment recycling and provenance studies. *Earth and Planetary Science Letters* **237**(3), 402–432.

Carrapa, B., DeCelles, P. G., Reiners, P. W., Gehrels, G. E., and Sudo, M. (2009) Apatite triple dating and white mica ^{40}Ar/^{39}Ar thermochronology of syntectonic detritus in the Central Andes: A multiphase tectonothermal history. *Geology* **37**(5), 407–410.

Carslaw, H. S. and Jaeger, J. C. (1959) *Conduction of Heat in Solids*, 2nd edn. Clarendon Press, Oxford.

Charlier, B. and Zellmer, G. (2000) Some remarks on U–Th mineral ages from igneous rocks with prolonged crystallisation histories. *Earth and Planetary Science Letters* **183**(3), 457–469.

Cherniak, D. J. and Watson, E. B. (2011) Helium diffusion in rutile and titanite, and consideration of the origin and implications of diffusional anisotropy. *Chemical Geology* **288**(3), 149–161.

Cherniak, D. J., Watson, E. B., and Thomas, J. B. (2009) Diffusion of helium in zircon and apatite. *Chemical Geology* **268**(1), 155–166.

Cherniak, D. J., Amidon, W., Hobbs, D., and Watson, E. B. (2015) Diffusion of helium in carbonates: Effects of mineral structure and composition. *Geochimica et Cosmochimica Acta* **165**, 449–465.

Cooper, F. J., van Soest, M. C., and Hodges, K. V. (2011) Detrital zircon and apatite (U–Th)/He geochronology of intercalated baked sediments: a new approach to dating young basalt flows. *Geochemistry, Geophysics, Geosystems* **12**(7).

Copeland, P., Watson, E. B., Urizar, S. C., Patterson, D., and Lapen, T. J. (2007) Alpha thermochronology of carbonates. *Geochimica et Cosmochimica Acta* **71**(18), 4488–4511.

Costantini, J. M., Trocellier, P., Haussy, J., and Grob, J. J. (2002) Nuclear reaction analysis of helium diffusion in britholite. *Nuclear Instruments and Methods in Physics Research Section B: Beam Interactions with Materials and Atoms* **195**(3), 400–407.

Cros, A., Gautheron, C., Pagel, M., *et al.* (2014) ^{4}He behavior in calcite filling viewed by (U–Th)/He dating, ^{4}He diffusion and crystallographic studies. *Geochimica et Cosmochimica Acta* **125**, 414–432.

Crowley, P. D., Reiners, P. W., Reuter, J. M., and Kaye, G. D. (2002) Laramide exhumation of the Bighorn Mountains, Wyoming: an apatite (U–Th)/He thermochronology study. *Geology* **30**(1), 27–30.

Damon, P. E. (1957) Determination of radiogenic helium in zircon by stable isotope dilution technique. *Eos, Transactions American Geophysical Union* **38**(6), 945–953.

Damon, P.E. and Kulp, J. L. (1957) Determination of radiogenic helium in zircon by stable isotope dilution technique. *Eos, Transactions American Geophysical Union* **38**(6), 945–953.

Danišík, M., Shane, P., Schmitt, A. K., *et al.* (2012) Re-anchoring the late Pleistocene tephrochronology of New Zealand based on concordant radiocarbon ages and combined $^{238}U/^{230}Th$ disequilibrium and (U–Th)/He zircon ages. *Earth and Planetary Science Letters* **349**, 240–250.

Dickinson, W. R. and Gehrels, G. E. (2003) U–Pb ages of detrital zircons from Permian and Jurassic eolian sandstones of the Colorado Plateau, USA: paleogeographic implications. *Sedimentary Geology* **163**(1), 29–66.

Dobson, K. J., Stuart, F. M., and Dempster, T. J. (2008) U and Th zonation in Fish Canyon Tuff zircons: implications for a zircon (U–Th)/He standard. *Geochimica et Cosmochimica Acta* **72** (19), 4745–4755.

Dodson, M. H. (1986) Closure profiles in cooling systems. *Materials Science Forum* **7**, 145–154.

Dunai, T. J. and Roselieb, K. (1996) Sorption and diffusion of helium in garnet: implications for volatile tracing and dating. *Earth and Planetary Science Letters* **139**(3), 411–421.

Edgar, C. J., Wolff, J. A., Olin, P. H., *et al.* (2007) The late Quaternary Diego Hernandez Formation, Tenerife: volcanology of a complex cycle of voluminous explosive phonolitic eruptions. *Journal of Volcanology and Geothermal Research* **160** (1), 59–85.

Ehlers, T. A., Farley, K. A., Rusmore, M. E., and Woodsworth, G. J. (2006) Apatite (U–Th)/He signal of large-magnitude accelerated glacial erosion, southwest British Columbia. *Geology* **34** (9), 765–768.

Evans, N. J., Wilson, N. S. F., Cline, J. S., McInnes, B. I. A., and Byrne, J. (2005) Fluorite (U–Th)/He thermochronology: constraints on the low temperature history of Yucca Mountain, Nevada. *Applied Geochemistry* **20**(6), 1099–1105.

Evenson, N. S., Reiners, P. W., Spencer, J. E., and Shuster, D. L. (2014) Hematite and Mn oxide (U–Th)/He dates from the Buckskin-Rawhide detachment system, western Arizona: gaining insights into hematite (U–Th)/He systematics. *American Journal of Science* **314**(10), 1373–1435.

Fanale, F. and Kulp, J. L. (1961) Helium in limestone and marble. *American Mineralogist* **46**.

Fanale, F. P. and Kulp, J. L. (1962) The helium method and the age of the Cornwall, Pennsylvania magnetite ore. *Economic Geology* **57**(5), 735–746.

Fanale, F. P. and Schaeffer, O. A. (1965) Helium–uranium ratios for Pleistocene and Tertiary fossil aragonites. *Science* **149**(3681), 312–316.

Farley, K. A. (2000) Helium diffusion from apatite: General behavior as illustrated by Durango fluorapatite. *Journal of Geophysical Research: Solid Earth (1978–2012)* **105**(B2), 2903–2914.

Farley, K. A. (2002) (U–Th)/He dating: techniques, calibrations, and applications. *Reviews in Mineralogy and Geochemistry* **47** (1), 819–844.

Farley, K. A. (2007) He diffusion systematics in minerals: evidence from synthetic monazite and zircon structure phosphates. *Geochimica et Cosmochimica Acta* **71**(16), 4015–4024.

Farley, K. A. and Flowers, R. M. (2012) (U–Th)/Ne and multidomain (U–Th)/He systematics of a hydrothermal hematite from eastern Grand Canyon. *Earth and Planetary Science Letters* **359**, 131–140.

Farley, K. A. and Stockli, D. F. (2002) (U–Th)/He dating of phosphates: Apatite, monazite, and xenotime. *Reviews in Mineralogy and Geochemistry* **48**(1), 559–577.

Farley, K. A., Wolf, R. A., and Silver, L. T. (1996) The effects of long alpha-stopping distances on (U–Th)/He ages. *Geochimica et Cosmochimica Acta* **60**(21), 4223–4229.

Farley, K. A., Kohn, B. P., and Pillans, B. (2002) The effects of secular disequilibrium on (U–Th)/He systematics and dating of Quaternary volcanic zircon and apatite. *Earth and Planetary Science Letters* **201**(1), 117–125.

Farley, K. A., Shuster, D. L., Watson, E. B., Wanser, K. H., and Balco, G. (2010) Numerical investigations of apatite $^4He/^3He$ thermochronometry. *Geochemistry, Geophysics, Geosystems* **11**(10).

Farley, K. A., Shuster, D. L., and Ketcham, R. A. (2011) U and Th zonation in apatite observed by laser ablation ICPMS, and implications for the (U–Th)/He system. *Geochimica et Cosmochimica Acta* **75**(16), 4515–4530.

Fechtig, H., Gentner, W., and Lämmerzahl, P. (1963) Argonbestimmungen an kaliummineralien—XII: Edelgasdiffusionsmessungen an stein- und eisenmeteoriten. *Geochimica et Cosmochimica Acta* **27**(11), 1149–1169.

Fechtig, H., & Kalbitzer, S. (1966) The diffusion of argon in potassium-bearing solids. *Potassium-Argon Dating*, 68–106.

Ferreira, M. P., Macedo, R., Costa, V., Reynolds, J. H., Riley, J. E., and Rowe, M. W. (1975) Rare-gas dating, II. Attempted uranium-helium dating of young volcanic rocks from the Madeira Archipelago. *Earth and Planetary Science Letters* **25**(2), 142–150.

Fitzgerald, P. G., Baldwin, S. L., Webb, L. E., and O'Sullivan, P. B. (2006) Interpretation of (U–Th)/He single grain ages from slowly cooled crustal terranes: a case study from the Transantarctic Mountains of southern Victoria Land. *Chemical Geology* **225** (1), 91–120.

Flowers, R. M. (2009) Exploiting radiation damage control on apatite (U–Th)/He dates in cratonic regions. *Earth and Planetary Science Letters* **277**(1), 148–155.

Flowers, R. M. and Kelley, S. A. (2011) Interpreting data dispersion and "inverted" dates in apatite (U–Th)/He and fission-track datasets: an example from the US midcontinent. *Geochimica et Cosmochimica Acta* **75**(18), 5169–5186.

Flowers, R. M., Ketcham, R. A., Shuster, D. L., and Farley, K. A. (2009) Apatite (U–Th)/He thermochronometry using a radiation damage accumulation and annealing model. *Geochimica et Cosmochimica Acta* **73**(8), 2347–2365.

Garcia, P., Martin, G., Desgardin, P., *et al.* (2012) A study of helium mobility in polycrystalline uranium dioxide. *Journal of Nuclear Materials* **430**(1), 156–165.

Garver, J. I., Brandon, M. T., Roden-Tice, M., and Kamp, P. J. (1999) Exhumation history of orogenic highlands determined by detrital fission-track thermochronology. *Geological Society, London, Special Publications* **154**(1), 283–304.

Gautheron, C. and Tassan-Got, L. (2010) A Monte Carlo approach to diffusion applied to noble gas/helium thermochronology. *Chemical Geology* **273**(3), 212–224.

Gautheron, C., Tassan-Got, L., Barbarand, J., and Pagel, M. (2009) Effect of alpha-damage annealing on apatite (U–Th)/He thermochronology. *Chemical Geology* **266**(3), 157–170.

Gautheron, C., Tassan-Got, L., Ketcham, R. A., and Dobson, K. J. (2012) Accounting for long alpha-particle stopping distances in (U–Th–Sm)/He geochronology: 3D modeling of diffusion, zoning, implantation, and abrasion. *Geochimica et Cosmochimica Acta* **96**, 44–56.

Gerling, E. K. (1939) Diffusion temperature of helium as a criterion for the usefulness of minerals for helium age determination. *Comptes rendus de l'Académie des sciences de l'URSS* **24**, 570–573.

Gibbons, J.F. (1972) Ion implantation in semiconductors—Part II: Damage production and annealing. *Proceedings of the IEEE*, **60**, 1062–1096.

Green, P. F. and Duddy, I. R. (2006) Interpretation of apatite (U–Th)/He ages and fission track ages from cratons. *Earth and Planetary Science Letters* **244**(3), 541–547.

Green, P. F., Crowhurst, P. V., Duddy, I. R., Japsen, P., and Holford, S. P. (2006) Conflicting (U–Th)/He and fission track ages in apatite: enhanced He retention, not anomalous annealing behaviour. *Earth and Planetary Science Letters* **250**(3), 407–427.

Guenthner, W. R., Reiners, P. W., Ketcham, R. A., Nasdala, L., and Giester, G. (2013) Helium diffusion in natural zircon: Radiation damage, anisotropy, and the interpretation of zircon (U–Th)/He thermochronology. *American Journal of Science* **313**(3), 145–198.

Hansen, K. and Reiners, P. W. (2006) Low temperature thermochronology of the southern East Greenland continental margin: evidence from apatite (U–Th)/He and fission track analysis and implications for intermethod calibration. *Lithos* **92**(1), 117–136.

Harrison, T. M., Célérier, J., Aikman, A. B., Hermann, J., and Heizler, M. T. (2009) Diffusion of 40 Ar in muscovite. *Geochimica et Cosmochimica Acta* **73**(4), 1039–1051.

Hart, S. R. (1984) He diffusion in olivine. *Earth and Planetary Science Letters* **70**(2), 297–302.

Heim, J. A., Vasconcelos, P. M., Shuster, D. L., Farley, K. A., and Broadbent, G. (2006) Dating paleochannel iron ore by (U–Th)/He analysis of supergene goethite, Hamersley province, Australia. *Geology* **34**(3), 173–176.

Holland, H. D. (1954) Radiation damage and its use in age determination. In *Nuclear Geology*, Faul, H. (ed.), 175–179. John Wiley & Sons, New York.

Holmes, A. and Paneth, F. A. (1936) Helium-ratios of rocks and minerals from the diamond pipes of South Africa. *Proceedings of the Royal Society of London A: Mathematical, Physical and Engineering Sciences* **154**(882), 385–413).

Hourigan, J. K., Reiners, P. W., and Brandon, M. T. (2005) U–Th zonation-dependent alpha-ejection in (U–Th)/He chronometry. *Geochimica et Cosmochimica Acta* **69**(13), 3349–3365.

House, M. A., Wernicke, B. P., and Farley, K. A. (1998) Dating topography of the Sierra Nevada, California, using apatite (U–Th)/He ages. *Nature* **396**(6706), 66–69.

House, M. A., Farley, K. A., and Stockli, D. (2000) Helium chronometry of apatite and titanite using Nd-YAG laser heating. *Earth and Planetary Science Letters* **183**(3), 365–368.

House, M. A., Wernicke, B. P., and Farley, K. A. (2001) Paleogeomorphology of the Sierra Nevada, California, from (U–Th)/He ages in apatite. *American Journal of Science* **301**(2), 77–102.

Huber, C., Cassata, W. S., and Renne, P. R. (2011) A lattice Boltzmann model for noble gas diffusion in solids: the importance of domain shape and diffusive anisotropy and implications for thermochronometry. *Geochimica et Cosmochimica Acta* **75**(8), 2170–2186.

Huneke, J. C., Nyquist, L. E., Funk, H., Köppel, V., and Signer, P. (1969) The thermal release of rare gases from separated minerals of the Mocs meteorite. *Meteorite Research* **12**, 901–921.

Hurley, P. M. (1952) Alpha ionization damage as a cause of low helium ratios. *Transactions of the American Geophysical Union* **33**, 174–183

Hurley, P. M. (1954) The helium age method and the distribution and migration of helium in rocks. In *Nuclear Geology*. Faul, H. (ed.), 301–329. John Wiley & Sons, New York.

Hurley, P. M. and Fairbairn, H. W. (1953) Radiation damage in zircons: a possible age method. *Bull Geol Soc Am* **64**, 659–674

Hurley, P. M., Larsen, E. S. Jr., and Gottfried, D. (1956) Comparison of radiogenic helium and lead in zircon. *Geochimica et Cosmochimica Acta* **9**, 98–102.

Jonckheere, R. and Van den Haute, P. (2002) On the efficiency of fission-track counts in an internal and external apatite surface and in a muscovite external detector. *Radiation Measurements* **35**, 29–40.

Keevil, N. B., Larsen, E. S. Jr., and Wank, F. J. (1944) Distribution of helium and radioactivity in rocks. VI: The Ayer granite-migmatite at Chelmsford, Mass. *American Journal of Science* **242**, 345–353.

Ketcham, R. A. (2005) Forward and inverse modeling of low-temperature thermochronometry data. *Reviews in Mineralogy and Geochemistry* **58**(1), 275–314.

Ketcham, R. A., Gautheron, C., and Tassan-Got, L. (2011) Accounting for long alpha-particle stopping distances in (U–Th–Sm)/He geochronology: refinement of the baseline case. *Geochimica et Cosmochimica Acta* **75**(24), 7779–7791.

Ketcham, R. A., Guenthner, W. R., and Reiners, P. W. (2013) Geometric analysis of radiation damage connectivity in zircon, and its implications for helium diffusion. *American Mineralogist* **98**(2–3), 350–360.

Krogh, T. E. (1982) Improved accuracy of U–Pb zircon ages by the creation of more concordant systems using an air abrasion technique. *Geochimica et Cosmochimica Acta* **46**(4), 637–649.

Kurz, M. D. and Jenkins, W. J. (1981) The distribution of helium in oceanic basalt glasses. *Earth and Planetary Science Letters* **53**(1), 41–54.

Larsen, E. S. Jr. and Keevil, N. B. (1942) The distribution of helium and radioactivity in rocks. III: Radioactivity and petrology of some California intrusives. *American Journal of Science* **240** (204–215).

Leventhal, J. S. (1975) An evaluation of the uranium-thorium-helium method for dating young basalts. *Journal of Geophysical Research* **80**(14), (1911–1914).

Levskiy, L. K. and Aprub, S. V. (1970) Diffusion of helium, neon, and argon from meteorites. *Geochemistry International USSR* **7** (5), 908.

Lippolt, H. J. and Weigel, E. (1988) [4]He diffusion in [40]Ar-retentive minerals. *Geochimica et Cosmochimica Acta* **52**(6), 1449–1458.

Lippolt, H. J., Wernicke, R. S., and Boschmann, W. (1993) [4]He diffusion in specular hematite. *Physics and Chemistry of Minerals* **20** (6), 415–418.

Lippolt, H. J., Leitz, M., Wernicke, R. S., and Hagedorn, B. (1994) (Uranium + thorium)/helium dating of apatite: experience with samples from different geochemical environments. *Chemical Geology* **112**(1), 179–191.

Lott, D. E. and Jenkins, W. J. (1984) An automated cryogenic charcoal trap system for helium isotope mass spectrometry. *Review of Scientific Instruments* **55**(12), 1982–1988.

Meesters, A. G. C. A. and Dunai, T. J. (2002a) Solving the production–diffusion equation for finite diffusion domains of various shapes: Part I. Implications for low-temperature (U–Th)/He thermochronology. *Chemical Geology* **186**(3), 333–344.

Meesters, A. G. C. A. and Dunai, T. J. (2002b) Solving the production–diffusion equation for finite diffusion domains of various shapes: Part II. Application to cases with α-ejection and nonhomogeneous distribution of the source. *Chemical Geology* **186**(3), 347–363.

Meesters, A. G. C. A. and Dunai, T. J. (2005) A noniterative solution of the (U–Th)/He age equation. *Geochemistry, Geophysics, Geosystems* **6**(4).

Min, K. and Reiners, P. W. (2007) High-temperature Mars-to-Earth transfer of meteorite ALH84001. *Earth and Planetary Science Letters* **260**(1), 72–85.

Min, K., Farley, K. A., Renne, P. R., and Marti, K. (2003) Single grain (U–Th)/He ages from phosphates in Acapulco meteorite and implications for thermal history. *Earth and Planetary Science Letters* **209**(3), 323–336.

Min, K., Reiners, P. W., Nicolescu, S., and Greenwood, J. P. (2004) Age and temperature of shock metamorphism of Martian meteorite Los Angeles from (U–Th)/He thermochronometry. *Geology* **32**(8), 677–680.

Min, K., Reiners, P. W., Wolff, J. A., Mundil, R., and Winters, R. L. (2006) (U–Th)/He dating of volcanic phenocrysts with high-U–Th inclusions, Jemez volcanic field, New Mexico. *Chemical Geology* **227**(3), 223–235.

Min, K., Reiners, P. W., and Shuster, D. L. (2013) (U–Th)/He ages of phosphates from St. Séverin LL6 chondrite. *Geochimica et Cosmochimica Acta* **100**, 282–296.

Mitchell, S. G. and Reiners, P. W. (2003) Influence of wildfires on apatite and zircon (U-Th)/He ages. *Geology* **31**(12), 1025–1028.

Murray, K. E., Orme, D. A., and Reiners, P. W. (2014) Effects of U–Th-rich grain boundary phases on apatite helium ages. *Chemical Geology* **390**, 135–151.

Nakajima, K., Serizawa, H., Shirasu, N., Haga, Y., and Arai, Y. (2011) The solubility and diffusion coefficient of helium in uranium dioxide. *Journal of Nuclear Materials* **419**(1), 272–280.

Nardi, L. V. S., Formoso, M. L. L., Müller, I. F., Fontana, E., Jarvis, K., and Lamarão, C. (2013) Zircon/rock partition coefficients of REEs, Y, Th, U, Nb, and Ta in granitic rocks: Uses for provenance and mineral exploration purposes. *Chemical Geology* **335**, 1–7.

Nasdala, L., Reiners, P. W., Garver, J. I., *et al.* (2004) Incomplete retention of radiation damage in zircon from Sri Lanka. *American Mineralogist* **89**(1), 219–231.

Nicolaysen, L. O. (1957) Solid diffusion in radioactive minerals and the measurement of absolute age. *Geochimica et Cosmochimica Acta* **11**(1), 41–59.

Nicolescu, S. and Reiners, P. W. (2005) (U–Th)/He dating of epidote and andradite garnet. *Geochimica et Cosmochimica Acta* **69** (10), A26.

Nyquist, L. E., Huneke, J. C., Funk, H., and Signer, P. (1972) Thermal release characteristics of spallogenic He, Ne, and Ar from the Carbo iron meteorite. *Earth and Planetary Science Letters* **14**(2), (207–215).

Ouchani, S., Dran, J. C., and Chaumont, J. (1998) Exfoliation and diffusion following helium ion implantation in fluorapatite: implications for radiochronology and radioactive waste disposal. *Applied Geochemistry* **13**(6), 707–714.

Perry, S. E., Garver, J. I., and Ridgway, K. D. (2009) Transport of the Yakutat terrane, southern Alaska: evidence from sediment petrology and detrital zircon fission-track and U/Pb double dating. *The Journal of Geology* **117**(2), 156–173.

Pidgeon, R. T., Brander, T., and Lippolt, H. J. (2004) Late Miocene (U + Th)–^{4}He ages of ferruginous nodules from lateritic duricrust, Darling Range, Western Australia. *Australian Journal of Earth Sciences* **51**(6), 901–909.

Pik, R. and Marty, B. (2009) Helium isotopic signature of modern and fossil fluids associated with the Corinth rift fault zone (Greece): implication for fault connectivity in the lower crust. *Chemical Geology* **266**(1), 67–75.

Pik, R., Marty, B., Carignan, J., and Lavé, J. (2003) Stability of the Upper Nile drainage network (Ethiopia) deduced from (U–Th)/He thermochronometry: implications for uplift and erosion of the Afar plume dome. *Earth and Planetary Science Letters* **215**(1), 73–88.

Pramono, Y., Sasaki, K., and Yano, T. (2004) Release and diffusion rate of helium in neutron-irradiated SiC. *Journal of nuclear science and technology* **41**(7), 751–755.

Rahl, J. M., Reiners, P. W., Campbell, I. H., Nicolescu, S., and Allen, C. M. (2003) Combined single-grain (U–Th)/He and U/Pb dating of detrital zircons from the Navajo Sandstone, Utah. *Geology* **31**(9), 761–764.

Rahl, J. M., Ehlers, T. A., and van der Pluijm, B. A. (2007) Quantifying transient erosion of orogens with detrital thermochronology from syntectonic basin deposits. *Earth and Planetary Science Letters* **256**(1), 147–161.

Reich, M., Ewing, R. C., Ehlers, T. A., and Becker, U. (2007) Low-temperature anisotropic diffusion of helium in zircon: implications for zircon (U–Th)/He thermochronometry. *Geochimica et Cosmochimica Acta* **71**(12), 3119–3130.

Reiners, P. W. and Farley, K. A. (1999) Helium diffusion and (U–Th)/He thermochronometry of titanite. *Geochimica et Cosmochimica Acta* **63**(22), 3845–3859.

Reiners, P. W. and Farley, K. A. (2001) Influence of crystal size on apatite (U–Th)/He thermochronology: an example from the Bighorn Mountains, Wyoming. *Earth and Planetary Science Letters* **188**(3), 413–420.

Reiners, P. W., Brady, R., Farley, K. A., Fryxell, J. E., Wernicke, B., and Lux, D. (2000) Helium and argon thermochronometry of the Gold Butte block, south Virgin Mountains, Nevada. *Earth and Planetary Science Letters* **178**(3), 315–326.

Reiners, P. W., Farley, K. A., and Hickes, H. J. (2002) He diffusion and (U–Th)/He thermochronometry of zircon: Initial results from Fish Canyon Tuff and Gold Butte. *Tectonophys* **349**, 247–308.

Reiners, P. W., Spell T.L., Nicolescu S., and Zanetti, K. A. (2004) Zircon (U–Th)/He thermochronometry: He diffusion and comparisons with ^{40}Ar/^{39}Ar dating. *Geochimica et Cosmochimica Acta* **68**, 1857–1887

Reiners, P. W., Campbell, I. H., Nicolescu, S., *et al.* (2005) (U–Th)/(He-Pb) double dating of detrital zircons. *American Journal of Science* **305**(4), 259–311.

Reiners, P. W., Thomson, S. N., McPhillips, D., Donelick, R. A., and Roering, J. J. (2007) Wildfire thermochronology and the fate and transport of apatite in hillslope and fluvial environments. *Journal of Geophysical Research: Earth Surface (2003–2012)*, **112**(F4).

Reynolds, J. H. (1960) Rare gases in tektites. *Geochimica et Cosmochimica Acta* **20**(2), 101–114.

Ruhl, K. W. and Hodges, K. V. (2005) The use of detrital mineral cooling ages to evaluate steady state assumptions in active orogens: An example from the central Nepalese Himalaya. *Tectonics* **24**(4).

Rutherford, E. (1905) Present problems in radioactivity. *Popular Science Monthly* **May**, 1–34.

Rutherford, E. (1906) *Radioactive Transformations.* Charles Scribner's & Sons, New York.

Saadoune, I. and De Leeuw, N. H. (2009) A computer simulation study of the accommodation and diffusion of He in uranium- and plutonium-doped zircon (ZrSiO 4). *Geochimica et Cosmochimica Acta* **73**(13), 3880–3893.

Saadoune, I., Purton, J. A., and de Leeuw, N. H. (2009) He incorporation and diffusion pathways in pure and defective zircon ZrSiO 4: a density functional theory study. *Chemical Geology* **258**(3), 182–196.

Saylor, J. E., Stockli, D. F., Horton, B. K., Nie, J., and Mora, A. (2012) Discriminating rapid exhumation from syndepositional volcanism using detrital zircon double dating: Implications for the tectonic history of the Eastern Cordillera, Colombia. *Geological Society of America Bulletin* **124**(5–6), 762–779.

Schildgen, T. F., Balco, G., and Shuster, D. L. (2010) Canyon incision and knickpoint propagation recorded by apatite ^{4}He/^{3}He thermochronometry. *Earth and Planetary Science Letters* **293**(3), 377–387.

Schmitt, A. K., Stockli, D. F., and Hausback, B. P. (2006) Eruption and magma crystallization ages of Las Tres Vírgenes (Baja California) constrained by combined ^{230}Th/^{238}U and (U–Th)/He dating of zircon. *Journal of Volcanology and Geothermal Research*, **158**(3), 281–295.

Schmitt, A. K., Stockli, D. F., Lindsay, J. M., Robertson, R., Lovera, O. M., and Kislitsyn, R. (2010) Episodic growth and homogenization of plutonic roots in arc volcanoes from combined U–Th and (U–Th)/He zircon dating. *Earth and Planetary Science Letters* **295**(1), 91–103.

Schmitt, A. K., Danišík, M., Evans, N. J., *et al.* (2011) Acıgöl rhyolite field, Central Anatolia (part 1): high-resolution dating of eruption episodes and zircon growth rates. *Contributions to Mineralogy and Petrology* **162**(6), 1215–1231.

Setina, J. (1999) New approach to corrections for thermal transpiration effects in capacitance diaphragm gauges. *Metrologia* **36**(6), 623.

Shuster, D. L. and Farley, K. A. (2004) ^{4}He/^{3}He thermochronometry. *Earth and Planetary Science Letters* **217**(1), 1–17.

Shuster, D. L. and Farley, K. A. (2005) Diffusion kinetics of proton-induced ^{21}Ne, ^{3}He, and ^{4}He in quartz. *Geochimica et Cosmochimica Acta* **69**(9), 2349–2359.

Shuster, D. L. and Farley, K. A. (2009) The influence of artificial radiation damage and thermal annealing on helium diffusion kinetics in apatite. *Geochimica et Cosmochimica Acta* **73**(1), 183–196.

Shuster, D. L., Farley, K. A., Sisterson, J., and Burnett, D. S. (2003) 4He/3He thermochronometry. *Geochimica et Cosmochimica Acta* **67**(18), A436–A436.

Shuster, D. L., Farley, K. A., Sisterson, J. M., & Burnett, D. S. (2004) Quantifying the diffusion kinetics and spatial distributions of radiogenic 4 He in minerals containing proton-induced 3 He. *Earth and Planetary Science Letters*, **217**(1), 19–32.

Shuster, D. L., Ehlers, T. A., Rusmoren, M. E., and Farley, K. A. (2005a) Rapid glacial erosion at 1.8 Ma revealed by ^{4}He/^{3}He thermochronometry. *Science* **310**(5754), 1668–1670.

Shuster, D. L., Vasconcelos, P. M., Heim, J. A., and Farley, K. A. (2005b) Weathering geochronology by (U–Th)/He dating of goethite. *Geochimica et Cosmochimica Acta* **69**(3), 659–673.

Shuster, D. L., Flowers, R. M., and Farley, K. A. (2006) The influence of natural radiation damage on helium diffusion kinetics in apatite. *Earth and Planetary Science Letters* **249**(3), 148–161.

Shuster, D. L., Cuffey, K. M., Sanders, J. W., and Balco, G. (2011) Thermochronometry reveals headward propagation of erosion in an alpine landscape. *Science* **332**(6025), 84–88.

Shuster, D. L., Farley, K. A., Vasconcelos, P. M., *et al.* (2012) Cosmogenic ^{3}He in hematite and goethite from Brazilian "canga" duricrust demonstrates the extreme stability of these surfaces. *Earth and Planetary Science Letters* **329**, 41–50.

Spotila, J. A., Farley, K. A., Yule, J. D., and Reiners, P. W. (2001) Near-field transpressive deformation along the San Andreas fault zone in southern California, based on exhumation constrained by (U–Th)/He dating. *Journal of Geophysical Research: Solid Earth* **106**(B12), 30909–30922.

Spiegel, C., Kohn, B., Belton, D., Berner, Z., and Gleadow, A. (2009) Apatite (U–Th–Sm)/He thermochronology of rapidly cooled samples: the effect of He implantation. *Earth and Planetary Science Letters* **285**(1), 105–114.

Stock, J. D. and Montgomery, D. R. (1996) Estimating palaeorelief from detrital mineral age ranges. *Basin Research* **8**(3), 317–327.

Stockli, D. F. (2005) Application of low-temperature thermochronometry to extensional tectonic settings. *Reviews in Mineralogy and Geochemistry* **58**(1), 411–448.

Stockli, D. F. and Farley, K. A. (2004) Empirical constraints on the titanite (U–Th)/He partial retention zone from the KTB drill hole. *Chemical Geology* **207**(3), 223–236.

Stockli, D. F., Farley, K. A., and Dumitru, T. A. (2000) Calibration of the apatite (U–Th)/He thermochronometer on an exhumed fault block, White Mountains, California. *Geology* **28**(11), 983–986.

Stockli, D. F., Surpless, B. E., Dumitru, T. A., and Farley, K. A. (2002) Thermochronological constraints on the timing and magnitude of Miocene and Pliocene extension in the central Wassuk Range, western Nevada. *Tectonics* **21**(4), 10–1.

Stockli, D. F., Dumitru, T. A., McWilliams, M. O., and Farley, K. A. (2003) Cenozoic tectonic evolution of the White Mountains,

California and Nevada. *Geological Society of America Bulletin* **115**(7), 788–816.

Stockli, D. F., Wolfe, M. R., Blackburn, T. J., Zack, T., Walker, J. D., and Luvizotto, G. L. (2007) He diffusion and (U–Th)/He thermochronometry of rutile. In *AGU Fall Meeting Abstracts*, Vol. **1**, 1548.

Strutt, R. J. (1908a) Helium and radio-activity in rare and common minerals. *Proceedings of the Royal Society of London. Series A, Containing Papers of a Mathematical and Physical Character* **80**(542), 572–594.

Strutt, R. J. (1908b) On the accumulation of helium in geological time. *Proceedings of the Royal Society of London. Series A, Containing Papers of a Mathematical and Physical Character* **81**(547), 272–277.

Strutt, R. J. (1908c) On helium in saline minerals, and its probable connection with potassium. *Proceedings of the Royal Society of London A: Mathematical, Physical and Engineering Sciences* **81**(547), 278–279.

Strutt, R. J. (1910a) The accumulation of helium in geological time. III. *Proceedings of the Royal Society of London. Series A, Containing Papers of a Mathematical and Physical Character* **83**(562), 298–301.

Strutt, R. J. (1910b) Measurements of the rate at which helium is produced in thorianite and pitchblende, with a minimum estimate of their antiquity. *Proceedings of the Royal Society of London. Series A, Containing Papers of a Mathematical and Physical Character* **84**(571), 379–388.

Strutt, R. J. (1910c) The accumulation of helium in geological time. IV. *Proceedings of the Royal Society of London. Series A, Containing Papers of a Mathematical and Physical Character* **84**(569) 194–196.

Tagami, T., Farley, K. A., and Stockli, D. F. (2003) (U–Th)/He geochronology of single zircon grains of known Tertiary eruption age. *Earth and Planetary Science Letters* **207**, 57–67.

Thomson, S. N., Brandon, M. T., Reiners, P. W., Zattin, M., Isaacson, P. J., and Balestrieri, M. L. (2010) Thermochronologic evidence for orogen-parallel variability in wedge kinematics during extending convergent orogenesis of the northern Apennines, Italy. *Geological Society of America Bulletin* **122**(7–8), 1160–1179.

Thomson, S. N., Reiners, P. W., Hemming, S. R., and Gehrels, G. E. (2013) The contribution of glacial erosion to shaping the hidden landscape of East Antarctica. *Nature GeoScience* **6**(3) (203–207).

Trachenko, K., Dove, M. T., and Salje, E. K. (2002) Structural changes in zircon under α-decay irradiation. *Physical Review B* **65**(18), 180102.

Tripathy-Lang, A., Hodges, K. V., Monteleone, B. D., and Soest, M. C. (2013) Laser (U–Th)/He thermochronology of detrital zircons as a tool for studying surface processes in modern catchments. *Journal of Geophysical Research: Earth Surface* **118**(3), 1333–1341.

Trull, T. W. and Kurz, M. D. (1993) Experimental measurements of ^{3}He and ^{4}He mobility in olivine and clinopyroxene at magmatic temperatures. *Geochimica et Cosmochimica Acta* **57**(6), 1313–1324.

Trull, T. W., Kurz, M. D., and Jenkins, W. J. (1991) Diffusion of cosmogenic ^{3}He in olivine and quartz: implications for surface exposure dating. *Earth and Planetary Science Letters* **103**(1), 241–256.

Turekian, K. K., Kharkar, D. P., Funkhouser, J., and Schaeffer, O. A. (1970) An evaluation of the uranium-helium method of dating of fossil bones. *Earth and Planetary Science Letters* **7**(5), 420–424.

Turner, G. (1969) Thermal histories of meteorites by the ^{39}Ar–^{40}Ar method. *Meteorite Research* **12**, 407–417.

Valla, P. G., Shuster, D. L., and van der Beek, P. A. (2011) Significant increase in relief of the European Alps during mid-Pleistocene glaciations. *Nature GeoScience* **4**(10), 688–692.

Van Soest, M. C., Monteleone, B. D., Hodges, K. V., and Boyce, J. W. (2011) Laser depth profiling studies of helium diffusion in Durango fluorapatite. *Geochimica et Cosmochimica Acta* **75**(9), 2409–2419.

Vasconcelos, P. M., Heim, J. A., Farley, K. A., Monteiro, H., and Waltenberg, K. (2013) ^{40}Ar/^{39}Ar and (U–Th)/He–^{4}He/^{3}He geochronology of landscape evolution and channel iron deposit genesis at Lynn Peak, Western Australia. *Geochimica et Cosmochimica Acta* **117**, 283–312.

Vermeesch, P. (2004) How many grains are needed for a provenance study? *Earth and Planetary Science Letters* **224**(3), 441–451.

Vermeesch, P. (2010) HelioPlot, and the treatment of overdispersed (U–Th–Sm)/He data. *Chemical Geology* **271**(3), 108–111.

Vermeesch, P., Sherlock, S. C., Roberts, N. M., and Carter, A. (2012) A simple method for *in-situ* U–Th–He dating. *Geochimica et Cosmochimica Acta* **79**, 140–147.

Watson, E. B., Wanser, K. H., and Farley, K. A. (2010) Anisotropic diffusion in a finite cylinder, with geochemical applications. *Geochimica et Cosmochimica Acta* **74**(2), 614–633.

Wernicke, R. S. and Lippolt, H. J. (1994a) ^{4}He age discordance and release behavior of a double shell botryoidal hematite from the Schwarzwald, Germany. *Geochimica et Cosmochimica Acta* **58**(1), 421–429.

Wernicke, R. S. and Lippolt, H. J. (1994b) Dating of vein specularite using internal (U + Th)/^{4}He isochrons. *Geophysical Research Letters* **21**(5), 345–347.

Wiens, R. C., Lal, D., Rison, W., and Wacker, J. F. (1994) Helium isotope diffusion in natural diamonds. *Geochimica et Cosmochimica Acta* **58**(7), 1747–1757.

Willett, S. D. and Brandon, M. T. (2013) Some analytical methods for converting thermochronometric age to erosion rate. *Geochemistry, Geophysics, Geosystems* **14**(1), 209–222.

Wolfe, M. R. (2009) *He diffusion in rutile and calibration of rutile (U–Th)/He thermochronology on the KTB ultra-deep borehole.* Unpublished MS, University of Kansas, Lawrence, Kansas.

Wolfe, M. R. and Stockli, D. F. (2010) Zircon (U–Th)/He thermochronometry in the KTB drill hole, Germany, and its implications for bulk He diffusion kinetics in zircon. *Earth and Planetary Science Letters* **295**(1), 69–82.

Wolf, R. A., Farley, K. A., and Silver, L. T. (1996) Helium diffusion and low-temperature thermochronometry of apatite. *Geochimica et Cosmochimica Acta* **60**(21), 4231–4240.

Wolf, R. A., Farley, K. A., and Silver, L. T. (1997) Assessment of (U–Th)/He thermochronometry: the low-temperature history of the San Jacinto mountains, California. *Geology* **25**(1), 65–68.

Wolf, R. A., Farley, K. A., and Kass, D. M. (1998) Modeling of the temperature sensitivity of the apatite (U–Th)/He thermochronometer. *Chemical Geology* **148**(1), 105–114.

Wolfe, M. R., Stockli, D. F., Shuster, D. L., Walker, J. D., and Mac-Pherson, G. L. (2007) Assessment of the rutile (U–Th)/He thermochronometry on the KTB drill hole, Germany. In *AGU Fall Meeting Abstracts* **1**, 1549.

Zashu, S. and Hiyagon, H. (1995) Degassing mechanisms of noble gases from carbonado diamonds. *Geochimica et Cosmochimica Acta* **59**(7), 1321–1328.

Zeitler, P. K., Herczeg, A. L., McDougall, I., and Honda, M. (1987) U–Th–He dating of apatite: a potential thermochronometer. *Geochimica et Cosmochimica Acta* **51**(10), 2865–2868.

Ziegler, J. F., Biersack J. P., and Ziegler, M. D. (2008) *SRIM The Stopping and Range of Ions in Matter*. SRIM Co., Chester, MD.

CHAPTER 12

Uranium-series geochronology

12.1 INTRODUCTION

The uranium and thorium decay series have been the object of study for over 100 years, and long before their use in geochronology they played an important role in the discovery of radioactivity. For example, the short half-lives of some of the intermediate daughters in the chain combined with the long half-life of ^{238}U led to the observation that uranium ores emitted more radiation than did pure uranium, spurring the search for other elements in the decay chain and illuminating many aspects of radioactivity (as discussed in more detail in Chapter 2). The recognition that radioactivity is a property of the nucleus, combined with the discovery that in some cases two substances with different radioactive behavior could not be separated chemically, led Frederick Soddy to propose the existence of isotopes of a given element which had different nuclear properties [*Soddy*, 1910]. The fundamentals of the sequence of nuclides (see Box 12.1) in the U and Th decay chains was established within only a few years of that date (see Chapter 2), laying the foundation for U-series geochronology. However, techniques for the routine measurement of some of the nuclides in the chains were not established until several decades after that, and it was not until the 1950s that U-series disequilibrium began to be widely applied to geochronology. Since that time, the field has grown rapidly, particularly during times when progress in analytical techniques led to development of increasingly precise measurements, allowing different types of problems to be addressed and/or higher precision to be applied to existing areas [e.g., *Ivanovich and Harmon*, 1992; *Bourdon et al.*, 2003]. Today, U-series geochronology is being applied to problems as diverse as magma generation, transport and storage, reconstruction of paleosea-level and paleoclimate, anthropology, paleoseismology, groundwater dating, and paleoceanography, to name just a subset of problems that have been addressed using this technique. One unique aspect of U-series geochronology is the fact that it makes use of a decay chain rather than a single parent–daughter pair, and therefore allows investigation of processes covering a broad range of timescales. In addition, the timescales accessible through this technique overlap with the timescales of interest for human interactions with Earth, including timescales relevant to volcanic eruptions, earthquakes, anthropology, and the development of scarce resources such as groundwater and marine organisms.

There are three different naturally occurring decay series and one that no longer exists naturally in the solar system due to the short half-life of its longest-lived member (neptunium-237, half-life 2.2×10^6 years). In each of the U-series or Th-series decay systems, the parent decays to its ultimate stable daughter (each of which is an isotope of lead) through a series of intermediate daughter nuclides, and each of these is radioactive. This provides the basis for the use of intermediate parent–daughter pairs within the decay chains for geochronology. The most widely used decay chains in geochronology (Fig. 12.1) are the uranium series (parent ^{238}U, stable daughter ^{206}Pb) and the actinium series (parent ^{235}U, stable daughter ^{207}Pb), which leads to the common usage of "U-series geochronology" to describe decay-chain geochronology in general. The ^{232}Th decay chain is also used for geochronology, although to a lesser extent. The neptunium-237 series has been produced artificially, and it is of interest in geochronology because it serves as the source of some of the nuclides commonly used as spikes for isotope dilution measurements (^{233}U, ^{229}Th, and ^{231}Pa).

There are several interesting features of the U-series system compared to other systems where the parent decays to a daughter in a single decay event. Arguably the most important of these features is that any U-bearing (or Th-bearing) system will eventually reach a state known as secular or radioactive equilibrium, where the *activity* (the number of decay events per unit time) of each of the nuclides in the chain is the same. A useful corollary of this state is that the ratio of the number of atoms of any two nuclides in secular equilibrium will be fixed (and inversely related to the ratio of their half-lives), providing important information about initial conditions prior to a disturbance. Second, each of the three decay chains produces intermediate daughters which are isotopes of elements with highly variable chemical behavior, for example from a high field-strength element (thorium) to an alkaline earth

Geochronology and Thermochronology, First Edition. Peter W. Reiners, Richard W. Carlson, Paul R. Renne, Kari M. Cooper, Darryl E. Granger, Noah M. McLean, and Blair Schoene.
© 2018 John Wiley & Sons Ltd. Published 2018 by John Wiley & Sons Ltd.

element (radium) to a noble gas (radon). The differing chemical behaviors of these elements (along with recoil effects, discussed more below) act to separate the intermediate daughters from each other during a wide variety of geological processes. The resulting deviations from secular equilibrium, and the eventual return to equilibrium, provides the basis for U-series geochronology. Finally, each parent–daughter pair in the chain returns to

equilibrium over a duration proportional to the half-life of the daughter. Therefore, the enormous variation in half-life along the chain (from milliseconds to hundreds of ka) allows investigation of processes that occur on a wide variety of timescales, many of which occur on timescales of interest to human-Earth interactions.

The concept of secular equilibrium merits further explanation. For each intermediate daughter in the chain, any fractionation event affecting a given daughter sets up a competition between ingrowth from its immediate parent and decay to the next daughter in the chain, such that the abundance of a particular nuclide is determined not only by its initial abundance and the decay of the parent over time, but also by the decay of the daughter itself. Eventually these two processes will balance each other, and the system will return to secular equilibrium (Fig. 12.2). For example, consider the case where an intermediate daughter is initially absent from the decay chain (e.g., Fig. 12.2b). Decay of the immediate parent will produce atoms of this daughter over time at a rate dictated by the abundance of the parent and the half-life

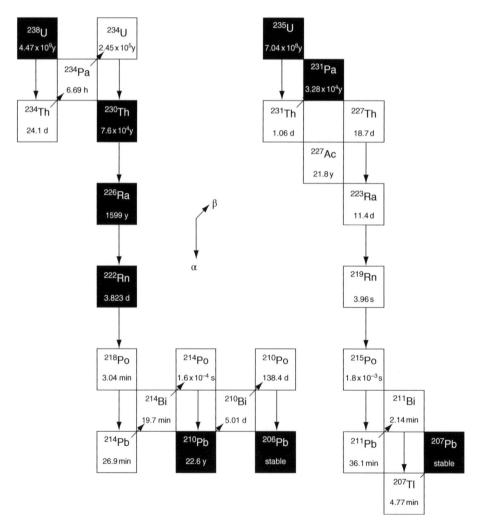

Fig. 12.1. Schematic decay chains for ^{238}U and ^{235}U. The isotopes used in the U-series dating discussed in this chapter are shown as black boxes. The half-lives shown are those compiled in *Bourdon et al.* [2003]. (Source: Adapted from *Bourdon et al.* [2003].)

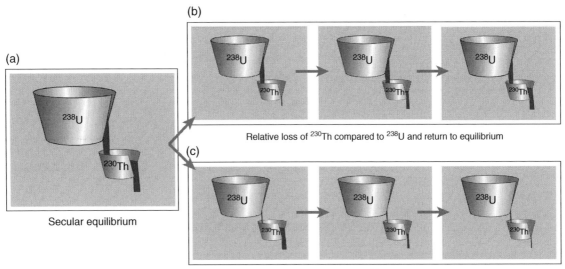

Fig. 12.2. Cartoon illustrating the return of the ^{238}U–^{230}Th parent–daughter pair to secular equilibrium after a disturbance (intermediate daughters are omitted in this diagram). Note that in these diagrams, the opening in each bucket is V-shaped in order to schematically represent the dependence of activity on the number of atoms present, and the different widths of the openings for ^{238}U and ^{230}Th represent the different half-lives. (a) The system at secular equilibrium, where the rate of decay of ^{230}Th matches that of ^{238}U. (b) Evolution of a system in which ^{230}Th has preferentially been lost (or excluded) relative to ^{238}U. (c) Evolution of a system in which ^{238}U has been preferentially lost (or excluded) relative to ^{230}Th. (Source: *Reid* [2008]. Reproduced with permission of Mineralogical Society of America.) (*See insert for color representation of the figure.*)

of the parent. However, as soon as these daughters are produced they will themselves begin to decay, at a rate dictated by the abundance of daughter at any given time and the half-life of the daughter. As the number of atoms of the daughter increases over time, so does the activity of the daughter, because the number of decay events per time is proportional to the number of atoms present and inversely proportional to the half-life. Eventually, the number of daughter atoms will increase to the point where the activity of the daughter will exactly match that of the parent, at which point the ingrowth and decay are precisely balanced. This state of balance is known as secular (or radioactive) equilibrium, and the parent–daughter pair will maintain equilibrium unless the system is further disturbed. A common analogy for the process of achieving secular equilibrium is that of a series of buckets representing each nuclide in the chain (Fig. 12.2), where water flows from one to the next at a rate determined both by the amount of water in each bucket (analogous to the number of atoms present) and by the size of the aperture feeding the next bucket (analogous to the half-life of the daughter). In this analogy, as long as the initial rate of supply of water is constant, a given bucket will fill at a rate proportional to the loss of water to the next stage, which is itself proportional to the amount of water in the bucket. This illustrates a fundamental property of the U-series decay chain: because the net rate of accumulation of the daughter nuclide is a function of its half-life, each intermediate parent–daughter pair will return to secular equilibrium at its own rate, and will therefore be sensitive to a different timescale. Once a system has returned to secular equilibrium, it can remain there indefinitely (as long as the supply of

the ultimate parent nuclide of the chain remains and the daughter is not removed) and therefore the amount of time elapsed since a prior fractionation event is indeterminate. The useful time range for geochronology for a given parent–daughter pair within the chain, therefore, is limited on the young end by how well initial conditions after fractionation can be estimated (and by analytical precision necessary to distinguish changes from this initial condition), and on the old end by how precisely differences from secular equilibrium can be measured analytically (Fig. 12.3; Table 12.1). The common rule of thumb, that each parent–daughter pair will return to secular equilibrium within five half-lives of the daughter nuclide, is based on an assumption that activity ratios can be measured to precisions of $\sim$3–5% (appropriate for alpha counting techniques), and the higher precision achievable with modern analytical techniques such as multicollector, inductively coupled plasma mass spectrometry (MC-ICP-MS) means that the timescales over which ages can be measured can be substantially longer. For example, for the ^{234}U–^{230}Th pair, the rule of thumb would predict return to secular equilibrium within $\sim$375 ka from fractionation, but ages of up to 800 ka can now be measured under favorable conditions [*Cheng et al.*, 2013].

12.2 THEORY AND FUNDAMENTALS

12.2.1 The mathematics of decay chains

Mathematically, the balance of ingrowth and decay in a decay chain requires some modification of the standard decay

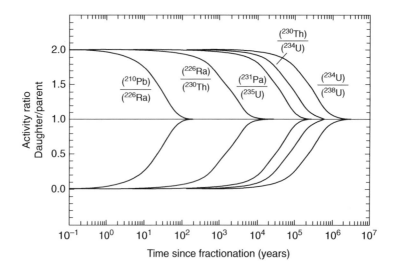

Fig. 12.3. Diagram showing timescale of return to secular equilibrium for different parent–daughter pairs in the decay chain. (Source: *Cooper* [2015].)

Table 12.1 Measurement techniques, sample size and preparation requirements, and typical analytical times

Technique	Typical sample size (whole-rock samples)	Sample preparation	Sample preparation time	Analytical time required for complete analysis of U and Th activity ratios (includes U and Th isotopic measurements, Th and U concentration measurements)
Alpha spectroscopy	~5–10 g	Wet chemistry, ion-exchange chromatography, plating	Weeks to months	~1–2 weeks
TIMS	0.1–1 g	Wet chemistry, ion-exchange chromatography, filament loading	Weeks	2 days
MC-ICP-MS	0.1–0.5 g	Wet chemistry, ion-exchange chromatography, solution preparation	Weeks	2–3 h
SIMS	40 μm × 5 μm spot (order of 1×10^{-8} g)	Polished thin section or grain mount	Days	~40 min
LA-MC-ICP-MS	100 μm × 25 μm spot (order of 1×10^{-6} g)	Polished thin section or grain mount	Days	~30 min

equations. The physics and mathematical description of the case of a single parent N decaying to a single daughter D (the standard decay equations) are described in Chapter 2 (e.g., equations 2.5–2.22). In the case of a decay chain, where $N_1 \rightarrow N_2 \rightarrow N_3 \ldots \rightarrow D$, expressions relating the abundance of the parent and daughter become more complicated because they also must consider ingrowth and decay of the intermediate daughters. Chapter 2 also summarizes the derivation of the fundamental decay-chain age equations (equations 2.25–2.31). For a full derivation of decay-chain equations, including those tracking the abundance of more than three nuclides in the chain, see *Ivanovich and Harmon* [1992]. We begin here with the most general equation describing the abundance of the first intermediate daughter, N_2 (equation 2.26):

$$N_2 = \frac{\lambda_1}{\lambda_2 - \lambda_1} N_1^0 \left(e^{-\lambda_1 t} - e^{-\lambda_2 t} \right) + N_2^0 e^{-\lambda_2 t}$$

where we can see that the number of atoms of the first daughter (N_2) present at time t is equal to the number of atoms produced by decay of the parent, minus the number of those that have since decayed (the first term on the right-hand side of the equation), plus the number of atoms of N_2 initially present minus the number of those that have decayed in time t (the second term on the right-hand side of the equation).

This is the most general form of the equation describing the abundance of N_2. However, for the parent–daughter pairs commonly used in U-series geochronology, two assumptions can simplify this expression. First, assuming that the time of interest is short with respect to the half-life of the parent, $\exp(-\lambda_1 t)$ is approximately equal to $e^0 = 1$. Second, if the half-life of the daughter is much shorter than the half-life of the parent (which is the case for all of the parent–daughter pairs commonly used in U-series geochronology), then $(\lambda_2 - \lambda_1)$ is approximately equal to λ_2.

Making these two substitutions into equation (2.26) and multiplying through by λ_2 gives

$$\lambda_2 N_2 = \lambda_1 N_1^0 \left(1 - e^{-\lambda_2 t}\right) + \lambda_2 N_2^0 e^{-\lambda_2 t} \tag{12.1}$$

or, expressed in terms of activities:

$$\left(N_2\right) = \left(N_1^0\right)\left(1 - e^{-\lambda_2 t}\right) + \left(N_2^0\right)e^{-\lambda_2 t} \tag{12.2}$$

Specific versions of the decay equations relevant to particular parent–daughter pairs are presented in the sections below.

A few observations are worth making explicit. First, equation (12.2) provides a mathematical description of why the timescale for return to secular equilibrium depends only on the half-life of the daughter nuclide. Note that the decay constant for the parent no longer appears in this equation, although the decay constant for the daughter does. This is a direct result of the assumptions made during simplification of equation (2.26)—that the half-life of the parent is long both with respect to the time of interest and with respect to that of the daughter. If either of those assumptions is not justified for a particular case, the half-life of the parent must be explicitly considered and a more general form of the equation must be used.

Second, in the case where the daughter is longer-lived than the parent, secular equilibrium may never be achieved after a disturbance. For example, if fractionation results in an excess of the longer-lived daughter (and the immediate parent is unsupported by decay of a longer-lived nuclide farther up the chain), the parent will completely decay before decay of the daughter can restore an equilibrium activity ratio.

Finally, there are some circumstances where the intermediate decay steps can be ignored, and the age of a system is measured by ingrowth of the stable daughter D (e.g., $^{238}\text{U} \rightarrow {}^{206}\text{Pb}$), in which case the more conventional decay equations apply (see Chapter 8). This requires (i) that the system is closed to all intermediate daughters, as well as closed to uranium and lead, and (ii) that that the time of interest is long enough that any initial disequilibrium in the decay chain and the resulting time that it took to regain secular equilibrium is insignificant compared to the total time. There are some notable cases in geochronology where these assumptions may not hold true and corrections must be made for initial disequilibrium; for further discussion, see Chapter 8.

12.2.2   Mechanisms of producing disequilibrium

Disequilibrium between nuclides in a decay chain can be produced by two fundamental mechanisms: chemical fractionation and recoil effects. The first of these is more intuitive to most geochronologists, considering that it is the same mechanism that segregates parent from daughter in many other radiogenic geochronology systems. Due to their differing chemical behavior, the many elements present in each of the decay chains can be separated from each other during a wide variety of geologic processes, including crystallization, partial melting, precipitation of minerals from a fluid, degassing, dissolution, oxidation/reduction, and in fact any other geologic process which involves chemical reactions. For example, during precipitation of a mineral from a melt, the $\left({}^{230}\text{Th}\right)/\left({}^{238}\text{U}\right)$ ratio (activity ratio) in the mineral will, in general, be different from that of the host melt due to different partitioning behavior of Th and U. This can be expressed mathematically as:

$$\left(\frac{{}^{230}\text{Th}}{{}^{238}\text{U}}\right)_{\text{mineral}} = \frac{D_{\text{Th}}}{D_{\text{U}}}\left(\frac{{}^{230}\text{Th}}{{}^{238}\text{U}}\right)_{\text{melt}} \tag{12.3}$$

where D is the Nernst partition coefficient for the parent or daughter relating the equilibrium distribution of an element between two phases (in this case, the mineral of interest compared to a melt), given for example by

$$D_{\text{Th}} = \frac{[\text{Th}]_{\text{mineral}}}{[\text{Th}]_{\text{melt}}} \tag{12.4}$$

where brackets indicate concentration. From equation (12.3) it is clear that if the mineral-melt partition coefficients for parent and daughter for any two nuclides in the chain are not equal, crystallization will produce an activity ratio different from that in the host melt, resulting in disequilibrium within the mineral (whether or not the host melt itself begins in secular equilibrium).

The second class of processes that produce disequilibrium is related to the decay events themselves, and can be particularly important in low-temperature applications. Many of the decay events in a decay chain are alpha decays, in which particles of substantial mass are ejected from the nucleus. Total momentum and kinetic energy are conserved in the decay reaction, and result in displacement of both the alpha particle and the daughter nucleus from the original position of the parent nucleus. The effects of this are threefold: first, the daughter atom is displaced from its original position, and may be ejected completely from the original lattice location into a neighboring mineral or fluid phase. The recoil distance will vary depending on the energy of the reaction and on the nature of the substrate, but is on the order of tens of nanometers. The effects of direct recoil can be modeled assuming a geometry of the mineral of interest, in a manner analogous to the recoil correction applied to the alpha particles themselves during (U–Th)/He dating (see Chapter 11). However, two other related effects are that the lattice site is damaged by the recoil, and that the daughter atom may be displaced from its lattice site, both of which make it more likely that the daughter will be mobilized during subsequent interactions (e.g., fluid–mineral interactions resulting in preferential leaching of the radiogenic daughter). This preferential leaching effect is most prominent during applications involving water–rock interactions.

12.3   METHODS AND ANALYTICAL TECHNIQUES

12.3.1   Analytical techniques

Analysis of samples for U-series dating involves a number of steps, with varying pathways depending on sample size and composition

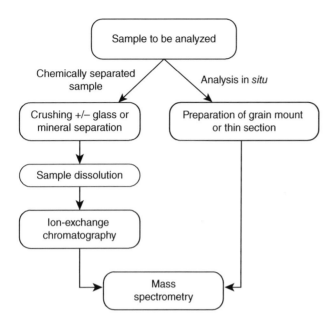

Fig. 12.4. Flow chart showing analytical protocols for analysis *in situ* and chemically separated samples.

and the analytical precision required for the specific geochronologic application. Two broad branches are determined by whether the analyses are performed on chemically separated fractions, where preconcentration of the elements to be analyzed allows analysis of samples with low abundances of U-series nuclides, or *in situ* on minimally processed samples, which allows preservation of textural context and/or analysis of different parts of the same crystals, but which requires samples with higher concentrations of the U-series elements. (Fig. 12.4).

12.3.1.1 Chemical separations

With the exception of ion-microprobe and laser-ablation techniques *in situ*, all U-series analyses require chemical separation of the elements of interest from the rock matrix. This is generally accomplished through wet chemistry using ion-exchange resins, which necessitates dissolution of the volume of rock or mineral to be analyzed (see Chapter 3). For pure carbonate materials, this is relatively easily accomplished using nitric acid (HNO_3), but breaking the stronger Si–O bonds in silicate minerals generally requires use of hydrofluoric acid, and (depending on the size and type of sample) may require multiple stages of dissolution in different acids. For example, a typical silicate dissolution procedure involves an initial stage of HF-HNO_3 digestion, followed by evaporation and subsequent dissolutions in HNO_3 and HCl.

Once the rock or mineral sample has been dissolved, the segregation of U, Th, and other elements of interest from the rock matrix is usually achieved through the use of organic ion-exchange resins, where the affinity of the resin for a particular element changes as a function of the type and strength of the acid solution in which the rock is dissolved. Thus, by initially loading the dissolved sample on a column in an acid for which the resin

has a high affinity for the element of interest (i.e., has a high K_D), other elements in the matrix can be removed by flushing the resin with acid, and finally the element of interest is removed by changing the acid on the column to one in which the resin has a low affinity for that element. Details of the procedures vary depending on the specific element to be analyzed, and on the rock or mineral matrix; for a review of some common approaches for silicate and carbonate rocks, see *Goldstein and Stirling* [2003].

U-series dating generally requires measuring both the isotopic composition and the concentration ratio of the parent and daughter. For example, in the case of ^{238}U–^{230}Th dating, calculating activity ratios and dates requires knowledge of the $^{238}U/^{234}U$ and $^{230}Th/^{232}Th$ ratios of the sample, as well as concentrations (or at least the concentration ratio) of U and Th in the sample. The isotopic composition measurements can be made directly on purified U and Th fractions, but measuring the concentrations at high precision requires use of isotope dilution techniques, where the natural sample is "spiked" with a known quantity of a nonnaturally occurring isotope of the same element, or a spike which is highly enriched in one isotope relative to natural abundances. After measuring the isotopic composition of the mixture of sample and spike, the abundance of the element in the sample can be calculated given the known amount, concentration, and isotopic composition of the spike present in the mixture (see Chapter 3 for more complete description of isotope dilution measurements).

12.3.1.2 Isotopic analyses
Chemically separated samples

Once the U-series elements to be measured have been purified, isotopic compositions (of spiked or unspiked samples) must be measured in order to calculate activities and activity ratios. U-series disequilibrium dating, like most other isotopic dating techniques, has gone through a series of stages of increasing analytical precision and decreasing sample size required for analysis, where these stages were driven by developments in the technology of mass spectrometry. In general, there are a number of factors which limit the precision of any isotopic analysis: the instrument background (which reflects both electronic noise in the instrument and any instrument blank consisting of atoms of the element that are present in the mass spectrometer itself), the ionization efficiency and transmission of a given element in the mass spectrometer (which combined dictate the yield, or proportion of ions entering the mass spectrometer that end up getting to the detector), and the sensitivity of the detectors. Over the decades during which U-series measurements have been made, significant improvements have been in all of these areas, resulting in progressively higher precision possible with analysis of increasingly smaller samples.

Early analyses of U and Th isotopes for U-series dating used alpha spectrometry [e.g., *Barnes et al.*, 1956], a technique that measures the alpha emissions from a sample. Precision of this technique is limited by the relatively long half-lives of ^{238}U and ^{230}Th and the statistics of counting small numbers of decay events, and the resulting 1σ precision was on the order of 1% and 3% for U and Th isotopic compositions, respectively, for samples on the order of 10 g of material. A major advance in U-series analyses was the development in the 1980s and 1990s of techniques for measurement of U and Th isotopic compositions by thermal ionization mass spectrometry (TIMS). Unlike alpha spectrometry, mass spectrometric techniques measure the isotopic composition of an element directly. Both techniques are limited by counting statistics, but whereas only a small proportion of U or Th atoms present will decay in any given time interval (e.g., over a counting time of one week, only one out of every five million atoms of ^{230}Th can be detected), individual ions can be measured by mass spectrometry. Even considering the relatively low efficiency of ionization of thorium by heating (approximately one ion in 1000–10,000 atoms), orders of magnitude more atoms of thorium are detectable by TIMS than by alpha spectrometry for the same sample size. Thus, smaller sample sizes are measureable to higher precision by TIMS, with the added advantage of shorter analytical time (order of 4–8 h per sample, compared to days to weeks). A similar step-function advance in analytical methods occurred in the late 1990s and early 2000s with the increasing availability of MC-ICP-MS. Ionization of almost all elements is extremely effective (>90%) in the plasma source of MC-ICP-MS, but on the other hand the transmission of ions is relatively low because going from a source at atmospheric pressures to the 10^{-9} mbar pressures required at the detectors results in loss of the majority of the ions created to the vacuum pumps. Transmission through the mass spectrometer for a TIMS instrument is nearly 100%. However, for elements like thorium and protactinium, which have relatively poor thermal ionization efficiency, the overall ion yield at the detector (the proportion of ions detected to the number of atoms present during sample introduction) is still more than an order of magnitude better for modern MC-ICP-MS instruments than for TIMS. An additional advantage is that the total analytical time per sample by MC-ICP-MS is on the order of 1 h (including washout between samples) compared to 4–8 h by TIMS. Under optimal circumstances epsilon-level precision can now be achieved for analyses of U and Th by MC-ICP-MS [*Cheng et al.*, 2013], and this is the analytical technique of choice. For elements that have higher ionization efficiencies by TIMS (e.g., Ra has thermal ionization efficiency of several percent), total ion yield may be comparable to TIMS, but the advantages in terms of sample loading time and analysis time make this technique attractive for all but the smallest Ra samples.

Analyses in situ

In some cases, abundances of U and other U-series elements is high enough that sufficient counts can be achieved by analysis techniques *in situ*, such as secondary ionization mass spectrometry (SIMS) or laser ablation (LA) MC-ICP-MS. For example, accessory minerals, such as zircon and allanite in young volcanic rocks, and pedogenic opal and carbonate have been analyzed by SIMS [e.g., *Reid et al.*, 1997; *Maher et al.*, 2007], and recently

there has been development of dating of some types of samples (carbonate, iron oxides, and zircon) by LA-MC-ICP-MS [*Bernal et al.*, 2005, 2014; *Potter et al.*, 2005]. For all these techniques, there is a trade-off between improved spatial resolution and lower analytical precision compared to TIMS or MC-ICP-MS. Therefore, application of the techniques *in situ* typically is limited to those where the spatial resolution is a critical factor in analysis or interpretation. For example, dating of pedogenic opal and carbonate [*Potter et al.*, 2005; *Maher et al.*, 2007] allows targeting of very fine-scale layers or the cleanest areas of a layer, which in many cases avoids corrections for the initial ^{230}Th that is often present in larger, mixed, samples. An additional example is that of dating of volcanic zircon by ion microprobe, which allows comparison of ages of multiple zones within the same crystal, illuminating the history of individual grains. One recent development in U-series dating of accessory minerals in volcanic rocks *in situ* is analysis of unpolished surfaces of individual crystals followed by remounting and polishing into the interiors of the same crystals (Fig. 12.5), which provides information about both the youngest stages of crystal growth (surfaces) and the older phases of crystal growth (interiors). Table 12.1 compares the sample size and sample preparation requirements, analytical time required, and typical precision of measurements of U and Th activity ratios by a variety of techniques.

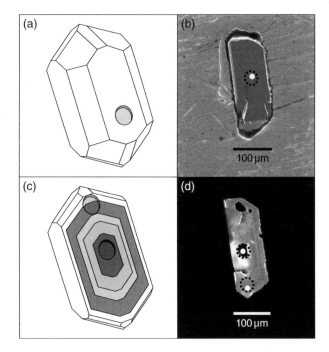

Fig. 12.5. Diagram showing analytical strategy for accessory phases (e.g. zircon, shown here). (a, b) First, unpolished grains are mounted in indium and surfaces are analyzed. (c, d) Next, grains are removed from indium and remounted in epoxy, and then polished to expose interiors. Note that the geometry of the ion microprobe spot (~40 μm diameter by ~5 μm deep) is oriented such that it captures the latest stages of crystallization better in surface analyses than in interior spots located near the rim (c), which are more likely to intersect multiple growth zones.

12.4 APPLICATIONS

12.4.1 U-series dating of carbonates

Dating of marine carbonates was among the first applications of U-series dating to geochronology [*Barnes et al.*, 1956] and has continued to be an important contribution to understanding sea-level changes related to climate change over geologic time and, more recently, local sea-level changes related to earthquakes. In addition, application of U-series dating to speleothems has provided an important temporal constraint on paleoclimate proxies within the speleothems, and also has provided critical temporal context for archeological and anthropological studies of artifacts or bones found in caves. More recently, U-series dating has also been applied to pedogenic carbonates and opal, dating geomorphic surfaces that can then be used in reconstruction of past fault movements or geomorphic feature development. In all of these applications, the most common dating method relies on the first few steps in the ^{238}U decay chain (Fig. 12.1). ^{234}Th and ^{234}Pa have half-lives that are short enough (hours to days) compared to the timescales of interest in carbonate dating that they can be assumed to be in secular equilibrium with ^{238}U and can be ignored in the decay equations. However, ^{234}U cannot be assumed to be in secular equilibrium with ^{238}U. Therefore, for carbonate dating the nuclides that need to be considered explicitly are ^{238}U, ^{234}U, and ^{230}Th. More recently, although less commonly, the first part of the ^{235}U decay chain has also been used, specifically ^{231}Pa (Fig. 12.1). Similar to ^{238}U–^{234}U–^{230}Th dating, the nuclide intermediate between ^{235}U and ^{231}Pa (^{231}Th) has a half-life short enough that it can be ignored in carbonate dating, and the relevant nuclides are ^{235}U and ^{231}Pa.

12.4.1.1 Generation of initial disequilibrium

The mechanism of generation of disequilibrium for carbonate dating is related to the hydrologic cycle and elemental behavior during weathering and precipitation. During weathering under oxidizing conditions, uranium is present mostly in the 6+ valence state, whereas protactinium is 5+ and thorium 4+. In the 6+ state, uranium forms soluble uranyl complex ions (UO_2^{2+}), which play an important role in uranium transport during weathering. In contrast, thorium and protactinium have low solubility in most natural waters and are transported primarily as insoluble restite minerals or adsorbed onto the surface of clay minerals. As a result, natural waters have ^{230}Th/^{238}U ratios on the order of 10^5 lower than at secular equilibrium, and ^{231}Pa/^{235}U ratios 10^4 times lower than at secular equilibrium. Therefore, weathering and soil formation lead to extreme fractionation of parents and daughters. Carbonates precipitated from natural waters inherit this fractionation signal, so that even though uranium and thorium or protactinium have similar partition coefficients in aragonite and calcite, pure carbonate (e.g., surface corals and some speleothems) will be precipitated with very low initial thorium or protactinium and this initial daughter can be neglected in the decay equations.

12.4.1.2 Age equations

Assuming that the assumptions discussed above apply, and further assuming that initial ^{230}Th/^{238}U = 0, the following age equation can be derived from equation 2.26:

$$\left(\frac{^{230}\text{Th}}{^{238}\text{U}}\right) - 1 = -e^{-\lambda_{230}t} + \left[\frac{\delta^{234}\text{U}_m}{1000}\right]\left[\frac{\lambda_{230}}{\lambda_{230}-\lambda_{234}}\right]\left(1 - e^{-(\lambda_{230}-\lambda_{234})t}\right)$$

(12.5)

where the parentheses around the first term on the left indicate an activity ratio (note that in the carbonate literature it is common to indicate activity ratios by square brackets or no brackets around the chemical symbols, but we use the convention of parentheses indicating activities in order to maintain consistency throughout the chapter). The second term on the right includes $\delta^{234}\text{U}_m$, which is the per thousand deviation of the ^{234}U/^{238}U activity ratio measured in the sample from that at secular equilibrium (i.e. 1):

$$\delta^{234}\text{U} = 1000\left[\frac{\left(^{234}\text{U}\right)/\left(^{238}\text{U}\right)_{sample}}{\left(^{234}\text{U}\right)/\left(^{238}\text{U}\right)_{equil}} - 1\right]$$

(12.6)

For example, if $(^{234}\text{U})/(^{238}\text{U}) = 1.005$, $\delta^{234}\text{U} = 5$. Or if $(^{234}\text{U})/(^{238}\text{U}) = 0.997$, $\delta^{234}\text{U} = -3$. This term did not appear in the original formulation of the age equation presented in *Barnes et al.* [1956] because at that time it was not known that seawater is not in ^{234}U/^{238}U radioactive equilibrium. In fact, seawater today has $(^{234}\text{U})/(^{238}\text{U}) = 1.146$ (or $\delta^{234}\text{U} = 146$), due to the preferential fractionation of U into the aqueous phase during weathering.

The combination of measured $\delta^{234}\text{U}_m$ and $(^{230}\text{Th})/(^{238}\text{U})$ also uniquely constrains the initial $\delta^{234}\text{U}$ for a given sample. Once equation (12.5) has been solved for t, the following equation gives initial $\delta^{234}\text{U}$:

$$\delta^{234}\text{U}_m = \left(\delta^{234}\text{U}_i\right)e^{-\lambda_{234}t}$$

(12.7)

The initial $\delta^{234}\text{U}$ of a sample can be used as a test of closed-system behavior in cases where the expected value can be constrained—for example, precipitation of carbonate by corals is expected to retain the seawater value for $\delta^{234}\text{U}$, and deviations from the seawater initial composition can indicate that fossil corals have had their U modified (e.g., by exchange with or uptake of U from groundwater after uplift onto land), which invalidates the closed-system assumption for calculating age.

Because t appears in both the left-hand and right-hand sides of equation (12.5), the equation must be solved either iteratively or graphically. Figure 12.6 shows a plot of $\delta^{234}\text{U}_m$ versus $(^{230}\text{Th})/(^{238}\text{U})$ [*Edwards et al.*, 2003b]. Two sets of contours show solutions to equation (12.5) (steep lines labeled with ages) and solutions to equation (12.7) (shallow lines labeled with initial $\delta^{234}\text{U}$ values). The position of a data point plotted on this diagram will thus uniquely define both age and initial $\delta^{234}\text{U}$. Isotopic compositions to the right of the heavy line in the area labeled "altered" cannot be achieved by closed-system decay and ingrowth, and therefore would indicate open-system behavior.

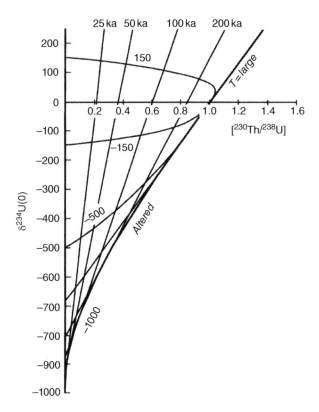

Fig. 12.6. δ^{234}U measured versus (^{230}Th)/(^{238}U). Contours with steep positive slopes indicate solutions to the age equation (equation 12.5) and contours with shallow slopes labeled "150", "–150", etc., indicate solutions to the δ^{234}U equation (equation 12.7). Seawater likely has values close to the modern value of ~149‰, so carbonates precipitated from seawater (like corals) should plot close to the "150" curve, unless they have been altered. In addition, isotopic compositions to the right of the bold steep line marked "Altered" cannot be achieved by closed-system decay and ingrowth, and therefore would indicate open-system behavior. (Source: *Edwards et al.* [2003]. Reproduced with permission of Mineralogical Society of America.)

In cases where the assumption of zero initial daughter does not hold, it is still possible to obtain an age if the amount of initial Th present is relatively small and its isotopic composition can be measured or estimated. Two approaches allow dating while accounting for this complication. First, in the case where initial Th is small but non-negligible, equation (12.5) can be modified to correct for the initial daughter present:

$$\left[\left(\frac{^{230}Th}{^{238}U}\right) - \left(\frac{^{232}Th}{^{238}U}\right)\left(\frac{^{230}Th}{^{232}Th}\right)_i\left(e^{-\lambda_{230}t}\right)\right] - 1$$
$$= -e^{-\lambda_{230}t}\left[\frac{\delta^{234}U_m}{1000}\right]\left[\frac{\lambda_{230}}{\lambda_{230}-\lambda_{234}}\right]\left(1-e^{-(\lambda_{230}-\lambda_{234})t}\right) \quad (12.8)$$

The second term on the left-hand side corrects for initial ^{230}Th, but requires knowing ^{230}Th/^{232}Th for the Th initially present (often referred to as the "detrital" component). In theory, the silicate (detrital) component could be physically separated from the carbonate component and its Th isotopic composition measured directly. However, in practice, physical separation can be difficult for fine grain sizes, and preferential

dissolution of a carbonate component can lead to differential leaching of U and Th. Instead, other approaches are typically used to estimate the Th isotopic composition of the detrital component. First, in cases where the amount of initial Th is small and therefore the correction is small, a broad estimate of Th isotopic composition (with appropriately large uncertainties) can be used. For example, using an average bulk-silicate ^{232}Th/^{238}U value of approximately 3.8 (atomic) and assuming secular equilibrium between ^{230}Th and ^{238}U in the detrital component gives ^{230}Th/^{232}Th of 4.4×10^{-6} (atomic). Even assigning relatively large uncertainties to these estimates (10–20%) will result in relatively small increases in uncertainty of calculated ages if the total correction is small (e.g., if measured (^{230}Th)/(^{232}Th) activity ratios are 10–20 or higher). A second approach is to assume that the sample represents a mixture of two components (detrital and authigenic), and to use an isochron approach. By physically separating and analyzing subsamples with the same age but different amounts of the detrital component, the composition of the end-member detrital component can be calculated by extrapolation to a detrital Th-free end-member. Over time, different subsamples with different initial ^{238}U/^{232}Th will evolve different isotopic compositions according to the laws of radioactive decay, and will therefore have different measured ^{238}U/^{232}Th, ^{234}U/^{238}U, and ^{230}Th/^{232}Th ratios. If the subsamples had the same initial ^{230}Th/^{232}Th, they will define a line in three-dimensional isotopic space.

Analogous age equations can be developed for the ^{235}U–^{231}Pa system. The equation appropriate for closed-system decay with no initial daughter is a simplified version of equation (12.9) (with the term relating to the initial daughter removed):

$$\left(\frac{^{231}Pa}{^{235}U}\right) - 1 = -e^{-\lambda_{231}t} \quad (12.9)$$

Unlike Th, there is no longer-lived isotope of Pa that could be used to infer the presence of initial ^{231}Pa. However, in cases where it is reasonable to assume zero initial Th, it is reasonable also to assume zero initial Pa given their geochemical similarity.

In cases where both systems have been measured in the same sample, the ^{238}U–^{234}U–^{230}Th and ^{235}U–^{231}Pa age equations can be combined:

$$\left(\frac{^{231}Pa}{^{230}Th}\right) = \frac{\left(1-e^{-\lambda_{231}t}\right)}{\left(\frac{^{238}U}{^{235}U}\right)\left\{1-e^{-\lambda_{230}t}+\left[\frac{\delta^{234}U_m}{1000}\right]\left[\frac{\lambda_{230}}{\lambda_{230}-\lambda_{234}}\right]\left(1-e^{-(\lambda_{230}-\lambda_{234})t}\right)\right\}} \quad (12.10)$$

(^{238}U/^{235}U) can be treated as constant because the magnitude of its variations in nature are negligible in this context. Because an isotope of U is the parent for both decay chains and because the ^{238}U/^{235}U ratio can be considered to be fixed, this equation removes the necessity for measuring abundance of U. ^{231}Pa–^{230}Th ages calculated in this way are analogous to ^{207}Pb/^{206}Pb ages, and are not sensitive to recent U gain or loss (although they are sensitive to older U mobility). Combined ^{231}Pa–^{230}Th dating can also be used to test for closed-system behavior, as the ages calculated using each system independently should be concordant (see below). In a manner analogous to comparison of $^{235}U-^{207}Pb$ and $^{238}U-^{206}Pb$ ages from the same sample in U–Pb dating, concordance between the two systems is expected if the carbonate has maintained a closed system. Because of the different half-lives involved, loss or gain of U or of its daughters will lead to discordance if sufficient time has elapsed after such a disturbance for decay to be measurable. However, $^{235}U-^{231}Pa$ dating is more technically challenging and therefore less common than $^{238}U-^{234}U-^{230}Th$ dating, which in practice limits the use of combined ^{230}Th and ^{231}Pa ages.

12.4.1.3 Application 1: dating marine carbonates

The most widely used parent–daughter pair in dating marine and lacustrine carbonates is $^{238}U-^{230}Th$, but this application has also seen the most work on combined $^{238}U-^{230}Th$ and $^{235}U-^{231}Pa$ dating.

One of the first applications of ^{230}Th dating of carbonates was to shallow corals as indicators of sea-level changes. For example, one of the most significant success stories of high-precision dating has been the construction of a detailed history of sea-level changes during the last deglaciation. The general approach is to measure ages of coral species that grow near the sea surface but which are no longer at sea level. An individual measurement then gives an age and an elevation, and many such points can be combined into a sea-level curve (Fig. 12.7)[*Edwards et al.*, 2003a]. A similar approach can be applied to speleothems (cave deposits), which grow only subaerially, providing a maximum elevation for sea level at a given time. This direct reconstruction technique has several challenges, including the discontinuous nature of the record, assessment of whether diagenesis has affected ages, corrections for tectonic uplift or subsidence, and difficulty recovering submerged samples—because we currently are in an interglacial period, most samples that record the recent history of sea level are now submerged. Nevertheless, through a combination of corals, speleothems, and careful assessment of the effects of diagenesis and tectonics, a great deal of information about the history of sea-level changes over the past ~500 ka has been acquired (Fig. 12.7). The curves often show correlations of sea level with high northern latitude summer insolation, which supports the idea that changes in orbital geometry are a primary driver for sea-level changes, although they also document that sea-level changes are responding to millennial-scale climate variations in addition to the longer (100 ka or 40 ka) Milankovich cycles. The history of sea-level change has important implications for our future. For example, the reconstructions show that the highest recorded sea level is ~20 m above present (prior to ~420 kyr; *Edwards et al.*, 2003a], which implies that the east Antarctic ice sheet must have been smaller than it is at present, as complete melting of the Greenland and west Antarctic ice sheets would lead to less than 20 m of sea-level rise. This finding could have important implications for climate in the near future, as the orbital conditions at that time were similar to those today. Another interesting time period in the sea level record is the last

deglaciation (Fig. 12.7), which is known in greater detail than older records and which spans a large total amount of sea-level change (over 120 m). This period also records some very rapid increases in sea level, for example rapid rises at ~19 ka, ~14.5 ka and 11–10 ka, reaching ~20 m/ka rates of rise (Fig. 12.7).

Corals can also record relative sea-level changes due to tectonic uplift, which can be applied to paleoseismology. Annual lowest tidal levels limit the highest level of growth of corals (highest level of survival, or HLS). Coral heads with surfaces that rise towards the edge (called microatolls) indicate slow relative sea-level rise during decades of growth. Emergence of the corals above sea level and their resulting death can be correlated with uplift due to earthquakes, thus dating of corals can provide a history of earthquakes over centuries. One such record for three sites in Sumatra is presented in Fig. 12.8, and shows a history of large earthquake events related to the Sunda megathrust offshore of Sumatra. The tip of each sawtooth in the relative sea-level record indicates emergence of the coral, and the ramp leading up to each tip records subsidence related to strain accumulation prior to the earthquake (which is typically recorded for only a few decades prior to earthquakes, when the coral head is resubmerged). From this record, *Sieh et al.* [2008] argued that the 2007 magnitude 8.4 earthquake was only the beginning of a cycle of failure of the Mentawi patch of the fault, and that the amount of potential slip remaining after that earthquake would be sufficient to generate a magnitude 8.8 earthquake in the next few decades.

12.4.1.4 Application 2: dating cave deposits (speleothems)

In addition to providing a constraint on sea-level changes that is complementary to the coral records, speleothems (cave deposits) serve as important archives of past climate changes and important dating constraints for archaeological specimens found in caves. Speleothems are most often composed of calcite, although aragonite is sometimes found as well, and depending on the structure of the speleothem (e.g., stalagmites vs. flowstones) may have regular layering that can be traced laterally across the deposit. As with corals, the initial high U/Th ratio in speleothems is due primarily to inheriting this ratio from the extreme fractionation of U from Th (and Pa) in the hydrosphere. Unlike near-surface corals, however, speleothems often have significant detrital component (likely derived from silicate dust within the cave), although the extent of contamination can vary widely based on the distance of the speleothem from the cave entrance, the growth rate of the speleothem, and other environmental factors, and may vary significantly over the history of the speleothem. Therefore, dating typically involves a correction for initial Th, as in equation (12.8). Furthermore, the initial $^{234}U/^{238}U$ ratio is dictated by the drip waters from which the calcite or aragonite precipitated, which can vary widely in their uranium isotopic composition, making it difficult to use $\delta^{234}U$ to assess the degree to which the closed-system assumption holds. These two factors can make dating speleothems more challenging than dating corals. Nevertheless, in many samples these limitations can be overcome

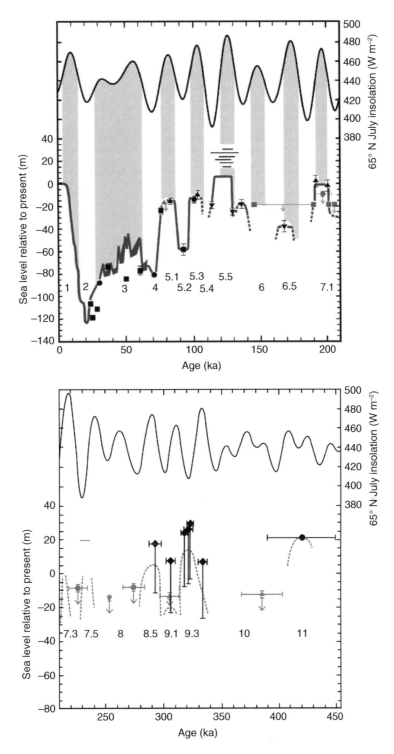

Fig. 12.7. Direct sea-level data from corals (solid symbols) and speleothems (gray symbols) for time periods of (a) 0–200 ka and (b) 200–450 ka. Numbers indicate marine isotope stages (MISs). The line segments above MIS 5.5 are different estimates of the timing and duration of peak sea level. July summer insolation at 65°N latitude is shown in the upper curves of each panel. Vertical gray bars in (a) indicate times of high summer insolation to facilitate comparison with the sea-level curves. (Source: *Edwards et al.* [2003a]. Reproduced with permission of Elsevier.)

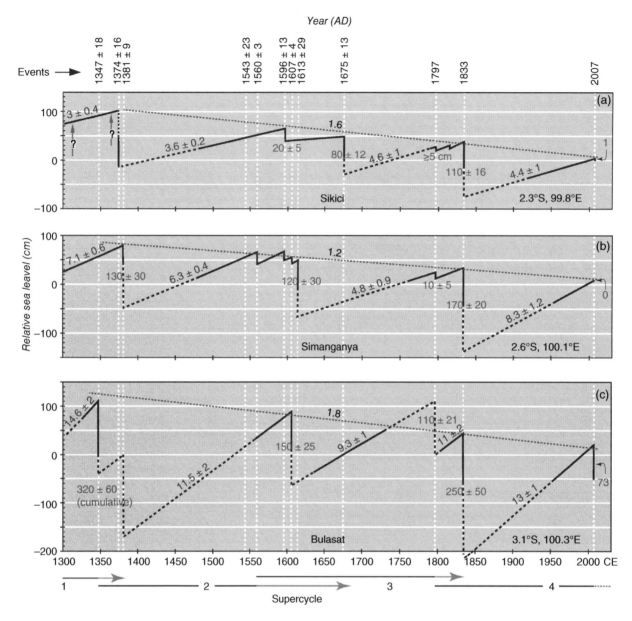

Fig. 12.8. Histories of interseismic submergence and coseismic emergence through seven centuries at three sites along the Sunda megathrust. Data constrain solid parts of the curves well; dotted portions are inferred. Emergence values (in centimeters ±2σ) are red. Interseismic submergence rates (in millimeters per year, ±2σ) are blue. Millennial emergence rates are black. Vertical dashed white lines mark dates of emergences. Red arrows at bottom highlight the timing of the failure sequence for each supercycle. (Source: *Sieh et al.* [2008]. Reproduced with permission of The American Association for the Advancement of Science.) (*See insert for color representation of the figure.*)

through correction or use of an isochron technique, and accurate and precise dates can be obtained.

Abundances of trace elements and isotopic compositions (especially of Sr, O, and C) of speleothem material contain information about past environmental conditions, and these proxy records can be compared to other paleoclimate information (e.g., marine-core records or ice-core records) if their absolute ages can be constrained—for example, through U-series dating. In particular, speleothem records provide an important complement to the marine records, because they can provide information about continental climate variations, whether local or

global, which can be combined with marine data for a more complete record of climate change. For example, one of the most widely cited $\delta^{18}O$ records is from the Devils Hole calcite, which covers the period from 568 ka to 4.5 ka [*Winograd et al.*, 1992, 2006]. Devils Hole is an open fissure (fault-related) that is lined with a > 30 cm thick deposit of calcite that grew apparently continuously from the calcite-supersaturated groundwater over the past ~500 ka. Several cores have been combined to form a continuous record ending at 4.5 ka. The Devils Hole $\delta^{18}O$ time series closely tracks sea-surface temperature records from marine sediments cored off the California coast (Fig. 12.9), suggesting

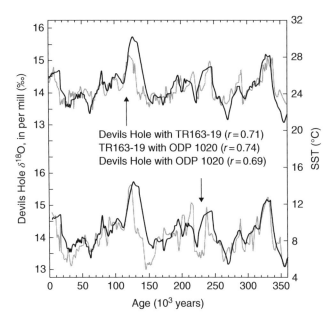

Fig. 12.9. Devils Hole $\delta^{18}O$ (in black) and sea surface temperature (SST; in gray) time series, for two different SST records: marine core TR 163-19 and ODP 1020. Correlation coefficients (*r*) calculated for the period 360,000 to 4500 years ago. (Source: *Winograd et al.* [2006]. Reproduced with permission of Elsevier.) (*See insert for color representation of the figure.*)

that climate in the continental interior is closely linked with that in the ocean. Interestingly, both the sea-surface temperature and Devils Hole records show warming that began ~5000–10,000 years prior to the start of the last two deglaciations, which in turn began at about 20,000 and 135,000 years ago. The causes of this early warming are still poorly understood, but could be related to warming in some other locations in the low to mid-latitudes in other ocean basins prior to the last two deglaciations [*Winograd et al.*, 2006].

Another high-resolution and long speleothem record of continental climate changes is provided by speleothems from several caves in China that cover the past 228,000 years: Hulu, Dongge, and Sanbao caves [*Wang et al.*, 2001, 2008; *Cheng et al.*, 2006]. The records document Chinese interstadial events (periods of relatively strong East Asian Monsoon (EAM)) that support the interpretation that tropical monsoons vary in intensity primarily due to changes in Northern Hemisphere summer radiation. An additional impact of these high-resolution well-dated records is that they have provided independent data for calibrating atmospheric ^{14}C, and therefore the ^{14}C age calibration curves.

Cave deposits have also provided important timing constraints for archeological studies, as many important artifacts and faunal remains are preserved in caves and rock shelters. For example, the mid-Pleistocene Sima de los Huesos ("pit of the bones") in Spain contains an important collection of human bones of at least 28 individuals, which represents more than 80% of the global record of the genus *Homo* in the middle Pleistocene [*Bischoff and Shamp*, 2003; *Bischoff et al.*, 2007]. Re-dating of speleothem deposits overlaying these bones using high-precision

MC-ICP-MS techniques [*Bischoff et al.*, 2007] improved the precision of the ages to the point where they could be reliably distinguished from secular equilibrium, yielding finite dates of ~600 ka, in contrast to the earlier TIMS dates which provided only a minimum age of ~350 ka [*Bischoff and Shamp*, 2003]. This also provides an illustration of the interpretive improvements that can accompany advances in analytical precision— because the Sima hominids are evolutionarily ancestral to Neanderthals, providing a finite age of these specimens is an important advance in understanding human evolution.

U-series dating can also be important in paleontology, for example, contributing to dating the Pleistocene megafauna extinction in Australia. Deposits in Naracoorte caves in southeastern Australia were used both to date faunal remains in the caves and as recorders of paleoclimate [*Prideaux et al.*, 2007]. The climate records indicate relatively cool, moist conditions for most of the last glacial, which should have favored megafauna. Therefore, the extinction of megafauna by ~45 ka was not due to climate changes, and may instead have been the result of human actions [*Prideaux et al.*, 2007].

12.4.1.5 Application 3: pedogenic carbonates

Pedogenic (soil) carbonate and opal deposits have also long been targets for U-series dating [e.g., *Ku et al.*, 1979; *Ludwig and Paces*, 2002; *Sharp et al.*, 2003; *Paces et al.*, 2004; *Maher et al.*, 2007]. Soils and geomorphic surfaces are difficult to date, and U-series dating of carbonate disseminated within soils or as rinds on clasts can provide much-needed age constraints. Early efforts using alpha spectrometry were hampered by the requirement for large sample sizes and relatively low precision. However, development of TIMS and MC-ICP-MS techniques has renewed interest in pedogenic carbonate dating [e.g., *Ludwig and Paces*, 2002; *Sharp et al.*, 2003; *Fletcher et al.*, 2011]. In general, correction for detrital ^{230}Th is required, but if samples are chosen carefully, a correction assuming average continental U/Th ratios can be used, removing the need to characterize the detrital component separately.

The first study to apply modern mass spectrometric techniques to dating pedogenic carbonates was *Ludwig and Paces* [2002], which analyzed multiple subsamples of mixed silicate–carbonate clast rinds from samples in Crater Flat, a semiarid, structurally controlled basin near Yucca Mountain, Nevada. They used a broad estimate of the composition of the detrital component, yielding a "single-sample, single date" rather than using an isochron approach. They found that the samples preserved stratigraphic order both within a single rind and for multiple samples from different deposits, and that most (>90%) of the samples were consistent with closed-system behavior despite some samples being hundreds of thousands of years old. This proof-of-concept study was followed by other studies, which applied the updated technique to dating soils in multiple contexts. For example, *Sharp et al.* [2003] dated rinds on clasts in coarse gravels capping Wind River fluvial terraces, using them to calculate incision rates for the Wind River of 0.26 ± 0.05 m/ka over the last glacial cycle, slower than that suggested by cosmogenic nuclide dating. *Fletcher et al.* [2010, 2011] dated alluvial fans offset by strands of the San Andreas fault in California, concluding that:

(1) the time lag between formation of the geomorphic surface and deposition of the pedogenic carbonate was no longer than ~10 ka;

(2) the U-series age was older than the age of the same surface determined from ^{10}Be dating, suggesting that some shielding of the cobbles dated by ^{10}Be occurred after deposition of the fan [*Fletcher et al.*, 2010].

A following study [*Fletcher et al.*, 2011] used U-series dating of alluvial fans offset by the Elsinore strand of the San Andreas to infer long-term slip rates of ~1.5–2 mm/year. A similar approach by *Gold et al.* [2011] applied to offset terraces along the Altyn Tagh fault in China (the major strike-slip system that forms the northern boundary of the Tibetan Plateau) are consistent with a long-term average slip rate of 8–12 mm/year.

12.4.2 U-series dating in silicate rocks

U-series dating has long been used as a tool for constraining timescales of magmatic processes, and has been applied to a broad variety of questions including dating eruptions and quantifying timescales of melt generation, transport, storage, and degassing [e.g., *Kigoshi*, 1967; *Allègre*, 1968; *Spiegelman and Elliott*, 1993; *Reid et al.*, 1997; *Reid*, 2003; *Cooper and Reid*, 2008; *Schmitt*, 2011; *Sims et al.*, 2013; *Cooper*, 2015]. The materials analyzed for these different approaches range from whole-rock or glass samples to mineral separates to gases emitted from fumaroles. In the following sections, we present examples of a variety of applications of U-series to geochronology of magmas and volcanic rocks.

12.4.2.1 Isochron and evolution diagrams
Many applications of U-series disequilibria to magmatic processes rely on use of an isochron diagram (Fig. 12.10). As with other isochron approaches (e.g., Chapter 6), the fundamental isochron approach makes use of a modification of the basic decay equation, where the abundances of parent and daughter are normalized to the abundance of a stable (or long-lived) isotope of the daughter. For the ^{238}U–^{230}Th pair, this involves modifying equation (12.2) by normalizing activities of parent and daughter by the activity of ^{232}Th, which, although not stable, has a half-life that is long enough (14 Ga) that decay of ^{232}Th can be neglected. The resulting equation is given here:

$$\left(\frac{^{230}\text{Th}}{^{232}\text{Th}}\right) = \frac{(^{230}\text{Th})}{(^{232}\text{Th})}e^{-\lambda_{230}t} + \frac{(^{238}\text{U})_0}{(^{232}\text{Th})}\left(1-e^{-\lambda_{230}t}\right) \quad (12.11)$$

As with other isochron approaches, this equation represents a straight line on a plot of normalized parent versus daughter abundance (in this case, $(^{238}\text{U})/(^{232}\text{Th})$ versus $(^{230}\text{Th})/(^{232}\text{Th})$;

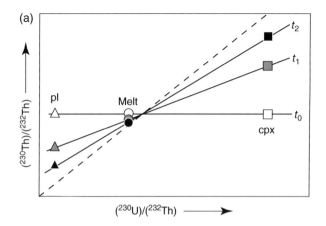

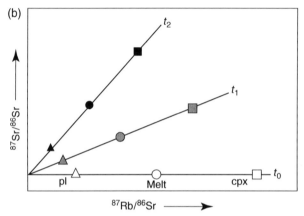

Fig. 12.10. Comparison of ^{238}U–^{230}Th isochron diagram to ^{87}Rb–^{87}Sr isochron diagram. See text for discussion of differences in evolution of the position of data points over time on the two diagrams.

Fig. 12.10). Extraction of a date from a group of data using the isochron approach requires the following standard set of assumptions.

(1) The materials to be dated must have formed at the same time—for example, in the case of dating crystals, the duration of crystallization must be short with respect to the half-life of ^{230}Th.

(2) The materials to be dated all began with the same initial ^{230}Th/^{232}Th ratio [i.e., $(^{230}\text{Th})_0/(^{232}\text{Th})$]. In the case of magmatic systems, this requires that the different samples must have formed at the same time from the same parent (e.g., coprecipitation of different minerals from a given melt at the same time).

(3) The materials to be dated must have acted as closed systems with respect to both parent and daughter.

In addition, the derivation of equation (12.2) implies the additional assumptions that the half-life of the parent is much longer than that of the daughter, and that the time of interest is short with respect to the half-life of the parent.

The first application of the isochron approach to U–Th dating of volcanic rocks was by *Allègre* [1968], who plotted data for individual minerals separated from a volcanic rock that were analyzed by *Kigoshi* [1967] in an attempt to date the eruption age of the volcanic rock. Cogenetic minerals with the same crystallization age will form an inclined array on an isochron diagram with a slope proportional to the crystallization age, similar to the arrays on more "conventional" isochron diagrams (Fig. 12.10). However, the ^{238}U–^{230}Th isochron has a number of distinctive features when compared to a "conventional" isochron diagram. For example, secular equilibrium is represented by a line with slope of unity (known as the equiline) and any phase that initially has disequilibrium between parent and daughter will plot off this line. As each phase approaches secular equilibrium through radioactive decay, the data points will move toward the equiline. However, unlike conventional isochrons, the trajectories of data points may be either upward or downward, depending on whether the mineral started out with an initial excess or deficit of the daughter. Although the trajectories are not technically vertical, decay of ^{232}Th and ^{238}U are slow enough that their activity ratio will not change measurably over time (consistent with the assumptions involved in deriving equation (12.2). Any phases that have aged enough to reach radioactive equilibrium will remain on the equiline unless, or until, some additional geological event fractionates parent from daughter; this means that a mineral array that lies along the equiline has an indeterminate age. In addition, as each phase exponentially approaches equilibrium the rate of vertical movement on an isochron diagram will slow; this leads to asymmetric age uncertainties as symmetric analytical uncertainty on the atomic ratio $^{230}Th/^{232}Th$ will represent a longer period of time on the side of the data point closest to the equiline than on the side of the data point farthest from the equiline. Finally, the initial $(^{230}Th)/(^{232}Th)$ for the array is given by the intersection of the mineral array with the equiline rather than by the intersection of the array with the *y*-axis.

Additional complications arise when using ^{230}Th–^{226}Ra disequilibria to date minerals. Because ^{226}Ra is the longest-lived isotope of radium, the isochron equation cannot be normalized to a stable isotope of Ra and must instead be normalized to barium as the closest chemical analog of radium. This implicitly assumes that Ra and Ba are exact chemical analogs, but because their ionic radii differ, theoretical and experimental models of trace-element partitioning indicate that in most situations their behavior will be similar but not the same [e.g., *Blundy and Wood*, 2003; *Fabbrizio et al.*, 2008, 2009, 2010]. As a result, data plotted on a ^{230}Th–^{226}Ra isochron diagram often does not follow the expected pattern for isochron diagrams. For example, minerals coprecipitating from a melt will generally have different initial Ra/Ba ratios as well as different initial Th/Ba ratios (Fig. 12.11), and the arrays of equal age will not form a straight line. This makes the use of a ^{230}Th–^{226}Ra isochron diagram cumbersome for calculating ages. Instead, an evolution diagram is a more intuitive method of visualizing the evolution of activity ratios over time. On this kind of a diagram, curves are plotted showing the evolution of $(^{226}Ra)/[Ba]$ over time (which

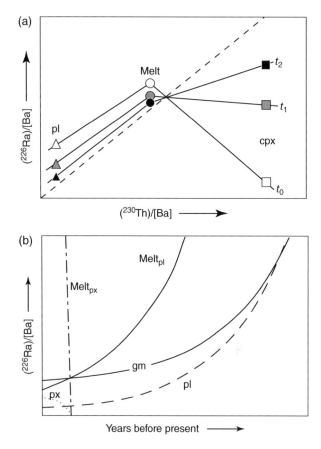

Fig. 12.11. (a) ^{230}Th–^{226}Ra isochron diagram: position of points for plagioclase (pl) and clinopyroxene (cpx) are shown schematically at different points in time after crystallization (t_0). Note that it is not until the minerals are close to secular equilibrium that the data array approximates a straight line. (b) Evolution diagram: curves showing the evolution of (^{226}Ra)/Ba in groundmass (gm), plagioclase (pl) and pyroxene (px) over time. Present day (measured compositions) is shown along the *y* axis and curves are calculated from $(^{230}Th)/(^{226}Ra)$ and (^{230}Th)/Ba ratios in the minerals. Also shown are curves for (^{226}Ra)/Ba in hypothetical melts that would be in equilibrium with plagioclase (Melt$_{pl}$) and pyroxene (Melt$_{px}$) at different times in the past, based on measured or calculated partition coefficients for Ra and Ba.

implicitly includes $(^{230}Th)/[Ba]$, as that controls the magnitude of change of $(^{226}Ra)/[Ba]$ over a given time interval) for cogenetic phases [*Cooper et al.*, 2001]. In the case of minerals crystallizing from a melt, the time at which curves for a given mineral intersect the curve for the evolution of the melt represents the time at which they had the same $(^{226}Ra)/[Ba]$ ratio; in other words, the isochron age assuming no initial fractionation of Ra from Ba. Because Ra/Ba fractionation during crystallization has been shown both theoretically and experimentally to be important in many silicate minerals [e.g., *Blundy and Wood*, 2003; *Fabbrizio et al.*, 2008, 2009, 2010], the effects of this initial fractionation during crystallization must be accounted for. On an evolution diagram, this can be undertaken by correcting the data for a given mineral phase using the ratio of the partition coefficients of Ra and Ba in that phase to calculate the $(^{226}Ra)/[Ba]$ expected in the liquid that would have been in equilibrium with that phase at a particular time. The intersection of this curve for a model melt in equilibrium with a mineral with the curve for the glass composition of a given rock will then represent the time at which they were in chemical equilibrium, i.e., the crystallization age (Fig. 12.11). The model crystallization age can thus be determined graphically. The model age can also be computed from the following equation [*Cooper and Reid*, 2003):

$$t_{GM-A} = \frac{1}{\lambda_{226}} \ln \left(\frac{\left[\frac{(^{230}\text{Th})_{GM}}{[\text{Ba}]_{GM}} \right] - \left(\frac{D_{Ba}}{D_{Ra}} \right)_A \left[\frac{(^{230}\text{Th})_A}{[\text{Ba}]_A} \right]}{\left(\frac{D_{Ba}}{D_{Ra}} \right)_A \left\{ \left[\frac{(^{236}\text{Ra})_A}{[\text{Ba}]_A} \right] - \left[\frac{(^{230}\text{Th})_A}{[\text{Ba}]_A} \right] \right\} - \left[\frac{(^{230}\text{Ra})_{GM}}{[\text{Ba}]_{GM}} \right] + \left[\frac{(^{230}\text{Th})_{GM}}{[\text{Ba}]_{GM}} \right]} \right)$$

where the subscripts for each term refer to the ratios measured in the groundmass or glass (GM) or a given mineral phase (A).

12.4.2.2 Whole-rock disequilibria in volcanic rocks

Measurements of U-series disequilibria in whole-rock samples can be used to constrain the timing of magmatic processes from melt generation to crustal storage and differentiation to degassing and eruption. For example, the interplay of parent–daughter fractionation, ingrowth of daughter nuclides, and melt transport during the melting process can lead to different degrees of disequilibrium in magmas produced during mantle melting. In addition, whole-rock disequilibria measured in a suite of lavas from a series of eruptions that periodically tap the same magma reservoir can be used to constrain the rates of chemical evolution and the residence time of magma within the reservoir.

Mantle melting rates and melt transport times
U-series disequilibria can be produced in magmas generated by partial melting. In particular, a great deal of attention has been focused on using U-series disequilibria to assess mantle melting processes [e.g., *Williams and Gill*, 1989; *McKenzie*, 1985, 2000; *Spiegelman and Elliott*, 1993; *Bourdon and Sims*, 2003; *Lundstrom et al.*, 2003; *Turner et al.*, 2003]. There are two broad types of effects that can contribute to disequilibrium in the melt. First, instantaneous chemical fractionation of parent from daughter (e.g., during batch or fractional melting) can create some disequilibrium in the melt if the partition coefficients for parent and daughter are different. However, this effect can only produce disequilibria if the melt fraction *F* is similar in magnitude or smaller than the bulk partition coefficients *D*. For mid-ocean ridge basalts, especially, this presents a problem because bulk partition coefficients for U and Th during melting of peridotite are on the order of 0.001, whereas melt fractions inferred from other trace elements are on the order of 0.15, yet observations typically show disequilibrium between ^{230}Th and ^{238}U, which can reach 20–40% excesses of ^{230}Th (i.e., $(^{230}\text{Th})/(^{238}\text{U}) = 1.2$–1.4). For this reason, other models of the behavior of parent and daughter during melting which explicitly consider decay that happens during the melting process are more successful at matching the observations.

Two conceptual models of this type are dynamic melting and equilibrium porous flow models (Fig. 12.12). Both lead to higher disequilibria in the melt than that produced by comparable melt fractions from models which do not consider the melting time, but they differ in important respects such as the depth at which the disequilibria is produced and the dependence on melting (or upwelling) rates versus porosity. Dynamic melting assumes that any melt produced up to a "critical porosity" remains in chemical equilibrium with the solid, and any melt produced above that porosity is instantaneously extracted from the solid, pooled and mixed elsewhere, and erupted [*Williams and Gill*, 1989]. The increase in disequilibria over instantaneous melting is related to the difference in residence time of the parent versus daughter in the residual solid. If the parent is more compatible in the solid than the daughter, it will remain in the solid longer during the melting process before it is "forced" into the liquid by progressive melting. During the time over which the parent is preferentially retained in the solid, it continues to decay, and the "excess" daughter produced during this time is added to that in the liquid. Equilibrium porous flow models [e.g., *Spiegelman and Elliott*, 1993] are similar in that differences in residence time of the parent and daughter enhance disequilibria compared to instantaneous melting. However, the difference in residence time in this case is produced by chromatographic flow. If the parent is more incompatible than the daughter, its effective velocity through the melting region is slower than that of the daughter, and if this effective velocity is slow relative to the half-life of the parent, continued decay during transport will produce "excess" daughter that is added to that already present in the melt. A range in both ^{238}U–^{230}Th and ^{230}Th–^{226}Ra disequilibria can be produced by dynamic melting and equilibrium porous flow, and the different parent–daughter pairs are sensitive to different melting parameters. For example, for melting rates and porosities that could be expected at mid-ocean ridges, $(^{226}\text{Ra})/(^{230}\text{Th})$ is primarily controlled by porosity of the melting region, whereas $(^{230}\text{Th})/(^{238}\text{U})$ is primarily controlled by melting rate. Within this context, for a given set of melting parameters, the disequilibria produced by dynamic melting will generally be lower than those produced by equilibrium porous flow. By combining data for multiple parent–daughter pairs, it is possible in some cases to constrain the melting parameters, including the melting rate

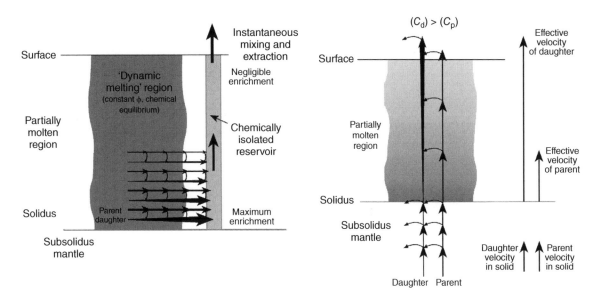

Fig. 12.12. Cartoons illustrating conceptual models of U-series disequilibria produced during mantle melting. (a) Dynamic melting model. This type of model assumes that melt remains in equilibrium with the solid in the "dynamic melting" region with a small, constant, porosity. Any melt that is produced in excess of this critical porosity is instantaneously extracted into a reservoir that is chemically isolated from the solid. Physically, this could be a network of channels where melt transport is too rapid to allow equilibration with the surrounding solid. Nearly all the ingrowth of the daughter nuclides occurs at the bottom of the melting column. (b) Equilibrium porous flow model. In this model, melt is transported through the solid with continuous melt–solid interaction in a steady-state, one-dimensional melting column. Elements can travel at different velocities due to the "chromatographic effect" and therefore have different residence times in the melting region. Below the melting region, parent and daughter travel at the same rate and maintain secular equilibrium. Within the melting region, the daughter is less compatible and travels faster than the parent, allowing additional time for decay of the parent to produce excess daughter. (Source: *Spiegelman and Elliott* [1993]. Reproduced with permission of Elsevier.)

[e.g., *Pietruszka and Garcia*, 1999; *Sims et al.*, 1999, 2002, 2003, 2008] (Fig. 12.13).

Further constraints on the transport rates of magma at mid-ocean ridges can be gained from the shorter-lived ^{226}Ra–^{210}Pb pair. Disequilibria between ^{210}Pb and ^{226}Ra have also been observed in mid-ocean ridge basalts (MORB) [*Rubin et al.*, 2005) and in Icelandic lavas [*Sigmarsson*, 1996]. ^{210}Pb activity in these studies is lower than expected for secular equilibrium (i.e., $(^{210}$Pb$)/(^{226}$Ra$) < 1$, or "^{210}Pb deficit"), and in these studies the most extreme disequilibrium between ^{226}Ra and ^{210}Pb occurred in the most primitive samples, suggesting that shallow crystallization did not produce the ^{210}Pb deficits. In contrast, the deficits are attributed to a mantle origin. ^{210}Pb has a half-life of 22.6 years, which means that ^{226}Ra–^{210}Pb disequilibria will persist only ~100 years after fractionation, which in turn would imply that the time for transport of the magma from the melting region to the surface was less than ~100 years.

Magma degassing rates

The observation that many subduction-related and oceanic island volcanic rocks show disequilibria between ^{210}Pb and ^{226}Ra (Fig. 12.14) [*Berlo and Turner*, 2010] has been attributed to the effects of degassing superimposed on the potential melting effects discussed above for MORB. Much of the work to date using U-series to study degassing processes has therefore focused on disequilibrium between ^{226}Ra and ^{210}Pb where ^{222}Rn (half-life 3.82 days) is the only intermediate daughter nuclide between

the pair that has a half-life longer than a few minutes. The short half-life of ^{210}Pb restricts these studies to very recent eruptions. Because radon will typically partition into the vapor phases during degassing of magmas, degassing can disrupt the decay series, producing disequilibrium. The difficulty of simultaneously measuring a gas phase and a solid or liquid phase in the same system makes use of the ^{226}Ra–^{210}Pb pair more attractive than direct measurement of ^{226}Ra–^{222}Rn disequilibria.

Degassing can lead to either ^{210}Pb deficits or ^{210}Pb excesses (i.e., $(^{210}$Pb$)/(^{226}$Ra$) < 1$ or >1, respectively) in volcanic rocks, depending on whether the volatile phase escapes from the magma that is eventually erupted (producing ^{210}Pb deficits) or accumulates in the magma (^{210}Pb excesses) [e.g., *Condomines et al.*, 2010]. The general principle is that ^{222}Rn will be produced from decay of ^{226}Ra in a melt phase, and will in turn decay to produce ^{210}Pb. However, because Rn is volatile, it will be strongly partitioned into a gas phase, whereas both Ra and Pb will be preferentially retained in the silicate melt relative to Rn. Therefore, if there is differential movement of the exsolved gas and the liquid, Rn will be transported from its source region, leaving the source depleted in ^{222}Rn—and as a consequence, its daughter ^{210}Pb—until secular equilibrium between ^{226}Ra and its daughters ^{222}Rn and ^{210}Pb is once again achieved. If the exsolved gas bubbles are transported to a different melt or a different region in the same magma body, decay of the ^{222}Rn within the bubbles will produce ^{210}Pb, which will partition into the coexisting silicate liquid, producing an excess of ^{210}Pb relative to the ^{226}Ra present in the

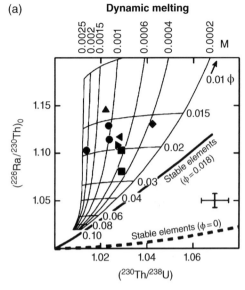

(a) **Dynamic melting**

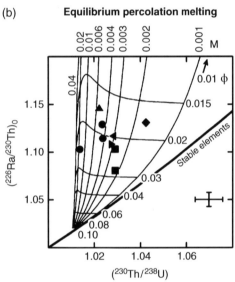

(b) **Equilibrium percolation melting**

Fig. 12.13. Calculations for (a) dynamic melting and (b) melting by equilibrium porous flow for data from Kilauea Volcano, Hawaii, contoured for different values of melting rate (M: kg/m³/year) and melt-zone porosity (φ: volume fraction). (Source: *Pietruszka et al.* [2001]. Reproduced with permission of Elsevier.)

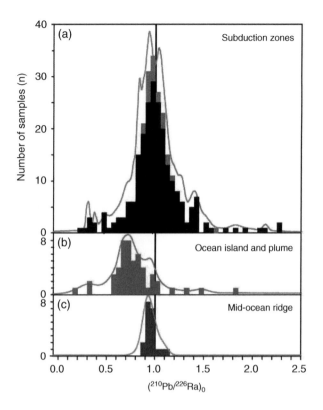

Fig. 12.14. Histograms of $(^{210}Pb)/(^{226}Ra)_0$ in magma erupted at (a) subduction zones, (b) ocean islands and (c) mid-ocean ridges. Vertical gray line shows secular equilibrium, and the curves along the histograms show the cumulative probability, which takes into account the standard deviations of each sample. (Source: *Berlo and Turner* [2010]. Reproduced with permission of Elsevier.)

melt. The extent to which ^{210}Pb will be depleted in a magma which has experienced gas loss and/or enriched in a magma which has experienced gas accumulation depends on the proportion of ^{222}Rn effectively extracted from the degassing magma and accumulated in a different magma, the mass ratio of degassing and accumulating magmas, and the residence time of the magma in the degassing part of the reservoir [*Condomines et al.*, 2010]. Early models required that the degassing magma be 100–1000 times the mass of the accumulating magma in order to explain the observed ratios [e.g., *Berlo et al.*, 2006; *Reagan et al.*, 2006], but more recent models suggest that steady fluxes within

a closed-system model (no loss or gain of Rn atoms from outside the system; implying that the gas moves from one region of a reservoir to another but does not leave the reservoir) can produce significant excesses without requiring extreme mass ratios (Fig. 12.15) [*Condomines et al.*, 2010]. Testing the extent to which these models reflect processes actually happening beneath volcanoes will require additional study and careful comparison of the patterns of disequilibria with other trace-element signatures [e.g., *Berlo and Turner*, 2010; *Condomines et al.*, 2010].

Magma differentiation and storage times

One approach to using U-series disequilibria to constrain magma storage and differentiation times is to compare disequilibria in magmas of different composition that are erupted from the same volcanic system. For example, disequilibria would be expected to decrease with time in a closed-system magma reservoir, which would predict that more evolved magmas would be closer to secular equilibrium than less evolved magmas. One example of this kind of an approach is shown in Fig. 12.16 [*Turner et al.*, 2004; *Turner and Costa*, 2007]. The decrease in $(^{226}Ra)/(^{230}Th)$ with increasing Th concentration (an index of differentiation) can be interpreted as closed-system evolution over thousands of years. However, caution must be used when interpreting whole-rock

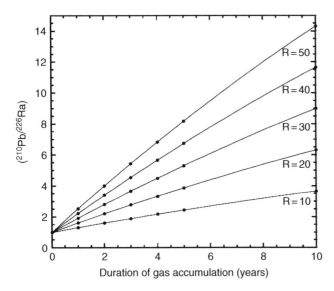

Fig. 12.15. Diagram showing the evolution of $(^{210}Pb)/(^{226}Ra)$ over time in a magma accumulating Rn due to gas movement in a closed system as a function of time. R represents the ratio of degassing magma to accumulating magma. Significant excesses of ^{210}Pb can develop on timescales of years in a closed system, even without requiring very large ratios of degassing to accumulating magma [*Condomines et al.*, 2010]. (Source: *Condomines et al.* [2010]. Reproduced with permission of Elsevier.)

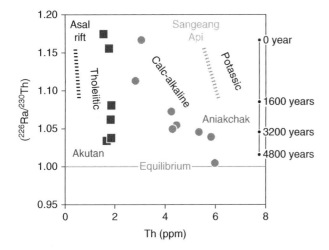

Fig. 12.16. Diagram showing $(^{226}Ra)/(^{230}Th)$ as a function of the differentiation of the magma. Here, concentration of thorium, an incompatible element, is shown as an index of differentiation, with higher concentrations of Th indicating more differentiated magma. This pattern of decreasing disequilbria with increasing differentiation can be interpreted as a timescale of differentiation (axis on right side of diagram), but only in the case of closed-system evolution. (Source: *Turner and Costa* [2007]. Reproduced with permission of Mineralogical Society of America.) (*See insert for color representation of the figure.*)

data in this way, as mixing between magmas of different compositions and/or assimilation can also result in arrays which look similar to those expected for closed-system differentiation, and which may have no age significance. Another way to assess magma storage times, albeit indirectly, is to date crystals in volcanic rocks, as is discussed in the following section.

12.4.2.3 U-series dating of volcanic crystals

Crystal compositions reflect those of the magmas in which they grew, providing a rich archive of chemical variation and chemical changes within magma reservoirs. However, accessing that archive in order to understand the timescales of magmatic processes requires combining crystal ages with crystal-scale measurements of chemical variations, and therefore has only recently become widely used as the analytical techniques necessary for such measurements have become more broadly available. Originally, crystals in volcanic rocks were viewed as belonging to two classes: *phenocrysts* (macroscopic crystals that formed from the liquid in which they erupted), and *xenocrysts* (crystals inherited by assimilation of the crust surrounding a magma reservoir). Recently, an intermediate category—that of crystals that did not form from the liquid which brought them to the surface, but which nevertheless formed within the broader magmatic system rather than being incorporated from preexisting crust—has been recognized. These crystals have been termed *antecrysts* (although the terminology is not universally accepted). In recent years, the importance of such antecrysts has become more apparent, as they often record a long history of processes within a given magmatic system, and different crystals (or different crystal populations) may have widely differing histories. Thus it is not surprising in retrospect that the earliest measurements of U-series crystal ages for a volcanic rock and construction of a ^{238}U–^{230}Th isochron, originally intended to provide information about the age of eruption, produced crystal ages that were significantly older than eruption [*Kigoshi*, 1967; *Allègre*, 1968].

In this section, we present two case studies to illustrate how U-series crystal ages are calculated, the factors that may affect the age calculations, the complexities that must be considered in interpreting the ages, and the implications of the data in terms of magma reservoir processes. The first case study, of Mount St. Helens, WA, focuses on major-phase dating with additional information provided by dating of accessory phases. The second case study, for Yellowstone Caldera, WY, focuses on zircon dating with additional information provided by major-phase dating. Finally, we discuss some general findings and implications for magmatic processes that have been derived from recent compilations of, and new approaches to using, U-series crystal age data.

Case study 1: Mount St. Helens

Crystals in volcanic rocks erupted from Mount St Helens were among the first to be dated using U–Th–Ra disequilibria analyzed by TIMS [*Volpe and Hammond*, 1991]. ^{238}U–^{230}Th dating of crystals in six volcanic rocks ranging in eruption age from ~2000 years before present to AD 1980 showed that the data for most samples formed linear arrays that were consistent with ages of 2–6 ka for the younger four samples, and ages of 27–34 ka for the older two samples (Fig. 12.17). An interpretation of magma residence of > 10 ka for the older magmas (assuming that the magma was coeval with the crystals), is not consistent with the observation of $(^{226}Ra)/(^{230}Th)$ disequilibria in the groundmass separates, and *Volpe and Hammond* [1991] interpreted the data

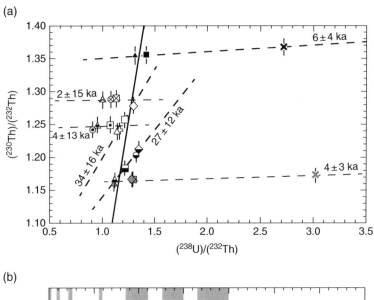

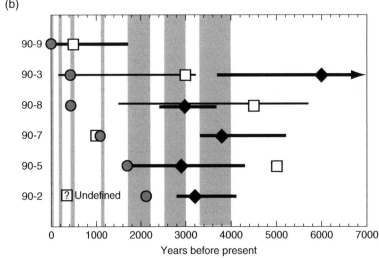

Fig. 12.17. (a) ^{238}U–^{230}Th isochron diagram and (b) summary of ^{230}Th–^{226}Ra ages for Mount St. Helens samples [*Volpe and Hammond*, 1991; *Cooper and Reid*, 2003]. In (b), filled diamonds represent "best-estimate" model ages for plagioclase considering the effects of Ra–Ba partitioning, and thick bars represent uncertainties including a range of temperatures and effects of inclusions in the mineral separates. Thin lines indicate model ages for pyroxene. Gray circles represent eruption ages and open squares represent mineral–mineral isochron ages that do not account for differences in partitioning.

for the two oldest eruptions to reflect mixing of young magma with older crystals. Volpe and Hammond also measured Th–Ra disequilibria and calculated isochron ages for the same mineral separates and obtained ages of 500–5000 years for most samples, with the Th–Ra ages of crystals in the two oldest eruptions being undefined. They also noted that ^{226}Ra/Ba ratios in the groundmass samples were higher than would be predicted with reference to the minerals (i.e., the groundmass point on a Ba-normalized Th–Ra isochron diagram plotted above the line connecting mineral phases for a given sample), which suggested open-system behavior of Ra, and in particular suggested a late-stage addition of Ra to the magma after crystal formation but prior to eruption. However, they did not account for those differences when calculating ages because the differences between partitioning of Ra and Ba had not yet been documented. These data were

reevaluated in light of the new information about Ra–Ba partitioning, and by accounting for both the effects of inclusions of other phases in the mineral separates and for the effects of differential partitioning of Ra and Ba, model Th–Ra ages of a few thousand years before present were obtained for all but one sample [*Cooper and Reid*, 2003] (Fig. 12.17). The correction for Ra/Ba fractionation fully accounted for the "anomalously" high Ra/Ba ratios in the groundmass, removing the need to call upon a late-stage addition of Ra to the system. However, obtaining finite Th–Ra ages (i.e. ages of <10 kyr) for crystals in the two older eruptions then presented a new interpretive challenge of explaining why the Th–Ra ages were apparently younger than the U–Th ages of the same mineral separates. Two models could explain this pattern of data: 1) mixing of young (2–4 ka) minerals with a melt with higher ^{230}Th/^{232}Th (and ^{226}Ra–^{230}Th

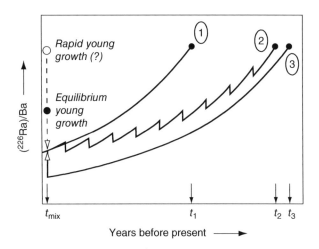

Fig. 12.18. Model showing how a given measured (^{226}Ra)/[Ba] ratio can reflect (1) a single stage of crystallization at t_1 (the model age) followed by decay, (2) multiple punctuated crystallization events that started at t_2, or (3) mixing of a single old population (crystallized at t_3) with young crystal population crystallized just prior to the mixing event. (Source: *Cooper and Reid* [2003]. Reproduced with permission of Elsevier.)

disequilibrium), or 2) mixing of minerals of different age in the mineral separates, whether through progressive crystallization or mixing of separate populations. The second model is illustrated in Fig. 12.18, which shows that a given measured ^{226}Ra/Ba ratio can be produced by either a single-stage crystallization event followed by ageing of the crystals, or by mixing of young crystallization with older crystals. Considering that trace-element data show considerable heterogeneity within the crystals in each eruption, and that further work on the 2004–2008 eruption showed ^{238}U–^{230}Th isotopic heterogeneity in crystals from different phases of a single eruption [*Cooper and Donnelly*, 2008], the interpretation of multi-stage crystallization seems the most likely.

^{238}U–^{230}Th ages of zircon have also been measured in samples erupted from Mount St. Helens over the past ~300 kyr [*Claiborne et al.*, 2010]. Most of the zircon grains in this suite (15 of 18 samples) lacked any zircon spot ages within error of the eruption age for a given samples, and in many cases the youngest spot ages were tens of kyr older than eruption. In each eruption, zircon of tens or even hundreds of thousands of years prior to eruption were scavenged and brought to the surface. These data suggest that the zircon capture a different (and longer-term) history of the magma reservoir, dominantly recording conditions relevant to the intrusive history of the magma system [*Claiborne et al.*, 2010]. Thus, the zircon and major phase records may provide complementary views on magma reservoirs.

Case study 2: Yellowstone

Yellowstone caldera, Wyoming, is a classic example of a large-volume, long-lived silicic magmatic system. It has generated three caldera-forming eruptions at 2.059 Ma, 1.285 Ma, and 639 ka, which produced the Huckleberry Ridge Tuff, the Mesa Falls Tuff, and the Lava Creek Tuff, respectively. In addition, the Yellowstone magmatic system has produced many intracaldera and postcaldera eruptions during its history [e.g., *Christiansen*, 2001; *Lanphere et al.*, 2002; *Christiansen et al.*, 2007]. Yellowstone caldera has been studied extensively, providing an excellent context for the crystal age data discussed here. The Yellowstone magmatic system has undergone systematic changes in oxygen-isotope composition over time, with abrupt drops in δ^{18}O immediately after caldera-forming eruptions followed by a rise in δ^{18}O of the erupted material between caldera-forming eruptions [e.g., *Hildreth et al.*, 1984, 1991; *Bindeman and Valley*, 2001; *Bindeman et al.*, 2008]. These data, together with other isotopic and trace-element data, indicate that the Yellowstone magmatic system incorporates varying amounts of low-δ^{18}O material, likely hydrothermally altered plutonic material and/or crustal material, mixed with a mantle-derived component. However, despite a wealth of age, chemical, and isotopic data for the rhyolites erupted over Yellowstone's history, the specific mechanism of producing rhyolites and the physical state of the magma reservoir remain debated. End-member models range from a dominantly subsolidus magma reservoir that experiences rapid episodes of remelting to produce the erupted rhyolites [e.g., *Bindeman et al.*, 2008] to a long-lived, chemically coherent crystal mush from which melts are occasionally extracted and erupted [*Hildreth et al.*, 1991]. In this case study we focus on the eruptions postdating the Lava Creek Tuff, as they are young enough to employ U-series dating.

Eruptions postdating the Lava Creek Tuff have been divided into three episodes: the early Upper Basin Member rhyolites (erupted 516–479 ka), the late Upper Basin Member rhyolites (erupted ~255 ka), and the Central Plateau Member (erupted 170–70 ka) [*Christiansen et al.*, 2007]. The early Upper Basin Member rhyolites have significant heterogeneity in δ^{18}O of the crystals, which has been interpreted to reflect generation of heterogeneous, independent batches of magma through remelting of volcanic and plutonic rocks predating Lava Creek Tuff and earlier-erupted Upper Basin Member rhyolites [*Bindeman et al.*, 2008; *Girard and Stix*, 2010; *Watts et al.*, 2012]. In contrast, the Central Plateau Member rhyolites show limited heterogeneity in δ^{18}O in the minerals but show systematic variation in whole-rock trace-element and isotopic compositions, compositions of major mineral phases, and geothermometry over time [*Hildreth et al.*, 1991; *Vazquez et al.*, 2009; *Girard and Stix*, 2010; *Watts et al.*, 2012], suggesting coherent chemical evolution of the magma reservoir over time. A more recent study combining age and trace-element data for zircon with age, trace-element, and Pb isotopic compositions of sanidine in Central Plateau Member rhyolites [*Stelten et al.*, 2015] produced several interesting observations, with implications for the structure of the magma reservoir and the processes of generation of the erupted bodies of magma.

First, ^{238}U–^{230}Th dating of zircon shows that spot ages of zircon interiors are both more diverse and on average older than ages of spots located on unpolished surfaces. Furthermore, trace-element compositions of the zircon interiors are also more diverse than those of the surfaces, yet averages of both interiors and surfaces show systematic changes in trace-element chemistry that mirror those in the glasses (Fig. 12.19). Second, bulk sanidine separates for a given eruption are younger than zircon interiors but coeval with zircon surfaces and within error of eruption age (Fig. 12.19). In addition, Pb isotopic compositions of individual sanidine crystals measured by LA-ICP-MS show that almost all the sanidine crystals in each eruption are in Pb isotopic equilibrium with the host glass, suggesting that these crystals

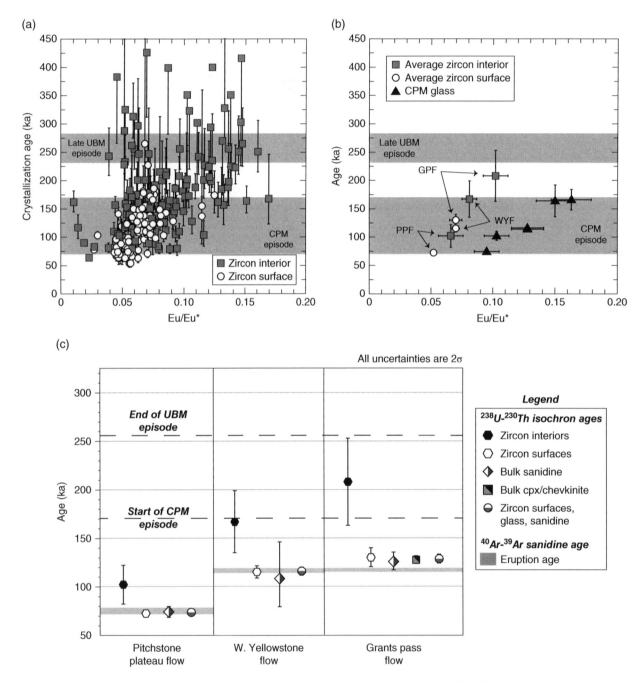

Fig. 12.19. Summary of U-series age data for Yellowstone case study; diagrams modified from *Stelten et al.* [2015]. (a) ^{238}U–^{230}Th zircon spot age versus Eu/Eu* for the same spots. Eu/Eu* is the europium anomaly (the deviation of Eu measured from that expected based on neighboring rare earth element concentrations), which is related to feldspar fractionation and magma differentiation. Spots located on zircon surfaces are shown as white circles, those located on polished interiors are shown as orange squares. (b) Average ages and Eu/Eu* of all zircon interior (orange squares) and surface (white circles) spot ages for individual eruptions studied, compared to glass data (black triangles) for the same eruptions. (c) Average zircon spot ages compared to sanidine ages and eruption ages (gray bars) for individual samples. (Source: *Stelten et al.* [2015]. Reproduced with permission of Oxford University Press.) (*See insert for color representation of the figure.*)

formed from the erupted melt. Liquids of the erupted bulk composition should be saturated in sanidine at all temperatures at which zircon would be saturated in the liquid, and the expectation is therefore that sanidine and zircon should on average be the same age. The lack of an older sanidine population indicates that the zircon interiors must have been separated from the coeval older sanidine population. Putting all of this evidence together, the data indicate a model where zircon started growing in a chemically heterogeneous, long-lived part of the reservoir, and then these zircon grains plus the rhyolitic liquid were separated from a crystal mush containing the sanidine that would have been cocrystallizing with the zircon interiors. This rhyolitic liquid then moved to a new location within the broader magma reservoir, where zircon surfaces and new sanidine crystals grew prior to eruption. The timescales of these different modes of storage are quite different—on average, zircon interiors are tens of thousands of years older than eruption and thus storage and crystallization of zircon interiors in a crystal mush occurs over tens of thousands of years, whereas the zircon surfaces and sanidine crystallized within error of eruption age and at most a few thousand years prior to eruption, indicating that the final stage of melt aggregation and storage prior to eruption was short.

General observations on U-series crystal ages

The two case studies presented above are by no means intended to represent a comprehensive view of all of the recent work using U-series crystal ages to study magmatic processes, and a number of review papers on the general topic of timescales of magmatic processes can provide more information for interested readers [*Hawkesworth et al.*, 2000; *Condomines et al.*, 2003; *Reid*, 2003, 2008; *Peate and Hawkesworth*, 2005; *Turner and Costa*, 2007; *Costa et al.*, 2008; *Schmitt*, 2011; *Cooper*, 2015]. However, these case studies illustrate several points that are representative of the U-series crystal data set as a whole.

(1) Individual spot ages for accessory phases (zircon or allanite) from a given eruption (and in some cases individual spots measured on single crystals) typically span a large range in age, from near eruption to hundreds of thousands of years prior to eruption (Fig. 12.20). Nevertheless, the range of ages from a given eruption is almost always within the age of activity of that magmatic center—in other words, few true xenocrysts (i.e. crystals derived from the country rock outside the magma reservoir) are present.

(2) The main peak (mode) in spot ages for accessory mineral interiors is typically tens of thousands of years prior to eruption for a given sample, whereas the main peak in ages for surfaces is always younger than the main peak in ages for interiors, and is typically within error of eruption age (Fig. 12.20).

(3) In contrast, less than half the ^{238}U–^{230}Th ages of bulk separates of major phases are more than 10 ka older than eruption age (Fig. 12.21).

(4) Most of the cases where ^{230}Th–^{226}Ra ages have been measured for major phases yield average ages of less than 10 ka

(Fig. 12.21). Although this is in some sense by necessity, since any crystals older than 10 ka will be in secular equilibrium with indeterminate ages, most of the measurements of ^{230}Th–^{226}Ra disequilibria are not within error of secular equilibrium (i.e., not shown as >10 ka in Fig. 12.21).

(5) In many cases where ^{238}U–^{230}Th ages and ^{230}Th–^{226}Ra ages have been measured in the same samples, they give discordant apparent ages (Fig. 12.21), indicating multistage crystallization and/or mixing of multiple crystal populations.

(6) There are relatively few cases where accessory phases and major phases have been dated in the same samples. In some cases the ages of accessory and major phases agree, and in some cases they do not [*Cooper*, 2015]. In most of these cases, as with each of the case studies discussed above, the major phases and the accessory phases appear to be recording different stages and/or locations of crystal and magma storage within the broader reservoir system.

These general observations provide a great deal of information about magmatic processes. Originally interpreted in terms of magma residence times, U-series ages of both major and accessory phases are now thought to be dominantly recording information about the longer-term history of the magma reservoir. Within this context, the crystals appear to be recording complex histories of storage and rejuvenation of largely crystalline parts of a reservoir, mixing of different magmas, and a much shorter history of final amalgamation and storage of melt-dominated bodies prior to eruption.

12.4.2.4 Dating eruptions

Dating the timing of volcanic eruptions has long been a goal of geochronology, partly because of its importance in understanding volcanic history, and partly because of the use of volcanic ash layers as stratigraphic markers. Two broad approaches have been applied to U-series dating of volcanic eruptions:

(1) measuring U-series disequilibria in whole-rock samples can provide constraints on eruption age, if the initial disequilibria at the time of eruption can be constrained;

(2) measuring U-series ages of minerals in volcanic rocks can provide the eruption age, if the minerals can be reasonably assumed to have crystallized essentially syneruptively (e.g., microlites formed during eruption).

Using measurements of U-series disequilibria in whole-rock samples to date eruptions requires knowledge of the initial disequilibria at the time of eruption. One example of this approach is studies that attempt to date MORB by assuming that $(^{230}$Th)/$(^{238}$U), $(^{231}$Pa)/$(^{235}$U), or $(^{226}$Ra)/$(^{230}$Th) ratios are relatively constant for a given segment of a mid-ocean ridge, and analyzing young, on-axis samples as representative of the "zero-age" average. This approach was applied to MORB from the Juan de Fuca and Gorda ridges [*Rubin and Macdougall*, 1990; *Goldstein et al.*, 1992, 1993, 1994; *Cooper et al.*, 2003], with good agreement between ages obtained from the different parent–daughter pairs. However, recent work analyzing multiple closely spaced samples along the East Pacific Rise (EPR) [*Sims et al.*, 2002] and

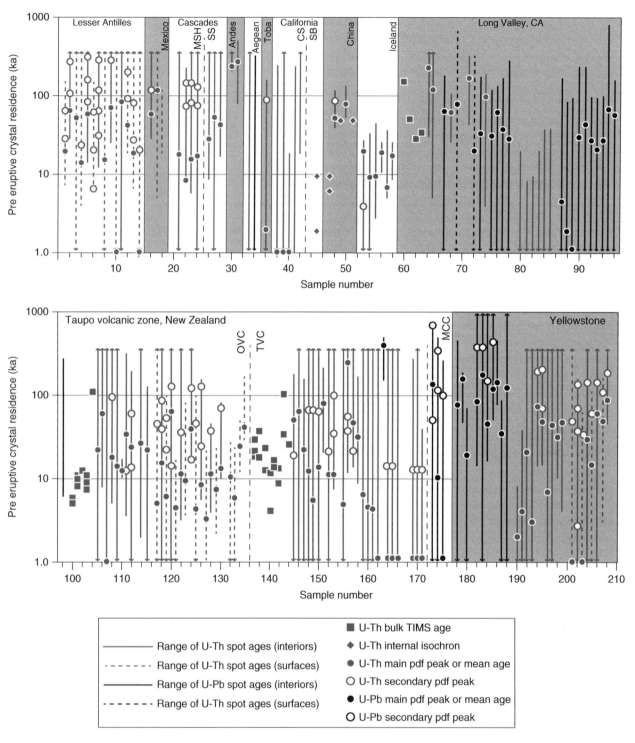

Fig. 12.20. Compilation of ^{238}U–^{230}Th and U–Pb ages of accessory minerals in volcanic rocks from *Cooper* [2015], expressed as preeruptive crystal residence ages (crystal age minus eruption age). Most ages shown are model ages based on ^{238}U–^{230}Th analyses *in situ* compared to host glasses (orange circles), with some U–Pb analyses *in situ* (green circles). Also shown are some internal isochron ages based on multiple spot analyses (orange diamonds), and a few bulk analyses of zircon by TIMS (orange squares). Each vertical array shows data for a single sample (or a group of closely related samples). Mean age or the dominant peak on probability density functions (PDF) is shown by solid circles; secondary PDF peaks are indicated by open circles. No circles are shown if no mean age or PDF is calculated in the original reference. Lines indicate range of ages for individual spot analyses; solid for interior analyses and dashed for unpolished surface analyses. Upward-pointing arrowheads on lines indicate analyses within error of secular equilibrium (or off-scale for U–Pb ages); downward-pointing arrowheads represent analyses within error of eruption age. Sample numbers are arbitrary. Geographic regions are separated by shaded bars, individual eruptive centers within these areas are separated by dashed black lines and labeled as follows: MSH, Mount St. Helens; SS, South Sister; CS, Coso; SB, Salton Buttes; OVC, Okataina Volcanic Center; TVC, Taupo Volcanic Center; MCC, Mangakino Caldera Center. (Source: *Cooper* [2015]. Reproduced with permission of The Geological Society.) (*See insert for color representation of the figure.*)

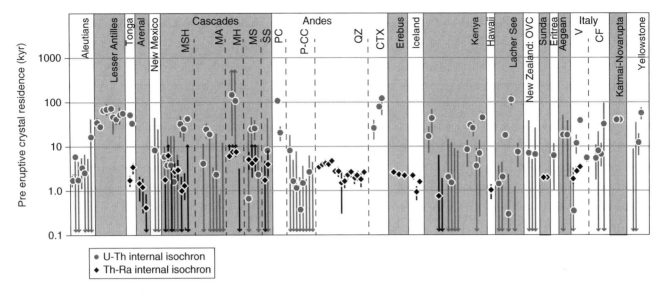

Fig. 12.21. Compilation of ^{238}U–^{230}Th and ^{230}Th–^{226}Ra ages of bulk mineral separates of major phases, expressed as the pre-eruptive residence age (bulk mineral age minus the eruption age). Error bars for U-series crystal residence ages indicate maximum and minimum residence ages calculated by combining the 2σ uncertainty on the crystal age with the 2σ uncertainty on the eruption. Arrows on error bars indicate samples with ages within error of eruption age (downward-pointing arrows) or indeterminate age maxima (upward-pointing arrows). Mean or best-estimate ages are shown by light blue filled circles (^{238}U–^{230}Th ages) or dark blue filled diamonds (^{230}Th–^{226}Ra ages); lines without symbols indicate samples for which only a maximum and minimum age estimate were available. Regions are labeled, and individual volcanic centers within each region are abbreviated as follows: MSH, Mount St. Helens; MA, Mound Adams; MH, Mount Hood; MS, Mount Shasta; SS, South Sister; PC, Parinacota; P-CC, Puyehue-Cordon Caulle; QZ, Quizapu; CTX, Cotopaxi; V, Vesuvius; CF, Campei Flegrei. (Source: *Cooper* [2015]. Reproduced with permission of The Geological Society.) (*See insert for color representation of the figure.*)

observations of disequilibria in off-axis EPR lavas [*Sims et al.*, 2003] have shown that the assumption of constant initial ratios is not always valid, even in a restricted geographic location.

For very young eruptions, repeated measurement of ^{210}Pb-^{210}Po disequilibria after sample collection can constrain the eruption age. Assuming that Po is completely degassed during eruption (a reasonable assumption given its volatility), measuring ingrowth of ^{210}Po at a sequence of times after eruption can produce an ingrowth curve which, when extrapolated back to zero initial ^{210}Po, will provide an eruption age. This method has been successfully applied to MORB [*Rubin et al.*, 1994], but is restricted in its application due to the very short (~2 years) time frame for return of the ^{210}Po–^{210}Pb pair to secular equilibrium.

Dating eruptions using U-series disequilibria is also possible through dating crystals that can be shown to have formed shortly prior to or during eruptions. For example, *Sims et al.* [2007] separated groundmass minerals from Quaternary volcanic rocks in the Zuni-Bandera volcanic field in New Mexico, and determined a ^{238}U–^{230}Th isochron age that was consistent with cosmogenic ^{3}He exposure age dating (after correction for erosion rates). Another recent approach to dating eruptions by using volcanic crystals is to date the unpolished surfaces of zircon or allanite crystals in rhyolitic lavas, either on its own or combined with (U–Th)/He dating of the same crystals [e.g., *Schmitt et al.*, 2006; *Schmitt, 2011; Danisik et al.*, 2012; *Vazquez and Lidzbarski*, 2012]. The surface analyses capture the most recent stage of crystallization, which is likely to have continued up to the

point of eruption in zircon-saturated magmas, (even in cases where the interiors of the zircon crystals are significantly older), and the addition of (U–Th)/He dating to the surface analyses provides an independent test of the crystallization ages representing eruption ages.

12.5 SUMMARY

As the diversity of the applications discussed above demonstrates, U-series geochronology is a versatile tool for late Pleistocene and Quaternary studies. U-series dating has provided critical age control for studies in fields as varied as paleoclimate, anthropology, paleoseismology, paleontology, soil pedogenesis, Quaternary fault slip histories, dating of Quaternary volcanic eruptions, magma generation and transport, and magma reservoir processes (among many other applications for which there was no room to discuss here). Many advances in analytical procedures since the technique was developed have resulted in orders of magnitude improvements in the precision of U-series dating and the sample sizes required for analysis, opening up fundamentally new applications and questions that can be addressed using this technique. For example, improvements in precision and sensitivity for *in situ* techniques have led to significant improvements in our ability to deconvolve multiple stages of growth of igneous minerals, and to better constraints on the detrital correction for dating of carbonates and opals. Furthermore, the increased efficiency of analyses from MC-ICP-MS techniques and *in situ* techniques can lead to much larger data sets being generated, which opens up possibilities for attacking problems with statistical techniques. U-series disequilibria have been, and will undoubtedly remain, a widely used and important geochronologic tool in years to come.

12.6 REFERENCES

Allègre, C. L. (1968) [230]Th dating of volcanic rocks: a comment. *Earth and Planetary Science Letters* **5** 209–210.

Barnes, J. W., Lang, E. J., and Potratz, H. A. (1956) The ratio of ionium to uranium in coral limestone. *Science* **124**, 175–176.

Berlo, K. and Turner, S. (2010) [210]Pb–[226]Ra disequilibria in volcanic rocks. *Earth and Planetary Science Letters* **296**(3–4), 155–164. DOI: https://dx.doi.org/10.1016/j.epsl.2010.05.023.

Berlo, K., Turner, S., Blundy, J., Black, S., and Hawkesworth, C. (2006) Tracing pre-eruptive magma degassing using (^{210}Pb/[226]Ra) disequilibria in the volcanic deposits of the 1980–1986 eruption of Mount St. Helens. *Earth and Planetary Science Letters* **249**, 337–349.

Bernal, J.-P., Eggins, S. M., and McCulloch, M. T. (2005) Accurate *in situ* [238]U–[234]U–[232]Th–[230]Th analysis of silicate glasses and iron oxides by laser-ablation MC-ICP-MS. *Journal of Analytical and Atomic Spectrometry* **20**, 1240–1249.

Bernal, J.-P., Solari, L. A., Gomez-Tuena, A., *et al.* (2014) *In-situ* [230]Th/U dating of Quaternary zircons using LA-MCICPMS. *Quaternary Geochronology* **23**, 46–55.

Bindeman, I. N. and Valley, J. W. (2001) Low-δ^{18}O rhyolites from Yellowstone: magmatic evolution based on analyses of zircons and individual phenocrysts. *Journal of Petrology* **42**, 1491–1517.

Bindeman, I. N., Fu, B., Kita, N. T., and Valley, J. W. (2008) Origin and evolution of silicic magmatism at Yellowstone based on ion microprobe analysis of isotopically zoned zircons. *Journal of Petrology* **49**(1), 163–193. DOI: 10.1093/petrology/egm075.

Bischoff, J. L. and Shamp, D. D. (2003) The Sima de los Huesos Hominids Date to Beyond U/Th Equilibrium (>350 kyr) and Perhaps to 400–500 kyr: New Radiometric Dates. *Journal of Archaeological Science* **30**, 275–280. DOI: 10.1006/jasc.2002.0834.

Bischoff, J. L., Williams, R. W., Rosenbauer, R. J., *et al.* (2007) High-resolution U-series dates from the Sima de los Huesos hominids yields: implications for the evolution of the early Neanderthal lineage. *Journal of Archaeological Science* **34**(5), 763–770. DOI: https://dx.doi.org/10.1016/j.jas.2006.08.003.

Blundy, J. and Wood, B. (2003) Mineral-melt partitioning of uranium, thorium and their daughters. In *Uranium-Series Geochemistry*, Bourdon, B., Henderson, G. M., Lundstrom, C. C., and Turner, S. P. (eds), 59–123. *Reviews in Mineralogy and Geochemistry*, Vol **52**. Mineralogical Society of America and Geochemical Society. Washington, DC.

Bourdon, B. and Sims, K. W. W. (2003) U-series constraints on intraplate basaltic magmatism. In *Uranium-Series Geochemistry*, Bourdon, B., Henderson, G. M., Lundstrom, C. C., and Turner, S. P. (eds), 215–254. *Reviews in Mineralogy and Geochemistry*, Vol **52**. Mineralogical Society of America and Geochemical Society. Washington, DC.

Bourdon, B., Henderson, G. M., Lundstrom, C. C., and Turner, S. P. (2003) *Uranium-Series Geochemistry*. In: *Reviews in Mineralogy and Geochemistry*, Vol. **52**. Mineralogical Society of America and Geochemical Society. Washington, DC.

Cheng, H., Edwards, R. L., Wang, Y., *et al.* (2006) A penultimate glacial monsoon record from Hulu Cave and two-phase glacial terminations. *Geology* **34**(3), 217–220. DOI: 10.1130/G22289.1.

Cheng, H., Edwards, R. L., Shen, C.-C., *et al.* (2013) Improvements in [230]Th dating, [230]Th and [234]U half-life values, and U–Th isotopic measurements by multi-collector inductively coupled plasma mass spectrometry. *Earth and Planetary Science Letters* **371–372**, 82–91.

Christiansen, R. L. (2001) The Quaternary and Pliocene Yellowstone Plateau Volcanic Field of Wyoming, Idaho, and Montana. *US Geological Survey Professional Paper* **729-G**,145.

Christiansen, R. L., Lowenstern, J. B., Smith, R. B., *et al.* (2007) Preliminary assessment of volcanic and hydrothermal hazards in Yellowstone National Park and vicinity. *US Geological Survey Open-File Report* **1071**, 94.

Claiborne, L. L., Miller, C. F., Flanagan, D. M., Clynne, M. A., and Wooden, J. L. (2010) Zircon reveals protracted magma storage and recycling beneath Mount St. *Helens. Geology* **38**(11), 1011–1014. DOI: 10.1130/G31285.1.

Condomines, M., Gauthier, P. J., and Sigmarsson, G. (2003) Time-scales of magma chamber processes and dating of young volcanic rocks. In *Uranium-Series Geochemistry*, Bourdon, B., Henderson, G. M., Lundstrom, C. C., and Turner, S. P. (eds), 125–174. *Reviews in Mineralogy and Geochemistry*, Vol **52**. Mineralogical Society of America and Geochemical Society. Washington, DC.

Condomines, M., Sigmarsson, O, and Gauthier, P. J. (2010) A simple model of [222]Rn accumulation leading to ^{210}Pb excesses in volcanic rocks. *Earth and Planetary Science Letters* **293**(3–4), 331–338. DOI: https://dx.doi.org/10.1016/j.epsl.2010.02.048.

Cooper, K. M. (2015) Timescales of crustal magma reservoir processes: insights from U-series crystal ages. In *Chemical, Physical and Temporal Evolution of Magmatic Systems*, Caricchi, L and Blundy, J. D. (eds), 141–174. Special Publications 422, Geological Society of London.

Cooper, K. M. and Donnelly, C. T. (2008) [238]U–[230]Th–[226]Ra Disequilibria in dacite and plagioclase from the 2004–2005 eruption of Mount St. Helens. In *A Volcano Rekindled: The Renewed Eruption at Mount St Helens (2004–2006)*, Sherrod, D. R., Scott, W. E., and Stauffer, P. H. (eds), 827–846. US Geological Survey Professional Paper 1750.

Cooper, K. M. and Reid, M. R. (2003) Re-examination of crystal ages in recent Mount St. Helens lavas: Implications for magma reservoir processes. *Earth and Planetary Science Letters* **213**(1–2), 149–167.

Cooper, K. M. and Reid, M. R. (2008) Uranium-series crystal ages. In *Minerals, Inclusions and Volcanic Processes*, Putirka, K. and Tepley III, F. J. (eds), 479–544. *Reviews in Mineralogy and Geochemistry*, Vol. **69**. Mineralogical Society of America and Geochemical Society, Washington, DC.

Cooper, K. M., Reid, M. R., Murrell, M. T., and Clague, D. A. (2001) Crystal and magma residence at Kilauea Volcano, Hawaii: [230]Th–[226]Ra dating of the 1955 east rift eruption. *Earth and Planetary Science Letters* **184**(3–4), 703–718.

Cooper, K. M., Goldstein, S. J., Sims, K. W. W., and Murrell, M. T. (2003) Uranium-series chronology of Gorda Ridge volcanism: new evidence from the 1996 eruption. *Earth and Planetary Science Letters* **206**(3–4), 459–475.

Costa, F., Dohmen, R., and Chakraborty, S. (2008) Time scales of magmatic processes from modeling the zoning patterns of crystals. In *Minerals, Inclusions and Volcanic Processes*, Putirka, K. and Tepley III, F. J. (eds), 545–594. *Reviews in Mineralogy and Geochemistry*, Vol. **69**. Mineralogical Society of America and Geochemical Society, Washington, DC.

Danisik, M., Shane, P., Schmitt, A. K., *et al.* (2012) Re-anchoring the late Pleistocene tephrochronology of New Zealand based on concordant radiocarbon ages and combined U-238/Th-230 disequilibrium and (U–Th)/He zircon ages. *Earth and Planetary Science Letters* **349**, 240–250. DOI: 10.1016/j.epsl.2012.06.041.

Edwards, R. L., Cutler, K. B., and Cheng, H. (2003a) Geochemical evidence for Quaternary sea-level changes. *Treatise on Geochemistry* **6**, 343–364.

Edwards, R. L., Gallup, C. D., and Cheng, H. (2003b) Uranium-series dating of marine and lacustrine carbonates. In *Uranium-Series Geochemistry*, Bourdon, B., Henderson, G. M., Lundstrom, C. C., and Turner, S. P. (eds), 363–405. *Reviews in Mineralogy and Geochemistry*, Vol **52**. Mineralogical Society of America and Geochemical Society. Washington, DC.

Fabbrizio, A., Schmidt, M. W., Gunther, D., and Eikenberg, J. (2008) Experimental determination of radium partitioning between leucite and phonolite melt and ^{226}Ra-disequilibrium crystallisation ages of leucite. *Chemical Geology* **255**, 377–387.

Fabbrizio, A., Schmidt, M. W., Gunther, D., and Eikenberg, J. (2009) Experimental determination of Ra mineral/melt partitioning for feldspars and ^{226}Ra-disequilibrium crystallization ages of plagioclase and alkali-feldspar. *Earth and Planetary Science Letters* **280**, 137–148.

Fabbrizio, A., Schmidt, M. W., Gunther, D., and Eikenberg, J. (2010) Ra-partitioning between plhogopite and silicate melt and ^{226}Ra/Ba-^{230}Th/Ba isochrons. *Lithos* **114**, 121–131.

Fletcher, K. E. K., Sharp, W. D., Kendrick, K. J., Behr, W. M., Hudnut, K. W., Hanks, T. C. (2010) Th-230/U dating of a late Pleistocene alluvial fan along the southern San Andreas fault. *Geological Society of America Bulletin* **122**(9–10), 1347–1359. DOI: 10.1130/B30018.1.

Fletcher, K. E. K., Rockwell, T. K., and Sharp, W. D. (2011) Late Quaternary slip rate of the southern Elsinore fault, Southern California: Dating offset alluvial fans via Th-230/U on pedogenic carbonate. *Journal of Geophysical Research-Earth* **116**. Artn F02006. DOI: 10.1029/2010jf001701.

Girard, G. and Stix, J. (2010) Rapid extraction of discrete magma batches from a large differentiating magma chamber: the Central Plateau Member rhyolites, Yellowstone Caldera, Wyoming. *Contributions to Mineralogy and Petrology* **160**(3), 441–465. DOI: 10.1007/s00410-009-0487-1.

Gold, R. D., Cowgill, E., Arrowsmith, J. R., *et al.* (2011) Faulted terrace risers place new constraints on the late Quaternary slip rate for the central Altyn Tagh fault, northwest Tibet. *Geological Society of America Bulletin* **123**(5–6), 958–978. DOI: 10.1130/B30207.1.

Goldstein, S. J. and Stirling, C. H. (2003) Techniques for measuring uranium-series nuclides: 1992–2002. In *Uranium-Series Geochemistry*, Bourdon, B., Henderson, G. M., Lundstrom, C. C., and Turner, S. P. (eds), 23–57. *Reviews in Mineralogy and Geochemistry*, Vol **52**. Mineralogical Society of America and Geochemical Society. Washington, DC.

Goldstein, S. J., Murrell, M. T., Janecky, D. R., Delaney, J. R., and Clague, D. A. (1992) Geochronology and petrogenesis of MORB from the Juan de Fuca and Gorda Ridges by ^{238}U-^{230}Th disequilibria. *Earth and Planetary Science Letters* **109**, 255–272.

Goldstein, S. J., Murrell, M. T., and Williams, R. W. (1993) ^{231}Pa and ^{230}Th chronology of mid-ocean ridge basalts. *Earth and Planetary Science Letters* **115**, 151–159.

Goldstein, S. J., Perfit, M. R., Batiza, R., Fornari, D. J., and Murrell, M. T. (1994) ^{226}Ra and ^{231}Pa systematics of axial MORB, crustal residence ages, and magma chamber characteristics at 9–10 N East Pacific Rise. *Mineralogical Magazine* **58A**(A-K), 335–336.

Hawkesworth, C. J., Blake, S., Evans P, *et al.* (2000) Time scales of crystal fractionation in magma chambers—integrating physical, isotopic and geochemical perspectives. *Journal of Petrology* **41**(7), 991–1006.

Hildreth, W., Christiansen, R. L., and O'Neil, J. R. (1984) Catastrophic isotopic modification of rhyolitic magma at times of caldera subsidence, Yellowstone Plateau Volcanic Field. *Journal of Geophysical Research* **89**(B10), 8339–8369.

Hildreth, W., Halliday, A. N., and Christiansen, R. L. (1991) Isotopic and chemical evidence concerning the genesis and contamination of basaltic and rhyolitic magma beneath the Yellowstone Plateau Volcanic Field. *Journal of Petrology* **32**(1), 63–138.

Ivanovich, M. and Harmon, R. S. (1992) *Uranium-series Disequilibria: Applications to Earth, Marine, and Environmental Sciences.* Oxford University Press, Oxford, 910 pp.

Kigoshi, K. (1967) Ionium dating of igneous rocks. *Science* **156**, 932–934.

Ku, T.-L., Bull, W. B., Freeman, S. T., and Knauss, K. G. (1979) Th230-U234 dating of pedogenic carbonates in gravelly desert soils of Vidal Valley Southeastern California. *Geological Society of America Bulletin* **90**, 1063–1073.

Lanphere, M. A., Champion, D. E., Christiansen, R. L., Izett, G. A., and Obradovich, J. D. (2002) Revised ages for tuffs of the Yellowstone Plateau volcanic field: assignment of the Huckleberry Ridge Tuff to a new geomagnetic polarity event. *Geological Society of America Bulletin* **114**(5), 559–568.

Ludwig, K. R. and Paces, J. B. (2002) Uranium-series dating of pedogenic silica and carbonate, Crater Flat, Nevada. *Geochimica et Cosmochimica Acta* **66**(3), 487–506.

Lundstrom, C. C., Hoernle, K., and Gill, J. (2003) U-Series disequilibri in volcanic rocks from the Canary Islands: plume versus lithospheric melting. *Geochimica et Cosmochimica Acta* **67**(21), 4153–4177.

Maher, K., Wooden, J. L., Paces, J. B., and Miller, D. M. (2007) ^{230}Th–U dating of surficial deposits using the ion microprobe

(SHRIMP-RG): a microstratigraphic perspective. *Quaternary International* **166**, 15–28.

McKenzie, D. (1985) ^{230}Th–^{238}U disequilibria and the melting processes beneath ridge axes. *Earth and Planetary Science Letters* **72**, 149–157.

McKenzie, D. (2000) Constraints on melt generation and transport from U-series activity ratios. *Chemical Geology* **162**, 81–94.

Paces, J. B., Neymark, L. A., Wooden, J. L., Persing, H. M. (2004) Improved spatial resolution for U-series dating of opal at Yucca Mountain, Nevada, USA, using ion-microprobe and microdigestion methds. *Geochimica et Cosmochimica Acta* **68**, 1591–1606.

Peate, D. W. and Hawkesworth, C. J. (2005) U series disequilibria: insights into mantle melting and the timescales of magma differentiation. *Reviews of Geophysics* **43**(1).

Pietruszka, A. J. and Garcia, M. O. (1999) A rapid fluctuation in the mantle source and melting history of Kilauea Volcano inferred from the geochemistry of its historical summit lavas (1790–1982). *Journal of Petrology* **40**(8), 1321–1342.

Pietruszka, A. J., Rubin, K. H., and Garcia, M. O. (2001) ^{226}Ra–^{230}Th–^{238}U disequilibria of historical Kilauea lavas (1790–1982) and the dynamics of mantle melting within the Hawaiian plume. *Earth and Planetary Science Letters* **186**, 15–31

Potter, E.-K., Stirling, C. H., Weichert, U. H., Halliday, A. N., and Spotl, C. (2005) Uranium-series dating of corals *in situ* using laser-ablation MC-ICPMS. *International Journal of Mass Spectrometry* **240**, 27–35.

Prideaux, G. J., Roberts, R. G., Megirian, D., Westaway, K. E., Hellstrom, J. C., and Olley, J. M. (2007) Mammalian responses to Pleistocene climate change in southeastern Australia. *Geology* **35**(1), 33–36. DOI: 10.1130/g23070a.1.

Reagan, M. K., Tepley III, F. J., Gill, J. B., Wortel, M., and Garrison, J. (2006) Timescales of degassing and crystallization implied by ^{210}Po–^{210}Pb–^{226}Ra disequilibria for andesitic lavas erupted from Arenal volcano. *Journal of Volcanology and Geothermal Research* **157**, 135–146.

Reid, M. R. (2003) Timescales of magma transfer and storage in the crust. In *The Crust*, Vol. 3, Rudnick, R. L. (ed.), 167–193. Elsevier, Oxford.

Reid, M. R. (2008) How long does it take to supersize an eruption? *Elements* **4**, 23–28.

Reid, M. R., Coath, C. D., Harrison, T. M., and McKeegan, K. D. (1997) Prolonged residence times for the youngest rhyolites associated with Long Valley Caldera: ^{230}Th–^{238}U ion microprobe dating of young zircons. *Earth and Planetary Science Letters* **150**, 27–39.

Rubin, K. H. and Macdougall, J. D. (1990) Dating of neovolcanic MORB using (^{226}Ra/^{230}Th) disequilibrium. *Earth and Planetary Science Letters* **101**, 313–322.

Rubin, K. H., Macdougall, J. D., and Perfit, M. R. (1994) ^{210}Po–^{210}Pb dating of recent volcanic eruptions on the sea floor. *Nature* **368**, 841–844.

Rubin, K. H., van der Zander I, Smith, M. C., and Bergmanis, E. C. (2005) Minimum speed limit for ocean ridge magmatism from ^{210}Pb–^{226}Ra–^{230}Th disequilibria. *Nature* **437**, 534–538.

Schmitt, A. K. (2011) Uranium series accessory crystal dating of magmatic processes. *Annual Review of Earth and Planetary Sciences* **39**, 321–349.

Schmitt, A. K., Stockli, D. F., and Hausback, B. P. (2006) Eruption and magma crystallization ages of Las Tres Virgenes (Baja California) constrained by combined Th-230/U-238 and (U–Th)/He dating of zircon. *Journal of Volcanology and Geothermal Research* **158**(3–4), 281–295.

Sharp, W. D., Ludwig, K. R., Chadwick, O. A., Amundson, R., and Glaser, L. L. (2003) Dating fluvial terraces by Th-230/U on pedogenic carbonate, Wind River Basin, Wyoming. *Quaternary Research* **59**(2), 139–150.

Sieh, K., Natawidjaja, D. H., Meltzner, A. J., *et al.* (2008) Earthquake supercycles inferred from sea-level changes recorded in the corals of West Sumatra. *Science* **322**, 1674–1678.

Sigmarsson, O. (1996) Short magma chamber residence time at an Icelandic volcano inferred from U-series disequilibria. *Nature* **382**(6590), 440–442.

Sims, K. W. W., DePaolo, D. J., Murrell, M. T., *et al.* (1999) Porosity of the melting zone and variations in solid mantle upwelling rate beneath Hawaii: Inferences from ^{238}U–^{230}Th–^{226}Ra and ^{235}U–^{231}Pa disequilibria. *Geochimica et Cosmochimica Acta* **63** (23/24), 4119–4138.

Sims, K. W. W., Goldstein, S. J., Blichert-Toft, J., *et al.* (2002) Chemical and isotopic constraints on the generation and transport of melt beneath the East Pacific Rise. *Geochimica et Cosmochimica Acta* **66**(19), 3481–3504.

Sims, K. W. W., Blichert-Toft, J., Fornari, D., *et al.* (2003) Aberrant youth: chemical and isotopic constraints on the young off-axis lavas of the East Pacific Rise. *Geochemistry, Geophysics, Geosystems* **4**(10), 8621. DOI: 8610.1029/2002GC000443.

Sims, K. W. W., Ackert Jr., R. P., Ramos, F. C., Sohn, R. A., Murrell, M. T., and DePaolo, D. J. (2007) Determining eruption ages and erosion rates of Quaternary basaltic volcanism from combined U-series disequilibria and cosmogenic exposure ages. *Geology* **35**, 471–474.

Sims, K. W. W., Hart, S. R., Reagan, M. K., *et al.* (2008) ^{238}U–^{230}Th–^{226}Ra–^{210}Pb–^{210}Po, ^{232}Th–^{228}Ra and ^{235}U–^{231}Pa constraints on the ages and petrogenesis of Vailulu and Malumalu Lavas, Samoa. *Geochemistry, Geophysics, Geosystems* **9** (Q04003):10.1029/2007GC001651.

Sims, K. W. W., Pichat, S., Reagan, M. K., *et al.* (2013) On the time scales of magma genesis, melt evolution, crystal growth rates and magma degassing in the Erebus volcano magmatic system using the U-238, U-235 and Th-232 decay series. *Journal of Petrology* **54**(2), 235–271. DOI: 10.1093/Petrology/Egs068.

Soddy, F. (1910) Radioactivity. In *Annual Reports on the Progress of Chemistry*, Vol. 7, 257–286. The Chemical Society, London.

Spiegelman, M. and Elliott, T. (1993) Consequences of melt transport for uranium series disequilibria. *Earth and Planetary Science Letters* **118**, 1–20.

Stelten, M. E., Cooper, K. M., Vazquez, J. A., Calvert, A. T., and Glessner, J. J. G. (2015) Mechanisms and timescales of generating eruptible rhyolitic magmas at Yellowstone Caldera from zircon and sanidine geochronology and geochemistry. *Journal of*

Petrology **56**(8), 1607–1642. DOI: 10.1093/petrology/egv047.

Turner, S. and Costa, F. (2007) Measuring timescales of magmatic evolution. *Elements* **3**(4), 267–272. DOI: 10.2113/Gselements.3.4.267.

Turner, S., Black, S., and Berlo, K. (2004) Pb-210-Ra-226 and Ra-228–Th-232 systematics in young arc lavas: implications for magma degassing and ascent rates. *Earth and Planetary Science Letters* **227**(1–2), 1–16.

Turner, S. P., Bourdon, B., and Gill, J. (2003) Insights into magma genesis at convergent margins from U-series isotopes. In *Uranium-Series Geochemistry*, Bourdon, B., Henderson, G. M., Lundstrom, C. C., and Turner, S. P. (eds). *Reviews in Mineralogy and Geochemistry*, Vol **52**. Mineralogical Society of America and Geochemical Society. Washington, DC.

Vazquez, J. A. and Lidzbarski, M. I. (2012) High-resolution tephrochronology of the Wilson Creek Formation (Mono Lake, California) and Laschamp event using U-238–Th-230 SIMS dating of accessory mineral rims. *Earth and Planetary Science Letters* **357**, 54–67. DOI: 10.1016/j.epsl.2012.09.013.

Vazquez, J. A., Kyriazis, S. F., Reid, M. R., Sehler, R. C., and Ramos, F. C. (2009) Thermochemical evolution of young rhyolites at Yellowstone: evidence for a cooling but periodically replenished postcaldera magma reservoir. *Journal of Volcanology and Geothermal Research* **188**(1–3), 186–196 DOI: 10.1016/j.jvolgeores.2008.11.030.

Volpe, A. M. and Hammond, P. E. (1991) ^{238}U–^{230}Th–^{226}Ra disequilibria in young Mount St. Helens rocks: time constraint for magma formation and crystallization. *Earth and Planetary Science Letters* **107**, 475–486.

Wang, Y., Cheng, H., Edwards, R. L., *et al.* (2008) Millenial- and orbital-scale changes in the East Asian monsoon over the past 224,000 years. *Nature* **451**, 1090–1093. DOI: 10.1038/nature06692.

Wang, Y. J., Cheng, H., Edwards, R. L., *et al.* (2001) A high-resolution absolute-dated late Pleistocene monsoon record from Hulu Cave, China. *Science* **294**, 2345–2348.

Watts, K. E., Bindeman, I. N., and Schmitt, A. K. (2012) Crystal scale anatomy of a dying supervolcano: an isotope and geochronology study of individual phenocrysts from voluminous rhyolites of the Yellowstone caldera. Contributions to Mineralogy and Petrology **164**(1), 45–67 DOI:. DOI 10.1007/S00410-012-0724-X.

Williams, R. W. and Gill, J. B. (1989) Effects of partial melting on the uranium decay series. *Geochimica et Cosmochimica Acta* **53**, 1607–1619.

Winograd, I. J., Coplen, T. B., Landwehr, J. M., *et al.* (1992) Continuous 500,000-year Climate Record from Vein Calcite in Devils Hole, Nevada. *Science* **258**, 255–260.

Winograd, I. J., Landwehr, J. M., Coplen, T. B., *et al.* (2006) Devils Hole, Nevada, δ18O record extended to the mid-Holocene. *Quaternary Research* **66**(2), 202–212. DOI: https://dx.doi.org/10.1016/j.yqres.2006.06.003.

CHAPTER 13

Cosmogenic nuclides

13.1 INTRODUCTION

Cosmogenic nuclides are rare particles, often radioactive, that are produced by nuclear reactions initiated by cosmic rays. They have a variety of applications in geochronology, including dating of meteorites, radiocarbon dating of organic matter, tracing and dating water and soil in the environment, surface-exposure dating of landforms, burial dating of sediments, and erosion-rate determinations for both rocks and watersheds. This chapter covers the fundamentals of cosmogenic nuclide production and discusses some of the many ways that cosmogenic nuclides can be used in geochronology.

13.2 HISTORY

The history of cosmogenic nuclide geochronology can be traced to the 1940s and 1950s, when *Bauer* [1947] realized that helium measured in meteorites could be a product of cosmic-ray interactions rather than strictly that of radioactive decay. *Paneth et al.* [1952] then confirmed this theory by measuring a high $^3He/^4He$ ratio as predicted for cosmic-ray interactions; *Paneth et al.* [1953] subsequently showed that the helium content systematically decreased towards the interior of the large iron meteorite 'Carbo' (see Fig. 13.6), proving that the helium was produced from external radiation. After the relative production rates in space were established by comparing the build up of stable ^3He with short-lived tritium [*Begemann et al.*, 1957; *Fireman and Schwarzer*, 1957], the door was opened for measuring the exposure ages of meteorites in space using ^3He and a host of other cosmogenic nuclides [see *Honda and Arnold*, 1967].

At the same time that cosmogenic nuclides were being measured in meteorites, *Libby* [1946] suggested that cosmic rays would produce measurable quantities of the cosmogenic nuclides ^{14}C and ^3H in Earth's atmosphere. *Anderson et al.* [1947] then detected cosmogenic ^{14}C in methane derived from sewage from the city of Baltimore, Maryland, proving the incorporation of cosmogenic ^{14}C in the biosphere and setting off the field of radiocarbon dating of organic material [*Arnold and Libby*, 1949]. Other meteoric cosmogenic nuclides were soon detected, including ^7Be [*Arnold and Al-Salih*, 1955] and ^{10}Be [*Goel et al.*, 1956] in rainwater, as suggested by *Peters* [1955], and ^{26}Al and ^{10}Be in glacier ice [*McCorkell et al.*, 1967].

The measurement of cosmogenic nuclides produced *in situ* within mineral grains on Earth began with the pioneering measurements of ^{36}Cl by *Davis and Schaeffer* [1955], who measured ^{36}Cl by decay-counting tens of grams of chlorine separated from ~10 kg of a chlorine-rich rock collected at high altitude. They calculated a rock exposure age to secondary cosmic ray neutrons of about 24,000 years. Twenty years later, *Hampel et al.* [1975] measured ^{26}Al by decay-counting a 402-gram disc of Al_2O_3 chemically separated and purified from tens of kilograms of chert. The extremely large quantities of rock and painstaking chemistry required, though, severely limited the numbers of measurements that could be made by conventional decay-counting methods and the method languished until the development of accelerator mass spectrometry (AMS) and improvements in noble gas mass spectrometry.

AMS was first developed by *Müller* [1977] using the University of California cyclotron to measure tritium, followed closely by independent measurements of ^{14}C at the University of Rochester and McMaster University [*Bennett et al.*, 1977; *Nelson et al.*, 1977]. The general principle of AMS is that ions are accelerated to high energy (MeV) and stripped of electrons to reduce molecular interferences. Particles are then separated by mass and charge in a magnetic sector and passed through a series of additional selective elements such as electrostatic deflectors, velocity selectors (i.e., Wien filters), and/or gas-filled-magnets. The particles are then typically separated from interferences by ranging them in a ΔE-E detector, either solid-state or a gas ionization detector. AMS often allows isotope ratios to be measured at ratios as low as 10^{-16} in samples that contain as little as 10^4 atoms of the rare nuclide. (For a review of the instrumentation

Geochronology and Thermochronology, First Edition. Peter W. Reiners, Richard W. Carlson, Paul R. Renne, Kari M. Cooper, Darryl E. Granger, Noah M. McLean, and Blair Schoene.
© 2018 John Wiley & Sons Ltd. Published 2018 by John Wiley & Sons Ltd.

see *Tuniz et al.* [1998].) What this implies is that AMS can analyze cosmogenic nuclides in chemically separated targets of under a milligram rather than the tens or hundreds of grams required for decay counting. Rocks from Earth's surface can now be routinely dated using only grams of minerals rather than tens of kilograms as before. AMS quickly revolutionized the application of cosmogenic nuclides to the earth surface, making measurements of meteoritic and meteoric cosmogenic nuclides far easier and making possible the measurement of cosmogenic nuclides produced *in situ*, such as ^{10}Be and ^{26}Al.

The application of *in situ*-produced cosmogenic nuclides to exposure dating and problems in earth surface processes exploded in the mid-1980s. *Kubik et al.* [1984] measured ^{36}Cl produced in limestone, and *Kurz* [1986] measured ^{3}He in Hawaii basalts. Other cosmogenic nuclides then rapidly followed. *Phillips et al.* [1986] measured ^{36}Cl in volcanic rocks in the western United States. *Klein et al.* [1986] reported ^{26}Al and ^{10}Be in Libyan Desert glass. *Lal and Arnold* [1985] first suggested the use of ^{10}Be and ^{26}Al in the mineral quartz due to its high purity and resistance to weathering, and *Nishiizumi* et al. [1986] then promptly reported measurements of both these nuclides. The calibration of ^{26}Al and ^{10}Be production rates in quartz from glacially polished rocks [*Nishiizumi et al.*, 1989] was a milestone that opened up the possibility of quantitative exposure dating and measurement of rock erosion rates.

The use of meteoric cosmogenic nuclides to address problems in earth surface processes also blossomed in the 1980s, when AMS methods were developed to measure ^{10}Be and ^{36}Cl. Beryllium-10 attaches to soil particles and was used to infer basin-wide erosion rates [*Valette-Silver et al.*, 1986; *Brown et al.*, 1988] by tracking the export of surface soils in river sediments. Chlorine-36, in contrast, stays in solution and was used to track groundwater [*Kubik et al.*, 1984; *Bentley et al.*, 1986] and the accumulation of evaporites [*Phillips et al.*, 1983].

Many of the geologic applications of *in situ*-produced cosmogenic nuclides used today, including exposure dating, burial dating, and the measurement of bedrock erosion rates were built upon theory that was originally developed for studying meteorites, but were adapted for earth surface processes. Seminal papers by *Lal and Arnold* [1985] and *Lal* [1991] laid out the theoretical framework for calculating exposure ages, erosion rates, and burial ages for quartz exposed anywhere on Earth, and introduced cosmogenic nuclides to a broader audience of geologists and geomorphologists. An especially important early application included the use of cosmogenic nuclides to exposure date glacial moraines [*Phillips et al.*, 1990; *Gosse et al.*, 1995], which remains a mainstay of cosmogenic nuclide dating [*Balco*, 2011].

There are several reviews available that cover the history and applications of cosmogenic nuclides. The textbook of *Dunai* [2010] and the article by *Gosse and Phillips* [2001] are notable for being particularly thorough. *Granger et al.* [2013] offers a historical overview; *Bierman* [1994] and *Nishiizumi et al.* [1993] are excellent earlier reviews for the application of ^{10}Be and ^{26}Al.

13.3 THEORY, FUNDAMENTALS, AND SYSTEMATICS

13.3.1 Cosmic rays

Cosmic rays were first described in the early 1900s when ionizing radiation was detected within closed electrometer chambers. *Hess* [1912] showed that the radiation increased with altitude, and thus came from space'. (The term 'cosmic ray' was later coined by *Millikan and Cameron* [1926], who believed gamma radiation to be responsible; the name stuck, even though the cosmic rays are actually particles.) We now know that there are two basic types of cosmic ray: high-energy galactic cosmic rays (GCRs) originating outside our solar system, and lower-energy solar cosmic rays (SCRs) emitted by our Sun (Chapter 2).

Galactic cosmic rays are energetic particles, mostly protons and alpha particles that are accelerated to very high energies by supernovae [*Ackermann et al.*, 2013]. Most have energies ranging from 0.1–10 GeV, although some ultrahigh energy cosmic rays—with energies $> 10^{20}$ eV—are the most energetic particles ever measured (Figure 2.14). These energies are much larger than nuclear binding energies of ~ 8 MeV/nucleon (Chapter 2, Figure 2.2), so when GCRs collide with nuclei they have more than enough energy to break them apart through spallation-type reactions. The GCRs thus instigate a cascade of nuclear reactions as they encounter meteoroids, planetary surfaces, or atmospheres such as Earth's (Fig. 13.1) [*Lal and Peters*, 1967]. We refer to galactic cosmic rays as *primary cosmic rays*, and to the high-energy products of the nuclear cascade as *secondary cosmic rays*. Secondary cosmic rays include a variety of particles such as neutrons and protons, as well as pions that decay to muons in the upper atmosphere, and electrons, positrons, and neutrinos. Because protons are slowed as they pass through matter due to their charge, the secondary cosmic-ray flux in the lower atmosphere is dominated by the least reactive particles with the longest penetration lengths, primarily neutrons and muons.

Solar cosmic rays are lower-energy particles, typically < 100 MeV, emitted by the Sun. Most lack sufficient energy to pierce through the geomagnetic field at low latitudes, so they are largely deflected towards the poles where they are responsible for the sweeping aurorae that illuminate the high-latitude skies. Solar cosmic rays do not produce many cosmogenic nuclides on Earth, although they do produce nuclides through low-energy reactions in the outermost few millimeters of meteoroids and the lunar surface, which are unprotected by atmospheres [*Reedy and Arnold*, 1972; *Reedy et al.*, 1983; *Nishiizumi et al.*, 2009].

13.3.2 Distribution of cosmic rays on Earth

The flux of cosmic rays to Earth depends on their energy and the strength of the solar-magnetic and geomagnetic fields. Because the incoming cosmic rays are charged, they are deflected by the geomagnetic field. Low-energy particles are deflected most, and are either spun away from Earth altogether or funneled towards the magnetic poles. Very high-energy particles have

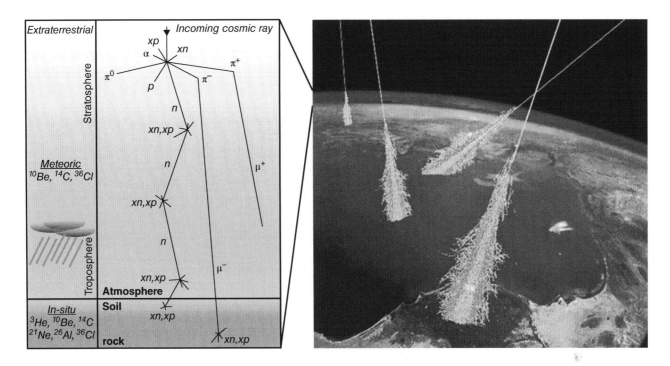

Fig. 13.1. A schematic cosmic-ray cascade generating cosmogenic nuclides in the atmosphere, soil, and rock. The incoming primary cosmic ray strikes an atmospheric nucleus and breaks it into smaller fragments. The high-energy fragments then travel through the atmosphere and generate further nuclear reactions, causing a cascade. Short-lived pions in the upper atmosphere decay into muons that travel long distances through the atmosphere and into rock. The products of these nuclear reactions are called cosmogenic nuclides. Those in the atmosphere are referred to as meteoric, those in rock soil are referred to as *in situ*-produced, and those produced in meteorites outside Earth's atmosphere are referred to as extraterrestrial. (*See insert for color representation of the figure.*)

sufficient momentum to penetrate the magnetic field and enter the atmosphere at lower latitudes. Both the flux and the energy spectrum of the secondary cosmic rays therefore vary with geomagnetic latitude due to blocking of the low-energy component at low latitudes. Whether or not a cosmic ray is able to penetrate the magnetic field at a particular point depends on its rigidity, defined as the ratio of momentum to charge. For each point on Earth there is a cutoff rigidity (R_C), below which cosmic rays are deflected. The cutoff rigidity depends on the local magnetic field strength, and changes over time as the geomagnetic field varies in strength and in pattern (Fig. 13.2). The implication is that cosmogenic nuclide production rates vary around the globe due to cutoff rigidity (which is often approximated simply as latitude because local and nondipole fluctuations are damped over periods of thousands to millions of years). For example, the production rate of ^{10}Be at sea level varies by about a factor of two from the equator to the pole [see *Lifton et al.*, 2014; *Phillips et al.*, 2016].

13.3.3 What makes a cosmogenic nuclide detectable and useful?

Primary and secondary cosmic rays produce a wide variety of cosmogenic nuclides in the atmosphere as well as in rock and soil. The vast majority of these, however, are undetectable. This is because cosmic-ray reactions are so rare (production rates at

Earth's surface are typically measured in atoms per gram of rock per year), that the cosmogenic nuclides are only detectable if they produce a particle that is otherwise essentially absent. In general there are two such types of nuclides:

(1) noble gases that are excluded from minerals during crystallization;

(2) parentless radionuclides that are not part of a decay chain and have a short enough half-life that they have completely decayed since nucleosynthesis [*Lal and Peters*, 1967].

Noble gas cosmogenic nuclides can be detected if they are either present at very low concentrations or are generally excluded from minerals. Examples include ^{3}He, one of the most abundant cosmogenic nuclides that is produced as a fragment during spallation of heavier nuclei, and ^{21}Ne, the least abundant stable isotope of neon. Heavier noble gases such as Ar, Kr, and Xe are occasionally analyzed for their cosmogenic component as well, mostly for extraterrestrial samples [e.g., *Wieler*, 2002] or in groundwater [*Collon et al.*, 2004].

Radioactive cosmogenic nuclides can be detected if they build up to sufficiently high concentrations, and if they are uniquely produced by cosmic-ray reactions. They must also have a half-life that is neither too long nor too short. Long-lived radionuclides that are still present from nucleosynthesis are useful for other types of geochronology, but due to their relatively high concentration in rocks today it is difficult or impossible to detect the small cosmogenic component. Likewise, radionuclides that are

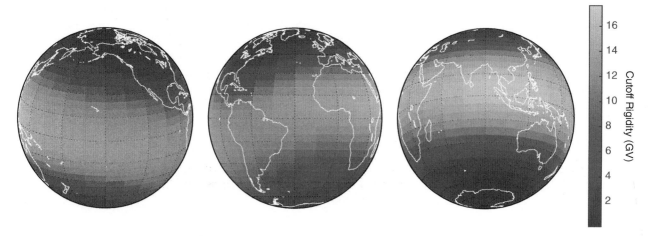

Fig. 13.2. Effective vertical cutoff rigidity (R_C). Lower energy cosmic rays are funneled to the poles, while only higher energy cosmic rays are able to penetrate the geomagnetic field to enter the atmosphere at lower latitudes. Cosmogenic nuclide-production rates are therefore partly a function of latitude. Variations in cutoff rigidity account for about a factor of two in cosmogenic nuclide production rates at the ground surface from high to low latitudes. (Source: *Sato et al.* [2008].) (*See insert for color representation of the figure.*)

daughter products of long-lived nuclides are also not suitable because they are too abundant. On the other end of the spectrum, radionuclides with very short half-lives are also unsuitable because they do not build up to sufficiently high concentrations before they reach secular equilibrium, at which point the concentrations reach a maximum due to radioactive decay. The "sweet spot" for cosmogenic nuclides is a radionuclide whose half-life is short enough that it has decayed since nucleosynthesis, but long enough that it can build up to measurable concentrations. Ideally, the radionuclide cannot be produced as a byproduct of radioactive decay, either directly or indirectly through capture of radiogenic neutrons or alpha particles, although in practice this may sometimes occur. A list of the main cosmogenic nuclides measured today is given in Table 13.1.

In addition to the commonly measured nuclides listed in Table 13.1, the much more intense cosmic-ray exposure experienced by some meteorites and lunar samples results in cosmogenic nuclide production at detectable levels in many other elements. For example, attempts to date early solar-system materials using the now extinct nuclides ^{53}Mn and ^{182}Hf must account for cosmogenic production or destruction of their daughter isotopes ^{53}Cr and ^{182}W, which alters the signal produced by radioactive decay (see Chapter 14). A small number of moderately abundant elements, for example Sm and Gd, have isotopes with sufficiently high thermal neutron capture cross-sections that their isotopic compositions can be modified substantially by capture of secondary cosmic-ray neutrons [e.g., *Hidaka et al.*, 2000]. These systems have proven useful to deduce the cosmic-ray irradiation history of the lunar surface and meteorites.

Finally, it is worth mentioning that secondary cosmic rays are important to consider in dosimetric dating methods such as luminescence and electron spin resonance (ESR) because they excite electrons as they pass through mineral grains. These are not cosmogenic nuclides, however, and will not be discussed in this chapter.

Table 13.1 List of commonly measured cosmogenic nuclides, their dominant production mechanisms and their radioactive half-lives

Nuclide	Main target elements	Half-life
^{3}He	Spallation of $Z > 2$ elements	Stable
	Li(n,α)^{3}H $\rightarrow$ ^{3}He	
^{10}Be	^{16}O(n,4p3n)^{10}Be	1.39 Ma
	^{16}O(μ^-,αpn)^{10}Be	
	^{28}Si(n,x)^{10}Be	
	^{28}Si(μ^-,x)^{10}Be	
	^{7}Li(α,p)^{10}Be	
	^{10}B(n,p)^{10}Be	
	^{13}C(n,α)^{10}Be	
	^{12}C(n,2pn)^{10}Be	
^{14}C	^{14}N(n,p)^{14}C	5730 years
	^{16}O(n,2pn)^{14}C	
^{21}Ne	Spallation: Na, Mg, Si, Al, Fe	Stable
	^{16}O(α,n)^{21}Ne	
	^{19}F(α,pn)^{21}Ne	
^{26}Al	^{28}Si(n,p2n)^{26}Al	702 ka
	^{28}Si(μ^-,2n)^{26}Al	
	^{23}Na(α,n)^{26}Al	
	^{40}Ar(n,x)^{26}Al	
^{36}Cl	^{40}Ca(n,3p2n)^{36}Cl	300 ka
	^{40}Ca(μ^-,α)^{36}Cl	
	^{39}K(n,α)^{36}Cl	
	^{39}K(μ^-,p2n)^{36}Cl	
	Fe(n,x)^{36}Cl	
	Ti(n,x)^{36}Cl	
	^{35}Cl(n,γ)^{36}Cl	
	^{40}Ar(n,p4n)^{36}Cl	
^{41}Ca	^{40}Ca(n,γ)^{41}Ca	102 ka
	Spallation of Fe, Ti	
^{53}Mn	Spallation of Fe, Co, Ni, Cu	3.7 Ma

13.3.4 Types of cosmic-ray reactions

Once a primary cosmic ray strikes matter, whether a meteoroid in space or Earth's atmosphere, it rapidly loses energy in a series of nuclear reactions (Fig. 13.1). Most cosmogenic

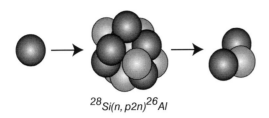

$^{28}Si(n, p2n)^{26}Al$

Fig. 13.3. An incoming neutron strikes a silicon nucleus and spalls a fragment containing a proton and two neutrons. The resulting particle is radioactive ^{26}Al, a cosmogenic nuclide that is commonly measured in quartz.

nuclides are made through a few basic types of nuclear reactions including nucleon spallation, negative muon capture and fast muon reactions, and neutron capture [e.g., *Lal and Peters*, 1967].

(1) Nucleon spallation is the most common reaction in the atmosphere and in rocks near the surface. This is simply a reaction in which an incoming neutron or a proton collides with enough energy to break apart, or spall, a nucleus into lighter fractions. An example is shown in Fig. 13.3. Most often the nucleus only loses a few neutrons, protons, or alpha particles, but occasionally it is broken into larger fragments. The product of a nuclear spallation is always a lighter nucleus or a byproduct such as helium nuclei (^3He). Depending on the number of neutrons and protons lost, the product could be stable or radioactive, common or very rare. The secondary particles released during a spallation reaction can carry sufficient energy to cause additional reactions when they collide with other nuclei, continuing the nuclear cascade.

(2) A muon is a type of secondary cosmic ray produced primarily in the upper atmosphere through pion decay. Muons are far less reactive than nucleons, and therefore penetrate deeper through the atmosphere and rock. Muons can carry charge, and negative muons can be captured into orbit around the nucleus. Because they are ~200 times heavier than electrons, their orbit interacts with the nucleus, where they can convert a proton to a neutron plus a neutrino. The energy of the reaction evaporates part of the nucleus, forming a cosmogenic nuclide [*Charalambus*, 1971]. Negative muon capture reactions account for less production than nucleon spallation in the atmosphere and near the surface, but they dominate production on Earth below a few meters depth [e.g., *Stone et al.*, 1998; *Heisinger et al.*, 2002a]. There are also reactions from high-energy muons that occur at even slower rates but at greater depths [*Heisinger et al.*, 2002b].

(3) Secondary cosmic ray neutrons are slowed as they pass through matter. Neutrons with very low thermal or epithermal energies can be captured into a nucleus, producing an isotope that is heavier by one atomic mass unit. Neutron capture reactions can be very important for certain cosmogenic nuclides, such as ^{36}Cl produced from ^{35}Cl or ^{41}Ca produced from ^{40}Ca (Table 13.1), but for most others this reaction is irrelevant. Thermal neutrons are very efficiently absorbed by

water as well as certain elements such as boron and gadolinium. Consequently, production rates by neutron capture can be highly variable depending on factors such as snow, soil moisture, and trace element concentrations [e.g., *Dep et al.*, 1994; *Liu et al.*, 1994; *Zweck et al.*, 2013].

13.3.5 Cosmic-ray attenuation

Secondary cosmic rays rapidly lose their energy as they pass through matter, whether in the atmosphere or in rock. The flux of secondary cosmic-ray neutrons through the atmosphere, for example, drops by approximately two orders of magnitude from the top of the atmosphere to sea level. The flux drops by another two orders of magnitude within the top 3 m of rock that the secondary cosmic-ray neutrons pass through.

There are several major implications for terrestrial cosmogenic nuclide production. Because ~99% of the secondary cosmic-ray neutrons are attenuated in the atmosphere, the bulk of cosmic-ray reactions happen there [*Lal and Peters*, 1967]. Of the remaining cosmic-ray flux, the vast majority of cosmogenic nuclide production happens within the topmost few meters of rock and soil at the ground surface. Cosmogenic nuclides produced in the atmosphere include ^{14}C and ^{10}Be (produced from N and O), and ^{36}Cl (produced from Ar) (Fig. 13.1 and Table 13.1). The ^{14}C is incorporated into CO_2 and enters the global carbon cycle, while ^{10}Be and ^{36}Cl tend to be scavenged by rainwater and aerosols. In either case, the concentrations of meteoric cosmogenic nuclides far exceed those found within mineral grains.

Another implication is that the secondary cosmic-ray flux depends strongly on atmospheric pressure, which varies primarily with altitude [*Lal and Peters*, 1967; *Lal*, 1991; *Stone*, 2000]. This is important for setting the production rate of cosmogenic nuclides in rocks and minerals at the ground surface. The production rate of ^{10}Be in quartz, for example, varies by a factor of 30 from sea level to high mountain altitudes of 5 km. This far outweighs the factor-of-two latitudinal variability from equator to pole due to the cutoff rigidity as illustrated in Fig. 13.2. The flux at any given point on Earth therefore depends on both altitude and latitude.

After the secondary cosmic rays traverse the atmosphere, they continue to pass through soil, water, and rock. In general the penetration of energetic nucleons and muons is nearly independent of composition, so the production rates by spallation and muon-capture scale with the mass traversed regardless of rock type, in units of density times length (g/cm^2). The same is not true for production by neutron capture, which requires composition-dependent calculations.

Production by nucleon spallation decreases rapidly below the ground surface, with an exponential penetration length of ~160 g/cm^2. The exact penetration length varies somewhat according to the energy spectrum, which in turn depends on latitude and elevation [see *Marrero et al.* [2016] and references therein]. The penetration length also varies somewhat with the slope of the ground surface due to topographic shielding of secondary cosmic rays that approach obliquely [*Dunne et al.*, 1999]. It is often easier to think about penetration lengths in terms of distance; for rock of density $2.6\,g/cm^3$ a penetration length of $160\,g/cm^2$ corresponds to about $62\,cm$, while for soil of density $1.5\,g/cm^3$ it corresponds to about $107\,cm$. In the case of rock, the production rate by neutron spallation decreases to 10% of its surface value at a depth of $1.4\,m$ and by 1% of the surface value by $2.8\,m$. Essentially all cosmogenic nuclide production by spallation reactions occurs within the uppermost few meters near the ground surface. This is very important for applications of *in situ*-produced cosmogenic nuclides such as surface exposure dating, measuring erosion rates, and for burial dating, as discussed in later sections.

Production rates by negative muon capture and by high-energy muon reactions decline far less rapidly with depth. Muons have a smaller reaction cross-section than neutrons, and so have lower production rates but much longer penetration lengths. Muogenic production is less important near the ground surface but it dominates at depths greater than a few meters. The end result is that although most cosmogenic nuclide production happens near the surface, there is a small but measurable production rate that continues at depths of tens or even hundreds of meters (Fig. 13.4).

13.3.6 Calibrating cosmogenic nuclide-production rates in rocks

The production rate of cosmogenic nuclides within a rock on Earth varies with *latitude* due to magnetic cutoff rigidity, with *altitude* due to atmospheric shielding, with *depth* due to attenuation of secondary cosmic rays, and with *time* due to changes in the geomagnetic field. Practically any application of *in situ*-produced cosmogenic nuclides requires first that the production rate be known. Cosmogenic-nuclide researchers must use calibration curves to scale production rates by latitude, longitude, altitude, and through time.

Production rates vary with latitude and longitude due to the cutoff rigidity imposed by the geomagnetic field. The first step in most scaling models is to determine the cutoff ridigity at a site. Some models (e.g., *Lifton et al.* [2014]; hereafter LSD scaling) account for location-specific nondipole patterns in the cutoff rigidity (see Fig. 13.2) while others (e.g., *Lal* [1991]; *Stone* [2000]; hereafter Lal/Stone scaling) assume that fluctuations in the nondipole components average out over long exposure times, and only consider latitude. In general the two methods agree over a large part of the globe although there are discrepancies, particularly at low latitudes and high altitudes [*Phillips et al.*, 2016].

The next step in scaling production rates is to model the atmospheric pressure. This is often taken from the standard atmospheric pressure equation, although there can be significant local deviations in air pressure [*Stone*, 2000]. There is currently no widely used model that accounts for variations in air pressure

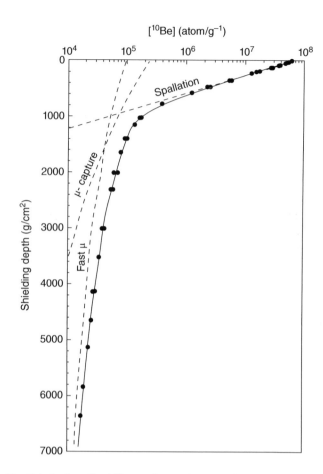

Fig. 13.4. Depth profile of ^{10}Be in a deep sandstone core from a slowly eroding site at Beacon Heights, Antarctica. The dashed lines represent production by spallation, negative muon capture, and fast muon reactions. Although muons account for only a few percent of production at the ground surface, they dominate at depths greater than $1000\,g/cm$, or about $3–4\,m$ in rock. Note semi-logarithmic scale. (Source: *Phillips et al.* [2016]. Reproduced with permission of Elsevier.)

over time (but see *Staiger et al.* [2007]). Local air pressure is then used to scale cosmogenic nuclide production, using either empirical scaling parameters (Lal/Stone) or nuclide-specific scaling from models of neutron propagation through the atmosphere (LSD; *Marrero et al.* [2016]).

The production rate is also known to vary as a function of time due to the changing geomagnetic field strength. Many models assume that variations in field strength are averaged out over long exposure times, and so ignore this variability. More complex time-dependent models such as LSD use magnetic field reconstructions to explicitly model the changing production rate over the exposure time. At low-latitude sites the instantaneous production rate at a given time can vary by a factor of two.

Although we know from theory how production rates vary with latitude, with elevation, and with time, the production rates must still be calibrated against measured samples of independently known age [e.g., *Nishiizumi et al.*, 1989]. This is most often done by dating glacial moraines whose age is constrained

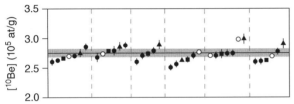

Fig. 13.5. Production-rate calibration site at the Bonneville shoreline. Photograph shows a wave-polished rock on a relict shoreline visible as the flat bench at Promontory Point, Utah. The shoreline dates to a lake highstand at 18.36 ± 0.30 ka. This and five other boulders were sampled for cosmogenic nuclides; ^{10}Be results are shown in the graph. The average ^{10}Be concentration is 2.75 ± 0.07 atom/g. Dashed lines separate individual samples, and symbols represent different chemistry and/or AMS laboratories. Adapted from (Source: *Lifton et al.* [2015]. Reproduced with permission of Elsevier.)

by bracketing radiocarbon ages, but it can also be done on pristine volcanic flows or on wave-cut benches (Fig. 13.5) that are independently dated [e.g., *Phillips et al.*, 2016]. At each calibration site the cosmogenic nuclide concentration is measured and then scaled using a model to a reference production rate at sea-level high-latitude (SLHL). It is important to recognize that the reference SLHL production rate is model-dependent. The same model can then be used to scale the reference SLHL production rate to any point on the globe.

There are many different production-rate scaling models in the literature, although the Lal/Stone and LSD models cited above are most widely used at this time. Recently, online calculators [e.g., *Balco et al.*, 2008; *Marrero et al.*, 2016] have become available that incorporate multiple scaling models and can account for time-dependent production rates when determining surface exposure times and erosion rates.

13.4 APPLICATIONS

13.4.1 Types of cosmogenic nuclide applications

Applications of cosmogenic nuclides can be broken down into three main classes. *Extraterrestrial* cosmogenic nuclides are

measured in meteorites and on lunar and planetary surfaces. *Meteoric* cosmogenic nuclides are produced in Earth's atmosphere and track production rates, atmospheric transport, and depositional processes. They serve as important tracers in the environment. *In situ-produced* cosmogenic nuclides are produced within mineral grains near the ground surface and are used to date landforms, measure erosion rates, and date sediment burial. These three different classes will be discussed separately below.

It is worth noting that there are several abbreviations for cosmogenic nuclides in the literature. These include CRN for cosmogenic radionuclide (excluding noble gases) and TCN for terrestrial cosmogenic nuclide (excluding extraterrestrial applications). In this chapter we will eschew the more restrictive acronyms and simply refer to cosmogenic nuclides *sensu lato*.

13.4.2 Extraterrestrial cosmogenic nuclides

Meteorites have very high concentrations of cosmogenic nuclides because they are exposed to intense cosmic-ray fluxes in space for many millions of years without shielding by magnetic fields or atmospheres. The concentrations are high enough that it was possible to measure the cosmogenic nuclides by decay counting and noble gas mass spectrometry long before the advent of accelerator mass spectrometry (AMS) [e.g., *Paneth et al.*, 1952]. (Today even very small samples such as individual cosmic spherules that weigh only micrograms can be measured by AMS; *Nishiizumi et al.* [1995].)

The build up of cosmogenic nuclides in a meteorite indicates the time that the meteoroid was exposed in space. Generally this represents the time since a larger parent body was broken up in a collision, exposing previously shielded material as roughly meter-sized objects that are then exposed to cosmic rays. To this point, thousands of meteorites have been measured, generating age distributions that inform us about the origin of meteorites and events in solar system history. In general, the range of cosmic-ray exposure ages varies with the class of meteorite. Stony meteorites tend to be younger, with ages up to tens of millions of years. Iron meteorites tend to be older, with ages ranging from hundreds of millions of years up to more than a billion years (Fig. 13.6). The grouping of exposure ages by meteorite class suggests that many meteorites are derived from a relatively small number of distinct collisions that each ejected a large number of similar asteroid and planitesmal fragments [*Eugster et al.*, 2006; *Herzog*, 2010].

Calculating the cosmic-ray exposure age of meteorites is not as simple as just measuring the concentration that has built up at the surface. This is because meteorites have been ablated during atmospheric entry, and they may have experienced break-up either before or during their fall. Because the production rate of cosmogenic nuclides is a function of depth, the exact production rate for a given meteorite is not known a priori. Instead, the production rate must be determined individually by measuring the concentration of a radionuclide that is short-lived with

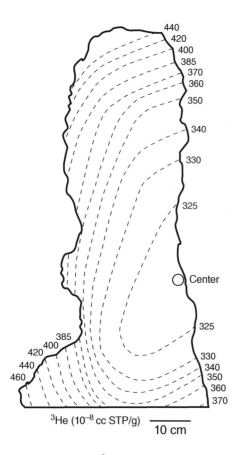

Fig. 13.6. Contours of cosmogenic ^{3}He in the iron meteorite Carbo measured by *Hoffman and Nier* [1959]. The center of the meteorite prior to ablation in the atmosphere on entry was near the circle shown. This meteorite has an exposure age of 850 ± 140 Ma [*Markowski et al.*, 2006]. (Source: *Hoffman and Nier* [1959]. Reproduced with permission of Elsevier.)

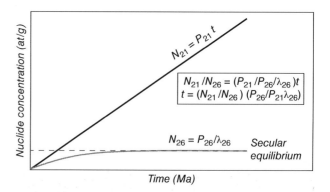

Fig. 13.7. The nuclide pair ^{21}Ne and ^{26}Al can be used to date the cosmic-ray exposure age (t) that a meteorite experienced in space. The radionuclide ^{26}Al builds up to a constant secular equilibrium concentration shown by the dashed line while stable ^{21}Ne continues to accumulate linearly over time. The cosmic-ray exposure age can be calculated from the measured nuclide ratio if the relative production rates and radioactive mean-lives are known. In this figure N_i refers to the concentration of a cosmogenic nuclide i, P_i refers to its production rate, and λ_i refers to its radioactive decay constant. For chondrites the ratio P_{21}/P_{26} is about 2.5. Axes not to scale.

production rates are much slower than in space) can yield the time of the fall [*Goel and Kohman*, 1962]. This is very similar to the burial dating method used for terrestrial *in situ*-produced cosmogenic nuclides, discussed in a later section.

Terrestrial ages of meteorites vary around the globe. In humid environments meteorites tend to weather quickly. However, in deserts or in ablating ice sheets (e.g., Antarctica) the meteorites can survive much longer (and are also easier to find). Terrestrial ages for most meteorites typically range up to a few tens of thousands of years, although a few have been dated at > 2 Ma (see *Herzog* [2010] and references therein).

13.4.3 Meteoric cosmogenic nuclides

Not to be confused with meteoritics (the study of meteorites), *meteoric* cosmogenic nuclides are those produced in Earth's atmosphere. They are sometimes referred to as "garden variety" cosmogenic nuclides, to distinguish them from *in situ*-produced cosmogenic nuclides that are present in mineral grains at much lower concentrations. Commonly measured meteoric cosmogenic nuclides include ^{14}C, ^{10}Be, and ^{36}Cl.

13.4.3.1 Carbon-14

By far the most commonly measured cosmogenic nuclide is meteoric ^{14}C, which forms the basis for radiocarbon dating of organic matter. It is produced in the atmosphere primarily from thermal neutron reactions on ^{14}N, and also from spallation of oxygen. Carbon-14 is rapidly oxidized to CO_2 in the atmosphere, where it is mixed over timescales of a few years. It is incorporated into the global carbon cycle, and can be used to date organic matter, to trace ocean circulation, and in many other ways [e.g., *Levin and Hesshaimer*, 2000]. With its relatively short half-life of 5730 years, ^{14}C can be measured by either decay

respect to the exposure age, and then assuming that its concentration is at secular equilibrium (i.e., $N = P/\lambda$). The inferred production rate can then be used to normalize the production rate of a stable nuclide that has been retained for the entire exposure history. There are several sets of carefully calibrated pairs of stable and radioactive cosmogenic nuclides, for which the relative production rates are largely independent of depth, or are a known function of depth [e.g., *Leya et al.*, 2000]. These include, for example, stable ^{21}Ne and radioactive ^{26}Al, illustrated conceptually in Fig. 13.7.

In addition to yielding the exposure age in space, cosmogenic nuclides in meteorites can also be used to determine their terrestrial age, or the time since they fell to earth. For meteorites recovered after observed falls the terrestrial age is obviously known. But for meteorites that are found in the field there is usually little indication of when they fell to Earth. A good way to date the terrestrial age is to measure the concentrations of short-lived nuclides such as ^{14}C, ^{36}Cl, ^{41}Ca, or ^{10}Be. By measuring a pair of nuclides whose relative production rates are known, the relative decay of the two after falling to Earth's surface (where counting for large samples, or most commonly by AMS for small samples.

There are many textbooks that deal with radiocarbon dating, especially within the context of archaeology [e.g., *Taylor and Bar-Yosef*, 2014], and so the treatment of radiocarbon dating here will be brief. Of special interest with respect to broader cosmogenic nuclide studies is the fact that ^{14}C provides an excellent record of variations in global cosmogenic nuclide production rates over time. Because ^{14}C is nearly homogenized in the atmosphere its concentration integrates over latitudinal and spatial variations in production rate. Because it is incorporated into the marine and terrestrial carbon cycle it is rapidly scrubbed from the atmosphere, and so represents globally averaged production rates over a period of years.

Radiocarbon dating uses the simple dating equation,

$$t = -(1/\lambda)\ln\left[\frac{\left(^{14}C/^{12}C\right)_{\text{meas}}}{\left(^{14}C/^{12}C\right)_{\text{init}}}\right] \tag{13.1}$$

where the isotope ratio $^{14}C/^{12}C$ in organic matter is measured with respect to that in the modern atmosphere. Here, the term "modern" refers to a reference atmospheric ratio in the year 1950 that predates atmospheric testing of nuclear weapons, which caused a spike in atmospheric ^{14}C concentrations [e.g., *Hua*, 2009]. One of the main complications with radiocarbon dating is that the initial $^{14}C/^{12}C$ ratio in the atmosphere varies over time due to changes in the solar and geomagnetic fields that modulate the cosmic-ray flux (see Box 13.2). Successful radiocarbon dating therefore requires that the initial $^{14}C/^{12}C$ ratio be calibrated as a function of time. This is carried out by measuring the $^{14}C/^{12}C$ ratio in samples whose age is known by other methods, and back-calculating the initial ratio [e.g., *Reimer et al.*, 2013]. The most reliable results are obtained by calibrating against dendrochronology, using tree rings that are carefully counted. Each tree's growth rings record a time series of $^{14}C/^{12}C$; the widths of the rings also reflect the local environment, with wide rings associated with favorable conditions and narrow rings associated with stresses such as drought. By comparing growth rings among trees in a given region and matching the pattern of widths a time series can be constructed that is much longer than the lifespan of any given tree (Fig. 13.8).

Beyond the timespan of dendrochronology, radiocarbon can be calibrated against calcite speleothem (e.g., stalagmites) that grow in caves, as well as corals and foraminifera [e.g., *Fairbanks et al.*, 2005]. A fraction of the carbon in the calcium carbonate minerals comes from atmospheric CO_2 that is dissolved in groundwater or ocean water. The calcium carbonate can then be dated using U-series (Chapter 8) to calibrate the initial $^{14}C/^{12}C$.

The radiocarbon calibration curve is constantly being refined by additional measurements and methods of cross-calibration. For consistency among dates, most researchers use an internationally accepted calibration curve, and clearly cite the calibration

Box 13.1   Cosmic-ray-flux reconstructions

The cosmic-ray flux on Earth is regulated by the geomagnetic field and solar activity. In addition, short-term spikes can occur due to local supernovae. Cosmogenic nuclides that are produced in the atmosphere can thus provide information about these past conditions, provided that a suitable record can be found.

Radiocarbon preserved in tree rings and speleothem provide a well-dated high-resolution record (see Figures 13.8 and 13.9). However, because radiocarbon has a multiyear residence time in the atmosphere and is exchanged in the global carbon cycle it may not record very short-term spikes due to supernovae. It is also limited by its short radioactive decay time.

Beryllium-10 is more rapidly scrubbed from the atmosphere, but on the other hand its fallout rate can fluctuate due to local variations in atmospheric circulation and precipitation. Beryllium-10 in ice cores provides a well-dated record with potentially subannual resolution [e.g., *Berggren et al.*, 2009; *Miyake et al.*, 2015]. Beryllium-10 in marine sediments can extend further back in time to constrain variations in the geomagnetic field [*Frank et al.*, 1997; *Ménabréaz et al.*, 2014].

Figure 13.B1.1 shows a reconstruction of the cosmic-ray flux over the past 9400 years determined by combining radiocarbon and ice-core ^{10}Be records [*Steinhilber et al.*, 2012]. Comparison with the geomagnetic field strength shows that the overall flux is regulated by the geomagnetic field, but that short-term fluctuations reflect solar forcing.

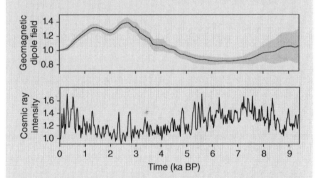

Fig. 13.B1.1. A record of cosmic-ray intensity inferred from ^{10}Be in ice cores and ^{14}C in tree rings. The top panel shows the relative geomagnetic dipole field over the past 9400 years, with uncertainty in gray. The bottom panel shows relative cosmic-ray intensity over the same time period. Note that the two curves are anti-correlated; high dipole field strength corresponds to lower cosmic ray intensity (as inferred from lower cosmogenic nuclide production rates). The long-term fluctuations in cosmic-ray intensity are regulated by the geomagnetic field, while solar forcing drives short-term fluctuations. (Source: *Steinhilber et al.* [2012]. Reproduced with permission of Proceedings of the National Academy of Science (PNAS).)

curve used. A section of the calibration curve from the IntCal13 calibration is shown Fig. 13.8 [*Reimer et al.*, 2013]. The calibrated age is referred to in units of calibrated (cal.) years. Uncalibrated radiocarbon dates based on raw measurements and equation (13.1) are reported in ^{14}C years BP, which should not be confused with true ages.

Fig. 13.8 shows a section of the IntCal13 calibration curve extending from 2000–4000 years BP (i.e., years before present,

Box 13.2   A test of the isochron burial-dating method

It is important to compare results from various dating methods to test the assumptions inherent in geochronology. For isochron burial dating (Section 13.4.3.16), these assumptions include production rates, production rate scaling, and decay constants. Fig. 13.B2.1 shows a nearly ideal case for comparing cosmogenic dating with $^{40}Ar/^{39}Ar$, at a place found on the north pediment of the Kunlun Mountains at the border of the Tarim basin in northern China. A gravel layer is sandwiched between two volcanic flows that cap the Xiyu conglomerate. The volcanic flows can be dated with $^{40}Ar/^{39}Ar$, while the gravel can be dated using cosmogenic nuclides.

Zhao et al. [2016] collected samples from near the bottom of the gravel layer for isochron burial dating. Individual clasts lie on an isochron indicating an age of 1.38 ± 0.07 Ma, assuming a $^{26}Al/^{10}Be$ production rate ratio of 6.8. This compares very well with previous $^{40}Ar/^{39}Ar$ dating of the volcanic flows, which yielded ages of 1.41 ± 0.04 Ma for the lower flow and 1.20 ± 0.05 Ma for the upper flow.

Intercomparisons such as these can be used to place constraints on the physical parameters used in geochronology. For example, after accounting for uncertainties in the radioactive half-lives of ^{26}Al and ^{10}Be, Zhao et al. [2016] were able to show that the $^{26}Al/^{10}Be$ production ratio at the intercomparison site is bracketed to $6.8^{+0.4}/_{-0.8}$ at 67% confidence, similar to the value of 6.8 that was assumed in the calculation. Additional intercomparisons will help to further constrain the production rates and other model assumptions.

Fig. 13.B2.1. Dating intercomparison site at Pulu, Xinjiang, northern China. Photograph on left shows two volcanic flows that sandwich a ~10 m thick gravel package. $^{40}Ar/^{39}Ar$ ages are shown for the two bracketing flows. Eight gravel clasts from the conglomerate were analyzed for ^{26}Al and ^{10}Be, shown in a burial dating isochron on the right as 1σ error ellipses. The best-fit isochron has an age of 1.38 ± 0.07 Ma, assuming an initial production rate ratio of 6.8, shown by the sloping line. The positive intercept of the isochron indicates postburial production. (Source: Zhao et al. [2016]. Reproduced with permission of Elsevier.)

where 'present' is defined as 1950). It shows some wiggles that have important implications for radiocarbon dating. Note that the section around 2450 ^{14}C years BP forms an exceptionally flat plateau, with steep sections to either side. This is because the ^{14}C inventory in the atmosphere was decreasing at that time for several hundred years due to a decreasing cosmic-ray flux to the atmosphere. The result is that a tree from the year 2700 cal. years BP and another one 300 years younger at 2400 cal. years BP will have almost exactly the same ^{14}C concentration. It is not possible to tell them apart from their radiocarbon concentrations alone. This leads to some interesting problems in calculating the calibrated radiocarbon dates, and for this reason radiocarbon ages are often expressed graphically as probability density functions (PDFs).

An example of radiocarbon calibration is shown in Fig. 13.9. Three radiocarbon determinations are shown on the ordinate, each with identical 1% uncertainties. The oldest measurement (2450 ^{14}C years BP) corresponds to a period of time in which ^{14}C was relatively stable, and so its calibrated date has a broad

ange from ~2350–2700 cal. years BP. The uncertainty in the calibrated date is much larger than the uncertainty in the radiocarbon determination. In contrast, the middle measurement (2325 ^{14}C years BP) corresponds to a time when atmospheric radiocarbon was changing rapidly. Its calibration is thus controlled very tightly, so that the calibrated date actually has a smaller relative uncertainty than the measurement itself. A more complex situation exists for the youngest measurement (2160 ^{14}C years BP). A wiggle in the calibration curve implies that there are two possible calibrated dates, each with roughly equal uncertainty. The calibrated date could either be near 2150 or 2300 cal. years BP; it is not possible to tell from the measurement alone.

Radiocarbon dating can be applied to many different types of materials that incorporate atmospheric CO_2 [Taylor and Bar-Yosef, 2014]. The most common and most reliable ages are applied to charcoal from plants, which is largely stable against diagenetic changes in the ^{14}C content. Dating can also be applied to the collagen from bone or to calcium carbonate shells, or many

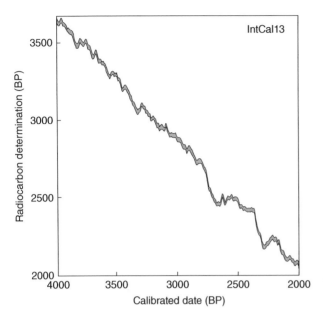

Fig. 13.8. The IntCal13 radiocarbon calibration curve from 2000 to 4000 cal. years BP [*Reimer et al.*, 2013]. This section of the curve is based on nearly 700 ^{14}C measurements of tree rings of known age. The curve represents the best fit and its uncertainties. Short-term wiggles are due to solar variability, while longer-term variations are due to changing geomagnetic field strength. This curve is used to determine the calibrated age (cal. BP) of samples for which the radiocarbon age (^{14}C BP) has been measured and determined using equation (13.1). (Source: *Reimer et al.* [2013]. Reproduced with permission of Cambridge University Press.)

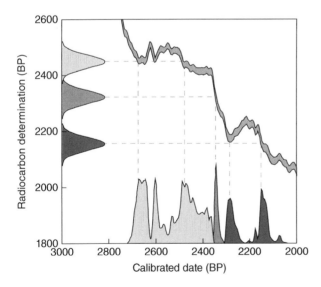

Fig. 13.9. An example of radiocarbon date calibration for three measurements. The oldest sample in light gray corresponds to a flat section of the calibration curve, and so has a broad calibrated date distribution. The center sample in medium gray corresponds to a steep section of the calibration curve, and so has a narrow calibrated date distribution. The youngest sample in dark gray falls near a reversal in the curve, and so exhibits a double peak in its calibrated date distribution. Calibrated date PDFs were generated using OxCal v. 4.2.4.

other materials that incorporate atmospheric CO_2. Caution must be used when dating marine fossils or nearshore deposits because carbon in the oceans can be thousands of years older than in the atmosphere (the reservoir effect). Likewise waters that contain dissolved carbonate rock have carbon that is depleted in ^{14}C so shells, plants, and animals within the water will appear older than their true age (the hard water effect). Another very important effect is the 'bomb pulse' that was produced from atmospheric nuclear testing that peaked in the 1960s and doubled atmospheric ^{14}C. There are such strong variations due to the bomb pulse that biological materials from that era can often be dated with an uncertainty as little as 1–2 years [*Hua*, 2009].

13.4.3.2 Meteoric ^{36}Cl

Chlorine-36 is produced in the atmosphere mostly by spallation of argon. Chlorine-36 production occurs mainly in the stratosphere, where ^{36}Cl remains on average only 2–3 years before deposition [*Elmore et al.*, 1982; *Synal et al.*, 1990]. Atmospheric circulation patterns lead to peak stratosphere-troposphere mixing at mid-latitudes, and a corresponding maximum in ^{36}Cl deposition rates [*Lal and Peters*, 1967]. The ^{36}Cl/Cl ratio in precipitation, however, is also controlled by atmospheric chloride. Because chloride is derived mostly from sea-salt aerosols and is removed by precipitation, its concentration decreases towards continental interiors. The ^{36}Cl/Cl ratio in rainwater at a particular location thus reflects a complex interplay of global geomagnetic field strength, atmospheric production and transport, atmospheric mixing, and the entrainment and precipitation of marine aerosols [e.g., *Phillips*, 2000; *Moysey et al.*, 2003].

After reaching the ground surface ^{36}Cl tends to remain in solution and follows surface water and groundwater flowpaths. It is largely flushed from soil reservoirs, except in arid or hyperarid environments that can accumulate halide salts. Meteoric ^{36}Cl is mostly used to trace and date groundwater flowpaths, although it has also been used in ice cores and even dried packrat urine [*Phillips*, 2000].

13.4.3.3 Meteoric ^{10}Be

Beryllium-10 in the atmosphere (^{10}Be$_{met}$) is produced mainly by spallation of oxygen and nitrogen. It is a particle-reactive species that is rapidly attached to atmospheric aerosols. Beryllium-10 then falls to the ground both in precipitation and as dry deposition of aerosols. It is produced mostly in the stratosphere, similar to ^{36}Cl. Deposition at the surface represents a combination of atmospheric production and stratosphere–troposphere exchange, with a maximum at midlatitudes [e.g., *Lal and Peters*, 1967; *Field et al.*, 2006; *Heikkilä et al.*, 2013].

Meteoric ^{10}Be is used to understand a wide spectrum of environmental and geological processes [*McHargue and Damon*, 1991; *Willenbring and von Blanckenburg*, 2010a]. Beryllium-10 that falls to the ground surface is strongly adsorbed to particles and is highly retentive in most fine-grained soils at neutral to alkaline pH, although it is imperfectly retained in acid or sandy soils [e.g., *E. Brown et al.*, 1992].

To first order it accumulates in a soil profile over time and is lost due to processes of physical erosion, where it is removed together with the soil particles [e.g., *L. Brown et al.*, 1988; *Graly et al.*, 2010; *Willenbring and von Blanckenburg*, 2010a]. Meteoric [10]Be accumulates in glacier ice where it records the local fallout rate [e.g., *Yiou et al.*, 1997]. It also accumulates in marine sediments, where the [10]Be/[9]Be ratio records the deposition of [10]Be from the atmosphere and the delivery of stable [9]Be in rivers from continental weathering [e.g., *Bourlès et al.*, 1989; *Willenbring and von Blanckenburg*, 2010b]. Beryllium-10 is present not only in marine sediments but is also concentrated in slowly growing manganese nodules on the sea floor [e.g., *Sharma and Somayajulu*, 1982; *von Blanckenburg et al.*, 1996]. Beryllium-10 can even be measured in many arc volcanoes due to incorporation of subducted marine sediments into magma generated by the downgoing slab [*Morris et al.*, 1990]. The very presence of [10]Be indicates that subduction recycling occurs over timescales short enough that [10]Be has not decayed, within about 10 Ma.

13.4.3.4 Ice cores

Ice cores provide an exceptionally high-fidelity record of meteoric [10]Be deposited with snow. In some cases the [10]Be deposition can be measured with annual resolution provided that sufficient ice is available [e.g., *Beer et al.*, 1990; *Berggren et al.*, 2009]. Its concentration in falling snow depends not only on the production rate in the atmosphere but also on details of atmospheric transport and precipitation rate. It therefore provides an important complement to the ^{14}C record (see Box 13.1).

Similar to ^{14}C in tree rings, the [10]Be in ice layers records the solar modulation of the cosmic-ray flux to Earth, with the 11-year solar cycle clearly evident. An advantage of [10]Be is that it has a much shorter residence time in the atmosphere [*McHargue and Damon*, 1991], and so the [10]Be record can be used to identify individual solar events and supernovae that produce transient spikes in global cosmogenic nuclide production [e.g., *Miyake et al.*, 2015]. The patterns of [10]Be deposition can be used to correlate ice cores [e.g., *Horiuchi et al.*, 2008], and transient spikes, once identified, can be used to anchor ice-core chronologies.

13.4.3.5 Ocean

Meteoric [10]Be is incorporated into marine sediments and manganese nodules. Its concentration primarily records the long-term deposition of [10]Be$_{met}$ on the oceans, with a small fraction coming from riverine inputs from the continents. Marine sediments thus provide a valuable long-term archive of the global production rate as modulated by the geomagnetic field. Beryllium-10 can be used to date marine sediments, assuming that the initial concentration is known, although more often it is used to explore variations in the geomagnetic field strength [e.g, *Frank et al.*, 1997; *Ménabréaz et al.*, 2014].

Beryllium isotopes in the ocean reflect input of cosmogenic [10]Be as fallout, and also riverine input of [9]Be weathered from minerals [*von Blanckenburg and Bouchez*, 2014]. Because beryllium is particle-reactive, the [10]Be/[9]Be ratio varies in ocean basins, with more [9]Be near continental weathering sources. This behavior allows the [10]Be/[9]Be ratio to be exploited as a proxy for subcontinental-scale to global-scale weathering. If weathering rates increase, and the fraction of [9]Be delivered from continents to the ocean in solution remains fixed, then the [10]Be/[9]Be ratio in the ocean waters will decrease. In this way, [10]Be/[9]Be in mid-ocean basins have been used to show that the global weathering flux has remained roughly constant over the past 12 million years [*Willenbring and von Blanckenburg*, 2010b], despite climate change and Quaternary glaciation.

13.4.3.6 Soils and sediments

Meteoric [10]Be deposited on land can be used to date soils and to estimate long-term erosion rates [*Graly et al.*, 2010; *Willenbring and von Blanckenburg*, 2010a]. As a first approximation the amount of [10]Be at a given site represents the influx due to local fallout, losses due to radioactive decay, losses to groundwater, and losses due to physical transport on eroded mineral grains (Fig. 13.10). This approach assumes a steady soil mass, i.e., that any sediment influx from upslope plus sediment generated by bedrock erosion is balanced by downslope transport. It also ignores any biogeochemical cycling of beryllium.

In this simple case the rate of change of the [10]Be inventory can be written as:

$$\frac{dN}{dt} = F_{met} - E\,N_{ero} - Q\,N_{gw} - \lambda\,N_{soil} \qquad (13.2)$$

where F_{met} is the local fallout rate (which itself can be a function of time), N_{soil} is the total concentration of [10]Be$_{met}$ in the profile (measured in units of atoms per gram), E is the physical erosion rate of soil, N_{ero} is the [10]Be$_{met}$ concentration in the eroded soil, Q is the flux to groundwater and N_{gw} is the [10]Be$_{met}$ concentration in groundwater.

The most common approach is to assume steady state ($dN/dt = 0$) and set the [10]Be$_{met}$ influx equal to outflux. If losses to groundwater are negligible, then the equation is simplified to

$$F_{met} = E\,N_{ero} - \lambda\,N_{soil} \qquad (13.3)$$

Using this simple approximation, if the local fallout rate is known then the erosion rate for a whole hillslope or watershed can be estimated from the concentration of [10]Be transported with the sediment [*Brown et al.*, 1988]. It is often assumed that radioactive decay is negligible, however, that may not always be the case. It must be recognized that there can be many complications, including a strong grain-size dependence to the [10]Be$_{met}$ concentration, isotopic exchange with river water, and violations of steady-state soil inventory due to increases or decreases in the [10]Be$_{met}$ concentration with depth. Some of these issues can be addressed by normalizing the [10]Be concentration by its stable isotope [9]Be, assuming that [9]Be is derived from weathering of a uniform mineral inventory in the watershed [*von Blanckenburg et al.*, 2012]. The use of [10]Be$_{met}$ for estimating erosion rates was popular in the 1980s and early 1990s, but waned with the

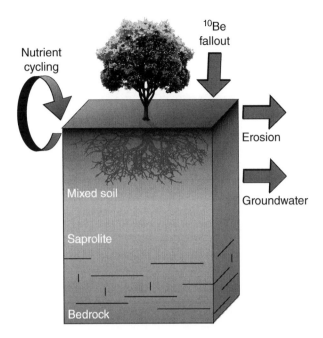

Fig. 13.10. A simplified ^{10}Be$_{met}$ budget in a soil column. Meteoric ^{10}Be accumulates from wet and dry deposition on the surface. It quickly adsorbs to soil particles and is removed over time by particulate erosion. Depending on local rainfall and soil pH, a fraction is also carried away in groundwater. Beryllium can also be passively incorporated into the nutrient cycle, although this remains a little studied aspect of cosmogenic nuclides. (*See insert for color representation of the figure.*)

Nutrient cycling

^{10}Be fallout

Erosion

Groundwater

Mixed soil

Saprolite

Bedrock

advent of methods based on *in situ*-produced ^{10}Be in quartz described in a later section. The method is experiencing a resurgence in popularity due partly to its simplicity but also because the method works over a wide variety of rock types and does not require a particular mineral such as quartz.

Meteoric ^{10}Be can also be used to estimate the minimum time that a soil or landform has been exposed to atmospheric fallout. In this case, equation (13.2) is solved explicitly, with losses to groundwater and erosion usually considered negligible. Because losses are minimized, the age determined in this way will always be a minimum. The accumulation age for ^{10}Be$_{met}$ has been used, for example, to show that soils in the Appalachian piedmont must have been exposed for at least 0.8 Ma [*Pavich et al.*, 1985].

Meteoric ^{10}Be has been used in many other ways, including in lake sediments [e.g., *Czymzik et al.*, 2016], in loess [e.g., *Zhou et al.*, 2014], in rivers [*Kusakabe et al.*, 1991], and in estuarine sediments [*Valette-Silver et al.*, 1986]. While in some cases ^{10}Be$_{met}$ seems to be quantitatively retained in soils, it is clearly not in others. Much work remains to be done to understand how ^{10}Be$_{met}$ is transported through the environment.

13.4.3.7 Cosmogenic nuclides produced *in situ*
Apart from radiocarbon dating, most applications of cosmogenic nuclides are made using nuclides that build up from secondary cosmic ray reactions *in situ* within mineral grains. With the notable exception of ^{3}He, *in situ*-produced cosmogenic nuclides are

retained in mineral grains as a closed system, which greatly simplifies their interpretation. Similar to other closed-system radionuclides, the concentration (N) of a cosmogenic nuclide is determined by a balance between production and decay.

$$\frac{dN}{dt} = P - \lambda N \qquad (13.4)$$

where N is generally in units of atoms per gram of target material, and P is the production rate in atoms per gram per year.

The key difference between cosmogenic nuclide dating and most other methods is that the production term P is not necessarily a constant. While in most parent–daughter dating systems the production rate is either constant or an exponential function of time, production rates of cosmogenic nuclides depend on many factors, including geomagnetic latitude (i.e., cutoff rigidity), altitude (i.e., air pressure), time, and also depth beneath the ground surface.

$$P = f(\text{latitude, altitude, time, depth}) \qquad (13.5)$$

The concentration of cosmogenic nuclides in a mineral grain represents the solution of equations (13.4) and (13.5) over time, which can in many cases be a complicated function of a changing production rate. Cosmogenic nuclide dating thus depends strongly on an individual sample's history. Successful cosmogenic nuclide dating requires finding samples with a sufficiently simple production-rate history such that the cosmogenic nuclides provide useful information about either the sample's exposure age, erosion rate, or the time since exposure in the case of buried sediments.

In the following sections some of the basic assumptions and mathematical treatments for three different scenarios will be discussed. *Surface exposure dating* relies on the assumption that a rock at the surface today was exposed from great depth and has been at the surface ever since. It is appropriate for landforms such as glacial moraines, landslides, or fault scarps. *Erosion rates* can also be measured, assuming that mineral grains at the surface today were exhumed from depth over time. Usually it is assumed that they were exhumed at a constant rate, so that production rates increased exponentially over time due to the exponential production-rate curve (Fig. 13.3). This method is suitable for outcrops, soil-mantled landscapes, and often for detrital minerals in stream sediment. *Burial dating* considers rocks that were first exposed near the surface and accumulated cosmogenic nuclides, but then were subsequently buried. It takes advantage of the relative decay of one nuclide relative to another to date when the rock or minerals were shielded. Burial dating can be applied to cave sediments, to alluvial or marine terraces, or to lake and marine sediments. All three cases are fundamentally based on equation (13.4), but exploit different production-rate histories.

13.4.3.8 Surface exposure dating
In surface exposure dating it is assumed that a rock has been continuously exposed to cosmic rays at the ground surface. The production rate is either considered constant, or it is a known

function of time given changes in geomagnetic field strength [*Lifton et al.*, 2014]. For a constant production rate, solving equation (13.4) yields

$$N(t) = N_{inh}e^{-\lambda t_{exp}} + \frac{P}{\lambda}\left(1 - e^{-\lambda t_{exp}}\right) \quad (13.6)$$

where N_{inh} represents an 'inherited' concentration that is present at the time the rock is first exposed and t_{exp} represents exposure time. If radioactive decay is not important, for example for stable nuclides or for exposure times much shorter than the radioactive half-life, and if there is no inheritance, then equation (13.6) simplifies to

$$N(t) = P\, t_{exp} \quad (13.7)$$

The concentration simply builds up proportionally to how long the rock has been exposed. The exposure age can be easily calculated as long as the production rate is known.

Exposure dating can be applied to any situation where a rock surface was either exposed from depth or, in the case of volcanic rock, was emplaced by an eruption. In either case, ideally the rock begins with no inherited cosmogenic nuclides and the concentration can be interpreted simply in terms of the buildup over time.

The most common application of exposure dating is the study of glacial moraine boulders and glaciated surfaces [*Balco*, 2011]. The method assumes that glaciers erode very rapidly, and that they have exhumed the rocks in their moraines and the bedrock beneath them from several meters depth. In this case, the cosmogenic nuclide inheritance in the rocks will be low. Most of the time this assumption holds true. The boulders within mountain moraines can be derived from rockfalls at the glacier head or from plucking of bedrock beneath the ice. In both cases the cosmogenic nuclide inheritance within the boulders should be low, especially for the majority of the boulder faces that were not exposed at the surface prior to removal. There is always a chance, however, that any particular boulder will have an inheritance; it is important to sample several rocks to be sure that their exposure ages agree.

Exposure dating of glacial moraines and glaciated surfaces can sometimes be ambiguous, especially when ages do not agree [e.g., *Putkonen and Swanson*, 2003; *Heyman et al.*, 2011]. Exposure ages can be younger than the true age if the rocks have been partially shielded by a cover of snow or glacial till that is now eroded away. Measured ages can also be too young if the boulder itself has eroded since deglaciation, or if the boulder rolled to a different orientation after deposition. On the other hand, the exposure age can be too old if the rock was not exhumed by glacial erosion from sufficient depth, so that it contains inherited cosmogenic nuclides. For this reason, researchers normally collect several boulders from a single moraine crest, preferably boulders in stable positions and protruding high enough from the surface that wind is likely to sweep them from snow cover. For young moraines the presence of glacial polish or striations on the rocks indicates with a high degree of confidence that there has been insignificant postglacial erosion. Even with careful

selection at some sites there can be outliers that must be interpreted as having inheritance or snow cover. Most of the time, however, the method yields reliable exposure ages up to a range of about 100,000 years, after which erosion tends to become significant [e.g., *Balco*, 2011; *Heyman et al.*, 2011].

A large boulder on a moraine crest is ideal for surface exposure dating (Fig. 13.11). Ideally the rock was deposited with no prior exposure to cosmic rays, and it is sufficiently tall or windswept that it protrudes above any snow cover. The boulder must also have experienced very little erosion so that its original surface is retained; any suspected erosion must be accounted for.

When dating glacial moraines the presence of inherited cosmogenic nuclides causes the exposure age to be too old. Sometimes, however, inheritance can be used to advantage to show that glaciated surfaces have *not* been eroded. If the deglaciation age is known independently, either from other dating methods or from exposure dating elsewhere, then the presence of excess cosmogenic nuclides indicates that glacial erosion must have been insufficient to remove the topmost few meters of rock [e.g., *Briner and Swanson*, 1998; *Stroeven et al.*, 2002].

One way that cosmogenic nuclide inheritance is used in glacial geomorphology is to diagnose the presence of cold-based glaciation. Cold-based glaciers that are frozen to their beds do not erode the bedrock beneath them. Even though the glacier is flowing, it does not slide across the ground and so fragile features such as weathered bedrock and even plants can be preserved intact, emerging effectively untouched as the ice melts away. Because cold-based glaciers do not leave many traces on the landscape, it can be very difficult to tell whether or for how long a particular area was covered by ice. Long-lived cosmogenic nuclides such as ^{10}Be are largely retained in a rock surface beneath an ice cover, while shorter-lived nuclides such as ^{26}Al or ^{36}Cl decay to a greater degree. *In situ*-produced ^{14}C is especially useful in this regard because it can decay completely away during a single glacial advance [*Goehring et al.*, 2011]. If the glacier is erosive, then all of the cosmogenic nuclides are removed during glaciation and they all indicate the same surface exposure age. If the glacier is cold-based, then the various cosmogenic nuclides are discordant, with the long-lived nuclides having the highest concentrations [e.g., *Briner et al.*, 2014].

Although exposure dating is used most often for glaciated surfaces and glacial moraines, it can be used for any rock that was rapidly exhumed from depth. It can be used to date landslides [e.g., *Ballantyne and Stone*, 2004; *Ivy-Ochs et al.*, 2009; *Barlow et al.*, 2016], impact craters [*Nishiizumi et al.*, 1991], or fault scarps [e.g., *Benedetti et al.*, 2002], as well as volcanic flows [*Phillips et al.*, 1986; *Schimmelpfennig et al.*, 2011]. Practically any rock that has been continuously exposed without erosion can be dated.

13.4.3.9 Profile dating

An important variation of exposure dating is *profile dating* of sedimentary surfaces such as fluvial terraces, marine terraces, or alluvial fans [*Repka et al.*, 1997; *Hidy et al.*, 2010]. The grains within

Fig. 13.11. Sampling a bouldery glacial moraine. It is a good idea to sample multiple boulders that protrude from the moraine surface to ensure that individual samples do not suffer from erosion, prior inheritance, or snow cover. For this boulder the researchers are recording the topographic shielding of low-angle secondary cosmic rays by measuring the angle to the horizon. Such topographic shielding must be applied as a correction factor to the cosmogenic nuclide-production rate. (Source: Courtesy of Marc Caffee.) (*See insert for color representation of the figure.*)

a sedimentary deposit come from an eroding watershed, so they are deposited with an inheritance. The key to profile dating is to assume that the mineral grains being dated all contain the *same* inheritance. While this is most definitely not the case for individual grains within the deposit, it can be true on average for amalgamated samples of many clasts, either sand or gravel. In theory, if enough grains are averaged together then the bulk cosmogenic nuclide concentration averages over the grain-to-grain variability and remains nearly constant within a single stratigraphic unit.

In profile dating the samples from the surface represent the total contribution of inheritance plus exposure, while samples from depths of several meters still contain an equal amount of inheritance but negligible additional buildup from exposure. By collecting samples in a profile, a production curve can be fit to the data to solve for both the exposure age and the average inheritance.

Profile dating has been very effective for dating alluvial terraces and alluvial fans [e.g., *Owen et al.*, 2014], especially in arid environments where the surfaces can be preserved intact with negligible erosion and little bioturbation by plant roots and burrowing animals. It is important to recognize that because the secondary cosmic-ray nucleon flux follows an exponential attenuation with depth below the ground surface, it can be difficult or impossible to distinguish whether a surface has been eroded or not without a very deep profile. That is, eroding the top from an exponential profile still leaves an exponential below. The exception is for a deep profile of several meters, in which case a "knee" in the

profile can be identified at the transition from nucleon spallation to negative muon capture [*Braucher et al.*, 2009].

13.4.3.10 Erosion rates

In contrast to surface exposure dating, which assumes that production rates remain constant, the determination of erosion rates usually assumes that production rates increase exponentially over time as a mineral grain is gradually carried from deep within the bedrock and approaches the surface [*Lal*, 1991]. Mathematically, this can be seen in equations (13.8) and (13.9), which consider production rates as a function of depth.

$$\frac{dN}{dt} = P(x) - \lambda N \tag{13.8}$$

$$P(x) = P_{surf} e^{\frac{-\rho x}{\Lambda}} \tag{13.9}$$

For a surface that is eroding at a steady rate, the depth x decreases steadily over time. In this case the cosmogenic nuclide concentration can obtained by integrating equation (13.8) back through time, beginning at the depth at which the sample was collected. For samples that are collected at the surface, and have no inherited concentration prior to being eroded from depth, the solution is given by equation (13.10).

$$N = \frac{P_{surf}}{\lambda + \rho E / \Lambda} \tag{13.10}$$

Equation (13.10) as written assumes that the cosmogenic nuclide concentration is entirely due to production by nucleon spallation, whose flux decreases exponentially over depth with a length constant of Λ/ρ. In practice, production by muons is also accounted for either by solving equation (13.8) analytically [e.g., *Marrero et al.*, 2016] or by using a multiexponential approximation to the muon depth profile. The interested reader is referred to *Granger and Riebe* [2014] for a more thorough treatment.

13.4.3.11 Erosion of bedrock

Equation (13.10) is strictly valid for steady-state erosion of bedrock. The cosmogenic nuclide concentration reflects an erosion rate E that can be thought of as the rate of mass removal above the sample. The cosmic rays do not record whether mass is removed by solution or by physical weathering; their penetration length is determined almost entirely by mass alone. (The exception is low-energy neutrons, which can contribute to production of ^{36}Cl by neutron capture; however they do not contribute to ^{10}Be or ^{26}Al that are more commonly used to determine erosion rates.)

One way to think about the erosion rate equation is to compare it to the exposure-dating equation. The cosmogenic nuclide concentration is proportional to an "effective exposure time" of $1/(\lambda + \rho E/\Lambda)$. For the case in which radioactive decay can be ignored (i.e., $\rho E/\Lambda >> \lambda$), the effective exposure time represents the amount of time that the mineral grain has spent within one penetration length of the surface.

$$t_{\text{eff}} = \Lambda/\rho E \qquad (13.11)$$

For a rock of density 2.65 g/cm^3 and a penetration length of $L = 160$ g/cm^2, the effective exposure age represents the time taken to erode about 60 cm.

The erosion rate of bedrock has been measured in many different rock types throughout the world [*Portenga and Bierman*, 2011]. It has also been measured not in exposed bedrock outcrops, but instead at the top of weathered bedrock (i.e., saprolite) below the mobile soil layer (Fig. 13.12). In this case the production rate in equation (13.10) is attenuated by the soil cover, but the mathematics and the interpretation remains the same as long as the soil thickness and density remain constant over time. In steady-state, the cosmogenic nuclides at the top of the saprolite record the rate of bedrock conversion to soil at a given site [*Heimsath et al.*, 1997].

13.4.3.12 Vertically mixed soil

In the case of bedrock erosion a mineral grain is gradually exhumed to the surface and then eroded away. Its trajectory through the rock is always upwards, and so equation (13.8) can be easily integrated to show that the cosmogenic nuclides reflect the erosion rate. In contrast, however, most landscapes are mantled by a mobile soil or regolith in which mineral grains are mixed vertically as they travel downslope. It would be impossible to know the history of any particular mineral grain as it is exhumed from bedrock and then carried downslope. How then can we interpret cosmogenic nuclides from vertically mixed soils and detrital sediment? The answer comes from averaging over many grains [*E. Brown et al.*, 1995; *Bierman and Steig*, 1996; *Granger et al.*, 1996].

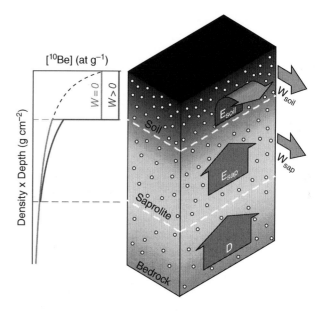

Fig. 13.12. Erosion and weathering convert bedrock to saprolite and soil. The total denudation rate (D) is equal to the sum of erosion (E) and chemical weathering (W). Weathering of minerals during the conversion of bedrock to soil leaves behind resistant minerals such as quartz, illustrated by the white circles. Additional weathering within the soil increases the quartz concentration. This results in a higher cosmogenic nuclide concentration in the quartz than would otherwise be present without chemical weathering. The curve at left shows the expected ^{10}Be concentration as a function of depth within the column, expressed in units of g/cm^2. The dashed line shows the ^{10}Be production rate. Vertical mixing within the soil homogenizes the upper part of the profile, leading to the gray curve for the case where weathering is negligible (equation (13.17)). Chemical weathering leads to higher concentrations in the profile, illustrated as the black curve (equation (13,18)). (See insert for color representation of the figure.)

In the simplest case of vertical mixing within the soil we can use a single box model to average cosmogenic nuclide concentration in the soil as if it were homogenized [*Granger et al.*, 1996]. Such a model is undoubtedly oversimplified for cases where the vigor of mixing depends on depth, but it serves to illustrate the power of analyzing bulk sediment. Figure 13.12 shows a column of rock with a weathering front that grows downward. Pristine bedrock at the bottom of the column is gradually weathered to saprolite as it approaches the surface, losing some of its mass in solution but remaining intact. The saprolite is then broken up and vertically mixed within the regolith or soil, where there is additional loss to weathering as well as losses by physical transport by erosion. Within the mixed zone mineral grains are added from below, they are lost to erosion, and they are also lost to chemical weathering.

A simple differential equation can be used to describe the total accumulation rate of cosmogenic nuclides within mineral grains in the soil. Ignoring chemical weathering, which will be dealt with later, the average concentration in the soil will build up according to equation (13.12)

$$\frac{d\langle N\rangle_{\text{soil}}}{dt} = \langle P\rangle + N_{\text{sap}}\frac{E}{h} - \lambda\langle N\rangle_{\text{soil}} - \langle N\rangle_{\text{soil}}\frac{E}{h} \qquad (13.12)$$

where $<N>_{soil}$ represents the average concentration of cosmogenic nuclides in the soil, $<P>$ represents the depth-averaged production rate, and N_{sap} represents the concentration of material introduced to the soil by erosion at the top of the saprolite. At steady-state $d<N>/dt = 0$ and the average concentration of cosmogenic nuclides in the soil may be represented by:

$$\langle N \rangle_{soil} = \frac{\langle P \rangle + N_{sap}\dfrac{E}{h}}{\lambda + \dfrac{E}{h}} \qquad (13.13)$$

The depth-averaged production rate can be found by integrating production over the soil depth h, and is given by

$$\langle P \rangle = P_{surf}\frac{\Lambda}{\rho h}\left(1 - e^{\frac{-\rho h}{\Lambda}}\right) \qquad (13.14)$$

and the concentration at the top of the saprolite is given by

$$N_{sap} = \frac{P_{surf} e^{\frac{-\rho h}{\Lambda}}}{\lambda + \dfrac{\rho E}{\Lambda}} \qquad (13.15)$$

Substituting these into equation (13.12) and solving for steady state,

$$\langle N \rangle_{soil} = \frac{P_{surf}\dfrac{\Lambda}{\rho h}}{\lambda + \dfrac{E}{h}}\left[1 - e^{\frac{-\rho h}{\Lambda}} + \left(\frac{\rho E}{\Lambda}\right)\frac{e^{\frac{-\rho h}{\Lambda}}}{\lambda + \dfrac{\rho E}{\Lambda}}\right] \qquad (13.16)$$

For the case in which the time to erode one penetration length is much shorter than the radioactive mean-life (i.e., $\rho E/\Lambda > 1/\lambda$), radioactive decay can be ignored and equation (13.16) simplifies to

$$\langle N \rangle_{soil} = \frac{P_{surf}\Lambda}{\rho E} \qquad (13.17)$$

This is the identical simple analytical solution as obtained for the surface erosion of bedrock. While this may seem counterintuitive, at steady-state the export of cosmogenic nuclides by erosion must balance the total cosmogenic nuclide-production rate, whether in bedrock or in soil. Consequently, the concentration in the entire mixed soil is equal to the concentration at the top of bedrock, and so the total inventory of cosmogenic nuclides in soil is higher than in rock (Fig. 13.12).

The treatment above is simplified in that it neglects losses to weathering. A more complete solution that includes chemical erosion can be calculated for simple cases [*Riebe et al.*, 2001; *Dixon et al.*, 2009; *Riebe and Granger*, 2013]. Figure 13.12 illustrates such a conceptual model. Usually the mineral quartz is analyzed because it is resistant to chemical weathering and it is known to remain free of contamination. During chemical weathering, however, the fractional concentration of quartz in the bedrock and soil increases as other minerals are dissolved away. This enrichment of quartz implies that that residence time

of quartz within the soil is longer than the residence time for other, more easily weathered minerals.

The concentration of cosmogenic nuclides in quartz can be expressed in terms of quartz enrichment factors. Ignoring radioactive decay, equation (13.17) is rewritten as

$$\langle N \rangle_{soil} = \frac{P_{surf}\Lambda}{\rho D}\left\{\frac{f_{sap}}{f_{bedrock}}\left[\frac{f_{soil}}{f_{sap}}\left(1 - e^{\frac{-\rho h}{\Lambda}}\right) + e^{\frac{-\rho h}{\Lambda}}\right]\right\} \qquad (13.18)$$

where f_{soil}, f_{sap}, and $f_{bedrock}$ represent the mass fraction of quartz in soil, saprolite, and bedrock, and D represents denudation, equal to the sum of chemical and physical erosion. The correction due to chemical erosion can range from a few percent in desert environments or where erosion is rapid, to a factor of two or more where chemical weathering is intense, for example in tropical environments or in calcareous or other soluble bedrock [*Riebe and Granger*, 2013].

13.4.3.13 Mixing across landscapes: basin-averaged erosion rates

If cosmogenic nuclide concentrations in soils represent the local erosion rate, regardless of the thickness of the mixed soil layer, then it follows that cosmogenic nuclide concentrations in stream sediment should represent the erosion rate of an entire watershed. This is true for the special case in which radioactive decay can be ignored (i.e., if erosion rates are significantly faster than about one penetration length per mean-life or about 1 m/Ma for ^{26}Al and ^{10}Be) and if quartz is present at equal abundance throughout the watershed.

The concentration of a cosmogenic nuclide in well-mixed stream sediment can be straightforwardly calculated as a flux-weighted average (Fig. 13.13). The flux of quartz from a particular part of the watershed should be proportional to its local erosion rate. Conversely, equation (13.17) shows that the concentration of cosmogenic nuclides should be *inversely* proportional to the erosion rate. These two effects balance each other out, so that the average concentration is simply equal to the spatially averaged erosion rate for the basin. This can be seen in equations (13.19)–(13.21), where the areal integrals are used to calculate the flux-weighted spatially averaged cosmogenic nuclide concentration in stream sediment.

$$\langle N \rangle_{stream} = \frac{A\int \rho E \langle N \rangle_{soil} dA}{A\int \rho E dA} \qquad (13.19)$$

$$\langle N \rangle_{stream} = \frac{A\int \rho_{surf} \Lambda dA}{A\int \rho E dA} \qquad (13.20)$$

$$\langle N \rangle_{stream} = \langle P \rangle_{surf}\frac{\Lambda}{\rho \langle E \rangle} \qquad (13.21)$$

Equation (13.21) shows that the average concentration in well-mixed stream sediment can be interpreted in terms of the spatially averaged production rate and the spatially averaged erosion rate.

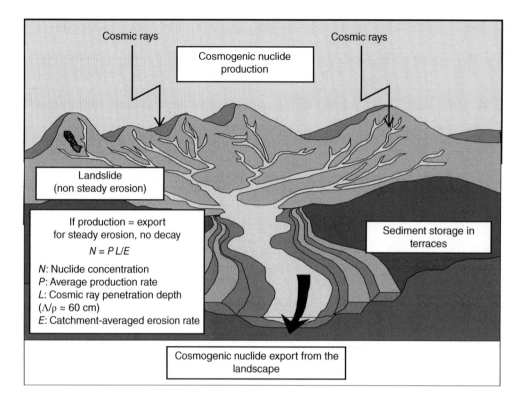

Fig. 13.13. Streams mix sediment that is eroded from throughout the landscape. The average ^{10}Be concentration in stream sediment can be used to estimate the spatially averaged erosion rate in the watershed, assuming steady state as in equation (13.21). Complications can arise due to nonsteady erosion processes such as landslides, or due to sediment storage and remobilization in terraces. Adapted from. (Source: *Granger and Schaller* [2014]. Reproduced with permission of Mineralogical Society of America.) (*See insert for color representation of the figure.*)

Many studies throughout the world have used equation (13.21) to estimate basin-averaged erosion rates as a function of climate, tectonics, and land use [see *Portenga and Bierman*, 2011; *Willenbring et al.*, 2013). There are many complications that can compromise the interpretation, including: uneven quartz distribution; landsliding that can bias the quartz content towards a small area; storage of sediment in valley bottoms or in stream terraces; chemical weathering; and radioactive decay for slowly eroding watersheds. Nonetheless, there is generally good agreement among replicate samples within a watershed and the cosmogenic nuclide method has become a standard way to estimate erosion rates over timescales of thousands to hundreds of thousands of years [*Portenga and Bierman*, 2011].

Just as for erosion of bedrock, we can think about erosion rates of landscapes as representing the time taken to erode one cosmic-ray penetration length (about 60 cm of bedrock). In a crude way this is analogous to the interpretation of thermochronometers such as (U–Th)/He or fission tracks. In the case of thermochronometry the age represents the time taken to exhume a mineral from its depth zone of partial annealing (PAZ) or partial retention (PRZ). In the case of cosmogenic nuclides the age represents the time taken to exhume a mineral grain through a cosmic ray penetration length. Much of the time, a cosmic ray penetration

length is comparable to the depth of the soil. The cosmogenic nuclides thus measure erosion rates over timescales relevant to soil formation.

13.4.3.14 Paleoerosion rates

If cosmogenic nuclides in the quartz of modern stream sediment measure the erosion rate of the watershed, then cosmogenic nuclides in quartz in sedimentary archives should represent the erosion rate at the time of sediment deposition. This is known as the paleoerosion rate. It can be calculated for sediment that was buried quickly and deeply, for example on a river terrace, within a sedimentary basin, or even in sediment washed into caves. If the age of the sediment is known, then the measured cosmogenic nuclide concentration can be corrected for radioactive decay and the concentration can be interpreted similarly to modern sediment [e.g., *Schaller et al.*, 2002].

Complications can arise if sedimentation is slow, in which case the sediment accumulates an additional dose of cosmogenic nuclides during gradual burial. Over long timescales in tectonically active environments the elevation of the source area may also change over time. Because production rates vary with altitude it can be difficult or impossible to distinguish long-term uplift or subsidence from a long-term shift in erosion rates. Paleoerosion rate methods have been used to examine changes in erosion rate over glacial–interglacial timescales as well as over millions of years [e.g., *Charreau et al.*, 2011; *Granger and Schaller*, 2014; *Hidy et al.*, 2014].

13.4.3.15   Burial dating

Every rock exposed near the ground surface accumulates cosmogenic nuclides. The longer the rock is exposed or the slower the erosion rate, the higher the concentration of cosmogenic nuclides within it. If a rock from the surface is subsequently buried and shielded from further cosmic-ray exposure, then the cosmogenic radionuclides that it inherited from exposure will decay away over time. This is the fundamental principle of burial dating. As discussed in the section on extraterrestrial applications, burial dating is an idea that was first used to date the terrestrial age of meteorites. It is now used to date terrestrial sediments in a variety of deposition settings ranging from caves and fluvial terraces to lakes, marine terraces, and oceans.

The concentration of a cosmogenic nuclide in a rock that was first exposed at the surface and then buried underground can be modeled as

$$N_i(t) = N_{i,\text{inh}} e^{-\lambda_i t} + \int P_i(t) e^{-\lambda_i t} dt \qquad (13.22)$$

where the subscript i refers to a particular cosmogenic nuclide and P refers to the production rate in the mineral grain after burial. The integral in equation (13.22) represents the total postburial production that has occurred from the time of burial to present. For deeply buried rocks the postburial production is very small and can often be ignored, but for rocks buried only several meters deep postburial production can be significant.

The key to burial dating is to solve equation (13.22) for two different cosmogenic nuclides simultaneously. If the inherited ratio of the two nuclides is known, and if the postburial production rates are known, then the differential decay of the two nuclides provides a clock that dates the time of burial [*Lal and Arnold*, 1985; *Klein et al.*, 1986]. The nuclide pair that is used almost exclusively for burial dating is ^{26}Al and ^{10}Be in the mineral quartz. This is because these particular nuclides have a production rate ratio that is nearly constant, and they have similar production rate profiles with depth. Their production rates are well known and they can both be measured in quartz, which is common in sediments.

If quartz is buried very deeply, for example in a cave or under a large column of water deep in a lake or marine setting, then postburial production can be ignored. In that case, equation (13.22) can be written for both ^{26}Al and ^{10}Be in a form that is known as simple burial dating.

$$\left(\frac{N_{26}}{N_{10}}\right)(t) = \left(\frac{N_{26}}{N_{10}}\right)_{\text{inh}} e^{-\lambda_{\text{bur}} t} \qquad (13.23)$$

where

$$\lambda_{\text{bur}} = \lambda_{26} - \lambda_{10} = 0.4808 \times 10^{-6} year^{-1} \qquad (13.24)$$

Solution of equation (13.23) requires that the initial ^{26}Al/^{10}Be ratio be known. It is most commonly assumed that the sediment is derived from a watershed that is eroding at steady-state. In this case the initial ratio can be estimated from equation (13.21), solved for the ratio N_{26}/N_{10}.

$$\left(\frac{N_{26}}{N_{10}}\right)_{\text{inh}} = \frac{P_{26}}{P_{10}} \frac{\left(\lambda_{10} + \dfrac{\rho E}{\Lambda}\right)}{\left(\lambda_{26} + \dfrac{\rho E}{\Lambda}\right)} \qquad (13.25)$$

Alternatively, if there is good reason to believe that the sediment is better represented by a case of constant exposure, then the initial ratio can be estimated from equation (13.26).

$$\left(\frac{N_{26}}{N_{10}}\right)_{\text{inh}} = \frac{P_{26}\lambda_{10}}{P_{10}\lambda_{26}} \frac{\left(1 - e^{-\lambda_{26} t}\right)}{\left(1 - e^{-\lambda_{10} t}\right)} \qquad (13.26)$$

There are two ways to visualize these equations graphically. The first is known as an exposure–burial diagram [*Klein et al.*, 1986; *Lal*, 1991], or sometimes informally as a 'banana diagram' due to its shape (Fig. 13.14). In an exposure–burial diagram the ^{26}Al/^{10}Be ratio is plotted as a function of ^{10}Be concentration. There are two lines at the top: one represents steady-state erosion and the other represents constant exposure. The space between those lines is known as the steady erosion island [*Lal*, 1991]. There is no physical way for samples to lie above these lines unless there is ^{26}Al contamination, which is unlikely. The position of a sample on these lines depends on the exposure time or the erosion rate. The higher the exposure time or the slower the production rate, the further to the right a sample will plot on the curves. The two curves terminate at a location that represents secular equilibrium, where production rates are balanced by decay. It is not possible for samples to lie to the right of this point, unless their production rates were higher in the past. After burial, both ^{26}Al and ^{10}Be will decay, so the measured ^{26}Al/^{10}Be ratio will lie below the curves. Decay lines are shown as well as the position expected at million year intervals.

A second way to visualize the equations is to plot ^{26}Al versus ^{10}Be (Fig. 13.15). It is helpful to plot the concentrations normalized by their secular equilibrium values so that the plot can be applied universally. These dimensionless concentrations are denoted by a star.

$$N_i^* = \frac{N_i \lambda_i}{P_i} \qquad (13.27)$$

On such a graph the concentrations due to steady erosion and constant exposure at the surface are seen to be two gentle curves that begin at the origin and terminate at secular equilibrium. Just as on a traditional exposure–burial diagram the position on the line represents the time of exposure or the rate of erosion.

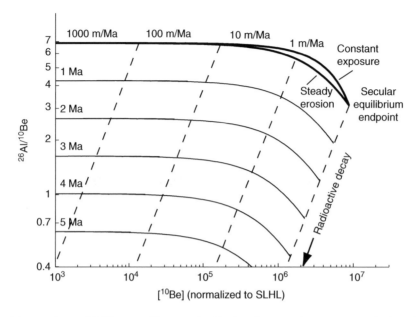

Fig. 13.14. An exposure-burial diagram showing $^{26}Al/^{10}Be$ versus ^{10}Be, normalized by the reference production rate at sea level and high latitude (SLHL). On this diagram samples at the surface are expected to fall between the steady erosion line predicted by equation (13.10) and the constant exposure line predicted by equation (13.6). Their position on the line is determined by the erosion rate labeled above the curve, or by the exposure time. The maximum ^{10}Be and ^{26}Al concentrations are reached at the secular equilibrium endpoint. If a sample is buried, then the $^{26}Al/^{10}Be$ ratio decreases parallel to the dashed lines. The location of a sample on this diagram can be used to infer the burial age, shown in millions of years, as well as the pre-burial erosion rate. (Source: Granger 2006. Reproduced with permission of Geological Society of America.)

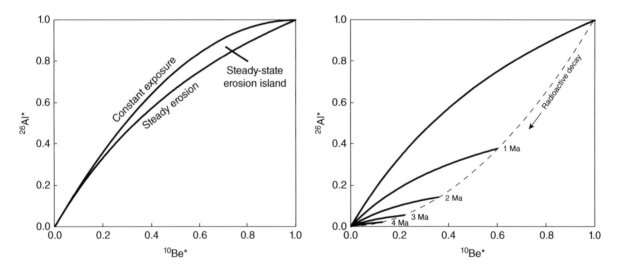

Fig. 13.15. The top panel shows a normalized plot of $^{26}Al^*$ versus $^{10}Be^*$ for samples that are eroding at steady state or continuously exposed. The "steady state erosion island" lies between these two curves, similar to that on Fig. 13.14. Samples with fast erosion rates or short exposure times plot near the origin, while those with very long exposure times or slow erosion rates approach secular equilibrium at the top right-hand corner. The bottom panel illustrates a family of curves for samples that begin on the steady erosion line but have been buried. As the burial time increases, the slope of the curve becomes more gentle and its length becomes shorter. The dashed line illustrates decay from the secular equilibrium endpoint.

After burial radioactive decay causes the curves to become shorter and to decrease systematically in slope. This representation of ^{26}Al and ^{10}Be is very useful for isochron burial-dating methods described in the next section [*Balco and Rovey*, 2008].

Simple burial dating often works very well for cave sediments and for deeply buried sediments [e.g., *Granger et al.*, 1997, 2001). It does suffer some important weaknesses, however. First, it must be recognized that when using one sample alone it is impossible to identify whether it has experienced more than one burial episode. If the sample was buried once and then reworked into another deposit prior to being sampled, then the burial age will be older than the true age. Also, if there is any postburial production due to insufficient burial depth then the burial age will be younger than the true age. For these reasons, it is often better to use the isochron burial-dating approach, which explicitly identifies and solves for these two complications.

13.4.3.16 Isochron burial dating

In isochron burial dating equation (13.22) is solved for multiple samples that share the same burial age but different inheritances [*Balco and Rovey*, 2008]. In some ways this is similar to other isochron methods in which different mineral fractions of the same age have variable amounts of a parent radionuclide that decay to a daughter. The slope of parent–daughter isochron indicates the age, even if the entire sample has excess daughter inherited from the time of crystallization. In important ways isochron burial dating is different though, because it is not a parent–daughter system but instead uses two radionuclides that begin at a known ratio. Still, the slope of the burial-dating isochron indicates the burial age, even if all of the samples have excess cosmogenic nuclides due to postburial production.

The basis of isochron dating is that equation (13.22) can be solved for a suite of samples, each of which has an identical amount of postburial production. No matter how complicated the burial history, all postburial production must be shared by all of the samples together. In this case the postburial production can be treated as a constant among all of the samples. Equation (13.22) can be reexpressed as

$$N_i(t) = N_{i,\text{inh}}e^{-\lambda_i t} + N_{i,pb} \tag{13.28}$$

where the subscript pb represents postburial and $N_{i,pb}$ is a constant that is shared among all of the samples.

Solving equation (13.28) for ^{26}Al and ^{10}Be yields the following equation.

$$N_{26} = \left(N_{10} - N_{10,pb}\right)\left(\frac{N_{26}}{N_{10}}\right)_{\text{inh}} e^{-\lambda_{\text{bur}}t} + N_{26,pb} \tag{13.29}$$

Substituting the solutions for $(N_{26}/N_{10})_{\text{inh}}$ for steady erosion, and using the star notation introduced above,

$$N_{26}^* = \left(N_{10}^* - N_{10,pb}^*\right)\frac{\left(1 + \dfrac{\rho E}{\lambda_{10}\Lambda}\right)}{\left(1 + \dfrac{\rho E}{\lambda_{26}\Lambda}\right)}e^{-\lambda_{\text{bur}}t} + N_{26,pb}^* \tag{13.30}$$

Equation (13.30) describes a gentle curve on a graph of N_{26}^* versus N_{10}^*, as illustrated in Fig. 13.15. The intercept of the curve depends on the values for postburial production, while the slope of the curve depends on the burial age.

As with all isochron methods, the burial-dating isochron approach requires that samples have a spread in their initial concentrations. In this case, the samples must have variable amounts of inherited cosmogenic nuclides so that they are distributed out on the steady erosion or constant exposure line prior to burial. There are several ways to ensure variability. One is to sample along a depth profile in a buried paleosol [*Balco and Rovey*, 2008]. In this case the variability in the inheritance at the time

of burial simply reflects the production-rate curve with depth. Alternatively, one can collect various grain-size fractions [*Balco et al.*, 2013], samples with varying degrees of weathering, or simply analyze individual gravel clasts within a conglomerate [e.g., *Erlanger et al.*, 2012]. In all of these cases it is difficult to know *a priori* how much variability in the inherited concentrations will occur, but it is usually the case that different samples have different amounts of inheritance. (The exception is for samples derived from glaciated watersheds, where all samples begin with very low inheritances.)

There are several advantages to the isochron approach for burial dating. First, it allows the identification of individual clasts that may have experienced prior burial, because they will lie below the line formed by the other clasts. Second, it allows the explicit solution of postburial production if there is any offset of the isochron from the origin. Isochron burial dating can be successfully achieved for samples buried only a few meters deep, as opposed to tens of meters often required for simple burial dating. Isochron burial dating is becoming an important tool for dating cave and terrace deposits not only in geomorphology but in archaeology and human evolution as well [e.g., *Granger et al.*, 2015].

13.5 CONCLUSION

Cosmogenic nuclide geochronology differs from most radiogenic nuclide dating in that production rates depend on external factors. For meteorites this is primarily the shielding depth within the parent body, but on Earth it also includes factors such as altitude, latitude, depth beneath the ground surface, and geomagnetic field strength over time. Consequently there has been a suite of applications developed that exploit variability in production rates to quantify things such as mineral exposure time to cosmic radiation, erosion rates of surfaces, and burial ages.

Cosmogenic nuclides produced *in situ* within mineral grains can be used to measure surface exposure, whether the exposure of a meteorite in space or the exposure time of boulders on a glacial moraine. In a real sense the cosmogenic nuclides reveal the time since the material that shielded that rock disappeared. This gives information about the rock and its environment that would be difficult or impossible to learn otherwise, whether it is the breakup time of a meteoroid in space or the time since a glacier melted.

In situ-produced cosmogenic nuclides can also be used to learn about erosion and weathering rates, whether of exposed bedrock, saprolite, soil, or detrital sediment. Cosmogenic nuclides measure erosion rates over the time taken to remove the uppermost meter or so of rock and soil. This is an important timescale for forming soils due to mechanical and chemical weathering and for depleting them by land use. The quantification of erosion rates has been important in geomorphology for understanding the factors that influence soil formation rates as well as how landscapes evolve in response to tectonic forcing or climate change.

A third use for *in situ*-produced cosmogenic nuclides is to date when a rock was shielded from cosmic radiation using the burial dating methods. This could be a meteorite that is 'buried' beneath Earth's atmosphere when it falls to the ground, or a grain of quartz that is buried in a cave or a sedimentary deposit. Applications of burial dating with ^{26}Al and ^{10}Be in quartz have ranged from landscape evolution to archaeology and human evolution.

Much higher quantities of cosmogenic nuclides are produced in the atmosphere than in mineral grains, and they form the basis of radiocarbon dating of organic matter. Once nuclides such as ^{10}Be and ^{36}Cl are removed from the atmosphere into the environment they also can be used to trace soil and sediment, to date groundwater, to date and trace marine sediments, and to date carbon-rich materials such as organic matter and carbonate minerals. They can be used to infer past changes in the geomagnetic field strength from sedimentary records in marine sediments and in ice sheets.

Improvements in cosmogenic nuclide geochronology are occurring rapidly. New methods and increased sensitivity of accelerator mass spectrometry is allowing smaller quantities of cosmogenic nuclides to be measured more precisely. Improved measurements and understanding of the variation in production rates over time and space is yielding more accurate surface exposure ages. These improvements will undoubtedly lead to new applications and a better understanding of earth surface and environmental processes.

13.6 REFERENCES

Ackermann, M., Ajello, M., Allafort, A., *et al.* (2013) Detection of the characteristic pion-decay signature in supernova remnants. *Science* **339**(6121), 807–811.

Anderson, E. C., Libby, W. F., Weinhouse, S., Reid, A. F., Kirshenbaum, A. D., and Grosse, A. V. (1947). Natural radiocarbon from cosmic radiation. *Physical Review* **72**(10), 931.

Arnold, J. R. and Libby, W. F. (1949) Age determinations by radiocarbon content: checks with samples of known age. *Science* **110** (2869), 678–680.

Arnold, J. R. and Al-Salih, H. A. (1955) Beryllium-7 produced by cosmic rays. *Science* **121**(3144), 451–453.

Balco, G. (2011) Contributions and unrealized potential contributions of cosmogenic-nuclide exposure dating to glacier chronology 1990–2010. *Quaternary Science Reviews* **30**(1), 3–27.

Balco, G. and Rovey, C. W. (2008) An isochron method for cosmogenic-nuclide dating of buried soils and sediments. *American Journal of Science* **308**(10), 1083–1114.

Balco, G., Stone, J. O., Lifton, N. A., and Dunai, T. J. (2008) A complete and easily accessible means of calculating surface exposure ages or erosion rates from ^{10}Be and ^{26}Al measurements. *Quaternary Geochronology* **3**(3), 174–195.

Balco, G., Soreghan, G. S., Sweet, D. E., Marra, K. R., and Bierman, P. R. (2013) Cosmogenic-nuclide burial ages for Pleistocene sedimentary fill in Unaweep Canyon, Colorado, USA. *Quaternary Geochronology* **18**, 149–157.

Ballantyne, C. K. and Stone, J. O. (2004) The Beinn Alligin rock avalanche, NW Scotland: cosmogenic ^{10}Be dating, interpretation and significance. *The Holocene* **14**(3), 448–453.

Barlow, J., Moore, R., and Gheorghiu, D. M. (2016) Reconstructing the recent failure chronology of a multistage landslide complex using cosmogenic isotope concentrations: St Catherine's Point, UK. *Geomorphology* **268**, 288–295.

Bauer, C. A. (1947) Production of helium in meteorites by cosmic radiation. *Physical Review* **72**(4), 354.

Begemann, F., Geiss, J., and Hess, D. C. (1957) Radiation age of a meteorite from cosmic-ray-produced He3 and H3. *Physical Review* **107**(2), 540.

Beer, J., Blinov, A., Bonani, G., Hofmann, H. J., and Finkel, R. C. (1990) Use of Be-10 in polar ice to trace the 11-year cycle of solar activity. *Nature* **347**, 164–166.

Benedetti, L., Finkel, R., Papanastassiou, D., *et al.* (2002) Postglacial slip history of the Sparta fault (Greece) determined by 36Cl cosmogenic dating: evidence for non-periodic earthquakes. *Geophysical Research Letters* **29**(8).

Bennett, C. L., Beukens, R. P., Clover, M. R., *et al.* (1977) Radiocarbon dating using electrostatic accelerators: negative ions provide the key. *Science* **198**(4316), 508–510.

Bentley, H. W., Phillips, F. M., and Davis, S. N. (1986) Chlorine-36 in the terrestrial environment. *Handbook of Environmental Isotope Geochemistry* **2**, 427–480.

Berggren, A. M., Beer, J., Possnert, G., *et al.* (2009) A 600-year annual 10Be record from the NGRIP ice core, Greenland. *Geophysical Research Letters* **36**(11).

Bierman, P. R. (1994) Using *in situ* produced cosmogenic isotopes to estimate rates of landscape evolution: A review from the geomorphic perspective. *Journal of Geophysical Research: Solid Earth* **99**(B7), 13885–13896.

Bierman, P. and Steig, E. J. (1996) Estimating rates of denudation using cosmogenic isotope abundances in sediment. *Earth Surface Processes and Landforms* **21**(2), 125–139.

Bourlès, D., Raisbeck, G. M., and Yiou, F. (1989) ^{10}Be and ^{9}Be in marine sediments and their potential for dating. *Geochimica et Cosmochimica Acta* **53**(2), 443–452.

Braucher, R., Del Castillo, P., Siame, L., Hidy, A. J., and Bourles, D. L. (2009) Determination of both exposure time and denudation rate from an *in situ*-produced ^{10}Be depth profile: a mathematical proof of uniqueness. Model sensitivity and applications to natural cases. *Quaternary Geochronology*, **4**(1), 56–67.

Briner, J. P. and Swanson, T. W. (1998) Using inherited cosmogenic ^{36}Cl to constrain glacial erosion rates of the Cordilleran ice sheet. *Geology* **26**(1), 3–6.

Briner, J. P., Lifton, N. A., Miller, G. H., Refsnider, K., Anderson, R., and Finkel, R. (2014) Using *in situ* cosmogenic ^{10}Be, ^{14}C, and ^{26}Al to decipher the history of polythermal ice sheets on Baffin Island, Arctic Canada. *Quaternary Geochronology* **19**, 4–13.

Brown, E. T., Edmond, J. M., Raisbeck, G. M., Bourlès, D. L., Yiou, F., and Measures, C. I. (1992) Beryllium isotope geochemistry in tropical river basins. *Geochimica et Cosmochimica Acta* **56**(4), 1607–1624.

Brown, E. T., Stallard, R. F., Larsen, M. C., Raisbeck, G. M., and Yiou, F. (1995) Denudation rates determined from the accumulation of *in situ*-produced [10]Be in the Luquillo Experimental Forest, Puerto Rico. *Earth and Planetary Science Letters* **129**(1) 193–202.

Brown, L., Pavich, M. J., Hickman, R. E., Klein, J., and Middleton, R. (1988) Erosion of the eastern United States observed with [10]Be. *Earth Surface Processes and Landforms* **13**(5), 441–457.

Charalambus, S. (1971) Nuclear transmutation by negative stopped muons and the activity induced by the cosmic-ray muons. *Nuclear Physics A* **166**(2), 145–161.

Charreau, J., Blard, P. H., Puchol, N., *et al.* (2011) Paleo-erosion rates in Central Asia since 9 Ma: a transient increase at the onset of Quaternary glaciations? *Earth and Planetary Science Letters* **304**(1), 85–92.

Collon, P., Kutschera, W., and Lu, Z. T. (2004) Tracing noble gas radionuclides in the environment. *Annual Review of Nuclear and Particle Science* **54**, 39–67.

Czymzik, M., Adolphi, F., Muscheler, R., *et al.* (2016) A varved lake sediment record of the [10]Be solar activity proxy for the Lateglacial-Holocene transition. *Quaternary Science Reviews* **153**, 31–39.

Davis, R. and Schaeffer, O. A. (1955) Chlorine-36 in nature. *Annals of the New York Academy of Sciences*, **62**(5), 107–121.

Dep, L., Elmore, D., Fabryka-Martin, J., Masarik, J., and Reedy, R. C. (1994) Production rate systematics of *in situ*-produced cosmogenic nuclides in terrestrial rocks: Monte Carlo approach of investigating [35]Cl (n, γ) [36]Cl. *Nuclear Instruments and Methods in Physics Research Section B: Beam Interactions with Materials and Atoms* **92**(1), 321–325.

Dixon, J. L., Heimsath, A. M., and Amundson, R. (2009) The critical role of climate and saprolite weathering in landscape evolution. *Earth Surface Processes and Landforms* **34**(11), 1507–1521.

Dunai, T. J. (2010) *Cosmogenic Nuclides: Principles, Concepts and Applications in the Earth Surface Sciences.* Cambridge University Press.

Dunne, J., Elmore, D., and Muzikar, P. (1999) Scaling factors for the rates of production of cosmogenic nuclides for geometric shielding and attenuation at depth on sloped surfaces. *Geomorphology* **27**(1), 3–11.

Elmore, D., Tubbs, L. E., Newman, D., *et al.* (1982) [36]Cl bomb pulse measured in a shallow ice core from Dye 3, Greenland. *Nature* **300**, 735–737.

Erlanger, E. D., Granger, D. E., and Gibbon, R. J. (2012) Rock uplift rates in South Africa from isochron burial dating of fluvial and marine terraces. *Geology* **40**(11), 1019–1022.

Eugster, O., Herzog, G. F., Marti, K., and Caffee, M. W. (2006) Irradiation records, cosmic-ray exposure ages, and transfer times of meteorites. *Meteorites and the Early Solar System* II, 829–851.

Fairbanks, R. G., Mortlock, R. A., Chiu, T. C., *et al.* (2005) Radiocarbon calibration curve spanning 0 to 50,000 years BP based on paired [230]Th/[234]U/[238]U and ^{14}C dates on pristine corals. *Quaternary Science Reviews* **24**(16), 1781–1796.

Field, C. V., Schmidt, G. A., Koch, D., and Salyk, C. (2006) Modeling production and climate-related impacts on [10]Be concentration in ice cores. *Journal of Geophysical Research: Atmospheres*, **111**(D15).

Fireman, E. L. and Schwarzer, D. (1957) Measurement of Li6, He3, and H3 in meteorites and its relation to cosmic radiation. *Geochimica et Cosmochimica Acta* **11**(4), 252–262.

Frank, M., Schwarz, B., Baumann, S., Kubik, P. W., Suter, M., and Mangini, A. (1997) A 200 kyr record of cosmogenic radionuclide production rate and geomagnetic field intensity from [10]Be in globally stacked deep-sea sediments. *Earth and Planetary Science Letters* **149**(1), 121–129.

Goehring, B. M., Schaefer, J. M., Schluechter, C., *et al.* (2011) The Rhone Glacier was smaller than today for most of the Holocene. *Geology* **39**(7), 679–682.

Goel, P. S., Jha, S., Lal, D., and Radhakrishna, P. (1956) Cosmic ray produced beryllium isotopes in rain water. *Nuclear Physics* **1**(8) 196–201.

Goel, P. S. and Kohman, T. P. (1962) Cosmogenic carbon-14 in meteorites and terrestrial ages of "finds" and craters. *Science* **136**(3519), 875–876.

Gosse, J. C. and Phillips, F. M. (2001) Terrestrial *in situ* cosmogenic nuclides: theory and application. *Quaternary Science Reviews* **20**(14), 1475–1560.

Gosse, J. C., Klein, J., Evenson, E. B., Lawn, B., and Middleton, R. (1995) Beryllium-10 dating of the duration and retreat of the last Pinedale glacial sequence. *Science* **268**(5215), 1329.

Graly, J. A., Bierman, P. R., Reusser, L. J., and Pavich, M. J. (2010) Meteoric [10]Be in soil profiles–a global meta-analysis. *Geochimica et Cosmochimica Acta* **74**(23), 6814–6829.

Granger, D. E. (2006) A review of burial dating methods using [26]Al and [10]Be. *Geological Society of America Special Papers* **415**, 1–16.

Granger, D. E. and Riebe, C. S. (2014) Cosmogenic nuclides in weathering and erosion. *Treatise on Geochemistry* **5**, 401–436.

Granger, D. E. and Schaller, M. (2014) Cosmogenic nuclides and erosion at the watershed scale. *Elements* **10**(5), 369–373..

Granger, D. E., Kirchner, J. W., and Finkel, R. (1996) Spatially averaged long-term erosion rates measured from *in situ*-produced cosmogenic nuclides in alluvial sediment. *The Journal of Geology* **104**, 249–257.

Granger, D. E., Kirchner, J. W., and Finkel, R. C. (1997) Quaternary downcutting rate of the New River, Virginia, measured from differential decay of cosmogenic [26]Al and [10]Be in cave-deposited alluvium. *Geology* **25**(2), 107–110.

Granger, D. E., Fabel, D., and Palmer, A. N. (2001) Pliocene – Pleistocene incision of the Green River, Kentucky, determined from radioactive decay of cosmogenic [26]Al and [10]Be in Mammoth Cave sediments. *Geological Society of America Bulletin* **113**(7), 825–836.

Granger, D. E., Lifton, N. A., and Willenbring, J. K. (2013) A cosmic trip: 25 years of cosmogenic nuclides in geology. *Geological Society of America Bulletin* **125**(9–10), 1379–1402.

Granger, D. E., Gibbon, R. J., Kuman, K., Clarke, R. J., Bruxelles, L., and Caffee, M. W. (2015) New cosmogenic burial ages for Sterkfontein Member 2 Australopithecus and Member 5 Oldowan. *Nature* **522**(7554), 85–88.

Hampel, W., Takagi, J., Sakamoto, K., and Tanaka, S. (1975) Measurement of muon-induced [26]Al in terrestrial silicate rock. *Journal of Geophysical Research* **80**(26), 3757–3760.

Heikkilä, U., Beer, J., Abreu, J. A., and Steinhilber, F. (2013) On the atmospheric transport and deposition of the cosmogenic radionuclides (^{10}Be): a review. *Space Science Reviews* **176**(1–4), 321–332.

Heimsath, A. M., Dietrich, W. E., Nishiizumi, K., and Finkel, R. C. (1997) The soil production function and landscape equilibrium. *Nature* **388**(6640), 358–361.

Heisinger, B., Lal, D., Jull, A. T., *et al.* (2002a) Production of selected cosmogenic radionuclides by muons: 2. Capture of negative muons. *Earth and Planetary Science Letters* **200**(3), 357–369.

Heisinger, B., Lal, D., Jull, A. T., *et al.* (2002b) Production of selected cosmogenic radionuclides by muons: 1. *Fast muons. Earth and Planetary Science Letters* **200**(3), 345–355.

Herzog, G. F. (2010) Cosmic-ray exposure ages of meteorites. *Treatise on Geochemistry* **1**, 1–36.

Hess, V. F. (1912) Observations in low level radiation during seven free balloon flights. *Zeitschrift für Physik* **13**, 1084–1091.

Heyman, J., Stroeven, A. P., Harbor, J. M., and Caffee, M. W. (2011) Too young or too old: evaluating cosmogenic exposure dating based on an analysis of compiled boulder exposure ages. *Earth and Planetary Science Letters* **302**(1), 71–80.

Hidaka, H., Ebihara, M., and Yoneda, S. (2000) Isotopic study of neutron capture effects on Sm and Gd in chondrites. *Earth and Planetary Science Letters* **180**(1), 29–37.

Hidy, A. J., Gosse, J. C., Pederson, J. L., Mattern, J. P., and Finkel, R. C. (2010) A geologically constrained Monte Carlo approach to modeling exposure ages from profiles of cosmogenic nuclides: An example from Lees Ferry, Arizona. *Geochemistry, Geophysics, Geosystems* **11**(9).

Hidy, A. J., Gosse, J. C., Blum, M. D., and Gibling, M. R. (2014) Glacial–interglacial variation in denudation rates from interior Texas, USA, established with cosmogenic nuclides. *Earth and Planetary Science Letters* **390**, 209–221.

Hoffman, J. H. and Nier, A. O. (1959) The cosmogenic He 3 and He 4 distribution in the meteorite Carbo. *Geochimica et Cosmochimica Acta* **17**(1), 32–36.

Honda, M. and Arnold, J. R. (1967) Effects of cosmic rays on meteorites. In *Kosmische Strahlung II/Cosmic Rays II*, 613–632. Springer, Berlin.

Horiuchi, K., Uchida, T., Sakamoto, Y., *et al.* (2008) Ice core record of ^{10}Be over the past millennium from Dome Fuji, Antarctica: a new proxy record of past solar activity and a powerful tool for stratigraphic dating. *Quaternary Geochronology* **3**(3), 253–261.

Hua, Q. (2009) Radiocarbon: a chronological tool for the recent past. *Quaternary Geochronology* **4**(5), 378–390.

Ivy-Ochs, S., Poschinger, A. V., Synal, H. A., and Maisch, M. (2009) Surface exposure dating of the Flims landslide, Graubünden, Switzerland. *Geomorphology* **103**(1), 104–112.

Klein, J., Giegengack, R., Middleton, R., Sharma, P. T., Underwood, J. R., and Weeks, R. A. (1986) Revealing histories of exposure using *in situ* produced ^{26}Al and ^{10}Be in Libyan desert glass. *Radiocarbon* **28**(2A), 547–555.

Kubik, P. W., Korschinek, G., Nolte, E., *et al.* (1984) Accelerator mass spectrometry of ^{36}Cl in limestone and some paleontological samples using completely stripped ions. *Nuclear Instruments and Methods in Physics Research Section B: Beam Interactions with Materials and Atoms* **5**(2), 326–330.

Kurz, M. D. (1986) Cosmogenic helium in a terrestrial igneous rock. *Nature* **320**, 435–439.

Kusakabe, M., Ku, T. L., Southon, J. R., *et al.* (1991) Be isotopes in rivers/estuaries and their oceanic budgets. *Earth and Planetary Science Letters* **102**(3–4), 265–276.

Lal, D. (1991) Cosmic ray labeling of erosion surfaces: *in situ* nuclide production rates and erosion models. *Earth and Planetary Science Letters* **104**(2–4), 424–439.

Lal, D. and Arnold, J. R. (1985) Tracing quartz through the environment. *Proceedings of the Indian Academy of Sciences-Earth and Planetary Sciences* **94**(1), 1–5.

Lal, D. and Peters, B. (1967) Cosmic ray produced radioactivity on the Earth. In *Kosmische Strahlung II/Cosmic Rays II*, 551–612. Springer, Berlin.

Levin, I. and Hesshaimer, V. (2000) Radiocarbon—a unique tracer of global carbon cycle dynamics.

Leya, I., Lange, H. J., Neumann, S., Wieler, R., and Michel, R. (2000) The production of cosmogenic nuclides in stony meteoroids by galactic cosmic-ray particles. *Meteoritics & Planetary Science* **35**(2), 259–286.

Libby, W. F. (1946) Atmospheric helium three and radiocarbon from cosmic radiation. *Physical Review* **69**(11–12), 671.

Lifton, N., Sato, T., and Dunai, T. J. (2014) Scaling *in situ* cosmogenic nuclide production rates using analytical approximations to atmospheric cosmic-ray fluxes. *Earth and Planetary Science Letters* **386**, 149–160.

Lifton, N., Caffee, M., Finkel, R., *et al.* (2015) *In situ* cosmogenic nuclide production rate calibration for the CRONUS-Earth project from Lake Bonneville, Utah, shoreline features. *Quaternary Geochronology* **26**, 56–69.

Liu, B., Phillips, F. M., Fabryka-Martin, J. T., Fowler, M. M., and Stone, W. D. (1994) Cosmogenic ^{36}Cl accumulation in unstable landforms, 1. Effects of the thermal neutron distribution. *Water Resources Research* **30**, 3115–3125.

Markowski, A., Quitté, G., Halliday, A. N., and Kleine, T. (2006) Tungsten isotopic compositions of iron meteorites: chronological constraints vs. cosmogenic effects. *Earth and Planetary Science Letters* **242**(1), 1–15.

Marrero, S. M., Phillips, F. M., Borchers, B., Lifton, N., Aumer, R., and Balco, G. (2016) Cosmogenic nuclide systematics and the CRONUScalc program. *Quaternary Geochronology* **31**, 160–187.

McCorkell, R., Fireman, E. L., and Langway, C. C. (1967) Aluminum-26 and beryllium-10 in Greenland ice. *Science* **158**(3809), 1690–1692.

McHargue, L. R. and Damon, P. E. (1991) The global beryllium-10 cycle. *Reviews of Geophysics* **29**(2), 141–158.

Ménabréaz, L., Thouveny, N., Bourlès, D. L., and Vidal, L. (2014) The geomagnetic dipole moment variation between 250 and 800 ka BP reconstructed from the authigenic ^{10}Be/^{9}Be signature in West Equatorial Pacific sediments. *Earth and Planetary Science Letters* **385** 190–205.

Millikan, R. A. and Cameron, G. H. (1926) High frequency rays of cosmic origin III. Measurements in snow-fed lakes at high altitudes. *Physical Review* **28**(5), 851.

Miyake, F., Suzuki, A., Masuda, K., *et al.* (2015) Cosmic ray event of AD 774–775 shown in quasi-annual ^{10}Be data from the Antarctic Dome Fuji ice core. *Geophysical Research Letters* **42**(1), 84–89.

Morris, J. D., Leeman, W. P., and Tera, F. (1990) The subducted component in island arc lavas: constraints from Be isotopes and B-Be systematics. *Nature* **344**(6261), 31–36.

Moysey, S., Davis, S. N., Zreda, M., and Cecil, L. D. (2003) The distribution of meteoric ^{36}Cl/Cl in the United States: a comparison of models. *Hydrogeology Journal* **11**(6), 615–627.

Müller, R. A. (1977) Radioisotope dating with a cyclotron. *Science* **196**, 489–494.

Nelson, D. E., Korteling, R. G., and Stott, W. R. (1977) Carbon-14: direct detection at natural concentrations. *Science* **198**(4316), 507–508.

Nishiizumi, K., Lal, D., Klein, J., Middleton, R., and Arnold, J. R. (1986) Production of ^{10}Be and ^{26}Al by cosmic rays in terrestrial quartz *in situ* and implications for erosion rates. *Nature* **319**, 134–136.

Nishiizumi, K., Winterer, E. L., Kohl, C. P., *et al.* (1989) Cosmic ray production rates of ^{10}Be and ^{26}Al in quartz from glacially polished rocks. *Journal of Geophysical Research: Solid Earth* **94** (B12), 17907–17915.

Nishiizumi, K., Kohl, C. P., Shoemaker, E. M., *et al.* (1991) *In situ* ^{10}Be–^{26}Al exposure ages at Meteor Crater, Arizona. *Geochimica et Cosmochimica Acta* **55**(9), 2699–2703.

Nishiizumi, K., Kohl, C. P., Arnold, J. R., *et al.* (1993) Role of *in situ* cosmogenic nuclides ^{10}Be and ^{26}Al in the study of diverse geomorphic processes. *Earth Surface Processes and Landforms* **18** (5), 407–425.

Nishiizumi, K., Arnold, J. R., Brownlee, D. E., Caffee, M. W., Finkel, R. C., and Harvey, R. P. (1995) Beryllium-10 and aluminum-26 in individual cosmic spherules from Antarctica. *Meteoritics*, **30**(6), 728–732.

Nishiizumi, K., Arnold, J. R., Kohl, C. P., Caffee, M. W., Masarik, J., and Reedy, R. C. (2009). Solar cosmic ray records in lunar rock 64455. *Geochimica et Cosmochimica Acta*, **73**(7), 2163–2176.

Owen, L. A., Clemmens, S. J., Finkel, R. C., and Gray, H. (2014) Late Quaternary alluvial fans at the eastern end of the San Bernardino Mountains, Southern California. *Quaternary Science Reviews* **87**, 114–134.

Paneth, F. A., Reasbeck, P., and Mayne, K. I. (1952) Helium 3 content and age of meteorites. *Geochimica et Cosmochimica Acta* **2** (5), 300–303.

Paneth, F. A., Reasbeck, P., and Mayne, K. I. (1953) Production by cosmic rays of helium-3 in meteorites. *Nature* **172**, 200–201.

Pavich, M. J., Brown, L., Valette-Silver, J. N., Klein, J., and Middleton, R. (1985) 10Be analysis of a Quaternary weathering profile in the Virginia Piedmont. *Geology* **13**(1), 39–41.

Peters, B. (1955) Radioactive beryllium in the atmosphere and on the earth. *Proceedings of the Indian Academy of Sciences—Section A* **41**(2), 67–71.

Phillips, F. M. (2000) Chlorine-36. In *Environmental Tracers in Subsurface Hydrology*, 299–348. Springer.

Phillips, F. M., Smith, G. I., Bentley, H. W., Elmore, D., and Gove, H. E. (1983) Chlorine-36 dating of saline sediments: preliminary results from Searles Lake, California. *Science* **222**(4626), 925–927.

Phillips, F. M., Leavy, B. D., Jannik, N. O., Elmore, D., and Kubik, P. W. (1986) The accumulation of cosmogenic chlorine-36 in rocks: A method for surface exposure dating. *Science* **231** (4733), 41–43.

Phillips, F. M., Zreda, M. G., Smith, S. S., Elmore, D., Kubik, P. W., and Sharma, P. (1990) Cosmogenic chlorine-36 chronology for glacial deposits at Bloody Canyon, eastern Sierra Nevada. *Science* **248**(4962), 1529.

Phillips, F. M., Argento, D.C., Balco, G., *et al.* (2016) The CRONUS-Earth project: a synthesis. *Quaternary Geochronology* **31**, 119–154.

Portenga, E. W. and Bierman, P. R. (2011) Understanding Earth's eroding surface with ^{10}Be. *GSA Today* **21**(8), 4–10.

Putkonen, J. and Swanson, T. (2003) Accuracy of cosmogenic ages for moraines. *Quaternary Research* **59**(2), 255–261.

Reedy, R. C. and Arnold, J. R. (1972) Interaction of solar and galactic cosmic-ray particles with the Moon. *Journal of Geophysical Research*, **77**(4), 537–555.

Reedy, R. C., Arnold, J. R., and Lal, D. (1983) Cosmic-ray record in solar system matter. *Annual Review of Nuclear and Particle Science* **33**(1), 505–538.

Reimer, P. J., Bard, E., Bayliss, A., *et al.* (2013) IntCal13 and Marine13 radiocarbon age calibration curves 0–50,000 years cal BP. *Radiocarbon* **55**(4), 1869–1887.

Repka, J. L., Anderson, R. S., and Finkel, R. C. (1997) Cosmogenic dating of fluvial terraces, Fremont River, Utah. *Earth and Planetary Science Letters* **152**(1), 59–73.

Riebe, C. S. and Granger, D. E. (2013) Quantifying effects of deep and near-surface chemical erosion on cosmogenic nuclides in soils, saprolite, and sediment. *Earth Surface Processes and Landforms* **38**(5), 523–533.

Riebe, C. S., Kirchner, J. W., and Granger, D. E. (2001) Quantifying quartz enrichment and its consequences for cosmogenic measurements of erosion rates from alluvial sediment and regolith. *Geomorphology* **40**(1), 15–19.

Sato, T., Yasuda, H., Niita, K., Endo, A., and Sihver, L. (2008) Development of PARMA: PHITS-based analytical radiation model in the atmosphere. *Radiation Research* **170**(2), 244–259.

Schaller, M., Von Blanckenburg, F., Veldkamp, A., Tebbens, L. A., Hovius, N., and Kubik, P. W. (2002) A 30 000 yr record of erosion rates from cosmogenic ^{10}Be in Middle European river terraces. *Earth and Planetary Science Letters* **204**(1), 307–320.

Schimmelpfennig, I., Benedetti, L., Garreta, V., *et al.* (2011) Calibration of cosmogenic ^{36}Cl production rates from Ca and K spallation in lava flows from Mt. Etna (38°N, Italy) and Payun Matru (36°S, Argentina). *Geochimica et Cosmochimica Acta* **75** (10), 2611–2632.

Sharma, P. and Somayajulu, B. L. K. (1982) ^{10}Be dating of large manganese nodules from world oceans. *Earth and Planetary Science Letters* **59**(2), 235–244.

Staiger, J., Gosse, J., Toracinta, R., Oglesby, B., Fastook, J., and Johnson, J. V. (2007) Atmospheric scaling of cosmogenic nuclide production: climate effect. *Journal of Geophysical Research: Solid Earth* **112**(B2).

Steinhilber, F., Abreu, J. A., Beer, J., *et al.* (2012) 9,400 years of cosmic radiation and solar activity from ice cores and tree rings. *Proceedings of the National Academy of Sciences* **109**(16), 5967–5971.

Stone, J. O. (2000) Air pressure and cosmogenic isotope production. *Journal of Geophysical Research: Solid Earth* **105**(B10), 23753–23759.

Stone, J. O. H., Evans, J. M., Fifield, L. K., Allan, G. L., and Cresswell, R. G. (1998) Cosmogenic chlorine-36 production in calcite by muons. *Geochimica et Cosmochimica Acta* **62**(3), 433–454.

Stroeven, A. P., Fabel, D., Harbor, J., Hättestrand, C., and Kleman, J. (2002) Quantifying the erosional impact of the Fennoscandian ice sheet in the Torneträsk–Narvik corridor, northern Sweden, based on cosmogenic radionuclide data. *Geografiska Annaler: Series A, Physical Geography* **84**(3–4), 275–287.

Synal, H. A., Beer, J., Bonani, G., Suter, M., and Wölfli, W. (1990) Atmospheric transport of bomb-produced ^{36}Cl. *Nuclear Instruments and Methods in Physics Research Section B: Beam Interactions with Materials and Atoms* **52**(3), 483–488.

Taylor, R. E. and Bar-Yosef, O. (2014) *Radiocarbon Dating: An Archaeological Perspective*, 2nd edn. Routledge, 404 pp.

Tuniz, C., Kutschera, W., Fink, D., Herzog, G. F., and Bird, J. R. (1998) *Accelerator Mass Spectrometry: Ultrasensitive Analysis for Global Science*. CRC Press.

Valette-Silver, J. N., Brown, L., Pavich, M., Klein, J. F., and Middleton, R. (1986) Detection of erosion events using ^{10}Be profiles: example of the impact of agriculture on soil erosion in the Chesapeake Bay area (USA). *Earth and Planetary Science Letters* **80**(1–2), 82–90.

Von Blanckenburg, F. and Bouchez, J. (2014) River fluxes to the sea from the ocean's ^{10}Be/^{9}Be ratio. *Earth and Planetary Science Letters* **387**, 34–43.

Von Blanckenburg, F., O'Nions, R. K., Belshaw, N. S., Gibb, A., and Hein, J. R. (1996) Global distribution of beryllium isotopes in deep ocean water as derived from Fe–Mn crusts. *Earth and Planetary Science Letters* **141**(1), 213–226.

Von Blanckenburg, F., Bouchez, J., and Wittmann, H. (2012) Earth surface erosion and weathering from the ^{10}Be (meteoric)/^{9}Be ratio. *Earth and Planetary Science Letters* **351**, 295–305.

Wieler, R. (2002) Cosmic-ray-produced noble gases in meteorites. *Reviews in Mineralogy and Geochemistry* **47**(1), 125–170.

Willenbring, J. K. and von Blanckenburg, F. (2010a) Meteoric cosmogenic Beryllium-10 adsorbed to river sediment and soil: Applications for Earth-surface dynamics. *Earth-Science Reviews* **98**(1), 105–122.

Willenbring, J. K. and von Blanckenburg, F. (2010b) Long-term stability of global erosion rates and weathering during late-Cenozoic cooling. *Nature* **465**(7295), 211–214.

Willenbring, J. K., Codilean, A. T., and McElroy, B. (2013) Earth is (mostly) flat: Apportionment of the flux of continental sediment over millennial time scales. *Geology* **41**(3), 343–346.

Yiou, F., Raisbeck, G. M., Baumgartner, S., *et al.* (1997) Beryllium 10 in the Greenland ice core project ice core at summit, Greenland. *Journal of Geophysical Research: Oceans* **102**(C12), 26783–26794.

Zhao, Z., Granger, D., Zhang, M., *et al.* (2016) A test of the isochron burial dating method on fluvial gravels within the Pulu volcanic sequence, West Kunlun Mountains, China. *Quaternary Geochronology* **34**, 75–80.

Zhou, W., Xian, F., Du, Y., Kong, X., and Wu, Z. (2014) The last 130 ka precipitation reconstruction from Chinese loess ^{10}Be. *Journal of Geophysical Research: Solid Earth* **119**(1) 191–197.

Zweck, C., Zreda, M., and Desilets, D. (2013) Snow shielding factors for cosmogenic nuclide dating inferred from Monte Carlo neutron transport simulations. *Earth and Planetary Science Letters* **379**, 64–71.

CHAPTER 14

Extinct radionuclide chronology

14.1 INTRODUCTION

All the radioactive isotopes that constitute geochronometers will eventually decay away. Many stellar-produced radioactive nuclei present when Earth formed, that have half-lives much shorter than the age of the solar system, are now effectively extinct, but the record of their presence can be found in abundance variations of their decay products. The most commonly used of these stellar-produced extinct radionuclides are listed in Table 14.1.

Some of these radioactive isotopes, for example ^{26}Al, ^{53}Mn, and ^{129}I, also are produced continuously through interaction of cosmic rays with Earth materials. The application of such cosmogenically produced radioactive isotopes is covered in Chapter 13. This chapter focuses on radioactive isotopes that are produced by stellar nucleosynthesis that have half-lives well less than that of the age of the Earth. These now extinct radionuclides were present in the mixture of elements from which the solar system and Earth formed. The extinct radionuclides offer at least two traits that make them important chronometers for events that occurred before the parent isotopes decayed away to undetectable levels.

- Their short half-lives provide high chronological resolution.
- The diverse chemical characteristics of the various elements involved allow these systems to date a variety of processes important in the differentiation of natural materials, from condensation and/or volatile loss, to metal–silicate separation, to igneous fractionation in silicate melts, to weathering on planetesimals.

Extinct radionuclide chronology, however, also presents a variety of technical difficulties that must be overcome in order to obtain accurate ages:

- These systems do not provide absolute ages, but only relative ages to some time when the abundance of the parent isotope has been determined by comparison with an absolute chronometer, for example, U–Pb. In most cases, time periods calculated from short-lived isotope systems are reported as time intervals referenced to the 4567 Ma age of the solar system, not to the present day as is the case for long-lived radiometric systems.
- Ages from short-lived radionuclides assume that the only process affecting the abundance of the parent isotope is radioactive decay. For this to be true requires that the parent isotope was homogeneously distributed within the solar system before planet formation began. This appears to have been the case, but should not be taken as a certainty, as there is still considerable discussion about the existence of variability in the initial abundance of the parent isotopes in the region of planet formation.
- All the extinct radionuclides are low abundance isotopes, so their contribution to increasing the abundance of the daughter isotopes also is small. Their use as chronometers thus requires either extremely high-precision isotope ratio measurements, or examination of mineral phases with extreme fractionation of parent from daughter isotope, or both.
- Given the small contribution of the extinct radionuclides to the daughter element, other processes, for example, reactions with high-energy cosmic rays or initial nucleosynthetic isotopic heterogeneity, can produce isotopic abundance variability in the daughter isotope that rivals, or exceeds, that produced by the decay of the radioactive parent. Obtaining an accurate age thus requires distinguishing the radioactive contribution from these other causes of isotopic variability.
- Given the rarity of some of these isotopes, their half-lives (e.g., Table 14.1) are not always determined to an accuracy that allows confidence in the ages they provide. The most obvious example of the problem is that both ^{60}Fe and ^{146}Sm have recent half-life determinations that differ by almost a factor of two compared to previous measurements.

Geochronology and Thermochronology, First Edition. Peter W. Reiners, Richard W. Carlson, Paul R. Renne, Kari M. Cooper, Noah M. McLean, and Blair Schoene.
© 2018 John Wiley & Sons Ltd. Published 2018 by John Wiley & Sons Ltd.

Table 14.1 Extinct radionuclides used as geochronometers and cosmochronometers

Parent	Initial abundance	Half-life (Ma)	Decay constant (year⁻¹)	Daughter
^{26}Al	^{26}Al/^{27}Al $= 5.1 \times 10^{-5}$	0.70	$(9.83 \pm 0.25) \times 10^{-7}$ [a]	^{26}Mg
^{60}Fe	^{60}Fe/^{56}Fe $= 3.7 \times 10^{-7}$	2.6	$(2.65 \pm 0.06) \times 10^{-7}$ [b]	^{60}Ni
^{53}Mn	^{53}Mn/^{55}Mn $= 9.1 \times 10^{-6}$	3.7	$(1.87 \pm 0.17) \times 10^{-7}$ [c]	^{53}Cr
^{107}Pd	^{107}Pd/^{108}Pd $= 2.8 \times 10^{-5}$	6.5	$(1.06 \pm 0.04) \times 10^{-7}$ [d]	^{107}Ag
^{182}Hf	^{182}Hf/^{177}Hf $= 9.7 \times 10^{-5}$	8.9	$(7.79 \pm 0.08) \times 10^{-8}$ [e]	^{182}W
^{129}I	^{129}I/^{127}I $= 1.0 \times 10^{-4}$	15.7	$(4.41 \pm 0.01) \times 10^{-8}$ [f]	^{129}Xe
^{244}Pu	^{244}Pu/^{238}U $= 0.0068$	81.2	$(8.54 \pm 0.05) \times 10^{-9}$ [g]	Fission Xe
^{146}Sm	^{146}Sm/^{144}Sm $= 0.0085$	103	$(6.73 \pm 0.31) \times 10^{-9}$ [h]	^{142}Nd

[a] *Nishizumi* [2004].
[b] *Rugel et al.* [2009].
[c] *Honda and Imamura* [1971].
[d] *Flynn and Glendenin* [1969].
[e] *Vockenhuber et al.* [2004].
[f] *Emery et al.* [1972].
[g] *Aggarwal* [2006].
[h] *Meissner et al.* [1987].

14.2 HISTORY

As discussed in Chapter 2, stellar nucleosynthesis produces a wide range of radioactive isotopes. Those with half-lives similar to, or longer than, the age of the Earth form the traditional radiometric systems K–Ar, Rb–Sr, Sm–Nd, Lu-Hf and U/Th–Pb. As Table 14.1 shows, there also are a number of radioactive species with half-lives long enough that they might have been present when Earth formed, depending on how long Earth took to form, but have since decayed away to undetectable levels. While the radioactive parents of these short-lived systems are extinct, their decay has left imprints in the isotopic composition of the daughter elements. The first of these systems to be put into use as a chronometer involved the detection of large excesses in ^{129}Xe in the Xe isotope composition of primitive stone meteorites [*Reynolds*, 1960]. The singular excess in ^{129}Xe compared to the other eight isotopes of xenon (Fig. 14.1) was interpreted as reflecting the decay of ^{129}I. Because the studied meteorite was formed of materials that had never condensed Xe, but had iodine, the meteorite had a much higher I/Xe ratio than solar. Some portion of this iodine was ^{129}I, which then decayed to create the observed excess in ^{129}Xe. The elevated abundances of the heavy Xe isotopes (Fig. 14.1) was later attributed to the fission decay of another short-lived radionuclide ^{244}Pu.

Following the discovery of a signal from the decay of now extinct ^{129}I and ^{244}Pu, 14 years passed until evidence was found for the decay of the next short-lived nuclide. In 1974, *Gray and Compston* [1974] measured excesses in ^{26}Mg that they attributed to the decay of ^{26}Al, but this interpretation was disputed by Lee and Papanastassiou [1974], who in their own data found no correlation between the ^{26}Mg excess and Al/Mg ratio, which would be expected if the excess ^{26}Mg was due to ^{26}Al decay. Later, with improved data and additional analyses, *Lee et al.* [1976] did find a correlation between ^{26}Mg excesses and Al/Mg ratios in samples of the Allende meteorite, testifying to the likely role of ^{26}Al decay in creating the observed excesses in ^{26}Mg.

The delay in the discovery of the first short-lived radionuclide that did not decay to a noble gas reflects the very small

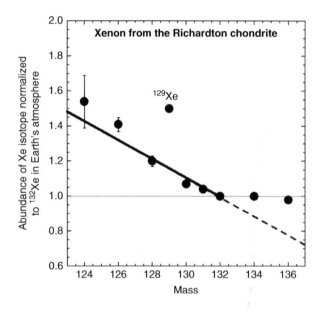

Fig. 14.1. Abundance of each xenon isotope in the xenon released by heating a sample of the Richardton chondritic meteorite. The chondrite data are normalized to the Xe isotopic composition of Earth's atmosphere and additionally by setting the relative abundance of ^{132}Xe in Richardton equal to that in Earth's atmosphere. (Source: *Reynolds* [1960]. Reproduced with permission of American Physical Society.)

contribution these short-lived radionuclides make to their daughter isotope. To detect this contribution, one either needs extremely high parent/daughter ratios, as is possible in a system involving a noble gas, like Xe, where the daughter concentration can become extremely low due to outgassing, or extremely high isotope ratio measurement precision. Going to high spatial resolution, for example through the use of the ion probe, can aid in analyzing mineral phases with very high parent/daughter ratios, which has proven particularly useful for Al–Mg [*MacPherson et al.*, 1995] and Fe-Ni systematics [*Tachibana and Huss*, 2003]. Improvements in mass spectrometry and the ability to measure isotope ratios to much higher precision were the primary factors that led to the addition of several new short-lived radionuclide systems to the chronological toolbox, including ^{53}Mn [*Birck and Allégre*, 1985], ^{60}Fe [*Birck and Lugmair*, 1988], ^{107}Pd [*Kelly and Wasserburg*, 1978; *Carlson and Hauri*, 2001], ^{146}Sm [*Lugmair and Marti*, 1977], and ^{182}Hf [*Lee and Halliday*, 1995]. The detection of the decay products of short-lived radionuclides is the first step in developing them into useful chronometers. As outlined in this chapter, these systems have already provided a much more precise chronology for events happening in the early solar system, but their full use will depend on further improvements in isotope-ratio measurement techniques that will allow full exploitation of the signals they provide.

14.3   SYSTEMATICS AND APPLICATIONS

14.3.1 ^{26}Al–^{26}Mg

Using ^{26}Al–^{26}Mg as an example, the standard radioactive decay equation takes the form:

$$\frac{^{26}Al}{^{27}Al_t} = \frac{^{26}Al}{^{27}Al_0} \times e^{-\lambda t} \tag{14.1}$$

where ^{27}Al is the only stable isotope of aluminum, t is a time interval after some starting time (0), and λ is the decay constant for ^{26}Al. As with other radioactive decay schemes, the buildup of the daughter product of ^{26}Al decay, ^{26}Mg, can be described as:

$$\left(^{26}Mg\right)_t = \left(^{26}Mg\right)_0 + \left(^{26}Al\right)_0 - \left(^{26}Al\right)_t \tag{14.2}$$

and dividing both sides by the most abundant stable isotope of Mg, ^{24}Mg, gives:

$$\left(\frac{^{26}Mg}{^{24}Mg}\right)_t = \left(\frac{^{26}Mg}{^{24}Mg}\right)_0 + \left(\frac{^{26}Al}{^{24}Mg}\right)_0 - \left(\frac{^{26}Al}{^{24}Mg}\right)_t \tag{14.3}$$

Unlike the case for long-lived radioactive isotopes, the ^{26}Al/^{24}Mg ratio cannot be measured in modern samples because ^{26}Al is now extinct. The present day ^{26}Al term in equation (14.3), however, can be expanded as:

$$\left(\frac{^{26}Al}{^{24}Mg}\right)_t = \left(\frac{^{26}Al}{^{27}Al}\right)_t \times \left(\frac{^{27}Al}{^{24}Mg}\right) \tag{14.4}$$

where ^{27}Al/^{24}Mg does not change with time, as both are stable isotopes, in lieu of some chemical process that fractionates Al from Mg. Combining these equations gives:

$$\left(\frac{^{26}Mg}{^{24}Mg}\right)_t = \left(\frac{^{26}Mg}{^{24}Mg}\right)_0 + \frac{^{27}Al}{^{24}Mg} \times \left(\frac{^{26}Al}{^{27}Al}\right)_0 \times \left(1 - e^{-\lambda t}\right) \tag{14.5}$$

This is similar to the standard radioactive daughter product buildup equation used for long-lived radioactive systems, with the exception that the initial abundance of the radioactive parent is included in the equation, rather than its present-day abundance.

A plot of the two measureable parameters in this equation, $(^{26}Mg/^{24}Mg)_t$ versus ^{27}Al/^{24}Mg, describes a line whose slope is equal to $\frac{^{26}Al}{^{27}Al_0} \times \left(1 - e^{-\lambda t}\right)$. The intercept gives the ^{26}Mg/^{24}Mg ratio present in the sample at the time of its formation. The slope is equivalent to the ^{26}Al/^{27}Al ratio in the sample at the time of its formation. With a few additional parameters, the initial ^{26}Al/^{27}Al ratio determined from the isochron can provide the age difference between two samples. For two samples (S_1, S_2) that provide slopes indicative of different ^{26}Al/^{27}Al ratios at the time of sample formation, the age difference can be calculated using the basic equation describing the reduction in ^{26}Al abundance with time:

$$\left(\frac{^{26}Al}{^{27}Al}\right)_{S2} = \left(\frac{^{26}Al}{^{27}Al}\right)_{S1} \times e^{-\lambda t} \tag{14.6}$$

the age difference, or Δt, is then:

$$\Delta t = -\frac{1}{\lambda} \times ln\left[\frac{\left(\frac{^{26}Al}{^{27}Al}\right)_{S2}}{\left(\frac{^{26}Al}{^{27}Al}\right)_{S1}}\right] \tag{14.7}$$

An example of the procedure follows. Table 14.2 lists data obtained for minerals separated from a calcium–aluminum-rich inclusion (CAI) from the Allende meteorite that was given the name "AJEF" [*Jacobsen et al.*, 2008]. CAIs are small, millimeter to centimeter sized inclusions in primitive meteorites that are made up of refractory minerals [*MacPherson*, 2003]. These minerals would be the first to condense from a gas of solar composition as the early solar nebula cooled, and hence are believed to be the first solids formed in the solar system. Table 14.2 also lists data for measurements of a whole rock and mineral separates from the meteorite D'Orbigny [*Spivak-Birndorf et al.*, 2009] that belongs to a class of meteorites called angrites. Angrites are interpreted to be melts that erupted onto, or intruded into, an asteroidal-sized object early in solar system history [*Keil*, 2012]. Aluminum and magnesium are both refractory elements, with aluminum having the higher condensation temperature. Both elements are lithophile, meaning that they occur in silicates and not metal or sulfide. As a result, the main process that causes fractionation of Al from Mg is partial melting of silicate rocks and fractional crystallization of magmas. Magnesium is compatible in many mantle minerals and hence stays in the residue of melting, whereas aluminum is incompatible in most mantle minerals and hence selectively partitions into the melt. The range in Al/Mg ratios in Table 14.2 thus primarily reflects the partitioning of Al and Mg between the various minerals during the crystallization of the magmas from which the CAI and angrite formed.

The mineral data from the Allende CAI define a slope on the Al–Mg isochron diagram of Fig. 14.2 equal to $(4.96 \pm 0.25) \times 10^{-5}$. This slope is equal to the ^{26}Al/^{27}Al ratio at the time the CAI formed. The slope on its own provides no age information. The same CAI, however, was also dated using the U–Pb system and provided an age of 4567.6 ± 0.4 Ma. The angrite data show very little variability in Mg isotopic composition in spite of the

Table 14.2 Al–Mg data for the Allende CAI AJEF and the D'Orbigny Angrite

Sample		^{27}Al/^{24}Mg	Uncertainty	^{26}Mg/^{24}Mg	Uncertainty
Allende	Whole rock	3.07	0.06	0.139503	0.000002
CAI-AJEF	Pyroxene	2.23	0.04	0.139458	0.000004
	Melilite-pyroxene	2.67	0.05	0.139481	0.000004
	Melilite	3.50	0.07	0.139524	0.000002
	Melilite	4.55	0.09	0.139575	0.000002
	Plagioclase	4.28	0.09	0.139559	0.000002
D'Obigny Angrite	Whole rock	1.68	0.03	0.139354	0.000004
	Olivine	0.04	0.0008	0.139349	0.000003
	Pyroxene	1.02	0.02	0.139354	0.000004
	Plagioclase	77.18	1.5	0.139386	0.000004
	Plagioclase	107.39	2.1	0.139407	0.000003

AJEF data from *Jacobsen et al.* [2008]; D'Orbigny data from *Spivak-Birndorf et al.* [2009].

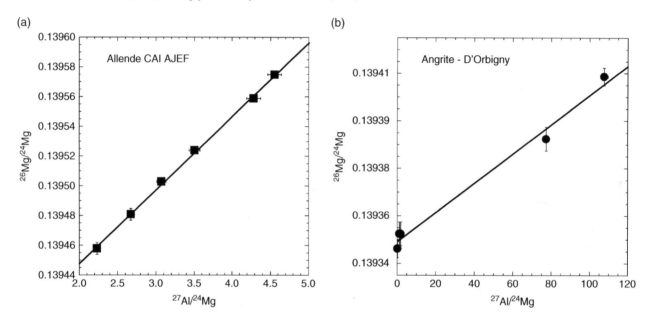

Fig. 14.2. Al–Mg isochrons for the data provided in Table 14.2.

much larger range in Al/Mg ratios in the mineral separates. The data define a slope on the Al–Mg isochron diagram of Fig. 14.2b equal to $(5.1 \pm 0.5) \times 10^{-7}$. This slope can be used to calculate the age difference between the Allende CAI and D'Orbigny by solving the equation:

$$\left(\frac{^{26}\text{Al}}{^{27}\text{Al}}\right)_{\text{D'Orbigny}} = \left(\frac{^{26}\text{Al}}{^{27}\text{Al}}\right)_{\text{AJEF}} \times e^{-\lambda t} \quad (14.8)$$

Solving for t gives:

$$\Delta t = -\frac{1}{9.49 \times 10^{-7}} \times \ln\left[\frac{5.1 \times 10^{-7}}{4.96 \times 10^{-5}}\right] \quad (14.9)$$

$\Delta t = 4.82 \pm 0.15\,\text{Ma}$

If the U–Pb age for the CAI is used to anchor the Al–Mg slope to an absolute age, then the age of D'Orbigny according to Al–Mg is 4562.8 ± 0.4 Ma. D'Orbigny also has been dated using the U–Pb system to provide an absolute age of 4563.37 ± 0.25 Ma

[*Brennecka and Wadhwa*, 2012] that overlaps within uncertainty the age determined by comparison of the Al–Mg systematics of D'Orbigny with those of the Allende CAI. The agreement in the time interval indicated by U–Pb and Al–Mg suggests that the lower ^{26}Al/^{27}Al ratio in D'Orbigny compared to the Allende CAI is due solely to decay of ^{26}Al. The alternative explanation is that the parent planetesimals of D'Orbigny and Allende formed in different parts of the solar nebula that might have had different initial amounts of ^{26}Al due to an imperfect homogenenization of ^{26}Al into the solar nebula prior to the start of planet formation. Although the homogeneous distribution of all the extinct radionuclides in the solar system remains a concern in the interpretation of the chronology they provide, an increasing number of age comparisons of the type discussed above allow more confidence that the extinct nuclides indeed initially were homogeneously distributed. If so, the variations in their abundance in different solar system objects can be used to provide a precise chronology for early solar system events.

The example above also illustrates a practical problem with extinct radionuclide chronology—the chronometer works only while the parent isotope is sufficiently abundant to create measurable increases in the daughter isotope. For D'Orbigny, the total range in the $^{26}Mg/^{24}Mg$ ratio between the mineral separates is only a bit more than a factor of ten larger than the uncertainty of any single measurement in spite of the very high Al/Mg ratios of the D'Orbigny plagioclase. In contrast, the total range in $^{26}Mg/^{24}Mg$ in the CAI minerals is about a factor of two larger than displayed by D'Orbigny mineral separates in spite of the nearly 50 times larger range in Al/Mg ratio measured in the D'Orbigny mineral separates. This reflects the fact that the 4.8 Ma age difference between the CAI and D'Orbigny corresponds to almost seven half-lives of ^{26}Al, by which time the abundance of ^{26}Al is so low that detecting its contribution to ^{26}Mg is becoming difficult.

Reflective of the generally small variation in daughter isotopic composition, a common reporting format expresses the daughter isotope ratio normalized to some standard. For example, in Al–Mg, Mg isotope data are usually reported as deviations in the measured $^{26}Mg/^{24}Mg$ ratio from the value of the DSM3 Mg isotope standard solution that has a $^{26}Mg/^{24}Mg = 0.139352$ [*Brand et al.*, 2014]. Mg isotope data are then commonly reported as:

$$\delta^{26}Mg = \left[\frac{\left(\frac{^{26}Mg}{^{24}Mg}\right)_{Sa}}{\left(\frac{^{26}Mg}{^{24}Mg}\right)_{DMS3}} - 1 \right] \times 1000 \qquad (14.10)$$

where "Sa" is the sample isotopic composition. In the case of D'Orbigny, the data shown in Table 14.2 represent a range in $\delta^{26}Mg$ from −0.200 to +0.396.

Thus, even though D'Orbigny is 4563 Ma old, its formation occurred near the end of the time interval during which the Al–Mg system was an effective chronometer. Given its 700,000 year half-life, Al–Mg dating is primarily confined to examination of events that occurred over 4.56 Ga ago [*Nyquist et al.*, 2009]. This includes the differentiation history of CAI's from different meteorites, the formation ages of the melt droplets, known as chondrules, that make up the majority of primitive meteorites, and the eruption ages of only the oldest magmatic meteorites—angrites and some members of another class of igneous meteorite known as eucrites.

14.3.2 ^{53}Mn–^{53}Cr chronometry

The 3.7 Ma half-life of ^{53}Mn increases the time period over which this system can serve as a chronometer compared to Al–Mg. ^{53}Mn was initially present in the early solar system at a ratio to the only other stable Mn isotope, ^{55}Mn, of 9.1×10^{-6} [*Lugmair and Shukolyukov*, 1998; *Nyquist et al.*, 2009]. ^{53}Mn β-decays to ^{53}Cr, which constitutes 9.5 atom% of normal Cr. Like Al and Mg, Mn and Cr are primarily fractionated from one another by igneous differentiation. In contrast to Al and Mg, Mn is significantly more volatile than Cr, and both are somewhat soluble in iron metal (siderophile), with Cr being significantly more siderophile than Mn. As a result, the Mn–Cr system potentially is sensitive to both volatile-loss processes and silicate-metal separation events.

The equations for Mn–Cr systematics normalize to the most abundant isotope of Cr, ^{52}Cr, which constitutes 83.8 atom% of natural Cr. The standard decay equation in Mn–Cr is thus:

$$\left(\frac{^{53}Mn}{^{55}Mn}\right)_t = \left(\frac{^{53}Mn}{^{55}Mn}\right)_0 \times e^{-\lambda t} \qquad (14.11)$$

and the buildup of the daughter ^{53}Cr is expressed as:

$$\left(\frac{^{53}Cr}{^{52}Cr}\right)_t = \left(\frac{^{53}Cr}{^{52}Cr}\right)_0 + \frac{^{55}Mn}{^{52}Cr} \times \left(\frac{^{53}Mn}{^{55}Mn}\right)_0 \times \left(1 - e^{-\lambda t}\right) \qquad (14.12)$$

Again, variations in the relative abundance of ^{53}Cr as a result of the decay of ^{53}Mn are quite small, so $^{53}Cr/^{52}Cr$ ratios are often expressed as:

$$\varepsilon^{53}Cr = \left[\frac{\left(\frac{^{53}Cr}{^{52}Cr}\right)_{Sa}}{\left(\frac{^{53}Cr}{^{52}Cr}\right)_{Std}} - 1 \right] \times 10000 \qquad (14.13)$$

Unfortunately, there is no consistently used standard (Std) for Cr isotope analysis although the NIST Cr standard solution SRM-3112a is used in many studies. SRM-3112a has a $^{53}Cr/^{52}Cr = 0.11346$, but some other normalizing values in use range to as low as 0.11345 (a difference of 0.9 $\varepsilon^{53}Cr$), so care should be exercised when comparing $\varepsilon^{53}Cr$ values reported in different publications [*Trinquier et al.*, 2008; *Qin et al.*, 2010].

As with the previous Al–Mg example, igneous meteorites, such as angrites, are a common target for Mn–Cr dating because their minerals provide a range of Mn/Cr ratios that allow for the creation of isochrons with precisely defined slopes. In addition, a quickly cooled volcanic rock like an angrite has a high probability of fulfilling the requirements for a radioactive clock to provide an accurate age in that all the minerals likely started with the same Cr isotopic composition that they inherited from the parent magma, and they cooled quickly below the Mn–Cr closure temperature as a result of the rapid cooling of the magma once it erupted onto the planetesimal surface.

Table 14.3 and Fig. 14.3 provide an example of the utility of Mn–Cr dating for deciphering the chronology of early igneous activity on the parent planetesmal of the angrite meteorites. Spinels, for example, chromite, have plentiful Cr, but little Mn and so are characterized by Mn/Cr ratios near zero. In contrast, olivine can have a high Mn/Cr ratio and hence, with time, high $^{53}Cr/^{52}Cr$. On the isochron diagram of Fig. 14.3, the whole rock and mineral separates from D'Orbigny define a line of slope equal to $(3.24 \pm 0.04) \times 10^{-6}$. The slope defined by the mineral separates from another angrite, LEW 86010 (meteorites found in Antarctica are given initials defined by their collection site, in this case Lewis Cliffs), is much shallower at $(1.25 \pm 0.07) \times 10^{-6}$.

Table 14.3 Mn–Cr data for the D'Orbigny Angrite and LEW 86010

Sample		$^{55}Mn/^{52}Cr$	Uncertainty	$e^{53}Cr$	Uncertainty
D'Orbigny Angrite	Whole rock	5.71	0.08	1.89	0.04
	Spinel	0.006	0.001	0.35	0.03
	Olivine 1	1.59	0.08	0.73	0.03
	Olivine 2	6.82	0.2	2.3	0.05
	Olivine 3	13.0	0.2	4.04	0.04
	Olivine 4	27.8	0.4	8.23	0.04
	Pyroxene	2.24	0.02	0.94	0.03
	Glass 1	5.83	0.25	2.02	0.08
	Glass 2	4.46	0.18	1.54	0.07
LEW86010	Whole rock	2.29	0.11	0.66	0.31
	Pyroxene 1	0.63	0.03	0.48	0.29
	Pyroxene 2	0.49	0.02	0.45	0.25
	Olivine	160	8	18.07	0.48

Source: *Glavin et al.* [2004]. Reproduced with permission of John Wiley & Sons.

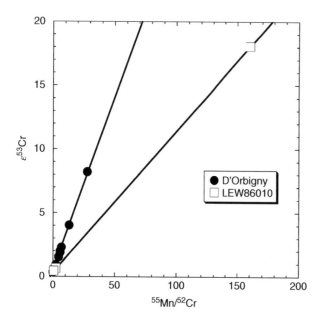

Fig. 14.3. Mn–Cr isochron diagram for the angrite data reported in Table 14.3.

The age difference between these two meteorites, according to Mn–Cr, is:

$$\Delta t = -\frac{1}{1.87 \times 10^{-7}} \times \ln\left[\frac{1.25 \times 10^{-6}}{3.24 \times 10^{-6}}\right] \qquad (14.14)$$

$$\Delta t = 5.1 \pm 0.3 \, \text{Ma}$$

Using the U–Pb age for D'Orbigny of 4563.37 Ma gives an age for LEW 86010 of 4558.3 Ma. The U–Pb age determined in an independent study of LEW 86010 is 4558.55 ± 0.15 Ma [*Amelin*, 2008], although this age drops to 4557.5 Ma using the nonterrestrial value of $^{235}U/^{238}U$ measured in angrites [*Brennecka and Wadhwa*, 2012]. U–Pb thus gives absolute ages for D'Orbigny and LEW 86010 that differ by 5.9 ± 0.4 Ma, some 800,000 years different than the age difference suggested by Mn–Cr. The explanation for this offset is not clear, but could

indicate: (i) an underestimation of the age uncertainties from either the U–Pb or Mn–Cr dates; (ii) different closure temperatures of U–Pb and Mn–Cr and the cooling history of the two angrites; or (iii) some heterogeneity in the initial abundance of ^{53}Mn, although this is unlikely given that the two angrites likely come from the same planetesimal.

This discrepancy in ages also brings up the question of processes, other than simple radioactive decay of ^{53}Mn, that can influence the isotopic composition of Cr. Two such processes are nucleosynthetic variability in Cr isotopic composition, and the production of Cr isotopes by interaction with cosmic rays. Recent studies [*Shukolyukov and Lugmair*, 2006; *Trinquier et al.*, 2007; *Qin et al.*, 2010] have identified Cr as one of an increasing number of elements whose stable isotopic composition is different between Earth and different types of meteorites. Carbonaceous chondrites show an excess in ^{54}Cr compared to Earth whereas ordinary chondrites and igneous meteorites like angrites and eucrites are deficient in ^{54}Cr compared to Earth (Fig. 14.4). The cause of this variability was tracked down to the presence of small Cr-rich grains in the meteorites that have Cr isotopic compositions very different from Earth, in one case enriched in ^{54}Cr by almost a factor of 50 [*Dauphas et al.*, 2010; *Qin et al.*, 2011]. The isotopic composition of these grains is so extreme that they most likely reflect grains coming from some other star. The strong excess in ^{54}Cr is expected from an exploding star known as a supernova (see Chapter 2). Although these grains have huge excesses in ^{54}Cr, their enrichments in ^{53}Cr are small, so they are unlikely to directly influence the abundance of ^{53}Cr in a way that would interfere with Mn–Cr chronometry.

A more significant problem for Mn–Cr dating comes from the modification of Cr isotopic composition caused by reactions with cosmic rays. Cr isotopes can be produced when an energetic cosmic-ray particle impacts an atom of Fe or Ni and knocks off a few nucleons in the process called spallation (Chapter 2), to leave behind a Cr atom. Spallation of iron produces Cr isotopes in declining abundance with mass away from the main iron mass at 56. Consequently, spallation produces more ^{54}Cr than ^{53}Cr, more ^{53}Cr than ^{52}Cr and more ^{52}Cr than ^{50}Cr (Fig. 14.5).

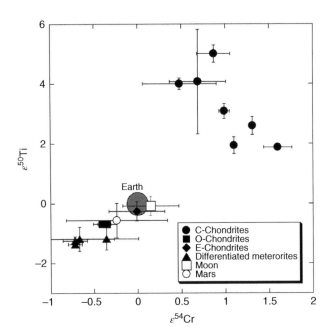

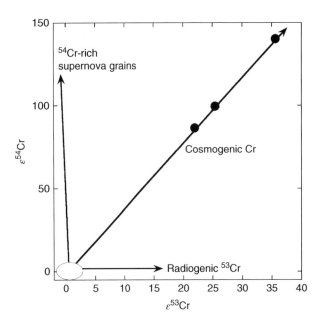

Fig. 14.4. Variation in the stable isotope composition of Cr and Ti between different groups of meteorites, Earth, Moon, and Mars. The axes show the range in $^{54}Cr/^{52}Cr$ and $^{50}Ti/^{47}Ti$ expressed in ε unit differences compared to the terrestrial standards shown by the gray oval. Data from *Trinquier et al.* [2007, 2009] and *Qin et al.* [2010]. Figure after *Carlson et al.* [2014] and *Warren* [2011].

Fig. 14.5. Deviations in $^{54}Cr/^{52}Cr$ and $^{53}Cr/^{52}Cr$ caused by radiogenic decay of ^{53}Mn, the production of Cr by spallation of Fe, and by addition of neutron-rich grains produced in supernovae. The isotope ratios are expressed as parts in 10,000 deviation from the isotopic composition of the terrestrial Cr standard. Most unirradiated meteorites, and Earth rocks, will plot in the oval in the bottom left of the figure. Decay of ^{53}Mn will produce only ^{53}Cr, not ^{54}Cr. The filled circles show three measurements of different splits of iron metal from the iron meteorite Carbo, which has a very long exposure to cosmic rays [*Qin et al.*, 2010]. Cosmic-ray spallation of Fe in Carbo has increased the $^{54}Cr/^{52}Cr$ and $^{53}Cr/^{52}Cr$ ratios in an approximate ratio of 4:1. Cr-rich supernova grains, likely responsible for the $^{54}Cr/^{52}Cr$ and $^{50}Ti/^{47}Ti$ variability shown in Fig. 14.4, contribute very high $^{54}Cr/^{52}Cr$ at slightly lower $^{53}Cr/^{52}Cr$ compared to terrestrial Cr [*Qin et al.*, 2011]. (Source: Adapted from *Qin et al.* 2010, 2011.)

The problem is particularly acute for any material with a high Fe/Cr ratio that has experienced significant exposure to cosmic rays. When cosmic-ray exposure is substantial, for example in the iron meteorite Carbo, $\varepsilon^{53}Cr$ can be increased by as much as 35 purely as a result of spallation-produced Cr [*Qin et al.*, 2010]. In Carbo, the increases in ^{53}Cr are well correlated with increases in ^{54}Cr (Fig. 14.5), so a correction can be made for the spallation component, but such secondary effects nevertheless have significant consequences for the use of Mn–Cr systematics in meteorites that have significant cosmic-ray exposure histories. In silicate meteorites, the problem is particularly pronounced for olivine due to the low Cr contents and high Fe/Cr ratio of olivine and the fact that olivine often forms the high Mn/Cr ratio end-member of chondrite isochrons [*Lugmair and Shukolyukov*, 1998].

While the evaluation of, and correction for, cosmogenic Cr must be made to ensure accurate Mn–Cr chronometry, the nucleosynthetic variability in ^{54}Cr also can influence Mn–Cr ages if ^{54}Cr is used for correcting the mass fractionation that occurs during mass spectrometry of Cr. In order to improve isotope-ratio precision and thereby enable development of the Mn–Cr chronometer, early Mn–Cr studies used a two-step fractionation correction that starts with $^{50}Cr/^{52}Cr$, but then adds an additional correction using $^{54}Cr/^{52}Cr$ if that ratio was different from the assumed value of the standard on the assumption that the $^{54}Cr/^{52}Cr$ ratio did not vary in meteorites or terrestrial samples [*Lugmair and Shukolyukov*, 1998]. In the mass spectrometer, preferential evaporation of the low-mass isotope causes the

measured isotopic composition to vary during the course of an analysis, as discussed in Chapter 3. This effect can be removed by assuming some mass dependency to the fractionation and correcting for the mass fractionation by using a fixed value for some stable isotope ratio. In most cases, the assumption of a fixed stable isotope composition is a good one, but for Cr, the variable nucleosynthetic contribution to ^{54}Cr causes the $^{54}Cr/^{52}Cr$ ratio to be different between different types of meteorites (Fig. 14.4). When $^{54}Cr/^{52}Cr$ is used to correct for instrumental mass fractionation, the variable abundance of ^{54}Cr can be propagated into the $^{53}Cr/^{52}Cr$ ratio via the mass fractionation correction. For example, carbonaceous chondrites have a $^{54}Cr/^{52}Cr$ ratio up to 1.5 parts in 10,000 higher than terrestrial Cr (e.g., $\varepsilon^{54}Cr$ = 1.5), whereas ordinary chondrites have $\varepsilon^{54}Cr$ of –0.5. If these two types of meteorite had the same $^{53}Cr/^{52}Cr$ ratio, but the measurement was made using $^{54}Cr/^{52}Cr$ for mass fractionation correction, then the mass fractionation correction would raise the $^{53}Cr/^{52}Cr$ ratio of the ordinary chondrite by $\varepsilon^{53}Cr$ = 0.25 and lower that of the carbonaceous chondrite by $\varepsilon^{53}Cr$ = 0.75, causing these two samples to have an apparent difference in $\varepsilon^{53}Cr$ of 1.00 instead of the 2.00 that they have when the measurement does not include ^{54}Cr for fractionation correction. If all the samples used to construct the isochron have the same $^{54}Cr/^{52}Cr$ ratio, the double fractionation correction will not affect the slope, and hence age, of the isochron, but if isochrons from different meteorite groups with different $^{54}Cr/^{52}Cr$ ratios are compared, for example carbonaceous chondrites versus ordinary chondrites, the double fractionation correction will produce a systematic difference in initial $^{53}Cr/^{52}Cr$ ratio that reflects the difference in the $^{54}Cr/^{52}Cr$ ratio between the meteorite groups, not a regional difference in the relative abundance of ^{53}Mn in the early solar nebula [*Lugmair and Shukolyukov*, 1998].

14.3.3 $^{107}Pd-^{107}Ag$

As mentioned previously, one of the strengths of extinct radionuclide chronology is that the elements involved have such diverse chemical characteristics that they can be used to trace a wide variety of processes involved in early planet formation and differentiation. One such process is the separation of iron metal from silicate that occurs during core formation, arguably the most significant chemical differentiation event that a planet experiences. Most long-lived radioactive systems, with the exception of Re–Os, are insoluble in iron metal and so are not sufficiently fractionated by metal–silicate separation to provide an accurate chronology of this process. Two short-lived systems, Pd–Ag and Hf–W are quite strongly affected by metal–silicate separation. The Pd–Ag system is based on the 6.5 million year half-life decay of ^{107}Pd to ^{107}Ag. Palladium is a so-called "highly siderophile element," meaning that when iron metal and silicate are together, the ratio of Pd in the metal to that in the silicate is of order 10^5 or higher. Silver also is siderophilic, but is more soluble in sulfides, and hence is a chalcophile element.

The relevant equations for the Pd–Ag system are:

$$\left(\frac{^{107}Pd}{^{108}Pd}\right)_t = \left(\frac{^{107}Pd}{^{108}Pd}\right)_0 \times e^{-\lambda t} \qquad (14.15)$$

$$\left(\frac{^{107}Ag}{^{109}Ag}\right)_t = \left(\frac{^{107}Ag}{^{109}Ag}\right)_0 + \frac{^{108}Pd}{^{109}Ag} \times \left(\frac{^{107}Pd}{^{108}Pd}\right)_0 \times \left(1-e^{-\lambda t}\right) \qquad (14.16)$$

$$\varepsilon^{107}Ag = \left[\frac{\left(\frac{^{107}Ag}{^{109}Ag}\right)Sa}{\left(\frac{^{107}Ag}{^{109}Ag}\right)Std} - 1\right] \times 10000 \qquad (14.17)$$

where "Sa" is sample and $(^{107}Ag/^{109}Ag)_{Std}$ is the value measured for the NIST978a Ag standard. Values reported for the $^{107}Ag/^{109}Ar$ ratio of this standard include 1.07916 ± 0.00052 [*Carlson and Hauri*, 2001], 1.07976 ± 0.00016 [*Schoenbaechler et al.*, 2007], 1.08048 ± 0.00042 [*Woodland et al.*, 2005], 1.08048 ± 0.00013 [*Matthes et al.*, 2015], and 1.0811 ± 0.0017 (TIMS measurement from *Chen and Wasserburg* [1983]). The range in standard values is 0.18% or $\varepsilon^{107}Ag$ of 18, and so is very large compared to the measurement precision. The $\varepsilon^{107}Ag$ reported for samples are calculated relative to the value measured for the NIST978a standard during the course

of measuring the samples, so one can directly compare $\varepsilon^{107}Ag$ values reported in different publications, but care should be taken when converting $\varepsilon^{107}Ag$ back to $^{107}Ag/^{109}Ag$ to ensure that the calculation is done using the same normalizing $^{107}Ag/^{109}Ag$ for NIST978a as was used to calculate the $\varepsilon^{107}Ag$ reported in the publication. In addition, this standard appears to be somewhat mass fractionated compared to Ag in the bulk silicate Earth as measurements of a number of terrestrial basalts provided an average $\varepsilon^{107}Ag = -2.2 \pm 0.7$ [*Schoenbaechler et al.*, 2010], which overlaps values measured in low Pd/Ag ratio chondrites [*Schoebaechler et al.*, 2008].

The Pd–Ag system has not seen much use, in part because the abundance of Ag is low in most materials, but mostly because with only two isotopes, distinguishing radiogenic ingrowth of ^{107}Ag from mass fractionation of the $^{107}Ag/^{109}Ag$ ratio is difficult. Until the advent of inductively coupled plasma mass spectrometry (ICP-MS), the inability to correct for instrument-induced mass fractionation limited precision on $^{107}Ag/^{109}Ag$ ratio determinations to about 0.1% [*Chen and Wasserburg*, 1983]. Using the ICP-MS with standard-sample comparison along with another element (usually Pd) to monitor and correct for instrumental mass fractionation has improved precisions on this ratio to about 0.005% [*Schoenbaechler et al.*, 2007]. The ICP approach, however, only corrects for the mass fractionation that occurs during analysis, not for the isotope mass fractionation that can occur during natural chemical processes that the sample may have experienced. Natural mass fractionation of Ag can cause variations in the $^{107}Ag/^{109}Ag$ ratio in the range of 0.1% that are not related to ^{107}Pd decay [*Schoebaechler et al.*, 2008]. What has enabled a good fraction of the applications of the Pd–Ag system is that Ag is a moderately volatile element under the conditions present in the early solar nebula, but Pd is refractory, so some high-temperature processes can produce extreme fractionation of Pd from Ag. The best example of this is the group of iron meteorites known as the IVA (four-A) iron meteorites.

Figure 14.6 shows the data obtained for one such IVA meteorite, Gibeon [*Chen and Wasserburg*, 1990]. With $^{108}Pd/^{109}Ag$ ratios of over 100,000 in the metal from Gibeon, compared to chondritic $^{108}Pd/^{109}Ag$ ratios that range from 1.5 to 20 [*Schoenbaechler et al.*, 2008], the $^{107}Ag/^{109}Ag$ ratio in Gibeon metal has been increased due to ^{108}Pd decay by factors of three to eight over the typical terrestrial value of 1.079. The slope of the isochron defined by the Gibeon data is $(2.4 \pm 0.5) \times 10^{-5}$ and corresponds to the $^{107}Pd/^{108}Pd$ ratio at the time the Pd–Ag system closed in Gibeon. In contrast, the Grant meteorite, which belongs to an iron meteorite group (the IIIB) that has not suffered the volatile element loss that characterizes the IVAs, has much lower Pd/Ag ratios, and as a result, much less variation in its $^{107}Ag/^{109}Ag$ ratio. This is where the improved precision allowed by modern Ag isotope analysis using an ICP-MS shows its worth compared to the lower precision data measured by thermal ionization mass spectrometry. Using the ICP-MS data alone, the slope of the Grant isochron is $(1.6 \pm 0.7) \times 10^{-5}$. Compared to the initial $^{107}Pd/^{108}Pd$ ratio defined by the Gibeon isochron,

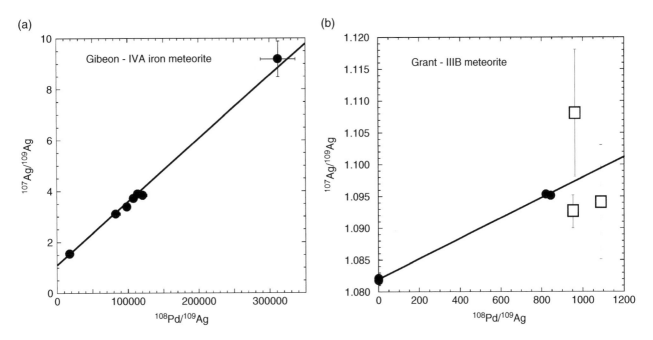

Fig. 14.6. Pd–Ag isochrons for two different types of iron meteorite. (a) Data for the Gibeon iron meteorite that is a representative of the very volatile-element depleted IVA group of iron meteorites. The volatile depletion is reflected in the extremely high Pd/Ag ratios of various splits of the iron from this meteorite shown by the filled circles. (Source: *Chen and Wasserburg* [1990]. Reproduced with permission of Elsevier.)(b) Thermal ionization (open squares) and ICP-MS (filled circles) data for the Grant IIIB iron meteorite that is much less volatile depleted, and hence characterized by lower Pd/Ag ratios. (Source: data from *Chen et al.* [2002] and *Carlson and Hauri* [2001].)

the result for Grant indicates that the Pd–Ag system in Grant closed 3.8 ± 5.6 Ma after that in Gibeon. The uncertainty on the two slopes allows these two meteorites to be the same age.

One IVA iron meteorite, Muonionalusta, has provided a precise absolute U–Pb age of 4565.3 ± 0.1 Ma [*Blichert-Toft et al.*, 2010]. The Pd–Ag systematics of this meteorite are somewhat scattered, but a fit through the majority of the data provides an isochron slope of $(2.14 \pm 0.3) \times 10^{-5}$ [*Horan et al.*, 2012]. Combining this isochron slope with the absolute U–Pb age allows calculation of the initial abundance of ^{107}Pd in the solar system using the equation:

$$\left(\frac{^{107}\text{Pd}}{^{108}\text{Pd}}\right)_t = \left(\frac{^{107}\text{Pd}}{^{108}\text{Pd}}\right)_0 \times e^{-\lambda t} \qquad (14.18)$$

For a solar system age of 4567 Ma ($\Delta t = 1.7$ Ma), the solar system initial $^{107}\text{Pd}/^{108}\text{Pd} = 2.56 \times 10^{-5}$. Using this initial abundance of ^{107}Pd, the slopes of the Gibeon and Grant isochrons then provide ages younger than 4567 Ma by 0.61 Ma (Gibeon) and 4.4 Ma (Grant).

Another piece of information provided by the isochron systematics is the initial Ag isotopic composition at the time of formation of the Grant and Gibeon meteorites. The initial $^{107}\text{Ag}/^{109}\text{Ag}$ ratios are 1.11 ± 0.03 for Gibeon and 1.08186 ± 0.00016 for Grant. The Grant initial $^{107}\text{Ag}/^{109}\text{Ag}$ ratio is particularly interesting because it is resolvably higher than the modern terrestrial $^{107}\text{Ag}/^{109}\text{Ag}$ of 1.07976. The bulk solar system has a low $^{108}\text{Pd}/^{109}\text{Ag}$ ratio near 1.6, as measured for primitive

carbonaceous chondrites [*Schoenbaechler et al.*, 2008], and hence would not evolve significant radiogenic ^{107}Ag. The high initial $^{107}\text{Ag}/^{109}\text{Ag}$ ratio measured for Grant implies that its parental materials must have been characterized by a high Pd/Ag ratio prior to the event that formed the material that would become the Grant meteorite. One can estimate the $^{108}\text{Pd}/^{109}\text{Ag}$ ratio of the source of Grant using the equation:

$$\left(\frac{^{107}\text{Ag}}{^{109}\text{Ag}}\right)_t = \left(\frac{^{107}\text{Ag}}{^{109}\text{Ag}}\right)_0 + \frac{^{108}\text{Pd}}{^{109}\text{Ag}} \times \left(\frac{^{107}\text{Pd}}{^{108}\text{Pd}}\right)_0 \times \left(1 - e^{-\lambda t}\right) \qquad (14.19)$$

To derive the Pd/Ag ratio from this equation requires assuming the starting Ag isotopic composition and the time interval over which the parent body of Grant grew in its radiogenic initial Ag isotopic composition. One choice for these parameters would be the starting $^{107}\text{Ag}/^{109}\text{Ag}$ of the solar system, which is 1.07972, and the 4.4 Ma age difference between Grant and the 4567 Ma age of the Solar system. Using these values, we have:

$$1.0816 = 1.07972 + \frac{^{108}\text{Pd}}{^{109}\text{Ag}} \times 2.56 \times 10^{-5}$$
$$\times \left(1 - e^{-\left(1.06 \times 10^{-7} \times 4.4 \times 10^6\right)}\right) \qquad (14.20)$$

which reduces to $^{108}\text{Pd}/^{109}\text{Ag} = 197$.

Given that typical chondrite $^{108}\text{Pd}/^{109}\text{Ag}$ ratios are between 1.5 and 20, the high inferred Pd/Ag for the Grant parent body implies that this meteorite did not form from segregation of metal from a body that had chondritic Pd/Ag, but instead formed from a body with much higher Pd/Ag, likely as a result of minor depletion in volatile Ag.

As in many short-lived radiometric systems, exposure to cosmic rays can have important consequences for Pd–Ag systematics. In the case of Pd–Ag, when an energetic galactic cosmic ray (GCR) impacts an iron meteorite, it creates a cascade of nuclear particles, some of which are neutrons. These neutrons can be captured by ^{108}Pd to form ^{109}Pd [*Leya and Masarik*, 2013], which then decays with a half-life of about 13 h to ^{109}Ag. The result of GCR irradiation is thus a decrease in the $^{107}Ag/^{109}Ag$ ratio that will be proportional to both the duration of the irradiation and the Pd/Ag ratio in the sample [*Matthes et al.*, 2015]. In their study of a number of iron meteorites, *Matthes et al.* [2015] calculated that the change in $^{107}Ag/^{109}Ag$ ratio due to cosmic ray irradiation should correlate with both Pd/Ag ratio and with Pt isotopic composition. In this case, Pt serves as a relatively abundant trace element in an iron meteorite whose isotopic composition also is modified by neutron capture, and hence can be used as a monitor for the neutron dose experienced by the sample. Their calculations show that the $^{107}Ag/^{109}Ag$ ratio can be reduced by up to 1% depending on magnitude of irradiation, as determined by the offset in $^{196}Pt/^{195}Pt$ from standard Pt, and the Pd/Ag ratio of the sample (Fig. 14.7a). Using this relationship to correct the Ag isotopic data they measured for a number of iron meteorites for the GCR exposure these meteorites experienced, they were able to significantly improve the Pd–Ag isochrons. An extreme example is the strongly irradiated meteorite Carbo, also discussed in the Mn–Cr section, where measured $\varepsilon^{107}Ag$ in a number of metal samples from Carbo ranged from −10 to −30. These values are lower than the initial $\varepsilon^{107}Ag = -3.1$ of the solar system [*Schoenbaechler et al.*, 2008], reflecting the reduction in $^{107}Ag/^{109}Ag$ ratio caused by neutron capture in this sample. Using the offset in Pt isotopic composition from the terrestrial Pt isotopic composition in these samples, and their model for GCR neutron correction, *Matthes et al.* [2015] found that these very low $\varepsilon^{107}Ag$ values corrected to a range from +17 to +34, and formed a respectable linear array on a Pd–Ag isochron diagram that gives a slope consistent with the slopes seen for a number of other iron meteorites (Fig. 14.7b).

14.3.4 $^{182}Hf–^{182}W$

In the Pd–Ag system, the parent isotope is concentrated in the metal, but in the Hf–W system, the parent is almost totally excluded from the metal. Both Hf and W are very refractory elements. Hf is a lithophile element, meaning that it is much more soluble in silicate than in metal. In contrast, W is a moderately siderophile element, meaning that it has higher solubility in metal than silicate. As a result, when core formation occurs on a planet, the core metal has a Hf/W ratio of essentially zero. The removal of the core metal from the mantle raises the Hf/W ratio of the mantle by simple mass balance. The Hf–W system thus has seen the majority of its use in defining the time of metal–silicate separation on planets and planetesimals, but it also can be used to date silicate-rich meteorites such as CAIs and angrites, an example of which is given in Table 14.4 and Fig. 14.8.

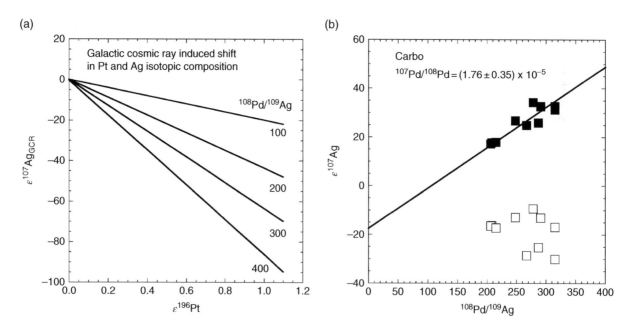

Fig. 14.7. Effect of cosmic-ray irradiation on Pd–Ag systematics. (a) Galactic cosmic rays (GCR) create neutrons that are captured by ^{108}Pd to make ^{109}Ag, and by ^{195}Pt to make ^{196}Pt, producing the correlations shown for different Pd/Ag ratios in the meteorite as indicated by the numbers at the right of each line. (b) Using these relationships to correct the measured Ag isotopic composition of metal fragments from the Carbo meteorite (open squares) for the long GCR exposure experienced by Carbo, produces the neutron-corrected Ag isotopic compositions shown by the filled squares, which form a respectable Pd–Ag isochron. (Source: *Matthes et al.* [2015]. Reproduced with permission of Elsevier.)

Table 14.4 Hf–W data for minerals from an Allende CAI and the D'Orbigny Angrite

Sample		$^{180}Hf/^{184}W$	Uncertainty	$^{182}W/^{184}W$	Uncertainty
Allende CAI	Melilite	1.07	0.001	0.864672	0.000039
	Mix	4.28	0.004	0.864982	0.000039
	Fassaite	12.8	0.2	0.865859	0.000104
D'Orbigny Angrite	Pyroxene	20.86	0.16	0.866234	0.000030
	Whole rock 1	6.85	0.11	0.865213	0.000022
	Whole rock 2	7.65	0.05	0.865259	0.000030
	Fine Fraction	4.12	0.03	0.864983	0.000025

CAI data from *Burkhardt et al.* [2008]; D'Orbigny data from *Markowski et al.* [2007].

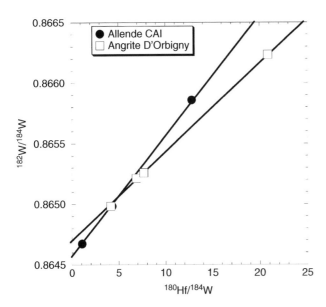

Fig. 14.8. Hf–W isochron diagram for the data in Table 14.4.

The equations for Hf–W dating include:

$$\left(\frac{^{182}Hf}{^{180}Hf}\right)_t = \left(\frac{^{182}Hf}{^{180}Hf}\right)_0 \times e^{-\lambda t} \tag{14.21}$$

$$\left(\frac{^{182}W}{^{184}W}\right)_t = \left(\frac{^{182}W}{^{184}W}\right)_0 + \frac{^{180}Hf}{^{184}W} \times \left(\frac{^{182}Hf}{^{180}Hf}\right)_0 \times \left(1 - e^{-\lambda t}\right) \tag{14.22}$$

As with many other isotope systems, the variability in the $^{182}W/^{184}W$ ratio is so small that it is convenient to express the deviations in the ratio from those of a standard, which, in this case, is a terrestrial W solution that has $^{182}W/^{184}W = 0.864863$ [*Kleine et al.*, 2009], although values used for this standard differ somewhat between different publications.

$$\varepsilon^{182}W = \left[\frac{\left(\frac{^{182}W}{^{184}W}\right)Sa}{\left(\frac{^{182}W}{^{184}W}\right)Std} - 1\right] \times 10000 \tag{14.23}$$

Table 14.4 shows Hf–W data for minerals separated from a single CAI from the Allende meteorite, and from the angrite

D'Orbigny. These data define the Hf–W isochrons shown in Fig. 14.8. For the CAI, the slope is $(1.0 \pm 0.4) \times 10^{-4}$ and the initial $^{182}W/^{184}W$ is 0.86456, which corresponds to $\varepsilon^{182}W = -3.5$ [*Burkhardt et al.*, 2008]. CAIs also display minor variability in the abundance of ^{183}W due to nucleosynthetic processes [*Burkhardt et al.*, 2012]. Correcting for the nucleosynthetic variability, and using a wider range of data for CAIs, results in a current best estimate of the starting solar system Hf–W values of $^{182}Hf/^{180}Hf = (9.72 \pm 0.44) \times 10^{-5}$ [*Burkhardt et al.*, 2008] and $\varepsilon^{182}W = -3.51 \pm 0.10$ [*Burkhardt et al.*, 2012], which corresponds to $^{182}W/^{184}W = 0.864559 \pm 0.000009$.

The slope for the D'Orbigny isochron is $(7.4 \pm 0.2) \times 10^{-5}$. As is the case for all the extinct radionuclide systematics, these slopes correspond to the ratio of parent isotope to the stable isotope of the element used in the denominator of the age equation at the time the system closed in these rocks. Using the two slopes and equation (14.21) gives the age difference between the CAI and D'Orbigny indicated by Hf–W as 3.5 ± 1.4 Ma, which compares well with the 4.8 ± 0.15 Ma age difference defined by Al–Mg and the 4.2 Ma age difference between the U–Pb ages of these two samples, as discussed previously.

14.3.4.1 Hf–W model ages for core formation

A major chronological application of the Hf–W system is the determination of the time of metal–silicate separation in planetesimals, which includes both the formation of iron meteorites and the cores of the terrestrial planets. Because Hf is essentially insoluble in iron metal, the Hf–W chronometry of iron meteorites works not by tracking the ingrowth of ^{182}W, but by noting the time that the ingrowth stopped because the W was placed into an environment—the iron core of a planetesimal—that no longer had Hf to decay to W. This application is similar to the model age approaches used in a number of other radioactive decay schemes (Chapters 6 and 7) in that it calculates the time when the $^{182}W/^{184}W$ ratio of some sample deviated from the W isotope evolution of some model reservoir. The typical reservoir used for Hf–W is the bulk composition of the solar system, as represented by primitive chondritic meteorites. The average present-day $^{182}W/^{184}W$ of chondrites is 0.864699 ($\varepsilon^{182}W = -1.9$ compared to the W isotopic composition of the modern terrestrial mantle) [*Kleine et al.*, 2009]. Using this isotopic composition for chondrites and the initial $^{182}W/^{184}W$ and $^{182}Hf/^{180}Hf$ ratios

(a)

(b)

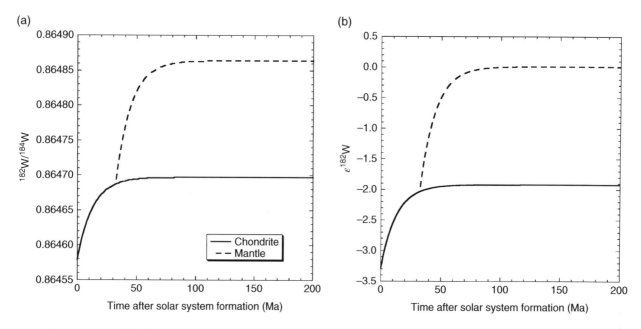

Fig. 14.9. (a) Evolution of the ^{182}W/^{184}W ratio in two reservoirs with different Hf/W ratios. The solid line describes the W isotope evolution of a reservoir with chondritic Hf/W ratio. The dotted line shows the evolution of a reservoir formed at 30 Ma that has a much higher Hf/W ratio, as expected in the mantle after core segregation has removed substantial W, but no Hf, from the mantle. (b) The same evolution in ε^{182}W space.

determined from the CAI isochron of Fig. 14.8 (0.864559 and 9.72×10^{-5}, respectively), allows calculation of the average Hf/W ratio of chondrites using equation (14.22). This equation describes the W isotope evolution shown in Fig. 14.9. With t being the age of the solar system, the term $e^{-\lambda t}$ becomes zero, so the average chondrite ^{180}Hf/^{184}W ratio is:

$$\frac{^{180}\text{Hf}}{^{184}\text{W}} = \frac{[0.864699 - 0.864559]}{9.72 \times 10^{-5}} = 1.4 \qquad (14.24)$$

A Hf–W model age can be calculated relative to a chondritic model by calculating the time when the ^{182}W/^{184}W ratio of the sample was the same as that of the model. This requires two equations, one for the model and one for the sample, with the measurable parameters being the present-day W isotopic composition and Hf/W ratio of both the chondrite model and the sample:

$$^{p}_{Ch}\left(\frac{^{182}\text{W}}{^{184}\text{W}}\right) = ^{0}_{Ch}\left(\frac{^{182}\text{W}}{^{184}\text{W}}\right) + _{Ch}\left(\frac{^{180}\text{Hf}}{^{184}\text{W}}\right) \times {}^{t}\left(\frac{^{182}\text{Hf}}{^{180}\text{Hf}}\right) \qquad (14.25)$$

and

$$^{p}_{Sa}\left(\frac{^{182}\text{W}}{^{184}\text{W}}\right) = ^{0}_{Sa}\left(\frac{^{182}\text{W}}{^{184}\text{W}}\right) + _{Sa}\left(\frac{^{180}\text{Hf}}{^{184}\text{W}}\right) \times {}^{t}\left(\frac{^{182}\text{Hf}}{^{180}\text{Hf}}\right) \qquad (14.26)$$

Where "p" is the present-day measured ratio, "Ch" is average chondrite, "Sa" is the sample, and t is the time when the sample's

W isotopic composition began to diverge from chondritic as a result of its formation with a nonchondritic Hf/W ratio. The sample's model age is thus the time when both the sample and average chondrite had the same ^{182}W/^{184}W ratio, or when:

$$^{0}_{Ch}\left(\frac{^{182}\text{W}}{^{184}\text{W}}\right) = ^{0}_{Sa}\left(\frac{^{182}\text{W}}{^{184}\text{W}}\right) \qquad (14.27)$$

With this requirement, subtracting the two equations above gives:

$$\left[^{p}_{Ch}\left(\frac{^{182}\text{W}}{^{184}\text{W}}\right) - ^{p}_{Sa}\left(\frac{^{182}\text{W}}{^{184}\text{W}}\right)\right] = {}^{t}\left(\frac{^{182}\text{Hf}}{^{180}\text{Hf}}\right)$$
$$\times \left[_{Ch}\left(\frac{^{180}\text{Hf}}{^{184}\text{W}}\right) - _{Sa}\left(\frac{^{180}\text{Hf}}{^{184}\text{W}}\right)\right] \qquad (14.28)$$

so:

$$^{t}\left(\frac{^{182}\text{Hf}}{^{180}\text{Hf}}\right) = \frac{\left[^{p}_{Ch}\left(\frac{^{182}\text{W}}{^{184}\text{W}}\right) - ^{p}_{Sa}\left(\frac{^{182}\text{W}}{^{184}\text{W}}\right)\right]}{\left[_{Ch}\left(\frac{^{180}\text{Hf}}{^{184}\text{W}}\right) - _{Sa}\left(\frac{^{180}\text{Hf}}{^{184}\text{W}}\right)\right]} \qquad (14.29)$$

The time can be calculated by noting that:

$$^{t}\left(\frac{^{182}\text{Hf}}{^{180}\text{Hf}}\right) = ^{0}\left(\frac{^{182}\text{Hf}}{^{180}\text{Hf}}\right) \times e^{-\lambda t} \qquad (14.30)$$

where $^{0}(^{182}\text{Hf}/^{180}\text{Hf})$ is the initial solar system ^{182}Hf/^{180}Hf ratio at 4567 Ma (1.07×10^{-4}).

This leads to the model age equation:

$$t = \frac{-1}{\lambda} \times \ln\left[\frac{\left[{}_{Ch}^{P}\left(\frac{{}^{182}W}{{}^{184}W}\right) - {}_{Sa}^{P}\left(\frac{{}^{182}W}{{}^{184}W}\right)\right]}{\left[{}_{Ch}\left(\frac{{}^{180}Hf}{{}^{184}W}\right) - {}_{Sa}\left(\frac{{}^{180}Hf}{{}^{184}W}\right)\right] \times {}^{0}\left(\frac{{}^{182}Hf}{{}^{180}Hf}\right)}\right] \quad (14.31)$$

If we want to apply this model age to calculate the time of core formation on Earth, for the sample parameters, we substitute the measured ${}^{182}W/{}^{184}W$ ratio of the modern mantle (0.864863) and an estimate of the average ${}^{180}Hf/{}^{184}W$ ratio (20) of the silicate portion of the Earth. Substituting these values and those appropriate for the average chondrite model, we obtain:

$$t = \frac{-1}{7.79 \times 10^{-8}} \times \ln\left[\frac{[0.864699 - 0.864863]}{[1.4 - 20] \times 9.72 \times 10^{-5}}\right] \quad (14.32)$$

These values result in a terrestrial core formation model age interval of 31 Ma after solar system formation (4567 Ma) or a model age of 4536 Ma.

As shown in Fig. 14.9, this type of model age assumes a single core-forming event. More realistic models have examined incremental core formation occurring in many discrete events or continuously throughout the accumulation of Earth [*Nimmo and Agnor*, 2006]. Although more complex, these models basically use the same data, and the same modeling approach, just with many separate core-formation events instead of the single event as depicted in Fig. 14.9.

This type of model age uses the W isotopic evolution of a planetary mantle. In contrast, measurements of the W isotopic composition of iron meteorites are dealing with the other side of the problem—the metal—whose Hf/W ratio dropped to zero on its segregation into the planetary core. In this case, the model age is calculated for when the ingrowth of ${}^{182}W$ stopped, using the equation:

$$\left(\frac{{}^{182}W}{{}^{184}W}\right)_t = \left(\frac{{}^{182}W}{{}^{184}W}\right)_0 + \left(\frac{{}^{180}Hf}{{}^{184}W}\right)_{Ch} \times \left(\frac{{}^{182}Hf}{{}^{180}Hf}\right)_0 \times \left(1 - e^{-\lambda t}\right) \quad (14.33)$$

and assuming that the parent body of the iron meteorite evolved with chondritic Hf/W ratio (${}^{180}Hf/{}^{184}W = 1.4$) up to the point where the metal segregated to form the planetesimal core. Solving for t gives:

$$t = \frac{-1}{\lambda} \times \ln\left[1 - \left\{\frac{\left[\left(\frac{{}^{182}W}{{}^{184}W}\right)_t - \left(\frac{{}^{182}W}{{}^{184}W}\right)_0\right]}{\left[\left(\frac{{}^{180}Hf}{{}^{184}W}\right)_{Ch} \times \left(\frac{{}^{182}Hf}{{}^{180}Hf}\right)_0\right]}\right\}\right] \quad (14.34)$$

The IIE group of iron meteorites have the highest measured $\varepsilon^{182}W = -2.3$ (${}^{182}W/{}^{184}W = 0.864664$) [*Markowski et al.*, 2006b]. Using equation (14.34) and substituting the appropriate parameters gives:

$$t = \frac{-1}{7.79 \times 10^{-8}} \times \ln\left[1 - \left\{\frac{[0.864664 - 0.864559]}{1.4 \times 9.72 \times 10^{-5}}\right\}\right] \quad (14.35)$$

The result provides a model age for the IIE irons of 19 Ma after solar-system formation.

Some meteorites, notably the IID iron meteorite Carbo with its measured $\varepsilon^{182}W = -4.29$ (${}^{182}W/{}^{184}W = 0.864492$) [*Qin et al.*, 2015] have $\varepsilon^{182}W$ even lower than the solar-system initial value. This would translate to an age 5 Ma older than the CAI if it reflected only the decay of ${}^{182}Hf$. While this result could be interpreted to mean that iron meteorites formed before CAIs, another means of modifying the W isotopic composition of a sample must be considered—the role of cosmic-ray irradiation. As discussed previously, Carbo is known to have a very long cosmic-ray exposure history. For W, the main consequence of cosmic-ray irradiation is the production of secondary neutrons, which are then captured by W isotopes, transforming them into a heavier W isotope. The net effect is that, when corrected for instrumental mass fractionation in the mass spectrometer using the ${}^{184}W/{}^{186}W$ ratio, cosmic-ray irradiation results in a lowering of the measured ${}^{182}W/{}^{184}W$ ratio that translates, erroneously, into apparently old ages in Hf–W systematics. A variety of correction schemes to the Hf–W systematics of meteorites that have suffered extensive cosmic-ray exposure histories are in the process of being developed. Most of these use either ${}^{3}He$ abundance [*Markowski et al.*, 2006a], or variations in the isotopic composition of elements such as Pt or Os [*Kruijer et al.*, 2013; *Qin et al.*, 2015], that are as, or more, sensitive to neutron exposure than is W. From the isotopic changes to He, Pt, or Os that must reflect only cosmic-ray interaction, empirical corrections to ${}^{182}W/{}^{184}W$ ratios in the same samples can account for the cosmic-ray-related modifications to W isotopic composition. The goal of such corrections is to isolate the changes in ${}^{182}W$ abundance due only to ${}^{182}Hf$ decay. The correction procedures, and the analysis of meteorites that have not suffered long cosmic-ray exposures, show that many iron meteorites have ${}^{182}W/{}^{184}W$ ratios only marginally more radiogenic than the CAI initial ${}^{182}W/{}^{184}W$ ratio, implying that many groups of iron meteorites formed within a million years, or less, of CAI formation (Fig. 14.10).

The very quick formation model ages for many iron meteorites appear to disagree with the few million year ages for the same meteorites indicated by the Pd–Ag system. The important difference here is that the Hf–W model ages of iron meteorites measure the time when the metal became chemically isolated from the silicate portion of the parent planetesimal. In contrast, the Pd–Ag ages provide the time when the Pd–Ag system closed in the iron meteorites following their crystallization and cooling. The magnitude of the difference between the Hf–W and Pd–Ag ages is consistent with iron meteorite cooling rates determined by mineralogic methods based on the composition and grain-sizes of the different alloys of iron present in the meteorite [*Matthes et al.*, 2015]. This example points out the importance of understanding exactly what process sets the "age" that is measured by the radioactive clock being used.

14.3.5 I–Pu–Xe

As an even Z element, overlapping the magic neutron number 82 (see Chapter 2 for why these parameters are important for

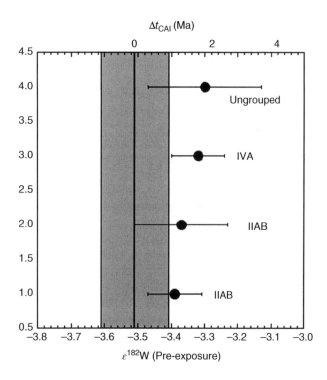

Fig. 14.10. Average W isotopic composition of various groups of iron meteorites after correction for cosmic-ray exposure history [*Kruijer et al.*, 2012]. The gray band shows the initial solar system $^{182}W/^{184}W$ ratio as determined from a CAI isochron [*Burkhardt et al.*, 2008]. (Source: adapted from *Kruijer et al.* [2012] and *Burkhardt et al.* [2008].)

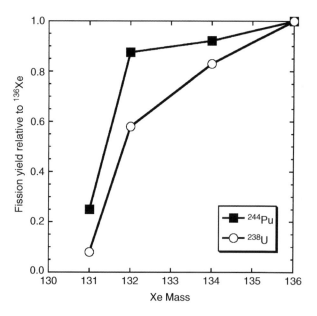

Fig. 14.11. Proportion of Xe isotopes produced by fission of ^{244}Pu and ^{238}U. (Source: data from *Ballentine and Burnard* [2002].)

determining the number of isotopes in an element), Xe has many stable isotopes, nine in all. Xenon receives radiogenic contributions from the β-decay of ^{129}I, and from the spontaneous fission of both ^{238}U and ^{244}Pu. Xenon, as a noble gas, is a highly volatile element meaning that it will be outgassed into a vapor phase, such as the atmosphere, during any significant heating of a rock. Iodine also is volatile, though nowhere near as volatile as xenon. Both uranium and plutonium are refractory lithophile elements. Fractionation of Xe from I, U, and Pu thus will be primarily driven by heating and volatile loss events.

Unlike the discrete isotope-to-isotope decay of α and β decay, fission produces a range of decay-product isotopes spanning several elements. So far, the spontaneous fission products of ^{238}U and ^{244}Pu in natural materials have been observed only in Xe because loss of the noble gas Xe from a material during heating can lead to very large U and Pu to Xe ratios. In the Xe mass range, the fission yields for both ^{244}Pu and ^{238}U peak at ^{136}Xe and decline as the mass decreases as illustrated in Fig. 14.11.

I–Xe systematics follow the same decay equations used in other extinct radionuclide systems:

$$\left(\frac{^{129}Xe}{^{130}Xe}\right)_t = \left(\frac{^{129}Xe}{^{130}Xe}\right)_0 + \left(\frac{^{129}I}{^{127}I}\right)_0 \times \left(\frac{^{127}I}{^{130}Xe}\right) \times \left(1 - e^{-\lambda t}\right)$$

$$(14.36)$$

^{130}Xe is used in the denominator of these equations because its direct production by U and Pu fission is small (yields < 0.01%) and its production from the decay of other neutron-rich isotopes is blocked by very long-lived ^{130}Te.

Plutonium and uranium fission decay follow similar equations, but with a number of additional complications. First, the primary route of radioactive transformations for both elements is α-decay, not fission. Only 0.125% of ^{244}Pu atoms decay by fission, and only 5.45×10^{-5} % of ^{238}U atoms decay by spontaneous fission. Next, only a small fraction of the fission products of either U or Pu end up as xenon. For ^{244}Pu, only 5.6% of fissioned Pu ends up as ^{136}Xe. For ^{238}U, the fission yield of ^{136}Xe is 6.3%. Combining the percentage of atoms of U and Pu that fission with the number of fission products that end up as ^{136}Xe, only 7×10^{-5} of ^{244}Pu atoms fission to ^{136}Xe and 3.5×10^{-8} ^{238}U atoms fission to ^{136}Xe. Finally, there is the additional problem that plutonium does not occur naturally, so there is no direct way to determine the Pu/Xe ratio at the time a sample formed. To overcome this problem, some other element with similar geochemical properties to Pu, often U or Nd [*Lugmair and Marti*, 1977], is taken as a proxy for Pu. As a result, using U as the geochemical proxy for Pu, the Pu–Xe dating equation takes on the form:

$$\left(\frac{^{136}Xe}{^{130}Xe}\right)_t = \left(\frac{^{136}Xe}{^{130}Xe}\right)_0 + \left(\frac{^{244}Pu}{^{238}U}\right)_0 \times \left(\frac{^{238}U}{^{136}Xe}\right)$$
$$\times \left(7 \times 10^{-5}\right) \times \left(1 - e^{-\lambda t}\right)$$

$$(14.37)$$

where the term 7×10^{-5} accounts for the number of ^{244}Pu atoms that end up as ^{136}Xe. $\left(\frac{^{136}Xe}{^{130}Xe}\right)_t$ and $\frac{^{238}U}{^{136}Xe}$ are the two measureable parameters in the equation.

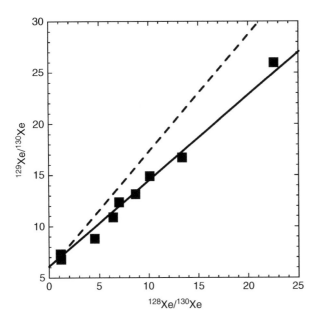

Fig. 14.12 Xe isotopic composition measured for apatite grains separated from the Acapulco meteorite. The dotted line shows the best fit line through data for various whole-rock splits of the meteorite Bjurbole that experienced the same neutron irradiation as the Acapulco apatites. (Source: data from *Nichols et al.* [1994].)

A complication in the I–Xe system is that iodine abundances in many types of samples often are at, or below, detection limits. This problem was overcome through much the same approach as used in $^{40}Ar/^{39}Ar$ dating (see Chapter 9)—irradiation of a sample with neutrons turns some of the ^{127}I in the sample into ^{128}I by neutron capture, which then β-decays with a 25 min half-life to ^{128}Xe. Through neutron irradiation, the $^{128}Xe/^{130}Xe$ ratio in the sample becomes proportional to the I/Xe ratio in the sample. Figure 14.12 shows an example of Xe isotope analyses of apatite grains extracted from the Acapulco meteorite that were subjected to neutron irradiation in a nuclear reactor powered by ^{235}U fission. The data show a good correlation between the measured $^{128}Xe/^{130}Xe$, which is proportional to the I/Xe ratio in the sample, and $^{129}Xe/^{130}Xe$ ratios that reflect the decay of ^{129}I. The slope of the line corresponds to the $^{129}I/^{127}I$ ratio at the time the apatites formed, multiplied by a constant that reflects the efficiency of the transformation of ^{127}I into ^{128}I during neutron irradiation. The best way to determine this irradiation constant is to include in the irradiation package a sample of known age; in this example, whole-rock samples of the meteorite Bjurbole were used. The difference in slope between these two lines is proportional to the difference in $^{129}I/^{127}I$ at the time of formation of the two samples because the irradiation factor is the same in both samples and hence divides out in the comparison of the two slopes. In this case, the difference in slope between the standard and the sample can be translated into an age difference through the equation:

$$\left(\frac{^{127}I}{^{129}I}\right)_{Sa} = \left(\frac{^{127}I}{^{129}I}\right)_{Std} \times e^{-\lambda t} \tag{14.38}$$

or

$$t = \frac{-1}{\lambda} \times \ln \left[\frac{\left(\frac{^{127}I}{^{129}I}\right)_{Sa}}{\left(\frac{^{127}I}{^{129}I}\right)_{Std}} \right] \tag{14.39}$$

Using these equations, the results shown in Fig. 14.12 indicate that the I–Xe system in the Acapulco apatites closed 8.1 ± 1.2 Ma after the formation of Bjurbole. These relative ages can then be converted to absolute ages by comparison with U–Pb ages measured on apatites from chondrites [*Göpel et al.*, 1994].

Xe is sufficiently rare in most materials that its isotopic composition can be modified by the production of Xe isotopes via spallation due to cosmic rays. The isotopic composition of the Xe produced by spallation depends on the composition of the material being irradiated because Ba and the rare-earth elements (REE) are primary targets for Xe production by spallation. When the Xe in a sample contains a combination of Xe produced by ^{129}I, ^{238}U, and ^{244}Pu decay, along with spallation Xe, these various components must be stripped sequentially from the Xe isotopic composition in order to get to the original Xe isotopic composition appropriate for calculating a radiogenic age from the Xe data. Extracting the proportion of the various contributions can be done by solving the series of linear mixing equations of the form:

$$^iXe_{Mix} = {}^iXe_S^* X_S + {}^iXe_A^* X_A + {}^iXe_{Sp}^* X_{Sp} + {}^iXe_U^* X_U$$
$$+ {}^iXe_{Pu}^* X_{Pu} + {}^iXe_I^* X_I \tag{14.40}$$

where "i" designates each isotope of Xe and "X" is the proportion of solar (S), air (A), spallation (Sp), U fission (U), Pu fission (Pu) and ^{129}I decay (I) in the measured mixture of Xe in the sample, where:

$$X_S + X_A + X_{Sp} + X_U + X_{Pu} + X_I = 1 \tag{14.41}$$

A simplified example of the component deconvolution procedure is shown in Table 14.5. In this example, the sample consists only of a mixture of solar Xe and U-fission Xe, so the equations above reduce to:

$$^iXe_{mix} = {}^iXe_S^* X_S + iXe_U^* X_U \text{ and } X_S + X_U = 1 \tag{14.42}$$

Solving these equations for X_U, and using ^{136}Xe because it is the most abundant isotope produced by U fission, gives:

$$X_U = \left({}^{136}Xe_{Mix} - {}^{136}Xe_S\right) / \left({}^{136}Xe_U - {}^{136}Xe_S\right) \tag{14.43}$$

Using the numbers from Table 14.5 for the mixture shows that this mixture is composed of 20% U-fission Xe and 80% solar Xe. Extracting multiple components from a natural Xe sample follows the same approach, though the length of the equations involved increase with the number of components. A recent example of this more complex deconvolution of Xe components is given in [*Parai and Mukhopadhyay*, 2015].

Table 14.5 Isotopic composition, in atom%, of various Xe components

	124	126	128	129	130	131	132	134	136
Solar	0.00125	0.00114	0.0221	0.276	0.0433	0.216	0.263	0.0976	0.0792
Planetary	0.00122	0.00109	0.0215	0.274	0.0425	0.215	0.262	0.0997	0.0837
Air	0.00095	0.000887	0.0192	0.264	0.0407	0.212	0.269	0.104	0.0886
Spallation	0.0612	0.103	0.156	0.101	0.101	0.388	0.0854	0.00454	0
U fission	0	0	0	0	0	0.0321	0.233	0.333	0.402
Pu fission	0	0	0	0	0	0.0823	0.287	0.302	0.328
^{129}I	0	0	0	1	0	0	0	0	0
Mixture	0.0010	0.000912	0.0176	0.221	0.0347	0.179	0.257	0.145	0.144

14.3.6 ^{146}Sm–^{142}Nd

Both neodymium and samarium have seven isotopes. Neodymium receives two radioactive contributions from Sm decay. The more commonly used is based on the 106 Ga half-life α-decay of ^{147}Sm to ^{143}Nd, which is covered in detail in Chapter 6. In this chapter, we will focus on the much shorter half-life α-decay of ^{146}Sm to ^{142}Nd. This system tracks the same type of geochemical differentiation processes as does the ^{147}Sm–^{143}Nd system, but it does so with a half-life a thousand times shorter. The short half-life of ^{146}Sm pushes this system into the extinct radionuclide category. Samarium and Nd are neighboring REE and share the group's chemical properties in that they have high condensation temperatures (refractory elements), are more soluble in silicates (lithophile elements) than metal or sulfide, and exist as +3 valence state ions in minerals. Under extremely reducing conditions, the REE do become soluble in sulfides, notably CaS (oldhamite), but such conditions are rare to nonexistent on the Earth, Moon, Mars, and most meteorites. An exception is a type of meteorite known as an enstatite chondrite, where a substantial portion of the whole-rock REE abundances can be present in CaS. With the exception of these very reducing conditions, the primary geochemical process that fractionates Sm from Nd is magmatic differentiation.

The ^{146}Sm–^{142}Nd clock relies on the decay of the low-abundance ^{146}Sm to the most abundant isotope of Nd, ^{142}Nd, and as a result, the changes to the ^{142}Nd/^{144}Nd ratio caused by ^{146}Sm decay are small. Consequently, variations in ^{142}Nd/^{144}Nd ratio are usually reported in either ε or μ units defined as:

$$\left[\left(\frac{\left(\frac{^{142}\text{Nd}}{^{144}\text{Nd}} \right)_{Sa}}{\left(\frac{^{142}\text{Nd}}{^{144}\text{Nd}} \right)_{Std}} \right) - 1 \right] \times \text{C} \qquad (14.44)$$

where C is a constant that is either 10^4 for ε, or 10^6 for μ.

The equation used for ^{146}Sm–^{142}Nd dating is similar to those used for other extinct-radionuclide systems:

$$\left(\frac{^{142}\text{Nd}}{^{144}\text{Nd}} \right)_t = \left(\frac{^{142}\text{Nd}}{^{144}\text{Nd}} \right)_0 + \left(\frac{^{144}\text{Sm}}{^{144}\text{Nd}} \right) \times \left(\frac{^{146}\text{Sm}}{^{144}\text{Sm}} \right)_0 \times \left(1 - e^{-\lambda t} \right)$$

$$(14.45)$$

For most reports involving the ^{146}Sm–^{142}Nd system, ^{144}Sm is used as the stable reference isotope of Sm rather than ^{147}Sm, which is the most commonly used stable Sm isotope for the ^{147}Sm–^{143}Nd system. This is partly just by convention, but also reflects the fact that both ^{144}Sm and ^{146}Sm are produced purely by the p-process route of nucleosynthesis (see discussion on nucleosynthesis in Chapter 2). If the solar nebula prior to the start of planet formation did not have a well-mixed distribution of s-, r-, and p-process contributions from surrounding stars, then normalizing the ^{146}Sm abundance to ^{144}Sm minimizes the impact of variability in the ratio of s-, to r-, to p-process components to ^{146}Sm–^{142}Nd dating.

The half-life of ^{146}Sm is currently in dispute. Measurements from the 1950s through 1980s resulted in the most commonly used value of 103 million years [*Friedman et al.*, 1966]. *Kinoshita et al.* [2012] recently reported a value of 68 ± 14 million years. A recent geological attempt to determine the ^{146}Sm half-life [*Marks et al.*, 2014] compared the ages of various meteorites dated using ^{146}Sm–^{142}Nd and a variety of other techniques to show that the 103 Ma half-life appears to provide more concordant ages than does the 68 Ma half-life. Remembering that half-lives are uncertain parameters subject to ongoing refinement is important in the evaluation of the uncertainties of ages, particularly when the ages determined by two or more independent systems are compared. For the remaining discussion in this chapter, the 103 Ma value for the ^{146}Sm half-life will be used.

Table 14.6 and Fig. 14.13 show an example of ^{146}Sm–^{142}Nd dating applied to the angrite meteorite NWA 4590. The whole rock and mineral separates for this angrite define good isochrons for both ^{147}Sm–^{143}Nd and ^{146}Sm–^{142}Nd. The slope of the ^{147}Sm–^{143}Nd isochron is 0.03033 ± 0.00018, from which one can calculate an age of 4562 ± 35 Ma. The ^{146}Sm–^{142}Nd data produce an isochron of slope of 0.00765 ± 0.00082, which corresponds to the ^{146}Sm/^{144}Sm ratio at the time NWA4801 formed. As with other extinct radionuclide systems, this slope cannot be translated directly into an age without additional information. One approach would to be to use an independently determined absolute age for this angrite in order to pin the ^{146}Sm/^{144}Sm ratio to some point in time. The most obvious age to use for this purpose is the ^{147}Sm–^{143}Nd age determined on exactly the same mineral separates as used for the ^{146}Sm–^{142}Nd measurements. Unfortunately, given the long half-life of ^{147}Sm, the uncertainty on the ^{147}Sm–^{143}Nd age is so large as to be of little use in constraining the short-lived ^{146}Sm–^{142}Nd system that is capable of age precisions of a few million years or less. NWA 4590, however, has been dated by the

Table 14.6 Sm–Nd results for the angrite NWA 4590

Material	[Nd] ppm	$^{147}Sm/^{144}Nd$	$^{143}Nd/^{144}Nd$	$^{142}Nd/^{144}Nd$	Error
Whole rock	6.665	0.2260	0.513531	1.141868	0.000006
Pyroxene	7.756	0.2365	0.513845	1.141882	0.000015
Plagioclase	0.261	0.1574	0.511494	1.141732	0.000017
Olivine	0.193	0.5037	0.521962	1.142297	0.000015
Phosphate	766	0.1473	0.511157	1.141751	0.000006

Data from *Sanborn et al.* [2015].

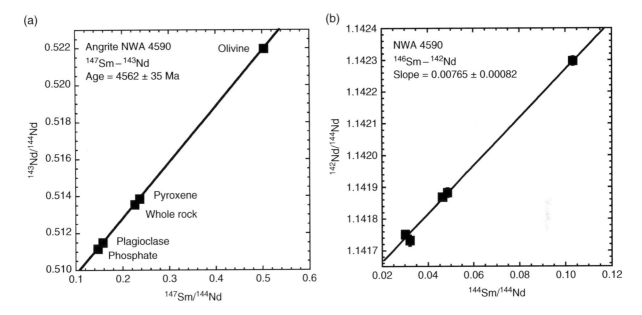

Fig. 14.13. (a)$^{147}Sm–^{143}Nd$ and (b) $^{146}Sm–^{142}Nd$ isochrons for the analyses of angrite NWA 4801 given in Table 14.6. The very high Sm/Nd ratio of the olivine separate is unusual and reflects the high Ca content of angrite olivines that ranges into the composition of kirschsteinite, the Ca–Fe-rich end member of the olivine group minerals. (*Source: Sanborn et al.* [2015]. Reproduced with permission of Elsevier.)

U–Pb system at an age of 4557.8 ± 0.4 Ma [*Amelin*, 2008; *Brennecka and Wadhwa*, 2012]. From this, one can calculate the $^{146}Sm/^{144}Sm$ ratio present at the start of solar-system formation at 4567 Ma using the equation:

$$\left(\frac{^{146}Sm}{^{144}Sm}\right)_{NWA} = \left(\frac{^{146}Sm}{^{144}Sm}\right)_0 \times e^{-\lambda \Delta t} \qquad (14.46)$$

where Δt is the time difference between the age of NWA 4590 and the time when solar-system initial $^{146}Sm/^{144}Sm$ is defined (4567 Ma in this case). Δt is thus 9.2 Ma for NWA 4590 on the basis of its U–Pb age. Substituting these values into equation (14.46) provides a value for $(^{146}Sm/^{144}Sm)_0 = 0.00814$ if the half-life of ^{146}Sm is 103 Ma. Using the 103 Ma half-life, other estimates of the solar-system initial $^{146}Sm/^{144}Sm$ ratio range from 0.00828 ± 0.00044 [*Marks et al.*, 2014] to 0.0085 ± 0.0007 [*Boyet et al.*, 2010], overlapping within uncertainty the value calculated above for NWA 4590.

14.3.6.1   Planetary isochrons

Another application of the $^{146}Sm–^{142}Nd$ system is the creation of so-called "planetary isochrons" that date major early events of igneous differentiation. The most successful application in this regard is to the Moon [*Nyquist et al.*, 1995; *Boyet and Carlson*, 2007; *Brandon et al.*, 2009] and Mars [*Foley et al.*, 2005; *Debaille et al.*, 2007], both of which appear to have gone through an initial planetary-scale differentiation that likely reflects the cooling and crystallization from a largely molten state—the so-called "magma ocean" phase of planet differentiation. This approach is analogous to whole-rock dating, but on a planetary scale, and is based on the assumption that the source region of all the lavas studied formed in a single differentiation event early in the history of the planet. Whether or not this is true will remain a topic of debate for some time to come, but the planetary isochron approach illustrates many of the requirements that must be met for an isochron to provide a meaningful age.

Table 14.7 lists Sm–Nd data for a variety of mare basalts that span the compositional spectrum of basalts erupted on the Moon. The eruption age of these basalts, determined by mineral isochrons from the individual samples, range from 2.99 to 3.9 Ga. The relatively large variability in their initial $^{143}Nd/^{144}Nd$ of almost 9 parts in 10,000 ($\epsilon^{143}Nd$ from -2.2 to $+6.7$) shows that the magmas formed from source materials in the lunar mantle

Table 14.7 Sm–Nd data for lunar basalts and meteorites

Sample	Age (Ga)	$^{147}Sm/^{144}Nd$	$^{143}Nd/^{144}Nd$	$(^{143}Nd/^{144}Nd)i$	$\varepsilon^{143}Nd$	$(^{147}Sm/^{144}Nd)_s$	$^{142}Nd/^{144}Nd$
SAU169	3.90	0.1691	0.511814	0.507445	−2.2	0.1690	1.141816
15386	3.85	0.1696	0.511827	0.507502	−2.2	0.1691	1.141821
LAP02205	2.99	0.1950	0.512635	0.508784	0.5	0.1983	1.141832
15555	3.32	0.1988	0.512839	0.508475	3.0	0.2136	1.141836
70017	3.69	0.2575	0.514372	0.508082	4.8	0.2366	1.141856
74275	3.72	0.2483	0.514252	0.508137	6.7	0.2546	1.141859

Source: *Brandon et al.* [2009]. Reproduced with permission of Elsevier.

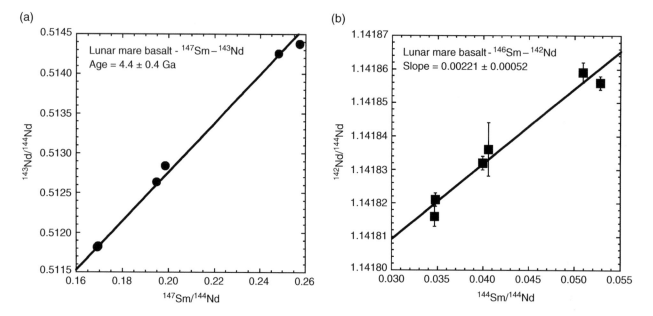

Fig. 14.14. (a) ^{147}Sm–^{143}Nd and (b) ^{146}Sm–^{142}Nd isochrons for the lunar mare basalt data listed in Table 14.7.

that had very different Sm/Nd ratios as a result of lunar differentiation events that occurred long before the eruption of the individual lavas. These samples thus clearly do not meet the requirement of simultaneous formation with constant initial $^{143}Nd/^{144}Nd$ that would allow them to form a whole-rock isochron that provides a meaningful age. Nevertheless, the data do align along a somewhat scattered correlation on the ^{147}Sm–^{143}Nd isochron diagram (Fig. 14.14) that gives a slope of 0.0295 ± 0.0027, from which an "age" of 4.4 ± 0.4 Ga can be calculated. What could this older age mean for a series of rocks with known crystallization ages from 3.0 to 3.9 Ga? The steep slope on the ^{147}Sm–^{143}Nd isochron diagram reflects the fact that the initial $^{143}Nd/^{144}Nd$ ratios of these samples correlate, at least in a rough way, with the samples' Sm/Nd ratios. This means that the compositional distinction of the basalts, as reflected in their different Sm/Nd ratios, reflects a chemical differentiation event that predates the eruption of the individual samples. In other words, basalts with high Sm/Nd ratios came from source materials that also had high Sm/Nd ratios, whereas those with low Sm/Nd ratio came from sources with low Sm/Nd ratios. This conclusion is reinforced by the observation that the samples show a correlation

between their Sm/Nd and $^{142}Nd/^{144}Nd$ ratios. Given its 103 Ma half-life, ^{146}Sm was extinct by the sub-4 Ga eruption ages of these basalts. The Sm/Nd–$^{142}Nd/^{144}Nd$ correlation thus implies that the distinct sources of these basalts formed while ^{146}Sm was still extant.

To this point, we have ignored the fractionation of Sm from Nd that almost certainly occurred during the melt generation and differentiation that accompanied the formation and eruption of these basalts. One can make an attempt to correct for this fractionation by using the initial $^{143}Nd/^{144}Nd$ ratio of each sample to calculate the Sm/Nd ratio of its source. The model one uses to calculate source Sm/Nd evolution prior to the basalt-forming event can be simple, or complex. An example of a simple model is to assume that the mare basalt sources started with chondritic $^{143}Nd/^{144}Nd$ (0.506686) at 4567 Ma and evolved with a constant Sm/Nd ratio up until the time of eruption of any given sample. The $^{147}Sm/^{144}Nd$ ratio in the source of the sample can then be calculated using the equation:

$$\left(\frac{^{143}Nd}{^{144}Nd}\right)_t = \left(\frac{^{143}Nd}{^{144}Nd}\right)_0 + \left(\frac{^{147}Sm}{^{144}Nd}\right)_t \times \left(e^{\lambda t} - 1\right) \quad (14.47)$$

where t in this case is the time difference between 4567 Ma and the eruption age of the sample. Solving for $^{147}Sm/^{144}Nd$ gives:

$$\frac{^{147}Sm}{^{144}Nd} = \frac{\left[\left(\frac{^{143}Nd}{^{144}Nd}\right)_t - \left(\frac{^{143}Nd}{^{144}Nd}\right)_0\right]}{(e^{\lambda t} - 1)} \qquad (14.48)$$

But this gives the $^{147}Sm/^{144}Nd$ ratio at the eruption age of the sample, which needs to be corrected for the additional ^{147}Sm decay that has occurred to the present day (p) by using the equation:

$$\left(\frac{^{147}Sm}{^{144}Nd}\right)_p = \left(\frac{^{147}Sm}{^{144}Nd}\right)_0 \times e^{-\lambda t} \qquad (14.49)$$

Doing so provides the source Sm/Nd ratios (($^{147}Sm/^{144}Nd$)s) shown in Table 14.7. Using the measured $^{144}Sm/^{144}Nd$ and $^{142}Nd/^{144}Nd$ ratios (the $^{144}Sm/^{144}Nd$ ratio can be obtained from the $^{147}Sm/^{144}Nd$ ratio by multiplying by the constant $^{144}Sm/^{147}Sm$ ratio of 0.20503), the $^{146}Sm-^{142}Nd$ isochron (Fig. 14.14b) provides a slope of 0.00221 ± 0.00052, which corresponds to an age of 200^{+40}_{-46} Ma after the 4567 Ma beginning of solar system formation. The errors on the isochron age are not symmetrical about the age because of the exponentially changing $^{146}Sm/^{144}Sm$ ratio. Using the modeled source Sm/Nd ratios derived from the $^{147}Sm-^{143}Nd$ results instead of the measured ratios results in more scatter on the $^{146}Sm-^{142}Nd$ isochron diagram, with a best fit slope of 0.00244 ± 0.00057, which corresponds to an age of 185^{+59}_{-30} Ma. The fact that the scatter on the isochron increases when using the source, instead of measured, Sm/Nd ratios suggests that the model used to calculate source Sm/Nd ratios does not well reflect the true source history of these lavas. One can thus explore more complicated models, for example a Moon that evolves with chondritic Sm/Nd ratio until 4.4 Ga before forming the compositionally distinct reservoirs, with different Sm/Nd ratios, that will eventually melt to produce the mare basalts. This type of model assumes that Moon formation was delayed until 4.4 Ga [*Borg et al.*, 2011]. Applying this model to the data of Table 14.7 improves the linearity of the $^{146}Sm-^{142}Nd$ isochron to provide a best fit line with slope of 0.00187 ± 0.00027, which corresponds to an age of 225^{+23}_{-20} Ma. That this model improves the linearity of the $^{146}Sm-^{142}Nd$ isochron may suggest that it is a better approximation of the true source history of these lavas than either of the other models considered. In the end, however, all these approaches are only models that attempt to explain the fact that the basalts, when they were erupted, had different $^{143}Nd/^{144}Nd$ and $^{142}Nd/^{144}Nd$ ratios, and hence do not meet the requirements for these whole rocks to provide a valid whole-rock isochron.

14.3.6.2 Relative sensitivity to metamorphic resetting
One of the advantages of extinct nuclide chronometry is that the isochrons they provide are less sensitive to disturbance of the system at any time after the parent becomes extinct compared to

Table 14.8 Nd isotopic evolution for a group of rocks formed at 4.4 Ga that experience metamorphic changes to their Sm/Nd ratios at 2.7 Ga

Age	Sample	$^{147}Sm/^{144}Nd$	$^{143}Nd/^{144}Nd$	$^{142}Nd/^{144}Nd$
At 4.4 Ga	1	0.24	0.506909	1.141709
	2	0.22	0.506909	1.141709
	3	0.20	0.506909	1.141709
	4	0.18	0.506909	1.141709
	5	0.16	0.506909	1.141709
	6	0.14	0.506909	1.141709
	7	0.12	0.506909	1.141709
At 2.7 Ga	1	0.2610	0.509562	1.141844
	2	0.2394	0.509342	1.141833
	3	0.2176	0.509120	1.141822
	4	0.1780	0.508899	1.141810
	5	0.1424	0.508678	1.141799
	6	0.1246	0.508456	1.141788
	7	0.1068	0.508236	1.141775
Present day	1	0.2564	0.514130	1.141844
	2	0.2352	0.513532	1.141833
	3	0.2138	0.512929	1.141822
	4	0.1749	0.512015	1.141810
	5	0.1399	0.510948	1.141799
	6	0.1224	0.510636	1.141788
	7	0.1049	0.510105	1.141775

long-lived radiometric systems that integrate through the entire history of a sample. This tendency is illustrated by the following example. Table 14.8 provides a demonstration of the relative effects of post-formation metamorphism on the long-lived $^{147}Sm-^{143}Nd$ and short-lived $^{146}Sm-^{142}Nd$ systems. This table follows the Nd isotope evolution of a group of rocks formed at 4.4 Ga, initially with the same Nd isotopic composition, but a range of Sm/Nd ratios. In this model, the rocks are allowed to grow ^{142}Nd and ^{143}Nd undisturbed until 2.7 Ga, when a hypothetical metamorphic event causes 10% changes in their Sm/Nd ratios, but no change in their $^{142}Nd/^{144}Nd$ and $^{143}Nd/^{144}Nd$ ratio. To somewhat unrealistically accentuate the effect in the model, the rocks with high Sm/Nd ratios have their Sm/Nd ratios increased by 10% in this event while those with low Sm/Nd ratio have their ratios lowered by 10%. Again arbitrarily, the sample with $^{147}Sm/^{144}Nd = 0.18$ experiences no change in its Sm/Nd ratio at 2.7 Ga. At 2.7 Ga, ^{146}Sm is extinct, so the $^{142}Nd/^{144}Nd$ ratios present in each sample at 2.7 Ga continue through to today unchanged. In contrast, the $^{143}Nd/^{144}Nd$ ratios of the samples continue to change with ingrowth of ^{143}Nd dictated by the post-metamorphism Sm/Nd ratios in each sample. When measured today, the Sm–Nd systematics of these rocks provide a very scattered $^{147}Sm-^{143}Nd$ isochron whose slope provides the meaningless age of 3.97 ± 0.16 Ga (Fig. 14.15). Although the isochron plot of the metamorphosed samples in Fig. 14.15a appears linear, the overall range in $^{143}Nd/^{144}Nd$ ratios of the samples is so large compared to the measurement precision on this ratio that the true divergence of the samples from the best fit line cannot be seen on a conventional isochron diagram. A common way to show the quality of the data fit to an isochron is the construction of what is known as the δY or εY diagram (Fig. 14.15b). The εY diagram compares the fractional

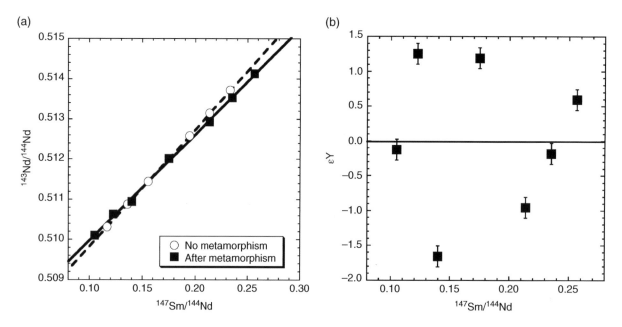

Fig. 14.15. (a) ^{147}Sm–^{143}Nd isochron diagram and (b) εY diagram for the data listed in Table 14.8. (a) The open symbols, and dotted line, show the 4.4 Ga isochron for the samples if they had not been affected by the hypothetical metamorphic event. The filled squares show the measured data for the samples that experienced the metamorphic changes to their Sm/Nd ratios at 2.7 Ga. (b) The offset of the individual data points from the solid best fit line of (a). The y-axis still represents the ^{143}Nd/^{144}Nd ratios of each sample, but shows the relative offset of this ratio for each sample from the best fit line magnified by a factor of 10,000.

offset of the measured ^{143}Nd/^{144}Nd ratio from the value calculated for the best fit line at the sample's Sm/Nd ratio. εY for each sample is thus calculated as:

$$\varepsilon Y = \left\{ \left[\frac{\left(\frac{^{143}\text{Nd}}{^{144}\text{Nd}} \right)_{Sa}}{\left\{ \left(\frac{^{143}\text{Nd}}{^{144}\text{Nd}} \right)_0 + \left(\frac{^{147}\text{Sm}}{^{144}\text{Nd}} \right)_{Sa} \times \left(e^{\lambda t} - 1 \right) \right\}} \right] - 1 \right\} \times 10000$$

(14.50)

Where "Sa" corresponds to the ratios measured in the sample and $(^{143}\text{Nd}/^{144}\text{Nd})_0$ is the initial Nd isotopic composition provided by the isochron whose slope provides the $e^{\lambda t} - 1$ term in equation (14.50). δY is calculated the same way except that the final constant is 1000 instead of 10000. In the εY diagram, the best fit isochron to all the data is a horizontal line at εY = 0, and the individual data points are shown with their offset from this best fit line at a level of "magnification" that allows the measurement uncertainties of the individual points to be seen. In this example, while the conventional isochron diagram for the metamorphosed samples (Fig. 14.15a) appears reasonably linear, the εY diagram (Fig. 14.15b) shows that most of the points lie off the best fit line within analytical uncertainty.

In contrast, in the ^{146}Sm–^{142}Nd isochron diagram (Fig. 14.16), the scatter of the data about the best fit line is clear, yet each point lies within analytical uncertainty of the best fit line. The slope of the line is 0.00204 ± 0.00019, which corresponds to an age of 4356^{+15}_{-13} Ma. While the ^{146}Sm–^{142}Nd age is inaccurate, and does not overlap the true 4.4 Ga age within uncertainty, it is much closer to the original formation age of the samples than is the ^{147}Sm–^{143}Nd result. The key difference here is that the

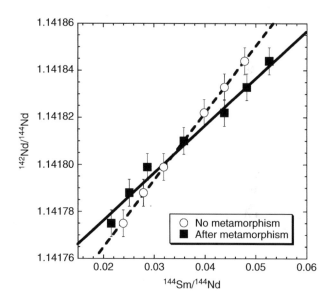

Fig. 14.16. ^{146}Sm–^{142}Nd isochron diagram for the data in Table 14.8. As in Fig. 14.15, the open symbols and dotted line show the data unaffected by 2.7 Ga metamorphism while the filled symbols and line shows the data after changing the Sm/Nd ratios at 2.7 Ga.

^{147}Sm–^{143}Nd system records both the pre-metamorphic and postmetamorphic evolution in the ^{143}Nd/^{144}Nd ratio of each sample, whereas the ^{146}Sm–^{142}Nd system does not record the 2.7 Ga change in Sm/Nd ratio in the ^{142}Nd/^{144}Nd ratios measured today. As a result, while both systems provide evidence for metamorphic resetting of the Sm–Nd system through the scatter of the data, the age offset experienced by the ^{147}Sm–^{143}Nd system as a result of metamorphism that occurs after the extinction of ^{146}Sm is much greater than reflected in the ^{146}Sm–^{142}Nd results.

14.4 CONCLUSIONS

Many of the short-lived radionuclides have been used for chronometry for only about 20–30 years. During this time, they have moved from the excitement of simply detecting the presence of the short-lived parent through its imprint in the isotopic composition of the daughter, to their use as chronometers capable of providing temporal resolution as high as 10^4 years for events occurring over 4.5 billion years ago. The systems with the shortest half-lives are useful for deciphering the timescales of formation of the first solids in the solar system and their assembly into planetesimals. The longer lived of the extinct radionuclides extend the time period of their use into the early stages of planetary differentiation, including that of the Earth. The detection of variability in both ^{142}Nd/^{144}Nd and ^{182}W/^{184}W in modern Earth materials [*Willbold et al.*, 2011; *Touboul et al.*, 2012; *Rizo et al.*, 2016] shows that Earth was never perfectly homogenized after its formation, so defining the extent of isotopic variability in the daughter products of these extinct radionuclides provides an opportunity to investigate the earliest processes that shaped the Earth. Given that most of these systems require the highest precision in isotope ratio measurement currently possible, as mass spectrometry continues to improve the number of uses of extinct radionuclide chronometers is likely to increase dramatically, because they provide access to a time, and temporal precision, window that few other systems can reach.

14.5 REFERENCES

Aggarwal, S. K. (2006) Precise and accurate determination of alpha decay half-life of ^{244}Pu by relative activity method using thermal ionization mass spectrometry. *Radiochimica Acta* **94**, 397–401.

Amelin, Y. (2008) U–Pb ages of angrites. *Geochimica et Cosmochimica Acta* **72**, 221–232.

Ballentine, C. J. and Burnard, P. G. (2002) Production, release and transport of noble gases in the continental crust. *Reviews in Mineralogy and Geochemistry* **47**, 481–538.

Birck, J.-L. and Allégre, C. J. (1985) Evidence for the presence of ^{53}Mn in the early solar system. *Geophysical Research Letters* **12**, 745–748.

Birck, J. L. and Lugmair, G. W. (1988) Nickle and chromium isotopes in Allende inclusions. *Earth and Planetary Science Letters* **90**, 131–143.

Blichert-Toft, J., Moynier, F., Lee, C.-T., Telouk, P., and Albarede, F. (2010) The early formation of the IVA iron meteorite parent body. *Earth and Planetary Science Letters* **296**, 469–480.

Borg, L. E., Connelly, J. N., Boyet, M., and Carlson, R. W. (2011) Chronological evidence that the Moon is either young or did not have a global magma ocean. *Nature* **477**, 70–73.

Boyet, M. and Carlson, R. W. (2007) A highly depleted moon or a non-magma ocean origin for the lunar crust? *Earth and Planetary Science Letters* **262**, 505–516.

Boyet, M., Carlson, R.W., and Horan, M. (2010) Old Sm–Nd ages for cumulate eucrites and redetermination of the solar system initial ^{146}Sm/^{144}Sm ratio. *Earth and Planetary Science Letters* **291**, 172–181.

Brand, W. A., Coplen, T. B., Vogl, J., Rosner, M., and Prohaska, T. (2014) Assessment of international reference materials for isotope-ratio analysis (IUPAC Technical Report). *Pure and Applied Chemistry* **86**, 425–467.

Brandon, A. D., Lapen, T. J., Debaille, V., Beard, B. L., Rankenburg, K., and Neal, C. (2009) Re-evaluating ^{142}Nd/^{144}Nd in lunar mare basalts with implications for the early evolution and bulk Sm/Nd of the Moon. *Geochimica et Cosmochimica Acta* **73**, 6421–6445.

Brennecka, G. A. and Wadhwa, M. (2012) Uranium isotope compositions of the basaltic angrite meteorites and the chronological implications for the early solar system. *Proceedings of the National Academy of Sciences* **109**, 9299–9303.

Burkhardt, C., Kleine, T., Bourdon, B., *et al.* (2008) Hf–W mineral isochron for Ca,Al-rich inclusions: age of the solar system and the timing of core formation in planetesimals. *Geochimica et Cosmochimica Acta* **72**, 6177–6197.

Burkhardt, C., Kleine, T., Dauphas, N., and Wieler, R. (2012) Nucleosynthetic tungsten isotope anomalies in acid leachates of the Murchison chondrite: implications for hafnium-tungsten chronometry. *The Astrophysical Journal Letters* **753**, L6.

Carlson, R. W. and Hauri, E. H. (2001) Extending the ^{107}Pd–^{107}Ag chronometer to low Pd/Ag meteorites with multicollector plasma-ionization mass spectrometry. *Geochimica et Cosmochimica Acta* **65**, 1839–1848.

Carlson, R. W., Garnero, E., Harrison, T. M., *et al.* (2014) How did early Earth become our modern world? *Annual Reviews in Earth and Planetary Science* **42**, 151–178.

Chen, J. H. and Wasserburg, G. J. (1983) The isotopic composition of silver and lead in two iron meteorites: Cape York and Grant. *Geochimica et Cosmochimica Acta* **47**, 1725–1737.

Chen, J. H. and Wasserburg, G. J. (1990) The isotopic composition of Ag in meteorites and the presence of ^{107}Pd in protoplanets. *Geochimica et Cosmochimica Acta* **54**, 1729–1743.

Chen, J. H., Papanastassiou, D. A., and Wasserburg, G. J. (2002) Re–Os and Pd–Ag systematics in Group IIIAB irons and in pallasites. *Geochimica et Cosmochimica Acta* **66**, 3793–3810.

Dauphas, N., Remusat, L., Chen, J. H., *et al.* (2010) Neutron-rich chromium isotope anomalies in supernova nanoparticles. *Astrophysical Journal* **720**, 1577–1591.

Debaille, V., Brandon, A. D., Yin, Q., and Jacobsen, B. (2007) Coupled ^{142}Nd–^{143}Nd evidence for a protracted magma ocean in Mars. *Nature* **450**, 525–528.

Emery, J. F., Reynolds, S. A., Wyatt, E. I., and Gleason, G. I. (1972) Half-lives of radionuclides—IV. *Nuclear Science and Engineering* **48**, 319–323.

Flynn, K. F. and Glendenin, L. E. (1969) Half-life of ^{107}Pd. *Physical Review* **185**, 1591–1593.

Foley, C. N., Wadhwa, M., Borg, L. E., Janney, P. E., Hines, R., and Grove, T. L. (2005) The early differentiation history of Mars from ^{182}W–^{142}Nd isotope systematics in the SNC meteorites. *Geochimica et Cosmochimica Acta* **69**, 4557–4571.

Friedman, A. M., Milsted, J., Metta, D., *et al.* (1966) Alpha decay half-lives of ^{148}Gd, ^{150}Gd and ^{146}Sm. *Radiochimica Acta* **5**, 192–194.

Glavin, D. P., A.Kubny, Jagoutz, E., and Lugmair, G. W. (2004) Mn–Cr isotope systematics of the D'Orbigny angrite. *Meteoritics and Planetary Science* **39**, 693–700.

Göpel, C., Manhès, G., and Allègre, C. J. (1994) U–Pb systematics of phosphates from equilibrated ordinary chondrites. *Earth and Planetary Science Letters* **121**, 153–171.

Gray, C. M. and Compston, W. (1974) Excess ^{24}Mg in the Allende meteorite. *Nature* **251**, 495–497.

Honda, M. and Imamura, M. (1971) Half-life of Mn53. *Physical Review C* **4**, 1182–1188.

Horan, M. F., Carlson, R. W., and Blichert-Toft, J. (2012) Pd–Ag chronology of volatile depletion, crystallization and shock in the Muonionalusta IVA iron meteorite and implications for its parent body. *Earth and Planetary Science Letters* **351–352**, 215–222.

Jacobsen, B., Yin, Q., Moynier, F., *et al.* (2008) ^{26}Al–^{26}Mg and ^{207}Pb–^{206}Pb systematics of Allende CAIs: canonical solar initial ^{26}Al/^{27}Al reinstated. *Earth and Planetary Science Letters* **272**, 353–364.

Keil, K. (2012) Angrites, a small but diverse suite of ancient, silica-undersaturated volcanic-plutonic mafic meteorites, and the history of their parent asteroid. *Chemie der Erde—Geochemistry* **72** 191–218.

Kelly, W. R. and Wasserburg, G. J. (1978) Evidence for the existence of ^{107}Pd in the early solar system. *Geophysical Research Letters* **5**, 1079–1082.

Kinoshita, N., Paul, M., Kashiv, Y., *et al.* (2012) A shorter ^{142}Nd half-life measured and implications for ^{146}Sm–^{142}Nd chronology in the solar system. *Science* **335**, 1614–1617.

Kleine, T., Touboul, M., Bourdon, B., *et al.* (2009) Hf–W chronology of the accretion and early evolution of asteroids and terrestrial planets. *Geochimica et Cosmochimica Acta* **73**, 5150–5188.

Kruijer, T. S., Sprung, P., Kleine, T., Leya, I., Burkhardt, C., and Wieler, R. (2012) Hf–W chronometry of core formation in planetesimals inferred from weakly irradiated iron meteorites. *Geochimica et Cosmochimica Acta* **99**, 287–304.

Kruijer, T. S., Fischer-Goedde, M., Kleine, T., Sprung, P., Leya, I., and Wieler, R. (2013) Neutron capture on Pt isotopes in iron meteorites and the Hf–W chronology of core formation in planetesimals. *Earth and Planetary Science Letters* **361**, 162–172.

Lee, D.-C. and Halliday, A. N. (1995) Hafnium–tungsten chronometry and the timing of terrestrial core formation. *Nature* **378**, 771–774.

Lee, T. and Papanastassiou, D. A. (1974) Mg isotopic anomalies in the Allende meteorite and correlation with O and Sr effects. *Geophysical Research Letters* **1**, 225–228.

Lee, T., Papanastassiou, D. A., and Wasserburg, G. J. (1976) Demonstration of ^{26}Mg excess in Allende and evidence for ^{26}Al. *Geophysical Research Letters* **3**, 109–112.

Leya, I. and Masarik, J. (2013) Thermal neutron capture effects in radioactive and stable nuclide systems. *Meteoritics and Planetary Science* **48**, 665–685.

Lugmair, G. W. and Marti, K. (1977) Sm–Nd–Pu timepieces in the Angra dos Reis meteorite. *Earth and Planetary Science Letters* **35**, 273–284.

Lugmair, G. W. and Shukolyukov, A. (1998) Early solar system time-scales according to ^{53}Mn–^{53}Cr systematics. *Geochimica et Cosmochimica Acta* **62**, 2863–2886.

MacPherson, G. J. (2003) Calcium–aluminum-rich inclusions in chondritic meteorites. In *Meteorites, Comets, and Planets*, Vol. **1**, Davis, A. M. (ed.). *Treatise on Geochemistry*, Elsevier, Amsterdam.

MacPherson, G. J., Davis, A. M., and Zinner, E. K. (1995) The distribution of aluminum-26 in the early solar system—a reappraisal. *Meteoritics* **30**, 365–386.

Markowski, A., Leya, I., Quitte, G., Ammon, K., Halliday, A. N., and Wieler, R. (2006a) Correlated helium-3 and tungsten isotopes in iron meteorites: quantitative cosmogenic corrections and planetesimal formation times. *Earth and Planetary Science Letters* **250**, 104–115.

Markowski, A., Quitte, G., Halliday, A., and Kleine, T. (2006b) Tungsten isotopic compositions of iron meteorites: Chronological constraints vs. cosmogenic effects. *Earth and Planetary Science Letters* **242**, 1–15.

Markowski, A., Quitte, G., Kleine, T., Halliday, A. N., Bizzarro, M., and Irving, A. J. (2007) Hafnium–tungsten chronometry of angrites and the earliest evolution of planetary objects. *Earth and Planetary Science Letters* **262**, 214–229.

Marks, N. E., Borg, L. E., Hutcheon, I. D., Jacobsen, B., and Clayton, R. N. (2014) Samarium-neodymium chronology and rubidium-strontium systematics of an Allende calcium-aluminum-rich inclusion with implications for ^{146}Sm half-life. *Earth and Planetary Science Letters* **405**, 15–24.

Matthes, M., Fischer-Godde, M., Kruijer, T. S., Leya, I., and Kleine, T. (2015) Pd–Ag chronometry of iron meteorites: Correction of neutron capture-effects and application to the cooling history of differentiated protoplanets. *Geochimica et Cosmochimica Acta* **169**, 45–62.

Meissner, F., Schmidt-Ott, W.-D., and Ziegeler, L. (1987) Half-life and α-ray energy of ^{146}Sm. *Zeitschrift für Physik A* **327**, 171–174.

Nichols, R. H., Hohenberg, C. M., Kehm, K., Kim, Y., and Marti, K. (1994) I–Xe studies of the Acapulco meteorite: absolute I-Xe ages of individual phosphate grains and the Bjurbole standard. *Geochimica et Cosmochimica Acta* **58**, 2553–2561.

Nimmo, F. and Agnor, C. B. (2006) Isotopic outcomes of N-body accretion simulations: constraints on equilibration processes during large impacts from Hf/W observations. *Earth and Planetary Science Letters* **243**, 26–43.

Nishizumi, K. (2004) Preparation of ^{26}Al, A.M.S. standards. *Nuclear Instruments and Methods* **223–224**, 388–392.

Nyquist, L. E., Kleine, T., Shih, C.-Y., and Reese, Y. D. (2009) The distribution of short-lived radioisotopes in the early solar system and the chronology of asteroid accretion, differentiation, and secondary mineralization. *Geochimica et Cosmochimica Acta* **73**, 5115–5136.

Nyquist, L. E., Wiesmann, H., Bansal, B., Shih, C.-Y., Keith, J. E., and Harper, C. L. (1995) ^{146}Sm–^{142}Nd formation interval for the lunar mantle. *Geochimica et Cosmochimica Acta* **59**, 2817–2837.

Parai, R. and Mukhopadhyay, S. (2015) The evolution of MORB and plume mantle volatile budgets: constraints from fission Xe isotopes in Southwest Indian Ridge basalts. *Geochemistry, Geophysics, Geosystems* **16**(3), 719–735.

Qin, L., Alexander, C. M. O. D., Carlson, R. W., Horan, M. F., and Yokoyama, T. (2010) Contributors to chromium isotope variation in meteorites. *Geochimica et Cosmochimica Acta* **74**, 1122–1145.

Qin, L., Nittler, L. R., Alexander, C. M. O. D., Wang, J., Stadermann, F. J., and Carlson, R. W. (2011) Extreme ^{54}Cr-rich nano-oxides in the CI chondrite Orgueil—implication for a late supernova injection into the solar system. *Geochimica et Cosmochimica Acta* **75**, 629–644.

Qin, L., Dauphas, N., Horan, M. F., Leya, I., and Carlson, R. W. (2015) Correlated cosmogenic W and Os isotopic variations in Carbo and implications for Hf–W chronology. *Geochimica et Cosmochimica Acta* **153**, 91–104.

Reynolds, J. H. (1960) Isotopic composition of primordial xenon. *Physical Review Letters* **4**, 351–354.

Rizo, H., Walker, R. J., Carlson, R. W., *et al.* (2016) Preservation of Earth-forming events in the tungsten isotopic composition of modern flood basalts. *Science* **352**, 809–812.

Rugel, G., Faestermann, T., Knie, K., *et al.* (2009) New measurement of the ^{60}Fe half-life. *Physical Review Letters* 072502.

Sanborn, M.E., Carlson, R.W., and Wadhwa, M. (2015) $^{147,146}Sm$–$^{143,142}Nd$, ^{176}Lu–^{176}Hf, and ^{87}Rb–^{87}Sr systematics in the angrites: Implications for chronology and processes on the angrite parent body. *Geochimica et Cosmochimica Acta* **171**, 80–99.

Schoenbaechler, M., Carlson, R. W., Horan, M. F., Mock, T., and Hauri, E. H. (2007) High precision Ag isotope measurements in geological materials by multiple collector ICPMS: an evaluation of dry- versus wet-plasma. *International Journal of Mass Spectrometry* **261**, 183–191.

Schoebaechler, M., Carlson, R. W., Horan, M. F., Mock, T. D., and Hauri, E. H. (2008) Silver isotope variations in chondrites: volatile depletion and the initial ^{107}Pd abundance of the solar system. *Geochimica et Cosmochimica Acta* **72**, 5330–5341.

Schoebaechler, M., Carlson, R. W., Horan, M. F., Mock, T. D., and Hauri, E. H. (2010) Heterogeneous accretion and the moderately volatile element budget of Earth. *Science* **328**, 884–887.

Shukolyukov, A. and Lugmair, G. W. (2006) Manganese-chromium isotope systemtatics of carbonaceous chondrites. *Earth and Planetary Science Letters* **250** 200–213.

Spivak-Birndorf, L., Wadhwa, M., and Janney, P. (2009) ^{26}Al–^{26}Mg systematics in D'Orbigny and Sahara 99555 angrites: implications for high-resolution chronology using extinct chronometers. *Geochimica et Cosmochimica Acta* **73**, 5202–5211.

Tachibana, S. and Huss, G. R. (2003) The initial abundancs of ^{60}Fe in the solar system. *Astrophysical Journal Letters* **588**, L41–L44.

Touboul, M., Puchtel, I. S., and Walker, R. J. (2012) ^{182}W evidence for long-term preservation of early mantle differentiation products. *Science* **335**, 1065–1069.

Trinquier, A., Birck, J.-L., and Allègre, C. J. (2007) Widespread ^{54}Cr heterogeneity in the inner solar system. *The Astrophysical Journal* **655**, 1179–1185.

Trinquier, A., Birck, J. L., Allègre, C. J., Göpel, C., and Ulfbeck, D. (2008) ^{53}Mn–^{53}Cr systematics of the early solar system revisited. *Geochimica et Cosmochimica Acta* **72**, 5146–5163.

Trinquier, A., Elliott, T., Ulfbeck, D., Coath, C., Krot, A. N., and Bizzarro, M. (2009) Origin of nucleosynthetic isotope heterogeneity in the solar protoplanetary disk. *Science* **324**, 374–6.

Vockenhuber, C., Oberli, F., Bichler, M., *et al.* (2004) New half-life measurement of ^{182}Hf: improved chronometer for the early solar system. *Physical Review Letters* **93**, 172501.

Warren, P. H. (2011) Stable-isotope anomalies and the accretionary assemblage of the Earth and Mars: A subordinate role for carbonaceous chondrites. *Earth and Planetary Science Letters* **311**, 93–100.

Willbold, M., Elliott, T., and Moorbath, S. (2011) The tungsten isotopic composition of the Earth's mantle before the terminal bombardment. *Nature* **477** 195–198.

Woodland, S. J., Rehkamper, M., Halliday, A. N., Lee, D.-C., Hattendorf, B., and Gunther, D. (2005) Accurate measurement of silver isotopic compositions in geologic materials including low Pd/Ag meteorites. *Geochimica et Cosmochimica Acta* **69**, 2153–2163.

Index

Geochronology and Thermochronology, First Edition. Peter W. Reiners, Richard W. Carlson, Paul R. Renne, Kari M. Cooper, Darryl E. Granger, Noah M. McLean, and Blair Schoene.
© 2018 John Wiley & Sons Ltd. Published 2018 by John Wiley & Sons Ltd.